T0399801

Medicinal Plants of Turkey

This book is part of the book series titled Natural Products Chemistry of Global Plants, and examines the rich plant diversity of Turkey, with descriptions of the plants and pharmacognosy properties. There is a focus on the chemistry of natural products and areas rich in folklore, and botanical medicinal uses are covered with a particular interest in the region of Anatolia. This book focuses on the chemistry of natural products, and where possible links these molecules to pharmacological modes of action. Students and professionals interested in the ethnobotany, chemistry, pharmacology, and biological activities of species used medicinally in Turkey will benefit from this book.

Features

- Addresses the rich chemistry of the natural products and their respective biosynthetic building blocks.
- Includes the association that many of the extracts have today with important drugs, nutrition products, beverages, perfumes, cosmetics, and pigments.
- Describes the key natural products and their extracts, with emphasis on sources, their complex molecules, and applications in science.
- Fills a gap in our understanding of medicinal plants, specifically in Turkey.
- Provides an in-depth understanding of medicinal plants from Turkey, and their complex chemistry and structures.

Natural Products Chemistry of Global Plants

Series Editor: Clara Bik-san Lau

Founding Editor: Raymond Cooper

This unique book series focuses on the natural products chemistry of botanical medicines from different countries such as Sri Lanka, Cambodia, Brazil, China, Africa, Borneo, Thailand, and Silk Road Countries. These fascinating volumes are written by experts from their respective countries. The series will focus on the pharmacognosy, covering recognized areas rich in folklore as well as botanical medicinal uses as a platform to present the natural products and organic chemistry. Where possible, the authors will link these molecules to pharmacological modes of action. The series intends to trace a route through history from ancient civilizations to the modern day showing the importance to man of natural products in medicines, foods, and a variety of other ways. With special emphasis on plant parts for medicinal uses, phytochemistry and biological activities, this book series will be of useful reference to scientists/pharmacognosists/pharmacists/chemists/graduates/undergraduates/researchers in the fields of natural products, herbal medicines, ethnobotany, pharmacology, chemistry and biology. Furthermore, pharmaceutical companies may also find valuable information on potential herbs and lead compounds for the future development of health supplements and western medicines.

Recent Titles in this Series:

Natural Products Chemistry of Botanical Medicines from Cameroonian Plants, *Xavier Siwe Noundou*

Medicinal Plants and Mushrooms of Yunnan Province of China, *Clara Lau and Chun-Lin Long*

Medicinal Plants of Borneo, *Simon Gibbons and Stephen P. Teo*

Natural Products and Botanical Medicines of Iran, *Reza Eddin Owfi*

Natural Products of Silk Road Plants, *Raymond Cooper and Jeffrey John Deakin*

Brazilian Medicinal Plants, *Luzia Modolo and Mary Ann Foglio*

Medicinal Plants of Bangladesh and West Bengal: Botany, Natural Products, and Ethnopharmacology, *Christophe Wiart*

Traditional Herbal Remedies of Sri Lanka, *Viduranga Y. Waisundara*

Medicinal Plants of Ecuador, Pablo A. Chong Aguirre, Patricia Manzano Santana, Migdalia Miranda Martínez (Eds)

Medicinal Plants of Laos, Djaja Djendoel Soejarto, Bethany Gwen Elkington and Kongmany Sydara

Edible and Medicinal Mushrooms of the Himalayas: Climate Change, Critically Endangered Species and the Call for Sustainable Development, *Ajay Sharma, Garima Bhardwaj and Gulzar Ahmad Nayik*

Medicinal Plants of Turkey, Ufuk Koca-Caliskan

Medicinal Plants of Turkey

Edited by
Ufuk Koca Çalışkan
Gazi University, Faculty of Pharmacy, Department of Pharmacognosy and Phytotherapy, Turkey

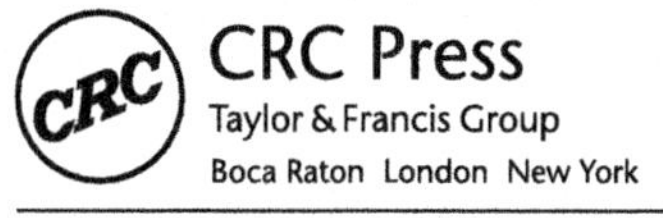

CRC Press is an imprint of the
Taylor & Francis Group, an **informa** business

Cover image: Shutterstock

First edition published 2024
by CRC Press
6000 Broken Sound Parkway NW, Suite 300, Boca Raton, FL 33487-2742

and by CRC Press
4 Park Square, Milton Park, Abingdon, Oxon, OX14 4RN

CRC Press is an imprint of Taylor & Francis Group, LLC

Library of Congress Cataloging-in-Publication Data
Names: Koca-Caliskan, Ufuk, editor. | Akkol, Esra Küpeli, editor.
Title: Medicinal plants of Turkey / edited by Ufuk Koca-Caliskan, Gazi University, Faculty of Pharmacy, Department of Pharmacognosy and Phytotherapy, Turkey, Esra Akkol, Gazi University, Faculty of Pharmacy, Department of Pharmacognosy and Phytotherapy, Turkey.
Description: First edition. | Boca Raton : CRC Press, 2024. | Series: Natural products chemistry of global plants | Includes bibliographical references and index
Identifiers: LCCN 2023009135 (print) | LCCN 2023009136 (ebook) | ISBN 9780367705671 (hardback) | ISBN 9780367705749 (paperback) | ISBN 9781003146971 (ebook)
Subjects: LCSH: Medicinal plants–Turkey. | Pharmacognosy–Turkey. | Materia medica–Turkey.
Classification: LCC RS180.T9 M43 2024 (print) | LCC RS180.T9 (ebook) DDC 615.3/21–dc23/eng/20230525
LC record available at https://lccn.loc.gov/2023009135
LC ebook record available at https://lccn.loc.gov/2023009136

ISBN: 978-0-367-70567-1 (hbk)
ISBN: 978-0-367-70574-9 (pbk)
ISBN: 978-1-003-14697-1 (ebk)

DOI: 10.1201/9781003146971

Typeset in Times
by Deanta Global Publishing Services, Chennai, India

Dedication

To Ray…

for his determination, constant source of knowledge, and encouragement in the preparation of this book.

I would have never thought about finishing this book without Ray three years ago. I should have talked to him, get his opinion and support… Unfortunately, we lost him in June 2022.

If you are watching us from somewhere, you probably would have seen our achievement without your presence but with your inspiration…

Contents

Preface

Turkey has an area of approximately 780,080 km^2 and is located at the midpoint of the continents of Europe and Asia between 36°–42° north latitude and 26°–45° east longitude. Turkey is the meeting point of the Mediterranean, Iran-Turanian, and Euro-Siberian flora regions, which are surrounded by seas on three sides. A map of the phytogeographical divisions of Turkey is given below. The land of the country has a huge plant potential with more than 9600 species, which reaches 12,000 with subspecies and varieties of which (approximately 3650, or about 30%) are endemic taxa. However, Turkey's endemic richness is not limited to this number, as it increases with the discovery of a new endemic taxon almost every month. The reasons for this richness can be explained by the climate differences (continental climate, oceanic climate, and Mediterranean climate), geological and geomorphological diversity, rich water resources (sea, lake, and stream), altitude differences, wide variety of habitat types and three phytogeographic regions, and ecological differences between its east and west and its reflection on floristic differences. The endemism rate of the plant taxa in the flora of Turkey is as follows: Asteraceae 15%, Fabaceae 14%, Lamiaceae 9%, Scrophulariaceae 8%, and Brassicaceae 7%. Moreover, Turkey is one of the central gene centers of plant diversity in the world. The country includes the eastern shores of the Mediterranean Sea, the southern slopes of the Taurus Mountains, and a large part of the "fertile crescent" covering the valleys of the Tigris and Euphrates rivers. Numerous ancient civilizations flourished and domestication of many nutrition and medicinal plants originated within these lands.

The 1st edition of this Turkish Medicinal Plants includes significant medicinal plants, which make up a very slight part of Turkey's flora, and has been selected with common usage and ethnobotanical research in awareness. The book contains 29 chapters, each comprising the botanical features, traditional uses, phytochemicals, and ethnopharmacological characteristics presented by the author expert in the field of pharmacy/pharmacognosy/pharmaceutical botany. Contributors provide detailed and fresh insights into selected significant plants and plant genus *Achillea* sp., *Arnebia* sp., *Astragalus* sp., *Centaurea* sp., *Ecbalium elaterium*., *Fumaria* sp., *Glycyrrhiza*, *Gundelia tournefortii*, *Hypericum perforatum*, *Juniperus* sp., *Lamium* sp., *Laurus nobilis*, *Lavandula* sp., *Malva silvestris*, *Origanum* sp., *Pistacia* sp, *Rosa damascena*, *Rosa canina*, *Salvia* sp., *Satureja* sp., *Scorzonera*., *Sideritis* sp., *Rhus coriaria*, *Teucrium* sp., *Papaver* sp., *Thymus* sp., *Trigonella foenum-graceum*, *Viburnum* sp., and *Viscum* sp. that grow in Turkey.

This book was in fact written during the challenging times of the global pandemic of coronavirus disease 2019 (COVID-19) caused by severe acute respiratory syndrome coronavirus 2 (SARS-CoV-2), with periods of lockdown in different parts of the world. The World Health Organization (WHO) declared the outbreak of a public health emergency of international concern on 30 January 2020, began referring to it as a pandemic on 11 March 2020, and ended its declaration of COVID-19 being a global health emergency on 5 May 2023, but continued to refer to it as a pandemic. Before the approval of the discovery of the COVID-19 vaccines in December 2020, preventive measures were taken that include social distancing, wearing masks, improving ventilation and air filtration, and quarantining those who have been exposed or are infected. Moreover, application of natural supplements, folk remedies, and medicinal plants against the symptoms of the virus came to the fore emphasizing the value of the medicinal plants.

This book is an attempt to provide scientific, reliable, and update information on commonly used medicinal plants grown in Turkey. The authors of the chapters sincerely recommend that this book will be found useful for scientists, pharmacists, pharmacognosists, chemists, undergraduate/graduate students, researchers in the fields of natural products, herbal medicines, ethnobotany, ethnopharmacology, chemistry, and biology, and those would like to acquire more scientific knowledge about the medicinal plants of Turkey.

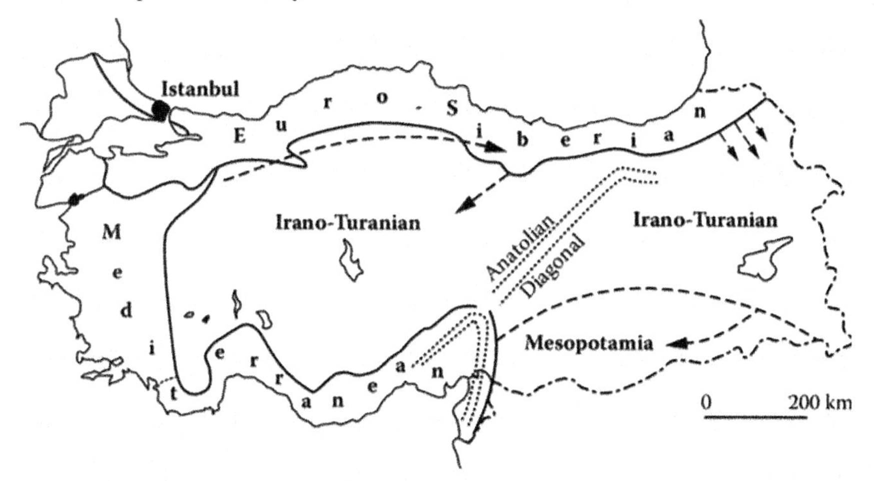

Editor Biography

PROFESSOR UFUK KOCA CALISKAN

Professor Ufuk Koca Caliskan received her diploma from Faculty of Pharmacy, Gazi University, in Turkey. She has a Master's degree in Biochemistry (Washington State University, USA) and a PhD degree in Plant Molecular and Cellular Biology (University of Florida, USA). She now holds the position of Professor in the Department of Pharmacognosy and Pharmaceutical Botany, at the Faculty of Pharmacy, Gazi University. She is an expert in Phytotherapy and Aromatherapy and has a long interest in plants and their healing properties, collecting folk medicine recipes at an early age from healers.

Ufuk studied the production and enhancing anticancer compound podophyllotoxin derivatives in *Linum flavum* and *L. usitatissimum* callus and suspension cultures, followed by a project dealing with the decreasing bitter taste of grapefruit via manipulation of the flavonoid pathway by means of gene transfer during her PhD at the University of Florida. After her PhD, she returned to the Faculty of Pharmacy, Gazi University, and established a plant tissue culture laboratory. She also served as an administrator at the University as a vice dean and dean of the Graduate School of Health Sciences, where she established nonthesis programs in Nursing and Phytotherapy, and worked to increase the success and the quality of education in Health Sciences. She also served as the ERASMUS constitutional coordinator for 4 years at the University. Currently, she is serving as the founder Dean of the Faculty of Pharmacy, Duzce University.

Ufuk served as an editorial or review board member on various international and national journals and published more than 100 papers, including 15 book chapters.

Contributors

Özlem Bahadır Acıkara
Department of Pharmacognosy, Faculty of Pharmacy, Ankara University, Turkey

Beste Atlı
Faculty of Pharmacy, Eastern Mediterranean University, Famagusta, Turkey

O. Tuncay Ağar
Department of Pharmacognosy, Faculty of Pharmacy, University of Suleyman Demirel, Isparta, Turkey

Dudu Altıntaş
Duzce University, Faculty of Pharmacy, Department of Pharmacognosy

Fatma Ayaz
Selçuk University, Faculty of Pharmacy, Department of Pharmacognosy, Konya, Turkey

Zehra Bektur
Gazi University, Faculty of Pharmacy, Department of Pharmacognosy Ankara, Turkey

Merve Yüzbaşıoğlu Baran
University of Health Sciences, Gulhane Faculty of Pharmacy, Department of Pharmacognosy, Ankara, Turkey

Erdal Bedir
Izmir Institute of Technology, Department of Bioengineering, Izmir, Turkey

Duygu Kaya Bilecenoğlu
Aydın Adnan Menderes University, Faculty of Health Sciences, Efeler, Aydın, Türkiye

Gülnur Ekşi Bona
Department of Pharmacognosy, School of Pharmacy, İstanbul Medipol University, İstanbul, Turkey

Ufuk Koca Çalişkan
Gazi University, Faculty of Pharmacy, Department of Pharmacognosy, Ankara, Turkey

I Irem Tatlı Çankaya
Hacettepe University, Faculty of Pharmacy, Department of Pharmaceutical Botany, Sıhhiye, Ankara, Turkey

L. Ömür Demirezer
Department of Pharmacognosy, Faculty of Pharmacy, University of Hacettepe, Ankara, Turkey

Ceylan Dönmez
Faculty of Pharmacy, Department of Pharmacognosy, Selçuk University, Konya, Turkey

Gökşen Dilşat Durbilmez Üstün
Ankara Medipol University, Faculty of Pharmacy, Department of Pharmacognosy, Ankara, Turkey

Sinem Aslan Erdem
Department of Pharmacognosy, Faculty of Pharmacy, Ankara University, Ankara, Turkey

Nuraniye Eruygur
Faculty of Pharmacy, Department of Pharmacognosy, Selçuk University, Konya, Turkey

Hasya Nazlı Gök
Department of Pharmacognosy, Faculty of Pharmacy, Gazi University, Turkey

Alper Gökbulut
Ankara University, Faculty of Pharmacy, Department of Pharmacognosy, Ankara, Turkey

Seren Gündoğdu
Hacettepe University, Faculty of Pharmacy, Department of Pharmacognosy, Ankara, Turkey

Tuğba Günbatan
Gazi University, Faculty of Pharmacy, Department of Pharmacognosy, Ankara, Turkey

İlhan Gürbüz
Gazi University, Faculty of Pharmacy, Department of Pharmacognosy, Ankara, Turkey

Burçin Ergene
Ankara University Faculty of Pharmacy, Pharmacognosy Department, Ankara, Turkey

Selen İlgün
Department of Pharmaceutical Botany, Faculty of Pharmacy, Erciyes University, Kayseri, Turkey

Gülçin Saltan İşcan
Ankara University Faculty of Pharmacy, Pharmacognosy Department, Ankara, Turkey

Methiye Mancak Karakuş
Gazi University, Faculty of Pharmacy, Department of Pharmacognosy, Ankara, Turkiye

Gökçe Şeker Karatoprak
Department of Pharmacognosy, Faculty of Pharmacy, Erciyes University, Kayseri, Turkey

Gülşen Kendir
Süleyman Demirel University, Faculty of Pharmacy, Department of Pharmaceutical Botany, Doğu Kampüsü Çünür, Isparta

Müberra Koşar
Faculty of Pharmacy, Eastern Mediterranean University, Famagusta, Turkey

Ayşegül Köroğlu
Ankara University, Faculty of Pharmacy, Department of Pharmaceutical Botany, Ankara, Türkiye

Günay Sarıyar
Faculty of Pharmacy, Cyprus International University, Nicosia, Northern Cyprus

Sıla Özlem Şener
University of Health Sciences, Gulhane Faculty of Pharmacy, Department of Pharmacognosy, Ankara, Turkey

Bilge Sener
Adjunct Professor, University of Karachi, International Center for Chemical and Biological Sciences, Karachi, Pakistan

Gizem Gülsoy Toplan
Faculty of Pharmacy, Istinye University, Istanbul, Turkey

Ayşe Uz
Hacettepe University, Faculty of Pharmacy, Department of Pharmacognosy, Ankara, Turkey

Nilgun Yakuboğulları
Izmir Institute of Technology, Department of Bioengineering, Izmir, Turkey

Funda N. Yalçın
Department of Pharmacognosy, Faculty of Pharmacy, Hacettepe University, Sıhhiye, Ankara, Türkiye

Erdem Yeşilada
Yeditepe University, Faculty of Pharmacy, Department of Pharmacognosy and Phytotherapy, İstanbul, Turkey

Erkan Yılmaz
Department of Pharmacognosy, Faculty of Pharmacy, Adıyaman University, Adıyaman, Turkey

Golshan Zare
Hacettepe University, Faculty of Pharmacy, Department of Pharmaceutical Botany, Sıhhiye, Ankara, Turkey

1 *Achillea* sp.

Bilge Şener

INTRODUCTION

The genus *Achillea* L. belongs to Asteraceae (Compositae), the largest family of vascular plants. Asteraceaeous plants are distributed throughout the world and are most common in the arid and semiarid regions of subtropical and lower temperate latitudes. *Achillea* consists of over 140 flowering and perennial species and occurs in Europe and temperate areas of Asia and a few grow in North America (Karamenderes, 2002). This genus is represented in Turkey with 6 sections 58 taxa, of which 31 are endemic (Huber-Morath, 1975; Duman, 2000; Guner *et al.*, 2012). From these species, *Achillea millefolium* L. is native to Europe and western Asia but is also widespread in most temperate regions including North America (Orav *et al.*, 2006). *A. millefolium* is known as common yarrow, milfoil, thousand weed, bloodwort, civanperçemi, and woundwort with common utilization in the traditional medicine of several cultures from Europe to Asia for the treatment of spasmodic gastrointestinal, hepatobiliary, and gynecological disorders as well as against inflammation and for wound healing. Various effects of these plants may be due to the presence of a broad range of secondary active metabolites such as flavonoids, phenolic acids, coumarins, terpenoids (monoterpenes, sesquiterpenes, diterpenes, triterpenes), and sterols which have been frequently reported from *Achillea* species (Karaalp, 2011). The botanical and medicinal properties of various species of *Achillea*, which have been examined on the basis of scientific *in vitro* and *in vivo* or clinical evaluations, have been reviewed.

BOTANICAL CHARACTERS

How Many Species in Turkey and Mainly Where They Were Located

Yarrow is an erect herbaceous perennial plant that produces one to several stems (0.2 to 1 m tall) and has a rhizomatous growth form. Leaves are evenly distributed along the stem, with the leaves near the middle and bottom of the stem being the largest. The leaves have varying degrees of hairiness (pubescence). The leaves are 5–20 cm long, bipinnate or tripinnate, almost feathery, and arranged spirally on the stems (Baytop, 2021). Yarrow grows up to 3500 m above sea level. The plant commonly flowers from May through June, and is a frequent component in butterfly gardens. Common yarrow is frequently found in the mildly disturbed soil of grasslands and open forests (Ağar, 2010). In North America, there are both native and introduced genotypes, and both diploid and polyploid plants. The basic chromosome number of this genus is X=9, and most of the species are diploid with great ecological ranges from desert to water-logged habitats (Inotai *et al.*, 2016; Uzay, 2017).

Plant Habitat

Yarrow is a common herb found in urban waste places throughout the temperate and boreal zones of the northern hemisphere and to a small extent in the southern hemisphere. *A. millefolium* is widespread in southern Europe, abundantly throughout central and northern Europe, and has been widely distributed in North America. Despite, most of the members of *Achillea* species are usually distributed in eastern, southeastern and middle Anatolia, some species such as *A. cretica* L., *A.*

DOI: 10.1201/9781003146971-1

nobilis L. subsp. densissima (O.Schwarz ex Bässler) Hub.-Mor., and *A. grandifolia* Friv. grow in the southwestern parts of Anatolia In India, it grows at an altitude of 1050–3600 m, especially in the temperate Himalayas. It is found in the western Himalayas from Kashmir to Kumaon. (Arabacı, 2006).

TRADITIONAL USES

Since *Achillea* L. genus is widespread all over the world, its species have been used by local people as folk or traditional herbal medicines since Middle Ages as natural remedies in several countries. The genus name *Achillea* is derived from the Greek myth of Achilles who was said to carry *A. millefolium* (also known in antiquity as herba militaris) into battle to treat wounds. The genus *Achillea* originates from the ancient use as a wound healing remedy by the Trojan hero Achilles, a powerful warrior in Greek mythology. *A. millefolium* is one of the oldest known botanicals used by humans (sensu lato): it is among the six medicinal plants whose pollen was found in a *Homo neanderthalensis* grave at Shanidar, dated to 65,000 BP (Karamenderes, 2002). Historically, this plant has been extensively used as a herbal remedy for numerous afflictions by many cultures on several continents. The use of yarrow in food and medicine dates back to at least 1200 BC. *A. millefolium* has been used in folk medicine in Pakistan, Iran, and Uzbekistan for wound healing as well. Bumadaran is a popular name for several species of *Achillea* in the Persian language. They are reported as tonic, anti-inflammatory, antispasmodic, diaphoretic, diuretic, and emmenagogic agents and have been used for the treatment of hemorrhage, pneumonia, rheumatic pain, and wound healing in Persian traditional literature (Sezik *et al.*, 2004). Native Americans and early settlers have used yarrow called as plumajillo, or "little feather", for its astringent property in wound healing and anti-bleeding. Unani medicine concerned the activity of *A. millefolium* and described its significant role in traditional medicine and successful administration in the treatment of inflammatory conditions and pain (wounds, cuts, and abrasions) that could be relieved by the application of lotions or ointments containing common yarrow extracts. In India, plants have been traditionally used for human and veterinary health care and also in the food and textile industries. Ninety percent of the local food resources known to indigenous people were undocumented to nutritional literature, trade, cosmetics, and perfumes, but India has a special position in the area of herbal medicines. In traditional Chinese medicine, *A. millefolium* is an antihemorrhagic, wound healing agent and an effective cosmetic soothing for sores, skin disorders (wounds), snakebites, and varicose veins (Ross, 2003). The oldest surviving texts to record the use of *A. millefolium* in the European classical medical tradition are by Pliny the Elder and Dioscorides, both during the first century AD. Dioscorides described the herb achilleios, or millefolium, as being useful to stop bleeding, including from wounds and abnormal menstrual bleeding, and reduce inflammation; a decoction could be used as a douche for menstrual bleeding and be drunk for dysentery (Osbaldeston & Wood, 2000). Pliny's natural history indicated that a plant probably identifiable as yarrow was known by names including achilleos, sideritis, and millefolia. Some people also used the former two names for several other species, some quite different in description, with all being considered valuable for wounds (Osbaldeston & Wood, 2000). *Achillea* species are the most important indigenous economic plants of Anatolia and are traditionally used as an herbal tea for abdominal pain and flatulence and the treatment of wounds in Turkey (Baytop, 2021; Fujita *et al.*, 1995; Honda *et al.*, 1996). Two *Achillea* species, namely *A. millefolium* and *A. ageratum*, are also described in the 18th century as natural remedies, also for skin disorders. Dried aerial parts of *A. millefolium* have been used as raw materials for the preparation of aqueous and alcoholic extracts applied externally in the form of compresses or baths for the treatment of skin and mucous membrane inflammations. Fresh or dried herbs or freshly squeezed juice from *A. millefolium* leaves has been used in traditional European medicine to stop wound bleeding and promote healing of minor wounds, ulcerations, and sores. *A. millefolium* oil macerate, prepared by a three-hour-long maceration of fresh or dried herb in vegetable oil, is recommended for the treatment of skin inflammation or as protection from sunburns. Many of these therapeutic usages have been confirmed by new experimental and clinical studies (Sezik *et al.*, 1997, 2001; Tabata *et al.*, 1994; Tanaka *et al.*, 1995; Tuzlacı *et al.*, 2010; Yeşilada *et al.*, 1993).

PHYTOCHEMICAL PROPERTIES

Phytochemical constituents of Yarrow (*A. millefolium*) mainly belong to several secondary metabolites, namely alkaloids (betonicine, stachydrine, trigonelline), coumarins, flavonoids (apigenin, luteolin, quercitin), sesquiterpene lactones (achillin, achillicine, achillifolin, millefin), polyacetylenes, volatile oil (linalool, camphor, sabinene, chamazulene), triterpenes, tannins, sterols and salicylic acid along with plant acids (Chandler *et al.*, 1982; Hosbas Coskun, 2015; Kachura, 2018; Kaneko *et al.*, 1971; Şabanoglu, 2016). *Achillea* species appear to be exceptionally rich in different types of secondary metabolites in comparison with other plant species. The genus *Achillea* comprises flavored species which produce intense essential oils. The essential oils of *Achillea* contain high levels of sesquiterpenes compared with monoterpenes (Bulatovic *et al.*, 1997; Feizbakhsh *et al.*, 2003; Simid *et al.*, 2000). There are several pharmacological actions that have been mostly attributed to the presence of azulenogenous sesquiterpene lactones in the essential oil of *Achillea* (Glasl *et al.*, 1999; Kaloshina & Neshta, 1973; Kastner *et al.*, 1992). The essential oil composition changes based on plant species and locations (Boscovic *et al.*, 2005; Farajpour *et al.*, 2017). The essential oils were obtained in yields of 0.9–9.5 mL/kg from *A. millefolium* growing in European countries. The major constituents of essential oil were sabinene, β-pinene, 1,8-cineole, artemisia ketone, linalool, α -thujone, β-thujone, camphor, borneol, fenchyl acetate, bornyl acetate, (*E*)-β-caryophyllene, germacrene D, caryophyllene oxide, β-bisabolol, δ-cadinol, and chamazulene. High amounts of monoterpenes and chamazulene were found in samples from European countries (Orav *et al.*, 2006). The chemical composition of essential oils from *Achillea crithmifolia, A. ligustica* All., and *A. umbellata* Thunb growing in Greece were analyzed (Tzakou *et al.*, 1993, 1995, 2009). The analysis of essential oils among Iranian *A. millefolium* accessions was performed. Based on their seven main components (1,8-cineole, β-thujone, camphor, germacrene-D, transnerolidol, isospathulenol, and cubenol) of the essential oils from the Iranian *A. millefolium*, accessions were categorized into five groups. Moreover, essential oils of other Iranian *Achillea* species, namely *A. biebersteinii* Afan, *A. eriophota, A. nobilis,* and *A. wilhelmsii* (Ghani *et al.*, 2008), were chemically examined. The chemical composition of essential oils obtained from *A. fragrantissima* (Forssk.) Sch. Bip. and *A. santolina* L. growing in Egypt were determined (el-Shazly *et al.*, 2004). In Turkey, the analysis of essential oils obtained from *Achillea* species including *A. biebersteinii* Afan (Chialva *et al.*, 1993; Kusmenoglu *et al.*, 1995), *A. boissieri* (Kucukbay *et al.*, 2010), *A. falcata* L. (Kürkçüoğlu *et al.*, 2003), *A. goniocephala* Boiss. (Başer *et al.*, 2001b), *A. ketenoglui, A. lycaonica* (Başer *et al.*, 2001a), *A. schischkinii* Sosn (Dönmez *et al.*, 2005; İşcan *et al.*, 2006), *A. phrygia* Boiss. et Bull. (Başer *et al.*, 2000), and *A. teretifolia* Willd. and *A. nobilis* (Demirci *et al.*, 2009) were investigated in detail. The chemical composition of essential oils from *Achillea abrotanoides, A. clypeolata* Sibth & Sm., *A. depressa* Janka, and *A. stricta* Schleicher et Koch. growing in Yugoslavia was analyzed (Chalchat *et al.*, 2005).

Besides essential oils, flavonoids (aglycons and glycosides), phenolic acids (mostly cinnamic and benzoic acid derivatives), terpenes (including guaianolides, diterpenes, sesquiterpenes, and their oxygenated forms), phytosterols, phenolic acids (caffeic and quinic acids), fatty acids, and alcohols were also determined as phytochemicals. Among them, flavonoids are the most widely represented in the form of aglycones and glucosides (Guedon *et al.*, 1993; Tzakou *et al.*, 1995). Keampferol and kaempferol-3-*O*-glycoside) are reported as the major components of *Achillea* extracts. Moreover, catechin, 5,7,3′-triacetoxy-3,6,4′-trimethoxyflavone, 7-*O*-methyl apigenin, apigenin, apigenin-7-*O*-glycoside, axillarin, centaureidin, chrysosplenol B, fisetin, galetin 3,6-dimethyl ether, hesperetin, hesperidin, hyperoside, isoquercetin, isovitexin, jaceidin, luteolin, luteolin-7-*O*-glycoside, luteolin-7-*O*-glucuronide, naringenin, nicotiflorin, penduletin, quercetin, and quercetin-3-*O*-glucuronide, isoharmnetin-3-*O*-glucoside and isoharmnetin-3-*O*-rutinoside, isoschaftoside, neoschaftoside, and schaftoside. In addition, the extracts from yarrow species are also rich sources of phytosterols such as stigmasterol and β-sitosterol were found in the extracts from the aerial parts of *Achillea* species (Karaalp, 2011).

BIOLOGICAL ACTIVITIES

Achillea species are well-known medicinal plants having an important role both in folk medicine and in modern phytotherapy. The main indications include loss of appetite, bloating, flatulence, minor menstrual spasms, and wound healing (Demirezer *et al.*, 2019). Common yarrow is also used to treat arthritis, colds, and fevers. Due to its antibacterial properties, common yarrow is used to treat dysentery and diarrhea. It is also used by women to ease menstruation (Demirci *et al.*, 2009). *A. millefolium* has recently been the subject of many *in vitro* and *in vivo* studies to determine its potential diverse physiological effects. Indeed, several reported bioactivities reflect yarrow's known ethnobotanical applications. Traditional indications of their use include digestive problems, liver and gallbladder conditions, menstrual irregularities, cramps, fever, and wound healing. The Commission E approves its internal use for loss of appetite and dyspeptic ailments (gastric catarrh, spastic discomfort), and externally it is used in the form of a sitz bath or as a compress against skin inflammation, slow healing wounds, and bacterial or fungal infections. Yarrow is traditionally used orally for the symptomatic treatment of gastrointestinal disorders. Traditionally, yarrow herb 4.5 g/day has been used for various conditions (EMA Monograph, 2020). However, there are no quality clinical studies to validate this dosing. Topically, the essential oil of *A. millefolium* is an emollient and itch-relieving adjunct in the treatment of skin diseases (Demirezer *et al.*, 2021). The use of yarrow is contraindicated in subjects allergic to Asteraceae depending on the sesquiterpene lactones.

ANTICHOLINESTERASE ACTIVITY

The anticholinesterase activity of *Achillea biebersteinii* Afan (Figure 1.1) and *A. millefolium* L. was investigated by using *in vitro* and docking studies. 6-OH-luteolin-7-*O*-β-D-glucoside isolated from *Achillea millefolium* L. was studied for anticholinesterase activity. This compound exhibited *in silico* and *in vitro* the strongest inhibition to AChE and BuChE and also has a strongest antioxidant capacity (Sevindik *et al.*, 2015a). In other studies, Patuletin-7-*O*-β-D-glucoside isolated from

FIGURE 1.1 *Achillea biebersteinii* AFAN

A. biebersteinii was determined to possess the strongest inhibition to acetylcholinesterase activity. Docking experiments revealed that this compound targeted both the catalytic and the peripheral active sites of acetylcholinesterase and butyrylcholinesterase with many hydrogen bond interactions (Sevindik *et al.*, 2015b).

Antihypertensive Activity

The control of hypertension is an important element in the management of cardiovascular diseases. Khan and Gilani reported that the aqueous–methanol extract of the aerial parts of *A. millefolium* caused a dose-dependent decrease in arterial blood pressure of rats under anesthesia. The extract also exhibited a high angiotensin-converting enzyme inhibition. These results indicated that *A. millefolium* exhibits hypotensive, cardiovascular inhibitory, and bronchodilatory effects, thus explaining its medicinal use in hyperactive cardiovascular and airway disorders, such as hypertension and asthma (Khan & Gilani, 2011).

Anti-Inflammatory Activity

Achillea millefolium and *A. nobilis* possess good anti-inflammatory activity and are used to treat rheumatism and muscle aches and heal inflamed cuts or wounds. Their phytochemical constituent flavonoids have an effect on prostaglandin production and possess anti-inflammatory activity. Another component of yarrow species as a powerful anti-inflammatory is azuline, which comprises almost half of yarrow's chemical composition (Küpeli *et al.*, 2007, 2011; Karabay-Yacasoğlu *et al.*, 2007).

Antimicrobial Activity

The antimicrobial activities of the extracts of *Achillea* species were evaluated (Karaalp *et al.*, 2009). The ether, hexane, and methanol extracts of Yarrow caused inhibition zones against *E. coli* in disc diffusion assays. The essential oil of *A. millefolium* showed antibacterial activity against *S. typhimurium* and *S. aureus* (Apel *et al.*, 2020). A number of studies on the essential oils of *Achillea* species have also been carried out for their antimicrobial properties (Başer *et al.*, 2002; Beckhechi *et al.*, 2011; Bezic *et al.*, 2003; Karamenderes *et al.*, 2003, 2007; Küçükbay *et al.*, 2011, 2012; Tabanca *et al.*, 2011). The essential oils obtained from *Achillea* species have potent antibacterial and antifungal activities.

Antinociceptive Activity

The antinociceptive activity of hydroalcoholic extracts of *A. millefolium* and *A. vulgaris* were evaluated by the hot plate, formalin, writhing, and intestinal transit tests for their use as anti-inflammatory, analgesic, and antispasmodic agents in traditional medicine. Rutin and caffeic acid derivatives were determined as a major constituent of both hydroalcoholic extracts of *A. millefolium* and *A. vulgaris*. Abdominal contractions were significantly inhibited by using these extracts due to the presence of flavonoid glycoside rutin (Pires *et al.*, 2009).

Antiulcer Activity

Cavalcanti *et al.* evaluated the antiulcer and antioxidant activity of the methanolic extract of *A. millefolium* leaves by using pylorus ligation-induced ulcers in rat stomach. Results showed that the pretreatment of *A. millefolium* extract at the dose of (100 mg/kg/*p.o.* and 125 mg/kg/*p.o.*) produced a dose-dependent decrease in ulcer index in pylorus ligation-induced ulcers in Wistar rats (Cavalcanti *et al.*, 2006).

Spasmolytic Activity

The effects of hydroalcoholic extract of *A. millefolium* on ileum contractions of Wister rats were investigated (Sedighi *et al.*, 2013). The hydroalcoholic extract of *A. millefolium* has a relaxation effect on ileum contraction activity; likely related effect could be attributed to flavonoids properties of plant, especially quercetin and apigenin. The antispasmodic activity was also determined on the rat-isolated duodenum by using the hydroalcoholic extract of *Achillea nobilis* L. subsp. *spylea* (O. Schwarz) Boiss. (Karamenderes & Apaydın, 2003).

Wound Healing Activity

Nowadays, the traditional usage of medicinal plants for wound healing has received attention from the scientific community. Wound healing is a complex process characterized by homeostasis, reepithelization, and granulation tissue formation and remodeling of the extracellular matrix. Medicinal plants may affect various phases of the wound healing process, coagulation, inflammation, and fibroplasia. Aqueous extract of the flowers of *A. millefolium* applied topically has shown significant wound healing activity in rats. The wound sizes of the test compared to control groups were reduced faster (Jalali *et al.*, 2007). The root extracts of *A. biebersteinii* growing in Turkey have been evaluated for wound healing activity by using Linear incision and circular excision wound models employed on mice and rats with the standard skin ointment Madecassol. The *n*-hexane extract-treated groups of animals were found to have better activity at 84.2% contraction, which was close to the contraction value of the reference drug Madecassol (100%). On the other hand, the same extract on the incision wound model demonstrated a significant increase (40.1%) in wound tensile strength as compared to other groups. According to the experimental data, *A. biebersteinii* was found to have remarkable wound healing activity (Küpeli *et al.*, 2011).

Cytostatic Activity

Recently, the cytostatic activity of *Achillea ageratum* was studied and the methanolic extract has shown cytotoxic effect against mouse Hep-2 and McCoy cells (Gomez *et al.*, 2001). Besides, *in vivo* antitumor activity of three sesquiterpenoids, achimillic acids A, B, and C, isolated from methanol extract of *A. millefolium* flower was studied against mouse P-388 leukemia cells. The compounds were found to be active against mouse P-388 leukemia cells in a concentration-dependent manner (Tozyo *et al.*, 1994).

CONCLUSIONS

Yarrow (*Achillea* L.) recognized as a powerful medicinal plant is widely distributed and has been medicinally used for thousands of years. Yarrow has attracted researchers' interest for a long time because of several biological and pharmacological properties. Popular indications of the several species of this genus include treatment of wounds, bleedings, headache, inflammation, pains, spasmodic diseases, flatulence, and dyspepsia. Phytochemical investigations of *Achillea* species have revealed that many secondary metabolites from this genus are highly bioactive. The ethnopharmacological approach, combined with chemical and biological methods, has provided and will continue to provide useful pharmacological leads. Recent findings have confirmed several traditional uses. The largest number of data was accumulated for antioxidant and anti-inflammatory effects. There are also positive results on the analgesic, antiulcer, choleretic, hepatoprotective, and wound healing activities. In the last decades, pharmacological studies became intensive, although human clinical investigations are still rare. Therefore, efforts should be made to further investigate standardized yarrow plant extracts using well-designed clinical trials in humans owing to their widespread use.

REFERENCES

Ağar, O.T. 2010. *Bazı Achillea L. türleri üzerinde farmasötik botanik araştırmalar*, YL tezi, Hacettepe Üniv. Sağlık Bilimleri Enstitüsü, Farmasötik Botanik programı, Ankara.

Apel, L., Lorenz, P., Urban, S., Sauer, S., Spring, O., Stintzing, F.C., Kammerer, D.R. 2020. Phytochemical characterization of different yarrow species (*Achillea* sp.) and investigations into their antimicrobial activity. *Zeitsch Naturforsch.*, 76(1–2), 55–65.

Arabacı, T. 2006. *Türkiye'de yetişen Achillea L. (Asteraceae) cinsinin revizyonu*, Doktara Tezi, İnönü Üniv. Fen Bilimleri Enstitüsü, Malatya.

Başer, K.H.C., Demirci, B., Demirci, F., Koçak, S., Akıncı, C., Malyer, H., Güleryüz, G. 2002. Composition and antimicrobial activity of the essential oil of *Achillea multifida*. *Planta Med.*, 68(10), 941–943.

Başer, K.H.C., Demirci, B., Duman, H. 2001. Composition of the essential oils of two endemic species from Turkey: *Achillea lycaonica* and *A. ketenoglui*. *Chem. Nat. Comp.*, 37(3), 245–252.

Başer, K.H.C., Demirci, B., Duman, H., Aytaç, Z., Adıgüzel, N. 2001. Composition of the essential oil of *Achillea goniocephala* Boiss. et Bal. from Turkey. *J. Ess. Oil Res.*, 13(4), 219–220.

Başer, K.H.C., Demirci, B., Kaiser, R., Duman, H. 2000. Composition of the essential oil of *Achillea phrygia* Boiss. et Ball. *J. Ess. Oil Res.*, 12(3), 327–329.

Baytop, T., 2021. *Türkiye'de Bitkiler İle Tedavi-Geçmişte ve Bugün*, s. 169–170, Ankara Nobel Kitabevleri, Ankara. (ISBN: 978-625-7146-86-9).

Bekhechi, C., Bekkara, F.A., Casanova, J., Tomi, F. 2011. Composition and antimicrobial activity of the essential oil of *Achillea odorata* L. subsp. *pectinata* (Lamk) var. *microphylla* (Willd.) Willk. from Northwestern Algeria. *J. Ess. Oil Res.*, 23(3), 42–46.

Bezic, N., Skocibusic, M., Dunkic, V., Radonic, A. 2003. Composition and antimicrobial activity of *Achillea clavennae* L. essential oil. *Phytother. Res.*, 17(9), 1037–1040.

Boskovic, Z., Radulovic, N., Stojanovic, G. 2005. Essential oil composition of four *Achillea* species from the balkans and its chemotaxonomic significance. *Chem. Nat. Comp.*, 41(6), 674–678.

Bulatovic, V.M., Savikin-Fodulovic, K.P., Menkovic, N.R., Grubisic, D.V., Dokovic, D.D., Vajs, V.E. 1997. Essential oil of *Achillea corabensis* (Heimerl) Micevski. *J. Ess. Oil Res.*, 9(5), 537–540.

Cavalcanti, A.M., Baggio, C.H., Freitas, C.S., Rieck, L., de Sousa, R.S., Da Silva-Santos, J.E., *et al.* 2006. Safety and antiulcer efficacy studies of *Achillea millefolium* L. after chronic treatment in Wistar rats. *J. Ethnopharmacol.*, 107(2), 277–284.

Chalchat, J.C., Petrovic, S.D., Maksimovic, Z.A., Gorunovic, M.S. 2005. Aromatic plants of Yugoslavia. III. Chemical composition of essential oils of *Achillea abrotanoides* (Vis.) Vis., *A. clypeolata* Sibth. & Sm., *A. depressa* Janka and *A. stricta* Schleicher et Koch. *J. Ess. Oil Res.*, 17(5), 549–552.

Chandler, R.F., Hooper, S.N., Harvey, M.J. 1982. Ethnobotany and phytochemistry of yarrow, *Achillea millefolium*, compositae. *Econ. Bot.*, 36(2), 203–223.

Chialva, F., Monguzzi, F., Manitto, P., Akgül, A. 1993. Essential oil constituents of *Achillea biebersteinii* Afan. *J. Ess. Oil Res.*, 5(1), 87–88.

Demirci, F., Demirci, B., Gürbüz, İ., Yeşilada, E., Baser, K.H.C. 2009. Characterization and Biological activity of *Achillea teretifolia* Willd. and *A. nobilis* L. subsp. *neilreichii* (Kerner) Formanek essential oils. *Turk. J. Biol.*, 33(2), 129–136.

Demirezer, L.Ö., Ersöz, T., Şener, B., Köroğlu, A. 2019. *A'dan Z'ye Tıbbi Bitkiler*, Hayykitap, İstanbul. (ISBN: 978-605-7674-34-0).

Demirezer, L.Ö., Ersöz, T., Şener, B., Köroğlu, A. 2021. *A'dan Z'ye Tıbbi Yağlar ve Aromatik Sular*, Hayykitap, İstanbul. (ISBN: 978-625-7685-15-3).

Dönmez, E., Tepe, B., Daferera, D., Polissiou, M. 2005. Composition of the essential oil of *Achillea schischkinii* Sosn. (Asteraceae) from Turkey. *J. Ess. Oil Res.*, 17(5), 575–576.

Duman, H. 2000. *Achillea* L. in: Güner, A., Özhatay, N., Ekim, T., Başer, K.H.C. (eds). *Flora of Turkey and the East Aegean Islands*, Vol. 11, pp. 158–159, Edinburgh University Press, Edinburgh.

el-Shazly, A.M., Hafez, S.S., Wink, M. 2004. Comparative study of the essential oils and extracts of *Achillea fragrantissima* (Forssk.) Sch. Bip. and *Achillea santolina* L. (Asteraceae) from Egypt. *Pharmazie*, 59(3), 226–230.

EMA Monographs. 2020. EMA/HMPC/509104/2019, Community herbal monograph on *Achillea millefolium* L.herba.

Farajpour, M., Ebrahimi, M., Baghizadeh, A., Aalifar, M. 2017. Phytochemical and yield variation among Iranian *Achillea millefolium* accessions. *Hortsci.*, 52(6), 827–830.

Feizbakhsh, A., Tehrani, M.S., Rustaiyan, A., Masoudi, S. 2003. Composition of the essential oil of *Achillea albicaulis* CA Mey. *J. Ess. Oil Res.*, 15(1), 21–22.

Fujita, T., Sezik, E., Tabata, M., Yesilada, E., Honda, G., Takeda, Y., *et al.* 1995. Traditional medicine in Turkey VII. Folk medicine in middle and west Black Sea regions. *Econ. Bot.*, 49(4), 406–422.

Ghani, A., Azizi, M., Hassanzadeh-Khayyat, M., Pahlavanpour, A.A. 2008. Essential oil composition of *Achillea eriophora, A. nobilis, A. biebersteinii* and *A. wilhelmsii* from Iran. *J. Ess. Oil Bearing Plants*, 11(5), 460–467.

Glasl, S., Presser, A., Werner, I., Wawrosch, C., Kastner, U., Jurenitsch, J., *et al.* 1999. Two proazulenes from *Achillea ceretanica* Sennen. *Phytochem.*, 50(4), 629–631.

Gomez, M.A., Garcia, M.D., Saenz, M.T., Ahumada, M.C., Aznar, J. 2001. Cytostatic activity of *Achillea ageratum* against cultured Hep-2 and McCoy cells. *Pharm. Biol.*, 39(1), 79–81.

Guedon, D., Abbe, P., Lamaison, J.L. 1993. Leaf and Flower Head Flavonoids of *Achillea millefolium* L. subspecies. *Biochem. Syst. and Ecol.*, 21(5), 607–611.

Güner, A., Aslan, S., Ekim, T., Vural, M., Babaç, M.T. (eds.). 2012. *Türkiye Bitkileri Listesi (Damarlı Bitkiler)*, pp. 108–112, Nezahat Gökyiğit Botanik Bahçesi ve Flora Araştırmaları Derneği Yayını, İstanbul.

Honda, G., Yeşilada, E., Tabata, M., Sezik, E., Fujita, T., Takeda, Y., Tanaka, T. 1996. Traditional medicine in Turkey VI. Folk medicine in west Anatolia: Afyon, Kütahya, Denizli, Muğla, Aydın provinces. *J. Ethnopharmacol.*, 53, 75–87.

Hoşbaş Coşkun, S. 2015. *Bazı Achillea türleri üzerinde farmakognozik araştırmalar*, DR tezi, Farmakognozi Anabilim Dalı, Gazi Üniv. Sağlık Bilimleri Enstitüsü, Ankara.

Huber-Morath, A. 1975. *Achillea* L. in: Davis, P.H. (ed). *Flora of Turkey and the East Aegean Islands*, Vol. 5, pp. 224–252, Edinburgh University Press, Edinburgh.

Inotai, K., György, Z., Kindlovits, S., Várady, G., Németh-Zámbori, E. 2016. Evaluation of yarrow (*Achillea*) accessions by phytochemical and molecular genetic tools. *J. Appl. Bot. Food Quality*, 89, 105–112.

İşcan, G., Kırımer, N., Kürkçüoğlu, M., Arabacı, T., Küpeli, E., Başer, K.H.C. 2006. Biological activity and composition of the essential oils of *Achillea schischkinii* Sosn. and *Achillea aleppica* DC. subsp. *aleppica*. *J. Agr. Food Chem.*, 54(1), 170–173.

Jalali, F.S.S., Tajik, H., Tehrani, A. 2007. Experimental evaluation of repair process of burn wound treated with aqueous extract of *Achillea millefolium* on animal model: Clinical and histopathological study. *J. Anim. Vet. Adv.*, 6(12), 1357–1361.

Kachura, A. 2018. An ethnobotanical, pharmacological, and phytochemical analysis of *Achillea millefolium* L. by parts. M.Sc. thesis, Biology Department of Biology, Faculty of Science University of Ottawa, Ottawa.

Kaloshina, N., Neshta, I. 1973. Flavonoids of *Achillea millefolium*. *Chem. Nat. Com.*, 9(2), 261–261.

Kaneko, H., Naruto, S., Takahashi, S. 1971. Sesquiterpenes of *Achillea sibirica*. *Phytochem.*, 10(12), 3305–3306.

Karaalp, C. 2011. *Achillea millefolium* L. Civanperçemi. in: Demirezer, L.Ö., Ersöz, T., Saraçoğlu, İ., Şener, B. (Editörler). *FFD Monografları- Tedavide Kullanılan Bitkiler*, 2. Baskı, s. 1–7, MN Medikal & Nobel Tıp Kitabevi, Ankara. (ISBN: 978-975-567-073-7).

Karaalp, C., Yurtman, A.N., Yavaşoğlu, N.U.K. 2009. Evaluation of antimicrobial properties of *Achillea* L. flower head extracts. *Pharm. Biol.*, 47(1), 86–91.

Karabay-Yavaşoğlu, N.U., Karamenderes, C., Baykan, S., Apaydın, S. 2007. Antinociceptive and anti-inflammatory activities and acute toxicity of *Achillea nobilis* subsp. *neilreichii* extract in mice and rats. *Pharm. Biol.*, 45(2), 162–168.

Karamenderes, C. 2002. *Achillea L. cinsinin Millefolium seksiyonu üzerinde Farmasötik Botanik yönünden araştırmalar*, DR Tezi, Ege Üniversitesi, İzmir.

Karamenderes, C., Apaydın, S. 2003. Antispasmodic effect of *Achillea nobilis* L. subsp. *sipylea* (O. Schwarz) Bässler on the rat isolated duodenum. *J. Ethnopharmacol.*, 84(2–3), 175–179.

Karamenderes, C., Karabay, N.U., Zeybek, U. 2003. Composition and antimicrobial activity of the essential oils of *Achillea setacea* Waldst. & Kit. collected from different localities of Turkey. *J. Fac. Pharm.*, 32(2), 113–120.

Karamenderes, C., Yavaşoğlu, N.U.K., Zeybek, U. 2007. Composition and antimicrobial activity of the essential oils of *Achillea nobilis* L. subsp. *sipylea* and subsp. *neilreichii*. *Chem. Nat. Comp.*, 43(5), 632–634.

Kastner, U., Jurenitsch, J., Glasl, S., Baumann, A., Robien, W., Kubelka, W. 1992. Proazulenes from *Achillea asplenifolia*. *Phytochem.*, 31(12), 4361–4362.

Khan, A.U., Gilani, A.U. 2011. Blood pressure lowering, cardiovascular inhibitory and bronchodilatory actions of *Achillea millefolium*. *Phytother. Res.*, 25, 577–583.

Küçükbay, F.Z., Kuyumcu, E., Arabacı, T. 2010. The essential oil of *Achillea boissieri*. *Chem. Nat Comp.*, 46(5), 824–825.

Küçükbay, F.Z., Kuyumcu, E., Bilenler, T., Yıldız, B. 2012. Chemical composition and antimicrobial activity of essential oil of *Achillea cretica* L. (Asteraceae) from Turkey. *Nat. Prod. Res.*, 26(18), 1668–1675.

Küçükbay, F.Z., Kuyumcu, E., Günal, S., Arabacı, T. 2011. Composition and antimicrobial activity of the essential oil of *Achillea formosa* subsp. *amanica*. *Chem. Nat. Comp.*, 47(2), 300–302.

Küpeli Akkol, E., Koca, U., Pesin, İ., Yılmazer, D. 2011. Evaluation of the wound healing potential of *Achillea biebersteinii* Afan. (Asteraceae) by *in vivo* excision and ıncision models. *Evidence-Based Compl. and Altern. Med.*, Article ID 474026, 7 pages. doi:10.1093/ecam/nep039.

Küpeli, E., Orhan, İ., Küsmenoğlu, S., Yeşilada, E. 2007. Evaluation of AntiInflammatory and antinociceptive activity of five Anatolian *Achillea* species. *Turk. J. Pharm. Sci.*, 4(2), 89–99.

Kürkçüoğlu, M., Demirci, B., Tabanca, N., Özek, T., Başer, K.H.C. 2003. The essential oil of *Achillea falcata* L. *Flavour & Frag J.*, 18(3), 192–194.

Küsmenoğlu, S., Başer, K.H.C., Özek, T., Harmandar, M., Goekalp, Z. (1995). Constituents of the essential oil of *Achillea biebersteinii* Afan. *J. Ess. Oil Res.*, 7(5), 527–528.

Orav, A., Arak, E., Raal, A. 2006. Phytochemical analysis of the essential oil of *Achillea millefolium* L. from various European Countries. *Nat. Prod. Res.*, 20(12), 1082–1088.

Osbaldeston, T.A., Wood, R.P.A. 2000. De Materia Medica*: Being* An Herbal *With Many* other *Medicinal* Materials*: Written* in Greek *In The First* Century *of the Common* Era*: A* New *Indexed* Version *in Modern* English, Ibidis, Johannesburg.

Pires, J.M., Mendes, F.R., Negri, G., Duarte-Almeida, J.M., Carlini, E.A. 2009. Antinociceptive peripheral effect of *A. millefolium* L. and *Artemisia vulgaris* L.: Both plants known popularly by brand names of analgesic drugs. *Phytother. Res.*, 23, 212–219.

Ross, J. 2003. *Combining Western Herbs and Chinese Medicine: Principles, Practice, and Materia Medica*, pp. 165–181, Greenfields Press, Seattle.

Şabanoğlu, S. 2016. Türkiye'de yetişen bazı *Achillea* L. türleri üzerinde farmakognozik araştırmalar, YL tezi, Farmakognozi Anabilim Dalı, Ankara Üniv. Sağlık Bilimleri Enstitüsü, Ankara.

Sedighi, M., Nasri, H., Rafieian-Kopaei, M., Mortazaei, S. 2013. Reversal effect of *Achillea millefolium* extract on ileum contractions. *J. Herb Med. Pharmacol.*, 2(1), 5–8.

Sevindik, H., Güvenalp, Z., Yerdelen, K.O., Yuca, H., Demirezer, L.Ö. 2015a. Research on drug candidate anticholinesterase molecules from *Achillea biebersteinii* Afan. using by molecular docking and *in vitro* methods. *Med. Chem. Res.*, 24(11), 3794–3802.

Sevindik, H.G., Güvenalp, Z., Yerdelen, K.O., Yuca, H., Demirezer, L.Ö. 2015b. The discovery of potential anticholinesterase compounds from *Achillea millefolium* L. *Ind. Crop. Prod.*, 76, 873–879.

Sezik, E., Yeşilada, E., Honda, G., Takaishi, Y., Takeda, Y., Tanaka, T. 2001. Traditional medicine in Turkey X. Folk medicine in Central Anatolia. *J. Ethnopharmacol.*, 75(2–3), 95–115.

Sezik, E., Yeşilada, E., Shadidoyatov, H., Kulivey, Z., Nigmatullaev, A.M., Aripov, H.N., Takaishi, Y., Takeda, Y., Honda, G. 2004. Folk medicine in Uzbekistan. I. Tashkent, Djizzax, and Samarqand provinces. *J. Ethnopharmacol.*, 92(2–3), 197–207.

Sezik, E., Yeşilada, E., Tabata, M., Honda, G., Takaishi Y., Fujita, T., Tanaka, T., Takeda, Y. 1997. Traditional medicine in Turkey VIII. Folk medicine in east Anatolia; Erzurum, Erzincan, Agri, Kars, Igdir province. *Econ. Bot.*, 51, 195–211.

Simid, N., Andjelkovid, S., Palid, R., Vajs, V., Milosavljevid, S. 2000. Volatile constituents of *Achillea serbica* Nym. *Flavour & Fragr. J.*, 15(3), 141–143.

Tabanca, N., Demirci, B., Gürbüz, I., Demirci, F., Becnel, J.J., Wedge, D.E., Başer, K.H.C. 2011. Essential oil composition of five collections of *Achillea biebersteinii* from central Turkey and their antifungal and insecticidal activity. *Nat. Prod. Comm.*, 6(5), 701–706.

Tabata, M., Sezik, E., Honda, G., Yeşilada, E., Fukui, H., Goto, K., *et al.* 1994. Traditional medicine in Turkey III. Folk medicine in East Anatolia, Van and Bitlis Provinces. *Int. J. Pharmacog.*, 32(1), 3–12.

Tanaka, T., Sezik, E., Yeşilada, E., Honda, G., Takaishi, Y., Takeda, Y. 1995. Traditional medicine in Turkey. V. Folk medicine in the inner Taurus mountains. *J. Ethnopharmacol.*, 46(3), 133–152.

Tozyo, T., Yoshimura, Y., Sakurai, K., Uchida, N., Takeda, Y., Nakai, H., Ishii, H. 1994. Novel antitumor sesquiterpenoids in *Achillea millefolium*. *Chem. Pharm. Bull.*, 42, 1096–1100.

Tuzlacı, E., İşbilen, D.F.A., Bulut, G. 2010. Turkish folk medicinal plants, VIII: Lalapaşa (Edirne). *Marmara Pharm. J.*, 14, 47–52.

Tzakou, O., Couladis, M., Verykokidou, E., Loukis, A. 1995. Leaf Flavonoids of *Achillea ligustica* and *Achillea holosericea*. *Biochem. Syst. and Ecol.*, 23(5), 569–570.

Tzakou, O., Loukis, A. 2009. Chemical composition of the essential oil of *Achillea umbellata* growing in Greece. *Nat. Prod. Res.*, 23(3), 264–270.

Tzakou, O., Loukis, A., Argyriadou, N. 1993. Volatile constituents of *Achillea crithmifolia* flowers from Greece. *J. Ess. Oil Res.*, 5(3), 345–346.

Tzakou, O., Loukis, A., Verykokidou, E., Roussis, V. 1995. Chemical constituents of the essential oil *Achillea ligustica* All. from Greece. *J. Ess. Oil Res.*, 7(5), 549–550.

Uzay, G. 2017. Türkiye'de yetişen *Achillea* L. (Asteraceae) türlerinin moleküler filogenetik analizi, YL tezi, İnönü Üniv. Sağlık Bilimleri Enstitüsü, Malatya.

Yeşilada, E., Honda, G., Sezik, E., Tabata, M., Goto, K., Ikeshiro, Y. 1993. Traditional medicine in Turkey IV. Folk medicine in the Mediterranean subdivision. *J. Ethnopharmacol.*, 39(1), 31–38.

2 *Arnebia* sp.

Merve Yüzbaşıoğlu Baran and Ayşe Uz

BOTANICAL FEATURES

The genus *Arnebia* belongs to the Boraginaceae family, which has about 2000 species in the world. In annual and perennial plants, purple dye is obtained from their rhizome. Stem simple or branched, pilose, hard or velvety hairy. Inflorescence a dense terminal head in perennials, numerously branched on upper stems in annuals; simos bract. Calyx deeply segmented; lobes linear to lanceolate. Corolla yellow, diffuse laminated tubular to funnel-shaped; corolla throat without scales and folds, ringed or absent. In flowers, the styli are of different shapes, the stamens are inset, protruding in a ring, or irregular at different levels. The stylus is a two-lobed biphyte with inset, straight, or biphyte branched. Nutlets flat ovoid or hemispherical to triangular, erect, acute to acuminate, with beak and ventral spine (Davis et al., 1965; Güner et al., 2000).

Arnebia decumbens (Vent.) Coss. and Kralik, *A. densiflora* Ledeb., *A. linearifolia A. DC., A. pulchra* (Roemer & Schultes) Edmondson, and *A. purpurea* S. Erik & H. Sümbül are among the five species included in the Flora of Turkey (Edmondson, 1978; Sümbül, 1988). *Arnebia pulchra*, a perennial herb, has been reassigned to the genus *Huynhia* as a result of anatomical, morphological, and phylogenetic research (Coppi et al., 2015). *A. purpurea* is distinguished by its purple flower and is the only rare endemic species (Yuzbasioglu et al., 2015).

Ethnobotanical Usage

In traditional Turkish medicine, the bark roots of *Arnebia densiflora* have been used for their ameliorative effects on wounds and burns (Baytop, 1999; Sezik et al., 1991). In an ethnobotanical study conducted in Kars, Iğdır region, it was reported that the ointment prepared by mixing the crushed wood of pine species and root bark of *A. densiflora* in butter was used in wound healing and burn treatment (Sezik et al., 1997). Also, the root barks of the plant, commonly known as "havaciva kökü, eyilik otu, kırmızı kök, eğnik, enlik", are used externally for the healing of calluses and burns and as a fabric dye in Turkey (Baytop, 1994; Baytop, 1999; Sezik et al., 1991).

The red colored root and root barks of *Arnebia densiflora* are used externally for wound healing and burn treatment among the people in Turkey. Another ointment is prepared from the roots of this plant are used topically for its wound healing activity. For preparing this balm, first, roots are cooked in butter and filtered, and then the drained oil is combined with pistachio gum, wax, and mistletoe over a fire to make a paste. The mixture is known as "eğlik paste" when honey is added. Internally, this material is used to treat tuberculosis and gastrointestinal ulcers. In the Eskişehir area, the boiled plant with wax is used to reduce arm and leg muscular tightness after removing a plaster bandage (Başer et al., 1986). The roots are a blue-purple dye source for woolen fabric in Aşağıkepen Village, Eskişehir, Turkey, according to a record in a PhD study (Yüzbaşıoğlu, 2016).

PHYTOCHEMICALS

The red colored (S)-alkannin and (R)-shikonin derivatives are the main compounds of the roots of these species growing in Turkey (Table 2.1). In addition to naphthoquinones, a benzoquinone derivative, 3,6-dihydroxy-2-isovaleryl 1,4 benzoquinone, a sterol type compound, stigmasterol and

DOI: 10.1201/9781003146971-2

TABLE 2.1
Alkannin/Shikonin Derivative Naphthoquinones

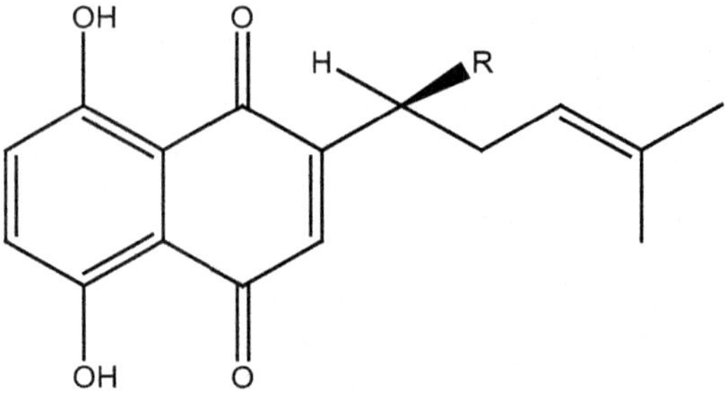

Names	R	Species	References
(S)-Alkannin	OH	*A. purpurea* *A. densiflora*	Yuzbasioglu et al. (2015); Yuzbasioglu Baran et al. (2020)
Deoxyalkannin	H	*A. decumbens*	Afzal and Al-Oriquat (1986)
(S)-Acetyl alkannin	$OCOCH_3$	*A. purpurea,* *A. densiflora*	Yuzbasioglu et al. (2015) Yuzbasioglu-Baran et al. (2020)
(S)-β,β-Dimethyl acryl alkannin	$OCOCH{=}C(CH_3)_2$	*A. densiflora*	Bozan et al. (1997,1999)
(S)-Isovaleryl alkannin	$OCOCH_2CH(CH_3)_2$	*A. purpurea,* *A. densiflora*	Yuzbasioglu et al. (2015) Yuzbasioglu-Baran et al. (2020)
(S)-β-Acetoxy isovaleryl alkannin	$OCOCH_2C(CH_3)_2$ OCH_3	*A. densiflora*	Yuzbasioglu-Baran et al. (2020)
Teracyrlalkannin	$OCOCH_2CH{=}C(CH3)_2$	*A. densiflora*	Bozan et al. (1997, 1999)
α-Methyl-*n*-butyl alkannin	$OCOCH(CH_3)CH_2CH_3$	*A. purpurea,* *A. densiflora*	Yuzbasioglu et al. (2015) Yuzbasioglu-Baran et al. (2020), Bozan et al. (1997, 1999)
(S)-Isobutyl alkannin	$OCOCH(CH_3)_2$	*A. purpurea,* *A. densiflora*	Yuzbasioglu et al. (2015) Bozan et al. (1997,1999)
2',3'-Epoxy alkannin	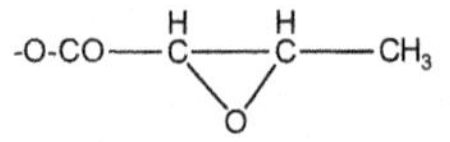 -O-CO—C(H)—C(H)—CH_3 (C–O–C epoxide)	*A. densiflora*	Bozan et al. (1997, 1999)
4-hydroxy-4-methyl valeryl alkannin	$OCOCH_2CH_2C(CH_3)_2$ OH	*A. densiflora*	Yuzbasioglu-Baran et al. (2020)
(R)-Isovaleryl shikonin	OCOCH2CH(CH3)2	*A. decumbens*	Afzal and Al-Oriquat (1986)
(R)-Acetyl shikonin	OCOCH3	*A. decumbens*	Afzal and Al-Oriquat (1986)

pyrrolizidine type alkaloids, O[9]-angeloyl retronesine, O[7]-angeloyl retronesine, 7-tigloyl retronesine, 9-tigloyl retronesine, supinine, heliothrine, likopsamine rinderine, and europine were yielded from *A. decumbens* (Hu et al., 2006; El-Dahmy and Ghani, 1995; El-Shazly and Wink, 2014).

3-*O*-Acetyl oleanolic acid, a triterpene, and a plant sterol β-sitosterol were isolated from the roots of *A. purpurea*. Moreover, three flavonoids, isorhamnetin-3-*O*-rutinoside, kaempferol-3-*O*-rutinoside, and kaempferol 3-*O*-β-(5''-acetyl) apiofuranoside-7-*O*-α-rhamnopyranoside, and a phenolic acid, rosmarinic acid, were obtained from the aerial parts of this endemic plant (Yuzbasioglu et al., 2015).

Many hydrocarbons, myristic acid, pentalene, nerolidolei heptadekan, bicyclo-3,1,1-heptan,2,6, 6-trimethyl-3(2-propenile), 3,3-dimethyl-hexane, dokozan, 2-methyl-undekan, hekzatraikontan, tetratetrakontan, phytol, and a naphthoquinone derivative, were isolated from the roots of *A. linearifolia* (Al-Mazroa et al., 2015; Yüzbaşıoğlu, 2016).

Also, a recent HPLC study revealed some phenolic compounds like rosmarinic acid, gallic acid, rutin, apigenin, quercetin, p-hydroxybenzoic acid, cinnamic acid, chlorogenic acid, luteolin, and benzoic acid in the methanol extract of the *Arnebia densiflora* aerial parts (Zengin et al., 2018).

Also, Çölgeçen et al. produced colored cultures in callus and suspension cultures of *Arnebia* and analyzed their chemical profiles spectrophotometrically (Çölgeçen et al., 2011).

BIOLOGICAL ACTIVITIES

In the previous studies on *Arnebia* and their isolated compounds, antitumoral, anti-inflammatory, prostaglandin synthesis inhibitory activity, antimicrobial, antiviral, platelet aggregation inhibitory, wound healing, and contraceptive effects were indicated (Yüzbaşıoğlu and Kuruüzüm-Uz, 2011).

Cytotoxic Effects

Isolated alkannin derivatives from *A. purpurea* roots were analyzed by MTT colorimetric assay which revealed significant cytotoxic activity on L929 cell line (Yuzbasioglu et al., 2015). The cytotoxic effects of the *n*-hexane extract of *Arnebia densiflora* roots and root barks (ADH) and its alkannin derivative compounds were examined against L929 murine fibrosarcoma, HEp-2 human laryngeal carcinoma, and HeLa human cervix carcinoma cell lines by *in vitro* MTT method. According to the results, ADH exhibited the strongest effect on L929 cell lines compared with the positive control doxorubicin (IC_{50} for ADH = 9.03 ζg/mL, IC_{50} for Dox = 6.26 ζg/mL). Alkannin derivatives showed low to moderate effects which suggests that they have synergistic effects while in the ADH (IC_{50} range 26.34–172.35 ζg/mL). On HEp-2 cells, ADH showed the strongest effect (IC_{50} = 4.96 ζg/mL) and its alkannin derivatives exhibited moderate-strong cytotoxic effects (IC_{50} value range = 8.48–34.33 ζg/mL). Also, ADH and alkannins demonstrated moderate-strong and concentration-dependent effects on HeLa cells (IC_{50} range: 14.35–30.72 ζg/mL). In the same research, cytotoxic activity mechanisms of the ADH and alkannins on HeLa cells were investigated. Using light microscope, DNA laddering, and caspase-3 activity assays, the apoptotic effects of ADH and alkannins were detected. All samples were administered at their IC_{50} concentrations to HeLa cells that had been grown for 24 hours. After 6 hours, all cells were scraped and morphological changes such as blebbing and the production of apoptotic bodies of each sample were noticed under the microscope. Another hallmark of apoptosis was observed with DNA laddering assay. According to the agarose gel image that was visualized under a UV illuminator, ADH and all alkannins have low molecular weight DNA fragments. After administering the IC_{50} dosage of all samples for 6 hours, the caspase 3 activity was increased for all samples. This activity has been shown to be induced at a maximum of three times that of ADH. The results revealed that all samples induced caspase-dependent apoptotic death in HeLa cells via increasing caspase-3 activity (Yuzbasioglu Baran et al., 2020).

Wound Healing Effects

A study was conducted to observe the wound healing effect of *A. densiflora* root extract on mouse palate mucosa. 10 mm thick mucosal wounds were created using a scalpel in the midline of the palate of mice. An ointment prepared with 10% *A. densiflora* root extract was applied topically once a day to the wounds in the experimental group. Histological examinations were performed by taking tissue samples from the wounds on the 4th, 7th, 14th, and 21st days. It was shown that 10% of *A. densiflora* root extract have wound healing effect in the experimental group (Kosger et al., 2009).

The *n*-hexane, chloroform, ethyl acetate, and methanol extracts and ointments containing 1% of each extract were prepared from *A. densiflora* roots. Wound healing effects of the extracts and ointments were investigated on the linear incision and excision wound models that were created on rats. The ointment containing 1% *n*-hexane extract showed comparably high efficacy to the reference ointment Madecassol in both models (Akkol et al., 2009). Orhan et al. (2018) investigated the effects of *A.*

purpurea extract on the viability of the skin flaps in rats. In this study, it was shown that the necrosis of skin flaps was reduced through topical application of *A. purpurea* extract (Orhan et al., 2018).

Antioxidant Activity

The methanol and water extracts of the *A. densiflora* roots exhibited a low-moderate 2,2-diphenyl-1-picrylhydrazyl (DPPH) radical scavenging effect at 2,000 μg/mL concentration (Orhan et al., 2011). The methanolic extract of the aerial parts of *A. densiflora* (ADM) was investigated for its *in vitro* antioxidant capacity using a range of assays, namely DPPH, ABTS, FRAP, CUPRAC, β-carotene/linoleic acid assay, phosphomolybdenum method, and metal chelating activity. The results of the assays revealed that *A. densiflora* has moderate-strong radical scavenging and antioxidant activities. While in the DPPH assay, 39.81 ± 1.21 mg Trolox equivalent (TE)/g ADM showed moderate activity, ABTS radical cation scavenging activity exhibited a strong effect with 115.15 ± 9.07 mg TE/g ADM. Antioxidant capacity tests, FRAP (68.14 ± 2.15 mg TE/g ADM) and CUPRAC (1.02 ± 2.35 mg TE/g ADM) showed moderate-strong activity, which is consisting with the phenolic content of the extract. ADM showed significant antioxidant activity (83.51% inhibition at 2 mg/mL concentration) in β-carotene/linoleic acid assay which is used to screen antioxidant activity. Butylated hydroxytoluene, a strong synthetic-lipophylic antioxidant, has a 97.32% inhibition value at the same concentration compared with the ADM. The phosphomolybdate method assay, which is another total antioxidant capacity test, exhibited low activity with the 1.41 ± 0.07 mmol TE/g ADM. Moreover, metal chelating activity test was conducted, which measures the antioxidant capacity by color reduction activity of the ADM that compete with ferrozine for the ferrous ions, as a result ADM showed weak chelating power (17.60 mg EDTAE/g ADM) (Zengin et al., 2018).

In Vitro Enzyme Inhibition Tests

Acetylcholinesterase (AChE) and Butrylcholinesterase Enzyme (BChE) Inhibition Tests

AChE and BChE are significant enzymes for the treatment of Alzheimer's disease, which is characterized by a cholinergic deficit. Orhan et al. (2011) investigated the AChE inhibition levels of the chloroform, ethyl acetate, methanol, and water extracts of *Arnebia densiflora* roots and root barks by using the Ellman method. Only *A. densiflora* chloroform extract exhibited mild inhibition value, 49.6 % ± 1.69 at 62.5 μg/mL. In another study, both AChE and BChE inhibition were tested on the methanol extract of *A. densiflora* aerial parts. The test results showed that *A. densiflora* has low activity against AChE with 1.63 ± 0.08 mg GALAE (Standard galantamine equivalent)/g extract and BChE with 1.11 ± 0.16 mg GALAE/g (Zengin et al., 2018).

Tyrosinase Enzyme Inhibitory Activity Test

Tyrosinase enzyme inhibitors are enormous potential drug candidates for hyperpigmentation diseases and melanogenesis. Therefore, the tyrosinase inhibitory activity of the *A. densiflora* methanol extract was assessed by comparing the standard kojic acid. The test results revealed that the extract showed low-moderate inhibitory activity against tyrosinase enzyme (20.98 ± 1.60 g kojic acid equivalent/g extract) (Zengin et al., 2018).

Carbohydrate Hydrolyzing Enzymes Inhibition Tests

Carbohydrate hydrolyzing enzymes such as α-amylase and α-glucosidase are key components of postprandial plasma glucose level. Natural compounds that inhibit these enzymes are potential regulators of hyperglycemia which by reducing the carbohydrate breakdown decrease the postprandial plasma glucose peaks. In a recent study, the methanolic extract of *A. densiflora* was investigated for its α-amylase and α-glucosidase inhibitory activities. The extract showed low carbohydrate hydrolyzing enzyme inhibitory activity (0.83 ± 0.04 mmol ACAE/g extract for α-amylase, 4.98 ± 0.17 mmol ACAE/g extract for α-glucosidase) compared with the standard acarbose (Zengin et al., 2018).

In Vitro DNA Protective Effects against Hydroxyl Radical

The hydroxyl radical may cause oxidative damage to DNA, which can result in a variety of negative consequences, such as genetic mutation, carcinogenesis, and cell death. The *in vitro* DNA protective effects of the *A. densiflora* methanolic extracts at various concentrations (25–400 µg/mL) against hydroxyl radical damage were studied. The methanolic extract of *A. densiflora* was shown to be dose-dependently efficient in preserving DNA when DNA bands were quantified. At a concentration of 100 ζg/mL, *A. densiflora* extract significantly reduced DNA damage (Zengin et al., 2018).

Antimicrobial Activity

The antimicrobial activity of petroleum ether (ADPE), hexane (ADH), chloroform (ADCHL), ethyl acetate (ADETOAC), *n*-butanol (ADBUT), and ethanol (ADETOH) extracts of the *A. densiflora* root barks was tested against different antibiotic-resistant strains of *Klebsiella pneumoniae* by using broth microdilution method. All types of extracts were active either at the concentration of 32 ζg/mL or 64 ζg/mL with similar results for the standard antibiotics – ampicilline and ofloxacin (Koca et al., 2010). In another research, the methanol extract of the *A. densiflora* aerial parts (ADM) was tested against different bacteria, yeast, and fungi strains. It has been found that ADM has low-moderate antibacterial activity against *Agrobacterium tumefaciens* (MIC 0.625 mg/mL) and *Azotobacter chroococum* (MIC 1.25 mg/mL). Also, the ADM was effective against *Fusarium oxysporum* and *Phialophora fastigiata* with MIC 2.5 mg/mL (Zengin et al., 2018).

CONCLUSION

The genus of *Arnebia* belongs to Boraginaceae family and is represented by four species, one of which is endemic to Turkey. *Arnebia* in Anatolia has been known as "eyilik otu, kırmızı kök, eğnik, enlik, havaciva" in Turkish. Because of their active ingredients such as naphthoquinones, benzoquinones, steroids, triterpenes, and flavonoids, they have been the subject of many studies over the years. This genus is commonly used in Eastern medicine and was included in many patented preparations as registered in the Chinese Pharmacopoeia. Additionally, naphthoquinones are important as a natural colorant in the cosmetic industry. The root and root bark of *A. densiflora* known as red root soaked in olive oil is used externally in wound healing and burn treatment in Anatolia for years. Besides the ethnobotanical usage of *A. densiflora* in wound and burn treatment, it was found that its naphthoquinones also have enormous cytotoxic activity. However, the information on the phytochemical and biological activity of the other species is still restricted. It is important to carry out studies on the large-scale production of *Arnebia* species, a valuable source of naphthoquinone derivatives used in the fields of medicine and cosmetics, with biotechnological techniques or good agricultural and collection practices.

Obtaining effective components and derivatives by synthesis will also enable medicinal and cosmetic product preparation in the future.

REFERENCES

Afzal, M., Al-Oriqat, G. (1986). Shikonin derivatives. V. Chemical investigations of *Arnebia decumbens*. *Agricultural and Biological Chemistry*, 50(3), 759–760.

Akkol, E., Koca, U., Pesin, I., Yılmazer, D., Toker, G., Yesilada, E. (2009). Exploring the wound healing activity of *Arnebia densiflora* (Nordm.) Ledeb. by in vivo models. *Journal of Ethnopharmacology*, 124(1), 137–141.

Al-Mazroa, S.A., Al-Wahaibi, L.H., Mousa, A.A., Al-Khathlan, H.Z. (2015). Essential oil of some seasonal flowering plants grown in Saudi Arabia. *Arabian Journal of Chemistry*, 8(2), 212–217.

Başer, K.H.C., Honda, G., Miki, W. (1986). *Herb Drugs and Herbalists in Turkey.* Tokyo: Institute for the Study of Languages and Cultures of Asia and Africa 27:83.

Baytop, T. (1994). *Türkçe Bitki Adları Sözlüğ ü.* Ankara: Atatürk Kültür, Dil Ve Tarih Yüksek Kurumu.

Baytop, T. (1999). *Geçmişte ve Bugün Türkiye'de Bitkiler ile Tedavi.* İstanbul: Nobel Tıp Kitabevi Ltd Şti.

Bozan, B., Başer, K., Kara, S. (1997). Quantitative determination of naphthaquinones of *Arnebia densiflora* (Nordm.) Ledeb. by an improved high-performance liquid chromatographic method. *Journal of Chromatography A*, 782(1), 133–136.

Bozan, B., Baser, K.H.C., Kara, S. (1999). Quantitative determination of naphthoquinones of *Arnebia densiflora* by TLC-densitometry. *Fitoterapia*, 70(4), 402–406.

Çölgeçen, H., Koca Çalışkan, U., Toker, G. (2011). Influence of different sterilization methods on callus initiation and production of pigmented callus in *Arnebia densiflora* Ledeb. *Turkish Journal of Biology*, 35(4), 513–520.

Coppi, A., Cecchi, L., Nocentini, D., Selvi, F. (2015). *Arnebia purpurea* a new member of formerly monotypic genus Huynhia (Boraginaceae-Lithospermeae). *Phytotaxa*, 204(2), 123.

Davis, P.H., Cullen, J., Coode, M.J.E. (1965). *Flora of Turkey and the East Aegean Islands.* Edinburgh: University Press.

Edmondson, J.R. (1978). *Arnebia* Forssk. P.H. Davis (Ed.). *Flora of Turkey and the Aegean Islands.* Edinburgh: Edinburgh University Press.

El-Dahmy, S., Ghani, A.A. (1995). Alkaloids of *Arnebia decumbens. Al-Azhar Journal of Pharmaceutical Sciences*, 15, 24–34.

El-Shazly, A., Wink, M. (2014). Diversity of pyrrolizidine alkaloids in the Boraginaceae structures, distribution, and biological properties. *Diversity*, 6(2), 188–282.

Güner, A., Özhatay, N., Ekim, T., Başer, K. (2000). *Flora of Turkey and the East Aegean Islands.* First Supplement. Edinburgh: University Press.

Hu, Y., Jiang, Z., Leung, K.S., Zhao, Z. (2006). Simultaneous determination of naphthoquinone derivatives in Boraginaceous herbs by high-performance liquid chromatography. *Analytica Chimica Acta*, 577(1), 26–31.

Koca, U., Ozcelik, B., Ozgen, S. (2010). Comparative *in vitro* activity of medicinal plants *Arnebia densiflora* and *Ecballium elaterium* against isolated strains of *Klebsiella pneumoniae. Turkish Journal of Pharmaceutical Sciences*, 7(3), 197–204.

Kosger, H.H., Ozturk, M., Sokmen, A., Bulut, E., Ay, S. (2009). Wound healing effects of *Arnebia densiflora* root extracts on rat palatal mucosa. *European Journal of Dentistry*, 3(2), 96–99.

Orhan, E., Yüzbaşıoğlu, M., Erol, Y.R., Bilensoy, E., Erdoğar, N., Demirezer, L.Ö., Kuruüzüm-Uz, A. (2018). The effect of *Arnebia purpurea* extract on the survival of random pattern skin flaps in rats. *The European Research Journal*, 4(3), 152–156.

Orhan, I., Şenol, F.S., Koca, U., Erçetin, T., Toker, G. (2011). Evaluation of the antioxidant and acetylcholinesterase inhibitory activities of *Arnebia densiflora* Ledeb. *Turkish Journal of Biology*, 35(1), 111–115.

Sezik, E., Tabata, M., Yesilada, E., Honda, G., Goto, K., Ikeshiro, Y. (1991). Traditional medicine in Turkey. 1. Folk medicine in Northeast Anatolia. *Journal of Ethnopharmacology*, 35(2), 191–196.

Sezik, E., Yesilada, E., Tabata, M., Honda, G., Takaishi, Y., Fujita, T., Fujita, T., Tanaka, T., Takeda, Y. (1997). Traditional medicine in Turkey. 8. Folk medicine in East Anatolia; Erzurum, Erzincan, Ağrı, Kars, Igdir provinces. *Economic Botany*, 51(3), 195–211.

Sümbül, S.E.H. (1988). *Arnebia purpurea.* P.H. Davis, R.R. Mill, ve K. Tan (Eds.). *Flora of Turkey and the East Aegean Islands.* Edinburgh: Edinburgh University Press.

Yüzbaşıoğlu, M., Kuruüzüm-Uz, A. (2011). *Arnebia* Forssk. türlerinin kullanışları ve biyolojik aktiviteleri. *Hacettepe University Journal of The Faculty of Pharmacy*, 32(1), 91–106.

Yüzbaşıoğlu, M., Kuruüzüm-Uz, A., Güvenalp, Z., Simon, A., Tóth, G., Harput, U.S., Kazaz, C., Bilgili, B., Duman, H., Saraçoğlu, İ., Demirezer, L.Ö. (2015). Cytotoxic compounds from endemic *Arnebia purpurea. Natural Product Communications*, 10(4), 595–596.

Yüzbaşıoğlu, M. (2016). *Pharmacognostical Researches on Arnebia Species Growing in Turkey.* Hacettepe University, Institute of Health Sciences, Pharmacognosy Programme, PhD Thesis, Ankara, Turkey.

Yuzbasioglu Baran, M., Guvenalp, Z., Saracoglu, I., Kazaz, C., Salih, B., Demirezer, L.O., Kuruuzum-Uz, A. (2020). Cytotoxic naphthoquinones from *Arnebia densiflora* (Nordm.) Ledeb and determining new apoptosis inducers. *Natural Product Research*, 34(2), 1669–1677.

Zengin, G., Ceylan, R., Katanić, J., Aktumsek, A., Matić, S., Boroja, T., Stanić, S., Mihailović, V., Seebaluck-Sandoram, R., Mollica, A., Mahomoodally, M.F. (2018). Exploring the therapeutic potential and phenolic composition of two Turkish ethnomedicinal plants–*Ajuga orientalis* L. and *Arnebia densiflora* (Nordm.) Ledeb. *Industrial Crops and Products*, 116, 240–248.

3 *Astragalus* sp.

Nilgun Yakuboğulları and Erdal Bedir

INTRODUCTION

Astragalus is one of the largest genera in Turkey and is widely distributed worldwide. One of the properties of *Astragalus* species is their well-known use in traditional medicine for the treatment of diabetes, nephritis, cold, cardiovascular diseases, leukemia, uterine cancer as well as fatigue. The main active constituents of *Astragalus* are saponins, flavonoids, and polysaccharides [1]. As *Astragalus* polysaccharides originating from the roots have been used in Chinese medicine for their strong immunomodulatory activities [2], numerous papers have been published on this topic over the last 50 years. However, a few articles have revealed the immunostimulatory properties of *Astragalus* saponins. Here, we present the summary of 25 years of research on Turkish *Astragalus* species from the isolation of saponins to the evaluation of their immunomodulatory activities, and finally bring forth a compound to be used as a vaccine adjuvant.

PHYTOCHEMISTRY OF ASTRAGALUS SPECIES

As the largest genus in the family of Fabaceae, *Astragalus* is widely distributed throughout the world, located primarily in Asia, North and South America, and Africa. In the flora of Turkey, *Astragalus* sp. is represented by 476 species in 63 sections with 51% endemism [3,4]. Thirty-one out of 476 *Astragalus* species, *A. amblolepis Fischer*, *A. angustifolius*, *A. aureus*, *A. baibutensis*, *A. brachypterus*, *A. campylosema* Boiss. ssp. *Campylosema*, *A. cephalotes* var. *Brevicalyx*, *A. elongatus*, *A. erinaceus*, *A. flavescens*, *A. gilvus* Boiss, *A. halicacabus* LAM, *A. hareftae* (NAB.) SIRJ., *A. icmadophilus*, *A. lycius* Boiss, *A. melanophrurius*, *A. microcephalus*, *A. oleifolius*, *A. pennatulus*, *A. plumosus* var., *A. prusianus*, *A. ptilodes* Boiss. var. *cariensis* Boiss., *A. pycnocephalus* var. *pycnocephalus*, *A. schottianus*, *A. stereocalyx* Bornm, *A. tauricolus*, *A. tmoleus* Boiss. var. *tmoleus*, *A. trojanus*, *A. vulneraria*, *A. wiedemannianus* Fischer, *A. zahlbruckneri*, and 14 different sections were investigated for their secondary metabolite profiles [5]. These studies revealed the presence of 17 new oleanane- [6–9] and 87 new cycloartane-type [5,7,9–24] triterpenoid glycosides, five new phenolic glycosides [25,26], a new maltol glycoside [12], a new flavonol glycoside [24], a new tryptophan derivative [6], apart from 63 known compounds. In the group of cycloartanes, three major sapogenin skeletons, viz. cycloastragenol, cyclocanthagenol, and cyclocephalogenol possessing 20,24-epoxy-, acyclic-, and 20,25-epoxy side-chains, respectively, were encountered [5].

THE JOURNEY FROM USE IN TRADITIONAL MEDICINE TO VACCINE ADJUVANT DISCOVERY

Astragalus extracts or saponins have been demonstrating several pharmacological activities such as antiprotozoal [13], antiviral [27], cardiotonic [28], neuroprotective [29], wound healing [30], etc. In this chapter, we will be focusing on the immunomodulatory activities of *Astragalus* saponins and their potential use in vaccines as immunological adjuvants.

DOI: 10.1201/9781003146971-3

PRELIMINARY IMMUNOMODULATORY ACTIVITY SCREENING STUDIES

In the Southeastern part of Turkey, the water extracts of *Astragalus* roots are used to cure leukemia in traditional medicine. To discover new pharmacologically active compounds inspired by its traditional use, *Astragalus* extracts and their saponins were investigated for their cytotoxic activities on cancer cell lines. The preliminary cytotoxicity panels revealed that neither *Astragalus* extracts nor their saponins had cytotoxicity. At this point, it was thought that the activity to rationalize the traditional use of *Astragalus* might be the stimulation of immune response. Thus, further studies were carried out to prove this hypothesis.

Firstly, Çalış et al. investigated the immunostimulatory properties of eight saponins isolated from the aqueous extracts of *A. melanophrurius* roots on human lymphocytes [21] as *Quillaja* and *Panax* saponins were reported for their immunomodulatory activities [31]. Based on their spectral data, these compounds were identified as Astrasieversianin, Astragalosides I and II, Astrasieversianin X, Astragaloside IV, Cyclocanthoside E, Astragaloside VI, and Cyclocanthoside G. To screen their immunostimulatory activities, human lymphocytes were isolated from heparinized human blood by Ficoll gradient centrifugation and treated with various concentrations of test compounds (0–200 μg/mL) and concanavalin A (ConA, 2 or 8 μg/mL). After 70 h incubation, the lymphocyte proliferation was assessed by liquid scintillation of radiolabeled thymidine, and an increase in the number of lymphocytes was presented as a percentage. Each compound can stimulate lymphocyte proliferation in a dose-dependent manner, especially at concentrations of 0.01–1 μg/mL. At higher concentrations up to 200 μg/mL, lymphocyte proliferation was inhibited. Based on structure-activity relationship analysis, cyclocanthogenol with an acyclic side-chain was more prominent in lymphocyte proliferation than 20,24-epoxy side-chain cycloastragenol. In detail, β-D-glucopyranose (β-D-Glc) moiety at C-6 on the aglycone backbone was essential for the stimulation of lymphocytes compared to β-D-xylopyranose (β-D-Xyl) [Astragaloside I (128 %, β-D-Glc) vs Astrasieversianin II (44 %, β-D-Xyl) and Astragaloside IV (71 %, β-D-Glc) vs Astrasieversianin X (53 %, β-D-Xyl)]. Moreover, aglycone-bearing monosaccharide or disaccharide substituent at C-3 and glucose moiety at C-6 altered the lymphocyte proliferation based on aglycone type and the number of sugar moieties at C-3. In acyclic-aglycone, Cyclocanthoside E (141 %, 3-β-D-Xyl) and Cyclocanthoside G (129 %, 3-β-D-Glc→ β-D-Xyl) were comparable to each other; however, monosaccharide nature increased the lymphocyte proliferation slightly compared to disaccharide substituent at 0.1 μg/mL concentration. In the case of aglycone deriving from 20,24-epoxy side-chain, Astragaloside VI (105 %, 3-β-D-Glc→ β-D-Xyl) augmented the lymphocyte proliferation in contrast to Astragaloside IV (32 %, 3-β-D-Xyl) at the concentration of 0.01 μg/mL. Interestingly, when the concentration was adjusted to 0.1 μg/mL, there was no difference in lymphocyte proliferation between Astragaloside IV and Astragaloside VI treatments, demonstrating the importance of treatment concentration.

The promising immunostimulatory activity results encouraged researchers to investigate 19 cycloartane-type triterpenoid glycosides in terms of NF-κB activation and pro-inflammatory cytokine production in macrophages [2]. The compounds and their origin were: Macrophyllosaponin B, Macrophyllosaponin C, and Macrophyllosaponin D from *A. oleifolius*; Askendoside G from *A. prusianus*; Cyclocanthoside E, Astragaloside IV, Cycloastragenol, Brachyoside B from *A. microcephalus*; Astrasieversianin II, Astragaloside I, Astragaloside II, Astragaloside VII, Trojanoside A, Trojanoside H from *A. trojanus*; Cyclocanthoside D, Cephalatoside A from *A. cephalotes*; Astrasieversianin X, Astragaloside VI from *A. melanophrurius*. Human leukemia monocytic cell line (THP-1) transfected with NF-κB reporter plasmid were treated with 50 ng/mL lipopolysaccharide (LPS) and 19 compounds at the concentration of 0.1 μg/mL, but none was active. When the concentration of compounds was changed and applied in the absence of LPS, only Astragaloside I (100 μg/mL) increased NF-κB-directed luciferase expression levels up to 65% compared to LPS alone (10 μg/mL). When the mRNA levels of IL-1β and TNF-α production were examined, Astragaloside I (200 μg/mL) was the only compound that augmented the mRNA expression of IL-1β and TNF-α in THP-1 cells. IL-1β is one of the pro-inflammatory cytokines, mainly produced by innate immune cells. It is also a key mediator of host immunity by influencing innate and adaptive immune responses [32]. On the other hand, TNF-α is a pleiotropic cytokine, and the inhibition of pathogenic effects of TNF and the preservation of its homeostatic activities are new research concepts in TNF-mediated diseases [33]. In assessed 19 compounds, two structural features are prominent for NF-κB activation and relevant cytokine production: (i) the presence of two acetyl groups on C-2(*O*) and C-3(*O*) of the β-D-xylopyranose moiety, and (ii) the presence of β-D-glucopyranose unit extending from C-6. The previous study carried out by Calis et al. (1997) demonstrated Cyclocanthoside E as one of the promising compounds for lymphocyte activation, but the same compound did not induce macrophage activation, implying different action mechanism(s) had a role in the immunomodulatory activities of *Astragalus* saponins.

The pro-inflammatory cytokine induction through Astragaloside I treatment led the researchers to investigate the cytokine release profiles of *Astragalus* species in human whole blood (hWB) [34]. As hWB contains diverse immune cells such as T cells, B cells, monocytes, dendritic cells, etc., this assay is a simple and effective approach to evaluate the immunomodulatory activities of the test compounds [35]. The test compounds and extracts were: Brachyoside A, Brachyoside C, Cyclocanthoside E, Cycloastragenol, Brachyoside B, Cyclocephaloside II, Astragaloside I, Astragaloside II, Astragaloside IV, Astragaloside VII, Trojanoside A, Trojanoside H, Cyclocephaloside I, Astrojanoside A (from *A. brachypterus*, *A. cephalotes*, *A. microcephalus*, and *A. trojanus*), and also the methanol extracts from the roots of *A. cephalotes*, *A. oleifolius*, and *A. trojanus*. hWB were treated with test materials (3 μg/mL) and phorbolacetate (PHA, 10 μg/mL) for IL-2, IFN-γ, IL-4, or LPS (1-30 μg/mL) for IL-1β, TNF-α, and IL-8 detection by using ELISA, and the cytokine production was presented as a percentage. IL-2 is a key growth factor for lymphocytes and is produced by mainly $CD4^+$ T cells, and to a small extent by $CD8^+$ T cells, natural killer (NK), natural killer T (NKT) cells, and at some point, by activated dendritic cells (DCs) under steady-state conditions. Upon exposure to an antigen, $CD4^+$ and $CD8^+$ T cells strongly produce IL-2 to transduce a further signal for the clearance of the pathogen [36]. When the cytokine induction was evaluated at the extract level, all methanol extracts increased IL-2 production, especially *A. oleifolius* being the most remarkable one (141.2%). On the other hand, the triterpene saponins were analyzed in detail revealing that all triterpene saponins had IL-2-inducing activity at different levels (35.9–139.6%). Based on aglycone type, 20,24-epoxy cycloartane-type saponins were more potent in IL-2 production than 20,25-epoxy cycloartane and acyclic-cycloartanes. Among triterpene saponins, the most promising compound was Astragaloside VII (139.6%), possessing a tridesmosidic structure and 20,24-epoxy side-chain. As the tridesmosidic structure of Astragaloside VII made a difference in IL-2 production, another tridesmosidic glycoside having an acyclic side-chain in cycloartane backbone (Brachyoside C) was not effective, signifying the importance of sugar moiety position and aglycone framework for IL-2 activity. One of the lowest scores for IL-2 production was obtained by Trojanoside A, which had C-16 acetyl functionality on Astragaloside IV structure, suggesting that the presence of free hydroxyl group at C-16 was also important for IL-2 inducing activity.

IL-4 mediates Th2 type immune response and has a role in fighting against helminth parasites and inactivation of toxins [37]. IL-4 was slightly augmented with the treatment of only Astragaloside VII, and it was concluded to be a special case for the compound. Other cytokines induced by the treatment of triterpene saponins were IL-1β and IL-8. IL-8 is a pro-inflammatory chemokine produced by macrophages, lymphocytes, mast cells as well as epithelial cells, and fibroblasts upon exposure to the inflammatory signals. IL-8 secretion in the microenvironment provides the activation and migration of neutrophils from peripheral blood to the site of infection that facilitates the clearance of pathogens and controls the maintenance of homeostatic balance [38,39]. Overall, almost all triterpene saponins induced the production of IL-1β and IL-8 at different levels except for a few compounds. In particular, the glycosides bearing an acyclic cycloartane-type aglycone [Brachyoside A (60.4% and 46%), Brachyoside C (52% and 40.8%), Cyclocanthoside E (64.8% and 50.6%)], the glycoside having 20,24-epoxy cycloartane type aglycone (Cyclocephaloside II, 48.9% and 32.7%) and aglycone itself (Cycloastragenol, 62%, and 46.6%) demonstrated substantially

Aglycone	Compound	R1	R2	R3	Activity
A 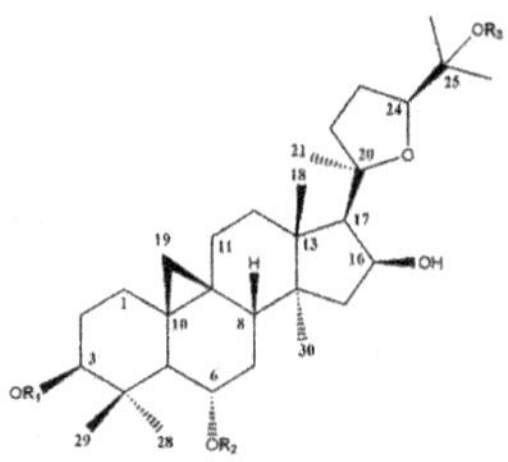	**Cycloastragenol**	H	H	H	• IL-1β ↑, IL-8 ↑, TNF-α ↓
	Astragaloside I	2',3'-*O*-acetyl-β-D-Xyl	β-D-Glc	H	• Macrophage Activation NF-κB ↑, IL-1β and IL-8 (mRNA) ↑ Moderate IL-2 and IL-8 ↑ Lymphocyte Proliferation ↑
	Astragaloside II	2'-*O*-acetyl-β-D-Xyl	β-D-Glc	H	• TNF-α ↓, IFN-γ ↓ Moderate IL-2 ↑ Lymphocyte Proliferation ↑
	Astragaloside IV	β-D-Xyl	β-D-Glc	H	• Moderate IL-1β ↑, IL-2 ↑ Moderate Lymphocyte Proliferation ↑ IL-1β, TNF-α, iNOS and IL-6 (mRNA) ↑ Expression of CD86 and CD40 ↑ Phosphorylation of p38, ERK and JNK ↑ Protein expression of p38, ERK and JNK ↓ Tregs ↓, Cytotoxic T Lymphocytes (CTLs) ↑
	Astragaloside VII	β-D-Xyl	β-D-Glc	β-D-Glc	• IL-2 and IL-4 ↑ Moderate IL-1β and IL-8 ↑ IFN-γ ↑, TGF-β ↑, IL-17A ↑ Splenocyte Proliferation ↑
	Brachyoside B	H	β-D-Glc	H	• TNF-α ↓
	Cyclocephaloside II	4'-*O*-acetyl-β-D-Xyl	β-D-Glc	H	• IL-2 ↑ Moderate IL-1β and IL-8 ↑
B 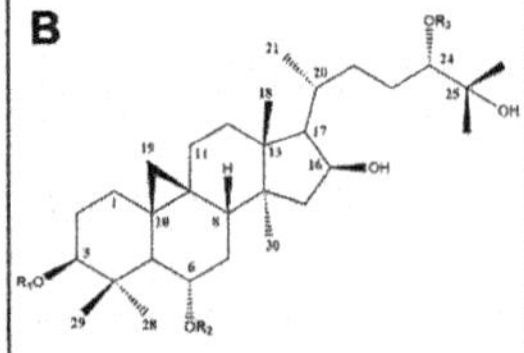	**Brachyoside A**	3'-*O*-β-D-Xyl → β-D-Xyl	β-D-Glc	H	• IL-1β ↑, IL-8 ↑
	Brachyoside C	H	β-D-Glc	β-D-Glc	• IL-1β ↑, IL-8 ↑
	Cyclocanthoside E	H	β-D-Glc	H	• IL-1β ↑, IL-8 ↑ Moderate IL-2 ↑ Lymphocyte Proliferation ↑
	Cyclocanthoside G	2'-*O*-β-D-Glc → β-D-Xyl	β-D-Glc	H	• Lymphocyte Proliferation ↑

FIGURE 3.1 The most potent *Astragalus* saponins isolated from Turkish *Astragalus* species, their structures and relevant immunomodulatory activities. **A)** 20,24-epoxy- [2,21,34,40–42]; **B)** acyclic side-chain [21,34] cycloartanes.

higher IL-1β and IL-8 secretion than other test compounds. Moreover, all triterpene saponins inhibited the production of TNF-α. In general, IFN-γ was not produced through the treatment with test compounds. A summary describing the most potent *Astragalus* saponins and their preliminary immunomodulatory activities is given in Figure 3.1.

In conclusion, a pro-inflammatory immune response was induced by the treatment of *Astragalus* saponins. A successful host immune response is obtained with balanced pro- and anti-inflammatory elements of the immune system that are organized to clear the pathogen and limit host damage [43]. As this balance is dynamic and responds to continuous feedback coming from pathogen and host, the type and magnitude of each cytokine will be altered to a different extent to maintain a steady-state condition.

POTENCY TO BE UTILIZED AS VACCINE ADJUVANT

Adjuvants are biological or chemical substances that increase antigen-specific immune responses by triggering and modulating both innate and adaptive immunity. Apart from their immunostimulatory effects, adjuvants also decrease the required dose of antigen and the number of immunizations, provide more rapid and long-lasting immunity, and increase the effectiveness of vaccines in poor responders [44].

Preliminary immunostimulatory activity screening studies carried out on *Astragalus* saponins and promising activity results led the researchers to investigate the potential of *Astragalus* saponins to be used as a vaccine adjuvant. For this purpose, Nalbantsoy et al. (2011) selected two potent cycloartane-type saponins, viz. Astragaloside VII (AST VII) and Macrophyllosaponin B (Mac B, the major saponin of the most active species *A. oleifolius*) in terms of their ability to produce IL-2.

Swiss albino mice were subcutaneously immunized with bovine serum albumin (BSA-100 μg) alone or in combination with AST VII (30 to 240 μg) and Mac B (30 to 120 μg). The immunostimulatory activity was evaluated in terms of cytokine and BSA-specific antibody production as well as stimulation of splenocyte proliferation. AST VII (120 and 240 μg) and Mac B (90 μg) stimulated the ConA-, LPS-, and BSA-induced splenocyte proliferation. BSA-specific IgG, IgG1, and IgG2b antibodies were increased by AST VII (120 μg) and Mac B (90 μg) treatments compared to BSA alone. The antibody levels were similar to Freund's adjuvant, a widely used oil-based adjuvant. Moreover, AST VII and Mac B augmented IFN-γ production over BSA alone. Interestingly, these compounds demonstrated minor hemolytic activity on rabbit red blood cells [45]. As the lysis of erythrocytes is one of the biological characteristics of saponins [46], AST VII and Mac B provide an advantage over other saponins for utilization in clinics. Overall, this study shows that AST VII and Mac B induce cellular and humoral immune responses and have the potential to be developed as a new class of saponin-based adjuvants.

A further attempt was made by Nalbantsoy et al. (2012) to investigate in detail the immunomodulatory and anti-inflammatory activities of AST VII and Mac B. Swiss albino mice were immunized subcutaneously and intraperitoneally by AST VII (120 μg) and Mac B (90 μg) alone or combination with LPS (12.5 μg) + AST VII (60 μg) and LPS (12.5 μg) + Mac (60 μg), respectively. The cytokines (IL-1β, TGF-β, TNF-α, IL-2, IFN-γ, and IL-4) and antibody (IgG, IgG1, IgG2b) responses were evaluated through the detection in mouse sera by ELISA. IgG1 is an indicator of Th2 whereas IgG2a antibody represents Th1-mediated immune response. AST VII (120 μg) and Mac B (90 μg) administration alone provided the production of IgG, IgG1, and IgG2b compared to physiological saline treatment, demonstrating Th1/Th2 balanced antibody response. Based on cytokine results, AST VII and Mac B alone did not alter IFN-γ or IL-4 levels but combining these compounds with LPS boosted the production of IFN-γ, IL-2, and TGF-β compared to LPS alone. There was no statistical difference in TNF-α, IL-1β, and IL-4 levels between saponin compounds and LPS. Th1-mediated cytokines such as IFN-γ and IL-2 are crucial for obtaining effector T cell function against viral infections or cancer, where the cellular immune response is needed. Moreover, the robust production of TGF-β by AST VII and Mac B treatments also highlighted their possible roles in immune regulation since TGF-β is one of the cytokines that controls innate and adaptive immunity [47]. To assess the anti-inflammatory response, NF-κB transcription, iNOS (inducible nitric oxide synthase), and NAG-1 (non-steroidal anti-inflammatory drug activated gene-1) activity were analyzed on cell-based assays. None of the compounds activated NAG-1 in human chondrosarcoma (SW1353) and colon adenocarcinoma cells (Caco-2 and HT-29) or NF-κB transcription in SW1353 cells. Only Mac B moderately inhibited iNOS activity (IC_{50}: 156 μg/mL) in murine macrophages (RAW 264.7), demonstrating no cytotoxicity on these cell lines [48].

Pathogens can enter the body through many routes and cause an infection in any part of the body, but eventually, antigens and lymphocytes will encounter each other in peripheral lymphoid organs such as lymph nodes, spleen, and mucosal lymphoid tissues [49]. In light of robust induction of Th1-mediated cytokines, early and later time activation markers (CD69 and CD25) were analyzed in spleen and lymph node tissue sections by immunohistochemistry staining. CD69 and CD25 showed moderate or weak cytoplasmic staining patterns at subcapsular zones and medulla, especially around the vessels by LPS (12.5 μg) + AST VII (60 μg) treatment. The subcapsular sinus allows the transportation of lymphatic fluid by receiving the afferent vessels and joining the medulla [50]. Moreover, LPS (12.5 μg) + AST VII (60 μg) and LPS (12.5 μg) + Mac B (60 μg) treatments demonstrated CD69 and CD25 immunostainings at spleen, especially in the subcapsular zone. As subcapsular zone and medulla sections are rich in $CD169^+$ macrophages [51], AST VII and Mac B may have a role in the activation of $CD169^+$ macrophages, similar to one of the *Quillaja* saponins (QS-21) [52].

Only a few adjuvants have been approved for use in human vaccines due to their limitations including lack of efficacy, unacceptable local or systemic toxicity, poor stability, and high manufacturing cost. Therefore, there is a huge need for novel, potent, and less toxic adjuvants. By the late

1980s, the adjuvant system concept was designed by simply combining different immunostimulatory agents into a delivery system, aiming to potentiate the immune response against a specific antigen [53]. In this perspective, an initial attempt was made to develop novel adjuvant systems, combining AST VII and its root partner *Astragalus* polysaccharide (APS) known as "gum tragacanth". To evaluate the adjuvant potential of AST VII along with APS, seasonal influenza antigen (H3N2) was selected as a model antigen. Adjuvant systems, which were 20–50 nm-sized nanovesicles, and AST VII (10, 25, 75 μg) simply admixture with inactivated H3N2 antigen (256 HA) and administered intramuscularly to Swiss albino mice. Four weeks after the immunization, the immune response was evaluated in terms of antibodies (IgG, IgG1, IgG2a) and cytokines (IL-2, IFN-γ, and IL-17A) responses as well as stimulation of splenocytes by comparing with QS-21, a golden standard of saponin-based adjuvant. Nanovesicles containing APS alone demonstrated significant Th2-driven (IgG1) and neutralizing antibody responses as well as modest protection in the embryos challenged with H3N2. The adjuvant system produced through the combination of APS plus AST VII elicited Th1/Th2 balanced antibody response along with remarkable proliferation on ConA- and LPS-restimulated splenocytes. AST VII alone (25 μg) showed a significant increase in IgG1 levels as well as a slight increase in IL-17A production. These results indicated that newly developed adjuvant systems have the potential to be used as an adjuvant in influenza vaccines in terms of mediating cellular and humoral immune responses [54]. Also, APS could be effectively used in an adjuvant system either as a delivery vehicle or an immunostimulating agent, besides its encapsulation properties. As a further study, the lipid nanoparticles (LNPs) were developed to deliver AST VII, aiming to increase the immunostimulatory activity of AST VII on the seasonal influenza (H3N2) vaccine *in vitro*. A novel adjuvant system containing polyethylene glycol (PEG300), cholesterol, and AST VII was prepared by solvent-assisted nanoprecipitation. The immunomodulatory activity of PEG300:Chol:AST VII LNPs was assessed in terms of cytokines release (IL-2, IFN-γ, and IL-17A) on phorbol 12-myristate 13-acetate (PMA)-ionomycin stimulated hWB and hemolytic activity on human erythrocytes. Eighty nm-sized spherical LNPs (3 and 6 μg/mL) alone or mixed with H3N2 (4 HA) were treated with hWB in parallel to AST VII and QS-21. PEG300:Chol:AST VII LNPs (3 μg/mL) alone increased IL-2 production, while AST VII (6 μg/mL) triggered the production of IL-2 and IFN-γ as compared to PMA-ionomycin. Co-treatment of adjuvant/adjuvant systems with H3N2 substantially decreased IL-2, IFN-γ, and IL-17A, but PEG300:Chol:AST VII LNPs (3 μg/mL) increased IFN-γ production in H3N2 administration. As IFN-γ is one of the crucial Th1 type cytokines and has a role in the clearance of viral antigens [55], an increase in IFN-γ level is worth developing PEG300:Chol:AST VII LNPs for viral vaccines. Only PEG300:Chol-AST-VII LNPs (250 μg/mL) showed hemolysis on human erythrocytes, indicating its immunostimulatory activity was not dependent on hemolysis [56].

After an initial attempt was made to develop a novel adjuvant system, further AST VII-based adjuvant systems were designed by formulating monophosphoryl lipid A (MPL), squalene, or *Astragalus* polysaccharide, and their immunomodulatory activities were assessed against Newcastle Disease vaccine (NDV). Swiss albino mice were immunized subcutaneously with NDV vaccines in the presence/absence of AST VII or developed adjuvant systems. In this study, AST VII (120 μg) administration both in live and inactivated vaccines induced the production of neutralizing and NDV-specific IgG, IgG1, and IgG2b antibody responses compared to AST VII administration only in inactivated NDV vaccine. These results have also revealed the effectiveness of adjuvant administration in live vaccines besides inactivated vaccines. Moreover, AST VII both in live and inactivated vaccines enhanced the splenocyte proliferation along with a slight increase in IL-2 and IL-4 production. The APS-based delivery system alone or its formulation with AST VII was prominent in neutralizing antibody production and provided the minor augmentation of IFN-γ and IL-2 levels. Squalene emulsion alone or its combination with AST VII was especially effective in NDV re-stimulated splenocyte proliferation. However, AST VII plus MPL formulation did not demonstrate advanced immune response in NDV vaccinations. Overall, AST VII alone or in an adjuvant system provided Th1/Th2 balanced antibody and cellular immune responses for NDV vaccines [57]. The

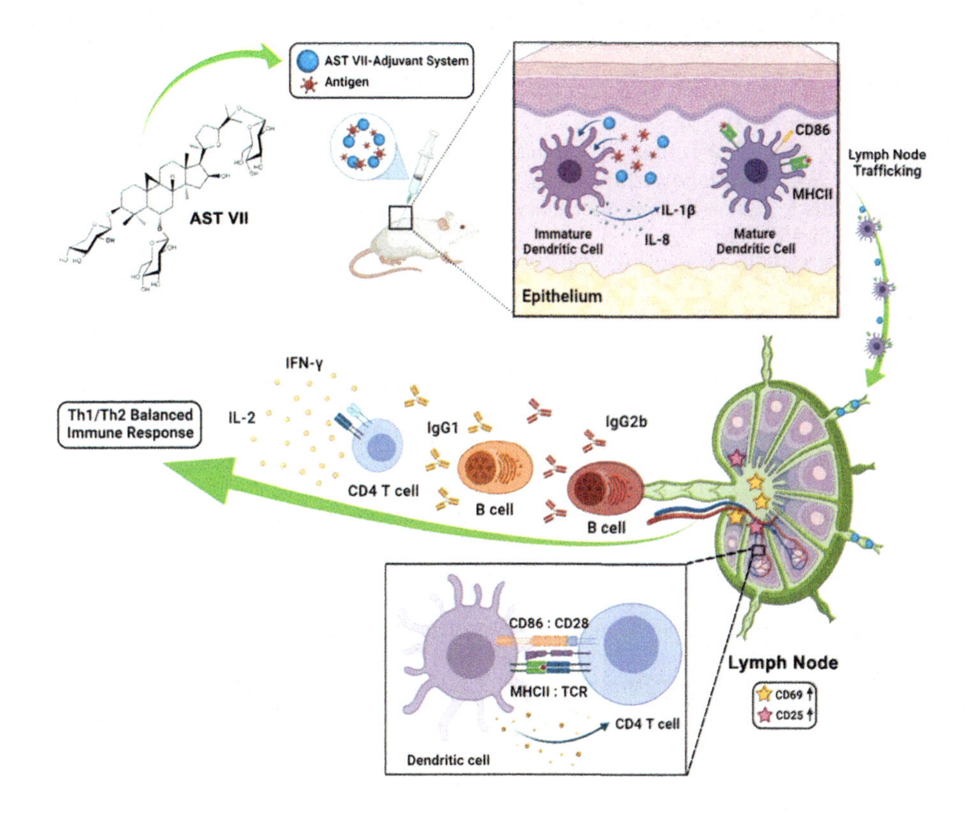

FIGURE 3.2 The overview of the immunomodulatory activities of AST VII and AST VII containing adjuvant systems. Created with BioRender.com.

summary of the immunomodulatory activities of AST VII alone or the AST VII-adjuvant system is illustrated in Figure 3.2.

CONCLUSION

Astragalus is one of the largest genera in the flora of Turkey including many endemic species. Up to now, the studies on Turkish *Astragalus* species gave rise to the isolation of 112 new compounds apart from 63 known compounds. The overriding basis for biological activity studies was the traditional use of *Astragalus* roots in the Southeastern region of Turkey to cure leukemia. As the isolated compounds did not demonstrate cytotoxicity, a hypothesis that the biological activity of *Astragalus* saponins might result from the activation of the immune system arises. Twenty-five years of research on the immunomodulatory effects of *Astragalus* saponins have shown that cycloartane-type triterpene saponins have balanced pro- and anti-inflammatory immune responses. Astragaloside VII, one of the most active compounds, and its formulation with other immunostimulating agents demonstrated humoral and cellular immune responses signifying its potential to be further developed as vaccine adjuvant/adjuvant systems for viral/bacterial/cancer vaccines as a new alternative to saponin-based adjuvants.

REFERENCES

1. Tang W, Eisenbrand G. *Chinese Drugs of Plant Origin*. Berlin: Springer-Verlag; 1992.
2. Bedir E, Pugh N, Calis I, et al. Immunostimulatory effects of cycloartane-type triterpene glycosides from astragalus species. *Biol Pharm Bull*. 2000;23:834–837.

3. Aytaç ZL. *Astragalus, Flora of Turkey and the East Aegean Islands*. Edinburgh: Edinburgh University Press; 2000.
4. Tunçkol B, Zeki A, Aksoy N, et al. Astragalus bartinense (Fabaceae), a new species from Turkey. *Acta Bot Croat*. 2020;79:131–136.
5. Gülcemal D, Aslanipour B, Bedir E. Secondary metabolites from Turkish Astragalus species. *Plant Hum Heal Phytochem Mol Asp*. 2019:43–97.
6. Bedir E, Çalis I, Aquino R, et al. Secondary metabolites from the roots of Astragalus trojanus. *J Nat Prod*. 1999;62:563–568.
7. Gülcemal D, Masullo M, Bedir E, et al. Triterpene glycosides from Astragalus angustifolius. *Planta Med*. 2012;78:720–729.
8. Horo I, Bedir E, Masullo M, et al. Saponins from Astragalus hareftae (NAB.) SIRJ. *Phytochemistry*. 2012;84:147–153.
9. Gülcemal D, Masullo M, Napolitano A, et al. Oleanane glycosides from Astragalus tauricolus: Isolation and structural elucidation based on a preliminary liquid chromatography-electrospray ionization tandem mass spectrometry profiling. *Phytochemistry*. 2013;86:184–194.
10. Horo İ, Bedir E, Perrone A, et al. Triterpene glycosides from Astragalus icmadophilus. *Phytochemistry*. 2010;71:956–963.
11. Çalış İ, Dönmez AA, Perrone A, et al. Cycloartane glycosides from Astragalus campylosema Boiss. ssp. campylosema. *Phytochemistry*. 2008;69:2634–2638.
12. Djimtombaye BJ, Alankuş- Çalışkan Ö, Gülcemal D, et al. Unusual secondary metabolites from Astragalus halicacabusLam. *Chem Biodivers*. 2013;10:1328–1334.
13. Özipek M, Dönme AA, Çalış İ, et al. Leishmanicidal cycloartanetype triterpene glycosides from Astragalus oleifolius. *Phytochemistry*. 2005;66:1168–1173.
14. Polat E, Caliskan-Alankus O, Perrone A, et al. Cycloartane-type glycosides from Astragalus amblolepis. *Phytochemistry*. 2009;70:628–634.
15. Çaliş I, Zor M, Saracoğlu I, et al. Four novel cycloartane glycosides from Astragalus oleifolius. *J Nat Prod*. 1996;59:1019–1023.
16. Çalış İ, Koyunoğlu S, Yeşilada A, et al. Antitrypanosomal Cycloartane glycosides from Astragalus baibutensis. *Chem Biodivers*. 2006;3:923–929.
17. Bedir E, Calis I, Khan IA. Macrophyllosaponin E: A novel compound from the roots of Astragalus oleifolius. *Chem Pharm Bull*. 2000;48:1081–1083.
18. Bedir E, Calis I, Aquino R, et al. Cycloartane triterpene glycosides from the roots of Astragalus brachypterus and Astragalus microcephalus. *J Nat Prod*. 1998;61:1469–1472.
19. Bedir E, Calis I, Dunbara C, et al. Two novel cycloartane-type triterpene glycosides from the roots of Astragalus prusianus. *Tetrahedron*. 2001;57:5961–5966.
20. Gülcemal D, Alankuş-Çalışkan O, Perrone A, et al. Cycloartane glycosides from Astragalus aureus. *Phytochemistry*. 2011;72:761–768.
21. Çaliş I, Yürüker A, Taşdemir D, et al. Cycloartane triterpene glycosides from the roots of Astragalus melanophrurius. *Planta Med*. 1997;63:183–186.
22. Polat E, Bedir E, Perrone A, et al. Triterpenoid saponins from Astragalus wiedemannianus Fischer. *Phytochemistry*. 2010;71:658–662.
23. Denizli N, Horo I, Gülcemal D, et al. Cycloartane glycosides from Astragalus plumosus var. krugianus and evaluation of their antioxidant potential. *Fitoterapia*. 2014;92:211–218.
24. Bedir E, Çalıs I, Piacente S, et al. A new flavonol glycoside from the aerial parts of Astragalus vulneraria. *Chem Pharm Bull*. 2000;48:1994–1995.
25. Calis I, Gazar HA, Piacente S, et al. Secondary metabolites from the roots of Astragalus zahlbruckneri. *J Nat Prod*. 2001;64:1179–1182.
26. Horo İ, Kocabaş F, Alankuş-Çalışkan Ö, Özgökçe F, et al. Secondary metabolites from Astragalus lycius and their cytotoxic activities. *Nat Product Commun*. 2016;11:1847–1850.
27. Abdallah RM, Ghazy NM, El-Sebakhy NA, et al. Astragalosides from Egyptian Astragalus spinosus Vahl. *Pharmazie*. 1993;48:452–454.
28. Khushbaktova ZA, Agzamova MA, Syrov VN, et al. Influence of cycloartanes from plants of the genus Astragalus and their synthetic analogs on the contractive function of the myocardium and the activity of Na K-ATPase. *Chem Nat Compd*. 1994;30:469–473.
29. Luo Y, Qin Z, Hong Z, et al. Astragaloside IV protects against ischemic brain injury in a murine model of transient focal ischemia. *Neurosci Lett*. 2004;363:218–223.
30. Sevimli-Gür C, Onbaşılar İ, Atilla P, et al. In vitro growth stimulatory and in vivo wound healing studies on cycloartane-type saponins of Astragalus genus. *J Ethnopharmacol*. 2011;134:844–850.

31. Hostettmann K, Marston A. *Saponins, Chemistry & Pharmacology of Natural Products.* Cambridge, University Press; 1995, pp. 277–279, 327, 332.
32. Dinarello CA. Overview of the IL-1 family in innate inflammation and acquired immunity. *Immunol Rev.* 2018;281:8–27.
33. Kalliolias GD, Ivashkiv LB. TNF biology, pathogenic mechanisms and emerging therapeutic strategies. *Nat Rev Rheumatol.* 2016;12:49–62.
34. Yesilada E, Bedir E, Çalış İ, et al. Effects of triterpene saponins from Astragalus species on in vitro cytokine release. *J Ethnopharmacol.* 2005;96:71–77.
35. Parameswaran N, Patial S. Tumor necrosis factor-α signaling in macrophages. *Crit Rev Eukaryot Gene Expr.* 2011;20:87–103.
36. Boyman O, Sprent J. The role of interleukin2 during homeostasis and activation of the immune system. *Nat Rev Immunol.* 2012;12:180–190.
37. Heeb LEM, Egholm C, Boyman O. Evolution and function of interleukin-4 receptor signaling in adaptive immunity and neutrophils. *Genes Immun.* 2020;21:143–149.
38. Baggiolini M, Walz A, Kunkel SL. Neutrophil-activating peptide-1/interleukin 8, a novel cytokine that activates neutrophils. *J Clin Invest.* 1989;84:1045–1049.
39. Benakanakere MR, Finoti LS, Tanaka U, et al. Investigation of the functional role of human Interleukin-8 gene haplotypes by CRISPR/Cas9 mediated genome editing. *Sci Rep.* 2016;6:31180.
40. Li Y, Meng T, Hao N, et al. Immune regulation mechanism of Astragaloside IV on RAW264.7 cells through activating the NF-κB/MAPK signaling pathway. *Int Immunopharmacol.* 2017;49:38–49.
41. Wang YP, Li XY, Song CQ, et al. Effect of astragaloside IV on T, B lymphocyte proliferation and peritoneal macrophage function in mice. *Acta Pharmacol Sin.* 2002;23:263–266.
42. Zhang A, Zheng Y, Que Z, et al. Astragaloside IV inhibits progression of lung cancer by mediating immune function of Tregs and CTLs by interfering with IDO. *J Cancer Res Clin Oncol.* 2014;140:1883–1890.
43. Cicchese JM, Evans S, Hult C, et al. Dynamic balance of pro- and anti-inflammatory signals controls disease and limits pathology. *Immunol Rev.* 2018;285:147–167.
44. Pifferi C, Fuentes R, Fernández-Tejada A. Natural and synthetic carbohydrate-based vaccine adjuvants and their mechanisms of action. *Nat Rev Chem.* 2021;5:197–216.
45. Nalbantsoy A, Nesil T, Erden S, et al. Adjuvant effects of Astragalus saponins Macrophyllosaponin B and Astragaloside VII. *J Ethnopharmacol.* 2011;134:897–903.
46. Podolak I, Galanty A, Sobolewska D. Saponins as cytotoxic agents: A review. *Phytochem Rev.* 2010;9:425–474.
47. Batlle E, Massague J. Transforming growth Factor-b signaling in immunity and cancer. *Immunity.* 2019;50:924–940.
48. Nalbantsoy A, Nesil T, Yilmaz-Dilsiz Ö, et al. Evaluation of the immunomodulatory properties in mice and in vitro anti-inflammatory activity of cycloartane type saponins from Astragalus species. *J Ethnopharmacol.* 2012;139:574–581.
49. Murphy K, Travers P, Walport M, et al. *Janeway's Immunobiology.* 6th ed. New York: Garland Science; 2008.
50. von Andrian UH, Mempel TR. Homing and cellular traffic in lymph nodes. *Nat Rev Immunol.* 2003;3:867–878.
51. Chávez-Galán L, Olleros ML, Vesin D, et al. Much more than M1 and M2 macrophages, there are also CD169+ and TCR+ macrophages. *Front Immunol.* 2015;6:263.
52. Detienne S, Welsby I, Collignon C, et al. Central role of CD169+ lymph node resident macrophages in the adjuvanticity of the QS-21 component of AS01. *Sci Rep.* 2016;6:39475.
53. Garçon N, Di Pasquale A. From discovery to licensure, the Adjuvant System story. *Hum Vaccines Immunother.* 2017;13:19–33.
54. Yakuboğulları N, Genç R, Çöven F, et al. Development of adjuvant nanocarrier systems for seasonal influenza A (H3N2) vaccine based on Astragaloside VII and gum tragacanth (APS). *Vaccine.* 2019;37:3638–3645.
55. Kang S, Brown HM, Hwang S. Direct Antiviral Mechanisms of Interferon-Gamma. *Immune Netw.* 2018;18:e33.
56. Genç R, Yakuboğulları N, Nalbantsoy A, et al. Adjuvant potency of Astragaloside VII embedded cholesterol nanoparticles for H3N2 influenza vaccine. *Turkish J Biol.* 2020;44:304–314.
57. Yakubogullari N, Coven FO, Cebi N, et al. Evaluation of adjuvant activity of Astragaloside VII and its combination with different immunostimulating agents in Newcastle Disease vaccine. *Biologicals.* 2021;70:28–37.

4 *Centaurea* sp.

Beste Atlı, Özlem Bahadır Acıkara and Müberra Koşar

INTRODUCTION

Centaurea species, which belongs to Asteraceae family, are widely grown all over the world and especially in the Mediterranean and Western Asia with approximately 500–900 species (Forgo *et al.*, 2012), 221 of which are endemic to Europe. The genus *Centaurea* gets its name from Greek mythology, where it is said to have cured the wound of Chiron, a Centaur (half-man, half-horse beast), thus the generic name (Sokovic *et al.*, 2017). The genus *Centaurea* is represented by 168 species in Turkey (Arif *et al.*, 2004), and six species in Northern Cyprus (Hand, Hadjikyriakou and Christodoulou, 2011). *Centaurea* is the third largest genus in Turkey, after *Astragalus* and *Verbascum*. Turkey is one of the gene centers of this genus, as evidenced by its high endemism ratio (61.6%) (Formisano *et al.*, 2008). The plant is known by Turkish names such as "peygamber çiçeği", "zerdali dikeni", "çoban kaldıran", and "Timur dikeni" (Arif *et al.*, 2004). *Centaurea* species have been used widely in folk medicines for hundreds of years. Many ethnobotanical uses have been proven by scientific studies. However, there are still many chemical constituents and biological activities that have not been discovered yet. In the few studies conducted, it is seen that the chemical content of these species is very rich. For instance, sesquiterpene lactones, flavonoids, and lignans have been identified as the major constituents of the genus *Centaurea* and their medicinal properties are substantially attributed to these unsaturated lactones. Some of *Centaurea* species are used (Figure 4.1) for therapeutic purposes among the people. Species used include *C. iberica*, *C. kurdica*, *C. virgata*, *C. cyanus*, and *C. calcitrapa*. Significant biological activities including anti-inflammatory, antipyretic, analgesic, antiplatelet, wound healing, anti-ulcerogenic, hepatoprotective, anti-plasmodial, cytotoxic, antioxidant, antibacterial, antifungal, have also been revealed by several studies. It is believed that more studies will be carried out on these species with the development of technology and facilitation of working opportunities.

TRADITIONAL USE

Centaurea species are annual, biennial, or perennial, and rarely small shrubs with spiny branches or larger shrubs with evergreen leaves. Rarely glabrous, usually range from tomentos or scabrous to multicellular hirsute hairs. They are often sessile gland hairy. Leaves alternate, sometimes all radical, very variable, but never spiny found in Turkey (except spinulose in *C. odyssei*); often pinnatifid or pinnatipartite, sometimes decurrent. Capitula are heterogamous, disciform, or radiant. Involucre are ovoid, subglobose, hemispherical, nearly cylindrical, oblong, or fusiform; phyllaries have pluriseriate, imbricate, with or without rigid, nearly always with a scarious, straw-textured, or coriaceous appendage of very variable form: entire or fringed to ciliate, orbicular, lanceolate or triangular, blunt or ending in a mucro, spinule, or rigid spine. Receptacles have smooth bristles. Flowers are pink, purple to blackish-purple, blue, yellow, or whitish; marginal ones neuter (sometimes with staminodes), funnel-shaped with 5–8 or more segments or nearly filiform and very inconspicuous with 4–5 linear segments, central ones are hermaphrodite. Achenes are usually glabrous when ripe, with or without laterally compressed, apex rounded or truncate, hilum lateral near base, often with elaiosome. Pappus of several series of unequal scabrous, barbellate, or plumose bristles, gradually

DOI: 10.1201/9781003146971-4

(A) *C. iberica* flower (Photo taken by Hüseyin Cahid Doğan)

(B) *C. kurdica* flower (Photo taken by Serdar Ölez)

(C) *C. virgata* flower (Photo taken by Hüseyin Cahid Doğan)

(D) *C. glastifolia* flower (Photo taken by Serdar Aslan) (Uysal, 2012)

FIGURE 4.1 Pictures of *Centaurea* species from different parts of Turkey.

elongated toward center, but innermost row often short and more scale-like, pappus persistent or rarely caducous, sometimes absent (Wagenitz, 1975).

Centaurea genus has been widely used for different purposes in folk medicine since ancient times. Species belonging to this genus are used as diuretics, tonic (Aktumsek *et al.*, 2011), antirheumatic (Koca *et al.*, 2009), anti-inflammatory (Garbacki *et al.*, 1999), antipyretic (Akkol *et al.*, 2009), antidiabetic (Ozturk *et al.*, 2018), antimalarial (Tagarelli, Tagarelli and Piro, 2010), antihypertensive (Amel, 2013), anticancer, stomachic, cough suppressant (Van Wyk and Gorelik, 2017), ophthalmic, wound healing, gynecological, treating skin problems (Forgo *et al.*, 2012), and antibacterial (Formisano *et al.*, 2008). Some *Centaurea* species used in folk medicine in Turkey have been reported (Table 4.1).

CHEMISTRY

Phytochemical studies on *Centaurea* species show that it is a characteristically rich genus, especially in terms of sesquiterpene lactones (Çelik *et al.*, 2006) (Table 4.2), of which many new compounds* have been isolated. In addition, *Centaurea* species contain flavonoids (Table 4.3), triterpenes (Table 4.4), phytosterols (Table 4.5), coumarins (Formisano *et al.*, 2008) (Table 4.6), lignans (Çelik *et al.*, 2006) (Table 4.7), and alkaloids (Sarker *et al.*, 2001) (Table 4.8), as well as other important secondary metabolites; for instance, phenolic acids (Erel *et al.*, 2011) (Table 4.9).

TABLE 4.1
List of Some *Centaurea* Species Used in Folk Medicine in Turkey

Species	Traditional usages	Parts used	References
C. iberica	Antipyretic, wound healing, kidney stone dropper, snake bite remedial, and antidiabetic	Aerial parts	(Cakilcioglu and Turkoglu, 2010), (Güzel, Güzelşemme and Miski, 2015), (Kaval, Behçet and Cakilcioglu, 2014), (Uysal *et al.*, 2010)
C. kurdica	Sedative	Capitulums	(Cakilcioglu and Turkoglu, 2010)
C. virgata	Antiallergic, wound healing, stomachic, antiepileptic, and antituberculosis	Aerial parts	(Cakilcioglu and Turkoglu, 2010)
C. lycophila	Wound healing	Aerial parts	(Demirci and Özhatay, 2012)
C. solstitialis subsp. *solstitialis*	Antimalarial, boils, and warts remover	Aerial parts	(Bulut and Tuzlacı, 2015), (Bulut *et al.*, 2017)
C. glastifolia	Prostate treatments, kidney stone reducing, wound healing, and antidiabetic	Aerial parts	(Dalar, 2018), (Karakaya *et al.*, 2019), (Kaval, Behçet and Cakilcioglu, 2014), (Mükemre, Behçet and Çakilcioglu, 2015)
C. pterocaula	Antidiarrheal and wound-healing	Aerial parts	(Kaval, Behçet and Cakilcioglu, 2014), (Mükemre, Behçet and Çakilcioglu, 2015)
C. cyanus	Gastrointestinal problems	Flowers and leaves	(Sargin, Akçicek and Selvi, 2013)
C. saligna	Astringent	Leaves	(Polat, 2019)
C. calcitrapa	Cardiac disorders and embolism	Aerial parts	(Polat, 2019)
C. calcitrapa subsp. *calcitrapa*	Antidiabetic	Leaves and capitulums	(Güzel, Güzelşemme and Miski, 2015)
C. urvillei subsp. *armata*	Wound care, anti-swelling, antitumoral, and skin cancer remover	Aerial parts	(Ahmet Sargin, 2015)
C. balsamita	Abscess relief	Leaves	(Altundag and Ozturk, 2011)
C. hyalolepis	Immunostimulant	Flowers	(Altundag and Ozturk, 2011)

Sesquiterpene lactones

TABLE 4.2
Sesquiterpene Llactones Isolated from Some *Centaurea* Species

Species	Compounds	References
C. ptosimopappa	Zaluzanin D*, zaluzanin C, janerin, chlorojanerin, cynaropicrin, deacylcynaropicrin, 11α,13-dihydro-deacylcynaropicrin, 11β,13-dihydro-deacylcynaropicrin, 4β,15-dihydro-3-dehydro-solstitialin A	(Çelik *et al.*, 2006)
C. behen	Cynaripicrin, arguerin B, deacylcynaropicrine, grosshemin, and 4β,15-dihydro-3-dehydrosolstitialin A	(Mosaddegh, Tavakoli and Behzad, 2018)
C. virgata	8α-hydroxysonchucarpolide, 8α-(3,4-dihydroxy-2-methylene-bu tanoyloxy)-dehydromelitensine, cnicin	(Tuzun *et al.*, 2017)
C. cuneifolia	Cnicin, dehydromelitensin, dehydromelitensin-8-[(2'-α-hydroxy-β-hydroxyethyl)-acryloyl)]	(Aslan and Öksüz, 1999)
C. solstitialis ssp. *solstitialis*	Centaurepensin (chlorohyssopifolin A), chlorojanerin, solstitialin A, 13-Acetylsolstitialin A	(Özçelik *et al.*, 2009), (Gürbüz and Yesilada, 2007)
C. calolepis	Cnicin	(Erel *et al.*, 2011)
C. depressa	Solstitialin A, acetyl solstitialin	(Akkol *et al.*, 2009)
C. hierapolitana	Hierapolitanins A–D*	(Karamenderes *et al.*, 2007)

Flavonoids

TABLE 4.3
Flavonoids Isolated from Some *Centaurea* Species

Species	Compounds	References
C. behen	Crisimaritin, jeceosidin, salvigenin, pectolinarigenin	(Mosaddegh, Tavakoli and Behzad, 2018)
C. amanicola	Apigenin	(Formisano *et al.*, 2008)
C. virgata	Apigenin, hispidulin, salvigenin, eupatorin, 3'-methyleupatorin), isokaempferide	(Tuzun *et al.*, 2017)
C. calolepis	Lucenin-2, schaftoside, vicenin-2,, vitexin, isovitexin, homoorientin, rutin, orientin, luteolin-7-*O*-glycoside	(Erel *et al.*, 2011)
C. cuneifolia	Salvigenin, eupatilin, jaceosidin, acacetin, kaemferol, 5,7,4'-trihydroxy-3-methylether, 5-hydroxy-3',',6,7-tetramethoxyflavone, 5-hydroxy-3',4',7,8-tetramet hoxyflavone	(Aslan and Öksüz, 1999)
C. hierapolitana	Hispidulin, jaceosidin, shaftoside, kaempferol-3-O-rutinoside	(Karamenderes *et al.*, 2007)

Triterpenes

TABLE 4.4
Triterpenes Isolated from *Centaurea* Species

Species	Compounds	References
C. behen	ψ-Taraxasterol, oleanolic acid	(Mosaddegh, Tavakoli and Behzad, 2018)
C. amanicola	α-Amyrin	(Formisano *et al.*, 2008)
C. cuneifolia	α-Amyrin	(Aslan and Öksüz, 1999)

Phytosterols

TABLE 4.5
Phytosterols Isolated from Some *Centaurea* Species

Species	Compounds	References
C. behen	β-Sitosterol, daucosterol	(Mosaddegh, Tavakoli and Behzad, 2018)
C. amanicola	β-Sitosterol	(Formisano *et al.*, 2008)
C. cuneifolia	β-Sitosterol	(Aslan and Öksüz, 1999)

Coumarins

TABLE 4.6
Coumarins Isolated from *Centaurea* Species

Species	Compound	Reference
C. amanicola	Scopoletin	(Formisano *et al.*, 2008)

Lignans

TABLE 4.7
Lignans Isolated from *Centaurea* Species

Species	Compounds	References
C. ptosimopappa	Arctigenin, matairesinol	(Çelik *et al.*, 2006)
C. cuneifolia	Arctigenin	(Aslan and Öksüz, 1999)

Alkaloids

TABLE 4.8
Alkaloids Isolated from *Centaurea* Species

Species	Compounds	Reference
C. cyanus	Moschamine, *cis*-moschamine, centcyamine, *cis*-centcyamine	(Sarker *et al.*, 2001)

Phenolic acids

TABLE 4.9
Phenolic Acids Isolated from Some *Centaurea* Species

Species	Compounds	References
C. calolepis	Chlorogenic acid	(Erel *et al.*, 2011)
C. amanicola	Vanillic acid, 4-Hydroxybenzoic acid	(Formisano *et al.*, 2008)
C. triumfetti	Syringic acid, trans-cinnamic acid, caffeic acid, ferulic acid, p-coumaric acid, chlorogenic acid	(Acet, 2021)

BIOLOGICAL ACTIVITIES

It is known that *Centaurea* species have been used in folk medicine for many years. Because of their widespread use, several studies have been performed to elucidate the various biological activities of *Centaurea* species. Among the determined effects of *Centaurea* species are antibacterial, antifungal, antiviral, antimalarial, anti-inflammatory, anti-ulcerogenic, antipyretic, antidiabetic, antioxidant, and cytotoxic activities.

Antibacterial Activity

Antibacterial activities of the methanol extracts of 12 *Centaurea* species collected from Turkey were investigated against four bacterial strains. Antibacterial activity was assessed by using the disc microdilution method. Eight *Centaurea* species have shown antibacterial activities against several bacterial strains. *C. cariensis* subsp. *microlepis*, *C. cariensis* subsp. *maculiceps*, *C. balsamita*, *C. kotschyi* var. *kotschyi*, *C. solsitialis* subsp. *solsitialis*, *C. urvillei* subsp. *Urvillei*, and *C. virgata* have shown antibacterial activities on *Escherichia coli*. *C. cariensis* subsp. *maculiceps* and *C. cariensis* subsp. *microlepis* have shown antibacterial activities on *Bacillus cereus*. *C. calolepis*, *C. cariensis* subsp. *Microlepis*, and *C. virgata* have shown antibacterial activities on *Salmonella enteritidis*.

C. balsamita, *C. cariensis* subsp. *maculiceps*, *C. cariensis* subsp. *microlepis*, *C. solsitialis* subsp. *Solsitialis*, and *C. urvillei* subsp. *urvillei* have shown antibacterial activities on *Staphylococcus aureus* (Tekeli *et al.*, 2011).

The *in vitro* antibacterial activities of nine *C. tchihatcheffii* extracts were investigated by liquid microdilution method. According to the data obtained, all the extracts were antibacterial effective against Gram-negative bacteria (*Eschericia coli*, *Pseudomonas aeruginosa*, *Proteus mirabilis*, *Klebsiella pneumoniae*, *Acinetobacter baumannii*) and Gram-positive bacteria (*Staphylococcus aureus*, *Enterococcus faecalis*), in the minimum concentration range of 2-16 µg/mL, while they were active against isolated strains of *E. coli*, *P. aeruginosa*, *P. mirabilis*, *K. pneumoniae*, *A. baumannii*, *S. aureus*, *E. faecalis*, and *Bacillus subtilis* the concentration range of 32–128 µg/mL (Koca and Özçelik, 2009).

The antibacterial activity of *C. amanicola*, *C. consanguinea*, and *C. ptosimopappa* essential oils was tested on some Gram-positive strains. *Bacillus subtilis* and *Staphylococcus aureus* were the bacteria that were most affected by *C. amanicola* essential oil, with Minimum Inhibitory Concentration (MIC) of 25 µg/mL^{-1} and Minimum Bactericidal Concentration (MBC) of 50 µg/mL^{-1}, respectively. *Bacillus cereus* and *B. subtilis*, the bacteria most affected by *C. consanguinea* essential oil, had the same MIC and MBC values. The essential oil of *C. ptosimopappa* had the least amount of action against the bacteria tested *B. cereus*, *B. subtilis*, *S. aureus*, *Staphylococcus epidermidis*, *Streptococcus faecalis*, *Escherichia coli*, *Proteus mirabilis*, *Proteus vulgaris*, *Pseudomonas aeruginosa, and Salmonella typhi* (Formisano *et al.*, 2008).

Antifungal Activity

The antifungal activities of nine *C. tchihatcheffii* extracts were evaluated against *Candida albicans* and *Candida parapsilosis* by liquid microdilution method. These extracts are water-chloroform interphase extract, chloroform, ethyl acetate, water, butanol fractions, ethyl acetate precipitate of aerial parts and 80% ethanol extracts of leaf, flower, and stem parts. *C. tchihatcheffii* extracts were found effective against *C. albicans* and *C. parapsilosis* at concentrations of 4 µg/mL and 8 µg/mL, respectively (Koca and Özçelik, 2009).

The antifungal effects of 10 *Centaurea* species (*C. calolepis*, *C. cariensis* subsp. *maculiceps*, *C. cariensis* subsp. *microlepis*, *C. hierapolitana*, *C. cadmea*, *C. reuterana* var. *reuterana*, *C. cyanus*, *C. depressa*, *C. urvillei* subsp. *urvillei* and *C. ensiformis*) were investigated on *C. albicans*, *Candida glabrata*, and *Candida krusei*. Antifungal activity has been observed in hexane extracts of *C. depressa* and *C. urvillei* subsp. *urvillei* against *C. krusei* (IC_{50} = 15 and 45 µg/ml, respectively) (Karamenderes *et al.*, 2006).

In vitro antifungal activity of three sesquiterpene lactones (chlorohyssopifolin A, chlorojanerin, and 13-acetyl solstitialin A) isolated from *C. solstitialis* ssp. *solstitialis* were investigated on *C. albicans* and *C. parapsilosis* using the microdilution method. These sesquiterpene lactones showed antifungal activity in these strains with 64 µg/ml MIC (Özçelik *et al.*, 2009).

Antiviral Activity

The antiviral activities of nine *C. tchihatcheffii* extracts were evaluated against *Herpes simplex* virus type-1 (HSV-1) and *Parainfluenza*-3 virus (PI-3 virus). According to the antiviral activity results, water-chloroform interphase extract, chloroform, and ethyl acetate fractions were found to be active against both DNA (HSV-1) and RNA (PI-3 virus) viruses (Koca and Özçelik, 2009).

Three sesquiterpene lactones (chlorohyssopifolin A, chlorojanerin, and 13-acetyl solstitialin A) obtained from *C. solstitialis* ssp. *solstitialis* were tested on *Herpes simplex* type-1 and *Parainfluenza* and 13-acetyl solstitialin A which showed significant antiviral activity on HSV-1 (Özçelik *et al.*, 2009).

Antimalarial Activity

In a study conducted on ten *Centaurea* species (*C. calolepis*, *C. cariensis* subsp. *maculiceps*, *C. cariensis* subsp. *microlepis*, *C. hierapolitana*, *C. cadmea*, *C. reuterana* var. *reuterana*, *C. cyanus*, *C. depressa*, *C. urvillei* subsp. *urvillei*, and *C. ensiformis*), the antimalarial effect was tested on *Plasmodium falciparum*. *C. hierapolitana* chloroform extract showed activity on chloroquine-sensitive and chloroquine-resistant *P. falciparum* clones with half maximal inhibitory concentration (IC_{50}) values of 7 and 7.3 μg/ml, respectively (Karamenderes *et al.*, 2006).

In an *in vivo* study, antimalarial activities of *C. hierapolitana*, *C. polyclada*, and *C. lydia* were examined on mice infected with *Plasmodium yoelii*. A dose of250 to 500 mg/kg dose of plant extract was given to mice as a single daily dose for 4 days. *C. lydia* chloroform extract and *C. polyclada* methanol extract showed antimalarial activities with reducing parasitemia (Ozbilgin *et al.*, 2014).

Antiinflammatory Activity

The *in vivo* anti-inflammatory activities of n-hexane, chloroform, ethyl acetate, and 85% aqueous methanol extracts of the aerial parts of *C. iberica* were investigated. The inhibitory effect on acetic acid-induced increase in capillary permeability was investigated for anti-inflammatory effect. Only methanol extract had a dose-dependent inhibitory efficacy at a dose of 200 mg/kg, with the greatest inhibitory value of 31.6% (Koca *et al.*, 2009).

In vitro anti-inflammatory effects of aerial part extracts of *C. aphrodisea*, *C. athoa*, *C. hyalolepis*, *C. iberica*, and *C. polyclada* plants were tested by NF-κB and iNOS inhibition tests, and *in vivo* anti-inflammatory effects were determined in rats by the carrageenan-induced paw edema test. NF-κB inhibition of the chloroform extract of *C. athoa* was considered equivalent to the positive control parthenolide (IC_{50} = 6 μg/ml). Also the carrageenan-induced paw edema test in rats revealed that this extract had anti-inflammatory activity at all hours at a dose of 50 mg/kg as compared to the control group (Erel *et al.*, 2014).

In vitro anti-inflammatory activity of cnicin obtained from aerial parts of *C. calolepis* was investigated by NF-κB and iNOS inhibition tests. Cnicin inhibited NF-κB activity and iNOS activity, with IC_{50} values of 1.8 and 6.5 μM, respectively (Erel *et al.*, 2011).

Antiulcerogenic Activity

The *in vivo* anti-ulcerogenic activity of the guaianolide-type sesquiterpene lactones chlorojanerin, 13-acetyl solstitialin A, and solstitialin A isolated from the chloroform extract of the aerial parts of *C. solstitialis* ssp. *solstitialis* was investigated on rats and mice. Chlorojanerin, 13-acetyl solstitialin A, and the mixture of 13-acetyl solstitialin A (95%) and solstitialin A (5%) have been shown to be effective in ulcer lesions (Gürbüz and Yesilada, 2007).

Antipyretic Activity

Freund's complete adjuvant-induced pyrexia model was used to evaluate the antipyretic activity of extracts, fractions, and subfractions of the root and aboveground portions of *C. solstitialis* subsp. *solstitialis* in mice. The ethanolic extract from the plant's aerial parts was found to have substantial antinociceptive ($p<0.01$) and antipyretic ($p<0.01$) properties. Solstitialin A and acetyl solstitialin were identified as active compounds (Akkol *et al.*, 2009).

Antidiabetic Activity

The phenolic content and antidiabetic effects of the hydrophilic extracts of the roots, stems, leaves, and flowers of *C. karduchorum*, an endemic plant growing in Eastern Anatolia, were examined.

Leaf extract showed the highest α-glucosidase inhibitory activity (IC_{50} < 1 mg/ml). The leaf extract inhibited glucosidase with a strong inhibitory activity (IC_{50} = 0.63 mg/ml) and amylase with a weak inhibitory activity (IC_{50}: 14.63±0.67 mg/ml) (Dalar *et al.*, 2015).

Ethanol, methanol, and ethyl acetate extracts of the aerial parts (flower and stem) of *C. triumfetti* were investigated for antidiabetic activity. To determine the activity, alpha-glucosidase and alpha-amylase enzyme inhibitory activities were examined by colorimetric methods. The highest inhibitory activity was observed in the ethyl acetate extract of the stem. In addition, stem extracts were found to be more active than flower extracts (Acet, 2021).

The antidiabetic activity of the essential oil obtained from the aerial parts of *C. pterocaula* was investigated. This activity was found by determining *in vitro* alpha-amylase inhibitory activity. The essential oil showed inhibitory activity on alpha-amylase with an IC_{50} value of 79.66 μg/mL (Sen *et al.*, 2021).

Antioxidant Activity

In vitro antioxidant effects of aerial part extracts of *C. aphrodisea*, *C. athoa*, *C. hyalolepis*, *C. iberica*, and *C. polyclada* plants were tested by 2-diphenyl-1-picrylhydrazyl (DPPH) and 2,20-azino -bis(3-ethylbenzothiazoline-6-sulphonicacid (ABTS) screening assays. Among the methanolic extracts of plants, *C. hyalolepis* showed the strongest activity in the DPPH test with an IC_{50} value of 147.8 μg/ml. In the ABTS test, the most potent activity was the methanol extract of *C. aphrodisea* with Trolox equivalent antioxidant capacity value (5.923±0.47) at 1000 μM concentration (Erel *et al.*, 2014).

The Folin–Ciocalteu method and the ferric reducing antioxidant power (FRAP) assay were used to estimate total reducing capacities, and the oxygen radical absorbance capacity (ORAC) assay was used to estimate oxygen radical scavenging capacities of the extracts *C. karduchorum*. The extracts' total reducing capacities ranged from 24.1 (root) to 38.5 (flower) mg gallic acid Eq./g DW (Folin–Ciocalteu assay) and from 274.0 (root) to 441.0 mmol Fe^{2+}/g DW (Folin–Ciocalteu assay) (FRAP assay). The oxygen radical absorbance capacity (ORAC assay) ranged from 930.5 μmol trolox Eq./g DW (root) to 1853.5 μmol trolox Eq./g DW (stem) (Dalar *et al.*, 2015).

Cytotoxic Activity

In vitro cytotoxic activities of aerial part extracts of *C. aphrodisea*, *C. athoa*, *C. hyalolepis*, *C. İberica*, and *C. polyclada* plants were tested against a panel of human solid tumor cell lines (SK-MEL: malignant melanoma, KB: oral epidermal carcinoma, BT-549: breast ductal carcinoma and SK-OV-3: ovary carcinoma) as well as non-cancerous kidney fibroblast (Vero) and kidney epithelial cells (LLC-PK1) by Neutral Red assay. The chloroform extract of *C. polyclada* had the greatest effect on the BT-549, KB, and SK-OV-3 cell lines (IC_{50} (μg/ml) values of 30, 33, and 47 mg/ml, respectively) (Erel *et al.*, 2014).

In vitro cytotoxic activity of cnicin obtained from aerial parts of *C. calolepis* was investigated against a panel of four human solid tumor cell lines SK-MEL, KB, BT-549, and SK-OV-3 and two non-cancerous kidney cell lines Vero and LLC-PK11. Cnicin was found to have cytotoxic activity against LLC-PK11, SK-MEL, and BT-549 cells, with IC_{50} values of 23.3, 14.0, and 18.3 μM, respectively (Erel *et al.*, 2011).

The cytotoxic properties of *C. behen* essential oil were investigated in human blood cell cultures. To investigate cytotoxic effects, researchers used 3-(4,5-dimethylthiazol-2-yl)-2,5 diphenyltetrazolium bromide (MTT) and lactate dehydrogenase (LDH) release assays. To measure oxidative potentials, researchers looked at total antioxidant capacity (TAC) and total oxidative status (TOS). The results showed that *C. behen* essential oil was cytotoxic and reduced cell viability (Çelikezen *et al.*, 2019).

CONCLUSION

Centaurea species have a great importance in Turkey due to their diversity and their traditional use. The presence of high amount of endemic *Centaurea* species in Turkey increases the importance of Turkey in this regard. Many studies show that its effects on folks have also been scientifically proven. Many new chemicals have been isolated in studies over the years. These chemically rich species can have a leading role in providing a natural source of new drugs.

REFERENCES

Acet, T. (2021) 'Determining the phenolic components by using HPLC and biological activity of *Centaurea triumfetti*', *Plant Biosystems*, 155(1), pp. 159–164. doi: 10.1080/11263504.2020.1722275.

Ahmet Sargin, S. (2015) 'Ethnobotanical survey of medicinal plants in Bozyazi district of Mersin, Turkey', *Journal of Ethnopharmacology*, 173, pp. 105–126. doi: 10.1016/j.jep.2015.07.009.

Akkol, E. K. *et al.* (2009) 'Sesquiterpene lactones with antinociceptive and antipyretic activity from two *Centaurea* species', *Journal of Ethnopharmacology*, 122(2), pp. 210–215. doi: 10.1016/j.jep.2009.01.019.

Aktumsek, A. *et al.* (2011) 'Screening for in vitro antioxidant properties and fatty acid profiles of five *Centaurea* L. species from Turkey flora', *Food and Chemical Toxicology*, 49(11), pp. 2914–2920. doi: 10.1016/j.fct.2011.08.016.

Altundag, E. and Ozturk, M. (2011) 'Ethnomedicinal studies on the plant resources of east Anatolia, Turkey', *Procedia – Social and Behavioral Sciences*, 19, pp. 756–777. doi: 10.1016/j.sbspro.2011.05.195.

Amel, B. (2013) 'Traditional treatment of high blood pressure and diabetes in Souk Ahras district', *Journal of Pharmacognosy and Phytotherapy*, 5(1), pp. 12–20. doi: 10.5897/JPP11.065.

Arif, R., Küpeli, E. and Ergun, F. (2004) 'The biological activity of *Centaurea* L. species', *Gazi University Journal of Science*, 17(4), pp. 149–164.

Aslan, Ü. and Öksüz, S. (1999) 'Chemical constituents of *Centaurea cuneifolia*', *Turkish Journal of Chemistry*, 23, pp. 15–20. doi: 10.1055/s-2007-971441.

Bulut, G. and Tuzlacı, E. (2015) 'Bayramiç (çanakkale-Türkiye) tıbbi bitkileri üzerinde etnobotanik çalışmalar', *Marmara Pharmaceutical Journal*, 19(3), pp. 268–282. doi: 10.12991/mpj.201519392830.

Bulut, G. *et al.* (2017) 'An ethnobotanical study of medicinal plants in Acipayam (Denizli-Turkey)', *Journal of Herbal Medicine*, 10(August), pp. 64–81. doi: 10.1016/j.hermed.2017.08.001.

Cakilcioglu, U. and Turkoglu, I. (2010) 'An ethnobotanical survey of medicinal plants in Sivrice (Elazig-Turkey)', *Journal of Ethnopharmacology*, 132(1), pp. 165–175. doi: 10.1016/j.jep.2010.08.017.

Çelik, S. *et al.* (2006) 'Guaianolides and lignans from the aerial parts of *Centaurea ptosimopappa*', *Biochemical Systematics and Ecology*, 34(4), pp. 349–352. doi: 10.1016/j.bse.2005.10.007.

Çelikezen, F. Ç. *et al.* (2019) 'Cytotoxic and antioxidant properties of essential oil of *Centaurea behen* L. in vitro', *Cytotechnology*, 71(1), pp. 345–350. doi: 10.1007/s10616-018-0290-9.

Dalar, A. *et al.* (2015) '*Centaurea karduchorum* Boiss. from Eastern Anatolia: Phenolic composition, antioxidant and enzyme inhibitory activities', *Journal of Herbal Medicine*, 5(4), pp. 211–216. doi: 10.1016/j.hermed.2015.09.006.

Dalar, A. (2018) 'Plant taxa used in the treatment of diabetes in van province, Turkey', *International Journal of Secondary Metabolite*, 5(3), pp. 170–184. doi: 10.21448/ijsm.430703.

Demirci, S. and Özhatay, N. (2012) 'An ethnobotanical study in Kahramanmaras (Turkey); wild plants used for medicinal purpose in Andirin, Kahramanmaraş', *Turkish Journal of Pharmaceutical Sciences*, 9(1), pp. 75–92.

Erel, S. B. *et al.* (2011) 'Secondary metabolites of *Centaurea calolepis* and evaluation of cnicin for anti-inflammatory, antioxidant, and cytotoxic activities', *Pharmaceutical Biology*, 49(8), pp. 840–849. doi: 10.3109/13880209.2010.551538.

Erel, S. B. *et al.* (2014) 'Bioactivity screening of five *Centaurea* species and *in vivo* anti-inflammatory activity of *C. athoa*', *Pharmaceutical Biology*, 52(6), pp. 775–781. doi: 10.3109/13880209.2013.868493.

Forgo, P. *et al.* (2012) 'Bioactivity-guided isolation of antiproliferative compounds from *Centaurea jacea* L.', *Fitoterapia*, 83(5), pp. 921–925. doi: 10.1016/j.fitote.2012.04.006.

Formisano, C. *et al.* (2008) 'Volatile constituents of aerial parts of three endemic *Centaurea* species from Turkey: *Centaurea amanicola* Hub.-Mor., *Centaurea consanguinea* DC. and *Centaurea ptosimopappa* Hayek and their antibacterial activities', *Natural Product Research*, 22(10), pp. 833–839. doi: 10.1080/14786410701218259.

Garbacki, N. *et al.* (1999) 'Anti-inflammatory and immunological effects of *Centaurea cyanus* flower- heads', *Journal of Ethnopharmacology*, 68(1–3), pp. 235–241. doi: 10.1016/S0378-8741(99)00112-9.

Gürbüz, I. and Yesilada, E. (2007) 'Evaluation of the anti-ulcerogenic effect of sesquiterpene lactones from *Centaurea solstitialis* L. ssp. *solstitialis* by using various *in vivo* and biochemical techniques', *Journal of Ethnopharmacology*, 112(2), pp. 284–291. doi: 10.1016/j.jep.2007.03.009.

Güzel, Y., Güzelşemme, M. and Miski, M. (2015) 'Ethnobotany of medicinal plants used in Antakya: A multicultural district in Hatay Province of Turkey', *Journal of Ethnopharmacology*, 174, pp. 118–152. doi: 10.1016/j.jep.2015.07.042.

Hand, R., Hadjikyriakou, G. N. and Christodoulou, C. S. (2011) *Flora of Cyprus – A Dynamic Checklist*. Available at: http://www.flora-of-cyprus.eu.

Karakaya, S. *et al.* (2019) 'An ethnobotanical investigation on medicinal plants in south of Erzurum (Turkey)', *Ethnobotany Research and Applications*, 18(December 2020), pp. 1–13. doi: 10.32859/era.18.13.1-18.

Karamenderes, C. *et al.* (2006) 'Antiprotozoal and antimicrobial activities of *Centaurea* species growing in Turkey', *Pharmaceutical Biology*, 44(7), pp. 534–539. doi: 10.1080/13880200600883080.

Karamenderes, C. *et al.* (2007) 'Elemanolide sesquiterpenes and eudesmane sesquiterpene glycosides from *Centaurea hierapolitana*', *Phytochemistry*, 68(5), pp. 609–615. doi: 10.1016/j.phytochem.2006.10.013.

Kaval, I., Behçet, L. and Cakilcioglu, U. (2014) 'Ethnobotanical study on medicinal plants in Geçitli and its surrounding (Hakkari-Turkey)', *Journal of Ethnopharmacology*, 155(1), pp. 171–184. doi: 10.1016/j.jep.2014.05.014.

Koca, U. *et al.* (2009) 'In vivo anti-inflammatory and wound healing activities of *Centaurea iberica* Trev. ex Spreng', *Journal of Ethnopharmacology*, 126(3), pp. 551–556. doi: 10.1016/j.jep.2009.08.017.

Koca, U. and Özçelik, B. (2009) 'Antiviral, antibacterial, and antifungal activities of *Centaurea tchihatcheffii* extracts', *Turkish Journal of Pharmaceutical Sciences*, 6(2), pp. 125–134.

Mosaddegh, M., Tavakoli, M. and Behzad, S. (2018) 'Constituents of the aerial parts of *Centaurea behen*', *Chemistry of Natural Compounds*, 54(5), pp. 1015–1017. doi: 10.1007/s10600-018-2539-0.

Mükemre, M., Behçet, L. and Çakilcioglu, U. (2015) 'Ethnobotanical study on medicinal plants in villages of Çatak (Van-Turkey)', *Journal of Ethnopharmacology*, 166, pp. 361–374. doi: 10.1016/j.jep.2015.03.040.

Ozbilgin, A. *et al.* (2014) 'Assessment of *in vivo* antimalarial activities of some selected medicinal plants from Turkey', *Parasitology Research*, 113(1), pp. 165–173. doi: 10.1007/s00436-013-3639-1.

Özçelik, B. *et al.* (2009) 'Antiviral and antimicrobial activities of three sesquiterpene lactones from *Centaurea solstitialis* L. ssp. *solstitialis*', *Microbiological Research*, 164(5), pp. 545–552. doi: 10.1016/j.micres.2007.05.006.

Ozturk, M. *et al.* (2018) *A Comparative Analysis of the Medicinal Plants Used for Diabetes Mellitus in the Traditional Medicine in Turkey, Pakistan, and Malaysia, Plant and Human Health, Volume 1*. doi: 10.1007/978-3-319-93997-1.

Polat, R. (2019) 'Ethnobotanical study on medicinal plants in Bingöl (City center) (Turkey)', *Journal of Herbal Medicine*, 16(February 2016), p. 100211. doi: 10.1016/j.hermed.2018.01.007.

Sargin, S. A., Akçicek, E. and Selvi, S. (2013) 'An ethnobotanical study of medicinal plants used by the local people of Alaşehir (Manisa) in Turkey', *Journal of Ethnopharmacology*, 150(3), pp. 860–874. doi: 10.1016/j.jep.2013.09.040.

Sarker, S. D. *et al.* (2001) 'Indole alkaloids from the seeds of *Centaurea cyanus* (Asteraceae)', *Phytochemistry*, 57(8), pp. 1273–1276. doi: 10.1016/S0031-9422(01)00084-X.

Sen, A. *et al.* (2021) 'Chemical composition, antiradical, and enzyme inhibitory potential of essential oil obtained from aerial part of *Centaurea pterocaula* Trautv', *Journal of Essential Oil Research*, 33(1), pp. 44–52. doi: 10.1080/10412905.2020.1839585.

Sokovic, M. *et al.* (2017) 'Biological activities of Sesquiterpene Lactones isolated from the genus *Centaurea* L. (Asteraceae)', *Current Pharmaceutical Design*, 23(19), pp. 2767–2786. doi: 10.2174/1381612823666170215113927.

Tagarelli, G., Tagarelli, A. and Piro, A. (2010) 'Folk medicine used to heal malaria in Calabria (southern Italy)', *Journal of Ethnobiology and Ethnomedicine*, 6, pp. 1–16. doi: 10.1186/1746-4269-6-27.

Tekeli, Y. *et al.* (2011) 'Antibacterial activities of extracts from twelve *Centaurea* species from Turkey', *Archives of Biological Sciences*, 63(3), pp. 685–690. doi: 10.2298/ABS1103685T.

Tuzun, B. S. *et al.* (2017) 'Isolation of chemical constituents of *Centaurea virgata* Lam. and Xanthine Oxidase inhibitory activity of the plant extract and compounds', *Medicinal Chemistry*, 13(5), pp. 498–502. doi: 10.2174/1573406413666161219161946.

Uysal, T. (2012) *Centaurea, Bizimbitkiler (2013).* Available at: http://www.bizimbitkiler.org.tr.

Wagenitz, G. (1975) '*Centaurea* L.', in: P.H. Davis (editor) *Flora of Turkey and the East Aegean Islands.* Vol. 5. Edinburgh University Press, Edinburgh, pp. 465–585.

Van Wyk, B. E. and Gorelik, B. (2017) 'The history and ethnobotany of Cape herbal teas', *South African Journal of Botany*, 110, pp. 18–38. doi: 10.1016/j.sajb.2016.11.011.

5 *Ecballium elaterium* (L.) A.Rich.

Hasya Nazlı Gök

INTRODUCTION

Ecballium elaterium (L.) A. Rich. (Figure 5.1) belongs to the Cucurbitaceae. It grows in Southern Europe, the Mediterranean region, Caucasus, Crimea, Turkey, Azores, and Southern Russia (Baytop, 1999; Jeffrey, 1972). It is a perennial, monoecious plant. When the fruits ripen, they separate from the stem. In this way, the seeds from the hole open at the bottom of the fruit, and the juice squirts out. Therefore, it is called "squirting cucumber". It has strong hydragog and cathartic effects. It increases the flow of urine and reduces edema in the body. (Grieve, 1970). Fruit juice is effective in the treatment of sinusitis. This effect has also been demonstrated in clinical studies (Sezik et al., 1982). The drug called "Elaterium" is obtained by condensing and drying the fruit juice and the extract obtained from the dried roots of the plant is used as a laxative (Baytop, 1999). Poultices and ointments prepared from the roots are used externally against tumors, chronic skin wounds, and eczema. The juice obtained by scraping the inner part of the root with a knife is wrapped on the skin as a pain reliever and wound healer. When the plant extract is taken internally in high amounts, it especially irritates the large intestine, so bloody diarrhea can be seen (Baytop, 1999). As a result of intranasal use of fruit juice without dilution, allergic reactions such as Quincke's edema, uvula edema, shortness of breath, difficulty in swallowing, and hypoxia may occur (Apostolos et al., 2012; Doğan et al., 2013; Plouvier et al., 1981). There may also be fatal cardiac and renal failure (Vlachos et al., 1994), therefore it should be used with caution. The phytochemical content of *E. elaterium* is formed especially from cucurbitacins (cucurbitacin B, C, D, E, G, H, I, L, R, etc.), sterols, flavonoids, terpenoids, fatty acids, and phenolic compounds.

TRADITIONAL USE

Ecballium elaterium has been used as a potent purgative since ancient times. Dioscorides reported that the roots of *Ecballium* are used for the treatment of the edemas, and the juice of the leaves is used for earache. Also, it is used for the treatment of gout, sciatica, toothache, the skin disorders such as leprosy, vitiligo, impetigo, and for cleaning black spots and scars on the face. Elaterium, which was precipitated from fruit juice, powdered, and mixed with milk, can be used in nasal cleansing in sinusitis and in relieving long-term headaches (Dioscórides, 2000; Gunther, 1968). The fruit juice of the plant was dropped into the nostrils in the treatment of sinusitis (Kultur, 2007; Tabata et al., 1988), 1–2 drops in the ear to relieve ear pain (Ugulu et al., 2009), it is used internally in the treatment of jaundice and hepatitis, for nocturia (Toker et al., 2003), and hemorrhoids by forming a pill with barley flour (Kultur, 2007).

Fruits are crushed and used externally for rheumatism and sciatica treatment, together with olive oil as a pain reliever. Its roots are used internally in the treatment of abdominal pain (Bulut & Tuzlaci, 2013), rheumatism (Bulut, 2011), hemorrhoids, uroclepsy (Kultur, 2007), and externally in eczema (Bulut, 2011). In Palestine, the juice of one fruit is given orally five times a day and is used for throat and liver cancer (Jaradat et al., 2016).

DOI: 10.1201/9781003146971-5

FIGURE 5.1 Fruits, flowers, and leaves of *Ecballium elaterium*. (Photo by H.N. Gök.)

BIOLOGICAL ACTIVITIES

Anti-sinusitis Activity

In a study conducted on 49 patients to determine the effect of *E. elaterium* (EE) fruit juice on sinusitis, the fruit juice was separated from the seeds and diluted with saline. The solution was applied to both nasal cavities of the patients. In a single-low dose administration, 3–8 drops of the solution were applied to the nostrils. Then, a double high dose was administered. In the double high dose, the patients were administered eight drops of the solution on the first day and four drops on the second day. Radiological and rhinological controls were done 10 days after the application of the solution. Medical and surgical intervention was not applied to the patients during the application of the solution. The solution improved aeration reduction by 71% in patients diagnosed with sinusitis. The double high dose provided 28% more improvement compared to the single low dose. The solution did not affect the mucosal thickening, cystic or polypoid formations (Sezik et al., 1982).

In another study, 0.1% benzoic acid was added to the filtrate as a preservative, the pH of the solution was adjusted to 6.4–6.8 with sodium acetate solution, and when used, it was diluted with saline at a rate of 50%. The solution was administered to patients intranasally three times a day for 10 days. In the first three days after application of the solution, hypersecretion occurred in the nose and an increase in nasal obstruction occurred. Hypersecretion stopped around the 4th day, and nasal obstruction returned to normal on the 8thor 9th day (Sezik, 1984).

Anti-inflammatory Activity

The anti-inflammatory activity of flower, fruit, and leaf ethanol extracts of *E. elaterium* was evaluated by inhibition of albumin denaturation, hypotonic solution-induced hemolysis methods *in vitro,* and inhibition of mouse paw edema induced by carrageenan method *in vivo*. It is shown that all of the extracts inhibited heat-induced protein denaturation effectively. Also, the fruit and flower extracts demonstrated high stability (78.70 ± 1.05%, 73.36 ± 1.19% of protection, respectively) with membrane stabilization test at 1 mg/mL. All of the extracts inhibited Paw edema induced by Carrageenan at 100 and 200 mg/kg *p.o.* The fruit extract (59.49% of edema reduction) at 100 mg/kg dose and flower extract (82.93% of edema reduction) at 200 mg/kg dose demonstrated the highest activity 3 h after inflammation initiation than the others (Bourebaba et al., 2020).

The aqueous extract of *E. elaterium* fruits was applied topically to rabbits with rhinosinusitis and its anti-inflammatory effects were investigated by measuring NO (nitric oxide) metabolites. It was observed that nitric oxide synthase (NOS) enzyme and NO radical metabolites were significantly decreased in the treatment group that received *E. elaterium* extract compared to the control group. Thus, it has been shown that *E. elaterium* extract can be used in the treatment of rhinosinusitis due to its anti-inflammatory effect (Uslu et al., 2006).

The effects on acetic acid-induced vascular permeability were investigated. Freeze-dried *E. elaterium* juices were dissolved in water and extracted with chloroform ($CHCl_3$), ethyl acetate (EtOAc), and *n*-Buthanol (*n*-BuOH), after that the fractions were administered intraperitoneally to mice. The active ingredient was determined to be cucurbitacin B isolated from the $CHCl_3$ extract (Yesilada et al., 1988). The anti-inflammatory activity of *E. elaterium* fruit juice and cucurbitacin B was tested on edema induced by serotonin and bradykinin in mice. Both the *E. elaterium* fruit juice and cucurbitacin B dose-dependently inhibited edema (Yesilada et al., 1989).

Antioxidant Activity

The antioxidant activity of flower, fruit, and leaf ethanol extracts of *E. elaterium* was evaluated by 2,2′-Azinobis-(3-Ethylbenzthiazolin-6-Sulfonic Acid) (ABTS), 2,2-Diphenyl-1-picrylhydrazyl (DPPH), ferric reducing antioxidant power (FRAP), and 2,2'-Azobis(2-amidinopropane) dihydrochloride (AAPH)-mediated inhibition of human erythrocyte hemolysis. The flower and leaf extract showed more activity against ABTS and DPPH radicals than the fruit extract, and the flower and fruit extract demonstrated more ferric reduced activity than the leaf extract. The extracts exerted moderate activity against the oxidative damage induced by AAPH (max $32.97 \pm 0.41\%$ with flower extract at 5 mg/ml) (Bourebaba et al., 2020).

The antioxidant activities of ethanolic fruit, root, and aerial part extracts, and fruit juice of *E. elaterium* were studied by using NO, DPPH, and DMPD radical scavenging activities, metal chelation capacity, and ferric- (FRAP) and phosphomolibdenum-reducing power (PRAP) analysis. The highest activities were shown in fruit ethanol extract by DPPH, NO radical scavenging, and FRAP activities ($12.21 \pm 2.34\%$, $40.26 \pm 2.21\%$, and 0.217 ± 0.005 OD, respectively) and were demonstrated in fruit juice by PRAP and metal-chelation activities (0.233 ± 0.016 OD, and $55.86 \pm 4.11\%$, respectively). Only fruit juice showed DMPD radical scavenging activity with $15.18 \pm 3.07\%$ value (Orhan et al., 2016).

The DPPH scavenging activity of the whole plant ethyl acetate extract of the *E. elaterium* was investigated, and as a result, it is found that the extract showed activity with 8.4 μg/ml SC_{50} value while the reference compound ascorbic acid SC50 value was 7.4 μg/ml (Elsayed & Badr, 2012).

Analgesic and Antipyretic Activity

The analgesic and antipyretic activity of Elaterium was investigated on male Wistar rats and Swiss male mice. It was determined that Elaterium (11.90 mg/kg) showed higher activity than Aspirin (200 mg/kg) in acetic acid-induced pain in mice. Unlike this, both Aspirin and Elaterium have not demonstrated analgesic activity with hot plate test in rats. In addition, a dose-dependent increase in antipyretic activity was determined in the study. An antipyretic effect of approximately 35 % was observed 4 hours after Elaterium was administered to rats at a dose of 11.90 mg/kg. This is equivalent to the effect of aspirin at a dose of 200 mg/kg (Agil et al., 1995).

Antimicrobial Activity

n-Hexane, dichloromethane, ethyl acetate, and methanol extracts of flower, fruit, leaf, and stem parts of *E. elaterium* were studied by using the disk diffusion method on *B. cereus* ATCC 14759, *E. coli* ATCC 25922, *Klebsiella* sp. CIP 104727, *L. monocytogenes* ATCC 19195, *Pseudomonas aeruginosa*

ATCC 27853, *Salmonella enteritidis* ATCC DMB560, and *S. aureus* ATCC 25923 strains. The hexane extract of *E. elaterium* fruit demonstrated moderate activity (1000 g/disk<MIQ< 125 g/disk) against *S. aureus* ATCC 25923 (Benzekri et al., 2016). In the other study, hexane, dichloromethane, and methanol extract of the seeds were studied by using the disc diffusion method against *Staphylococcus aureus, Staphylococcus epidermidis, Bacillus subtilis, Listeria monocytogenes, Escherichia coli, Salmonella typhi,* and *Candida albicans*. Only dichloromethane extract showed activity with 7–9 mm inhibitor zone against *Bacillus subtilis* (Karimi et al., 2019).

The antifungal activity of the whole plant ethyl acetate extract of *E. elaterium* was tested against *Penicillium* sp. *Saccharomyces cerevisiae*, *Aspergillus oryzae*, and *Aspergillus niger*. It is found that the extract showed significant activity against *A. niger* and *Aspergillus oryzae*, and moderate activity against *Penicillium* sp. The extract was not found to be effective against *Saccharomyces cerevisiae* (Elsayed & Badr, 2012).

Antimicrobial and antifungal activities of ethanol extract of *E. elaterium* fruits were investigated. The microdilution method was used to determine the minimum inhibitory concentration of extract and penicillin-extract combination against *Staphylococcus aureus*. Bifonazole was used as a positive control against *Candida albicans*. As a result, it has been shown that *E. elaterium* fruit extract has an antimicrobial effect against methicillin-resistant *S. aureus* (MRSA), methicillin-susceptible *S. aureus* (MSSA), and *Candida albicans* (Adwan et al., 2011). In another study, the antimicrobial activity of ethanolic fruit and leaf extracts of *E. elaterium* was evaluated by using the microdilution method on *E. coli* ATCC 25922 and clinical strains. As a result, the MIC values of both of the extracts on both *E. coli* strains were found to be 25 mg/ml (Abu-Hijleh et al., 2018).

Antitrypsin Activity

The antitrypsin activities of extract obtained from two different types of the seed of *E. elaterium* were investigated *in vitro*. Strong and specific antitrypsin activity was observed in three fractions separated by chromatographic methods from each extract (Attard & Attard, 2008).

Effect on the Cardiovascular System

The cardiovascular effect of *E. elaterium* juice was investigated in Langendorff perfused rabbit hearts. Eleven rabbit hearts were perfused with Krebs solution containing 200- to 1000-fold diluted fruit juice. There was a significant increase in left ventricular pressure and coronary flow ($p<0.05$) when *Ecballium* juice was applied at low concentrations (800- to 1000-fold dilution), but no significant increase in heart rate was observed. At higher concentrations (200-, 300-, 400-fold, etc. dilutions), there was a significant reduction in left ventricular pressure and heart rate ($p<0.005$). there was no change in coronary flow after 200-fold dilution. These results suggest that some components of *Ecballium* juice exert a positive inotropic effect with little change in heart rate, while some other components have a negative effect and decrease heart rate (Khatib et al., 1993).

Cytotoxic and Antiproliferative Activity

The antiproliferative activities of the *n*-hexane, dichloromethane, and methanol extracts of the aerial parts and rhizome on HepG2, SW480 (colon adenocarcinoma), and MCF7 (human breast adenocarcinoma) as Cancerous Cell Lines and HFFF2 (human fibroblast cell line) as normal cell lines during 48 hours were evaluated by Molavi et al. (2020). It is found that *n*-hexane extracts of the aerial parts and rhizomes demonstrated potent antiproliferative activity on the MCF7 cell line with 264.3 ± 5.2 and 321.1 ± 40.2 µg/mL IC_{50} values, respectively.

The activity of *E. elaterium* seed oil on the growth of fibrosarcoma (HT1080) and human colonic adenocarcinoma (HT29) cell lines was studied. As a result, it is found that the seed oil of *E.*

elaterium showed activity on the cell lines with 4.16 μg/ml and 4.86 μg/ml IC_{50} values, respectively. On the other hand, the cytotoxic activity of the seed oil was evaluated by using a lactate dehydrogenase (LDH) assay. It is found that the seed oil did not show cytotoxic activity on HT29 and HT1080 cells up to 5 μg/mL concentration. Also, in MTT analysis the seed oil did not affect the viabilities of HT29 and HT1080 cells up to 5 μg/mL and slightly affected up to 25 μg/mL concentrations (Touihri et al., 2019).

The activity of fruit juice and chloroform extract of *E. elaterium* and Cucurbitacin I, one of the compounds of *E. elaterium*, was investigated on MCF-7 and MDA-MB-231 breast cancer lines. Cucurbitacin I, *E. elaterium* fruit juice, and chloroform extract inhibited cell proliferation of MCF-7 cells in a time-dependent manner with 35.1 nM, 0.88 μg/mL, and 0.025 μg/mL IC_{50} values at 72 hours, respectively. Additionally, Cucurbitacin I, *E. elaterium* fruit juice, and chloroform extract inhibited cell proliferation of MDA-MB-231 cells with 17.7 nM, 4.21 mg/mL, and 2.5 μg/mL IC_{50} values at 72 hours, respectively. In the study, the anti-metastatic activities of cucurbitacin I, *E. elaterium* fruit juice, and chloroform extract were evaluated by using wound healing assay, and it is found that all samples showed anti-metastatic activity when compared to the control group. Also, it is found that Cucurbitacin I, *E. elaterium* fruit juice, and chloroform extract inhibited the colony-forming ability of the cells in both cancer cell lines (Yılmaz et al., 2018).

The inhibitory activity of *E. elaterium* seed oil on proliferation, adhesion, and migration of human brain cancer cell line (U87) was investigated. The seed oil inhibited the adhesion of U87 cell line to fibronectin (IC_{50} = 34.1 μg/mL) and fibrinogen (IC_{50} = 9.2 μg/mL), also the seed oil led to inhibition of cell migration to fibronectin (IC_{50} = 34 μg/mL) and fibrinogen (IC_{50} = 9 μg/mL). The seed oil showed antitumoral activity via the mediation of avb3 and a5b1 integrins (Touihri-Barakati et al., 2016).

The cytotoxic activity of ethyl acetate extract of the whole plant of *E. elaterium* was conducted on Hep-G2 (liver adenocarcinoma) cells, and it was found that as a result, the extract showed cytotoxic activity with 19.12 μg/ml IC_{50} value on Hep-G2 cells (Elsayed & Badr, 2012).

The immunomodulatory effect of cucurbitacin E isolated from *E. elaterium* was investigated on human peripheral lymphocytes. Lymphocytes were cultured with PC-3 (prostate cancer) and ZR-75-1 (breast cancer) cell lines and lymphocyte-mediated cytotoxicity was studied. Human peripheral lymphocytes stimulated by cucurbitacin E or phytohemagglutinin were tested on human cancer cell lines by the microtiter colorimetric method of WST-1 tetrazolium salt. Four different treatment groups were formed. After 48 hours of incubation of lymphocytes, cells were washed and then cultured with cancer cell lines for 48 hours. The proliferation of the cells was then evaluated. It was observed that more cytotoxicity occurred on both cancer cell lines in the treatment group compared to the control group. Thus, the potential cell-mediated effect of cucurbitacin E on prostate and breast cancer cell lines has been demonstrated (Attard et al., 2005).

Inhibitory effects of water, and methanol extracts of *E. elaterium* roots and *n*-hexane, chloroform, and *n*-butanol fractions of methanol extract on interleukin 1α (IL-1α), interleukin 1β (IL-1β), and tumor necrosis factor (TNFα) biosynthesis were investigated. Only *n*-hexane fraction was found to be active. In higher concentrations of the hexane fraction, higher activity was detected (Yeşilada et al., 1997).

The dried and powdered fruit juice of *E. elaterium* suppresses the mutagenic and carcinogenic effect of aflatoxin B1, and this effect is caused by the antitumoral activity of the plant (Başaran et al., 1993).

Anti-hepatotoxic Activity

To investigate the antioxidant and hepatoprotective effects of *E. elaterium* fruit juice, 400 mg/kg acetaminophen was administered to male albino rats and hepatotoxicity was induced. A single dose was administered orally every 48 hours for 22 days. Alkaline phosphatase (ALP), alanine aminotransferase (ALT), aspartate transaminase (AST), total bilirubin, total protein, antioxidant enzymes (catalase, glutathione peroxidase, glutathione reductase, and superoxide dismutase), and histopathological changes were examined. It was observed that liver function markers increased significantly with acetaminophen administration, but on the contrary, antioxidant enzymes decreased. It was observed that liver function markers increased significantly with acetaminophen administration, but on the contrary, antioxidant enzymes decreased. When *E. elaterium* fruit juice was administered orally at a dose of 1 mL/kg before acetaminophen administration, serum ALT, AST, and ALP levels were decreased. Histopathological data also supported the protective effect of *E. elaterium* fruit juice. A significant protective effect was observed with a decrease in necrotic areas and hepatocellular degeneration in the prophylactic and therapeutic groups (Elmhdwi et al., 2014).

The protective and curative activity of cucurbitacin B and Elaterium derived from the fruit juice of *E. elaterium* were investigated against hepatotoxicity induced by CCl_4 on male Swiss-OFFI mice. The ALT (alanine aminotransferase) enzyme level was significantly decreased with the use of Elaterium and cucurbitacin B before and after CCl_4 administration. It was also observed that fatty liver was significantly reduced in the test groups. Thus, it has been determined that elaterium and cucurbitasin B have preventive and therapeutic effects against CCl_4-induced hepatotoxicity (Agil et al., 1999).

To investigate the effect of *E. elaterium* (EE) fruit juice on serum bilirubin levels in male rats, jaundice was formed by surgically ligating the main bile duct. By intravenous administration of 0.05 mL fruit juice for 2 days after the surgical intervention, a 54% decrease in bilirubin level was observed 24 hours after the injection. A more significant decrease in bilirubin levels (64%, 52%, and 58%, respectively) was observed in 2, 4, and 24 hours with 0.1 mL application. There was a significant decrease in bilirubin level when the juice was administered intranasally at a dose of 0.1 mL. A more significant decrease was observed when two doses were given (Elayan et al., 1988).

Enzyme Inhibitory Activity

The acetyl (AChE) and butrylcholine esterase (BChE) and tyrosinase inhibitory activities of ethanolic fruit, root and aerial part extracts, and fruit juice of *E. elaterium* were studied. All of the extracts were not found effective significantly against AChE, BChE, and tyrosinase (Orhan et al., 2016).

The β-amylase, caseinolytic, L-asparaginase, lipoxygenase, lipase, pepsin, polyphenol oxidase, and tyrosinase enzyme activities (U/mg protein) of aqueous/ethanol extracts (30%) and latex of *E. elaterium* were evaluated. Both of the samples showed all of the enzyme activities (Imen & Yotova, 2016).

Genotoxic and Mutagenic Activity

The genotoxic activity of fruits and leaf ethanol extracts of *E. elaterium* was studied on the DNA genome and protein profile of *E. coli* ATCC 25922 and clinical strain by using SDS-PAGE, and ERIC-PCR analytical methods. In conclusion, the appearance or disappearance of some bands or changes in intensity was observed by the ERIC-PCR method in treated *E. coli* strains with extracts. Additionally, the appearance or disappearance of some bands or changes in intensity in protein profiles of treated *E. coli* strains were detected using SDS-PAGE gels (Abu-Hijleh et al., 2018).

The mutagenic and antimutagenic activity of *E. elaterium* fruit juice was determined. To investigate the mutagenic effect of *E. elaterium* fruit juice, the juice was applied on human peripheral lymphocytes at doses of 18, 36, and 72 μL/L for 24 and 48 hours. To investigate the antimutagenic effect of *E. elaterium* juices, human lymphocytes were treated with a mixture of *E. elaterium* juice and 0.25 μg/mL mitomycin C (MMC). When the juice was applied alone, it induced total chromosomal anomaly (especially the percentage of a structural anomaly rather than the percentage of numerical anomaly). When used as a mixture with Mitomycin-C, they induced the total chromosome anomaly synergistically. *E. elaterium* juice did not affect the frequency of sister chromatid exchange (SCE) during 24 and 48 hours of treatment. However, when the *E. elaterium* juice and MMC mixture was applied at the highest dose, it synergistically affected the frequency of sister chromatid exchange (SCE) after 48 hours of treatment. The use of a mixture of *E. elaterium* juice and MMC produced more cytotoxic effects than the use of *E. elaterium* juice alone. As a result, it was determined that *E. elaterium* fruit juice had mutagenic and cytotoxic effects on human peripheral lymphocytes, but no antimutagenic effect (Rencüzogullari et al., 2006).

Wound Healing Activity

The ethanol extract of fruits of *E. elaterium* 5 mg/kg was administered subcutaneously to the rats after the skin was incised 2.5 cm. The *E. elaterium* extract reduces fibrosis when compared to the control group, but there wasn't observed a significant difference in staining rates of caspase-3 and cd-34 in the immunohistochemical assay (İbiloglu et al., 2021).

CHEMISTRY

The substances contained in *E. elaterium* are cucurbitacins, phenolic compounds, sterols, amino acids, fatty acids, and some other substances. The isolated cucurbitacins, sterols, flavonoids, and terpenoids from *E. elaterium* are given in Tables 5.1, 5.2, 5.3, and 5.4, respectively.

OTHER COMPOUNDS

Fatty acids such as linoleic acid (48.64%), punicic acid (22.38%), oleic acid (15.58%), stearic acid, palmitic acid, myristic acid, arachidic acid were found in *E. elaterium* seed oil. Tocopherols such as γ-tocopherol (represents 71% of total tocopherols), α-tocopherol, β-tocopherol, and δ-tocopherol were found in the seed oil. Desmosterol, campesterol, stigmasterol, delta-7-campesterol, β-sitosterol, sitostanol, delta-5-avenasterol, delta-7-stigmastenol, and delta-7-avenasterol were also found in the seed oil by using GC-FID (Touihri et al., 2019). Additionally, heavy metals such as iron, chromium, zinc, copper, lead, nickel, cadmium, and copper were determined in the seed oil by using atomic absorption spectroscopy. Fe (23.46 mg/L) was found to be the major heavy metal in seed oil (Shahid & Arora, 2019).

In the other study, some volatile compounds were determined in the aerial part of *E. elaterium*, which are *n*-hentriacontane, loliolide, 1-allyl-1-but-3-enyl-1-silacyclobutane, neophytadiene, 3,7,11,15-tetramethyl-2-hexadecen-1-ol, hexadecanoic acid, linolenic acid methyl ester, phytol, propylhexedrine, *n*-pentacosane, alpha-tocohpherol, heptane, 3,4-dimethyl, thymol, carvacrol, 5-eocosene(e), limonene dioxide, and 7,10-pentadecadiynoic acid (Molavi et al., 2020). Also, lauric, myristic, and palmitic acids were determined as major fatty acids of fruits and roots, and linoleic and palmitic acids were determined as major fatty acids of leaves (Rodriguez Gonzalez & Martin Panizo, 1966). Hexadecanoic acid, octadecanoic acid, nonanoic acid, and tetradecanoic acid were isolated from the aerial part of *E. elaterium* (Lazaris et al., 1997).

Phenolic components such as hydroquinone, 4-hydroxy-3-metoxy-acetophenone, 4-hydroxy acetophenone, 2-nitroquinol, and ligballinol (4-hydroxyphenyl-bisepoxy lignan) were isolated from fresh fruit juice of *E. elaterium* (Rao & Lavie, 1974).

m-Carboxyphenylalanine and β-pyrazol-1-ylalanine were isolated from seeds of *E. elaterium* (Dunnill & Fowden, 1965). Allantoin was determined in the roots of *E. elaterium* (Constantinescu et al., 1969). Also, *Ecballium elaterium* trypsin inhibitor II (EETI-II), a 28-residue peptide, was isolated from seed of *E. elaterium* (Favel et al., 2009).

TABLE 5.1
Chemical Structures of the Isolated Cucurbitacins from *E. elaterium*

Compounds	Plant Parts	References
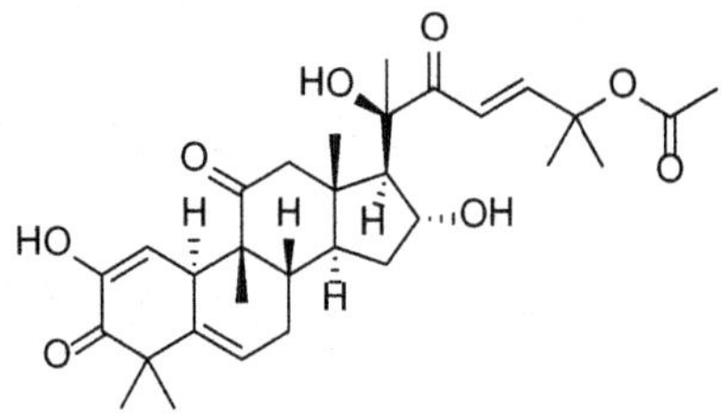 Cucurbitacin B	Fruit	Rao et al. (1974)
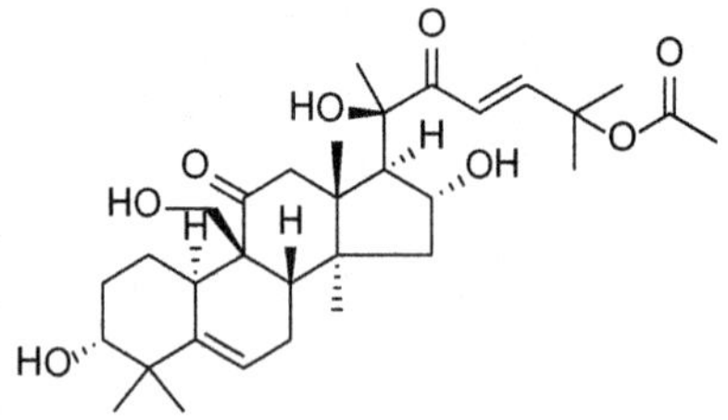 Cucurbitacin C	Leaves, fruits, and roots	Rodriguez Gonzalez and Martin Panizo (1967a)
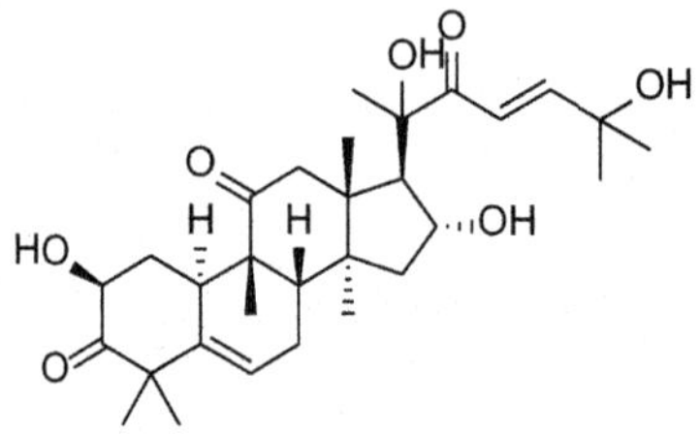 Cucurbitacin D (Elatericin A)	Leaves, fruits, and roots	Rodriguez Gonzalez and Martin Panizo (1967a), Lazaris et al. (1997), Seger, Sturm, Haslinger, et al. (2005)
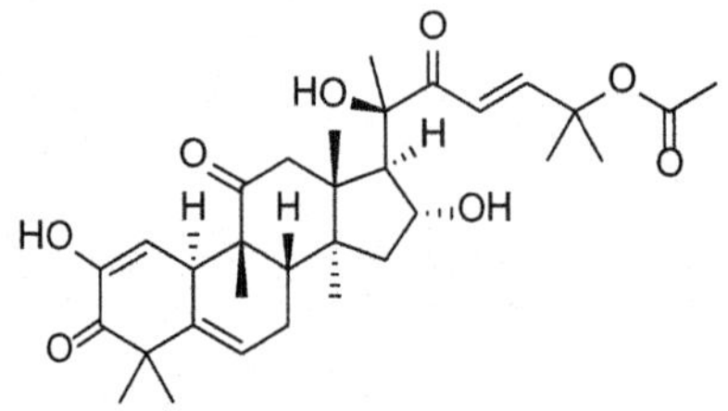 Cucurbitacin E (α-elaterin)	Leaves, fruits, and roots	Rodriguez Gonzalez and Martin Panizo (1967a), Seger, Sturm, Mair, et al. (2005)
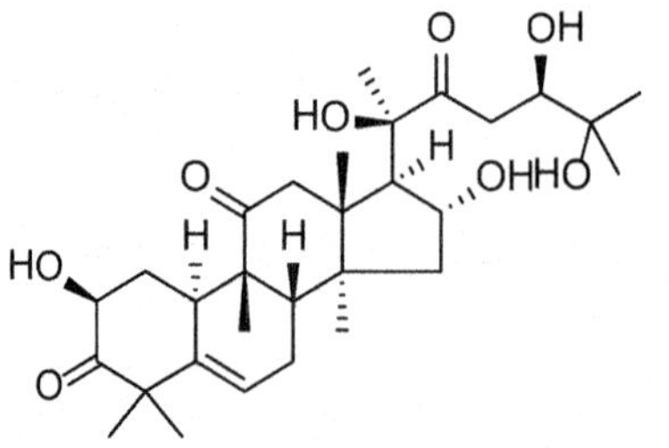 Cucurbitacin G	Leaves, fruits, and roots	Rodriguez Gonzalez and Martin Panizo (1967a)

(Continued)

TABLE 5.1 (CONTINUED)
Chemical Structures of the Isolated Cucurbitacins from *E. elaterium*

Compounds	Plant Parts	References
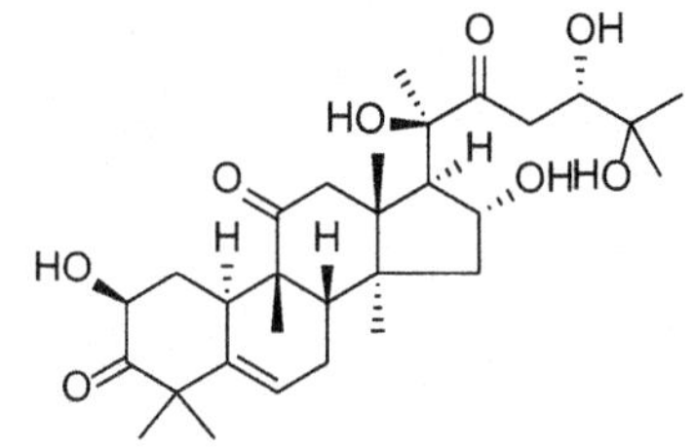 Cucurbitacin H	Leaves, fruits, and roots	Rodriguez Gonzalez and Martin Panizo (1967a)
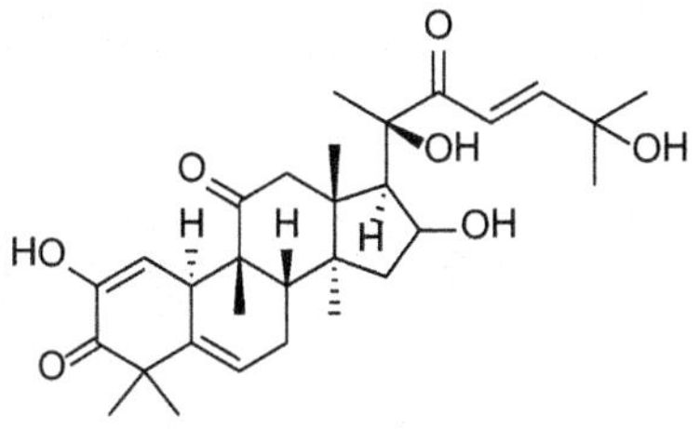 Cucurbitacin I (Elatericin B)	Leaves, fruits, and roots	Rodriguez Gonzalez and Martin Panizo (1967a), Lazaris et al. (1997), Seger, Sturm, Mair, et al. (2005)
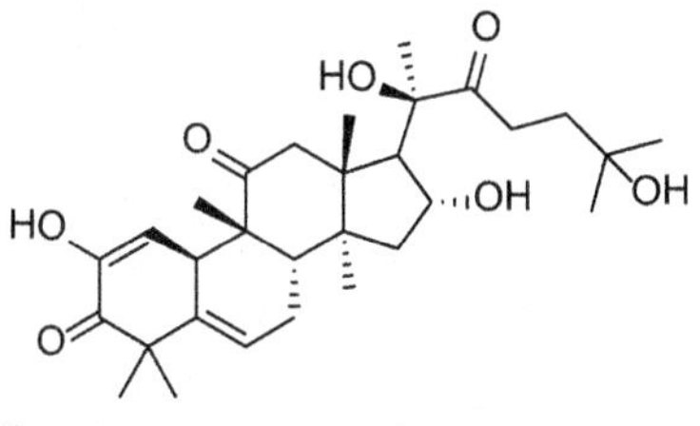 Cucurbitacin L	Fruit	Rao et al. (1974)
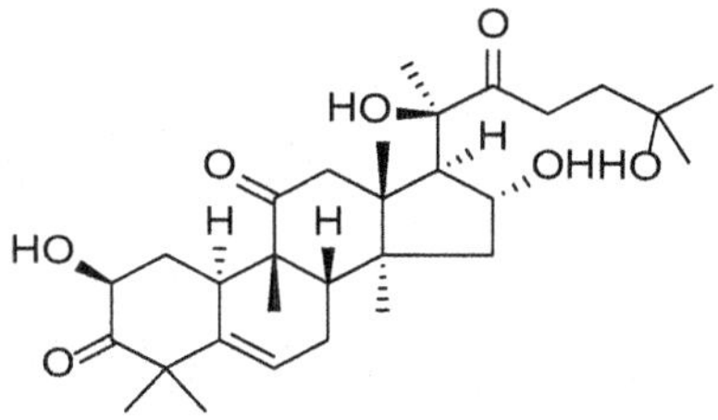 Cucurbitacin R	Fruit	Rao et al. (1974)
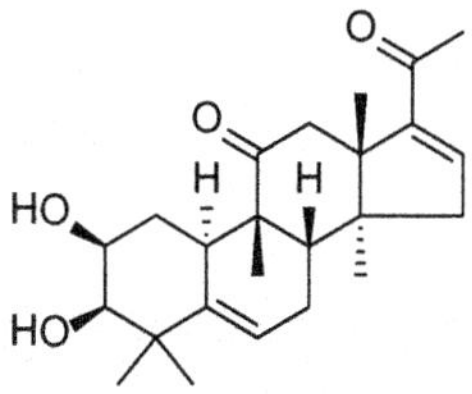 16-deoxy-Δ^{16}-hexanorcucurbitacin O	Fruit	Rao et al. (1974)

(Continued)

TABLE 5.1 (CONTINUED)
Chemical Structures of the Isolated Cucurbitacins from *E. elaterium*

Compounds	Plant Parts	References
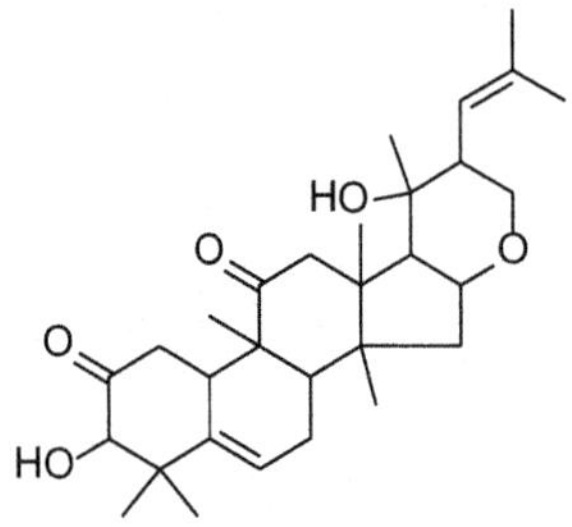 Anhidro 22-deoxo-3-epi-isocucurbitacin D	Fruit	Rao et al. (1974)
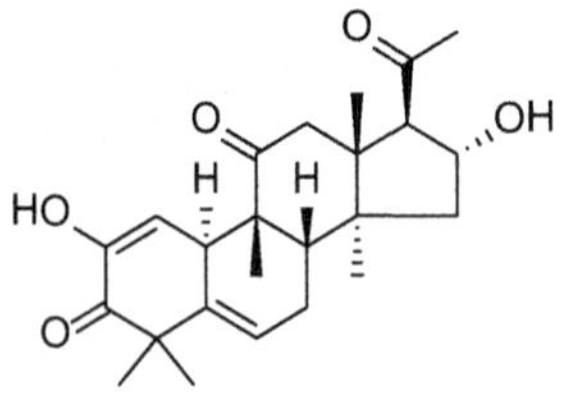 Hexanorcucurbitacin I	Fruit	Rao et al. (1974)
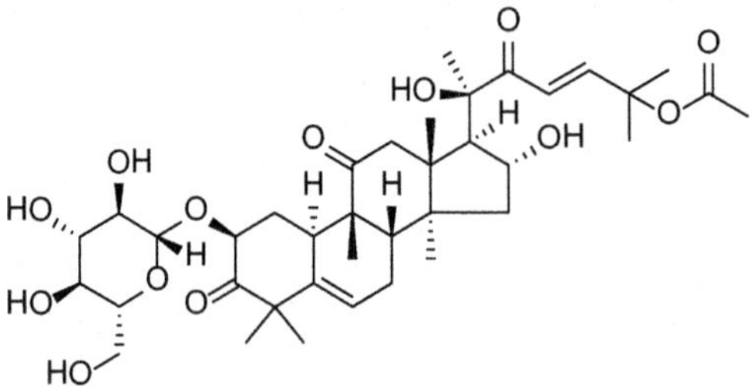 2-*O*-β-D-Glucopyranosylcucurbitacin B (Arvenin I)	Fruit	Seifert and Elgamal (1977)
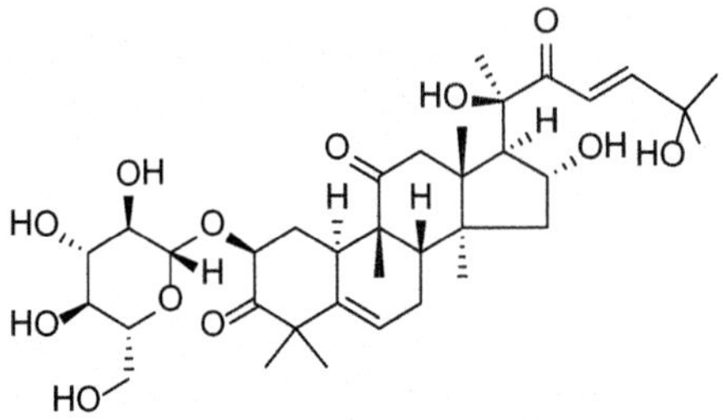 2-*O*-β-D-Glucopyranosylcucurbitacin D (Arvenin III)	Fruit	Seifert and Elgamal (1977)
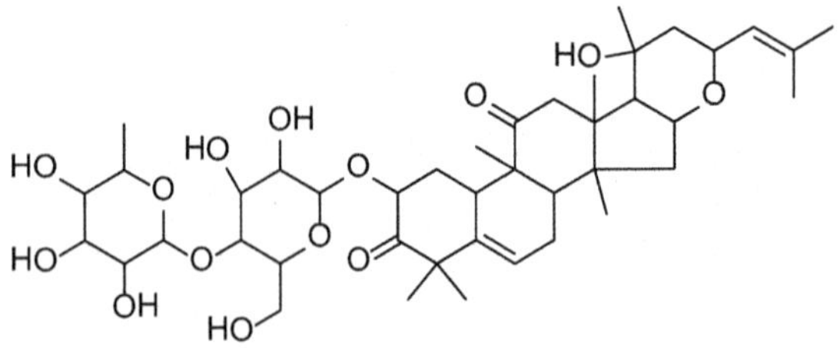 22-deoxocucurbitoside A	Fruit	Panosyan et al. (1985)

(*Continued*)

TABLE 5.1 (CONTINUED)
Chemical Structures of the Isolated Cucurbitacins from *E. elaterium*

Compounds	Plant Parts	References
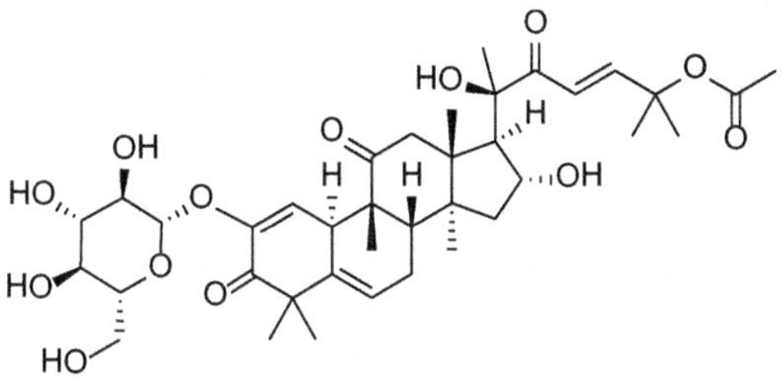 2-*O*-β-D-glucopyranosyl cucurbitacin E	Whole plant	Elsayed and Badr (2012)
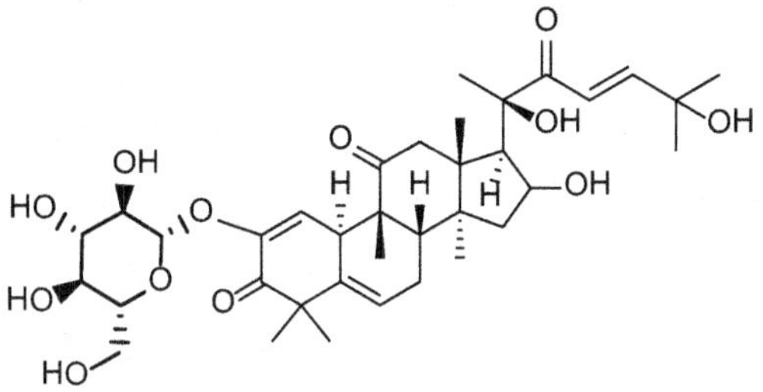 2-*O*-β-D-glucopyranosyl cucurbitacin I	Whole plant	Elsayed and Badr (2012)
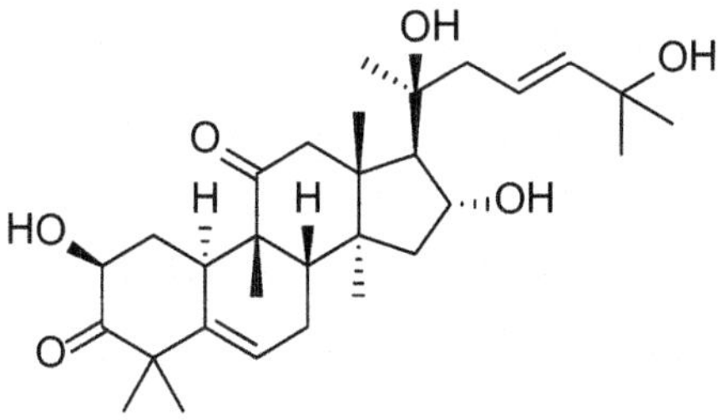 22-deoxocucurbitacin-D	Fruit	Seger, Sturm, Haslinger, et al. (2005)
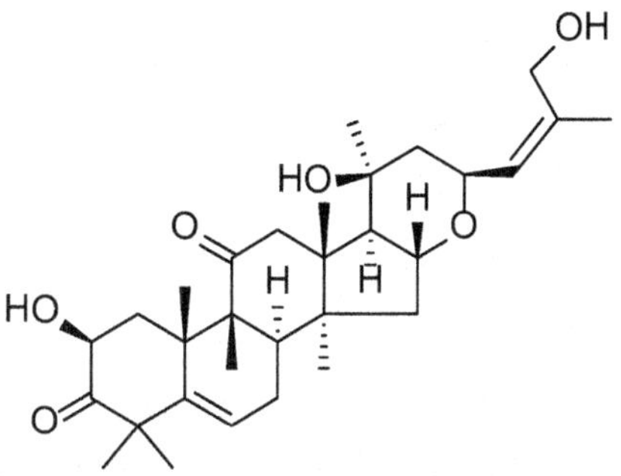	Fruit	Seger et al. (2004)

TABLE 5.2
Chemical Structures of the Isolated Sterols from *E. elaterium*

Compounds	Plant Parts	References
Elasterol	Aerial part, root and leaves	Hylands and Oskoui (1979b); Lazaris et al. (1997) Rodriguez Gonzalez and Martin Panizo (1967b)
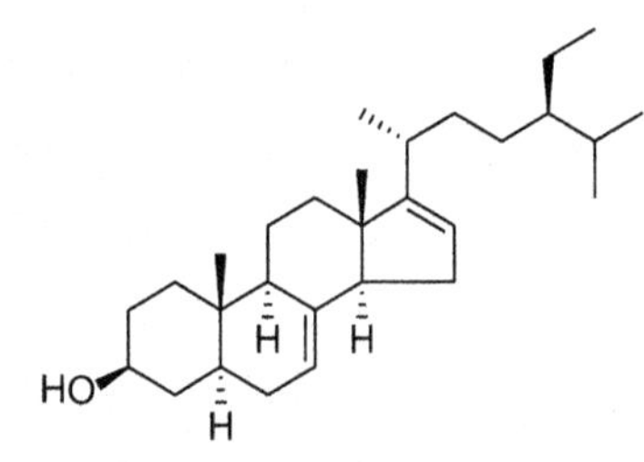 25,26-Dihydroelasterol		Rodriguez Gonzalez and Martin Panizo (1969)
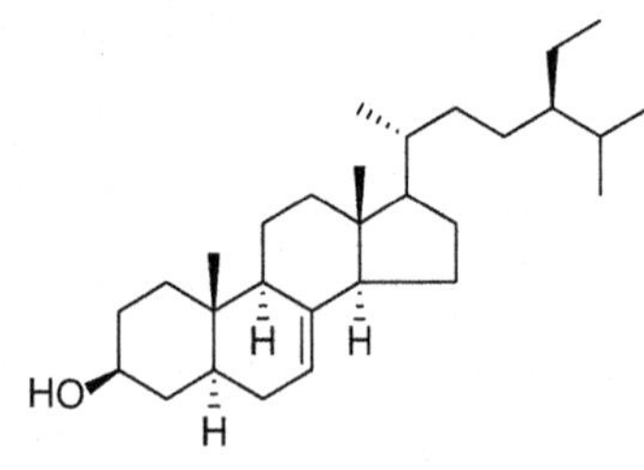 16,17,25,26-Tetrahydroelasterol		Rodriguez Gonzalez and Martin Panizo (1969)
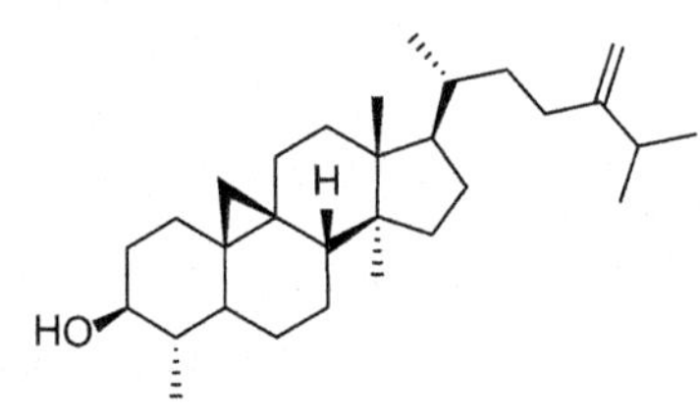 Cycloeucalenol	Leaves	Oskoui (1986)
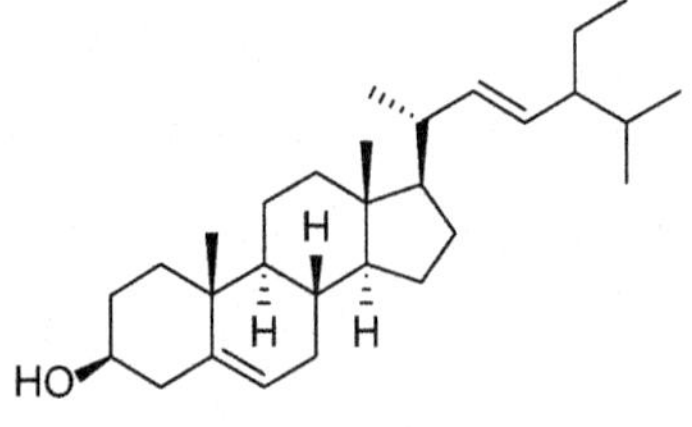 Stigmasterol	Whole plant	Hylands and Oskoui (1979a)

TABLE 5.3
Chemical Structures of the Isolated Flavonoids from *E. elaterium*

Compounds	Plant Parts	References
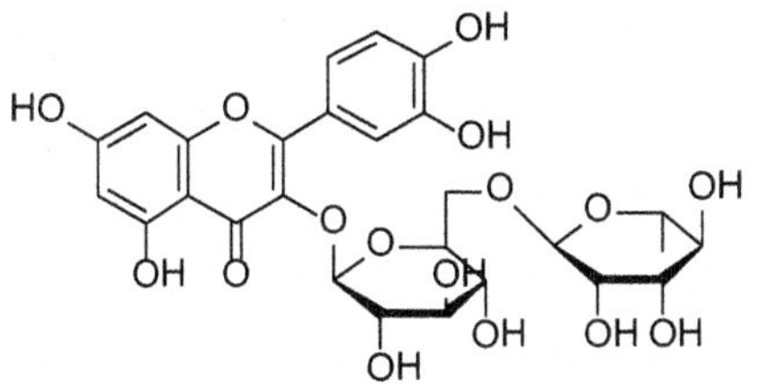 Rutin	Aerial part, whole plant, pollen, and stigma	Elsayed and Badr (2012); Imperato (1980); Krauze-Baranowska and Cisowski (1995)
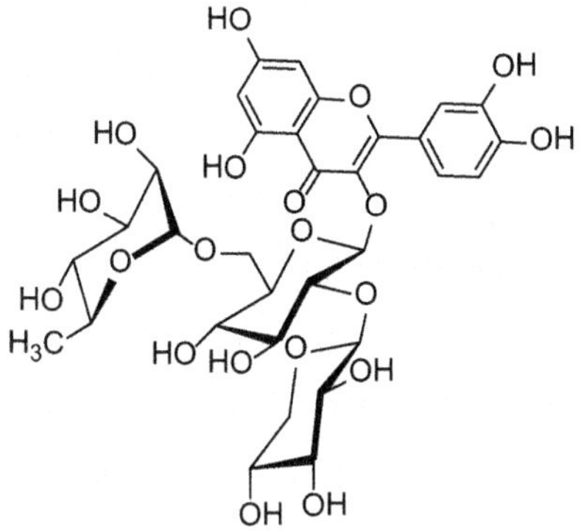 Quercetin 3-*O*-β-(2[G]-O-β-xylopyranosyl-6[G]-*O*-α-rhamnopyranosyl)glucopyranoside	Aerial part	Touil (2017)
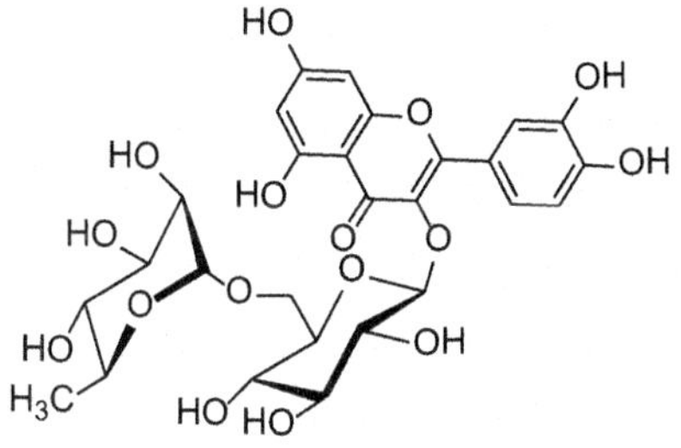 Quercetin 3-*O*-β-(6-*O*-α-rhamnopyranosyl)glucopyranoside	Aerial part	Touil (2017)
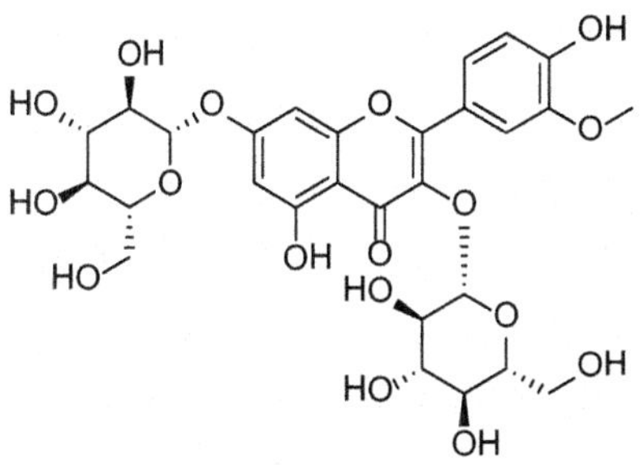 Isorhamnetin 3-*O*-β-D-glucopyranosyl-β-D-glucopyranoside -7-*O*-glucosyl	Whole plant	Elsayed and Badr (2012)

(Continued)

TABLE 5.3 (CONTINUED)
Chemical Structures of the Isolated Flavonoids from *E. elaterium*

Compounds	Plant Parts	References
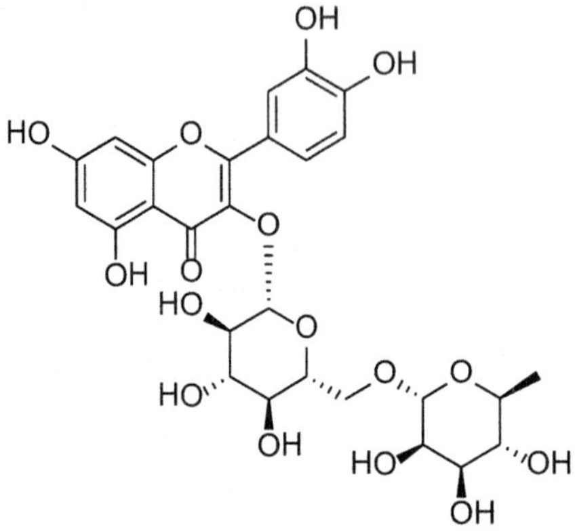 Quercetin 3-galactosylrhamnoside	Aerial part	Krauze-Baranowska and Cisowski (1995)
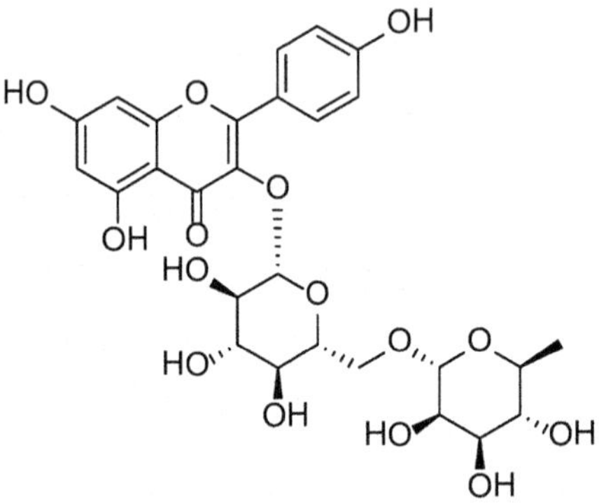 Kaempferol 3-*O*-rutinoside	Stigma and pollen	Imperato (1980)

E. elaterium is a plant that has been used from history to the present and has been included in various sources in the literature. Although it has many biological effects, especially anti-sinusitis and anti-inflammatory activity, its internal use in high doses causes dangerous poisoning. If it is applied too often, the effect on the stomach and intestines can be severe, inflammation occurs and can have fatal results (Baytop, 1999; Grieve, 1970). Quincke edema, fatal cardiac and renal failure may occur as a result of intranasal use of fruit juice (Plouvier et al., 1981; Vlachos et al., 1994). Therefore, it should be used with caution. Considering the clinical studies that determined the effect of the plant against sinusitis, this plant can be brought to the pharmaceutical industry. For this purpose, firstly, the plant can be cultivated and produced with good agricultural practices (GAP), and the extract standardized on cucurbitacins content of *Ecballium elaterium* can be prepared with standardization studies. Subsequently, new formulations containing these extracts in appropriate proportions can be developed.

TABLE 5.4
Chemical Structures of the Isolated Terpenoids from *E. elaterium*

Compounds	Plant Parts	References
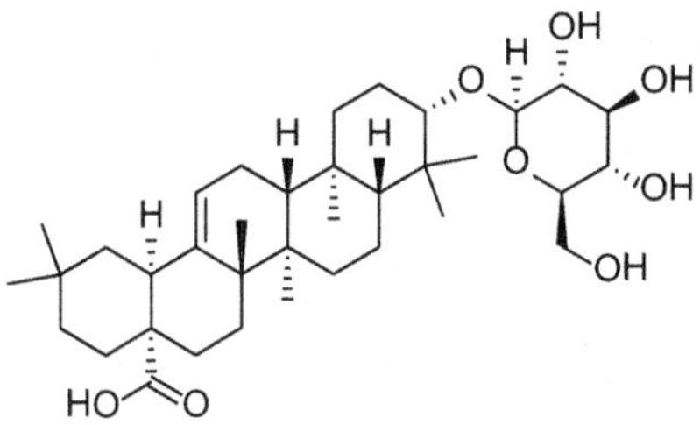 Oleanolic acid 3-*O*-β-D-glucopyranoside	Fruit	Chakravarty et al. (1997)
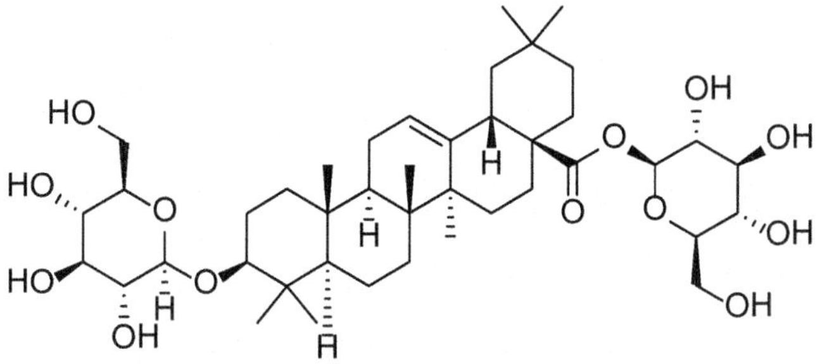 3-*O*-β-D-glucopyranosyl-3β-hydroxyolean-12-en-28-oic acid 28-*O*-β-D-glucopyranoside	Fruit	Chakravarty et al. (1997)
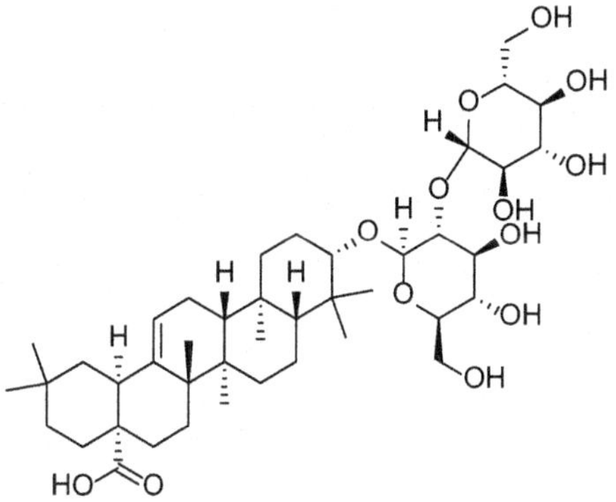 Oleanolic acid 3-O-β-D-glucopyranosyl (1→2)-β-D-glucopyranoside	Fruit	Chakravarty et al. (1997)
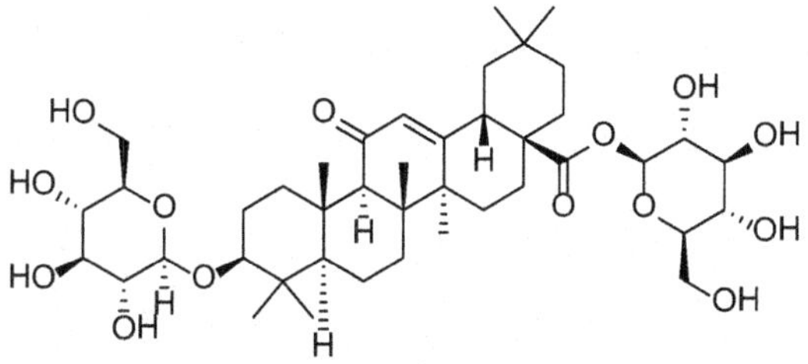 Elateroside A	Fruit	Chakravarty et al. (1997)

(Continued)

TABLE 5.4 (CONTINUED)
Chemical Structures of the Isolated Terpenoids from *E. elaterium*

Compounds	Plant Parts	References
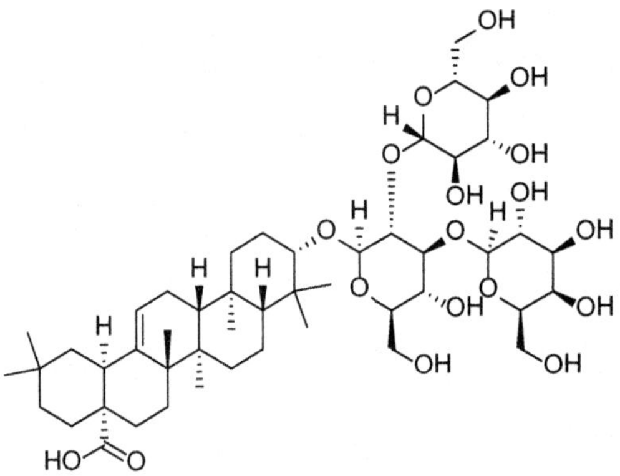 Elateroside B	Fruit	Chakravarty et al. (1997)

REFERENCES

Abu-Hijleh, A., Adwan, G., & Abdat, W. (2018). Biochemical and molecular evaluation of the plant *Ecballium elaterium* extract effects on *Escherichia coli*. *Journal of Advances in Biology & Biotechnology*, *19*(2), 1–11. https://doi.org/10.9734/jabb/2018/43882

Adwan, G., Salameh, Y., & Adwan, K. (2011). Effect of ethanolic extract of *Ecballium elaterium* against *Staphylococcus aureus* and *Candida albicans*. *Asian Pacific Journal of Tropical Biomedicine*, *1*(6), 456–460. https://doi.org/10.1016/s2221-1691(11)60100-7

Agil, A., Miró, M., Jimenez, J., Aneiros, J., Caracuel, M., García-Granados, A., & Navarro, M. (1999). Isolation of an anti-hepatotoxic principle from the juice of *Ecballium elaterium*. *Planta Medica*, *65*(7), 673–675. https://doi.org/10.1055/s-2006-960847

Agil, M. A., Risco, S., Miró, M., Navarro, M. C., Ocete, M. A., & Jiménez, J. (1995). Analgesic and antipyretic effects of *Ecballium elaterium* (L.) A. Richard. Extract in rodents. *Phytotherapy Research*, *9*(2), 135–138. https://doi.org/10.1002/ptr.2650090211

Apostolos, P., Athanasios, P., Georgios, G., Charalambos, S., Emmanouil, L., Ioannis, D., Ioannis, C., Anastasia, K., & Georgios, A. (2012). Severe uvular edema and resulting hypoxemia due to single use of *Ecbalium elaterium* extract: A case report. *American Journal of Case Reports*, *13*, 11–13. https://doi.org/10.12659/ajcr.882292

Attard, E., & Attard, H. (2008). Antitrypsin activity of extracts from *Ecballium elaterium* seeds. *Fitoterapia*, *79*(3), 226–228. https://doi.org/10.1016/j.fitote.2007.11.020

Attard, E., Brincat, M. P., & Cuschieri, A. (2005). Immunomodulatory activity of cucurbitacin E isolated from *Ecballium elaterium*. *Fitoterapia*, *76*(5), 439–441. https://doi.org/10.1016/j.fitote.2005.02.007

Başaran, A., Cakmak, E. A., Degirmenci, I., Başaran, N., Artan, S., Başer, K., & Kırımer, N. (1993). Chromosome aberrations induced by Aflatoxin B1 in rat bone marrow cells *in vivo* and their suppression by *Ecballium elaterium*. *Fitoterapia*, *64*(4), 310–313.

Baytop, T. (1999). *Türkiye'de Bitkiler ile Tedavi (Geçmişte ve Bugün), İlaveli 2. Baskı*. Nobel Tıp Kitabevleri.

Benzekri, R., Bouslama, L., Papetti, A., Snoussi, M., Benslimene, I., Hamami, M., & Limam, F. (2016). Isolation and identification of an antibacterial compound from *Diplotaxis harra* (Forssk.) Boiss. *Industrial Crops and Products*, *80*, 228–234. https://doi.org/10.1016/j.indcrop.2015.11.059

Bourebaba, L., Gilbert-Lopez, B., Oukil, N., & Bedjou, F. (2020). Phytochemical composition of *Ecballium elaterium* extracts with antioxidant and anti-inflammatory activities: Comparison among leaves, flowers and fruits extracts. *Arabian Journal of Chemistry*, *13*, 3286–3300.

Bulut, G. (2011). Folk medicinal plants of Silivri (Istanbul, Turkey). *Marmara Pharmaceutcal Journal*, *1*(15), 25–29. https://doi.org/10.12991/201115441

Bulut, G., & Tuzlaci, E. (2013, October 7). An ethnobotanical study of medicinal plants in Turgutlu (Manisa-Turkey). *Journal of Ethnopharmacology*, *149*(3), 633–647. https://doi.org/10.1016/j.jep.2013.07.016

Chakravarty, A. K., Sarkar, T., Das, B., Masuda, K., & Shiojima, K. (1997). Triterpenoid glycosides from fruits of *Ecballium elaterium*. *Journal of the Indian Chemical Society*, *74*(11–12), 858–863.

Constantinescu, E., Pislarasu, N., Istudor, V., & Forstner, S. (1969). Allantoin, an important factor in the therapeutic action of some plant products. *Herba Hungarica*, *8*(3), 101–106.

Dioscórides. (2000). *De materia medica: Being an herbal with many other medicinal materials: Written in Greek in the first century of the common era: A new indexed version in modern English* (T. A. Osbaldeston, Trans.; T. A. Osbaldeston & R. P. Wood, Eds.). Ibidis Press.

Doğan, N. Ö., Pamukçu Günaydın, G., Tekin, M., & Çevik, Y. (2013). Afflictive effect of Squirting Cucumber: Isolated uvular oedema due to complication of a herbal medicine. *Journal of Academic Emergency Medicine Case Reports*, *4*, 84–86.

Dunnill, P. M., & Fowden, L. (1965). The amino acids of seeds of the Cucurbitaceae. *Phytochemistry*, *4*(6), 933–944. https://doi.org/10.1016/s0031-9422(00)86271-8

Elayan, H. H., Gharaibeh, M. N., Zmeili, S. M., & Salhab, A. S. (1988). Effects of *Ecballium elatrium* juice on serum bilirubin concentration in male rats. *International Journal of Crude Drug Research*, *27*(4), 227–234. https://doi.org/10.3109/13880208909116908

Elmhdwi, M. F., Muftah, S. M., El tumi, S. G., & Elslimani, F. A. (2014). Hepatoprotective effect of *Ecballium elaterium* fruit juice against paracetamol induced hepatotoxicity in male albino rats. *International Current Pharmaceutical Journal*, *3*(5), 270–274. https://doi.org/10.3329/icpj.v3i5.18535

Elsayed, Z. I. A., & Badr, W. H. (2012). Cucurbitacin glucosides and biological activities of the ethyl acetate fraction from ethanolic extract of egyptian *Ecballium Elaterium*. *Journal of Applied Sciences Research*, *8*(2), 1252–1258.

Favel, A., Mattras, H., Coletti-Previero, M. A., Zwilling, R., Robinson, E. A., & Castro, B. (2009). Protease inhibitors from *Ecballium elaterium* seeds. *International Journal of Peptide and Protein Research*, *33*(3), 202–208. https://doi.org/10.1111/j.1399-3011.1989.tb00210.x

Grieve, M. (1970). *A modern herbal*. Hafner Publishing Co.

Gunther, R. T. (1968). *The greek herbal of dioscorides*. Hafner Publishing Co.

Hylands, P. J., & Oskoui, M. T. (1979a). Seasonal sterol variations in *Ecballium elaterium*. *Planta Medica*, *37*(9), 37–44. https://doi.org/10.1055/s-0028-1097292

Hylands, P. J., & Oskoui, M. T. (1979b). The structure of elasterol from *Ecballium elaterium*. *Phytochemistry*, *18*(9), 1543–1545. https://doi.org/10.1016/s0031-9422(00)98493-0

İbiloglu, İ., Alabalik, U., Keles, A. N., Aydogdu, G., Basuguy, E., & Buyukbayram, H. (2021). *Ecballium elaterium* extract reduces fibrosis during wound healing in rats. *Biotechnology & Biotechnological Equipment*, *35*(1), 696–703. https://doi.org/10.1080/13102818.2021.1920847

Imen, L., & Yotova, L. (2016). Investigation of different enzyme activities from *Pergularia tomentosa* L. and *Ecballium elaterium* L. *Journal of Chemical Technology and Metallurgy*, *51*(3), 263–270.

Imperato, F. (1980). Five plants of the family Cucurbitaceae with flavonoid patterns of pollens different from those of corresponding stigmas. *Experientia*, *36*(10), 1136–1137. https://doi.org/10.1007/bf01976084

Jaradat, N. A., Al-Ramahi, R., Zaid, A. N., Ayesh, O. I., & Eid, A. M. (2016). Ethnopharmacological survey of herbal remedies used for treatment of various types of cancer and their methods of preparations in the West Bank-Palestine. *BMC Complementary and Alternative Medicine*, *16*(1). https://doi.org/10.1186/s12906-016-1070-8

Jeffrey, C. (1972). *Ecballium* A. Rich. In P. H. Davis (Ed.), *Flora of Turkey and the east Aegean Islands* (Vol. 4). Edinburgh at the University Press.

Karimi, S., Farzaneh, F., Asnaashari, S., Parina, P., Sarvari, Y., & Hazrati, S. (2019, Fall). Phytochemical analysis and anti-microbial activity of some important medicinal plants from North-west of Iran. *Iranian Journal of Pharmaceutical Research*, *18*(4), 1871–1883. https://doi.org/10.22037/ijpr.2019.1100817

Khatib, S. Y., Mahmoud, I. I., & Hasan, Z. A. (1993). Effects of crude *Ecballium elaterium* juice on isolated rabbit heart. *International Journal of Pharmacognosy*, *31*(4), 259–268. https://doi.org/10.3109/13880209309082951

Krauze-Baranowska, M., & Cisowski, W. (1995). Flavonoid compounds of *Ecballium elaterium* (L.) A. Richard herb. *Herba Polonica*, *41*(1), 5–10.

Kultur, S. (2007, May 4). Medicinal plants used in Kirklareli Province (Turkey). *Journal of Ethnopharmacology*, *111*(2), 341–364. https://doi.org/10.1016/j.jep.2006.11.035

Lazaris, D., Chinou, I., Roussis, V., Vayias, C., & Roussakis, C. (1997). Chemical constituents from Ecballium elaterium and their effects on a non-small-cell bronchial carcinoma line. *Pharmaceutical and Pharmacological Letters*, *7*(4), 50–51.

Molavi, O., Torkzaban, F., Jafari, S., Asnaashari, S., & Asgharian, P. (2020). Chemical compositions and anti-proliferative activity of the aerial parts and rhizomes of Squirting Cucumber, Cucurbitaceae. *Jundishapur Journal of Natural Pharmaceutical Products*, *15*(1), e82990.

Orhan, I. E., Senol, F. S., Haznedaroglu, M. Z., Koyu, H., Erdem, S. A., Yılmaz, G., Cicek, M., Yaprak, A. E., Ari, E., Kucukboyaci, N., & Toker, G. (2016). Neurobiological evaluation of thirty-one medicinal plant extracts using microtiter enzyme assays. *Clinical Phytoscience*, *2*(1). https://doi.org/10.1186/s40816-016-0023-6

Oskoui, M. T. (1986, April). Cycloeucalenol from the leaves of *Ecballium elaterium*. *Planta Medica*, (2), 159–159. <Go to ISI>://WOS:A1986C046100025

Panosyan, A. G., Nikishchenko, M. N., & Avetisyan, G. M. (1985). Structure of 22-deoxocucurbitacins isolated from *Bryonia alba* and *Ecbalium elaterium*. *Chemistry of Natural Compounds*, *21*(5), 638–645. https://doi.org/10.1007/bf00579070

Plouvier, B., Trotin, F., Deram, R., De Coninck, P., & Baclet, J. L. (1981). Concombre d'ane (*Ecbalium elaterium*) une cause peu banale d'oed`eme de Quincke. *Nouvelle Presse Médicale*, *10*(31), 2590.

Rao, M. M., & Lavie, D. (1974). The constituents of Ecballium elaterium L.—XXII: Phenolics as minor components. *Tetrahedron*, *30*(18), 3309–3313. https://doi.org/10.1016/s0040-4020(01)97506-4

Rao, M. M., Meshulam, H., & Lavie, D. (1974). The constituents of *Ecballium elaterium* L. Part XXIII. Cucurbitacins and hexanorcucurbitacins. *Journal of the Chemical Society, Perkin Transactions 1*. https://doi.org/10.1039/p19740002552

Rencüzogullari, E., İla, H. B., Kayraldiz, A., Diler, S. B., Yavuz, A., Arslan, M., Funda Kaya, F., & Topaktas, M. (2006). The mutagenic and antimutagenic effects of *Ecballium elaterium* fruit juice in human peripheral lymphocytes. *Russian Journal of Genetics*, *42*(6), 623–627.

Rodriguez Gonzalez, B., & Martin Panizo, F. (1966). Composition of Ecballium elaterium. I. Introduction. *Anales de la Real Sociedad Espanola de Fisica y Quimica, Serie B: Quimica*, *62*(4–5), 553–562.

Rodriguez Gonzalez, B., & Martin Panizo, F. (1967a). Composition of *Ecballium elaterium*. II. Separation of free cucurbitacins. *Anales de la Real Sociedad Espanola de Fisica y Quimica, Serie B: Quimica*, *63*(9–10), 959–964.

Rodriguez Gonzalez, B., & Martin Panizo, F. (1967b). Structure of new sterol from Ecballium elaterium. *Anales de la Real Sociedad Espanola de Fisica y Quimica, Serie B: Quimica*, *63*(12), 1123–1136.

Rodriguez Gonzalez, B., & Martin Panizo, F. (1969). Sterols of Cucurbitaceae. II. Sterols of *Bryonia dioica*, *Bryonia verrucosa*, and *Ecballium elaterium*. *Anales de Quimica*, *65*(12), 1139–1152.

Seger, C., Sturm, S., Haslinger, E., & Stuppner, H. (2004). A new Cucurbitacin D related 16,23-epoxy derivative and its isomerization products. *Organic Letters*, *6*(4), 633–636. https://doi.org/10.1021/ol036469r

Seger, C., Sturm, S., Haslinger, E., & Stuppner, H. (2005). NMR signal assignment of 22-Deoxocucurbitacin D and Cucurbitacin D from *Ecballium elaterium* L. (Cucurbitaceae). *Monatshefte für Chemie – Chemical Monthly*, *136*(9), 1645–1649. https://doi.org/10.1007/s00706-005-0347-2

Seger, C., Sturm, S., Mair, M. E., Ellmerer, E. P., & Stuppner, H. (2005, Jun). 1H and 13C NMR signal assignment of cucurbitacin derivatives from *Citrullus colocynthis* (L.) Schrader and *Ecballium elaterium* L. (Cucurbitaceae). *Magnetic Resonance in Chemistry*, *43*(6), 489–491. https://doi.org/10.1002/mrc.1570

Seifert, K., & Elgamal, M. H. (1977). New cucurbitacin glucosides from *Ecballium elaterium* L. *Pharmazie*, *32*(10), 605–606.

Sezik, E. (1984). Acı hıyar sinüzite karşı kullanılan bir halk ilacının değerlendirilmesi. *Yeni Tıp Dergisi*, *1*(4), 61–64.

Sezik, E., Kaya, S., & Aydan, N. (1982). *Ecbalium elaterium* meyvalarının sinüzite etkisi. *Eczacılık bülteni*, *24*, 33–36.

Shahid, M., & Arora, A. K. (2019). *Ecballium elaterium* seed oil: Heavy metals and physicochemical analysis from Arid zone of Rajasthan. *Journal of Applicable Chemistry*, *8*(2), 844–849.

Tabata, M., Honda, G., & Sezik, E. (1988). *A Report on Traditional Medicine and Medicinal Plants in Turkey*. Faculty of Pharmaceutical Sciences, Kyot, Japan.

Toker, G., Memisoglu, M., Toker, M. C., & Yesilada, E. (2003, December). Callus formation and cucurbitacin B accumulation in Ecballium elaterium callus cultures. *Fitoterapia*, *74*(7–8), 618–623. https://doi.org/10.1016/S0367-326x(03)00165-5

Touihri-Barakati, I., Kallech-Ziri, O., Boulila, A., Khwaldia, K., Marrakchi, N., Hanchi, B., Hosni, K., & Luis, J. (2016). Targetting $\alpha v\beta 3$ and $\alpha 5\beta 1$ integrins with *Ecballium elaterium* (L.) A. Rich. seed oil. *Biomedicine and Pharmacotherapy*, *84*, 1223–1232. https://doi.org/10.1016/j.biopha.2016.10.035

Touihri, I., Kallech-Ziri, O., Boulila, A., Fatnassi, S., Marrakchi, N., Luis, J., & Hanchi, B. (2019). *Ecballium elaterium* (L.) A. Rich. seed oil: Chemical composition and antiproliferative effect on human colonic adenocarcinoma and fibrosarcoma cancer cell lines. *Arabian Journal of Chemistry*, *12*, 2347–2355.

Touil, A. (2017). A flavonol triglycoside from Ecballium elaterium (Cucurbitaceae). *Der Pharma Chemica, 9*(10), 59–61.

Ugulu, I., Baslar, S., Yorek, N., & Dogan, Y. (2009, May). The investigation and quantitative ethnobotanical evaluation of medicinal plants used around Izmir province, Turkey. *Journal of Medicinal Plants Research, 3*(5), 345–367.

Uslu, C., Karasen, R. M., Sahin, F., Taysi, S., & Akcay, F. (2006). Effect of aqueous extracts of *Ecballium elaterium* rich, in the rabbit model of rhinosinusitis. *International Journal of Pediatric Otorhinolaryngology, 70*(3), 515–518. https://doi.org/10.1016/j.ijporl.2005.07.020

Vlachos, P., Kanitsakis, N. N., & Kokonas, N. (1994). Fatal cardiac and renal failure due to *Ecbalium elaterium* (Squirting Cucumber). *Journal of Toxicology: Clinical Toxicology, 32*(6), 737–738. https://doi.org/10.3109/15563659409017981

Yesilada, E., Tanaka, S., Sezik, E., & Tabata, M. (1988). Isolation of an anti-inflammatory principle from the fruit juice of *Ecballium elaterium. Journal of Natural Products, 51*(3), 504–508. https://doi.org/10.1021/np50057a008

Yesilada, E., Tanaka, S., Tabata, M., & Sezik, E. (1989). Antiinflammatory effects of the fruit juice of *Ecballium elaterium* on edemas in mice. *Phytotherapy Research, 3*(2), 75–76. https://doi.org/10.1002/ptr.2650030210

Yeşilada, E., Üstün, O., Sezik, E., Takaishi, Y., Ono, Y., & Honda, G. (1997). Inhibitory effects of Turkish folk remedies on inflammatory cytokines: Interleukin-1α, interleukin-1β and tumor necrosis factor α. *Journal of Ethnopharmacology, 58*(1), 59–73. https://doi.org/10.1016/s0378-8741(97)00076-7

Yılmaz, K., Karakuş, F., Eyol, E., Tosun, E., Yılmaz, İ., & Ünüvar, S. (2018). Cytotoxic effects of Cucurbitacin I and *Ecballium elaterium* on breast cancer cells. *Natural Product Communications, 13*(11), 1445–1448.

6 *Fumaria* sp.

Nuraniye Eruygur

INTRODUCTION

The genus *Fumaria* L. (fumitory or fumewort, from Latin fumus terrae, "smoke of the earth") is a genus of about 60 species of annual flowering plants which belong to Fumariaceae (Papaveraceae) family. The genus is native to Europe, Africa, and Asia, most diverse in the Mediterranean region, and introduced to North and South America, and Australia. *Fumaria* species are sometimes used in herbal medicine (*Fumaria - Wikipedia*, 2021). Among the *Fumaria* species, the medicinal properties of *F. officinalis* (Figure 6.1) are worldwide recognized, and the plant is included in the European Pharmacopoeia, British Pharmacopoeia, and Turkish Pharmacopoeia. In Germany, this plant is approved for the indication of colicky pain affecting the gallbladder and biliary system, together with the gastrointestinal tract (Hentschel et al., 1995). Literature reviews show that there are many reports on pharmacological, immunological, biological, and other therapeutic activities of these valuable herbs which are reviewed in this article.

BOTANICAL FEATURES/ETHNOBOTANICAL USAGE

Botanical Features

The genus *Fumaria* (Fumariaceae, Papaverales) consists of 60 species widely distributed all over the world and especially in the Mediterranean region (Vrancheva et al., 2016). *Fumaria* species are mostly annual herbs that are scattered and branched, with the ability to climb on occasion. Compound leaves are divided into narrow, oblong–filiform segments (Al-Ghazzawi et al., 2020). In his book Species Plantarum, Linnaeus described the genus *Fumaria* in 1753. He got the name from the Latin fumus terrae, which means "earth's smoke". Earth smoke is the name given to Fumaria species, which comes from an early European legend that fumitory was formed by vapor rising from the earth. Beggary, fume-of-the-earth, fumiterre, fumusterre, God's fingers and thumbs, snapdragon, and wax dolls are some of the other common names for *Fumaria* species. Despite their delicate look, *Fumaria* taxa are typically regarded as noxious weeds capable of strangling entire crop fields and, as their common term beggary implies, thrive in nutrient-deficient soils. Since the fifth century AD, *Fumaria* species have been linked to witchcraft and superstition throughout Europe. The leaves were burned for their smoke, which was thought to have the ability to ward off evil spirits and spells (Zhang et al., 2020). Fumitory species are mostly found in temperate and cooler parts of the ancient world, where they commonly grow as weeds. Fumariaceae is mainly a temperate family that embraces 16 genera and 400 species. Economically, its use is limited to garden ornamentals (Mitich, 1997). Identification of the species belonging to this Fumaria genus is based on several specific morphological characteristics: presence or absence of the sepals, their length, shape of the fruit, length of the fruit pedicel, and length of the fruit pedicel bracteole (Păltinean et al., 2016).

Ethnobotanical Usage

Fumaria officinalis L., (Figure 6.1) named "smoke of the earth", "fumitory" is the most common plant of Papaveraceae family, and has been used as Asian folk medicine in many inflammatory and painful ailments like conjunctivitis and rheumatism (Raafat & El-Zahaby, 2020). Several *Fumaria*

DOI: 10.1201/9781003146971-6

FIGURE 6.1 Habitat photo of *Fumaria officianlis* L. (https://kocaelibitkileri.com/fumaria-officinalis/).

species, including *F. capreolata*, *F. densiflora*, *F. indica*, *F. officinalis*, *F. parviflora*, and *F. vaillantii*, have long been utilized in folk medicine. In Ayurvedic, Unani, and other traditional medical systems in South Asia, *Fumaria indica* is one of the most often utilized plants. Plants of the genus *Fumaria* have been used in traditional medicine as antihypertensives, diuretics, hepatoprotectants, and laxatives (to treat gastrointestinal ailments), as well as to treat several skin ailments (R. Suau et al., 2002; Zhang et al., 2020). Many of these therapeutical usages have been confirmed by *in vitro*, *in vivo*, and clinical studies. The consumption of *Fumaria* species especially for gastrointestinal disorders is common in folk medicine. However, there are still several unknown aspects of *Fumaria* plants that need more investigation and attention.

PHYTOCHEMICALS

Phytochemical investigations of *Fumaria* species have revealed that many components from this genus are highly bioactive. Here are the phytochemicals analyzed and isolated from different species of *Fumaria* genus.

ALKALOIDS

The Fumariaceae family is represented by two genera in the flora of Turkey, namely *Fumaria* L. and *Corydalis* Medik., which are known to produce isoquinoline alkaloids derived from phenylalanine and tyrosine (Orhan et al., 2007). Norfumaritine was isolated from *Fumaria kralikii* (Colton et al., 1985). Simple isoquinolines, (seco-)phthalide isoquinolines, aporphines, protopines (Figure 6.2), protoberberines, spirobenzylisoquinolines, benzophenanthridines, and indenobenzazepines have all been isolated from *Fumaria* species in previous phytochemical research (Chlebek et al., 2016). The biological activity of *Fumaria* is mostly associated with the presence of isoquinoline alkaloids (Sousek et al., 1999). Seger *et al.* isolated a set of ten alkaloids from *F. officinalis* (Seger et al., 2004). Protopine is an isoquinoline alkaloid, biosynthesized in *F. rostellata* and *F. officinalis* cell suspensions (Georgieva et al., 2015). Protopine and adlumidiceine were isolated from *Fumaria parviflora* LAM. as the major alkaloids, while protopine and cryptopine have been isolated from *F. kralikii* (Popova et al., 1982).

PHENOLICS

Organic acids, namely citric, coumaric, ferulic, fumaric, malic, 3-hydroxybenzoic, protocatechuic acid and caffeic acid (and its methylester), were identified by GC-MS in *Fumaria agraria*, *F. capreolata*, *F. densiflora*, *F. muralis*, *F. officinalis*, *F. parviflora*, and *F. vaillantii* (Sousek et al., 1999). Citric acid and malic acid, amino acids, and carbohydrates were determined in *F. officinalis*, *F.*

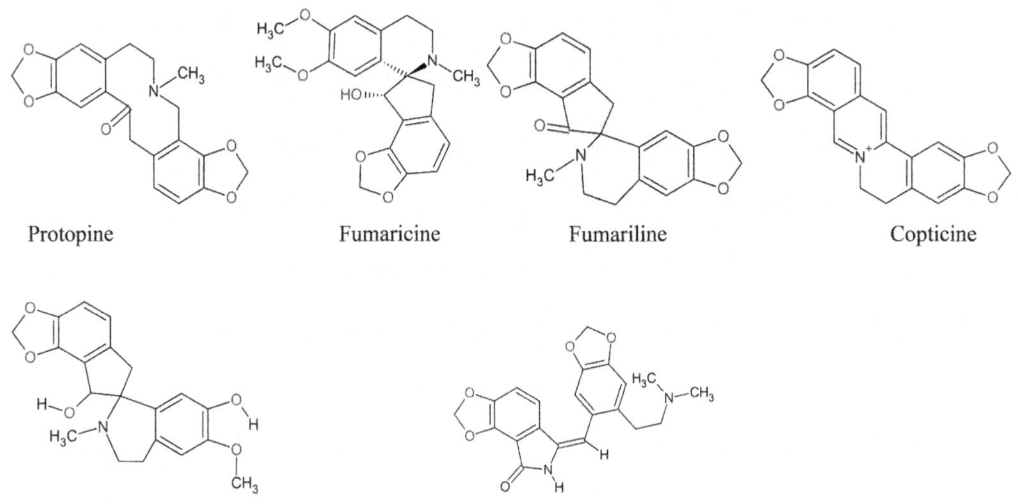

Protopine Fumaricine Fumariline Copticine

Fumaritine Fumaranine

FIGURE 6.2 Chemical structures and names of compounds reported from *Fumaria* species.

thuretii, *F. kralikii*, *F. rostellata*, and *F. schrammii* (Vrancheva et al., 2014). Myricetin, kaempferol, quercetin, rutin, hyperoside, hesperidin, apigenin, *p*-coumaric acid, ferulic acid, and sinapic acid were detected in *Fumaria officinalis*, *Fumaria thuretii*, *Fumaria kralikii*, *Fumaria rostellata*, and *Fumaria schrammii* (Ivanov et al., 2014). In a study, some phenol carboxylic acids including cynarine, chlorogenic, isochlorogenic, and ferulic acids as well as some flavonoids such as isovitexin, rutin, isoquercitrin, and quercitrin were determined in *Fumaria officinalis* by Paltinean et al. (2016). In another study by these researchers, some phenolic compounds such as *p*-coumaric acid, ferulic acid, isoquercitrin, rutin, quercitrin, quercetin, and kaempferol were determined by LC-MS method in rutin and isoquercetin were determined in *F. jankae*, *F. vailantii*, *F. schleicheri*, *F. officinalis*, *F. rostellata*, and *F. capreolata* (Paltinean et al., 2017).

BIOLOGICAL ACTIVITIES

ANTIOXIDANT

In rat mitochondrial membranes, the alkaloid extracts of eight *Fumaria* species showed higher antioxidant activities against tert-butyl hydroperoxide (tBH)-induced lipoperoxidation; however, the phenolic extracts were two- to threefold more effective as DPPH radical scavenger than the alkaloid extracts (Sousek et al., 1999). Methanol extract of *F. officinalis* was also investigated for radical scavenging and glycation inhibition activity, and results suggest that there was a significant correlation between antioxidant and anti-glycation activity (Safari et al., 2018). In fact, the IC_{50} value of DPPH-radical scavenging activity was found as 62.91± 1.85 mg/mL for *F. indica* methanol extract (Brice Landry et al., 2021).

ANTIMICROBIAL

A novel compound, n-octacosan-7β-ol (OC) isolated from methanol extract of whole plant of *F. parviflora* was shown to have significant antimicrobial activity against tested microorganisms *in vitro* with the minimum inhibitory concentration (MIC) of 250–500 µg/mL (Jameel, Islamuddin, et al., 2014). The silver nanoparticles using plant extract of *Fumaria officinalis* L. (AgNPs-E) showed strong antibacterial activity (Cakic et al., 2018). The methanolic extract of *F. indica* was evaluated for *in vitro* antimicrobial activity using agar disk diffusion and broth dilution method against bacterial and fungal strains, and results show that methanol extract possess greater antibacterial activity than antifungal activity, with the MIC and MBC values being 150 and 250 microL/mL against *Escherichia coli*, respectively (Khan et al., 2014). *F. asepala* extract showed antimicrobial activity, among the test strains, *Bacillus cereus* was the most susceptible, and *Escherichia coli* was the least sensitive (Jafari-Sales et al., 2019). The petroleum ether, methanol, and aqueous extract of *F. indica* showed antibacterial activity with MIC values of 1.65 mg/mL, 1.065 mg/mL, and 18.36 mg/mL, respectively (Safdar et al., 2017). The methanolic extract of *F. indica* was fractioned with n-hexane, $CHCl_3$, EtOAc, and n-BuOH sequentially, and antibacterial activity was checked against *Escherichia coli*, *Pasturella multocida*, *Bacillus subtilis*, and *Staphylococcus aureus* by the disc diffusion method. Results show that aqueous fraction of *F. indica* showed very good antibacterial activity against *P. multocida* with a zone of inhibition of 26 mm and MIC of 98 μg/mL; while chloroform and n-BuOH fractions showed good antifungal activity against *Ganoderma lucidum* with a zone of inhibition 24 mm and MIC of 115 μg/mL (Riaz et al., 2019). The 32 isoquinoline alkaloids from Turkish *Fumaria* and *Corydalis* species were tested for their antimicrobial activity against several microorganisms, among them 23 alkaloids have significant antibacterial activity against Gram-positive and Gram-negative bacteria at a concentration of 1 mg/mL (Abbasoglu et al., 1991).

ANTIVIRAL

Thirty-three isoquinoline alkaloids obtained from *Fumaria* and *Corydalis* species growing in Turkey have been evaluated for their *in vitro* antiviral activities. Both DNA virus Herpes simplex (HSV) and RNA virus Parainfluenza (PI-3) were employed for antiviral assay. Results show that the alkaloids were found to have selective inhibition against the PI-3 virus ranging between 0.5 and 64 μg/mL as minimum and maximum cytopathogenic effects (CPE) inhibitory concentrations, whereas they were completely inactive toward HSV (Orhan et al., 2007).

ENZYME INHIBITION

Aqueous and methanolic extracts of *F. officinalis* exhibited significant hypoglycemic effects at 0.5 and 0.75 g/kg dosages due to its antioxidant and alpha-amylase inhibitory activities (Fatima et al., 2019). The chloroform:methanol (1:1) extracts of a number of the plant species belonging to eight families were screened for their anticholinesterase activity by *in vitro* Ellman method at 10 μg/mL and 1 mg/mL concentrations, and among the extracts screened, all of the *Fumaria* extracts displayed highly potent inhibition against acetylcholinesterase and butyrylcholinesterase at 1 mg/mL concentration compared to the standard (Mukherjee et al., 2007; Orhan et al., 2004).

GASTROINTESTINAL SYSTEM

F. officinalis contains a variety of phytopharmaceuticals and is mostly used to treat hepatobiliary system functional problems such as colicky symptoms in the gallbladder, biliary system, and gastrointestinal tract. The presence of isoquinoline alkaloids in Fumariae herbs is primarily responsible for these biological actions (Sturm et al., 2006). The efficacy of *F. officinalis* has been studied in irritable bowel syndrome (IBS) patients due to its antispasmodic properties. Results showed that IBS-related pain decreased more in the fumitory group than in the placebo group in a randomized, double-blind, placebo-controlled experiment (Rahimi & Abdollahi, 2012). The EtOH extract of *F. asepala* was tested against a pentobarbital-induced hypnosis model in mice treated with carbon tetrachloride (CCl_4) as hepatotoxin and found to cause a significant hepatoprotective effect as compared to *Cynara scolymus* stem extract used as the reference drug (Deliorman et al., 1999). The leaves and stems of *Fumaria* spp. have been used as hepatoprotectant in Egyptian folk medicine by brewed and fresh herb consumption (Ghanadi et al., 2019).

TABLE 6.1
Ethnopharmacological Uses of *Fumaria* Species

Species	Region	Common Name	Used Plant Parts	Medicinal Uses	References
F. asepala	Turkey	Şahtere	Herb	Abdominal pain, headache, itching antiseptic	(Altundag & Ozturk, 2011)
F. microcarpa					
F. officinalis				Toothache, oral diseases	
F. vaillantii	Turkey	Şahtere	Extract	Blood purifier, skin diseases	(Şener, 2002)
Fumaria sp.	Romania	-	Aerial parts	Hepatobiliary diseases, diuretic	(Paltinean et al., 2017)
F. officinalis	Algeria	-	-	Hepatobiliary diseases	(Khamtache-Abderrahim et al., 2016)
F. indica	Pakistan	Pitpapra Shahtrah	Whole plant	Diuretic, laxative, colic, migraine, skin complaints	(Gulshan et al., 2012)
F. officinalis	Bulgaria	-	-	Antihypertensive, diuretic, hepatoprotectant, for skin rashes	(Ognyanov et al., 2018)
F. officinalis	İraq	-	-	Blood-related diseases and cancer	(Adham et al., 2021)
F. vaillantii/F. indica	Iran, India	Parpata, Pitpapra, Parpatakam	-	Hepatobiliary disorders, dermatological diseases, dysfunction and gastrointestinal disorders and as a blood purifier	(Srivastava & Choudhary, 2014)
F. indica	India	-	-	Laxative, diuretic, beneficial in dyspepsia and liver complaints	(Pandey et al., 2008)
F. parviflora	Greco-Arab, Saudi Arabia	Homaira	-	Hyperactive gut and respiratory disorders including diarrhea, abdominal cramps, indigestion, and asthma	(Bashir et al., 2012)
F. capreolata F. bastardii	Algeria	-	-	Hepatobiliary dysfunction and gastrointestinal disorders	(Maiza-Benabdesselam et al., 2007)

TABLE 6.2
Phytochemical Studies on *Fumaria* Species

Species Name	Phytochemical Constituents	Phytochemical's Name	References
Fumaria asepala	İsoquinoline alkaloid	N-methylcorydaldine, corydaldine, oxyhydrastinine	(Şener, 2002)
F. bastardii	Benzylisoquinolines		
F. boissieri	Protopines		
F. bracteosa	Protoberberines		
F. capreolata	Phthalideisoquinolines		
F. cilicica	Secophthalideisoquinolines		
F. densiflora	Aporphines		
F. flabellate	Benzophenanthridines		
F. gaillardotii	Spirobenzylisoquinolines		
F. Judaica	Secospirobenzyliseoquinolines		
F. kralikii	İndenobenzapzepines		
F. macrocarpa			
F. officinalis			
F. parviflorum			
F. petteri			
F. rostellata			
F. scheicheri			
F. vaillantii			
F. officinalis	Alkaloid		(Şener, 1982)
F. densiflora			
F. cilicica			
F. kralikii			
F. gaillardotii			
F. macrocarpa			
F. capreolata			
F. Judaica			
F. parviflora			
F. vaillantii			
F. macrocarpa			
F. asepala			
F. petteri subsp. *thuretii*			
F. officinalis	Alkaloid	Isoquinolines, (seco-)phthalide isoquinolines, aporphines, protopines, protoberberines, spirobenzylisoquinolines, benzophenanthridines, and indenobenzazepines	(Chlebek et al., 2016)
		Adlumine, corlumine, corydamine, cryptopine, fumarophycine, O-methylfumarophycine, hydrastine, parfumine, protopine, sinactine	(Seger et al., 2004)
		N-methylsinactine, dihydrofumariline	(Z.H.Mardirossian et al., 1983)
		Fumaricine, fumaritine, fumariline	(MacLean et al., 1969)
		Protopine, dl-tetrahydro-coptisine, cryptocavine, aurotensine	(Manske, 1938)

(Continued)

TABLE 6.2 (CONTINUED)
Phytochemical Studies on *Fumaria* Species

Species Name	Phytochemical Constituents	Phytochemical's Name	References
F. indica	Alkaloid	Fuyuziphine, (±)-a-hydrastine	(Pandey et al., 2008)
		Papracinine, paprazine, fumaritine N-oxide, parfumine, lastourvilline, feruloyl tyramine, fumariflorine, and N-methyl corydaldine	(Atta-Ur-Rahman et al., 1992)
		Paprafumine, paprarine, papraline	(Ahmad et al., 1995)
F. parviflora LAM	Alkaloid	Protopine, adlumidiceine, parfumine, fumariline, dihydrofumariline, cryptopine, (-)-stylopine, 8-oxocoptisine, sanguinarine, oxysanguinarine	(Popova et al., 1982)
F. kralikii JORDAN	Alkaloid	Protopine, fumarophycine, O-methylfumarophycine, adlumidiceine, berberine, coptisine, cryptopine, (-)-stylopine, (-)-canadine	
Fumaria kralikii	Alkaloid	Norfumaritine	(Colton et al., 1985)
F. densiflora DC	Alkaloid	Bicuculline, densiflorine, fumaritine, fumariline, scoulerine, coptisine, cryptopine, parfumine, protopine, sinactine, adlumine, fumaricine, fumaritridine, fumarophycine, fumarofine, corytuberine, cis-N-methylstylopinium iodide, stylopine, and two new 1-benzylisoquinoline alkaloids, viz. fumaflorine and fumaflorine methyl ester	(Taborska et al., 1996)
F. officinalis	Isoquinoline alkaloids	(-)-Fumaricine, fumaranine, (-)-dihydrofumariline, (-)-fumaritine	(Chlebek et al., 2016)
		Protopine	(Rakotondramasy-Rabesiaka et al., 2007)
F. parviflora	Alkaloid	Adlumidicein, copticine, fumariline, perfumine, protopine, fumaranine, fumaritine, paprafumicin and paprarine	(Bashir et al., 2012; Jameel, Ali, et al., 2014)
		Protopine, adlumidiceine, parfumine, fumariline, dihydrofumariline, cryptopine, stylopine, 8-oxocoptisine, sanguinarine, oxysanguinarine	(Popova et al., 1982)
F. kralikii	Alkaloid	Protopine, fumarophycine, 0-methylfumarophycine, adlumidiceine, berberine, coptisine, cryptopine, (-)-stylopine, (-)-canadine	

(Continued)

TABLE 6.2 (CONTINUED)
Phytochemical Studies on *Fumaria* Species

Species Name	Phytochemical Constituents	Phytochemical's Name	References
F. rostellata *F. officinalis*	Alkaloid	Protopine	(Georgieva et al., 2015)
F. sepium *F. agraria*	Alkaloid	Demeticine, (+)-Isoboldine, dihydrosanguinarine, noroxyhydrastinine, oxyhydrastinine, coptisine, (-)-stylopine, protopine, densiflorine, (-)-Fumaritine,N-oxide, (+)-parfumine	(Rafael Suau et al., 2002)
Fumaria agraria *F. capreolata* *F. densiflora* *F. muralis* *F. officinalis* *F. parviflora* *F. vaillantii*	Organic acid	Citric, coumaric, ferulic, fumaric, malic, 3-hydroxybenzoic, protocatechuic acid, and caffeic acid	(Sousek et al., 1999)
F. officinalis	Phenol carboxylic acids	Cynarine, chlorogenic, isochlorogenic and ferulic acids	(Paltinean et al., 2016)
	Flavonoid	Isovitexin, rutin, isoquercitrin and quercitrin	
	isoquinoline alkaloids	Allocryptopine, chelidonine, protopine, bicuculline, sanguinarine, cheleritrine, stylopine and hydrastine	
F. jankae *F. vailantii* *F. schleicheri* *F. officinalis* *F. rostellata* *F. capreolata*	Flavonoid	Rutin, isoquercetin	(Paltinean et al., 2017)
Fumaria officinalis *Fumaria thuretii* *Fumaria kralikii* *Fumaria rostellata* *Fumaria schrammii*	Phenolic	Myricetin, kaempferol, quercetin, rutin, hyperoside, hesperidin, apigenin, p-coumaric acid, ferulic acid, sinapic acid	(Ivanov et al., 2014)
F. officinalis *F. thuretii* *F. kralikii* *F. rostellata* *F. schrammii*	Phenolic acid Amino acid Carbohydrates	Citric acid, malic acid	(Vrancheva et al., 2014)

TABLE 6.3
***In Vitro, In Vivo* Studies on *Fumaria* sp.**

Species Name	Biological Activity	Sample	Effective Doses	References
İn vivo				
F. asephalae	Hepatoprotective	Ethanol extract	Pentobarbital-induced sleeping time in mice, 500 mg/kg per os	(Deliorman et al., 1999)
F. asepalae *F. vailantii*	Hepatoprotective	80% aqueous EtOH extract	CCl_4-induced hepatotoxicity in rats, 500 mg/kg per os	(Aktay et al., 2000)
Fumaria indica	Hepatoprotective	Protopine	10–20 mg/kg (p.o.)	(Gupta et al., 2012; Rathi et al., 2008)
Fumaria indica	Antihepatotoxic	Monomethyl fumarate	Carbon tetrachloride-, paracetamol-, and rifampicin-induced hepatotoxicity	(Rao & Mishra, 1998)
Fumaria parviflora	Hepatoprotective	Aqueous-methanol extract	500 mg/kg, paracetamol- and CCl_4-induced hepatic damage	(Gilani et al., 2005)
F. cilicica *F. densiflora* DC. *F. kralikii* Jordan *F. parviflora*	Hepatoprotective	Ethanol extract	400 mg/kg b.w.	(Orhan et al., 2012)
Fumaria indica	Hepatoprotective	50% aqueous ethanol extract	200 and 400 mg/kg	(Hussain et al., 2012)
F. officinalis	Antidiabetic	Aqueous extract	10% extract feeding for 8 weeks in diabetes-induced male Wistar rats	(Goodarzi et al., 2011)
F. parviflora Lam.	Antihypertensive, antispasmodic	-	-	(Hilal et al., 1989)
In vitro				

(Continued)

TABLE 6.3 (CONTINUED)
***In Vitro, In Vivo* Studies on *Fumaria* sp.**

Species Name	Biological Activity	Sample	Effective Doses	References
F. indica	Antioxidant	Methanol extract	IC_{50} values of DPPH-radical scavenging activity: 62.91± 1.85 mg/mL	(Brice Landry et al., 2021)
F. officinalis *F. thuretii* *F. kralikii* *F. rostellata* *F. schrammii*	Antioxidant	Ethanol extract	DPPH, ABTS, FRAP, CUPRAC	(Ivanov et al., 2014)
F. capreolata *F. bastardii*	Antioxidant	Alkaloid extract	Reducing power, linoleic acid peroxidation, DPPH	(Maiza-Benabdesselam et al., 2007)
F. asepala *F. capreolata* *F. cilicica* *F. densiflora* *F. flabellate* *F. judaica* *F. kralikii* *F. macrovarpa* *F. parviflora* *F. petteri* *F. vaillantii*	Anticholinesterase	Chloroform:methanol (1:1) Extract of whole plant	At 1 mg/mL concentration	(Mukherjee et al., 2007), (Orhan et al., 2004; Sener & Orhan, 2004)
F. officinalis	Antibacterial	Silver nanoparticles		(Cakic et al., 2018)
F. densiflora	Antiplasmodial antitrypanosomal activity	Ethanol extract	*Plasmodium falciparum, Trypanosoma bruceirhodesiense* at 0.81–4.85 µg/mL	(Orhan et al., 2015)
F. indica	Antimicrobial	Petroleum ether, methanol, and aqueous extracts	MIC: 1.65, 1.065, and 18.36 mg/mL	(Safdar et al., 2017)
F. densiflora DC. *F. officinalis* L	Cytotoxicity	Hexane and ethyl acetate extract	In 10–1000 ppm concentration	(Erdoğan, 2009)

Cytotoxicity

F. officinalis has been shown to have cytotoxic action against human pharyngeal squamous carcinoma cells FaDu, human tongue squamous carcinoma cells SCC-25, and human breast adenocarcinoma cells MCF-7 (Petruczynik et al., 2019). The cytotoxic activity of *F. officinalis* extracts on two leukemia and nine multiple myeloma cell lines was investigated. Results show that ethyl acetate fraction induced autophagic cell death, while chloroform fraction stimulated iron-dependent cell death (Adham et al., 2021). The methanolic extract of *Fumaria indica* aerial parts was evaluated for cytotoxic activity against HepG2 cell line by MTT and trypan blue assays, showing dose-dependent cytotoxic activity (Brice Landry et al., 2021).

Hepatoprotective Activity

Fumaria indica 50% aqueous ethanol extract exerted a chemopreventive effect at a dose of 200 and 400 mg/kg on the experimental animals by reversing the oxidant-antioxidant imbalance during hepatocarcinogenesis induced by N-nitrosodiethylamine and CCl_4 (Hussain et al., 2012).

CONCLUSION

Fumaria has been used as popular medicine for its diuretic, anthelmintic, anti-inflammatory, and laxative properties in various regions throughout the world. It was used by northern European and North American native people as a cholagogue, stomachic, diaphoretic, and to purify blood. Some of these traditional and folk usages have been evaluated showing the potential medicinal use of the plant. The biological activity of *Fumaria* is mostly associated with the presence of isoquinoline alkaloids in the plant. In the last few years, many scientific reports have described the properties of *Fumaria*. As it is reviewed in this chapter, the antioxidant and hepatoprotective activity of various species of *Fumaria* is reported frequently. This is due to the high content of alkaloids in these plants. Among the medicinal properties of *Fumaria*, it is important to have hepatoprotective effects, especially since *Fumaria* species contain isoquinoline alkaloid constituents. Finally, the presence of anti-inflammatory compounds such as protopine and fumarine is another reason why these plants may be used as potential sources of medicinal compounds and drugs in the future.

REFERENCES

Abbasoglu, U., Sener, B., Günay, Y., & Temizer, H. (1991). Antimicrobial activity of some isoquinoline alkaloids. *Archiv der Pharmazie*, *324*(6), 379–380.

Adham, A. N., Naqishbandi, A. M., & Efferth, T. (2021). Cytotoxicity and apoptosis induction by *Fumaria officinalis* extracts in leukemia and multiple myeloma cell lines. *Journal of Ethnopharmacol*, *266*, 113458, Article 113458. https://doi.org/10.1016/j.jep.2020.113458

Ahmad, S., Bhatti, M. K., & Choudhary, M. I. (1995). Alkaloidal constituents of *Fumaria indica*. *Phytochemistry*, *40*(2), 593–596.

Aktay, G., Deliorman, D., Ergun, E., Ergun, F., Yeşilada, E., & Cevik, C. (2000). Hepatoprotective effects of Turkish folk remedies on experimental liver injury. *Journal of Ethnopharmacology*, *73*(1–2), 121–129.

Al-Ghazzawi, A. M., Abu Zarga, M. H., & Abdalla, S. S. (2020). Chemical constituents of *Fumaria densiflora* and the effects of some isolated spirobenzylisoquinoline alkaloids on murine isolated ileum and perfused heart. *Natural Product Research*, *34*(8), 1180–1185.

Altundag, E., & Ozturk, M. (2011). Ethnomedicinal studies on the plant resources of east Anatolia, Turkey. *Procedia-Social and Behavioral Sciences*, *19*, 756–777.

Atta-Ur-Rahman ,Bhatti, M. K., Akhtar, F., & Choudhary, M. I. (1992). Alkaloids of Fumaria indica. *Phytochemistry*, *31*(8), 2869–2872.

Bashir, S., Al-Rehaily, A. J., & Gilani, A.-H. (2012). Mechanisms underlying the antidiarrheal, antispasmodic and bronchodilator activities of Fumaria parviflora and involvement of tissue and species specificity. *Journal of Ethnopharmacology*, *144*(1), 128–137.

Brice Landry, K., Tariq, S., Malik, A., Sufyan, M., Ashfaq, U. A., Ijaz, B., & Shahid, A. A. (2021). Berberis lyceum and Fumaria indica: In vitro cytotoxicity, antioxidant activity, and in silico screening of their selected phytochemicals as novel hepatitis C virus nonstructural protein 5A inhibitors. *Journal of Biomolecular Structure and Dynamics*, *40*, 7829–7851.

Cakic, M., Glisic, S., Cvetkovic, D., Cvetinov, M., Stanojevic, L., Danilovic, B., & Cakic, K. (2018). Green synthesis, characterization and antimicrobial activity of silver nanoparticles produced from plant extract. *Colloid Journal*, *80*(6), 803–813. https://doi.org/10.1134/S1061933x18070013

Chlebek, J., Novak, Z., Kassemova, D., Safratova, M., Kostelnik, J., Maly, L., Locarek, M., Opletal, L., Host'alkova, A., Hrabinova, M., Kunes, J., Novotna, P., Urbanova, M., Novakova, L., Macakova, K., Hulcova, D., Solich, P., Perez Martin, C., Jun, D., & Cahlikova, L. (2016). Isoquinoline alkaloids from *Fumaria officinalis* L. and their biological activities related to Alzheimer's Disease. *Chemistry & Biodiversity*, *13*(1), 91–99. https://doi.org/10.1002/cbdv.201500033

Colton, M. D., Guinaudeau, H., Shamma, M., Önür, M. A., & Gözler, T. (1985). (-)-Norfumaritine: A new spirobenzylisoquinoline alkaloid from Fumaria kralikii. *Journal of Natural Products*, *48*(5), 846–847.

Deliorman, D., Ergun, F., & Yesilada, E. (1999). A preliminary study on hepatoprotective activity of some Turkish folk remedies by using Pentobartibal-induced hypnosis model in mice. *Gazi University Journal of the Faculty of Pharmacy*, *16*(2), 77–81.

Erdoğan, T. F. (2009). Brine shrimp lethality bioassay of *Fumaria densiflora* Dc. and *Fumaria officinalis* L. extracts. *Hacettepe University Journal of the Faculty of Pharmacy* (2), 125–132.

Fatima, S., Akhtar, M. F., Ashraf, K. M., Sharif, A., Saleem, A., Akhtar, B., Peerzada, S., Shabbir, M., Ali, S., & Ashraf, W. (2019). Antioxidant and alpha amylase inhibitory activities of *Fumaria officinalis* and its antidiabetic potential against alloxan induced diabetes. *Cell Mol Biol (Noisy-le-grand)*, *65*(2), 50–57. https://doi.org/10.14715/cmb/2019.65.2.8

Fumaria - Wikipedia. (2021). Retrieved 23 June 2021 from https://en.wikipedia.org/wiki/Fumaria

Georgieva, L., Ivanov, I., Marchev, A., Aneva, I., Denev, P., Georgiev, V., & Pavlov, A. (2015). Protopine production by fumaria cell suspension cultures: Effect of light. *Applied Biochemistry and Biotechnology*, *176*(1), 287–300. https://doi.org/10.1007/s12010-015-1574-6

Ghanadi, K., Rafieian-Kopaei, M., Karami, N. & Abbasi, N. (2019). An ethnobotanical study of hepatoprotective herbs from shahrekord, chaharmahal and bakhtiari province, southwest of Iran Egypt. *J. Vet. Sci*, *50*,(2), 129–134.

Gilani, A. H., Bashir, S., Janbaz, K. H., & Khan, A. (2005). Pharmacological basis for the use of *Fumaria indica* in constipation and diarrhea. *Journal of Ethnopharmacology*, *96*(3), 585–589.

Goodarzi, M. T., Tootoonchi, A. S., Karimi, J., & Panah, M. H. (2011). Improvement of hyperinsulinemia in type 2 diabetic rats by Trigonell Foenum compare to *Fumaria officinalis. Clinical Biochemistry*, *44*(13), S335–S335. https://doi.org/10.1016/j.clinbiochem.2011.08.830

Gulshan, A. B., Dasti, A. A., Hussain, S., & Atta, M. I. (2012). Indigenous uses of medicinal plants in rural areas of Dera Ghazi Khan, Punjab, Pakistan. *Journal of Agricultural and Biological Science*, *7*(9), 750–762.

Gupta, P. C., Sharma, N., & Rao, C. V. (2012). A review on ethnobotany, phytochemistry and pharmacology of *Fumaria indica* (Fumitory). *Asian Pacific Journal of Tropical Biomedicine*, *2*(8), 665–669.

Hentschel, C., Dressler, S., & Hahn, E. (1995). *Fumaria officinalis* (Fumitory) – Clinical applications. *Fortschritte der Medizin*, *113*(19), 291–292.

Hilal, S., Aboutabl, E., Youssef, S., Shalaby, M., & Sokkar, N. (1989). Alkaloidal content and certain pharmacological activities of *Fumaria parviflora* Lam. growing in Egypt. *Plantes Medicinales et Phytotherapie*, *23*(2), 109–123.

Hussain, T., Siddiqui, H. H., Fareed, S., Sweety, K., Vijayakumar, M., & Rao, C. V. (2012). Chemopreventive effect of *Fumaria indica* that modulates the oxidant-antioxidant imbalance during N-nitrosodiethylamine and CC14-induced hepatocarcinogenesis in Wistar rats. *Asian Pacific Journal of Tropical Biomedicine*, *2*(2), S995–S1001.

Ivanov, I., Vrancheva, R., Marchev, A., Petkova, N., Aneva, I., Denev, P., Georgiev, V. G., & Pavlov, A. (2014). Antioxidant activities and phenolic compounds in Bulgarian *Fumaria* species. *International Journal of Current Microbiology and Applied Sciences*, *3*(2), 296–306.

Jafari-Sales, A., Mobaiyen, H., Jafari, B., & Sayyahi, J. (2019). Assessment of antibacterial effect of alcoholic extract of *Centaurea depressa* MB, *Reseda lutea* L. and *Fumaria asepala* on selected standard strains *in vitro. Scientific Journal of Nursing, Midwifery and Paramedical Faculty*, *5*(3), 63–73.

Jameel, M., Ali, A., & Ali, M. (2014). Phytochemical investigation of the aerial parts of *Fumaria parviflora* Lam. *Journal of Pharmaceutical and Biosciences B*, *2*, 1–8.

Jameel, M., Islamuddin, M., Ali, A., Afrin, F., & Ali, M. (2014). Isolation, characterization and antimicrobial evaluation of a novel compound N-octacosan 7 β ol, from *Fumaria parviflora* Lam. *BMC Complementary and Alternative Medicine*, *14*(1), 1–9.

Khamtache-Abderrahim, S., Lequart-Pillon, M., Gontier, E., Gaillard, I., Pilard, S., Mathiron, D., Djoudad-Kadji, H., & Maiza-Benabdesselam, F. (2016). Isoquinoline alkaloid fractions of Fumaria officinalis: Characterization and evaluation of their antioxidant and antibacterial activities. *Industrial Crops and Products*, *94*, 1001–1008.

Khan, A., Tak, H., Nazir, R., Lone, B. A., & Parray, J. A. (2014). In vitro anthelmintic and antimicrobial activities of methanolic extracts of *Fumaria indica*. *Clinical Microbiology*, *3*, 161. https://doi.org/10.4172/2327-5073.1000161

MacLean, D., Bell, R., Saunders, J., Chen, C.-Y., & Manske, R. (1969). Structures of three minor alkaloids of *Fumaria officinalis* L. *Canadian Journal of Chemistry*, *47*(19), 3593–3599.

Maiza-Benabdesselam, F., Khentache, S., Bougoffa, K., Chibane, M., Adach, S., Chapeleur, Y., Henry, M., & Laurain-Mattar, D. (2007). Antioxidant activities of alkaloid extracts of two Algerian species of *Fumaria*: *Fumaria capreolata* and *Fumaria bastardii*. *Records of Natural Products*, *1*(2–3), 28.

Manske, R. H. (1938). The alkaloids of Fumariceous plants:XVIII. *Fumaria officinalis* L. *Canadian Journal of Research*, *16*(12), 438–444.

Mardirossian, Z. H., Kiryakovt, H. G., Ruder, J. P., & MacLean, D. B. (1983). Alkaloids of fumaria officinalis. *Phytochemistry*, *22*(3), 759–761.

Mitich, L. W. (1997). Intriguing world of weeds .59. Fumitory (*Fumaria officinalis* L). *Weed Technology*, *11*(4), 843–845.

Mukherjee, P. K., Kumar, V., Mal, M., & Houghton, P. J. (2007). Acetylcholinesterase inhibitors from plants. *Phytomedicine*, *14*(4), 289–300.

Ognyanov, M., Georgiev, Y., Petkova, N., Ivanov, I., Vasileva, I., & Kratchanova, M. (2018). Isolation and characterization of Pectic polysaccharide fraction from *In Vitro* suspension culture of *Fumaria officinalis* L. *International Journal of Polymer Science*, *2018*, Article 5705036. https://doi.org/10.1155/2018/5705036

Orhan, I. E., Ozturk, N., & Sener, B. (2015). Antiprotozoal assessment and phenolic acid profiling of five *Fumaria* (fumitory) species. *Asian Pacific Journal of Tropical Medicine*, *8*(4), 283–286.

Orhan, I., Özçelik, B., Karaoğlu, T., & Şener, B. (2007). Antiviral and antimicrobial profiles of selected isoquinoline alkaloids from *Fumaria* and *Corydalis* species. *Zeitschrift für Naturforschung C*, *62*(1–2), 19–26.

Orhan, I., Şener, B., Choudhary, M., & Khalid, A. (2004). Acetylcholinesterase and butyrylcholinesterase inhibitory activity of some Turkish medicinal plants. *Journal of Ethnopharmacology*, *91*(1), 57–60.

Orhan, I. E., Şener, B., & Musharraf, S. G. (2012). Antioxidant and hepatoprotective activity appraisal of four selected *Fumaria* species and their total phenol and flavonoid quantities. *Experimental and Toxicologic Pathology*, *64*(3), 205–209.

Paltinean, R., Mocan, A., Vlase, L., Gheldiu, A. M., Crisan, G., Ielciu, I., Vostinaru, O., & Crisan, O. (2017). Evaluation of polyphenolic content, antioxidant and diuretic activities of six fumaria species. *Molecules*, *22*(4), Article 639. https://doi.org/10.3390/molecules22040639

Paltinean, R., Toiu, A., Wauters, J. N., Frederich, M., Tits, M., Angenot, L., Tamas, M., & Crisan, G. (2016). Phytochemical analysis of *Fumaria Officinalis* L. (Fumariaceae). *Farmacia*, *64*(3), 409–413.

Pandey, M., Singh, A., Singh, J., Singh, V., & Pandey, V. (2008). Fuyuziphine, a new alkaloid from *Fumaria indica*. *Natural Product Research*, *22*(6), 533–536.

Petruczynik, A., Plech, T., Tuzimski, T., Misiurek, J., Kapron, B., Misiurek, D., Szultka-Mlynska, M., Buszewski, B., & Waksmundzka-Hajnos, M. (2019). Determination of selected Isoquinoline alkaloids from *Mahonia aquifolia*; *Meconopsis cambrica*; *Corydalis lutea*; *Dicentra spectabilis*; *Fumaria officinalis*; *Macleaya cordata* extracts by HPLC-DAD and comparison of their cytotoxic activity. *Toxins (Basel)*, *11*(10), Article 575. https://doi.org/10.3390/toxins11100575

Popova, M., Šimanek, V., Dolejš, L., Smysl, B., & Preininger, V. (1982). Alkaloids from *Fumaria parviflora* and *F. kralikii*. *Planta Medica*, *45*(6), 120–122.

Raafat, K. M., & El-Zahaby, S. A. (2020). Niosomes of active *Fumaria officinalis* phytochemicals: Antidiabetic, antineuropathic, anti-inflammatory, and possible mechanisms of action. *Chinese Medicine*, *15*(1), 40. https://doi.org/10.1186/s13020-020-00321-1

Rahimi, R., & Abdollahi, M. (2012). Herbal medicines for the management of irritable bowel syndrome: A comprehensive review. *World Journal of Gastroenterology*, *18*(7), 589–600. https://doi.org/10.3748/wjg.v18.i7.589

Rakotondramasy-Rabesiaka, L., Havet, J. L., Porte, C., & Fauduet, H. (2007). Solid-liquid extraction of protopine from *Fumaria officinalis* L. – Analysis determination, kinetic reaction and model building. *Separation and Purification Technology*, *54*(2), 253–261. https://doi.org/10.1016/j.seppur.2006.09.015

Rao, K., & Mishra, S. (1998). Antihepatotoxic activity of monomethyl fumarate isolated from *Fumaria indica*. *Journal of Ethnopharmacology*, *60*(3), 207–213.

Rathi, A., Srivastava, A. K., Shirwaikar, A., Rawat, A. K. S., & Mehrotra, S. (2008). Hepatoprotective potential of *Fumaria indica* Pugsley whole plant extracts, fractions and an isolated alkaloid protopine. *Phytomedicine*, *15*(6–7), 470–477.

Riaz, T., Abbasi, M. A., Shazadi, T., & Shahid, M. (2019). Assessment of *Fumaria indica*, *Dicliptera bupleuroides* and *Curcuma zedoaria* for their antimicrobial and hemolytic effects. *Pakistan Journal of Pharmaceutical Sciences*, *32*(2), 697–702.

Safari, M. R., Azizi, O., Heidary, S. S., Kheiripour, N., & Ravan, A. P. (2018). Antiglycation and antioxidant activity of four Iranian medical plant extracts. *Journal of Pharmacopuncture*, *21*(2), 82–89. https://doi.org/10.3831/KPI.2018.21.010

Safdar, N., Yaqeen, N., Kazmi, Z., & Yasmin, A. (2017). Antimicrobial potential of *Mazus japonicus* and *Fumaria indica* extracts: Individual vs. synergistic effect. *Journal of Herbs, Spices & Medicinal Plants*, *23*(4), 272–283.

Şener, B. (1982). Turkish species of Fumaria L. and their alkaloids. *Journal of Faculty of Pharmacy of Ankara University*, *12*(1), 83–104.

Şener, B. (2002). Molecular diversity in the alkaloids of Turkish *Fumaria* L. species. *Acta Pharmaceutica Sciencia*, *44*(3), 205–212.

Seger, C., Sturm, S., Strasser, E. M., Ellmerer, E., & Stuppner, H. (2004). ^{1}H and ^{13}C NMR signal assignment of benzylisoquinoline alkaloids from *Fumaria officinalis* L. (Papaveraceae). *Magnetic Resonance in Chemistry*, *42*(10), 882–886. https://doi.org/10.1002/mrc.1417

Sener, B., & Orhan, I. (2004). Molecular diversity in the bioactive compounds from Turkish plants evaluation of acetylcholinesterase inhibitory activity of Fumaria species. *Journal- Chemical Society of Pakistan*, *26*(3), 313–315.

Sousek, J., Guedon, D., Adam, T., Bochorakova, H., Taborska, E., Valka, I., & Simanek, V. (1999). Alkaloids and organic acids content of eight Fumaria species. *Phytochemical Analysis*, *10*(1), 6–11.

Srivastava, S., & Choudhary, G. (2014). Pharmacognostic and pharmacological study of *Fumaria vaillantii* Loisel: A review. *Journal of Pharmacognosy and Phytochemistry*, *3*(1), 194–197.

Sturm, S., Strasser, E. M., & Stuppner, H. (2006). Quantification of *Fumaria officinalis* isoquinoline alkaloids by nonaqueous capillary electrophoresis-electrospray ion trap mass spectrometry. *Journal of Chromatography A*, *1112*(1–2), 331–338. https://doi.org/10.1016/j.chroma.2005.12.008

Suau, R., Cabezudo, B., Rico, R., López-Romero, J. M., & Nájera, F. (2002). Alkaloids from *Fumaria sepium* and *Fumaria agraria*. *Biochemical Systematics and Ecology*, *30*(3), 263–265.

Taborska, E., Bochorakova, H., Sousek, J., Sedmera, P., Vavreckova, C., & Simanek, V. (1996). *Fumaria densiflora* DC. alkaloids. *Collection of Czechoslovak Chemical Communications*, *61*(7), 1064–1072. https://doi.org/10.1135/cccc19961064

Vrancheva, R. Z., Ivanov, I. G., Aneva, I. Y., Dincheva, I. N., Badjakov, I. K., & Pavlov, A. I. (2014). GC-MS based metabolite profiling of five Bulgarian *Fumaria* species. *Journal of BioScience & Biotechnology*, *3*(3), 195–201.

Vrancheva, R. Z., Ivanov, I. G., Aneva, I. Y., Dincheva, I. N., Badjakov, I. K., & Pavlov, A. I. (2016). Alkaloid profiles and acetylcholinesterase inhibitory activities of *Fumaria* species from Bulgaria. *Zeitschrift für Naturforschung. C, A Journal of Biosciences*, *71*(1–2), 9–14. https://doi.org/10.1515/znc-2014-4179

Zhang, R., Guo, Q., Kennelly, E. J., Long, C., & Chai, X. (2020). Diverse alkaloids and biological activities of Fumaria (Papaveraceae): An ethnomedicinal group. *Fitoterapia*, *146*, 104697, Article 104697. https://doi.org/10.1016/j.fitote.2020.104697

7 *Glycyrrhiza* sp.

Merve Yüzbaşıoğlu Baran and Sıla Özlem Şener

BOTANICAL FEATURES/ETHNOBOTANICAL USAGE

Glycyrrhiza genus contains 30 species worldwide. The genus is represented by six taxa named as *Glycyrrhiza iconica* Hub.-Mor, *Glycyrrhiza glabra* L., *Glycyrrhiza asymmetrica* Hub.-Mor, *Glycyrrhiza aspera* Pall., *Glycyrrhiza echinata L.*, and *Glycyrrhiza flavescens* Boiss. in Turkey.

G. glabra is sparsely pubescent perennial herbs, growing to 30–60 cm in height. Elliptic leaves contain 5–9-paired, 15–45 mm sized leaflets. Inflorescence is elongate, 5–9 cm sized. Calyx teeth are 3 mm. Blue to violet corolla is 9–11 mm long. 15–25 × 4–5 mm legume with six seed is red-brown color. *G. echinata* is erect, glabrescent, bushy perennial herb, growing to 1–2 m. 4–8-paired, 15–30 x 7–15 mm longed leaflets are varied from lanceolate to broadly elliptic. Inflorescence is 2–5 cm, dense, globose to elongate. Calyx teeth are 1 mm. 5–7 mm sized corolla is ranged from white to mauve color. Legume contained 1–3-seeded is red-brown color and 12–16 x 6–8 mm long. *G. iconica* is a sparsely pubescent perennial herb, growing to 20–30 cm. Leaflets are 2–4-paired, 15–23 x 10–17 mm long. Calyx teeth and lilac corolla are 2.5–4 mm and 18–20 mm sized, respectively. The endemic taxa are distinguished from *G. aspera* by its larger flowers and shorter calyces. *G. flavescens* is 30–85 cm longed perennial herb. Elliptic leaflets are 5–8-paired, 25–40 x 10–15 mm long. Many-flowered inflorescence is 8–20 cm sized. 12–18 mm long corolla is golden-yellow color. The other endemic taxa *G. asymmetrica* is bristly perennial herbs, growing to 30–70 cm. Leaflets are 2–4-paired, 30–40 x 20–30 mm, widely obovate to orbicular-cuneate. Calyx teeth are 3–5 mm sized. 15–16 mm longed Corolla is yellow. Fruits contain brown 2-seeds 4 x 1.5 cm sized and oblong (Davis et al., 1965).

The roots of *G. glabra* have been traditionally used for laxative, antitussive effective agents, and the treatment of ulcer and respiratory disorders in Kahramanmaraş, Turkey (Uzun et al., 2020). A traditional cold mixture known as "ava susê" is prepared from the roots of *G. glabra*. Washed roots are mixed with water, and then kneaded like dough to prepare ava susê. Extra water is added, after the amount of water decreased in mixture, several times. Obtained water from the liquorice root is expressed as yeast. Liquorice syrup is prepared by adding a little more water to the yeast. The foam of the syrup is separated to eliminate the bitter taste. In addition to this cultural use, *Glycyrrhiza glabra* has been used for treatment of colds, respiratory tract diseases, flu, bronchitis, gastrointestinal diseases, smoking addiction, and diabetes in traditional Turkish medicine, for a long time (Yesil et al., 2019).

CHEMICAL CONSTITUENTS

Phytochemical studies showed that triterpenoid saponins derivatives are main components for *Glycyrrhiza* species growing in Turkey. Oleanane-type triterpenoid glucuronosides, yunganoside G, yunganoside G, yunganoside M, yunganoside E-2, and macedonoside C were isolated from hydro-alcoholic (3:7) extract of roots of *G. echinata* (Ali et al., 2020). Oleanane-type triterpenoid saponins, Glabasaponin A-G, and macedonoside A were yielded from ethanolic extract of *G. glabra* roots (Shou et al., 2019) (Table 7.1). Glycyrrhizin is the most common triterpenoid saponin obtained from *G. glabra* roots. Also, different groups of secondary metabolites were isolated from *G. glabra* roots, like flavonoids liquiritin, rhamnoliquirilin, liquiritigenin, prenyllicoflavone

DOI: 10.1201/9781003146971-7

TABLE 7.1
Some Important Triterpenoid Saponins from *Glycyrrhiza* Species

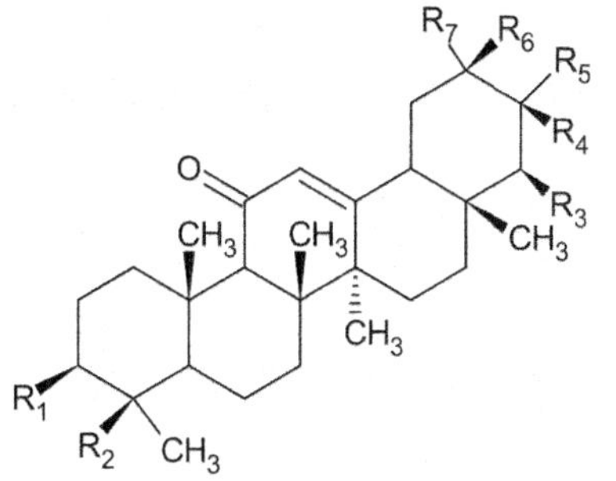

Compounds	R1	R2	R3	R4	R5	R6	R7	Species	References
Glabasaponin A	GlcA-GlcA	CH_2OH	H	OH	H	CH_3	COOH	*G. glabra*	(Shou et al., 2019)
Glabasaponin B	GlcA-GlcA	CH_2OH	H	H	OH	CH_3	COOH	*G. glabra*	(Shou et al., 2019)
Glabasaponin C	GlcA-GlcA	CH_3	H	Glu	H	CH_3	CH_2OH	*G. glabra*	(Shou et al., 2019)
Glabasaponin D	GlcA-GlcA	CH_3	H	Glu	H	CH_2OH	CH_3	*G. glabra*	(Shou et al., 2019)
Glabasaponin E	Rha	CH_3	H	Glu	H	CH_3	CH_2OH	*G. glabra*	(Shou et al., 2019)
Glabasaponin F	Rha	CH_3	Glu	H	H	CH_3	CH_2OH	*G. glabra*	(Shou et al., 2019)
Glabasaponin G	Rha	CH_3	H	Glu	H	CH_2OH	CH_3	*G. glabra*	(Shou et al., 2019)
Macedonoside A	Glu	CH_3	H	H	OH	CH_3	COOH	*G. glabra*	(Shou et al., 2019)
Yunganoside G_1	GlcA-GlcA-Rha	CH_2OH	H	OH	H	COOH	CH_3	*G. echinata*	(Ali et al., 2020)
Yunganoside G_2	GlcA-GlcA	CH_2OH	H	OH	H	COOH	CH_3	*G. echinata*	(Ali et al., 2020)
Glycyrrhizin	GlcA-GlcA	CH_3	H	H	H	COOH	CH_3	*G. glabra*	(Batiha et al., 2020)

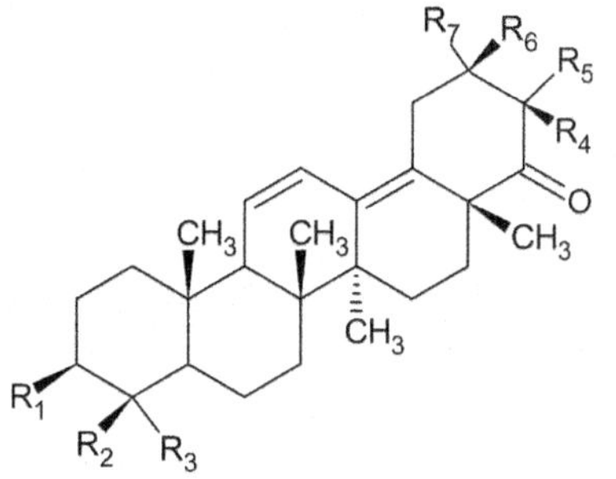

	R1	R2	R3	R4	R5	R6	R7	Species	References
Yunganoside M	GlcA-GlcA	CH_2OH	CH_3	H	H	COOH	CH_3	*G. echinata*	(Ali et al., 2020)
Yunganoside E_2	GlcA-GlcA	CH_3	CH_3	H	H	COOH	CH_3	*G. echinata*	(Ali et al., 2020)

GlcA: Glucuronic acid, Glc: Glucose, Rha: Rhamnose

A, glucoliquiritin, apioside, 1-methoxyphaseolin, shinpterocarpin, shinflavanone, licopyranocoumarin, glisoflavone, licoarylcoumarin, and coumarin-GU-12, and isoprenoid type phenolic constituents, isoangustone A, semilicoisoflavone B, licoriphenone, and 1-methoxyficifolinol (Batiha et al., 2020). Also, iconichalcone, hydroxymethoxychalcone, 2′-O-methylisoliquiritigenin, 3-O-methylkaempferol, topazolin, violanthin, glycyrrhisoflavone, licocoumarone, glyasperin C, licoricidin, licorisoflavan A, dehydroglyasperin C, iconisoflaven, glycycoumarin, 1-methoxyficifolinol, and edudiol of phenolic constituents were obtained from methanolic extract of *G. iconica* roots (Cakmak et al., 2016).

BIOLOGICAL ACTIVITIES

ANTIOXIDANT EFFECTS

The methanolic extract of *G. iconica* and *G. flavescens* aerial parts and roots were tested for their antioxidant effects through different antioxidant capacity tests. According to β-carotene/linoleic acid test system, Inhibition (%) values of *G. flavescens* aerial parts and roots were detected as 91.95 ± 0.15 and 95.40 ± 0.63, while the values were found as 85.58 ± 0.21 and 96.16 ± 0.07 for G. *iconica* and *G. flavescens* aerial parts and roots. Besides, free radical scavenging (DPPH) test showed that *G. iconica* roots (IC_{50} = 90.34 ± 1.64 µg/mL) and *G. flavescens* roots (IC_{50} = 127.15 ± 5.3) roots exhibited moderate antioxidant effects (Cakmak et al., 2016).

The total phenolic content and antioxidant activity of *G. glabra roots* cultivated in the wild are described. Total phenolics in *G. glabra* roots were 12.88 mg gallic acid equivalent (GAE)/g dry weight extract (DW). At a concentration of 800 µg/mL, root extract had the maximum antioxidant activity of 88.7% (Ercisli et al., 2008).

CYTOTOXIC EFFECTS

The methanol (MeOH) extract and chloroform ($CHCl_3$), ethyl acetate (EtOAc), *n*-buthanol (*n*-BuOH), and remaining water (rH_2O) subextracts of *G. iconica* roots and its isolated phenolic compounds were tested to determine their cytotoxic effects against hepatocellular (Huh7), breast (MCF7), and colorectal (HCT116) cancer cell lines by sulforhodamine B assay. The results showed that $CHCl_3$, EtOAc, and n-buOH subextracts, and hydroxymethoxychalcone, 2′-O-methylisoliquiritigenin, 3-O-methylkaempferol, topazolin, glycyrrhisoflavone, licocoumarone, glyasperin C, licoricidin, licorisoflavan A, and dehydroglyasperin of isolated compounds inhibited cancer cells with IC_{50} values in the range of 2.4 to 33 µM. Licoricidin, dehydroglyasperin C, iconisoflaven, and methoxyficifolinol caused apoptosis through caspase activation in Huh7 cells. Dehydroglyasperin C induced cytotoxic effects against Huh7 cells by activation of p53 expression (Cevik et al., 2019).

The MeOH extract, $CHCl_3$, and EtOAc subextracts of *G. glabra* roots and its isolated secondary metabolites exhibited cytotoxic effects against hepatocellular (Huh7), breast (MCF7), and colorectal (HCT116) cancer cells with IC_{50} values ranging from 5.6 to 33.6 µg/mL using sulforhodamine B assay. The compounds responsible for cytotoxic effects were determined as glabridin, 4′-O-methylglabridin, β-amyrin, kanzonol U, glabrene, and tetrahydroxymethoxychalcone (Cevik et al., 2018).

The least cytotoxic effect was observed in the sample collected from Calabria, Italy. The samples obtained from Dagestan and Uzbekistan showed less than 50% cell death at the 250 µg/mL test concentration. The sample collected from Afghanistan exhibited the most cytotoxic effect against A549 cells (IC_{50} = 191.3 µg/mL), HepG2 cells (IC_{50} = 248.5 µg/mL), and HaCaT cells (IC_{50} = 158.8 µg/mL). The sample obtained from Afghanistan showed cytotoxic effect against A549 cells (IC_{50} = 205.6 µg/mL). The samples collected from Syria and Turkey exhibited cytotoxic effects against HaCaT cells with IC_{50} values of 193.8 µg/mL and 196.6 µg/mL, respectively (Basar et al., 2015).

The cytotoxic effect of enriched ethanolic extract of *G. glabra* roots was evaluated alone and in combination with Adriamycin which was known as chemotherapeutic agent against PC-3 cells by MTT assay. The enriched extract treatment exhibited cytotoxic effect against PC-3 cells with 35.7 ± 2.0 µg/mL of IC_{50} value. The treatment in combination with extract and Adriamycin (IC_{50} = 11.6 ± 0.6 nM) increased the cytotoxic effect against PC-3 cells compared with administration of Adriamycin alone (IC_{50} = 24.0 ± 2.5 nM) (Gioti et al., 2020).

The cytotoxic effect of the water extract of *G. glabra* root was assessed against MCF-7 cells using sulforhodamine B (SRB) assay and the extract showed a cytotoxic effect at 0.1–100 µg/mL concentration ranges (Dong et al., 2007).

The water extract of *G. glabra* inhibited cell proliferation in C666-1 cells, according to results obtained in MTT assay. The inhibition of cell proliferation started at low concentration (0.2 mg/mL) of the extract. Also, the extract increased the activation of caspase-3 and caspase-9 in C666-1 cells, so the study has proven the anticancer effect of *G. glabra* root extract through apoptosis (Zheng et al., 2019).

Respiratory System

The extract of *G. glabra* was assessed to determine the effect of lung pathology, bronchoalveolar lavage oxidative stress markers, plasma immunoglobulin E (IgE), and Th2 cell cytokine levels for rats induced bronchial asthma with ovalbumin (OVA). The results showed that 10 mg/kg of dosage of the extract had a protective effect against OVA-originated mucus secretion and lung inflammation. The extract caused a reducing effect in the levels of interleukin (IL)-5 and IL-13. Moreover, the extract alleviated the levels of increased malondialdehyde and nitric oxide and resulted in antioxidant effect by enhancing the activity of superoxide dismutase and catalase (Khattab et al., 2018).

Glycyrrhizin has proven to be responsible for antitussive effect of *G. glabra* by releasing congestion in the upper respiratory tract and decreasing tracheal mucus secretion. Liquiritin apioside is another active component of *G. glabra* which exhibited antitussive effect through inhibition of capsaicin (Pastorino et al., 2018).

Anti-ulcerative Activity

G. glabra is used for gastric and duodenal ulcers and benefits as an adjuvant agent for the treatment of spasmodic pains of chronic gastritis. Glycyrrhizin is the major compound of *G. glabra* responsible for antiulcer activity by increasing the concentration of prostaglandins in the digestive tract and promoting stomach mucus secretion. Besides, *G. glabra* cause an antipepsin effect by expanding lifespan of stomach surface cells. Carbenoxolone is glycyrrhetinic acid analogue, which is obtained from *G. glabra*, induce increasing effect on prostaglandin levels and leading to therapeutic effect for gastric and duodenal ulcers (19Pastorino et al., 2018).

The antiulcer activity of *G. glabra* was tested in adult male albino mice. Male adult albino mice were given a high dosage of aspirin to create a gastric ulcer. A seven-day treatment plan was devised. Group 1 was given a regular diet, group 2 was given aspirin, and group 3 was given omeprazole plus aspirin as conventional ulcer therapy. Groups 4, 5, and 6 received *G. glabra* root powder in doses of 250, 500, and 750 mg/kg body weight, respectively, along with aspirin. The gastroprotective action of *G. glabra* was determined by histopathological examination. The findings imply that at the maximum dose, *G. glabra* has a significant gastroprotective effect (Awan et al., 2015).

Hepatoprotective Activity

The hydromethanolic root extract of *G. glabra* showed a significant protective effect at the dosages of 300 and 600 mg/kg against carbon tetrachloride-induced hepatotoxicity in the liver tissue of mice. The mice serum glutathione (GSH) levels of *G. glabra* extract groups and positive control Vitamin "C" groups were determined to be 12.07%, 30.36%, and 65.51%, respectively. Besides, the mice serum catalase levels of *G. glabra* extract groups and positive control Vitamin "C" groups were found to be 21.52%, 35.46%, and 70.65%, respectively. Thus, the study revealed that hepatoprotective activity of the extract may be related to antioxidant capacity as determined by GSH and CAT serum levels of mice (Sharma and Agrawal, 2014).

Neuroprotective Activity

The oral administration of *G. glabra* extract at 75–300 mg/kg concentration ranges for 7 days gave rise to an improving effect on learning and memory in mice. Similarly, the oral administration of aqueous extract of *G. glabra* root at 75 and 300 mg/kg concentration ranges for 6 weeks enhanced memory and learning capacity in mice. The memory-enhancing effects of *G. glabra* may be related to the neuroprotective role produced by antioxidant and anti-inflammatory activity (Pastorino et al., 2018).

Antidepressant Activity

The antidepressant effects of *G. glabra* were proven with forced swim test (FST) and tail suspension test (TST). The aqueous extracts of *G. glabra* (75, 150, and 300 mg/kg) were administered orally for 7 days in mice. The dosage of 150 mg/kg caused decreasing effect for the immobility times in both FST and TST, compared with positive control imipramine (15 mg/kg) and fluoxetine (20 mg/kg). The extract increased the levels of norepinephrine and dopamine in the mice brain by interaction with α1-adrenoceptors and dopamine D2 receptors (Dhingra et al., 2006).

Immunomodulatory Activity

The immunomodulatory activity of aqueous methanolic extract of *G. glabra* roots at 100, 200, and 300 mg/kg dosages was examined against *Eimeria* species infection in broiler chickens. The results showed that 300 mg/kg of dosage had a dose-dependent cell-mediated and humoral immune response similar to positive control vitamin E (Hussain et al., 2017).

Liquiritin, a compound obtained from *G. glabra*, was investigated for its immunomodulatory effects. The results of this study revealed that Liquiritin could be a candidate of a potential immune stimulator because it activated granulocytes in the range of 100–800 μg/mL concentration and NK cells in the range of 12–100 μg/mL concentration (Cheel et al., 2010).

Antimicrobial Activity

The ether–water extracts of *G. glabra* were proven to be effective on *Escherichia coli*, *B. subtilis*, *E. aerogenes*, *K. pneumoniae*, and *S. aureus*. The mechanism of antibacterial effect may be related to the decrease of bacterial gene expression, the inhibition of bacterial growth, and the impairment of bacterial toxin production. The antibacterial activity of *G. glabr*a was associated with glabridin, glabrol, glabrene, hispaglabridin A, hispaglabridinB, 40-methylglabridin, and 3-hydroxyglabrol of isolated compounds from *G. glabra. In vitro* studies showed that aqueous and ethanolic extracts of *G. glabra* had antibacterial effects on *Streptococcus pyogenes*. *G. glabra* has been found to have antibacterial activity against *Helicobacter pylori*. The effective compounds were determined as glabridin and glabrene. These compounds caused antibacterial effect on *H. pylori* via inhibition of the protein synthesis, DNA gyrase, and dihydrofolate reductase (Pastorino et al., 2018).

Antiviral Activity

The antiviral activity of *G. glabra* extracts was proven against different viruses *Herpes simplex*, *Varicella zoster*, *Japanese encephalitis*, *Influenza*, and *Vesicular stomatitis*. The responsible compounds were detected as glycyrrhizin and 18β -glycyrrhetinic acid for antiviral activity. These compounds exhibited antiviral ability by inhibiting virus gene replication and reducing the adhesion force and stress (Pastorino et al., 2018)

Glycyrrhiza species and its components have antibacterial and antiviral effects as well as antiprotozoal effects. Table 7.2 summarized some of the antimicrobial, antiviral, and antiprotozoal effects of *Glycyrrhiza* species and its components (Hosseinzadeh et al., 2015).

The COVID-19 illness outbreak has prompted a hunt for suitable vaccinations and medicines. However, because of the lack of vaccination supplies to match worldwide demand and a lack of

TABLE 7.2
Summary of the antimicrobial, antiviral, and antiprotozoal effects of *Glycyrrhiza* species and its components

Compound	Microorganism
Antibacterial effects	
Glycyrrhiza sp.	MRSA; *Escherichia coli, Bacillus subtilis, Enterobacter aerogenes, Klebsiella pneumoniae, Staphylococcus aureus, Enterococcus faecalis*
Glycyrrhiza glabra	*Helicobacter pylori*
Leaves	*S. aureus*
Roots	*S. aureus, Campylobacter jejuni*
Glycyrrhizin	*E. coli*
18β-Glycyrrhetinic acid	MRSA
Deglycyrrhizinated licorice	*Streptococcus mutans* UA159
Antiviral effects	
Glycyrrhiza uralensis	Enterovirus type 71, rotavirus
Glycyrrhiza	*Varicella zoster virus*, rotavirus, human respiratory syncytial virus, Newcastle disease virus
Glycyrrhizin	Influenza A virus, Glycyrrhizic acid HSV1, Coxsackievirus A16, enterovirus 71
Glycyrrhetinic acid	HIV
18β-Glycyrrhetinic acid	Rotavirus
Diammonium glycyrrhizin	PrV
Antifungal effects	
G. glabra	*Candida albicans*
Glycyrrhizic acid	*C. albicans*
18β-Glycyrrhetinic acid	*C. albicans*
Antiprotozoal effects	
G. glabra	
Aerial parts	*Plasmodium falciparum*
Roots	*P. falciparum*
Glycyrrhetinic acid	Filaria
18β-Glycyrrhetinic acid	*P. falciparum* *Leishmania donovani*

effective COVID-19 prescription pharmaceuticals, some patients are considering using complementary medicines, such as traditional herbal therapy. The therapeutic capabilities of medicinal plants are dependent on the active chemicals they contain. Herbal therapy, particularly in Asian cultures, has clearly played an important role in the treatment and prevention of the COVID-19 pandemic. Based on case reports, community surveys, and guidelines available in academic databases, the researchers described the important families and species that are sources of antiviral drugs against COVID-19 (Jalali et al., 2021).

According to the literature, Lamiaceae family members, *Zingiber officinale* and *Glycyrrhiza* sp. are frequently used as medicinal sources for COVID-19 treatment. The aqueous extract of *G. glabra* root has been shown to prevent SARS-CoV-2 infection in Vero E6 cells. The extract has antiviral properties at a concentration of 2 mg/mL, which is lower than the hazardous threshold. Glycyrrhizin, the major triterpenoid glycoside of the plant, has likewise exhibited viral inhibiting effectiveness at 0.5 mg/mL and 1 mg/mL, respectively. Furthermore, no cytotoxic effect has been observed at concentrations below 4 mg/mL. The EC_{50} value of the compound is 0.44 mg/mL. Glycyrrhizin has also been reported to lower SARS-CoV-2 RNA levels considerably. Also, this compound binds to the ACE2 receptor, potentially blocking viral entry, according to a computer simulation (Liana et al., 2021).

Anti-inflammatory Activity

Glycyrrhiza extracts and its isolated compounds were investigated to reveal anti-inflammatory activity by *in vivo* studies. The results of the studies showed that *G. glabra* and glabridin exhibited strong anti-inflammatory activity via inhibition of prostaglandin E2 (PGE2) and leukotriene B production. Another component isoliquiritigenin caused anti-inflammatory activity through reduction effect of the levels of PGE2 and thromboxane 2. *G. glabra* exhibited anti-inflammatory effect via reduction of paw swelling and histopathological changes for TPA and collagen-induced arthritis in rats. Also, hydroalcoholic extract of *G. glabra* root displayed anti-inflammatory effect by preventing leukocyte migration in carrageenan-induced rat paw (Batiha et al., 2020).

CONCLUSION

Glycyrrhiza genus consists of 30 species worldwide, six of which are naturally growing in Turkey. *G. glabra*, the most common species, is used for its beneficial effects on gastrointestinal and respiratory system disorders like in traditional Anatolian medicine. In particular, triterpenoid-type saponins like glycyrrhizin and other components such as glycyrrhizinic acid, glabridin, isoliquiritin, and licochalcone A are yielded from the *Glycyrrhiza* species which are responsible for their medicinal and pharmacological properties. These species have demonstrated antioxidant, anti-inflammatory, antidepressant, anti-ulcer, cytotoxic, and antiviral effects in *in vitro* and *in vivo* experiment models. Their potential curative effects on the treatment of COVID-19 disease have increased interest in this plant. Except for *G. glabra*, other species grown in Turkey have been studied very little in terms of pharmacological and phytochemicals. It is of great importance for drug development studies to conduct further research on species belonging to this genus, which have important secondary metabolites.

REFERENCES

Ali, Z., Srivedavyasasri, R., Zhao, J., Avula, B., Chittiboyina, A.G., Khan, I.A. (2020). Oleanane-type triterpenoid glucuronosides from *Glycyrrhiza echinata* L. root. *Biochemical Systematics and Ecology*, 92, 1–3.

Awan, T., Aslam, B., Javed, I., Khaliq, T., Ali, A., Sindhu, Z.U.D. (2015). Histopathological evaluation of *Glycyrrhiza glabra* on aspirin induced gastric ulcer in mice. *Pakistan Journal of Agricultural Sciences*, 52(2), 563–568.

Basar, N., Oridupa, O.A., Ritchie, K.J., Nahar, L., Osman, N.M.M., Stafford, A., Kushiev, H., Kan, A., Sarker, S. (2015). Comparative cytotoxicity of *Glycyrrhiza glabra* roots from different geographical origins against immortal human keratinocyte (HaCaT), lung adenocarcinoma (A549) and liver carcinoma (HepG2) cells. *Phytotherapy Research*, 29(6), 944–948.

Batiha, GE-S., Beshbishy, A.M., El-Mleeh, A., Abdel-Daim M.M., Devkota H.P. (2020). Traditional uses, bioactive chemical constituents, and pharmacological and toxicological activities of *Glycyrrhiza glabra* L. (Fabaceae). *Biomolecules*, 10(3), 352–371.

Cakmak, Y.S., Aktumsek, A., Duran, A., Cetin, O. (2016). Antioxidant activity and biochemical screening of two *Glycyrrhiza* L. species. *British Journal of Pharmaceutical Research*, 11(1), 1–11.

Cevik, D., Kan, Y., Kirmizibekmez, H. (2019). Mechanisms of action of cytotoxic phenolic compounds from *Glycyrrhiza iconica* roots. *Phytomedicine*, 58, 1–8.

Cevik, D., Yilmazgoz, S.B., Kan, Y., Guzelcan, E.A., Durmaz, I., Cetin-Atalay, R., Kirmizibekmez, H. (2018). Bioactivity-guided isolation of cytotoxic secondary metabolites from the roots of *Glycyrrhiza glabra* and elucidation of their mechanisms of action. *Industrial Crops and Products*, 124, 389–396.

Cheel, J., Van Antwerpen, P., Tumova, L., Onofre, G., Vokurkova, D., Zouaoui-Boudjeltia, K., Vanhaeverbeek, M., Nève, J. (2010). Free radical-scavenging, antioxidant and immunostimulating effects of a licorice infusion (*Glycyrrhiza* glabra L.). *Food Chemistry*, 122(3), 508–517.

Davis, P.H., Cullen, J., Coode, M.J.E. (1965). *Flora of Turkey and the East Aegean Islands*. Edinburgh: University Press.

Dhingra, D., Sharma, A. (2006). Antidepressant-like activity of *Glycyrrhiza glabra* L. in mouse models of immobility tests. *Progress in Neuro-Psycho-pharmacology and Biological Psychiatry*, 30(3), 449–454.

Dong, S., Inoue, A., Zhu, Y., Tanji, M., Kiyama, R. (2007). Activation of rapid signaling pathways and the subsequent transcriptional regulation for the proliferation of breast cancer MCF-7 cells by the treatment with an extract of *Glycyrrhiza glabra* root. *Food and Chemical Toxicology*, 45(12), 2470–2478.

Ercisli, S., Coruh, I., Gormez, A., Sengul, M., Bilena, S. (2008). Total phenolics, mineral contents, antioxidant and antibacterial activities of *Glycyrrhiza glabra* L. Roots grown wild in Turkey. *Italian Journal of Food Science*, 20(1), 91–99.

Gioti, K., Papachristodoulou, A., Benaki, D., Beloukas, A., Vontzalidou, A., Aligiannis, N., Skaltsounis, A.L., Mikros, E., Tenta, R. (2020). *Glycyrrhiza glabra* enhanced extract and adriamycin antiproliferative effect on PC-3 prostate cancer cells. *Nutrition and Cancer*, 72(2), 320–322.

Hosseinzadeh, H., Nassiri-Asl, M. (2015). Pharmacological effects of *Glycyrrhiza* spp. and its bioactive constituents: Update and review. *Phytheraphy Research*, 29(12), 1868–1886.

Hussain, K., Iqbal, Z., Abbas, R.Z., Khan, M.K., Saleemi, M.K. (2017). Immunomodulatory activity of *Glycyrrhiza glabra* extract against mixed Eimeria infection in chickens. *International Journal of Agriculture and Biology*, 19(4), 928–932.

Jalali, A., Dabaghian, F., Akbrialiabad, H., Foroughinia, F., Zarshenas, M.M. (2021). A pharmacology-based comprehensive review on medicinal plants and phytoactive constituents possibly effective in the management of COVID-19. *Phytheraphy Research*, 35(4), 1925–1938.

Khattab, H.A.-R.H., Abdel-Dayem, U.A., Jambi, H.A., Abbas, A.T., Abdul-Jawad M.T.A., El-Shitany N.A.E-A.F. (2018). Licorice (*Glycyrrhizza glabra*) extract prevents production of Th2 cytokines and free radicals induced by ova albumin in mice. *International Journal of Pharmacology and Pharmaceutical Sciences*, 14(8), 1072–1079.

Liana, D., Phanumartwiwath, A. (2021). Leveraging knowledge of Asian herbal medicine and its active compounds as COVID-19 treatment and prevention. *Journal of Natural Medicines*, 1–18.

Pastorino, G., Cornara, L., Soares, S., Rodrigues, F., Oliveira, M.B.P.P. (2018). Liquorice (*Glycyrrhiza glabra*): A phytochemical and pharmacological review. *Phytotherapy Research*, 32(12), 2323–2339.

Sharma, V., Agrawal, R.C. (2014). *In vivo* antioxidant and hepatoprotective potential of *Glycyrrhiza glabra* extract on carbon tetra chloride (CCl4) induced oxidative-stress mediated hepatotoxicity. *International Journal of Research in Medical Sciences*, 2(1), 314–320.

Shou, Q., Jiao, P., Hong, M., Jia, Q., Prakash, I., Hong, S., Wang, B., Bechman, E., Ma, G. (2019). Triterpenoid saponins from the roots of *Glycyrrhiza glabra*. *Natural Product Communications*, 14(1), 19–22.

Uzun, S.P., Koca, C. (2020). Ethnobotanical survey of medicinal plants traded in herbal markets of Kahramanmaras. *Plant Diversity*, 42(6), 443–454.

Yesil, Y., Celik, M., Yilmaz, B. (2019). Wild edible plants in Yesilli (Mardin-Turkey), a multicultural area. *Journal of Ethnobiology and Ethnomedicine*, 15(1), 1–19.

Zheng, C., Han, L., Wu, S. (2019). A metabolic investigation of anticancer effect of G. glabra root extract on nasopharyngeal carcinoma cell line, C666-1. *Molecular Biology Reports*, 46(4), 3857–3864.

8 *Gundelia tournefortii* L.

Gülsen Kendir

INTRODUCTION

Gundelia tournefortii belongs to Asteraceae family. It is of Iranian-Turanian origin and is widely distributed in the countries such as Cyprus, Egypt, Iran, Palestine, Israel, Jordan, Turkey, Azerbaijan, Turkmenistan, and Armenia (Asadi-Samani *et al.*, 2013; Çoruh *et al.*, 2007; Kupicha, 1975). Known as "Kangar" in Iran, this plant is known as "Akub" in Mediterranean countries such as Jordan, Syria, and Palestine. It is commonly known as Galgal, tumble thistle, or Tumbleweed in English (Asadi-Samani *et al.*, 2013). In Turkey, it is known as "Kengel, Kenger, Kenger tiken, Kengiotu, Kengir" (Baytop, 2007). Its different parts are consumed as food in different ways in some countries such as Iranian, Turkey, Palestine, and Syria (Baytop, 1999; Jamshidzadeh *et al.*, 2005; Prance ve Nesbitt, 2005). The product obtained by crushing and sifting the fruits in stone mortars after roasting in Turkey is used as a coffee substitute with the name "Kenger coffee" (Baytop, 1999). This coffee is also sold in Lebanese and Syrian markets (Duke *et al.*, 2008).

The plant, which is a hemicryptophyte, is perennial and herbaceous at a height of 20–100 cm. The stem is thick, erect, branched, glabrous, or with sparsely arachnoid hairs. The leaves are large, stiff, oblong-lanceolate, sessile or decurrent and somewhat drooping downward, glabrous or arachnoid hairy, very thickly veined, pennate lobed, lobes acuminate, pinnatiphyte-sub-2-pinnatisect, toothed margins, and teeth hard-thorny. The lower leaves 7–30 x 4.5–16 cm, sessile, and narrowed. The upper leaves are smaller in size. The inflorescence is in the form of a head; about 15–20 sessile capitulums and numerous bracts are frequently located on a short and non-fleshy axis, forming an ovoid-globose state, a spike. Bracts in the inflorescence vary in shape and size, thin. The margins are finely toothed towards the top and the teeth are softly spiny, longer and stronger at the apex. At the base of the spike, there are three brackets surrounding the inflorescence. They are longer than the inner bracts, coriaceous, sessile, and lanceolate. It contains capitulums in the form of an inverted pyramid. Capitulum 7–17 mm long, yellowish-green at maturity, 4–8 flowers, only the middle flowers are fertile. Corolla purplish-red, funnel-shaped, exceeding the length of the capitulum. In the period of capitula dispersal, obovate- quadrangular 10–16 x 5–9 mm (Fig. 8.1). Fertile achenes smooth, ovoid (Baytop, 1963; Kupicha, 1975; Tanker and Tanker, 1967).

TRADITIONAL USE

In a study on natural raw materials that were used medicinally in the region including Palestine, Syria, Lebanon, Israel, Jordan, Sinai peninsula, and Hatay in the period covering the Medieval and Ottoman periods (a period of 1100 years), it was found that the plant was used in stomach and intestinal diseases and also in psychiatric and antiepileptic use as mentioned in the literature records (Lev, 2002).

In Iran, its stems are employed as hepatoprotective and blood purifier (Jamshidzadeh ve ark., 2005). In the ancient books of Islamic traditional Iranian medicine poultice prepared from its latex was indicated to remove cancerous swellings (Emami *et al.*, 2012). Its roots are used orally for antiparasitic purposes for the digestive system (Mosaddegh *et al.*, 2012). Its leaves and stem are employed internally in indigestion, as a tonic, laxative, against tartar formation, and in diabetes (Ghasemi Pirbalouti *et al.*, 2013). Its leaves are consumed as fresh for kidney stones (Bahmani *et al.*, 2016).

DOI: 10.1201/9781003146971-8

In Lebanon, its latex is recommended for burning warts, drying wounds, emetic, and treating snake bites (Duke *et al.*, 2008). It is used for chest pain and heart ailments among Bedouins in Jordan (Halabi *et al.*, 2005). There is also information that the herbal hormones contained in the plant strengthen sexual libido (Nooraei, 2005). The aerial parts of *G. tournefortii* are eaten cooked against anemia by pregnant women and infants or children (<2 years old) in Palestine (Ali-Shtayeh *et al.*, 2015).

In Turkey, gum obtained from its root is called "Kenger sakızı veya Kanak sakızı". This gum is chewed as strengthening gums and is appetizing (Baytop, 1999; Baytop, 2007). For mumps, the sap obtained by beating the stem and thorny part of the fresh plant is applied to the swollen part of the neck and left for a day (Sarper *et al.*, 2009). The thorny part of the plant is separated and used raw as a blood sugar reducer (Çakılcıoğlu and Türkoğlu, 2009). To treat eczema, the roots are pounded and combined with crushed wheat to produce a mush. This mush is applied to the infected region and left to sit for one day. The decoction of roots is consumed to treat stomach ailments. The roots, which have been removed from the bark, are eaten to treat high blood pressure (Arasan and Kaya, 2015). *G. tournefortii* is included in the composition of various ointments prepared in Turkey. While preparing the ointment for skin cracks and rheumatism, *Rosmarinus officinalis* L. flowers, *Sinapis arvensis* L. leaves, *Hamamelis virginiana* L. leaves, *G. tournefortii* leaves, and tomato paste are boiled. The boiled mixture is filtered. The resulting filtrate is kneaded with broad bean flour and egg. The mixture is softened with cottonseed oil. For insect bites and inflammatory wounds, *Vicia faba* L. leaves, *Crataegus monogyna* Jacq. leaves, *G. tournefortii* leaves, and *Spinacia oleraceae* L. leaves are boiled and filtered. The resulting filtrate is kneaded with *Punica granatum* L. fruit peel, henna, and yogurt. The mixture is softened with *Ficus carica* L. latex. For scabies, itching, and inflamed wounds, *Peganum harmala* L. leaves, *Lycopersicum esculentum* Miller leaves, *Urtica dioica* L. leaves, *G. tournefortii* root, and *Linum usitatissimum* L. seeds are boiled. The mixture is filtered. The filtrate is kneaded with poplar tree ash and the resulting mixture is softened with milk (Uğulu and Başlar, 2010).

G. tournefortii var. *tournefortii* seeds are utilized in the form of coffee for vitiligo disease (Özgökçe and Özçelik, 2004). A decoction of the seeds is employed for colds and flu (Tabata *et al.*, 1994). The seeds are pounded, boiled, and drunk for liver ailments (Vural *et al.*, 1997). Its fresh seeds are also employed for its diuretic effect (Çoruh *et al.*, 2007). It is used in diabetes and diarrhea as an infusion of roots with latex or ingestion of a piece of latex (Hayta *et al.*, 2014). The seed, latex, root, and stem are utilized internally or externally in colds, catarrh, toothache, diabetes, edema, kidney pains, and vitiligo disease and as an appetite stimulant in the form of decoction (Altundağ and Öztürk, 2011). The latex and its roots are employed as an aphrodisiac and digestive regulator (Polat *et al.*, 2013; Tetik *et al.*, 2013). The latex obtained from the roots is chewed and used for thirst and removal of water in the spleen. The latex is also used externally for cuts (Sezik *et al.*, 2001). Its latex and seed are employed for dental abscesses, halitosis, and epilepsy. For this purpose, the latex is sucked or the seed is eaten raw (Dalar *et al.*, 2018). *Gundelia tournefortii var. armata* Freyn & Sint. latex is applied externally in the treatment of edema. The seed is prepared as a decoction and drunk for toothache (Altundağ and Öztürk, 2011). Its fruit is used orally for stomach ulcers (Tuzlacı, 2016). *G. tournefortii* var. *armata* latex obtained from the stem base is chewed extensively and used as a refreshing agent (Tuzlacı and Şenkardeş, 2011).

Activity

In Vitro

Antioxidant Effect

In the study on Jordanian plants, antioxidant activities and total phenol contents of water and methanol extracts of plants were determined. A significant linear relationship was observed between antioxidant activity and total phenol content of water and methanol extracts. The antioxidant activities of *G. tournefortii* water and methanol extract were determined as 63.6 and 53.0 μmol trolox equivalents g^{-1} dry weight, and total phenol contents were determined as 16 and 11 mg gallic acid equivalent g^{-1} dry weight, respectively (Alali *et al.*, 2007).

Antioxidant activities of the aerial parts and seeds were evaluated using DPPH radical scavenging and lipid peroxidation inhibition methods. In the DPPH assay, the seeds exhibited a higher antioxidant capacity compared to the aerial parts, with an IC_{50} of 0.07 mg/mL and a value of 0.15 mg/mL in the lipid peroxidation inhibition assay. The total phenol content of the extracts, especially the seed extracts (105.1 ± 8.7 μg/mL gallic acid equivalent), was associated with high antioxidant activity. The effect of the extracts on cytosolic glutathione-S-transferase was also investigated, and the seed extracts showed an effective inhibition effect with an IC_{50} of 97.51 μg/mL (Çoruh *et al.*, 2007).

Antioxidant activities of the water and methanol extracts by ABTS radical cation scavenging method were detected as 57.3 ± 2.7 and 63.9 ± 2 μmol trolox equivalents g^{-1} dry weight, respectively (Tawaha *et al.*, 2007).

G. tournefortii ethanol extract (200 mg/mL) inhibited iron-fructose-phosphate-induced lipid peroxidation (lecithin liposome 0.216 absorbance value at 533 nm and linoleic acid emulsion 0.57 absorbance value at 508 nm) in two different lipid systems (lecithin liposome and linoleic acid emulsion) (Mavi *et al.*, 2011).

The methanol extract prepared from its leaves was shown to have antioxidant potential by *in vitro* methods of scavenging DPPH and ABTS radicals, metal chelating, and reducing power assays. DPPH; 11.85 ± 0.5 mmol Trolox equivalent g^{-1} dry weight, ABTS; 5.79 ± 1.36 mmol Trolox equivalent g^{-1} dry weight, metal chelating activity; 2.97 ± 0.22 mmol EDTA equivalent g^{-1} dry weight, antioxidant power reducing activity; 21.72 ± 2.58 mmol ascorbic acid equivalent g^{-1} dry weight were detected (Özkan *et al.*, 2011).

Antioxidant activities of ethanol extracts of several herbal coffees traditionally consumed in Anatolia, including *G. tournefortii*, and of seed samples that form the source of coffee were determined through radical scavenging activity tests and metal-related tests including metal chelation capacity, ferric-reducing antioxidant power (FRAP), and phosphomolybdenum reducing antioxidant power (PRAP). While the seed extracts showed a significant radical scavenging effect against DPPH (27.74 ± 0.83 – 78.52 ± 3.97% inhibition value), the extract of its coffee was also found to have the best metal chelating activity (33.57 ± 1.75% at 3000 μg/mL) (Şekeroğlu *et al.*, 2012).

The DPPH assay was used to test the ability of the various plant extracts (chloroform, ethyl acetate, methanol, and aqueous extracts) to scavenge free radicals. Also, the total phenolics content of the plant extracts was determined. It was observed that the methanol extract had the highest antioxidant activity (IC_0 = 40.3 μg/mL) and total phenolic content (103.4 mgGA/g extract) (Dastan and Yousefzadi, 2016).

Anticancer Effect

Methanol and hexane extracts of the aerial parts were found to have potent anticancer activity against the HCT-116 cancer cell line. The half-maximal effective concentrations by the MTT assay were determined as 303.3 ± 12 μg/mL for the methanol extract and 313.3 ± 18.6 μg/mL for the hexane extract (Abu-Lafi *et al.*, 2019).

Antimicrobial Effect

In a study on Jordanian plants, methanol extract of *G. tournefortii* was combined with seven different antibiotics (penicillin G, chloramphenicol, gentamicin, cephalexin, erythromycin, tetracycline, nalidixic acid). The inhibitory effect of these combinations on the standard strain of *Pseudomonas aeruginosa* (ATCC 9027) and the resistant strain isolated from the patient were investigated. In combination with penicillin G and erythromycin, it did not inhibit the growth of the standard strain, but in combination with other antibiotics, it inhibited its growth. While it did not inhibit the growth of resistant strains in combination with cephalexin and nalidixic acid, it was observed that it inhibited the growth of resistant strains in combination with other antibiotics (Aburjai *et al.*, 2001).

In another study conducted in Jordan, the inhibitory effect of the combination of *G. tournefortii* methanol extract with seven different antibiotics (penicillin G, chloramphenicol, gentamicin, cephalexin, erythromycin, tetracycline, nalidixic acid) on methylicin-resistant *Staphylococcus aureus* (MRSA) isolated from a patient and standard strain of *S. aureus* (ATCC 25923) was evaluated. While the methanol extract of *G. tournefortii* did not inhibit MRSA growth in penicillin G, cephalexin, erythromycin, tetracycline, and nalidixic acid combinations, it was determined that it inhibited the growth of standard strain only in tetracycline (75.7 ± 3.6%) and nalidixic acid (98.4 ± 9.1%) combinations (Darwish *et al.*, 2002).

In another study conducted in Jordan, the inhibitory activity of combinations of *G. tournefortii* methanol extract with antibiotics (amoxicillin, chloramphenicol, neomycin, cephalexin, clarithromycin, doxycycline, and nalidixic acid) against standard strain of *Escherichia coli* (ATCC 8739) and resistant strain was evaluated. It was observed that the combinations generally increase inhibitory activity against the resistant strain. *G. tournefortii* increased the inhibitory activity of chloramphenicol, neomycin, doxycycline, cephalexin, and nalidixic acid against the standard strain (Darwish and Aburjai, 2010).

In a study conducted in Iran, it was determined that the essential oil obtained from the leaves had a bacteriostatic effect on the *Staphylococcus epidermis*. It was stated that the essential oil has an effect on some Gram-positive bacteria at a concentration of 30 µg/mL (Talei *et al.*, 2007).

G. tournefortii leaves showed low potency (17–25 mm inhibition zone) against seven different clinical isolates of *Helicobacter pylori* (Nariman *et al.*, 2009).

Antimicrobial activities of ethanol and water extracts of leaves and roots were evaluated against methicillin-resistant *S. aureus* (MRSA) (resistant to ampicillin, erythromycin, penicillin G, vancomycin), *E. coli* (resistant to ampicillin, chloramphenicol, erythromycin, novobiocin, penicillin G), *P. aeruginosa* (resistant to ampicillin, chloramphenicol, erythromycin, penicillin G, vancomycin), and *C. albicans* (resistant to ampicillin, erythromycin, penicillin G, vancomycin) which are skin pathogens resistant to multiple antibiotics. It was determined that the ethanol extract of its roots showed significant activity against *E. coli*, *P. aeruginosa*, MRSA, and *C. albicans* with MIC values of 64, 32, 32, and 16 mg/mL, respectively (Obeidat, 2011).

Antimicrobial activity of the ethanol extract was investigated against acne-causing *Propionibacterium acnes* ATCC 6919, ATCC 6921 strains, and eight different strains of this bacterium isolated from other clinical specimens, *E. coli* ATCC 25922, *Klebsiella pneumonia* ATCC 13883, *Proteus vulgaris* ATCC 13315, *Pseudomonas aeruginosa* ATCC 27853 and *S. aureus* ATCC 25923 by the disc diffusion method. While it did not show activity against *S. aureus*, it showed very low activity against other bacteria (Ali-Shtayeh *et al.*, 2013).

The ethyl acetate extract displayed antibacterial activity against *Bacillus cereus*, *Bacillus pumilus*, and *Bacillus subtilis* strains with 7.5 mg/mL MIC value (Dastan and Yousefzadi, 2016).

Antibacterial activities against *Salmonella typhimurium*, *E. coli*, *P. aeruginosa*, *S. aureus*, *Streptococcus pneumoniae*, and *Bacillus subtilis*, and antifungal activities against *Candida albicans*, *C. glabrata*, *C. guilliermondii*, and *C. krusei* of gold nanoparticles using aqueous extract of *G. tournefortii* leaves (AuNPs@GT) were assessed. AuNPs@GT showed greater antibacterial and antifungal activity than any other antibiotics ($p \leq 0.01$). At 2–4 mg/mL concentrations, AuNPs@GT inhibited the growth of all bacteria and fungi and completely eliminated them ($p \leq 0.01$) (Zhaleh *et al.*, 2019).

Antidiabetic Effect

The effectiveness of the methanol and hexane aerial extracts in enhancing glucose transporter 4 (GLUT4) translocation to the plasma membrane was investigated utilizing a cell-ELISA test in L6 muscle cells stably expressing myc-tagged GLUT4 (L6-GLUT4myc). In MTT and LDH leakage assays, it was detected that methanol and hexane extracts were safe up to 250 g/mL. The methanol extract was shown to be the most effective in improving GLUT4 translocation. In the absence and presence of insulin, it boosted GLUT4 translocation 1.5- and 2-fold compared to the control at 63 µg/mL (Kadan *et al.*, 2018).

Antiplatelet Effect

For the antiplatelet activity of the chloroform extract obtained from the aerial parts of *G. tournefortii*, its essential oil and the isolated mixture of scopoletin, isoscopoletin, esculin, β-sitosterol, and stigmasterol were studied using adenosine-5′-diphosphate and arachidonic acid. Chloroform extract showed a moderate inhibitory effect on platelet aggregation caused by adenosine-5′-diphosphate (IC_{50}: 780 ± 59.9 μg/mL) and arachidonic acid (IC_{50}: 600 ± 16 μg/mL). Isolated pure compounds and essential oil did not show an inhibitory effect on platelet aggregation (Halabi *et al.*, 2005).

Hepatoprotective Effect

The effect of different concentrations of hydroalcoholic extract of the footstalk on isolated rat liver cells was investigated. Freshly isolated liver cells were incubated in Krebs–Henseleit buffer under 95% O_2 and 5% CO_2 flow for 20 minutes, then different concentrations of extract and 10 mM CCl_4 were added to the incubation mixture. At concentrations of 0.2–0.8 mg/mL, the extract showed a protective effect against CCl_4-induced cytotoxicity, and the maximum protective effect was found at approximately 0.5 mg/mL. However, an increase in cytotoxicity was observed at 1 mg/mL and higher concentrations (Jamshidzadeh *et al.*, 2005).

Insecticidal Activity

The chloroform extract of seed displayed the highest toxicity against *Drosophila melanogaster* Meigen larvae with the lowest LC_{50} value (119.95 ppm), the chloroform leaf extract, as well as the ether extracts of leaf and seed, comes next. Terpinyl acetate and oleic acid from *G. tournefortii* were shown to be responsible for 80% and 70% of larval deaths, in turn, with LC_{50} values of 22.98 and 23.15 ppm. After the acetone and chloroform seed extracts were applied, mortality percentages for fly larvae due to carbaryl and imidacloprid insecticides were enhanced from 10–20% to 40–70% (Ghabeish, 2015).

Effect on Neurodegeneration

The effects of ethanol extracts of several herbal coffees traditionally consumed in Anatolia, including *G. tournefortii*, and of the seed samples that form the source of coffee against acetylcholinesterase, butyrylcholinesterase, and tyrosinase enzymes related to neurodegeneration were tested. While *G. tournefortii* seed extracts inhibited enzymes to some extent, it was determined as one of the most effective extracts that inhibited butyrylcholinesterase enzyme (41.07 ± 0.55% at 3000 μg/mL). The inhibition of tyrosinase enzyme (34.70 ± 3.59 μg/mL at 200 μg/mL) of its coffee extract was found to be remarkable (Şekeroğlu *et al.*, 2012).

Molluscicidal Effect

Oleanolic acid-derived saponins isolated from the roots showed potent molluscicidal activity against schistosomiasis spread by the snail *Biomphalaria glabrata* (Wagner *et al.*, 1984).

In Vivo

Anti-inflammatory and Antinociceptive Effects

The antinociceptive and anti-inflammatory effects of the aerial parts were evaluated with experiments on mice. After the lethal dose (LD_{50}) of the total extract was determined, different doses of the total extract were injected intraperitoneally into mice and then its antinociceptive and anti-inflammatory effects were determined by formalin and xylene tests. Compared with the control group, it was observed that it caused antinociceptive and anti-inflammatory effects at doses of 0.3, 0.6, 1.2, and 2.4 g/kg. While the most effective dose was found to be 2.4 g/kg, the LD_{50} value was found to be 6.3 g/kg (Oryan *et al.*, 2011).

Essential oil obtained from the aerial parts displayed significant antinociceptive activity. The essential oil at test doses of 10, 31.6, 100, 316, and 1000 mg/kg orally remarkably reduced pain response in a dose-dependent manner in the male Wistar rats ($p < 0.05$). Potassium channels sensitive to ATP and adrenergic receptors have been shown to be involved in this effect (Qnais *et al.*, 2016).

Antidiabetic Effect

In alloxan-induced diabetic mice, the nephroprotective and antidiabetic effects of *G. tournefortii* aqueous extract were investigated. Glibenclamide (10 mg/kg) and *G. tournefortii* aqueous extract (5, 10, 20, and 40 mg/kg doses) were given by gavage to the mice for 20 days. When compared to diabetic mice who were not given any treatment, *G. tournefortii* aqueous extract at all doses and glibenclamide significantly ($p \leq 0.05$) lowered elevated blood glucose, creatinine, and urea levels. When compared to the diabetes untreated group, multiple doses of the aqueous extract and glibenclamide considerably ($p \leq 0.05$) lowered the volume and length of renal structures (Mohammadi *et al.*, 2018).

Hypoglycemic and Hypolipidemic Effects

The effects of decoction of the roots were evaluated in male albino mice induced with hyperglycemia and hyperlipidemia with 1 mg/kg dexamethasone. The mice were administered with extract of *G. tournefortii* at doses of 75, 150, and 300 mg/kg of body weight orally. While dexamethasone treatment did not affect the total protein level (5.45 ± 0.186 g/dL), it significantly increased glucose (261.76 ± 8.123 mg/dL), cholesterol (334.11 ± 14.04 mg/dL), and triglyceride (188.03 ± 2.971 mg/dL) levels and decreased body weight (27.06 ± 0.537 g). The extract treatment at a dose of 75 mg/kg resulted in a significant reduction in glucose level (170.24 ± 8.399 mg/dL) and body weight (24.51 ± 0.723 g). At 300 mg/kg dose, it caused a decrease in glucose (184.87 ± 6.502 mg/dL), triglyceride (133.2 ± 6.186 mg/dL), and cholesterol (226.02 ± 9.449 mg/dL) levels (Azeez and Kheder, 2012).

The effect of *G. tournefortii* seed oil on blood total lipid and lipoproteins in rats with atherosclerosis was investigated. For this purpose, two main groups were formed from adult albino male rats. The first group consisted of five healthy rats (healthy control group), and the second group consisted of 25 rats that underwent atherosclerosis with the help of hydrogen peroxide and cholesterol and were equally divided into five subgroups. Sunflower oil was given orally to the first subgroup (atherosclerosis rat control group). In the second subgroup, 300 mg/kg clofibrate (hypolipidemic reference group), and 60, 90, and 120 mg/kg seed oil were given orally to the third, fourth, and fifth subgroups, respectively. It has been observed that the seed oil has a hypolipidemic effect on total lipid, total cholesterol, very low-density lipoprotein cholesterol, and low-density lipoprotein cholesterol in plasma, together with an improvement in the nutritional status of rats and a decrease in atherogenic symptoms. High-density lipoprotein cholesterol level (45.67 ± 4.62 mg/dL) increased and liver total cholesterol level (5.12 ± 0.41 mg/100 g) decreased in rats with atherosclerosis. The seed oil at a dose of 90 mg/kg was found to have higher efficacy compared to other doses. Hypolipidemic effect at this dose [total lipid (609.26 ± 62.5 mg/dL), total cholesterol (107.14 ± 6.52 mg/dL), very low-density lipoprotein cholesterol, and low-density lipoprotein cholesterol (61.47 ± 4.32 mg/dL)] were determined to be lower than clofibrate [total lipid (544.92 ± 34.5 mg/dL), total cholesterol (100.38 ± 5.81 mg/dL), very low-density lipoprotein cholesterol and low-density lipoprotein cholesterol (53.08 ± 4.69 mg/dL)] ($p < 0.05$) (Sharaf *et al.*, 2004).

The effect of *G. tournefortii* on some cardiovascular risk factors was investigated by the analysis of biochemical factors in an animal model. For this purpose, 20 male rabbits were randomly divided into four groups: normal diet, normal diet supplemented with *Gundelia*, high cholesterol diet, and high cholesterol diet supplemented with *Gundelia*. *Gundelia* decreased the amount of cholesterol, triglyceride, low-density lipoprotein cholesterol (LDL), very low-density lipoprotein cholesterol (VLDL), apolipoprotein B, oxidized LDL, and factor VII. It has also been observed to increase high-density lipoprotein (HDL) cholesterol and apolipoprotein A levels compared to a high-cholesterol diet (Asgary *et al.*, 2008).

Hepatoprotective Effect

The effects of different concentrations of hydroalcoholic extract of dried footstalks of *G. tournefortii* on CCl_4-induced hepatotoxicity were investigated. Different concentrations (200, 300, and 400 mg/kg) of plant extract were injected intraperitoneally for three days and then CCl_4 was injected into the rats. The extract showed a protective effect against liver damage caused by CCl_4 at doses of 200 and 300 mg/kg, and decreased serum levels of liver enzymes such as serum aminotransaminase, alanine aminotransferase, and alkaline phosphatase. However, it was found to be less effective at doses higher than 300 mg/kg (Jamshidzadeh *et al.*, 2005).

Aphrodisiac Effect

The effects of hydroalcoholic extracts prepared from *G. tournefortii* leaf and stem on sperm count and motility and serum concentration of testosterone were investigated. In the study, the male mice were divided into five groups as a control group and four experimental groups (100, 200, 400, and 800 mg/kg doses of the extracts administered intraperitoneally for 14 days). The sperm count increased significantly at 400 mg/kg dose. It was observed that the rate of sperm motility and testicular weight increased significantly at 200 and 400 mg/kg doses. When compared with the control group, it was determined that the serum level of testosterone increased significantly at doses of 100 ($p = 0.05$) and 400 mg/kg ($p = 0.001$) (Tabibian *et al.*, 2013).

The Iron Chelating Activity

The iron chelating activity of *G. tournefortii* methanolic extract was assessed in iron-overloaded rats. *G. tournefortii* methanolic extract dramatically lowered blood levels of iron, ferritin, liver biomarkers, and cardiac biomarkers, while also improving the lipid profile. *G. tournefortii*'s chelating activity, as well as its hepatoprotective and cardioprotective effects in iron-overloaded rats, was demonstrated in this research (Mansi *et al.*, 2020).

Wound Healing Effect

Cutaneous wound healing activities of gold nanoparticles using aqueous extract of *G. tournefortii* leaves (AuNPs@GT) were demonstrated. The use of AuNPs@GT ointment significantly reduced ($p \leq 0.01$) the wound area, total cells, neutrophil, and lymphocyte, while significantly increased ($p \leq 0.01$) the wound contracture, hydroxyl proline, hexosamine, hexuronic acid, fibrocyte, fibroblast, and fibrocytes/fibroblast rate in the rats (Zhaleh *et al.*, 2019).

The wound healing effect of the aerial parts extract with milkcream was evaluated on the healing of second-degree burns in a rat model. On the 14th and 21st days, macroscopic examination of wound sizes revealed that the wound surface after the aerial parts extract with milkcream administration was significantly reduced ($p < 0.001$). Also, histological data revealed that burn recovery was also significantly enhanced (Javanmardi *et al.*, 2020).

Clinical Studies

α-Amylase Enzyme Inhibitory Effect

The urinary α-amylase activity of 50 type-1 diabetes patients aged 1–16 years was investigated. The inhibitor dose required for the inhibition of enzyme activity of the petroleum ether extract prepared from *G. tournefortii* leaves was used. The information obtained was compared with the findings obtained from healthy individuals (30 individuals aged 1–16 years) used as the control group. α-Amylase activity in the serum of patients (281.70 ± 10.03 IU/24 hours) was increased compared to the control samples (43.38 ± 3.33 IU/24 hours). Enzyme inhibition was observed in patients using 15 mg/mL *Gundelia* extract (Hamad and Hasan, 2010).

Acute Toxicity

It has been stated that the aqueous alcoholic extract prepared from the petioles can cause hepatotoxic effects at high concentrations (Jamshidzadeh *et al.*, 2005).

Cardiovascular Effect

The effects of the aerial parts extract on total antioxidant capacity and lipid profile in patients with coronary artery disease were assessed. It is observed that the extract-consumed group displayed considerably lower energy intake compared to the placebo group ($p = 0.04$). In addition, the extract intake dropped body mass index by 3%. Total cholesterol levels were significantly reduced in the extract consumer group (from 151 ± 23.8 mg/dL at baseline to 131.1 ± 25.9 mg/dL at the end of the study). The mean lipoprotein cholesterol level fell from 86 ± 26 to 60.58 ± 29.9 mg/dL ($p = 0.001$). However, total antioxidant capacity increased considerably (Hajizadeh-Sharafabad *et al.*, 2016).

CHEMISTRY

Seven different saponins, the aglycone of which are oleanolic acid derivatives, were determined from their roots (Wagner *et al.*, 1984). In addition, the presence of alkaloid, flavonoid, phenol, tannin, glycoside, carbohydrate, and free amine groups in the water and alcohol extracts of the root (hot and cold) was determined qualitatively (Al-Younis and Argushy, 2009). In a study conducted in Jordan, a very high tannin content (39172 mg/kg dry weight) was found in naturally grown *G. tournefortii* (Alkurd *et al.*, 2008). A mixture of scopoletin, isoscopoletin, esculin, β-sitosterol, and stigmasterol was isolated from the aerial parts of the plant (Halabi *et al.*, 2005; Kery and Hussein, 1985).

The contents of chlorogenic acid and caffeic acid from phenolic acids were determined (Haghi and Hatami, 2010). Caffeic acid and caffeic acid derivatives such as neochlorogenic acid (*3-O*-caffeoylquinic acid), cryptochlorogenic acid (*4-O*-caffeoylquinic acid), chlorogenic acid (*5-O*-caffeoylquinic acid) were identified from the leaves and seeds, Total phenol content and chlorogenic acid content of leaf, seed coat, and peeled seed were determined (Haghi *et al.*, 2011).

High amounts of quinic acid, chlorogenic acid, and protocatechuic acid were detected in mature disseminules (fruit-seeds) of *G. tournefortii* and *G. tournefortii* var. *tenuisecta* (Ertas *et al.*, 2021).

In a study conducted in Jordan, α-terpinyl acetate (36.21%), methyl eugenol (12.6%), eugenol (6.7%), β-caryophyllene (5.9%), and zingiberene (5.8%) were determined as the main components in the essential oil obtained from the aerial parts (Halabi *et al.*, 2005). In another study, terpinyl acetate (15.2%), eugenol (9.5%), pcyme (9.5%), β-caryophyllene (7.1%), and estragole (6.1%) were found as the major constituents in essential oil from the aerial parts (Qnais *et al.*, 2016).

In a study conducted in Iran, thymol (11.2%), γ-terpinene (9.8%), germacrene D (6.6%), and *p*-cymene (6.3%) were determined as the major components in the essential oil from the aerial parts (Dastan and Yousefzadi, 2016). In another study conducted in Iran, the essential oil of the aerial parts was analyzed, and palmitic acid (12.48%), lauric acid (10.59%), alpha ionene (6.68%), myristic acid (4.45%), 1-hexadecanol, 2-methyl (3.61%), phytol (3.6%), and beta turmerone (3.4%) were detected as main compounds (Farhang *et al.*, 2016). In a study conducted in Israel, malic acid, methyl alpha-D-glucoside, lyxose, asparagine, inositol, glucopyranose, and D-Ribofuranose were in methanol extracts of the aerial parts and hexacosan-1-ol, α-linolenic acid, octacosan-1-ol, tetradecanoic acid, stigmasterol, (E)-3,7,11,15-tetramethylhexadec-2-en-1-ol, 12-Oleanen-3-yl acetate in hexane extracts of the aerial parts were detected as main components. In both extracts, stigmasterol was found (Kadan *et al.*, 2018).

Chemical compositions of essential oils of *G. tournefortii* var. *tournefortii* and var. *armata* dried aerial parts were analyzed. In *G. tournefortii* var. *tournefortii*, monoterpenes formed the main group, while sesquiterpenes were determined in *G. tournefortii* var. *armata* as the main group. Thymol (24.5%), γ-terpinene (10.7%), α-terpineol (8.7%), and *p*-cymene (7.3%) were detected as the main compounds in *G. tournefortii* var. *tournefortii*, while germakren D (21.6%), β-caryophyllene (7%), and bicyclogermacrene (5.2%) were detected in *G. tournefortii* var. *armata* (Bağcı *et al.*, 2010).

The oil of *G. tournefortii* mature flower buds was found to be rich in linoleic (57.8%), oleic (28.5%), and palmitic acid (8.1%), and the presence of stearic, vaksenic, arachidic, and linolenic acids in the oil. The total sterol content of the oil is 3.766 mg/kg, and β-sitosterol content (51.6%)

is dominant, while the other sterols 18.5% stigmasterol, 9.8% 5-avenasterol, 6% campesterol, 3.7% 7-stigmasterol, and 2.6% 7-avenasterol were also detected. The total vitamin E content in oil was found to be 51.9 mg/100 g, and the dominant isomers were α-tocopherol (48.9 mg/100 g) and γ-tocopherol (1 mg/100 g) (Matthäus and Özcan, 2011). The major components in the aerial parts of methanol extract were determined as (9Z)-9,17-octadecadienal, linoleic acid, palmitic acid, dodecanoic acid, hop-22(29)-en-3.β.-ol, and (Z)-11-hexadecenoic acid. Also, anticancer components were detected such as artemisinin and gitoxigenin. The major components in the aerial parts of hexane extract were detected as thunbergol, hop-22(29)-en-3.β.-ol, olean-12-en-3-yl acetate, A'-neogammacer-22(29)-en-3-ol, acetate, (3.β.,21.β.)-, and anticancer components such as lupeol, stigmasterol, β-sitosterol, α-amyrin (Abu-Lafi *et al.*, 2019). β-Sitosterol, α-amyrin, and 3-acetyl lupeol were identified in mature disseminules (fruit-seeds) of *G. tournefortii* and *G. tournefortii* var. *tenuisecta* (Ertas *et al.*, 2021).

It was demonstrated that the seed oil contains palmitic acid (12.1%), stearic acid (2.5%), oleic acid (23.4%), and linoleic acid (62%) (Erciyes *et al.*, 1989). Linoleic acid methyl ester (43.98%), oleic acid methyl ester (28.29%), palmitic acid methyl ester (13.42%), and 8-octadecenoic acid methyl ester (6.89%) were detected as major components in the fatty acid methyl ester contents of the seed oil (Al-Saadi *et al.*, 2017). It was indicated that mature disseminules (fruit-seeds) of *G. tournefortii* and *G. tournefortii* var. *tenuisecta* consisted predominantly oleic acid, palmitic acid, stearic acid, and linoleic acid (Ertas *et al.*, 2021).

The physicochemical properties of the oil obtained from its seeds and used in cooking were examined, the oil content was 22.8%, the saponification value was found as 166.1 mg/g, and it was observed that it contains a high level of unsaturated fatty acids and is soluble oil at ambient temperature. By specifying the oleic and linoleic fatty acid content, β-sitosterol and stigmasterol were determined as the main unsaponifiable ingredients of the oil, and color analysis revealed that the dominant color was yellow. The physicochemical properties of the oil have shown that the oil has high edible quality (Khanzadeh *et al.*, 2012). Cr was determined as 5.072 ± 0.5 ppm/g dry weight and Cu as 5.855 ± 0.22 ppm/g dry weight in the leaves of the plant grown in Amanos Mountains (Ergün *et al.*, 2012). P, K, S, Ca, Mg, Na, Fe, Mn, Zn, and Cu contents of the plant grown in the Eastern Anatolia Region (Turkey) were revealed and low Mn content was observed (Turan *et al.*, 2003). In a study on some edible wild plants consumed in Elazig (Turkey), high nitrate content was found in the plant (Çakılcıoğlu and Khatun, 2011). Fe content (3020 mg/kg dry weight) was found in high concentration in the plant grown in the Negev desert in Israel. The Cd content was also found to be above the toxic level (13 mg/kg dry weight) (Sathiyamoorthy *et al.*, 1997). Mature flower buds were found to be rich in crude oil, protein, and fiber and as well as P (1.8 mg/kg), K (412.1 mg/kg), Ca (316.3 mg/kg), Mg (266.6 mg/kg), Na (32.4 mg/kg), Zn (6.8 mg/kg), and Fe (5.8 mg/kg) (Matthäus ve Özcan, 2011).

The chemical composition and biological activities of *Gundelia tournefortii* have been extensively studied up to this point. Biological activities and clinical studies conducted on *G. tournefortii* have revealed that it is typically safe to use for the treatment of different ailments and diseases. However, it has been stated that it can become toxic at very high doses. Considering its traditional use, biological activity studies to be carried out on it and isolation studies of bioactive compounds will increase.

REFERENCES

Abu-Lafi, S., Rayan, B., Kadan, S., Abu-Lafi, M., Rayan, A. Anticancer activity and phytochemical composition of wild *Gundelia tournefortii*, *Oncol. Lett.*, 17, 713–717 (2019).

Aburjai, T., Darwish, M.R., Al-Khalil, S., Mahafza, A., Al-Abbadi, A. Screening of antibiotic resistant inhibitors from local plant materials against two different strains of *Pseudomonas aeruginosa*, *J. Ethnopharmacol.*, 76, 39–44 (2001).

Alali, F.Q., Tawaha, K., El-Elimat, T., Syouf, M., El-Fayad, M., Abulaila, K., Nielsen, S.J., Wheaton, W.D., Falkinham III, J.O., Oberlies, N.H. Antioxidant activity and total phenolic content of aqueous and methanolic extracts of Jordanian plants: An ICBG project, *Nat. Prod. Res.*, 21 (12), 1121–1131 (2007).

Ali-Shtayeh, M.S., Al-Assali, A.A., Jamous, R.M. Antimicrobial activity of Palestinian medicinal plants against acne-inducing bacteria, *Afr. J. Microbiol. Res.*, 7 (21), 2560–2573 (2013).
Ali-Shtayeh, M.S., Jamous, R.M., Jamous, R.M. Plants used during pregnancy, childbirth, postpartum and infant healthcare in Palestine, *Compl. Ther. Clin. Pr.*, 21, 84–93 (2015).
Alkurd, R.A., Takruri, H.R., Al-Sayyed, H. Tannin contents of selected plants used in Jordan, Jordan, *J. Agri. Sci.*, 4 (3), 265–274 (2008).
Al-Saadi, S.A.M., Qader, K.O., Hassan, T.O. Variations in fatty acid methyl ester contents and composition in oil seeds *Gundelia tournefortii* L. (Asteraceae), *Adv. Plants Agric. Res.*, 6 (6), 00236 (2017).
Altundağ, E., Öztürk, M. Ethnomedicinal studies on the plant resources of east Anatolia, Turkey, *Procedia Soc. Behav. Sci.*, 19, 756–777 (2011).
Al-Younis, N.K., Argushy, Z.M. Antibacterial evaluation of some medicinal plants from Kurdistan region, *J. Duhok Univ.*, 12 (1) (Special Issue), 256–261 (2009).
Arasan, S., Kaya, I. Some important plant belonging to Asteraceae Family used in folkloric medicine in Savur (Mardin/Turkey) area and their application areas, *J. Food Nutr. Res.*, 3, 337–340 (2015).
Asadi-Samani, M., Rafieian-Kopaei, M., Azimi, N. *Gundelia*: A systematic review of medicinal and molecular perspective, *Pak. J. Biol. Sci.*, 16 (21), 1238–1247 (2013).
Asgary, S., Movahedian, A., Badiei, A., Naderi, G.A., Amini, F., Hamidzadeh, Z. Effect of Gundelia tournefortii on some cardiovascular risk factors in animal model, *J. Med. Plants.*, 7 (28), 112–119 (2008).
Azeez, O.H., Kheder, A.E. Effect of *Gundelia tournefortii* on some biochemical parameters in dexamethasone-induced hyperglycemic and hyperlipidemic mice, *Iraqi J. Vet. Sci.*, 26 (2), 73–79 (2012).
Bağcı, E., Hayta, Ş., Kılıç, Ö., Koçak, A. Essential oils of two varieties of Gundelia tournefortii L. (*Asteraceae*) from Turkey, *Asian J. Chem.*, 22 (8), 6239–6244 (2010).
Bahmani, M., Baharvand-Ahmadi, B., Tajeddini, P., Rafieian-Kopaei, M., Naghdi, N. Identification of medicinal plants for the treatment of kidney and urinary stones, *J. Ren. Inj. Prev.* 5 (3), 129–133 (2016).
Baytop, T. *Türkiye'nin Tıbbi ve Zehirli Bitkileri*, İstanbul Üniv. Yay. No. 1039, Tıp Fak. Yay. No. 59, İstanbul (1963).
Baytop, T. *Türkiye'de Bitkiler ile Tedavi (Geçmişte ve Bugün)*, İlaveli 2. Baskı, Nobel Tıp Kitabevleri, İstanbul (1999).
Baytop, T. *Türkçe Bitki Adları Sözlüğ ü, Atatürk Kültür, Dil ve Tarih Yüksek Kurumu, Türk Dil Kurumu Yayınları: 578*, 3. Baskı, Türk Tarih Kurumu Basımevi, Ankara (2007).
Çakılcıoğlu, U., Türkoğlu, İ. Plants used to lower blood sugar in Elazığ central district, ISHS, in *Proc. Ist IC on Culinary Herbs*, Turgut, K. et al., (Eds.), *Acta Hort.*, 826, 97–104 (2009).
Çakılcıoğlu, U., Khatun, S. Nitrate, moisture and ash contents of edible wild plants, *J. Cell & Plant Sci.*, 2 (1), 1–5 (2011).
Çoruh, N., Sağdıçoğlu Celep, A.G., Özgokçe, F., İşcan, M. Antioxidant capacities of *Gundelia tournefortii* L. extracts and inhibition on glutathione-S-transferase activity, *Food Chem.*, 100, 1249–1253 (2007).
Dalar, A., Mukemre, M., Unal, M., Ozgokce, F. Traditional medicinal plants of Agri province, Turkey, *J. Ethnopharmacol.*, 226, 56–72 (2018).
Darwish, R.M., Aburjai, T., Al-Khalil, S., Mahafzah, A. Screening of antibiotic resistant inhibitors from local plant materials against two different strains of *Staphylococcus aureus*, *J. Ethnopharmacol.*, 79, 359–364 (2002).
Darwish, R.M., Aburjai, T.A. Effect of ethnomedicinal plants used in folklore medicine in Jordan as antibiotic resistant inhibitors on Escherichia coli, *BMC Complement. Altern. Med.*, 10 (9), 1–8 (2010).
Dastan, D., Yousefzadi, M. Volatile oil constituent and biological activity of *Gundelia tournefortii* L. from Iran, *J. Rep. Pharm. Sci.*, 5 (1), 18–24 (2016).
Duke, J.A., Duke, P.-A.K., du Cellie, J.L. *Duke's Handbook of Medicinal Plants of the Bible*, CRC Press, Taylor & Francis Group, Boca Raton, FL (2008).
Emami, S.A., Sahebkar, A., Tayarani-Najaran, N., Tayarani-Najaran, Z. Cancer and its treatment in main ancient books of islamic iranian traditional medicine (7th to 14th Century AD), *Iran Red. Crescent Med. J.*, 4 (12), 747–757 (2012).
Erciyes, A.T., Karaosmanoğlu, F., Civelekoğlu, H. Fruit oils of four plant species of Turkish origin, *JAOCS*, 66 (10), 1459–1464 (1989).
Ergün, N., Yolcu, H., Özçubukçu, S. Amanos dağlarındaki bazı tıbbi bitki türlerinde ağır metal birikimi, *Biyoloji Bilimleri Araştırma Dergisi*, 5 (1), 21–23 (2012).
Ertas, A., Fırat, M., Yener, I., Akdeniz, M., Yigitkan, S., Bakir, D., Çakir, C., Yilmaz, M.A., Ozturk, M., Kolak, U. Phytochemical fingerprints and bioactivities of ripe disseminules (fruit-seeds) of seventeen *Gundelia* (Kenger-Kereng Dikeni) species from Anatolia with chemometric approach, *Chem. Biodivers.*, 18, e2100207 (2021).

Farhang, H.R., Vahabi, M.R., Allafchian, A.R. Chemical compositions of the essential oil of *Gundelia tournefortii* L. (Asteraceae) from Central Zagros, Iran, *J. Herb Drugs*, 6 (4), 227–233 (2016).

Ghabeish, I.H. Insecticidal activity and synergistic effect of *Gundelia tournefortii* L. (Asteraceae: Compositae) extracts and some pure constituents on *Drosophila melanogaster* Meigen (Drosophilidae: Diptera), Jord, *J. Agric. Sci.*, 11, 353–366 (2015).

Ghasemi Pirbalouti, A., Momeni, M., Bahmani, M. Ethnobotanical study of medicinal plants used by kurd tribe in Dehloran and Abdanan districts, Ilam province, Iran, *Afr. J. Tradit. Complement Altern. Med.*, 10, 368–385 (2013).

Haghi, G., Hatami, A. Simultaneous quantification of flavonoids and phenolic acids in plant materials by a newly developed isocratic high-performance liquid chromatography approach, *J. Agric. Food Chem.*, 58, 10812–10816 (2010).

Haghi, G., Hatami, A., Arshi, R. Distribution of caffeic acid derivatives in Gundelia tournefortii L., *Food Chem.*, 124, 1029–1035 (2011).

Hajizadeh-Sharafabad, F., Alizadeh, M., Mohammadzadeh, M.H.S., Alizadeh-Salteh, S., Kheirouri, S. Effect of *Gundelia tournefortii* L. extract on lipid profile and TAC in patients with coronaryartery disease: A double-blind randomized placebo controlled clinicaltrial, *J. Herb. Med.*, 6, 59–66 (2016).

Halabi, S., Battah, A.A., Aburjai, T., Hudaib, M. Phytochemical and antiplatelet investigation of Gundelia tournifortii, *Pharm. Biol.*, 43 (6), 496–500 (2005).

Hamad, N.S., Hasan, H.G. Inhibitory effect of *Gundelia* extract on urinary α-amylase activity of type-I diabetes mellitus, *Zanco J. Med. Sci.*, 14 (3), 1–6 (2010).

Hayta, Ş., Polat, R., Selvi, S. Traditional uses of medicinal plants in Elazığ (Turkey), *J. Ethnopharmacol.*, 154, 613–624 (2014).

Jamshidzadeh, A., Fereidooni, F., Salehi, Z., Niknahad, H. Hepatoprotective activity of *Gundelia tourenfortii*, *J. Ethnopharmacol.*, 101, 233–237 (2005).

Javanmardi, S., Safari, I., Aghaz, F., Khazaei, M. Wound healing activities of *Gundelia tournefortii* L extract and milkcream ointment on second-degree burns of rat skin, *Int. J. Low. Extrem. Wounds*, 20 (3), 272–281 (2020).

Kadan, S., Sasson, Y., Saad, B., Zaid, H. *Gundelia tournefortii* antidiabetic efficacy: Chemical composition and GLUT4 translocation, *Evid. Based Complement. Alternat. Med.*, 8294320 (2018).

Kery, A., Hussein, A.M. Isolation of scopoletin in *Gundelia tournefortti* by DCCC, *Fitoterapia*, 56, 42–44 (1985).

Khanzadeh, F., Haddad Khodaparast, M.H., Elhami Rad, A.H., Rahmani, F. Physiochemical properties of Gundelia tournefortii L. seed oil, *J. Agr. Sci. Tech.*, 14, 1535–1542 (2012).

Kupicha, F.K. Gundelia, in *Flora of Turkey and the East Aegean Islands*, Davis, P.H. (Ed.), Vol. 5, Edinburgh University Press, Edinburgh, pp. 325–326 (1975).

Lev, E. Reconstructed materia medica of the Medieval and Ottoman al-Sham, *J. Ethnopharmacol.*, 80, 167–179 (2002).

Mansi, K., Tabaza, Y., Aburjai, T. The iron chelating activity of *Gundelia tournefortii* in iron overloaded experimental rats, *J. Ethnopharmacol.*, 263, 113114 (2020).

Mavi, A., Lawrence, G.D., Kordali, Ş., Yıldırım, A. Inhibition of iron-fructose-phosphate-induced lipid peroxidation in lecithin liposome ana linoleic acid emulsion systems by some edible plants, *J. Food Biochem*, 35, 833–844 (2011).

Matthäus, B., Özcan, M.M. Chemical evaluation of flower bud and oils of tumbleweed (*Gundelia tourneforti* L.) as a new potential nutrition sources, *J. Food Biochem.*, 35, 1257–1266 (2011).

Mohammadi, G., Zangeneh, M.M., Rashidi, K., Zangeneh, A. Evaluation of nephroprotective and antidiabetic effects of *Gundelia tournefortii* aqueous extract on diabetic nephropathy in male mice, *Res. J. Pharmacog.*, 5, 65–73 (2018).

Mosaddegh, M., Naghibi, F., Moazzeni, H., Pirani, A., Esmaeili, S. Ethnobotanical survey of herbal remedies traditionally used in Kohghiluyeh va Boyer Ahmad province of Iran, *J. Ethnopharmacol.*, 141, 80–95 (2012).

Nariman, F., Eftekhar, F., Habibi, Z., Massarrat, S., Malekzadeh, R. Antibacterial activity of twenty Iranian plant extracts against clinical isolates of Helicobacter pylori, *Iran. J. Basic Med. Sci.*, 12 (2), 105–111 (2009).

Nooraei, M. *The Great Encyclopedia of Islamic Medicine, Persian*, Fakhredin Press, Tehran (2005).

Obeidat, M. Antimicrobial activity of some medicinal plants against multidrug resistant skin pathogens, *J. Med. Plants Res.*, 5 (16), 3856–3860 (2011).

Oryan, S., Nasri, S., Amin, G., Kazemi-Mohammady, S. Antinociceptive and anti-inflammatory effects of aerial parts of *Gundelia tournefortii* L. on NMRI male mice, *J. Shahrekord Univ. Med. Sci.*, 12 (4) (Suppl 1), 8–15 (2011).

Özgökçe, F., Özçelik H. Ethnobotanical aspects of some taxa in East Anatolia (Turkey), *Eco. Bot.*, 58 (4), 697–704 (2004).

Özkan, A., Yumrutaş, Ö., Saygıdeğer, S.D., Kulak, M. Evaluation of antioxidant activities and phenolic contents of some edible and medicinal plants from Turkey's flora, *Adv. Environ. Biol.*, 5 (2), 231–236 (2011).

Polat, R., Çakılcıoğlu, U., Satıl, F. Traditional uses of medicinal plants in Solhan (Bingöl-Turkey), *J. Ethnopharmacol.*, 148, 951–963 (2013).

Prance, G., Nesbitt, M. (Eds.). *The Cultural History of Plants*, Routledge, New York (2005).

Qnais, E., Bseiso, Y., Wedyan, M., Al-Omari, M., Alkhateeb, H. Chemical composition and antinociceptive effects of essential oil from aerial parts of *Gundelia tournefortii* L Asteraceae (Compositae) in rats, *Trop. J. Pharm. Res.*, 15 (10), 2183–2190 (2016).

Sarper, F., Akaydın, G., Şimşek, I., Yeşilada, E. An ethnobotanical field survey in the Haymana district of Ankara province in Turkey, *Turk. J. Biol.*, 33, 79–88 (2009).

Sathiyamoorthy, P., Van Damme, P., Oven, M., Golan-Goldhirsh, A. Heavy metals in medicinal and fodder plants of the negev desert, *J. Environ. Sci. Health, A*, 32 (8), 2111–2123 (1997).

Şekeroğlu, N., Şenol, S., Erdoğan Orhan, İ., Gülpınar, A.R., Kartal, M., Şener, B. *In vitro* prospective effects of various traditional herbal coffees consumed in Anatolia linked to neurodegeneration, *Food Res. Int.*, 45, 197–203 (2012).

Sezik, E., Yeşilada, E., Honda, G., Takaishi, Y., Takeda, Y., Tanaka, T. Traditional medicine in Turkey X. Folk medicine in Central Anatolia, *J. Ethnopharmacol.*, 75, 95–115 (2001).

Sharaf, K.H., Ali, J.S. Hypolipemic effect of Kuub (*Gundelia tournefotii* A.) oil and clofibrate on lipid profile of atherosclerotic rats, *Vet. Arhiv.*, 74 (5), 359–369 (2004).

Tabata, M., Sezik, E., Honda, G., Yeşilada, E., Fukui, H., Goto, K., Ikeshiro, Y. Traditional medicine in Turkey III. Folk medicine in East Anatolia,Van and Bitlis provinces, *Int. J. Pharmacog.*, 32, 3–12 (1994).

Tabibian, M., Nasri, S., Kerishchi, P., Amin, G. The effect of *Gundelia tournefortii* hydro-alcoholic extract on sperm motility and testosterone serum concentration in mice, *Zahedan J. Res. Med. Sci.*, 15 (8), 18–21 (2013).

Talei, G.R., Meshkatalsadat, M.H., Mousavi, Z. Antibacterial activity and chemical composition of essential oils from four medicinal plants of Lorestan, Iran, *J. Med. Plants*, 6 (1), 45–52 (2007).

Tanker, M., Tanker, N. Kenger Kahvesini Veren Bitki: Gundelia tournefortii L., *İstanbul Üniv. Ecz. Fak. Mec.*, 3 (2), 63–74 (1967).

Tawaha, K., Alali, F.Q. Gharaibeh, M., Mohammad, M., El-Elimat, T. Antioxidant activity and total phenolic content of selected Jordanian plant species, *Food Chem.*, 104, 1372–1378 (2007).

Tetik, F., Civelek, Ş., Çakılcıoğlu, U. Traditional uses of some medicinal plants in Malatya (Turkey), *J. Ethnopharmacol.*, 146, 331–346 (2013).

Turan, M., Kordali, S., Zengin, H., Dursun, A., Sezen, Y. Macro and micro mineral content of some wild edible leaves consumed in eastern Anatolia, *Acta Agric. Scand., Sect. B, Soil and Plant Sci.*, 53, 129–137 (2003).

Tuzlacı, E., Şenkardeş, İ. Turkish folk medicinal plants, X: Ürgüp (Nevşehir), *Marmara Pharm. J.*, 15, 58–68 (2011).

Tuzlacı, E. *Türkiye Bitkileri Geleneksel İlaç Rehberi*, İstanbul Tıp Kitabevleri, İstanbul (2016).

Uğulu, İ., Başlar, S. The determination and fidelity level of medicinal plants used to make traditional Turkish salves, *J. Altern. Complement. Med.*, 16 (3), 313–322 (2010).

Vural, M., Karavelioğulları, F.A., Polat, H. Çiçekdağı (Kırşehir) ve çevresinin etnobotanik özellikleri, Ot Sistematik Botanik, *Dergisi*, 4 (1), 117–124 (1997).

Wagner, H., Nickl, H., Aynechi, Y. Molluscicidal saponins from *Gundelia tournefortii. Phytochem.*, 23 (11), 2505–2508 (1984).

Zhaleh, M., Zangeneh, A., Goorani, S., Seydi, N., Zangeneh, M.M., Tahvilian, R., Pirabbasi, E. *In vitro* and *in vivo* evaluation of cytotoxicity, antioxidant, antibacterial, antifungal, and cutaneous wound healing properties of gold nanoparticles produced via a green chemistry synthesis using *Gundelia tournefortii* L. as a capping and reducing agent, *Appl. Organomet. Chem.*, 33, e5015 (2019).

9 *Hypericum perforatum* L.

Erdem Yeşilada

INTRODUCTION

The *Hypericum* L (Hypericaceae) genus has a worldwide distribution comprising about 484 species (Meseguer and Sanmartin, 2012). Turkey has been considered one of the gene centers of this genus, with 82 species and 98 taxa, among which 45 taxa are reported to be endemic (45.92%) (Bingöl *et al.*, 2009). Despite such a rich diversity of the genus in Turkey, only a few species have found utilization in Turkish folkloric medicine. In a reference survey on published ethnobotanical reports, only 16 species have been recorded as folkloric remedies (Table 1). This book chapter aimed to discuss the utilization described by the local people in those ethnobotanical records based on the available scientific evidence to justify their validity.

St. John's wort is the well-known name of *Hypericum perforatum* and is among the most popular herbal remedies in the world. The flowering aerial part is used as an antidepressant, and its efficiency has been confirmed scientifically by experimental and clinical studies. However, such utilization is not a matter of common knowledge in Turkey. Although infusion has been noted as a "sedative" tea for calming the nervous system in three papers, possibly such an effect would not be correct since the active components, hypericin and hyperforin derivatives, are not soluble in water. Moreover, hyperforin is rapidly decomposed by heat and atmospheric oxygen once extracted.

As shown in Table 1, flowering aerial parts of *Hypericum* species are mainly prepared as aqueous, (decoction or infusion), and as oily extracts. The wound healing effect of its olive oil extract (oleate) received the highest scores in ethnobotanical publications related to Turkish traditional medicine. The flowering aerial parts are cut into a jar of raw olive oil and left for at least a month or two under sunshine to yield a red-colored extract. People keep this oleate as a first-aid treatment when necessary as a home remedy. The oleate is also used internally against stomach disorders, i.e., peptic ulcers, gastritis, or reflux. On the other hand, decoction or infusion of the flowering aerial parts has also been reported to treat gastric disorders, i.e., gastric ulcers and reflux. However, this information may not be realistic since the naphthodianthrones, i.e., hypericin, etc., known as active components, are not extractable with water (Saçıcı and Yesilada, 2021).

DERMATOLOGICAL PROBLEMS

Naphthoquinones, hypericin derivatives, have been the main active ingredients for wound healing, burns, cuts, lip chaps, and inflammatory skin conditions such as pimples, acne, diaper rash and eczema, and infectious cases, i.e., *Herpes labialis* infection (Süntar Peşin *et al.*, 2010). They are lipophilic and yield a red-colored oleate when left maceration under sunshine for one or two months. However, in two publications, infusion/decoction of *H. androsaemum* and *H. perforatum* is noted as a wound healing remedy, which needs further confirmation. On the other hand, oleates from the flowering aerial parts of *H. perforatum* and six other *Hypericum* species, i.e., *H. bithynicum*, *H. lydium*, *H. montbretii*, *H. perfoliatum*, *H. scabrum*, and *H. triquetrifolium*, are reported as a wound healing remedy. Interestingly similar use was also given for an oleate for the fruit of *H. perforatum* in one study. However, hypericin and pseudohypericin are not available in all *Hypericum* species. Ayan *et al.* (2004) investigated the hypericin contents of 18 *Hypericum*

 DOI: 10.1201/9781003146971-9

species of Turkey. According to the results of this study, hypericin contents of the abovementioned species *H. bithynicum* (0.174%), *H. montbretii* (0.0224%), and *H. perforatum* (0.303%) seem sufficient to support the folkloric use. While in *H. androsaemum* (0.00%), *H. scabrum* (0.00%), and *H. perfoliatum* (0.005 %), hypericin was absent or traced and thus would not be effective in wound healing if we consider naphthoquinones are the active components. However, in a recent study, the phenolic contents of *H. perfoliatum* have demonstrated a significant antioxidant and anti-inflammatory activity, and wound healing effect may be attributed to the contribution of these components (Celep *et al.*, 2017). Ayan and Çırak (2008), in the following study, have quantified the hypericin and pseudohypericin contents in the aerial parts. Total contents of these two components were found, respectively, 3.90% and 1.75% in *H. perforatum*, 2.9% and 4.16% in *H. triquetrifolium*, and 0.12% and 0.1% in *H. scabrum* (mg/g; dry weight). These results have supported the folkloric claim on *H. perforatum* and *H. triquetrifolium* oleates for wound healing, but this would not be a case for *H. scabrum. In vivo* experimental studies confirmed the wound healing activity of *H. perforatum* by Süntar-Peşin *et al.* (2010). In a clinical study carried out on intensive care unit patients, *H. perforatum* oleate was found to be significantly effective in the treatment of decubitus ulcers (Yücel *et al.*, 2017). Although naphthoquinone content was a trace in *H. scabrum*, in a recent study, its essential oil was reported to be effective on diabetic wounds in rats, possibly by regulating the antioxidant parameters in the wounded area (Ibaokurgil *et al.*, 2022). The ratio of its essential oil was about 0.96 % and was composed of monoterpenoids, mainly α-pinene (40-45 %). Since essential oils are lipophilic, the wound healing activity of its oleate formulation may be attributed to this content.

Since naphthoquinones increase sensitivity to sunshine for the external use of *Hypericum* species, their contents are important. In a recent study, only a few species have been reported to contain these components (Saçıcı and Yesilada, 2021). Eventually, the flowering aerial part of *H. olympicum* L. has been reported to act as an anti-inflammatory against sunburn due to a lack of hypericin derivatives (Kurt-Celep *et al.*, 2020). There are also a few quotations for *H. perforatum* oleate as an external remedy against diaper rash and lip chaps and *Herpes labialis* infection on lips. On the other hand, infusions from the flowering aerial parts of *H. androsaemum* for acne and allergy and the *H. scabrum* fruit for eczema were noted, which require further scientific evidence to confirm their efficacy.

GASTROINTESTINAL PROBLEMS

Among the gastric disorders, stomachache, gastritis, reflux, and gastric ulcers are the most frequently referred problems that *Hypericum* species have been used. If we consider gastritis, ulcers, and reflux as internal wounds, oral use of the oleate type preparations would be reasonable due to the lipophilic characteristics of naphthoquinones. However, only two receipts were oleate types among the traditional notes: *H. montbretii* and *H. perforatum*. The potent anti-ulcerogenic activity of the oily extract or lipophilic fractions of *H. perforatum* has been evidenced by Yesilada and Gürbüz (1998). Sofi *et al.* (2020) recently reported its possible activity mechanism by H+/K+ ATPase α inhibition. On the other hand, due to sufficient hypericin content in *H. montbretii* (0.224%), its oleate formulation would be expected to be highly effective. However, decoctions or infusions from the flowering aerial parts of eight *Hypericum* species were also reported for such gastric problems: *H. atomarium*, *H. bithynicum*, *H. empetrifolium*, *H. montbretii*, *H. polyphyllum*, *H. scabrum*, *H. thymifolium*, and *H. triquetrifolium*. Currently, there is no supporting scientific evidence for these species if used for gastric complications as such aqueous preparations.

Decoctions or infusions of various *Hypericum* species were also noted as a remedy for hemorrhoidal complaints, either externally (*H. androsaemum*) or orally (*H. lydium*, *H. montbretii*, *H. perforatum*, *H. scabrum*). The oleate of *H. perforatum* is used externally or internally for hemorrhoids. Among the other folkloric uses, decoctions/infusions of *H. montbretii* and *H. perforatum* for constipation; *H. cerastoides* and *H. perforatum* for diarrhea; and *H. perforatum* for colitis were recorded.

UROLOGIC PROBLEMS

Among the urinary problems, the decoctions of several *Hypericum* species (*montbretii, perforatum, tetrapterum,* and *thymifolium*) are referred for urinary infections, cystitis, and kidney stones. However, the effect of only *H. perforatum* on the urological problems has been investigated experimentally. With its hydroalcoholic extract (0.3% hypericin), pretreatment protected cisplatin-induced nephrotoxicity in rats (Shibayama *et al.,* 2007). Another study confirmed its nephroprotective potential in diabetic nephropathy rats (Abd El Motteleb and Abd El Aleem, 2017). The hydroalcoholic extract of *H. perforatum* leaves was shown to exert inhibitory effect on kidney stones in rats (Khalili *et al.*, 2012).

MENSTRUAL PROBLEMS

The herbal tea (decoctions/infusions) of *H. bitynicum, H. lydium, H. perforatum,* or *H. scabrum* is noted to act as an emmenagogue for menstrual cramps and pain. The effect of *H. perforatum* alcoholic extract was confirmed clinically to inhibit hot flush score and improve quality of life in perimenopausal women (Al-Akoum *et al.,* 2009). However, the possible contribution of the other *Hypericum* species requires further evidence.

DIABETES MELLITUS

H. empetrifolium, H. montbretii, H. perforatum, and *H. venustum* are used as herbal tea to control blood sugar levels. Can *et al.* (2011) reported that *H. perforatum* improved the impaired metabolic parameters and restored hyperalgesia in diabetic rats. Ku *et al.* (2014) isolated hyperoside, a flavonoid, as the antioxidant and antihyperlycemic component of *Hypericum* species, which would be beneficial against diabetic complications, particularly of diabetic vascular inflammatory conditions. Hyperoside has a widespread distribution in *Hypericum* species with water-soluble characteristics; using these *Hypericum* species in folkloric practice as herbal tea seems confirmed. However, in a recent review paper, the possible effect of *H. perforatum* was associated with the anti-inflammatory effects of hypericin and hyperforin contents on pancreatic ß cells (Novelli *et al.*, 2020).

RHEUMATISM

The oleate of *H. empetrifolium* externally and the herbal tea of *H. perforatum* internally were reported against rheumatic complaints. Hypericin was determined as the active anti-rheumatic principle in *H. perforatum* acting by interleukin-12 inhibition (Kang *et al.,* 2001). Jin *et al.* (2016) reported that hyperoside exerted anti-inflammatory activity by suppressing the NFκB signaling pathway in experimental animals. Based on the available references, the use of herbal tea from *H. perforatum* may be considered effective due to its hyperoside content. However, there is no report on the hyperoside content of *H. empetrifolium*; the efficiency of its oleate needs further evidence.

CONCLUSION

Despite a rich diversity of *Hypericum* species in Turkey, only 16 species have been reported to be used as a folkloric remedy. However, among them, *H. perforatum* has a unique value. Aqueous extracts (decoctions or infusions) or oleate prepared from its flowering aerial parts have been suggested as a remedy against a wide range of health problems. It is noteworthy that many scientific studies have evidenced its efficacy.

BIBLIOGRAPHY

Abd El Motteleb D.M., Abd El Aleem D.I., 2017. Renoprotective effect of *Hypericum perforatum* against diabetic nephropathy in rats: Insights in the underlying mechanisms. *Clin. Exp. Pharmacol. Physiol.* 44, 509–21.

Akaydın G., Şimşek I., Arıtuluk Z.C., Yeşilada E., 2013. An ethnobotanical survey in selected towns of the Mediterranean subregion (Turkey). *Turk. J. Biol.* 37, 230–47.

Akbulut S., Özkan Z.C., 2014. An ethnobotanical study of medicinal plants in Marmaris (Muğla, Turkey). *Kastamonu Univ. J. Forestry Faculty* 14, 135–45.

Akyol Y., Altan Y., 2013. Ethnobotanical studies in the Maldan Village (Province Manisa, Turkey). *Marmara Pharm. J.* 17, 21–5.

Al-Akoum M., Maunsell E., Verreault R., Provencher L., Otis H., Dodin S., 2009. Effects of *Hypericum perforatum* (St. John's wort) on hot flashes and quality of life in perimenopausal women: A randomized pilot trial. *Menopause: The Journal of The North American Menopause Society* 16, 307–14.

Altundağ E., Öztürk M., 2011. Ethnomedicinal studies on the plant resources of east Anatolia, Turkey. *Procedia Soc. Behav. Sci.* 19, 756–77.

Ayan A.K., Çırak C., Kudret K., Özen T., 2004. Hypericin in some *Hypericum* species from Turkey. *Asian J. Plant Sci.* 3, 200–2.

Ayan A.K., Çırak C., 2008. Hypericin and Pseudohypericin contents in Some *Hypericum* species Growing in Turkey. *Pharm. Biol.* 46, 288–91.

Bingöl Ü., Cosge B., Gürbüz B., 2011. *Hypericum* species in the flora of Turkey. *Med. Aromat. Plant Sci. Biotechnol.* 5 (Special Issue 1), 86–90.

Bulut G., 2011. Folk medicinal plants of Silivri (Istanbul, Turkey). *Marmara Pharm. J.* 15, 25–9.

Bulut G., Bozkurt M.Z., Tuzlacı E., 2017. The preliminary ethnobotanical study of medicinal plants in Uşak (Turkey). *Marmara Pharm. J.* 21/2, 305–10.

Bulut G., Haznedaroğlu M.Z., Doğan A., Koyu H., Tuzlacı E., 2017. An ethnobotanical study of medicinal plants in Acipayam (Denizli-Turkey). *J. Herbal Med.* 10, 64–81.

Bulut G., Tuzlacı E., 2013. An ethnobotanical study of medicinal plants in Turgutlu (Manisa—Turkey). *J. Ethnopharmacol.* 149, 633–47.

Bulut G., Tuzlacı E., 2015. An ethnobotanical study of medicinal plants in Bayramic (Canakkale-Turkey). *Marmara Pharm. J.* 19, 268–82.

Çakılcıoğlu U., 2011. Ethnopharmacological survey of medicinal plants in Maden (Elazig-Turkey). *J. Ethnopharmacol.* 137, 469–86.

Can Ö.D., Öztürk Y., Öztürk N., Sagratini G., Ricciutelli M., Vittori S., Maggi F., 2011. Effects of treatment with St. John's Wort on blood glucose levels and pain perceptions of streptozotocin-diabetic rats. *Fitoterapia* 82, 576–84.

Celep E., İnan Y., Akyüz S., Yesilada E., 2017. The bioaccessible phenolic profile and antioxidant potential of *Hypericum perfoliatum* L. after simulated human digestion. *Ind. Crops Prod.* 109, 717–23.

Dalar A., Mukemre M., Unal M., Özgökçe F., 2018. Traditional medicinal plants of Ağrı Province, Turkey. *J. Ethnopharmacol.* 226, 56–72.

Demirci S., Özhatay N., 2012. An ethnobotanical study in Kahramanmaraş (Turkey); Wild plants used for medicinal purpose in Andirin, Kahramanmaraş. *Turk. J. Pharm. Sci.* 9, 75–92.

Ecevit Genç G., Özhatay N., 2006. An ethnobotanical study in Çatalca (European part of Istanbul) II. *Turk. J. Pharm. Sci.* 3, 73–89.

Güler B., Manav E., Uğurlu E., 2015. Medicinal plants used by traditional healers in Bozüyük (Bilecik–Turkey). *J. Ethnopharmacol.* 173, 39–47.

Gürbüz İ., Özkan Gençler A.M., Akaydın G., Salihoğlu E., Günbatan T., Demirci F., Yeşilada E., 2019. Folk medicine in Düzce Province (Turkey). *Turk. J. Bot.* 43, 769–84.

Gürdal B., Kültür E., 2013. An ethnobotanical study of medicinal plants in Marmaris (Muğla, Turkey). *J. Ethnopharmacol.* 146, 113–26.

Güzel Y., Güzelşemme M., Miski M., 2015. Ethnobotany of medicinal plants used in Antakya: A multicultural district in Hatay Province of Turkey. *J. Ethnopharmacol.* 174, 118–52.

Ibaokurgil F., Yıldırım B.A., Yıldırım S., 2022. Effects of *Hypericum scabrum* L. essential oil on wound healing in streptozotocin-induced diabetic rats. *Cutan. Ocul. Toxıcol.* https://doi.org/10.1080/15569527.2022.2052890.

Jin X.N., Yan E.Z., Wang H.M., Sui H.J., Liu Z., Gao W., Jin Y., 2016. Hyperoside exerts anti-inflammatory and anti-arthritic effects in LPS-stimulated human fibroblast-like synoviocytes in vitro and in mice with collagen-induced arthritis. *Acta Pharmacologica Sinica* 37, 674–86.

Kang B.Y., Chung S.W., Kim T.S., 2001. Inhibition of Interleukin-12 production in lipopolysaccharide-activated mouse macrophages by hypericin, an active component of *Hypericum perforatum*. *Planta Med* 67, 364±6.

Khalili M., Jalali M.R., Mirzaei-Azandaryani M., 2012. Effect of hydroalcoholic extract of Hypericum perforatum L. leaves on ethylene glycol-induced kidney calculi in rats. *Urol J.* 9(2), 472–9.

Kılıçarslan Ç., Özhatay N., 2012. Wild plants used as medicinal purpose in the south part of Izmit (northwest Turkey). *Turk J. Pharm. Sci.* 9, 199–218.

Ku S.K., Kwak S., Kwon O.J., Bae J.S., 2014. Hyperoside inhibits high-glucose-induced vascular inflammation in vitro and in vivo. *Inflammation* 37, 1389–400.

Kurt-Celep İ., Celep E., Akyüz S., İnan Y., Barak T.H., Akaydın G., Telci D., Yesilada E., 2020. *Hypericum olympicum* L. recovers DNA damage and prevents MMP–9 activation induced by UVB in human dermal fibroblasts. *J. Ethnopharmacol.* 246, 112202.

Kültür Ş., 2007. Medicinal plants used in Kırklareli Province (Turkey). *J. Ethnopharmacol.* 41, 341–64.

Meseguer A.S., Sanmartín I., 2012. Paleobiology of the genus *Hypericum* (Hypericaceae): A survey of the fossil record and its palaeogeographic implications. *Anales del Jardín Botánico de Madrid* 69, 97–106.

Novelli M., Masiello P., Be P., Mene M., 2020. Protective role of St. John's wort and its components hyperforin and hypericin against diabetes through inhibition of inflammatory signaling: Evidence from in vitro and in vivo studies. *Int. J. Mol. Sci.* 21, 8108.

Özüdoğru B., 2011. Inferences from an ethnobotanical field expedition in the selected locations of Sivas and Yozgat provinces (Turkey). *J. Ethnopharmacol.* 137, 85–98.

Polat R., Çakılcıoğlu U., Kaltalioğlu K., Ulusan M.D., Türkmen Z., 2015. An ethnobotanical study on medicinal plants in Espiye and its surrounding (Giresun-Turkey). *J. Ethnopharmacol.* 163, 1–11.

Polat R., Çakılcıoğlu U., Satıl F., 2013. Traditional uses of medicinal plants in Solhan (Bingöl—Turkey). *J. Ethnopharmacol.* 148, 951–63.

Polat R., Satıl F., 2012. An ethnobotanical survey of medicinal plants in Edremit Gulf (Balıkesir–Turkey). *J. Ethnopharmacol.* 139, 626–41.

Saçıcı E., Yesilada E., 2021. Development of new and validated HPTLC methods for the qualitative and quantitative analysis of hyperforin, hypericin and hyperoside contents in *Hypericum* species. Phytochem. Anal. 33(1). https://doi.org/10.1002/pca.3093.

Sargın S.A., Selvi S., Büyükcengiz M., 2015. Ethnomedicinal plants of Aydıncık District of Mersin, Turkey. *J. Ethnopharmacol.* 174, 200–16.

Şenkardeş İ., Tuzlacı E., 2014. Some ethnobotanical notes from Gündoğmuş District (Antalya/Turkey). *MÜSBED* 4, 63–75.

Sezik E., Yesilada E., Honda G., Takaishi Y., Takeda Y., Tanaka T., 2001. Traditional medicine in Turkey X. Folk medicine in Central Anatolia. *J. Ethnopharmacol* 75, 95–115.

Sofi S.H., Nuraddin S.M., Amin Z.A., Al-Bustany H.A., Nadir M.Q., 2020. Gastroprotective activity of *Hypericum perforatum* extract in ethanol-induced gastric mucosal injury in Wistar rats: A possible involvement of H+/K+ ATPase α inhibition. *Heliyon* 6, e05249.

Shibayama Y., Kawachi A., Onimaru S., Tokunaga J., Ikeda R., Nishida K., Kuchiiwa S., Nakagawa S., Takamura N., Motoya T., Takeda Y., Yamada K., 2007. Effect of pre-treatment with St John's Wort on nephrotoxicity of cisplatin in rats. *Life Sciences* 81, 103–8.

Süntar Peşin İ., Küpeli Akkol E., Yılmazer D., Baykal T., Kırmızıbekmez H., Alper M., Yeşilada E., 2010. Investigations on the in vivo wound healing potential of *Hypericum perforatum* L. *J. Ethnopharmacol.* 127, 468–77.

Tetik F., Civelek Ş., Çakılcıoğlu U., 2013. Traditional uses of some medicinal plants in Malatya (Turkey). *J. Ethnopharmacol.* 146, 331–46.

Tuzlacı E., Alpaslan D.F., 2007. Turkish folk medicinal plants, Part V: Babaeski (Kırklareli). *J. Fac. Pharm. Istanbul* 39, 11–22.

Tuzlacı E., Aymaz P.E., 2001. Turkish folk medicinal plants, Part IV: Gönen – Balıkesir. *Fitoterapia* 72, 323–43.

Tuzlacı E., Bulut G.E., 2007. Turkish folk medicinal plants, Part V: Ezine (Çanakkale). *J. Fac. Pharm. Istanbul* 39, 39–51.

Tuzlacı E., İşbilen Alpaslan D.F., Bulut G., 2007. Turkish folk medicinal plants, VIII: Lalapaşa (Edirne). *Marmara Pharm. J.* 14, 47–52.

Tuzlacı E., Sadıkoğlu E., 2007. Turkish folk medicinal plants, Part VI: Koçarlı (Aydın). *J. Fac. Pharm. Istanbul* 39, 25–37.

Tuzlacı E., Şenkardeş İ., 2011. Turkish folk medicinal plants, X: Ürgüp (Nevşehir). *Marmara Pharm. J.* 15, 58–68.

Tuzlacı E., Tolon E., 2000. Turkish folk medicinal plants, part III: Şile (Istanbul). *Fitoterapia* 71, 673–85.
Yesilada E., Honda G., Sezik E., Tabata M., Fujita T., Tanaka T., Takeda Y., Takaishi Y., 1995. Traditional medicine in Turkey V. Folk medicine in the Inner Taurus Mountains. *J. Ethnopharmacol.* 46, 133–52.
Yesilada E., Honda G., Sezik E., Tabata M., Goto K., Ikeshiro Y., 1993. Traditional medicine in Turkey IV. Folk medicine in the Mediterranean subdivision. *J. Ethnopharmacol.* 39, 31–8.
Yeşil Y., Akalın E., 2009. Folk medicinal plants in Kürecik area (Akçadağ/Malatya-Turkey). *Turk. J. Pharm. Sci.* 6, 207–20.
Yeşilada, E., Gürbüz, İ., 1998. Evaluation of the Anti-ulcerogenic effect of the flowering herbs of *Hypericum perforatum*. *J. Fac. Pharm. Gazi Univ.*15, 25–31.
Yeşilyurt E.B., Şimşek I., Tuncel T., Akaydın G., Yesilada E., 2017. Marmara Bölgesi'nin Bazı Yerleşim Merkezlerinde Halk İlacı Olarak Kullanılan Bitkiler. *Marmara Pharm. J.* 21, 132–48.
Yücel A., Kan Y., Akın O., Yesilada E., 2017. Effect of St.John's wort (*Hypericum perforatum*) oily extract for the care and treatment of pressure sores. *J. Ethnopharmacol.* 196, 236–41.

10 *Juniperus* sp.

Sinem Aslan Erdem and Gülnur Ekşi Bona

INTRODUCTION

Juniperus L. (Cupressaceae) has a wide distribution in the northern hemisphere from sea level to timberline (Maria Loureiro Seca et al., 2015; Van Auken and Smeins, 2008) with one exception, *J. procera* Hochst. ex Endl. from Africa (Adams et al., 2002). The genus is represented by 52 species worldwide (Kandemir, 2018; Maria Loureiro Seca et al., 2015; The Plant List, 2019) which are monophyletic, including three monophyletic sections: sect. *Caryocedrus*, sect. *Juniperus*, and sect. *Sabina* (Mao et al., 2010). *Juniperus* species are evergreen shrubs or trees, dioecious or monoecious. Stem is ascending or erect. Shoots are rounded or angular. Leaves on young branches are acicular and rigid; mature leaves are either acicular, rigid, jointed at the base, in whorls of three, or scale-like and decussate, rarely short, acicular, and not jointed. Male flowers consist of numerous stamens. Female flowers are surrounded by persistent bracts at the base, and the bracts are composed of 3–9 scales. Seeds are 1–12 and unwinged (Coode and Cullenn, 1965; Kandemir, 2018).

Juniperus L. is generally known as "ardıç" in Turkey and is represented by eight species and nine taxa in three sections: *J. communis* L. (ardıç), *J. macrocarpa* Sm. (deniz ardıcı), and *J. oxycedrus* L. (katran ardıcı) belong to sect. *Juniperus* L.; *J. drupacea* Labill. (andız) belongs to sect. *Caryocedrus* Endl.; and *J. sabina* L. (saçağacı), *J. excelsa* M.Bieb. (bozardıç), *J. phoenicea* L. (fenike ardıcı), and *J. foetidissima* Willd. (kokarardıç) belong to sect. *Sabina*. In Turkey, apart from naturally distributed taxa, culture forms of *J. chinensis* L., *J. horizontalis* Moench, and *J. virginiana* L. are widely grown in parks and gardens (Baytop, 1999; Güner et al., 2012; Kandemir, 2018).

The genus is used as medicinal plants due to their secondary metabolites, promising therapeutic sources, and inspiring for new drug developments (Maria Loureiro Seca et al., 2015). Seeking new, pharmacologically active natural compounds which are more effective than the ones still in use is always a continuing process (Cragg and Newman, 2013). *Juniperus* species are rich in essential oils, diterpenes, flavonoids, and lignans (Maria Loureiro Seca et al., 2015). In recent years, numerous biological activities have been detected for their natural compounds (Tavares and Seca, 2018). In this chapter, botanical properties, traditional medicinal usages, phytochemical constituents, and bioactivities of *Juniperus* species naturally grown in Turkey are summarized based on the existing literature.

TRADITIONAL USE

Traditional use of plants dates back to the early days of humankind for food and medicinal purposes (Najar et al., 2020b; Prinsloo et al., 2018; Tavares and Seca, 2018). Traditional knowledge of medicinal plants is still an important source of humanity worldwide (Falzon and Balabanova, 2017; Tavares and Seca, 2018). For the last decades, the essential oils and extracts obtained from different parts of *Juniperus* species are used for their biological and therapeutic properties, making them an eminent topic for modern medicine (Ghosh et al., 2018; Tavares and Seca, 2018) and food and cosmetics industries (Lascurain Rangel et al., 2018; Orhan, 2019). Although some *Juniperus* species have toxic effects, they have many uses in folk medicine all over the world. Dried berries of junipers are used as a spice to flavor the meat, and ripened berries are used as one of the main constituents of gin (Orhan, 2019).

DOI: 10.1201/9781003146971-10

For centuries, *Juniperus* species have had importance in Turkey due to their numerous usages in folk medicine. In recent years, several studies conducted on the genus *Juniperus* reveal its traditional use such as abortive, antiparasitic, antirheumatic, antiseptic, digestive diuretic, anthelmintic and expectorant, for abdominal pain, amenorrhea, flu, disorder anemia, aphrodisiac, aphrodisiac, arthrosis, asthma, bed-wetting, bloating, blood sugar lowering, bronchitis, cancer, cardiac diseases, gall stones, gastrointestinal pains, gout, hair loss, heart diseases, hemorrhoids, hypercholesterolemia, infection, inflammation, injury treatment, intestinal pain, joint calcification, kidney stones, loss of appetite, marsh fever, medicinal tea, menstrual pains, menstrual regulatory, miscarry, painkiller, pneumonitis, prostate, psoriasis, respiratory system disorder, rheumatism, scabies of livestock, sinusitis, skin care, stomach ache, tuberculosis, ulcer, urethritis, urinary tract infection, warts, wounds fewer, cardiotonic, cardiovascular health, chest pain, cholesterol, cold, cough, cystitis, diabetes, diarrhea, and eczema (Table 10.1).

PHYTOCHEMISTRY

The genus *Juniperus* is rich in bioactive natural compounds from different types of secondary metabolites. Essential oil and diterpenes can be accounted as the main type of secondary metabolites of the *Juniperus* species. Flavonoids, phenolic acids, lignans, coumarins, fatty acids, sterols, tannins, sesquiterpenes, and triterpenes have also been discovered, whether in the aerial parts, leaves, or berries of *Juniperus* species (Fathima Firdose and Anjum, 2020; Ghasemnezhad et al., 2020; Miceli et al., 2020; Seca et al., 2015; Stankov et al., 2020a; Tavares and Seca, 2018).

Sesquiterpenes and/or diterpenes are found frequently in conifers as well as juniper. *Juniperus* species have been known to contain diterpenoids, unspecific sesquiterpenoids, as well as specific conifer sesquiterpenoids, of which some of the members of these classes of compounds are accepted as chemosystematic markers for conifers (Otto and Wilde, 2001). Topçu et al. isolated a new (2α -acetoxy-labda-8(17),13(16),14-trien-19-oic acid) diterpene from the berries of *J. excelsa* from Turkey (Topçu et al., 1999). Due to their high molecular weight, diterpenes are practically not found in essential oils. Contrary to this statement, diterpenes have been recognized in essential oils of *J. macrocarpa*, *J. excelsa*, and *J. communis* (Başer and Demirci, 2007; Höferl et al., 2014; Lesjak et al., 2017; Najar et al., 2020; Sezik et al., 2005; Topçu et al., 2005). The secondary metabolites of *Juniperus* species have received a lot of attention recently. They are located in various parts of the plants in considerable amounts, such as leaves, berries, and wood. Generally, the active ingredients of the juniper are often found in the essential oil group. The *Juniperus* species are characterized by their high amount of essential oil in berries, wood, and seed. Current researches have established that juniper essential oil composition is altered by sexual differences, seasonal fluctuations, diurnal variations, and distillation time (Rostaefar et al., 2017, Najar et al., 2020, Ghasemnezhad et al., 2020). According to the majority of research, the predominant component of the essential oil of distinct *Juniperus* species derived from diverse sections of the plant is α -pinene (Ben Ali et al., 2015; Hayta and Bagci, 2014; Labokas and Ložiene, 2013; Najar et al., 2020a; Pandey et al., 2018; Rostaefar et al., 2017; Sahin Yaglioglu et al., 2020; Topçu et al., 2005a).

According to an analysis of the essential oils from the leaves, berries, and twigs of *J. oxycedrus* collected from Hekimdağ, Eskişehir in Turkey, manoyl oxide, a diterpene, was found as the major component in the leaf (32.8%) and twig (35.4%) oils, whereas myrcene (44.6%) was found higher in the berry oil (Alan et al., 2016). On the other hand, GC-MS analysis of the essential oil of leaves, bark, and berries of *J. oxycedrus* subsp. *oxycedrus* gathered from the natural habitats in Bursa province in Turkey yielded α -pinene in the essential oils of the leaves (42.9%) and the bark (72.8%) as the main component; therewithal limonene (17.8%) was found as the second component in leaves oil, while β-pinene was detected in significant amounts in the bark oil. The essential oil of berries had β-myrcene (21.7%) as the main component, followed by notable amounts of α -pinene (20.9%) and germacrene D (19.6%) (Hayta and Bagci, 2014). In another study, essential oil from the

TABLE 10.1
Traditional Use of *Juniperus* Species in Turkey

Species Names	Local Names	Plant Parts	Uses	References
J. communis L. (Sect. *Juniperus*)	ardıç	berries	diuretic, antiseptic, stomachic	(Baytop, 1999)
		berries, stem	menstrual pains, diuretic	(Behlül Güler et al., 2015)
		berries and aerial parts	calcification, heart diseases, wounds, asthma, rheumatism	(Güneş, 2018)
J. drupacea Labill. (Sect. *Caryocedrus*)	andız kökü	roots	psoriasis, eczema, anemia	(Yaşar et al., 2019)
	andız	berries	asthma, bronchitis	(Güneş et al., 2018)
	enek	fruit jam (pekmez)	well being	(Semiz et al., 2007)
	andızçamı, andızçamıkozağı	berries	cough, cold	(Honda et al., 1996)
	andız	berries	cough, bronchitis, fever, urinary tract infection, skin care, cardiovascular health, gout, expectorant	(Aksoy et al., 2016)
	andız	berries	cough, hair loss, as molasses	(Kocabaş and Gedik, 2016)
	karaandız, diken andız, andız	berries	asthma, bronchitis, anemia	(Sargin and Büyükcengiz, 2019)
	andız	berries	asthma, bronchitis, cough	(Demirci and Ozhatay, 2012)
	ayıgiliği	berries	asthma, hypercholesterolemia	(Senkardes and Tuzlaci, 2014)
	andız	berries, molasses	asthma, bronchitis	(Sargin et al., 2015)
	andız, selbandız, selbi andızı, pıt andız	berries, seeds	asthma, bronchitis, cold, flu, gastrointestinal disease, enuresis	(Sargin, 2015)
	andız	berries	tonic, aphrodisiac, anthelmintic	(Baytop, 1999)
	andız, andız gılıği, ardıç, ardıçgeliği, ardıçgiliği	berries, leaves, seeds	stomach ache, anthelmintic, urethritis, gout	(Yeşilada et al., 1993)
	andız	berries	arthrosis, tuberculosis, cardiotonic, emmenagogue, antirheumatic, aphrodisiac, asthma	(Altundağ Çakır, 2017; Everest and Ozturk, 2005)
	andız	seeds	bed-wetting, arthrosis, tuberculosis, cardiotonic, emmenagogue, antirheumatic, aphrodisiac	(Everest and Öztürk, 2005)

(Continued)

TABLE 10.1 (CONTINUED)
Traditional Use of *Juniperus* Species in Turkey

Species Names	Local Names	Plant Parts	Uses	References
	andız, ardıçgeliği, ardıçgiliği, selbandız, selbi andızı, dikenli andız, pıt andız	fruit	flu, gastrointestinal diseases, anthelmintic	(Sargın, 2015; Yeşilada et al., 1993)
	andız	gall	diuretic, prostrate	(Ertuğ, 2002)
	andız, andız gılıği	tar, fruit	diarrhea, abdominal pain	(Yeşilada et al., 1995)
	andız	tar	antirheumatic, skin disorder, hemorrhoids	(Altundağ Çakır, 2017; Özçelik, 1987)
	katran ardıcı	tar	respiratory system disorder, urethritis, scabies of livestock	(Baytop, 1999)
J. excelsa M.Bieb. (Sect. *Sabina*)	ardıç	berries	sinusitis, inflammation	(Yaşar et al., 2019)
			cold, bronchitis, joint calcification	(Orhan et al., 2012a)
			hemorrhoids	(Altundag and Ozturk, 2011a)
			diabetes, asthma, cough	(Akaydin et al., 2013)
		shoot	coughs and cold	(Orhan et al., 2012a)
		seeds	vulnerary	(Tetik et al., 2013)
			kidney stones	(Cakilcioglu et al., 2011b)
		young shoots, leaves	pneumonitis, marsh fever	(Sargin and Büyükcengiz, 2019)
	çitandız	berries	bronchitis	(Sargin et al., 2015)
	andız	young shoots	gastrointestinal diseases	(Sargin, 2015)
	katran ağacı	stem	digestive, infection	(Ari et al., 2015)
J. foetidissima Willd. (Sect. *Sabina*)	yağlı ardıç, sakız ardıçı	resin	gastrointestinal pains, vaginal white discharge	(Sargin and Büyükcengiz, 2019)
	ardıç, sakız ardıcı, sakızlı ardıç		wound care, diabetes	(Sargin, 2015)
	kara ardıç	berries	stomach ache, diabetes, arthrosis	(Tuzlacı and Erol, 1999)
	kokarardıç, kokarardıçpürçüğü	shoots	cough, cold	(Honda et al., 1996)
	kokulu ardıç	berries, leaves	cough, cold	(Öztürk et al., 2011)
	kokar ardıcı	leaves	skin disorder, warts	(Ari et al., 2015)
	kokar ardıç	shoots	diuretic	(Sayar et al., 1995)

(*Continued*)

TABLE 10.1 (CONTINUED)
Traditional Use of *Juniperus* Species in Turkey

Species Names	Local Names	Plant Parts	Uses	References
J. oxycedrus L. (Sect. *Juniperus*)	katran ardıcı	berries	painkiller	(Erbay et al., 2017)
	katran ardıcı	branch	skin disorder, eczema	(Özdemir and Alpinar, 2015a)
	ardıç dikeni, ardıç, ardıç gıligılisi, tiken ardıçı, sığanardıç, sığanardıçdüveni, dikenliardıç düdemi, hardıç	berries, volatile oil	asthma, stomach ache, bronchitis, cold, cough, hemorrhoids, hypoglycemic, abdominal pain, kidney inflammation, kidney stone	(Honda et al., 1996)
	dikenli ardic, cirti	berries, leaves	parasitic disease	(Sezik et al., 1997)
	ardıç	branch	injure treatment	(Uysal et al., 2010)
	ardıç	berries, leaves	blood sugar lowering	(Orhan et al., 2012a)
	ardıç	bark	cancer	(Akyol and Altan, 2013)
	andız	berries, seeds	cardiac diseases, arthrolith	(Kocabaş and Gedik, 2016)
	ardıç	bark	cancer	(Akyol and Altan, 2013)
	ardıç	berries, seeds, leaves	bronchitis, cold, flu, diuretic, gastrointestinal disease, pleasure and medicinal tea	(Sargin, 2015)
	ardıç	berries	prostrate, chest pain	(Sargin, 2015)
	ardıç	leaves	bronchitis, cold, flu, diuretic, gastrointestinal disorder	(Sargin, 2015)
	ardıç	berries	cystitis	(Güzel et al., 2015b)
	ardıç	stem	injury treatment	(Uysal et al., 2010)
	ardıç	tar	antiseptic	(Ertuğ, 2002)
	ardıç	tar	skin disorder	(Koçyiğit and Özhatay, 2006)
	ardıç	tar	gallstone	(Sargin et al., 2013)
	ardıç	leaves	injury treatment	(Uysal et al., 2010)
	ardıç	berries	menstrual regulatory, antiseptic, expectorant, diuretic, stomachic	(Uğulu et al., 2009)
	ardıç	seeds	diuretic, gastrointestinal disorder, bronchitis, cold, flu	(Sargin, 2015)
	ardıç	seeds	hemorrhoids	(Koçyiğit and Özhatay, 2006)
	ardıç	berries	bronchitis	(Koçyiğit and Özhatay, 2006)

(Continued)

TABLE 10.1 (CONTINUED)
Traditional Use of *Juniperus* Species in Turkey

Species Names	Local Names	Plant Parts	Uses	References
	ardıç	berries	stomach ache	(Han and Bulut, 2015b)
	ardıç, tiken ardıç, tikenli ardıç	berries, resin	bronchitis, cold, flu, diuretic, stomach ache, intestinal pain, gastrointestinal spasm	(Sargin et al., 2015a)
	ardıç, diken ardıç giliği, ardıç giliği	berries, bud, tar, seeds	asthma, skin blemishes, stomach ache, bronchitis, rheumatism, kidney gravels, gastrointestinal disorder, cold	(Sargın et al., 2015b)
	diken ardıcı	berries	kidney stone, asthma, cold, bronchitis, ulcer, tuberculosis, stomach ache, diabetes	(Tuzlacı and Erol, 1999)
	gılı gılı	berries	cholesterol, diabetes	(Ari et al., 2015)
	diken ardıcı	berries	cough	(Demirci and Özhatay, 2012)
	tiken ardıcı, ardıç üzümü, diken ardıç giliği, dikenli ardıç, ardıç giliği	berries	hemorrhoids	(Gürhan and Ezer, 2004; Yeşilada et al., 1993)
	katran ardıcı	berries	prostate cancer	(Tuzlacı, 2016)
	ardıç, dikenli ardıç, katran ardıcı	berries	antirheumatic	(Altundağ and Öztürk, 2011; Özgökçe and Özçelik, 2004)
	dikenli boz andız, tiken, tiken ardıç	berries	asthma, bronchitis, anemia	(Sargin and Büyükcengiz, 2019)
	ardıç	berries, pix	gall stones, asthma, rheumatism	(Sargin et al., 2013)
	dikenli ardıç, ardıç, diken ardıç giliği, ardıç giliği	berries	cough	(Altundag and Ozturk, 2011a; Uğulu et al., 2009)
	tiken ardıçı,	berries	abdominal pain, stomach ache, loss of appetite	(Yeşilada et al., 1995)
	ardıç, diken ardıç, tikenli ardıç	berries	bronchitis, cold, flu, diuretic, stomach ache, asthma, spasm, rheumatism, intestinal pain, gastrointestinal disorder	(Sargin, 2015)
	cicamuk, çıtımık, ardıç	berries	asthma	(Özkan and Koyuncu, 2005)
	kokar ardıç	berries	stomachic	(Vural and Karavelioğulları, 1997)
	ardıç, ardıç üzümü	berries	gall bladder disorder, hemorrhoids	(Ezer and Avcı, 2005)
	dikenli ardıç	berries	antiparasitic	(Altundag and Ozturk, 2011b)
	ardıç üzümü	berries	diabetes, hemorrhoids	(Yeşilyurt et al., 2017)

(Continued)

TABLE 10.1 (CONTINUED)
Traditional Use of *Juniperus* Species in Turkey

Species Names	Local Names	Plant Parts	Uses	References
	ardıç giliği, tiken ardıçı, ayı andızı giliği, gecimik, gecimik çeçiği, dikenli ardıç, eşek andızı, bozandız, ardıç türü, ardıç katranı	berries, resin, tar	anal fistula, diarrhea, amenorrhea, catarrh, urethritis, abdominal pain	(Yeşilada et al., 1995)
	dikenli ardıç, ardıç giliği	berries	kidney stone, bloating	(Yeşilada et al., 1993)
	gılı gılı	leaves	cholesterol, diabetes	(Ari et al., 2015)
	diken ardıcı	leaves	diabetes	(Tuzlacı and Erol, 1999)
	ardıç giliği	resin	vulnerary	(Yeşilada et al., 1993)
	katran ardıcı	stem	skin disorder, eczema	(Özdemir and Alpinar, 2015b)
	ardıç, diken ardıç giliği	tar	asthma, antirheumatic	(Sargin et al., 2013)
	dikenli ardıç	tar	antiparasitic	(Altundag and Ozturk, 2011a)
	katran ardıcı	tar	scabies of livestock	(Baytop, 1999)
	andız	tar	psoriasis	(Demirci and Özhatay, 2012)
	kokar ardıç	tar	scabies	(Vural and Karavelioğulları, 1997)
J. phoenicea L. (sect. *Sabina*)	ardeş, ardıç	berries, shoots	abdominal pain, diarrhea, kidney stone	(Altundağ Çakır, 2017; Ertuğ, 2002)
J. sabina L. (sect. *Sabina*)	kara ardıç	leaves	miscarry, diuretic, menstrual regulatory	(Baytop, 1999)
	-	berries	stomach disorders, eczema, wounds	(Orhan et al., 2012a)
	sabin ardıcı	herba	diuretic, emmenagogue, abortive, in the treatment of diabetes mellitus	(Öztürk et al., 2011)

leaves and berries of the same species from the Amasya region were analyzed, and unlike other studies, caryophyllene oxide (31.6%) and β-pinene (29.9%) were found as the primary compounds (Sahin Yaglioglu et al., 2020). The essential oil composition differentiations of *J. macrocarpa* from Çeşme-İzmir province in Turkey were searched based on the seasons, and manoyl oxide (7.7%, 21.9%) was found as the predominant compounds in the May and August samples, respectively, while α -pinene (11.1%) and manoyl oxide (9.1%) were in high amounts in October sample (Sezik et al., 2005). The differences in the content and the yield of the oils can be attributed to the age of the trees, geographical origin, edaphic factors, harvesting time, and the prevalence of chemotypes (Tunalier et al., 2002; Unlu et al., 2008). The essential oils of the berries and leaves of *J. excelsa* from Isparta province in Turkey were analyzed by GC-MS; α -pinene (29.7%, 34.3%) and cedrol (25.3%, 12.3%) were determined as the main components of both of the leaves and berries with the amounts given, respectively (Topçu et al., 2005a). The major components in the essential oils of the leaves and berries of *J. excelsa* from the Amasya region were found as β-terpinyl acetate (38%) and α -pinene (37.3%), respectively (Sahin Yaglioglu et al., 2020). Similarly, α -pinene (55.5%) was determined as the major component of the essential oil from the berries of *J. excelsa* (Unlu et al., 2008). α -Pinene (46.1%) was found as the major component in the essential oil obtained from the aerial parts of *J. excelsa* collected from Antalya province in Turkey (Atas et al., 2012). The essential oil composition from the berries, leaves, and branches of *J. foetidissima* from Turkey was analyzed. The berries were found to have a higher amount of essential oil; branches were found to be high in α -pinene (25%); sabinene was the main component of berries (23.7%), seedless berries (19.7%), and seed oils (15.4%); and β-thujone (26.5%) was the major compound of the leaves. The second major components of berries, seedless berries, seeds, leaves, and branches were found as limonene (13%), β-thujone (12%), abietal (12%), sabinene (9%), and cedrol (11.4%) (Tunalier et al., 2002). In another study, the essential oil of the leaves and berries obtained from *J. foetidissima* of Antalya were analyzed; the major components were determined as α -pinene (56.1%, 90.2%) for both leaves and the berries, respectively (Sahin Yaglioglu et al., 2020). α -Pinene was stated as the chief component of both essential oils of leaves (35.9%) and berries (29.3%) of *J. communis* collected from Amasya province in Turkey (Sahin Yaglioglu et al., 2020). In a study where the components of the essential oils of the foliages of *J. communis* were determined and compared according to the seasons (spring, summer, autumn) and sexual differences, α -pinene was found as the major component in male samples in three seasons – spring, summer, and autumn; 61.69%, 48.81%, and 37.50%, respectively. Nonetheless, in terms of female samples, α -pinene was the principal compound in spring (42.15%) and autumn (36%) samples, whereas sabinene (44.47%) was the main compound in the summer samples (Rostaefar et al., 2017). A literature survey resulted in limited data about the essential oil analysis of *J. drupacea* and *J. phoenicea*. α -Pinene was found as the predominant component of the volatiles of the berries of *J. drupacea* from Egypt, as well as essential oil of the aerial parts of *J. phoenicea* from Algeria (Boulanouar et al., 2013; El-Ghorab et al., 2008). Likewise, a literature search yielded minimal information on *J. sabina* essential oil analysis. The essential oil obtained from the leaves of female *J. sabina* species was analyzed, and sabinene was found as the main compound of the essential oil with a value of 54.3% (Nikolić et al., 2016).

Researches on the phenolic compounds and flavonoids of *Juniperus* species have been in an upsurge, especially in recent years (Chaouche et al., 2015; Miceli et al., 2020; Sahin Yaglioglu and Eser, 2017; Stankov et al., 2020b; Živić et al., 2019). Some of the studies focused on the total flavonoid and phenol contents (Elmastaş et al., 2006; Öztürk et al., 2011; Živić et al., 2019); on the other hand, some researches are based on the identification of the related compounds by using analytical chromatographic methods such as HPLC-DAD-ESI-MS and HPLC-TOF/MS (Table 10.2) (Miceli et al., 2011; Sahin Yaglioglu and Eser, 2017; Taviano et al., 2013).

Juniperus species were also recognized to contain coumarins, fatty acids, lignans, sterols, tannins, alkaloids, and other aliphatic and aromatic chemicals (Marino et al., 2010; Orhan, 2019).

TABLE 10.2
The Phenolic and Flavonoid Contents of Some Selected *Juniperus* Species

Plants	Compounds	Parts	Analysis	References
J. communis L.	Amentoflavone, cupressoflavone, apigenin	berries	HPLC-DAD-ESI-MS	(Miceli et al., 2009)
	Catechin, rutin, quercitrin	berries	HPLC-TOF/MS	(Sahin Yaglioglu and Eser, 2017)
	Protocatechuic acid, rutin, cupressuflavone, amentoflavone, hinokiflavone	leaves	HPLC-PDA/ESI-MS	(Miceli et al., 2020)
	Epigallocatechin, catechin, rutin, quercitrin, cosmosiin,	leaves	HPLC-TOF/MS	(Sahin Yaglioglu and Eser, 2017)
J. communis var. *saxatilis* Pall.	Amentoflavone, cupressuflavone, apigenin	berries	HPLC-DAD-ESI-MS	(Miceli et al., 2009)
	Cupressuflavone, amentoflavone, hinokiflavone	leaves	HPLC-PDA/ESI-MS	(Miceli et al., 2020)
J. excelsa M.Bieb.	Catechin, rutin	berries	HPLC-TOF/MS	(Sahin Yaglioglu and Eser, 2017)
	Catechin, rutin, quercitrin	leaves	HPLC-TOF/MS	(Sahin Yaglioglu and Eser, 2017)
J. oxycedrus L.	Catechin, rutin	berries	HPLC-TOF/MS	(Sahin Yaglioglu and Eser, 2017)
	Catechin, methyl robustone, rutin,	leaves	HPLC-TOF/MS	(Sahin Yaglioglu and Eser, 2017)
	Cupressuflavone, amentoflavone, rutin, apigenin	ripe berries	HPLC-DAD-ESI-MS	(Taviano et al., 2013)
	Shikimic acid, 4-*O*-β-D- glucopyranosyl ferulic acid, oleuropeic acid-8-*O*-β-D-glucopyranoside	berries	Isolation, structure elucidation NMR and LC-TOF/MS	(Orhan et al., 2012b)
	Protocatechuic acid, cupressuflavone, amentoflavone, hinokiflavone	leaves	HPLC-PDA/ESI-MS	(Miceli et al., 2020)
J. macrocarpa Sm.	Cupressuflavone, amentoflavone, rutin, apigenin	ripe berries	HPLC-DAD-ESI-MS	(Taviano et al., 2013)
	Protocatechuic acid, rutin, cupressuflavone, amentoflavone, hinokiflavone	leaves	HPLC-PDA/ESI-MS	(Miceli et al., 2020)
J. drupacea Labill.	Gallic acid, protocatechuic acid, tyrosol, chlorogenic acid, amentoflavone	berries	HPLC-DAD-MS	(Miceli et al., 2011)
	Cupressuflavone, amentoflavone, hinokiflavone	leaves	HPLC-PDA/ESI-MS	(Miceli et al., 2020)
J. foetidissima Willd.	Catechin, rutin, quercitrin	berries	HPLC-TOF/MS	(Sahin Yaglioglu and Eser, 2017)
	p-Methylphenol, catechin, rutin	leaves	HPLC-TOF/MS	(Sahin Yaglioglu and Eser, 2017)

PHARMACOLOGICAL PROPERTIES

Anticancer Activity

Li et al. (2021) proved that the extract of *J. communis* prevents the esophageal squamous cell carcinoma cells' growth due to stimulating G0/G1 arrest by regulating the cycle regulatory proteins. Furthermore, the extract of *J. communis* stimulates cell apoptosis in esophageal squamous cell carcinoma cells. *J. communis* extract increases the anticancer effect of the clinical drug 5-fluorouracil, proposing that it may be a promising source for the current chemotherapeutic treatment (Li et al., 2021). Fernandez and Cock (2016) measured the inhibitory activity of *J. communis* berry extracts against human carcinoma cells, CaCo2, and HeLa. The results demonstrated the inhibitory effects of *J. communis* extract against intestine and cervix cancer cells. In this experiment, the IC_{50} values were between 1300 and 2500 µg/mL (Fernandez and Cock, 2016). According to the results of Akbari et al. (2021), oral administration of *J. communis* seed extract inhibits testosterone-induced hyperplasia in rats. The effect is most likely through the formation of inflammatory responses by reducing oxidative stress and antiproliferative effect in compliance with the results, which demonstrated a decrease in malondialdehyde and protein carbonyl and an increase in glutathione. This study suggests *J. communis* seed extract has active components and is a promising source for benign prostatic hyperplasia for future studies (Akbari et al., 2021). Tundis et al. (2021) designed a comparative study of hexane and dichloromethane extracts of *J. macrocarpa* and *J. oxycedrus* as potential source of antioxidant and antiproliferative agents. The results showed a selective antiproliferative activity of both *J. oxycedrus* and *J. macrocarpa* extracts in a dose-dependent manner against lung cancer cell lines A549 and COR-L23 (Tundis et al., 2021). Yaman et al. (2021) monitored a significant decrease in tumor incidence and tumor multiplicity with *J. communis* oil. They suggested that *J. communis* oil is a chemopreventive dietary source that prevents cell proliferation and COX-2 expression and leads to apoptosis, causing significant reduction in colon tumor formation (Yaman et al., 2021). Najar et al. (2020a) investigated the phytochemical components of the essential oils of leaves, seeds, and berries of *J. oxcycedrus*. According to the results, the monoterpene hydrocarbons are the major components in studied oils, and the sesquiterpene hydrocarbons are just a high percentage in fruit essential oil. Furthermore, essential oil from leaves showed cytotoxic activity on MCF7, and the fruit essential oil was active on SH-SY5Y (Najar et al., 2020a). Sahin Yaglioglu et al. (2020) investigated the aerial parts of *J. oxycedrus* subsp. *oxycedrus*, *J. foetidissima*, *J. excelsa*, and *J. communis* for their antioxidant and antiproliferative activities. They observed that the leaf extracts of investigated *Juniperus* species have strong antiproliferative activities against C6 cells (Table 10.3) (Sahin Yaglioglu et al., 2020).

Antioxidant Activity

Meriem et al. (2021) studied the antioxidant activity of *Juniperus phoenicea* L. leaves essential oil. The GC-MS analysis showed that α -pinene is the major component with a percentage of 74.14% in essential oil. The results suggested that the antioxidant activity of *J. phoenicea* essential oil is an effective antioxidant source (Meriem et al., 2021). *J. oxycedrus* ssp. *oxycedrus* berry and wood essential oils were identified by GC and GC/MS. The major components are found as α -pinene and β-myrcene. *In vitro* antioxidant activity showed a significant activity for both oils with IC_{50} values of 1.45 µl/mL for wood and 7.42 µl/mL for berries (Loizzo et al., 2007). Miceli et al. (2020) investigated *J. communis*, *J. drupacea*, *J. oxycedrus*, and *J. macrocarpa*. The results showed that the leaf extracts of *Juniperus* species are effective antioxidants and anti-lipid peroxidation agents. They also represent an antimicrobial efficacy against *S. aureus*. The results signalized their potential bioactive compounds responsible for their antioxidant and antimicrobial activity (Miceli et al., 2020). The antioxidant, anti-inflammatory, and antimicrobial activities of *J. macrocarpa* extracts and essential oils were investigated by Lesjak et al. (2014). Considerable differences between leaves

TABLE 10.3
Bioactivities of Selected *Juniperus* Species

Species Names	Biological Activities	Plant Parts	References
J. communis L.	Abortifacient	leaves	(Gardner et al., 1998)
	Antifertility	berries	(Agrawal et al., 1980)
	Diuretic	berries, leaves, aerial parts, and roots	(Firdose, Kouser and Anjum, 2020)
		berries	(Stanić et al., 1998)
		leaves	(Yarnell, 2002)
	Against gynecological disorders	fruit, jeaf	(Firdose, Kouser and Anjum, 2020)
	Hepatoprotective	jeaf	(Ved et al., 2017)
	Antioxidant	leaves, branches	(Elshafie et al., 2020)
		berries	(Elmastaş et al., 2006; Höferl et al., 2014; Orhan et al., 2011; Öztürk et al., 2011)
		leaves	(Miceli et al., 2020)
		aerial parts	(Sahin Yaglioglu et al., 2020)
	Nephroprotective	leaves	(Bais et al., 2015)
	Antiaging	berries	(Pandey et al., 2018)
	Antiproliferative	aerial parts	(Sahin Yaglioglu et al., 2020)
	Anticancer	berries	(Fernandez and Cock, 2016)
	Antibacterial	berries	(Najar et al., 2020b)
		leaves	(Miceli et al., 2020)
	Antimycobacterial	aerial parts, leaves, bark	(Jimenez-Arellanes et al., 2003)
	Hypoglycemic Antidiabetic	berries	(De Medina et al., 1994; Orhan et al., 2012b)
	Antibacterial	galbuli	(Zheljazkov et al., 2018)
		branch	(Taviano et al., 2011)
		branch	(Taviano et al., 2011)
		berries	(Öztürk et al., 2011)
	Anticholinesterase	berries	(Öztürk et al., 2011)
	Anti-inflammatory	leaves	(Bais et al., 2017)
		stem, fruit, and leaves	(Akkol et al., 2009)
		stem, fruit, and leaves	(Akkol et al., 2009)
J. drupacea Labill.	Anti-inflammatory	stem, fruit, and leaves	(Akkol et al., 2009)
	Antioxidant	leaves	(Miceli et al., 2020)
	Antibacterial	branch	(Taviano et al., 2011)
		leaves	(Miceli et al., 2020)
		berries	(Miceli et al., 2011)
J. excelsa M. Bieb.	Antimicrobial	berries	(Öztürk et al., 2011; Venditti et al., 2021)
		galbuli	(Zheljazkov et al., 2018)
	Antioxidant	berries	(Öztürk et al., 2011)
		aerial parts	(Sahin Yaglioglu et al., 2020)
	Anticancer	aerial parts	(Sahin Yaglioglu et al., 2020)

(*Continued*)

TABLE 10.3 (CONTINUED)
Bioactivities of Selected *Juniperus* Species

Species Names	Biological Activities	Plant Parts	References
		berries	(Topçu et al., 2005b)
		branches and berries	(Sadeghi-aliabadi et al., 2009)
		aerial parts	(Darvishi et al., 2016)
		leaves, berries	(Yaman et al., 2021)
		leaves	(Khanavi et al., 2019)
	Anticholinesterase	berries	(Öztürk et al., 2011)
	Anti-inflammatory	aerial parts	(Orhan et al., 2012a; Tunón et al., 1995)
	Antinociceptive	leaves, berries	(Orhan et al., 2012a)
	Antihypertension	aerial parts	(Khan et al., 2012)
J. foetidissima Willd.	Antibacterial	berries	(Öztürk et al., 2011)
	Antioxidant	berries	(Orhan et al., 2011; Öztürk et al., 2011)
		aerial parts	(Sahin Yaglioglu et al., 2020)
	Antiproliferative	aerial parts	(Sahin Yaglioglu et al., 2020)
	Anticholinesterase	berries	(Öztürk et al., 2011)
	Anti-inflammatory	aerial parts	(Orhan et al., 2012a; Tunón et al., 1995)
	Antifungal	wood	(Balaban et al., 2003; Ohashi et al., 1994)
	Antinociceptive	leaves, berries	(Orhan et al., 2012a)
J. macrocarpa Sm.	Anti-inflammatory	leaves, seeds	(Lesjak et al., 2014)
	Anticancer	berries	(Taviano et al., 2013)
	Antibacterial	berries	(Najar et al., 2020b)
		leaves, seeds	(Lesjak et al., 2014)
	Antioxidant	branches, leaves, seeds	(Lesjak et al., 2014)
J.. oxycedrus L.	Antidiarrheic	leaves, stem	(Moreno et al., 1997)
	Antiprurigen		
	Wound healing	berries	(Tumen et al., 2013)
	Vasorelaxing	leaves, stem	(Bello et al., 1997)
	Antibacterial	branch	(Taviano et al., 2011)
		berries	(Öztürk et al., 2011)
		berries	(Najar et al., 2020b)
		seeds	(Ateş and Turgay, 2003)
		galbuli	(Zheljazkov et al., 2018)
	Antifungal	wood	(Balaban et al., 2003; Ohashi et al., 1994)
	Antioxidant	roots bark,	(Chaouche et al., 2015)
		leaves	(Miceli et al., 2020)
		berries	(Öztürk et al., 2011)
		aerial parts	(Sahin Yaglioglu et al., 2020)
	Antiproliferative	aerial parts	(Sahin Yaglioglu et al., 2020)
	Anticancer	berries	(Taviano et al., 2013)
	Anticholinesterase	berries	(Öztürk et al., 2011)
	Anti-inflammatory	stem, berries, and leaves	(Akkol et al., 2009)
		aerial parts	(Orhan et al., 2012a; Tunón et al., 1995)
	Antinociceptive	leaves, berries	(Orhan et al., 2012a)
J. phoenicea L.	Antibacterial	berries	(Öztürk et al., 2011; Venditti et al., 2021)
		leaves	(Meriem et al., 2021)
	Antioxidant	berries	(Öztürk et al., 2011)
		leaves	(Ben Ali et al., 2015)
	Anticholinesterase	berries	(Öztürk et al., 2011)

(*Continued*)

TABLE 10.3 (CONTINUED)
Bioactivities of Selected *Juniperus* Species

Species Names	Biological Activities	Plant Parts	References
	Antiulcer	leaves	(Ben Ali et al., 2015)
J. sabina L.	Antibacterial	berries	(Öztürk et al., 2011)
		galbuli	(Zheljazkov et al., 2018)
	Antifertility	berries	(Xie et al., 2017)
	Hepatoprotective	aerial parts, leaves	(Abdel-Kader et al., 2017)
	Antioxidant	berries	(Öztürk et al., 2011)
	Anticholinesterase	berries	(Öztürk et al., 2011)
	Anti-inflammatory	aerial parts	(Orhan et al., 2012a; Tunón et al., 1995)
	Antinociceptive	leaves, berries	(Orhan et al., 2012a)

and seed berry extracts were found. Rutin, catechin, quercitrin, epicatechin, and amentoflavone were observed as the dominant compounds in the extracts. The major group was monoterpenes for the essential oil, and it was rich in α-pinene. The leaves and seed cones of *J. macrocarpa* extract and essential oils expressed moderate antioxidant activity. Seed cones presented high anti-inflammatory activity. The essential oils of *J. macrocarpa* demonstrated noteworthy antimicrobial activity against Gram-positive bacteria (Lesjak et al., 2014). Fernandez and Cock (2016) demonstrated that *J. communis* berry extracts prevent the growth of the bacterial triggers of rheumatoid arthritis, ankylosing spondylitis, and multiple sclerosis. Additionally, the results indicated that *J. communis* berry extracts may be useful to cure some cancer types (Fernandez and Cock, 2016). Ben Ali et al. (2015) examined the ulceroprotective, and antioxidant activity of essential oil of *J. phoenicea* leaves against hydrogen chloride/ethanol-induced ulcers in rats. The essential oil of *J. phoenicea* significantly decreased malondialdehyde content and increased the activities of superoxide dismutase, catalase, and glutathione peroxidase. Oral use of *J. phoenicea* showed significant inhibition on induced gastric ulcers compared to those of the proton pump inhibitor omeprazole (Table 10.3) (Ben Ali et al., 2015).

Ethanol and ethyl acetate extracts of the berries of *J. oxycedrus* and *J. communis* possess significant antioxidant activity, which may be used as preservative agents to extend the shelf-life in the food industry. There was a strong relationship between antioxidant activity and both phenolic and flavonoid compounds (Živić et al., 2019). Öztürk et al. (2011) investigated the fruit extracts of six *Juniperus* species from Turkey for their antioxidant, anticholinesterase, and antimicrobial activities. According to the results, berry extracts of *J. excelsa*, *J. sabina*, *J. Phoenicea*, and *J. oxycedrus* showed notable antioxidant activity. Besides, the berry extract of *J.oxycedrus* was found to be a potential source for anti-Alzheimer drug developments (Öztürk et al., 2011). Orhan et al. (2011) have screened the leaves and berries' ethanolic and water extracts of *J. communis*, *J. oxycedrus*, *J. sabina*, *J. foetidissima*, and *J. excelsa* for their cholinesterase inhibitory and antioxidant activities. The unripe fruit ethanol and water extracts of *J. foetidissima* displayed a considerable inhibition against butyrylcholineesterase. The leaf ethanolic extract of *J. foetidissima* presented higher antioxidant activity in DPPH radical scavenging activity and ferric-reducing antioxidant power assays (Nilufer Orhan et al., 2011).

Antidiabetic and Anti-hyperlipidemic Activities

Orhan et al. (2014) investigated the *in vitro* antidiabetic effects of *J. oxycedrus* subsp. *oxycedrus* and *J. communis* var. *saxatilis*. According to this study, *J. oxycedrus* subsp. *oxycedrus* leaf extract showed a potent *α* -amylase inhibitory effect. The highest *α* -glucosidase inhibitory activity is detected for *J. communis* var. *saxatilis* fruit extract (IC_{50}: 4.4 µg/mL) (Orhan et al., 2014). In order to determine the antidiabetic effect of leaf and fruit extracts and to evaluate the renal lipid peroxidation in streptozotocin-induced diabetic rats treated with *Juniperus oxycedrus* extracts. The extracts have been administered to rats at 500 and 1000 mg/kg doses for 10 days. *J. oxycedrus* interrelationships between the levels of trace elements like Zinc (Zn) have also been tested, and the extracts were found to increase Zn levels in the liver. Leaf and fruit extracts have decreased the blood glucose levels and the levels of lipid peroxidation in both liver and kidney tissues. Thus, results have pointed out that *J. oxycedrus* fruit and leaf extracts might be beneficial for diabetes and its complications (Orhan et al., 2011). Keskes et al. (2014) examined the antioxidant, *in vitro* *α* -amylase, and pancreatic lipase inhibitory activities of *J. phoenicea* essential oil and extracts. The IC_{50} values of essential oil, hexane, and methanol extracts against *α* -amylase were 35.44, 30.15, and 53.76 µg/mL, respectively, and against pancreatic lipase enzyme were 66.15, 68.47, and 60.22 µg/mL, respectively. The methanol extract has been found the richest in total phenolic content (265 mg as gallic acid equivalent/g extract) and demonstrated the highest antioxidant activity (IC_{50} ¼ 2 µg/mL). The study demonstrated that *J. phoenicea* leaves have powerful antioxidant, antidiabetic, and antiobesity effects (Keskes et al., 2014). Orhan et al. (2017) have determined the *in vitro* antidiabetic activities of the berry and leaf ethanol extracts for *J. foetidissima* and *J. sabina*. The leaf extract of *J. foetidissima* has shown stronger *α* -glucosidase inhibitory (>100%–88.6%) and *α* -amylase activity (65.1% at 1 mg/mL) than the berry extracts. Furthermore, extracts have lowered blood glucose concentrations of streptozotocin-induced diabetic animals significantly after oral administration. Amentoflavone has been determined as the major constituent for the leaf and berry extracts, and umbelliferone has been observed in the leaf extracts. Orhan et al. (2017) have attributed the antidiabetic activity of the extracts to amentoflavone which has strong inhibitory activity on carbohydrate digestive enzymes as well as positive effects on insulin resistance (Table 10.3).

Antimicrobial Activity

Antioxidant, anthelmintic, and antibacterial activities of *J. phoenicea* were carried out by Meriem et al. (2021) using the essential oil of leaves. The major component was found as *α* -pinene (74.14%) in the essential oil. As a result, *J. phoenicea* essential oil was found as an effective antibacterial source against six human pathogenic bacteria (*Escherichia coli, Bacillus cereus, Listeria monocytogenes, Salmonella typhimurium, Staphylococcus aureus, Salmonella enteritidis*) (Meriem et al., 2021). Venditti et al. (2021) have examined the essential oil of *C. libani* and *P. pinea* woods and *J. excelsa* berries. *C. libani* wood oil was observed as the most effective for both antioxidant and cytotoxic tests. For the antimicrobial assay, different sensitivities of different bacteria were observed. All the oils expressed remarkable activity against the yeast *Candida albicans* (Venditti et al., 2021). Zheljazkov et al. (2018) investigated the antibacterial effect of some different *Juniperus* species. The results showed that the species *J. communis* is the most effective in inhibiting *Candida glabrata. J. communis* was studied using different extracts at different concentrations of 20, 40, 80, 100, 200, and 300 mg/mL against bacterial pathogens, such as *Bacillus subtilis* and *Escherichia coli*. The results of this study showed that *J. communis* methanolic extract had the greatest effect on both bacteria at the concentration of 300 mg/mL (Zheljazkov et al., 2018). Unlu et al. (2008) examined the antimicrobial activity of *J. excelsa* using essential oil from the berry of the plant against a number of important microorganisms. The results presented that the obtained essential oil of *J. excelsa* berry demonstrates a significant inhibitory activity against *Clostridium perfringens*. The study has confirmed the presence of *a* -pinene as a dominant compound in the plant essential oil that may be responsible for the activity against *C. perfringens* (Table 10.3) (Unlu et al., 2008).

TOXICITY

There are also reports of toxic effects, for example, cade oil (juniper tar), obtained by distillation from the branches and wood of *J. oxycedrus*, has a dark color, a strong smoky smell, and has toxic effects apparently due to its phenol content (Koruk et al., 2005). The poisoning of healthy newborns treated with a topical application of cade oil for atopic dermatitis was observed (Achour et al., 2011). The ingestion of a spoonful of cade oil causes, within an hour, fever, headache, myalgia, nausea, vomiting, dyspnea, and a productive cough in an adult (Koruk et al., 2005; Maria Loureiro Seca et al., 2015). These and other cases were the incentives to a study on the safety and possible side effects of cade oil (Maria Loureiro Seca et al., 2015).

FUTURE PERSPECTIVES

Juniperus species, belonging to Cupressaceae, is represented by 52 species worldwide, of which eight are growing in Turkey naturally: *J. communis*, *J. macrocarpa*, and *J. oxycedrus* belong to sect. *Juniperus*; *J. drupacea* belongs to sect. *Caryocedrus*; *J. sabina*, *J. excelsa*, *J. phoenicea*, and *J. foetidissima* belong to sect. *Sabina*. They are generally known as "ardıç" in Turkey. The genus is traditionally used extensively as abortive, antiparasitic antirheumatic, antiseptic, digestive, diuretic, anthelmintic, aphrodisiac, and expectorant; for pains (abdominal, gastrointestinal, menstrual), asthma, bed-wetting, bloating, blood sugar lowering, bronchitis, cancer, etc. Phytochemical studies on the *Juniperus* species are mainly focused on essential oils and diterpenes, the major secondary metabolites of the genus. Flavonoids, phenolic acids, lignans, coumarins, fatty acids, sterols, tannins, sesquiterpenes, and triterpenes are also known to be found in *Juniperus* species' aerial parts, leaves, and berries. Several bioactivity studies were also conducted on *Juniperus* species, especially on the ones with traditional use, and their activities on anticancer, antioxidant, antidiabetic, antimicrobial, etc. were tested, and the tested extracts are found to be actively moderate to significant amounts. This study has demonstrated that, despite their traditional use, phytochemistry and bioactivity studies on *Juniperus* species are limited. More studies are necessary in order to reveal the structure-activity relationships and/or to clarify the mechanisms of action.

REFERENCES

Abdel-Kader, M.S., Alanazi Bin, M.T., Saeedan, A.S., Al-Saikhan, F.I., Hamad, A.M., 2017. Hepatoprotective and nephroprotective activities of Juniperus sabina L. aerial parts. *J. Pharm. Pharmacogn. Res.* 5, 29–39.

Achour, S., Abourazzak, S., Mokhtari, A., Soulaymani, A., Soulaymani, R., Hida, M., 2011. Juniper tar (cade oil) poisoning in new born after a cutaneous application. *BMJ Case Rep.* https://doi.org/10.1136/bcr.07.2011.4427

Adams, R.P., Hsieh, C.F., Murata, J., Naresh Pandey, R., 2002. Systematics of Juniperus from eastern Asia based on random amplified polymorphic DNAs (RAPDS). *Biochem. Syst. Ecol.* 30, 231–241. https://doi.org/10.1016/S0305-1978(01)00087-4

Agrawal, O.P., Bharadwai, S., Mathur, R., 1980. Antifertility effects of fruits of Juniperus communis. *Planta Med.* 40, 98–101. https://doi.org/10.1055/s-2008-1075011

Akaydin, G., Şimşek, I., Arituluk, Z.C., Yeşilada, E., 2013. An ethnobotanical survey in selected towns of the mediterranean subregion (Turkey). *Turkish J. Biol.* 37, 230–247. https://doi.org/10.3906/biy-1010-139

Akbari, F., Azadbakht, M., Megha, K., Dashti, A., Vahedi, L., Barzegar Nejad, A., Mahdizadeh, Z., Abdi Sarkami, S., Sadati, M., 2021. Evaluation of Juniperus communis L. seed extract on benign prostatic hyperplasia induced in male Wistar rats. *African J. Urol.* 27, 48. https://doi.org/10.1186/s12301-021-00137-x

Akkol, E.K., Güvenç, A., Yesilada, E., 2009. A comparative study on the antinociceptive and anti-inflammatory activities of five Juniperus taxa. *J. Ethnopharmacol.* 125, 330–336. https://doi.org/10.1016/j.jep.2009.05.031

Aksoy, A., Çelik, J., Tunay, H., 2016. Gazipaşa (Antalya) İlçe Pazarında Satılan ve Halk Tarafından Kullanılan Bazı Bitkiler ve Kullanım Amaçları. *Biyol. Bilim. Araştırma Derg.* 9, 55–60.

Akyol, Y., Altan, Y., 2013. Ethnobotanical studies in the Maldan Village (Province Manisa, Turkey). *Marmara Pharm. J.* 17, 21–25. https://doi.org/10.12991/201317388

Alan, S., Kürkçüoğlu, M., Şener, G., 2016. Türkiye'de yetişen Juniperus oxycedrus L. subsp. oxycedrus'un uçucu yağ bileşikleri. *Turkish J. Pharm. Sci.* 13, 300–303. https://doi.org/10.4274/tjps.2016.03

Altundağ Çakır, E., 2017. A comprehensive review on ethnomedicinal utilization of gymnosperms in Turkey. *Eurasian Journal of Forest Science.* https://doi.org/10.31195/ejejfs.327364

Altundag, E., Ozturk, M., 2011a. Ethnomedicinal studies on the plant resources of east Anatolia, Turkey, in: *Procedia – Social and Behavioral Sciences.* Elsevier Ltd, pp. 756–777. https://doi.org/10.1016/j.sbspro.2011.05.195

Altundag, E., Ozturk, M., 2011b. Ethnomedicinal studies on the plant resources of east Anatolia, Turkey. *Procedia – Soc. Behav. Sci.* 19, 756–777. https://doi.org/10.1016/j.sbspro.2011.05.195

Ari, S., Temel, M., Kargioğlu, M., Konuk, M., 2015. Ethnobotanical survey of plants used in Afyonkarahisar-Turkey. *J. Ethnobiol. Ethnomed.* 11, 1–15. https://doi.org/10.1186/s13002-015-0067-6

Atas, A.D., Goze, I., Alim, A., Cetinus, S.A., Durmus, N., Vural, N., Cakmak, O., 2012. Chemical composition, antioxidant, antimicrobial and antispasmodic activities of the essential oil of *Juniperus excelsa* subsp. *excelsa. J. Essent. Oil-Bearing Plants* 15, 476–483. https://doi.org/10.1080/0972060X.2012.10644075

Ateş, D.A., Turgay, Ö., 2003. Antimicrobial activities of various medicinal and commercial plant extracts Tıbbi ve Ticari Amaçlı Kullanılan Bazı Bitki Ekstraktlarının Antimikrobiyal Etkileri. *Turk J Biol.*

Bais, S., Abrol, N., Prashar, Y., Kumari, R., 2017. Modulatory effect of standardised amentoflavone isolated from Juniperus communis L. agianst Freund's adjuvant induced arthritis in rats (histopathological and X Ray anaysis). *Biomed. Pharmacother.* 86, 381–392. https://doi.org/10.1016/j.biopha.2016.12.027

Bais, S., Gill, N.S., Kumar, N., 2015. Neuroprotective effect of Juniperus communis on Chlorpromazine induced Parkinson Disease in animal model. *Chinese J. Biol.* 2015, 1–7. https://doi.org/10.1155/2015/542542

Balaban, M., Atik, C., Uçar, G., 2003. Fungal growth inhibition by wood extracts from Juniperus foetidissima and J. oxycedrus. *Holz als Roh - und Werkst.* 61, 231–232. https://doi.org/10.1007/s00107-003-0377-6

Başer, K.H.C., Demirci, F., 2007. Chemistry of essential oils, in: Berger, R.G. (Ed.), *Flavours and Fragrances Chemistry, Bioprocessing and Sustainability. Springer-Verlag, Germany*, p. 45. https://doi.org/10.1007/b136889

Baytop, T., 1999. *Türkiye'de Bitkiler ile Tedavi, Geçmişte ve Bugün.* Nobel Tıp Kitapevi, Istanbul.

Bello, R., Moreno, L., Beltrán, B., Primo-Yúfera, E., Esplugues, J., 1997. Effects on arterial blood pressure of methanol and dichloromethanol extracts from Juniperus oxycedrus L. *Phytotherapy Research.* John Wiley & Sons. https://doi.org/10.1002/(SICI)1099-1573(199703)11:2<161::AID-PTR57>3.0.CO;2-C

Ben Ali, M.J., Guesmi, F., Harrath, A.H., Alwasel, S., Hedfi, A., Ncib, S., Landoulsi, A., Aldahmash, B., Ben-Attia, M., 2015. Investigation of antiulcer and antioxidant activity of Juniperus phoenicea L. (1753) essential oil in an experimental rat model. *Biological and Pharmaceutical Bulletin.* https://doi.org/10.1248/bpb.b15-00412

Boulanouar, B., Abdelaziz, G., Aazza, S., Gago, C., Miguel, M.G., 2013. Antioxidant activities of eight Algerian plant extracts and two essential oils. *Ind. Crops Prod.* 46, 85–96. https://doi.org/10.1016/J.INDCROP.2013.01.020

Cakilcioglu, U., Khatun, S., Turkoglu, I., Hayta, S., 2011a. Ethnopharmacological survey of medicinal plants in Maden (Elazig-Turkey). *J. Ethnopharmacol.* 137, 469–486. https://doi.org/10.1016/j.jep.2011.05.046

Cakilcioglu, U., Khatun, S., Turkoglu, I., Hayta, S., 2011b. Ethnopharmacological survey of medicinal plants in Maden (Elazig-Turkey). *J. Ethnopharmacol.* 137, 469–486. https://doi.org/10.1016/j.jep.2011.05.046

Chaouche, T.M., Haddouchi, F., Atik-Bekara, F., Ksouri, R., Azzi, R., Boucherit, Z., Tefiani, C., Larbat, R., 2015. Antioxidant, haemolytic activities and HPLC-DAD-ESI-MSn characterization of phenolic compounds from root bark of Juniperus oxycedrus subsp. oxycedrus. *Ind. Crops Prod.* 64, 182–187. https://doi.org/10.1016/j.indcrop.2014.10.051

Coode, M.J.E., Cullenn, J., 1965. Juniperus L. in: Davis, P.H. (Ed.), *Flora of Turkey and the East Aegean Islands.* Edinburgh University Press, Edinburgh, pp. 78–84.

Cragg, G.M., Newman, D.J., 2013. Natural products: A continuing source of novel drug leads. *Biochim. Biophys. Acta - Gen. Subj.* 1830, 3670–3695. https://doi.org/10.1016/j.bbagen.2013.02.008

Darvishi, M., Esmaeili, S., Dehghan-nayeri, N., Mashati, P., Gharehbaghian, A., 2016. The effect of extract from aerial parts of *Juniperus excelsa* plant on proliferation and apoptosis of acute lymphoblastic leukemia cell lines, Nalm-6 and Reh TT (*Juniperus excelsa*). *Blood-Journal* 13, 304–313.

De Medina, F.S., Gamez, M.J., Jimenez, I., Jimenez, J., Osuna, J.I., Zarzuelo, A., 1994. Hypoglycemic activity of juniper "berries." *Planta Med.* 60, 197–200. https://doi.org/10.1055/s-2006-959457

Demirci, S., Ozhatay, N., 2012a. Wild plants used for medicinal purpose in Andırın, Kahramanmaraş article in Turkish. *Journal of Pharmaceutical Sciences.*

Demirci, S., Özhatay, N., 2012b. An ethnobotanical study in Kahramanmaras (Turkey); wild plants used for medicinal purpose in Andirin, Kahramanmaraş. *Turkish Journal of Pharmaceutical Sciences.*

El-Ghorab, A., Shaaban, H.A., El-Massry, K.F., Shibamoto, T., 2008. Chemical composition of volatile extract and biological activities of volatile and less-volatile extracts of juniper berry (*Juniperus drupacea* L.) fruit. *J. Agric. Food Chem.* 56, 5021–5025. https://doi.org/10.1021/jf8001747

Elmastaş, M., Gülçin, I., Beydemir, Ş., Küfrevioğlu, Ö.I., Aboul-Enein, H.Y., 2006. A study on the *in vitro* antioxidant activity of juniper (*Juniperus communis* L.) fruit extracts. *Anal. Lett.* 39, 47–65. https://doi.org/10.1080/00032710500423385

Elshafie, H.S., Caputo, L., De Martino, L., Gruľová, D., Zheljazkov, V.Z., De Feo, V., Camele, I., 2020. Biological investigations of essential oils extracted from three *Juniperus* species and evaluation of their antimicrobial, antioxidant and cytotoxic activities. *J. Appl. Microbiol.* 129, 1261–1271. https://doi.org/10.1111/jam.14723

Erbay, M.Ş., Anil, S., Melikoğlu, G., 2017. Türkiye'de geleneksel halk ilacı olarak kullanılan ağrı kesici etkili bitkiler-I MİDE AĞRISI. *Marmara Pharm. J.* https://doi.org/10.12991/mpj.2017.0

Ertuğ, F., 2002. Bodrum Yöresinde Halk Tıbbında Yararlanılan Bitkiler, in: Başer, K.H.C., N. Kırımer (Eds.), *14. BİHAT Toplantısı*, Bildiriler, Eskişehir, pp. 76–93.

Everest, A., Ozturk, E., 2005. Focusing on the ethnobotanical uses of plants in Mersin and Adana provinces (Turkey). *J. Ethnobiol. Ethnomed.* 1, 1–6. https://doi.org/10.1186/1746-4269-1-6

Ezer, N., Avcı, K., 2004. Folk medicines of Çerkeş (Çankırı) in Turkey. *Hacettepe University J. Fac. Pharm.* 24, 67–80.

Falzon, C.C., Balabanova, A., 2017. Phytotherapy: An introduction to herbal medicine. *Prim. Care - Clin. Off. Pract.* 44, 217–227. https://doi.org/10.1016/j.pop.2017.02.001

Fathima Firdose, K., Anjum, A., 2020. A comprehensive review of Juniperus communis with special reference to gynecology and Unani medicine. *International Journal of Innovative Science and Research Technology.*

Fernandez, A., Cock, I.E., 2016. The therapeutic properties of juniperus communis L.: Antioxidant capacity, bacterial growth inhibition, anticancer activity and toxicity. *Pharmacogn. J.* 8, 273–280. https://doi.org/10.5530/pj.2016.3.17

Gardner, D.R., Panter, K.E., James, L.F., Stegelmeier, B.L., 1998. Abortifacient effects of lodgepole pine (Pinus contorta) and common juniper (Juniperus communis) on cattle. *Vet. Hum. Toxicol.* 40, 260–263.

Ghasemnezhad, A., Ghorbanzadeh, A., Khoshhal Sarmast, M., Ghorbanpour, M., 2020. A review on botanical, phytochemical, and pharmacological characteristics of Iranian junipers (Juniperus spp.), in: *Plant-Derived Bioactives: Production, Properties and Therapeutic Applications*, pp. 493–508. https://doi.org/10.1007/978-981-15-1761-7_19

Ghosh, N., Ghosh, R.C., Kundu, A., Mandal, S.C., 2018. Herb and drug interaction, in: *Natural Products and Drug Discovery: An Integrated Approach*, pp. 467–490. https://doi.org/10.1016/B978-0-08-102081-4.00017-4

Güler, B., Kümüştekin, G., Uğurlu, E., 2015. Contribution to the traditional uses of medicinal plants of Turgutlu (Manisa – Turkey). *J. Ethnopharmacol.* 176, 102–108. https://doi.org/10.1016/j.jep.2015.10.023

Güler, B., Manav, E., Uǀurlu, E., 2015. Medicinal plants used by traditional healers in Bozüyük (Bilecik-Turkey). *J. Ethnopharmacol.* 173, 39–47. https://doi.org/10.1016/j.jep.2015.07.007

Güner, A., Aslan, S., Ekim, T., Vural, M., Babaç, M.T., Guner, A., Aslan, S., Ekim, T., Vural, M., Babac, M., 2012. Türkiye bitkileri listesi (damarlı bitkiler). Nezahat Gökyiğit Bot. *Bahçesi ve Flora Araştırmaları Derneği Yayını. İstanbul* 1, 47–83.

Güneş, F., 2018. An ethnobotany study in Enez town from Edirne. *Curr. Pers.* 1, 28–35.

Güneş, S., Savran, A., Paksoy, M.Y., Çakılcıoğlu, U., 2018. Survey of wild food plants for human consumption in Karaisalı (Adana-Turkey). *Indian Journal of Traditional Knowledge.*

Gürhan, G., Ezer, N., 2004. Halk Arasında Hemoroit Tedavisinde Kullanılan Bitkiler-I. *Hac Univ J Fac Pharm* 2, 37–60.

Güzel, Y., Güzelşemme, M., Miski, M., 2015a. Ethnobotany of medicinal plants used in Antakya: A multicultural district in Hatay Province of Turkey. *J. Ethnopharmacol.* 174, 118–152. https://doi.org/10.1016/j.jep.2015.07.042

Güzel, Y., Güzelşemme, M., Miski, M., 2015b. Ethnobotany of medicinal plants used in Antakya: A multicultural district in Hatay Province of Turkey. *J. Ethnopharmacol.* 174, 118–152. https://doi.org/10.1016/j.jep.2015.07.042

Han, M.I., Bulut, G., 2015a. The folk-medicinal plants of Kadişehri (Yozgat – Turkey). *Acta Soc. Bot. Pol.* 84, 237–248. https://doi.org/10.5586/asbp.2015.021

Han, M.I., Bulut, G., 2015b. The folk-medicinal plants of Kadişehri (Yozgat - Turkey). *Acta Soc. Bot. Pol.* 84, 237–248. https://doi.org/10.5586/asbp.2015.021

Hayta, S., Bagci, E., 2014. Essential oil constituents of the leaves, bark and cones of Juniperus oxycedrus subsp. oxycedrus L. from Turkey. *Acta Bot. Gall.* 161, 201–207. https://doi.org/10.1080/12538078.2014.921642

Höferl, M., Stoilova, I., Schmidt, E., Wanner, J., Jirovetz, L., Trifonova, D., Krastev, L., Krastanov, A., 2014. Chemical composition and antioxidant properties of juniper berry (Juniperus communis L.) essential oil. action of the essential oil on the antioxidant protection of saccharomyces cerevisiae model organism. *Antioxidants* 3, 81–98. https://doi.org/10.3390/antiox3010081

Honda, G., Yeşilada, E., Tabata, M., Sezik, E., Fujita, T., Takeda, Y., Takaishi, Y., Tanaka, T., 1996. Traditional medicine in Turkey VI. Folk medicine in West Anatolia: Afyon, Kutahya, Denizli, Mugla, Aydin provinces. *J. Ethnopharmacol.* 53, 75–87. https://doi.org/10.1016/0378-8741(96)01426-2

Jimenez-Arellanes, A., Meckes, M., Ramirez, R., Torres, J., Luna-Herrera, J., 2003. Activity against multidrug-resistant Mycobacterium tuberculosis in Mexican plants used to treat respiratory diseases. *Phyther. Res.* 17, 903–908. https://doi.org/10.1002/ptr.1377

Kandemir, A., 2018. Juniperus L., in: Güner, A., Kandemir, A., Menemen, Y., Yıldırım, H., Aslan, S., Ekşi, G., Güner, I., Çimen, Ç.A. (Ed.), *Resimli Türkiye Florası*. Ali Nihat Gökyiğit Vakfı, İstanbul, pp. 382–408.

Keskes, H., Mnafgui, K., Hamden, K., Damak, M., El Feki, A., Allouche, N., 2014. *In vitro* antidiabetic, antiobesity and antioxidant proprieties of Juniperus phoenicea L. leaves from Tunisia. *Asian Pac. J. Trop. Biomed.* 4, S649–S655. https://doi.org/10.12980/APJTB.4.201414B114

Khan, M., Khan, A.U., Najeeb-Ur-Rehman, Zafar, M.A., Hazrat, A., Gilani, A.H., 2012. Cardiovascular effects of *Juniperus excelsa* are mediated through multiple pathways. *Clin. Exp. Hypertens.* 34, 209–216. https://doi.org/10.3109/10641963.2011.631651

Khanavi, M., Enayati, A., Ardekani, M.R.S., Akbarzadeh, T., Razkenari, E.K., Eftekhari, M., 2019. Cytotoxic activity of *Juniperus excelsa* M. Bieb. leaves essential oil in breast cancer cell lines. *Res. J. Pharmacogn.* 6, 1–7. https://doi.org/10.22127/rjp.2019.84313

Kocabaş, Y.Z., Gedik, O., 2016. An Ethnobotanical Study of wild plants sold in district bazaar in Kahramanmaras. *J. Inst. Sci. Technol.* 4, 41–41. https://doi.org/10.21597/jist.2016624154

Koçyiğit, M., Özhatay, N., 2006. Wild plants used as medicinal purpose in Yalova (Northwest Turkey). *Turkish Journal of Pharmaceutical Sciences.*

Koruk, S.T., Ozyilkan, E., Kaya, P., Colak, D., Donderici, O., Cesaretli, Y., 2005. Juniper tar poisoning. *Clin. Toxicol.* 43, 47–49. https://doi.org/10.1081/CLT-45072

Labokas, J., Ložiene, K., 2013. Variation of essential oil yield and relative amounts of enantiomers of α-pinene in leaves and unripe cones of Juniperus communis L. growing wild in Lithuania. *J. Essent. Oil Res.* 25, 244–250. https://doi.org/10.1080/10412905.2013.775678

Lascurain Rangel, M., Guerrero-Analco, J.A., Monribot-Villanueva, J.L., Kiel-Martínez, A.L., Avendaño-Reyes, S., Díaz Abad, J.P., Bonilla-Landa, I., Dávalos-Sotelo, R., Olivares-Romero, J.L., Angeles, G., 2018. Anatomical and chemical characteristics of leaves and branches of Juniperus deppeana var. deppeana (Cupressaceae): A potential source of raw materials for the perfume and sweet candies industries. *Ind. Crops Prod.* 113, 50–54. https://doi.org/10.1016/j.indcrop.2017.12.046

Lesjak, M., Beara, I., Orčić, D., Anačkov, G., Knežević, P., Mrkonjić, Z., Mimica–Dukić, N., 2017. Bioactivity and chemical profiling of the *Juniperus excelsa*, which support its usage as a food preservative and nutraceutical. *Int. J. Food Prop.* 20, 1652–1663. https://doi.org/10.1080/10942912.2017.1352598

Lesjak, M.M., Beara, I.N., Orčić, D.Z., Petar, K.N., Simin, N.D., Emilija, S.D., Mimica-Dukić, N.M., 2014. Phytochemical composition and antioxidant, anti-inflammatory and antimicrobial activities of Juniperus macrocarpa Sibth. et Sm. *J. Funct. Foods* 7, 257–268. https://doi.org/10.1016/j.jff.2014.02.003

Li, C.Y., Lee, S.C., Lai, W.L., Chang, K.F., Huang, X.F., Hung, P.Y., Lee, C.P., Hsieh, M.C., Tsai, N.M., 2021. Cell cycle arrest and apoptosis induction by Juniperus communis extract in esophageal squamous cell carcinoma through activation of p53-induced apoptosis pathway. *Food Sci. Nutr.* 9, 1088–1098. https://doi.org/10.1002/fsn3.2084

Loizzo, M.R., Tundis, R., Conforti, F., Saab, A.M., Statti, G.A., Menichini, F., 2007. Comparative chemical composition, antioxidant and hypoglycaemic activities of Juniperus oxycedrus ssp. oxycedrus L. berry and wood oils from Lebanon. *Food Chem.* 105, 572–578. https://doi.org/10.1016/j.foodchem.2007.04.015

Mao, K., Hao, G., Liu, J., Adams, R.P., Milne, R.I., 2010. Diversification and biogeography of Juniperus (Cupressaceae): Variable diversification rates and multiple intercontinental dispersals. *New Phytol.* 188, 254–272. https://doi.org/10.1111/j.1469-8137.2010.03351.x

Maria Loureiro Seca, A., Pinto, D.C., S Silva, A.M., 2015. The current status of bioactive metabolites from the Genus Juniperus, in: Gupta, V.K. (Ed.), *Bioactive Phytochemicals: Perspectives for Modern Medicine.* M/S Daya Publishing House, New Delhi, India, pp. 365–408.

Marino, A., Bellinghieri, V., Nostro, A., Miceli, N., Taviano, M.F., Güvenç, A., Bisignano, G., 2010. *In vitro* effect of branch extracts of Juniperus species from Turkey on Staphylococcus aureus biofilm. *FEMS Immunol. Med. Microbiol.* 59, 470–476. https://doi.org/10.1111/j.1574-695X.2010.00705.x

Meriem, A., Msaada, K., Sebai, E., Aidi Wannes, W., Salah Abbassi, M., Akkari, H., 2021. Antioxidant, anthelmintic and antibacterial activities of red juniper (Juniperus phoenicea L.) essential oil. *J. Essent. Oil Res.* 1–10. https://doi.org/10.1080/10412905.2021.1941338

Miceli, N., Marino, A., Köroğlu, A., Cacciola, F., Dugo, P., Mondello, L., Taviano, M.F., 2020. Comparative study of the phenolic profile, antioxidant and antimicrobial activities of leaf extracts of five Juniperus L. (Cupressaceae) taxa growing in Turkey. *Nat. Prod. Res.* 34, 1636–1641. https://doi.org/10.1080/14786419.2018.1523162

Miceli, N., Trovato, A., Dugo, P., Cacciola, F., Donato, P., Marino, A., Bellinghieri, V., La Barbera, T.M., Gljvenç, A., Taviano, M.F., 2009. Comparative analysis of flavonoid profile, antioxidant and antimicrobial activity of the berries of Juniperus communis L. var. communis and Juniperus communis L. var. saxatilis Pall, from Turkey. *J. Agric. Food Chem.* 57, 6570–6577. https://doi.org/10.1021/jf9012295

Miceli, N., Trovato, A., Marino, A., Bellinghieri, V., Melchini, A., Dugo, P., Cacciola, F., Donato, P., Mondello, L., Güvenç, A., De Pasquale, R., Taviano, M.F., 2011. Phenolic composition and biological activities of Juniperus drupacea Labill. berries from Turkey. *Food Chem. Toxicol.* 49, 2600–2608. https://doi.org/10.1016/j.fct.2011.07.004

Moreno, L., Bello, R., Primo-Yúfera, E., Esplugues, J., 1997. *In vitro* studies of methanol and dichloromethanol extract of Juniperus oxycedrus L. *Phytotherapy Research.* https://doi.org/10.1002/(SICI)1099-1573(199706)11:4<309::AID-PTR87>3.0.CO;2-6

Najar, B., Pistelli, L., Buhagiar, J., 2020a. Volatilomic analyses of Tuscan Juniperus oxycedrus L. and *in vitro* cytotoxic effect of its essential oils on human cell lines. *J. Essent. Oil-Bearing Plants* 23, 756–771. https://doi.org/10.1080/0972060X.2020.1823891

Najar, B., Pistelli, L., Mancini, S., Fratini, F., 2020b. Chemical composition and *in vitro* antibacterial activity of essential oils from different species of Juniperus (section Juniperus). *Flavour Fragr. J.* 35, 623–638. https://doi.org/10.1002/ffj.3602

Nikolić, B., Vasilijević, B., Mitić-Culafić, D., Lesjak, M., Vuković-Gačić, B., Dukić, N.M., Knežević-Vukčević, J., 2016. Screening of the antibacterial effect of Juniperus sibirica and Juniperus sabina essential oils in a microtitre platebased MIC assay. *Bot. Serbica* 40, 43–48. https://doi.org/10.5281/zenodo.48858

Ohashi, H., Asai, T., Kawai, S., 1994. Screening of main japanese conifers for antifungal leaf components, sesquiterpenes of juniperus chinensis var. pyramidalis. *Holzforschung* 48, 193–198. https://doi.org/10.1515/hfsg.1994.48.3.193

Orhan, N., 2019. Juniperus species: Features, profile and applications to diabetes, in: *Bioactive Food as Dietary Interventions for Diabetes.* Elsevier, pp. 447–459. https://doi.org/10.1016/b978-0-12-813822-9.00030-8

Orhan, N., Akkol, E., Ergun, F., 2012. Evaluation of antiinflammatory and antinociceptive effects of some juniperus species growing in Turkey. *Turkish J. Biol.* 36, 719–726. https://doi.org/10.3906/biy-1203-32

Orhan, Nilüfer, Akkol, E., Ergun, F., 2012a. Evaluation of antiinflammatory and antinociceptive effects of some juniperus species growing in Turkey. *Turkish J. Biol.* 36, 719–726. https://doi.org/10.3906/biy-1203-32

Orhan, Nilüfer, Aslan, M., Pekcan, M., Orhan, D.D., Bedir, E., Ergun, F., 2012b. Identification of hypoglycaemic compounds from berries of Juniperus oxycedrus subsp. oxycedrus through bioactivity guided isolation technique. *J. Ethnopharmacol.* 139, 110–118. https://doi.org/10.1016/j.jep.2011.10.027

Orhan, Nilüfer, Berkkan, A., Deliorman Orhan, D., Aslan, M., Ergun, F., 2011. Effects of Juniperus oxycedrus ssp. oxycedrus on tissue lipid peroxidation, trace elements (Cu, Zn, Fe) and blood glucose levels in experimental diabetes. *J. Ethnopharmacol.* 133, 759–764. https://doi.org/10.1016/j.jep.2010.11.002

Orhan, N., Hoşbaş, S., Orhan, D.D., Aslan, M., Ergun, F., 2014. Enzyme inhibitory and radical scavenging effects of some antidiabetic plants of Turkey. *Iran. J. Basic Med. Sci.* 17, 426–432. https://doi.org/10.22038/ijbms.2014.2927

Orhan, N., Orhan, D.D., Gökbulut, A., Aslan, M., Ergun, F., 2017. Comparative analysis of chemical profile, antioxidant, In-vitro and In-vivo antidiabetic activities of Juniperus foetidissima willd. and Juniperus sabina L. *Iranian Journal of Pharmaceutical Research.* https://doi.org/10.22037/ijpr.2017.1962

Orhan, Nilufer, Orhan, I.E., Ergun, F., 2011. Insights into cholinesterase inhibitory and antioxidant activities of five Juniperus species. *Food Chem. Toxicol.* 49, 2305–2312. https://doi.org/10.1016/j.fct.2011.06.031

Otto, A., Wilde, V., 2001. Sesqui-, Di-, and Triterpenoids as chemosystematic markers in extant conifers – a review. *Bot. Rev.* 67, 141–238.
Özçelik, H., 1987. Akseki yöresinde doğal olarak yetişen bazı faydalı bitkilerin yerel adları ve kullanılışları. *Doğa Türk Bot. Dergisi* 11, 316–321.
Özdemir, E., Alpinar, K., 2015a. An ethnobotanical survey of medicinal plants in western part of central Taurus Mountains: Aladaglar (Nigde - Turkey). *J. Ethnopharmacol.* 166, 53–65. https://doi.org/10.1016/j.jep.2015.02.052
Özdemir, E., Alpinar, K., 2015b. An ethnobotanical survey of medicinal plants in western part of central Taurus Mountains: Aladaglar (Nigde - Turkey). *J. Ethnopharmacol.* 166, 53–65. https://doi.org/10.1016/j.jep.2015.02.052
Özgökçe, F., Özçelik, H., 2004. Ethnobotanical aspects of some taxa in East Anatolia, Turkey. *Econ. Bot.* 58, 697–704. https://doi.org/10.1663/0013-0001(2004)058[0697:EAOSTI]2.0.CO;2
Özkan, A., Koyuncu, M., 2005. Tradional medicinal plants used in Pınarbaşı area (Kayseri-Turkey). *Turkish J. Pharm. Sci.* 2, 63–82.
Öztürk, M., Tümen, I., Uğur, A., Aydoğmuş-Öztürk, F., Topçu, G., 2011. Evaluation of fruit extracts of six Turkish Juniperus species for their antioxidant, anticholinesterase and antimicrobial activities. *J. Sci. Food Agric.* 91, 867–876. https://doi.org/10.1002/jsfa.4258
Pandey, S., Tiwari, S., Kumar, A., Niranjan, A., Chand, J., Lehri, A., Chauhan, P.S., 2018. Antioxidant and anti-aging potential of Juniper berry (Juniperus communis L.) essential oil in Caenorhabditis elegans model system. *Ind. Crops Prod.* 120, 113–122. https://doi.org/10.1016/j.indcrop.2018.04.066
Prinsloo, G., Marokane, C.K., Street, R.A., 2018. Anti-HIV activity of southern African plants: Current developments, phytochemistry and future research. *J. Ethnopharmacol.* https://doi.org/10.1016/j.jep.2017.08.005
Rostaefar, A., Hassani, A., Sefidkon, F., 2017. Seasonal variations of essential oil content and composition in male and female plants of Juniperus communis L. ssp. hemisphaerica growing wild in Iran. *Journal of Essential Oil Research* 29, 357–360. https://doi.org/10.1080/10412905.2017.1279990
Sadeghi-aliabadi, H., Emami, A., Sadeghi, B., Jafarian, A., 2009. *In Vitro* cytotoxicity of two subspecies of *Juniperus excelsa* on cancer cells. *Iran. J. Basic Med. Sci.* 11, 250–253. https://doi.org/10.22038/ijbms.2009.5189
Sahin Yaglioglu, A., Eser, F., 2017. Screening of some Juniperus extracts for the phenolic compounds and their antiproliferative activities. *South African J. Bot.* 113, 29–33. https://doi.org/10.1016/j.sajb.2017.07.005
Sahin Yaglioglu, Ayse, Eser, F., Yaglioglu, M.S., Demirtas, I., 2020. The antiproliferative and antioxidant activities of the essential oils of Juniperus species from Turkey. *Flavour Fragr. J.* 35, 511–523. https://doi.org/10.1002/ffj.3586
Sahin Yaglioglu, A., Eser, F., Yaglioglu, M.S., Demirtas, I., 2020. The antiproliferative and antioxidant activities of the essential oils of Juniperus species from Turkey. *Flavour Fragr. J.* 35, 511–523. https://doi.org/10.1002/ffj.3586
Sargin, S.A., 2015. Ethnobotanical survey of medicinal plants in Bozyazi district of Mersin, Turkey. *J. Ethnopharmacol.* 173, 105–126. https://doi.org/10.1016/j.jep.2015.07.009
Sargin, S.A., Akçicek, E., Selvi, S., 2013. An ethnobotanical study of medicinal plants used by the local people of Alaşehir (Manisa) in Turkey. *J. Ethnopharmacol.* 150, 860–874. https://doi.org/10.1016/j.jep.2013.09.040
Sargin, S.A., Büyükcengiz, M., 2019. Plants used in ethnomedicinal practices in Gulnar district of Mersin, Turkey. *J. Herb. Med.* 15, 100224. https://doi.org/10.1016/j.hermed.2018.06.003
Sargin, S.A., Selvi, S., Büyükcengiz, M., 2015. Ethnomedicinal plants of Aydincik district of Mersin, Turkey. *J. Ethnopharmacol.* 174, 200–216. https://doi.org/10.1016/j.jep.2015.08.008
Sargın, S.A., Selvi, S., López, V., 2015. Ethnomedicinal plants of Sarigöl district (Manisa), Turkey. *J. Ethnopharmacol.* 171, 64–84. https://doi.org/10.1016/j.jep.2015.05.031
Sayar, A., Güvensen, A., Özdemir, F., Öztürk, M., 1995. Muğla (Türkiye) İlindeki Türlerin Etnobotanik Özellikleri. *Herb J. Syst. Bot.* 2, 151–160.
Seca, A.M.L., Pinto, P.C.G.A., Silva, A.M.S., 2015. The current status of bioactive metabolites from the Genus Juniperus. *Bioact. Phytochem. Perspect. Mod. Med.* 3, 365–408.
Semiz, G., Isik, K., Unal, O., 2007. Enek Pekmez production from Juniper “fruits” by native people on the taurus mountains in southern Turkey. *Econ. Bot.* 61, 299–301. https://doi.org/10.1663/0013-0001(2007)61[299:NOEP]2.0.CO;2
Senkardes, I., Tuzlaci, E., 2014. Some ethnobotanical notes from Gundogmus District (Antalya/Turkey). *J. Marmara Univ. Inst. Heal. Sci.* 1. https://doi.org/10.5455/musbed.20140303070652

Sezik, E., Kocakulak, E., Baser, K.H.C., Ozek, T., 2005. Composition of the essential oils of Juniperus oxycedrus subsp. macrocarpa from Turkey. *Chem. Nat. Compd.* 41, 352–354. https://doi.org/10.1007/s10600-005-0149-0

Sezik, E., Yesilada, E., Tabata, M., Honda, G., Takaishi, Y., Fujita, T., Tanaka, T., Takeda, Y., 1997. Traditional medicine in Turkey .8. Folk medicine in east Anatolia; Erzurum, Erzincan, Agri, Kars, Igdir provinces. *Econ. Bot.* 51, 195–211.

Stanić, G., Samaržija, I., Blažević, N., 1998. Time-dependent diuretic response in rats treated with juniper berry preparations, *Phytotherapy Research.* https://doi.org/10.1002/(SICI)1099-1573(199811)12:7<494::AID-PTR340>3.0.CO;2-N

Stankov, S., Fidan, H., Petkova, Z., Stoyanova, M., Petkova, N., Stoyanova, A., Semerdjieva, I., Radoukova, T., Zheljazkov, V.D., 2020a. Comparative study on the phytochemical composition and antioxidant activity of Grecian juniper (*Juniperus excelsa* M. Bieb) unripe and ripe Galbuli. *Plants* 9, 1–18. https://doi.org/10.3390/plants9091207

Stankov, S., Fidan, H., Petkova, Z., Stoyanova, M., Petkova, N., Stoyanova, A., Semerdjieva, I., Radoukova, T., Zheljazkov, V.D., 2020b. Comparative study on the phytochemical composition and antioxidant activity of Grecian Juniper (*Juniperus excelsa* M. Bieb) unripe and ripe Galbuli. *Plants* 9. https://doi.org/10.3390/PLANTS9091207

Tavares, W., Seca, A., 2018. The current status of the pharmaceutical potential of Juniperus L. metabolites. *Medicines* 5, 81. https://doi.org/10.3390/medicines5030081

Taviano, M.F., Marino, A., Trovato, A., Bellinghieri, V., La Barbera, T.M., Gvenç, A., Hrkul, M.M., Pasquale, R. De, Miceli, N., 2011. Antioxidant and antimicrobial activities of branches extracts of five Juniperus species from Turkey. *Pharm. Biol.* 49, 1014–1022. https://doi.org/10.3109/13880209.2011.560161

Taviano, M.F., Marino, A., Trovato, A., Bellinghieri, V., Melchini, A., Dugo, P., Cacciola, F., Donato, P., Mondello, L., Güvenç, A., Pasquale, R. De, Miceli, N., 2013. Juniperus oxycedrus L. subsp. oxycedrus and Juniperus oxycedrus L. subsp. macrocarpa (Sibth. & Sm.) Ball. "berries" from Turkey: Comparative evaluation of phenolic profile, antioxidant, cytotoxic and antimicrobial activities. *Food Chem. Toxicol.* 58, 22–29. https://doi.org/10.1016/j.fct.2013.03.049

Tetik, F., Civelek, S., Cakilcioglu, U., 2013. Traditional uses of some medicinal plants in Malatya (Turkey). *J. Ethnopharmacol.* 146, 331–346. https://doi.org/10.1016/j.jep.2012.12.054

The Plant List, 2019. Version 1.1. http://www.theplantlist.org/ (accessed 1st January of 2021).

Topçu, G., Erenler, R., Çakmak, O., Johansson, C.B., Çelik, C., Chai, H.B., Pezzuto, J.M., 1999. Diterpenes from the berries of *Juniperus excelsa. Phytochemistry* 50, 1195–1199. https://doi.org/10.1016/S0031-9422(98)00675-X

Topçu, G., Gören, A.C., Bilsel, G., Bilsel, M., Çakmak, O., Schilling, J., Kingston, D.G.I., 2005. Cytotoxic activity and essential oil composition of leaves and berries of *Juniperus excelsa. Pharm. Biol.* 43, 125–128. https://doi.org/10.1080/13880200590919429

Tumen, I., Süntar, I., Eller, F.J., Keleş, H., Akkol, E.K., 2013. Topical wound-healing effects and phytochemical composition of heartwood essential oils of *Juniperus virginiana* L., juniperus occidentalis hook., and juniperus ashei J. buchholz. *J. Med. Food* 16, 48–55. https://doi.org/10.1089/jmf.2012.2472

Tunalier, Z., Kirimer, N., Baser, K.H.C., 2002. The composition of essential oils from various parts of Juniperus foetidissima. *Chem. Nat. Compd.* 38, 43–47. https://doi.org/10.1023/A:1015725630556

Tundis, R., Bonesi, M., Loizzo, M.R., 2021. A Comparative Study of Phytochemical Constituents and Bioactivity of n-Hexane and dichloromethane extracts of Juniperus oxycedrus subsp. macrocarpa and J. oxycedrus subsp. oxycedrus. https://doi.org/10.3390/iecps2020-08770

Tunón, H., Olavsdotter, C., Bohlin, L., 1995. Evaluation of anti-inflammatory activity of some Swedish medicinal plants. Inhibition of prostaglandin biosynthesis and PAF-induced exocytosis. *J. Ethnopharmacol.* 48, 61–76. https://doi.org/10.1016/0378-8741(95)01285-L

Tuzlacı, E., 2016. *Türkiye Bitkileri Geleneksel İlaç Rehberi.* İstanbul Tıp Kitabevleri, İstanbul.

Tuzlacı, E., Erol, M.K., 1999. Turkish folk medicinal plants. Part II: Egirdir (Isparta). *Fitoterapia* 70, 593–610. https://doi.org/10.1016/S0367-326X(99)00074-X

Uğulu, İ., Başlar, S., Yorek, N., Doğan, Y., 2009. The investigation and quantitative ethnobotanical evaluation of medicinal plants used around Izmir province, Turkey. *J. Med. Plants* 3, 345–367.

Unlu, M., Vardar-Unlu, G., Vural, N., Donmez, E., Cakmak, O., 2008. Composition and antimicrobial activity of *Juniperus excelsa* essential oil. *Chem. Nat. Compd.* 44, 129–131. https://doi.org/10.1007/s10600-008-0040-x

Uysal, I., Onar, S., Karabacak, E., Çelik, S., 2010. Ethnobotanical aspects of Kapıdağ Peninsula (Turkey). *Biol. Divers. Conserv.* 3, 15–22.

Van Auken, O.W., Smeins, F., 2008. *Western North American Juniperus Communities: Patterns and Causes of Distribution and Abundance*, pp. 3–18. https://doi.org/10.1007/978-0-387-34003-6_1

Ved, A., Gupta, A., Singh Rawat, A., 2017. Antioxidant and hepatoprotective potential ofphenol-rich fraction of *Juniperus communis* Linn. leaves. *Pharmacogn Mag* 13, 108–113.

Venditti, A., Maggi, F., Saab, A.M., Bramucci, M., Quassinti, L., Petrelli, D., Vitali, L.A., Lupidi, G., El Samrani, A., Borgatti, M., Bernardi, F., Gambari, R., Abboud, J., Saab, M.J., Bianco, A., 2021. Antiproliferative, antimicrobial and antioxidant properties of *Cedrus libani* and *Pinus pinea* wood oils and *Juniperus excelsa* berry oil. *Plant Biosyst. - An Int. J. Deal. with all Asp. Plant Biol.* 1–12. https://doi.org/10.1080/11263504.2020.1864495

Vural, M., Karavelioğulları, F.A., 1997. Çiçekdağı (Kırşehir) ve çevresinin etnobotanik özellikleri. *Ot Sist. Bot. Derg.* 4, 117–124.

Xie, S., Li, G., Qu, L., Zhong, R., Chen, P., Lu, Z., Zhou, J., Guo, X., Li, Z., Ma, A., Qian, Y., Zhu, Y., 2017. Podophyllotoxin extracted from *Juniperus sabina* fruit inhibits rat sperm maturation and fertility by promoting epididymal epithelial cell apoptosis. *Evidence-based Complement. Altern. Med.* 2017. https://doi.org/10.1155/2017/6958982

Yaman, T., Uyar, A., Kömüroğlu, A.U., Keleş, Ö.F., Yener, Z., 2021. Chemopreventive efficacy of juniper berry oil (*Juniperus communis* L.) on azoxymethane-induced colon carcinogenesis in rat. *Nutr. Cancer* 73, 133–146. https://doi.org/10.1080/01635581.2019.1673450

Yarnell, E., 2002. Botanical medicines for the urinary tract. *World J. Urol.* 20, 285–293. https://doi.org/10.1007/s00345-002-0293-0

Yaşar, H.İ., Koyuncu, O., Turan Koyuncu, F., Kuş, G., 2019. Sale of medicinal and aromatic plants and economic dimensions in Eskişehir herbalist. *Res. J. Biol. Sci.* 12, 25–28.

Yeşilada, E., Honda, G., Sezik, E., Tabata, M., Fujita, T., Tanaka, T., Takeda, Y., Takaishi, Y., 1995. Traditional medicine in Turkey. V. Folk medicine in the inner Taurus Mountains. *J. Ethnopharmacol.* 46, 133–152. https://doi.org/10.1016/0378-8741(95)01241-5

Yeşilada, E., Honda, G., Sezik, E., Tabata, M., Goto, K., Ikeshiro, Y., 1993. Traditional medicine in Turkey IV. Folk medicine in the Mediterranean subdivision. *J. Ethnopharmacol.* 39, 31–38. https://doi.org/10.1016/0378-8741(93)90048-A

Yeşilyurt, E.B., Şimşek, I., Akaydin, G., Yeşilada, E., 2017. An ethnobotanical survey in selected districts of the black sea region (Turkey). *Turk. J. Botany* 41, 47–62. https://doi.org/10.3906/bot-1606-12

Zheljazkov, V.D., Kacaniova, M., Dincheva, I., Radoukova, T., Semerdjieva, I.B., Astatkie, T., Schlegel, V., 2018. Essential oil composition, antioxidant and antimicrobial activity of the galbuli of six juniper species. *Ind. Crops Prod.* 124, 449–458. https://doi.org/10.1016/j.indcrop.2018.08.013

Zheljazkov, V.D., Kacaniova, M., Dincheva, I., Radoukova, T., Semerdjieva, I.B., Astatkie, T., Schlegel, V., 2018. Essential oil composition, antioxidant and antimicrobial activity of the galbuli of six juniper species. *Ind. Crops Prod.* 124, 449–458. https://doi.org/10.1016/j.indcrop.2018.08.013

Živić, N., Milošević, S., Dekić, V., Dekić, B., Ristić, N., Ristić, M., Sretić, L., 2019. Phytochemical and antioxidant screening of some extracts of *Juniperus communis* L. and *Juniperus oxycedrus* L. *Czech J. Food Sci.* 37, 351–358. https://doi.org/10.17221/28/2019-CJFS

11 *Lamium* sp.

Duygu Kaya Bilecenoğlu and Funda N. Yalçın

INTRODUCTION

The genus *Lamium* L. belongs to the family Lamiaceae, which comprises about 40 species worldwide, distributing from Western Europe to Eastern Asia, including subtropical regions of Africa. It is herbaceous and possesses annual or perennial life forms (Mennema, 1989).

The common English name of the genus *Lamium* L. is "deadnettle", referring to their resemblance to the unrelated nettles of *Urtica* sp.; however, they do not have stinging hairs and are harmless or apparently "dead" (Yalçın & Kaya, 2006). Some of the local names of *Lamium* is "ballık" in the middle Aegean region of Turkey (Kargıoğlu et al., 2010), while "ballıbaba" is widely used in Malatya and many different parts of Anatolia (Baytop, 1999; Güler et al., 2015; Tetik et al., 2013).

Research on the phytochemistry of *Lamium* sp. has shown significant progress during the last four decades, emphasizing the presence of several chemicals such as iridoids and secoiridoids, phenylpropanoids, flavonoids, anthocyanins, phenolic acids, phytoecdysteroids, betaines, benzoxazinoids, terpenes, and megastigmen compounds, as well as essential oils (Salehi et al., 2019; Yalçın & Kaya, 2006).

Representatives of the genus *Lamium* display antispasmodic, astringent, anti-proliferative, anti-inflammatory, and antiviral activities, which have also been used in folk medicine worldwide as a remedy in the treatment of several disorders, such as trauma, fracture, paralysis, hypertension, menorrhagia, leucorrhea, and uterine hemorrhage (Salehi et al., 2019; Yalçın & Kaya, 2006).

In this chapter, we have focused on phytochemistry and biological activity studies of *Lamium* sp. by compiling comprehensive reviews, recent studies, and unpublished thesis in order to present up-to-date information on the above-mentioned medicinal plant.

BOTANICAL FEATURES/ETHNOBOTANICAL USAGE

The genus *Lamium* belongs to the subfamily Lamioideae, and rocky mountain slopes are the dominating habitat, including waste ground, cultivated fields, gardens, and roadsides (Mennema, 1989; Mill, 1982). The leaves are ovate to reniform, crenate to dentate. Inflorescences are verticillasters from dense to remote and contain 2–12 flowers. The calyx is tubular or campanulate, 5-veined, with five equal or subequal teeth. Corolla is purple mauve, pink, cream, or rarely white, 2-lipped; upper lip hooded; lower lip obcordate or broadly obovate with or without small lateral lobes (Figure 11.1). Nutlets are triquetrous, usually truncate at the apex (Mill, 1982).

Extensive taxonomical research carried out to date has revealed the presence of 31 *Lamium* species in the flora of Turkey (Celep, 2017; Mill, 1982). Approximately 23 taxa, including varieties and subspecies, are endemic (Baran & Ozdemir, 2013).

The ethnobotanical uses of *Lamium* plants have been recorded worldwide, particularly as food and medicine (Salehi et al., 2019; Yalçın & Kaya, 2006). Traditional uses mainly were listed for *L. album* as utilization in menorrhagia, uterine hemorrhage, vaginal and cervical inflammation, and leukorrhea treatment and getting rid of kidney stones, besides prevention of menstrual and musculoskeletal disorders as an edible plant and food supplement (Kelayeh et al., 2019; Salehi et al., 2019; Yalçın & Kaya, 2006). *L. album* is also listed for reducing pain in rheumatism and arthritis in Western Anatolia (Baytop, 1999). Besides these uses of *L. album*, it is one of the edible medicinal

DOI: 10.1201/9781003146971-11

FIGURE 11.1 Wild population of *Lamium garganicum* L. subsp. *laevigatum* Arcangeli in Türkiye. (Photograph: Duygu Kaya Bilecenoğlu.)

plants consumed as a food alternative and an ingredient for food supplements claimed as detoxifying agent (Konarska et al., 2021).

According to their traditional uses throughout the Mediterranean region, locals take advantage of *Lamium* sp. for a variety of reasons; for example, the medicinal use of *L. amplexicaule* was reported for its hemostatic effect, while *L. maculatum* is consumed as food, and some other *Lamium* species are utilized against the evil eye (Mulas, 2006). As for the Anatolian region, *L. album*, *L. maculatum*, and *L. purpureum* have been recorded to be used as a tonic and in the treatment of constipation as home remedies (Baytop, 1999).

In an ethnobotanical study performed in Denizli Province in the middle Aegean region of Turkey, aerial parts of *L. amplexicaule* decoction and infusion of plant were consumed as a beverage and used to treat urinary tract infections (Kargıoğlu et al., 2010). It was determined that *L. tenuiflorum*, a species endemic to Türkiye, is used to treat colds and bronchitis and for urinary tract infections internally and for boils and abscesses externally (Dulger, 2009).

The aerial parts of *L garganicum* L. subsp. *striatum* are used for fodder (Kargıoğlu et al., 2010). A herbal tea prepared by decoction or infusion methods from the leaves of *L. album* was reported to be used for respiratory tract problems by drinking one tea glass of the plant before the meal (Tetik et al., 2013). Likewise, the flowering stems of *L. purpureum* were prepared as herbal tea with an infusion method, and people drink it three or four times a day for influenza and constipation in Bilecik, Türkiye (Güler et al., 2015).

PHYTOCHEMICALS

The phytochemical composition of *Lamium* species has always been of interest to several researchers especially due to the presence of biologically active molecules among them. Until now, essential oil components and several compounds including iridoids, secoiridoids, phenylpropanoids, flavonoids, anthocyanins, phytoecdysteroids, betaines, benzoxazinoids, terpenes, and megastigmen have been identified from *Lamium* species (Frezza et al., 2019; Salehi et al., 2019; Yalçın & Kaya, 2006). A brief list of essential oil components and the structures of the main molecules previously isolated from *Lamium* species are given in the previous literature (Alipieva et al., 2007; Salehi et al., 2019; Yalçın & Kaya, 2006).

A wide range of *in vitro*, *in vivo*, and clinical studies performed on the genus *Lamium* showed that especially iridoid-rich members of the genus exhibit anticancer, antioxidant, antibacterial, anti-viral, anti-inflammatory, antiarthritic, immunomodulatory, neuroprotective, and wound healing activities (Kelayeh et al., 2019; Yordanova et al., 2014).

The iridoids found in the genus can work as markers of *L. album*, *L. amplexicaule*, *L. garganicum*, *L. maculatum*, and *L. purpureum* (Yalçın & Kaya, 2006). The presence of iridoids from *Lamium* species was determined by LC-ESI-MS analysis, in which lamalbide, shanzhiside methyl ester, and barlerin were found in the extracts of the frequently studied *L. album*, *L. amplexicaule*, *L. garganicum*, *L. maculatum*, and *L. purpureum*, while lamalbide and shanzhiside methyl ester were also detected from the subspecies of *Lamium garganicum* and *L. purpureum* L. var. *purpureum* grown wildly in Turkey by the same method (Alipieva et al., 2007). It can be obviously seen that iridoid monoglucosides containing 11-$COOCH_3$ iridoids or only methyl groups at the position C4 can work as chemotaxonomic markers for particular *Lamium* species: *L. album*, *L. amplexicaule*, *L. garganicum*, *L. maculatum*, and *L. purpureum* (Frezza et al., 2019; Kikuchi et al., 2009; Salehi et al., 2019). The iridoids, lamiol, and lamioside are detected as the chemotaxonomic markers for the genus *Lamium* (Figure 11.2) (Frezza et al., 2019). According to the studies, given some phylogenetic relationships within *Lamium*, iridoids with 11-COOR substituent were found in various concentrations in all studied species, whereas iridoids with 11-CH_3 were not detected in *L. garganicum* and found in *L. album* only in traces (Table 11.1) (Alipieva et al., 2006).

L. album is considered the most prominent member of the genus because of its traditional uses and several therapeutic activities, resulting in many phytochemical investigations carried out to date (Czerwinska et al., 2020; Czerwinska et al., 2018; Czerwinska et al., 2017; Pereira et al., 2012; Yordanova et al., 2014). Czerwinska et al. (2020) reported that phenylpropanoids were the most diverse compounds in the flowers and herbs of *L. album*. Lamiuside A (Lamalboside), lamiusides B, C, E, phlinoside D, verbascoside, and isoverbascoside were determined as the major phenylpropanoids in the phenolic compounds containing extract of *L. album* (Salehi et al., 2019).

Other studies about the chemical composition of *L. album* flower and herb extracts resulted in the detection and isolation of iridoids (lamalbid (lamiridoside), lamiol, caryoptoside), phenolic acids (such as chlorogenic acid), and high quantities of flavonoids in the form of glycosides and aglycones (Alipieva et al., 2007; Kelayeh et al., 2019). Pereira et al. (2012) isolated glycosides of common flavones from the aerial parts of *L. album* as the major flavonoids. The ethanolic and methanolic extracts were obtained from the glandular trichomes of corolla parts of *L. album* subsp. *album* which resulted in the isolation of triterpenes (β-amyrin, β-amyrin acetate, oleanolic acid) and flavonoids (rutoside, quercetin, and isoquercetin) (Konarska et al., 2021; Sulborska et al., 2020).

L. album is considered the most popular species that contains a variety of compounds: phenols, iridoids, triterpenes, saponins, fatty acids, phytoecdysteroids, essential oils, tannins, amines (Alipieva et al., 2007). Valyova et al. (2011) showed that *L. album* could be a source of natural antioxidants with potential use in food supplements (Chipeva et al., 2013).

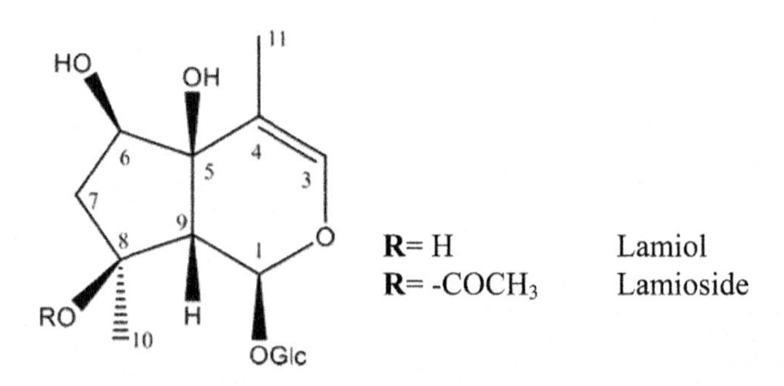

FIGURE 11.2 Molecular formula of lamiol and lamioside.

TABLE 11.1
Some Iridoids Detected from *Lamium* Species

	L. amplexicaule	*L. garganicum*	*L. album*	*L. maculatum*	*L. purpureum*
Iridoids with 11-CH3					
Lamiol	+		+	+	+
5-Deoxylamiol	+			+	+
Lamioside	+			+	+
5-Deoxylamioside	+				
6-Deoxylamioside	+				
Iridoids with 11-COOR					
Lamalbid	+	+	+	+	+
Caryoptoside	+		+		+
Penstemoside		+		+	
Ipolamiide	+				
Iridoids with 11-COOR					
Ipolamiidoside	+				
Lamiide	+				
Shanzhiside Methyl ester	+	+	+	+	+
Barlerine	+	+	+	+	+
Sesamoside		+		+	
Deacetyl asperulosidic acid	+				
Asperuloside	+				
Alboside A			+		
Alboside B			+		

(Alipieva et al., 2007; Alipieva et al., 2006; Ersöz et al., 2007; Ito et al., 2006; Kikuchi et al., 2009)

The phytochemical studies showed that *L. purpureum* has a slightly different phytochemical profile than *L. album. Lamium purpureum* contains mainly phenolic acids like caffeic acid and chlorogenic acid, as well as rutin, hyperoside, and quercetin derivatives (Bubueanu et al., 2013). Caryoptoside and lamalbide were the other isolated iridoids from the whole plants (aerial, leaf, stem) of *L. purpureum* (Ito et al., 2006).

Essential Oil–Bearing *Lamium* Plants

Since *Lamium* genus belongs to Lamioideae, a subfamily that possesses low levels of essential oils, only a small number of *Lamium* species bear essential oils with ratios between 0.01 and 0.31% (Flamini et al., 2005; Salehi et al., 2019). Terpenes present in essential oil were considered the responsible compounds for the pharmacological activity/medicinal properties and protective function in plants from the family Lamiaceae (Konarska et al., 2021).

Plenty of studies have been recorded to elucidate the composition of the essential oils obtained from *Lamium* species (Akkoyunlu & Dulger, 2019; Flamini et al., 2005; Grujić et al., 2020; Konarska et al., 2021; Morteza-Semnani et al., 2016; Sajjadi & Ghannadi, 2012). Essential oil composition studies were generally on *L. amplexicaule*, *L. purpureum*, *L. maculatum*, *L. garganicum*, *L. album*, *L. moschatum*, and *L. Garganicum*, which are also known to occur in Türkiye. Salehi et al. (2019) have gathered the essential oil composition of studied species in a table. According to the table, aldehydes, sesquiterpenes, alkanes, alcohols, and monoterpenes have been revealed from essential oils. The existence and percentages of the major compounds in essential oils vary because of different regions in which plants are grown, genetic and environmental factors (geographic, climatic, and altitude properties), and even the harvesting time (Konarska et al., 2021; Salehi et al., 2019). A pronounced abundance of germacrene D was prominent among the sesquiterpenes, followed by β-caryophyllene (Salehi et al., 2019). Other important substances for *L. purpureum* were β-pinene and α-pinene, while *L. amplexicaule* essential oil was mainly composed of α-pinene (Jones, 2012). The essential oil of *L. amplexicaule* was also characterized by high amounts of trans-chrysanthenyl acetate, a monoterpene derivative not detected in the other *Lamium* species (Flamini et al., 2005).

The essential oil (hydro-distillation) composition of the aerial parts of *L. amplexicaule* collected from two different regions of Iran was determined by GC-MS. The major components were germacrene-D, camphor, trans-phytol, octadecanol, hexadecanoic acid, and hexahydrofarnesyl acetone (Nickavar et al., 2008; Sajjadi & Ghannadi, 2012). The major component of the oil, trans-phytol is an abundant component of the essential oil of several Lamiaceae species. Essential oil composition variation depends on the soil or climatic conditions (Sajjadi & Ghannadi, 2012)

According to the recent studies conducted on the essential oil of Flos Lamii albi, several compounds have been found by histochemical methods (Konarska et al., 2021; Sulborska et al., 2020). These are triterpenes and iridoids besides the essential oil components. Konarska et al. (2021) showed the most important compounds isolated from *L. album* subsp. *album* essential oil. Sulborska et al. (2020) showed the presence of phenolic compounds which are tannins, phenolic acids, and flavonoids in trichomes of this species. *L. album* subsp. *album* flower oil was dominated by sesquiterpenes, alkanes, and aldehydes in different ratios, which depend on the flower's dry or fresh version (Konarska et al., 2021). A GC-MS study on *L. album* flowers has found the major compounds of the essential oil to be 6,10,14-trimethyl-2-pentadecanone (10.2%) and 4-hydroxy-4-methyl-2-pentanone (9.1%) (Morteza-Semnani et al., 2016).

BIOLOGICAL ACTIVITIES/EFFECTS

In general, *Lamium* species has some ethnobotanical uses which have been confirmed by studies around the world. These ethnobotanical uses are mostly based on the plant's biological activities tested *in vitro* and *in vivo* assays of uterotonic, astringent, antispasmodic, anti-inflammatory, antinociceptive, antioxidant, antimicrobial, cytotoxic, and antischistosomal activities, as well as for pain relief in rheumatism and arthritis and as tonic for constipation (Salehi et al., 2019).

In a recent study, hemostatic activity and acute toxicity of butanolic extracts of aerial parts of *L. album* and *L. purpureum* were compared, revealing that the toxicity of *L. album* was quite low, confirming applicability in phytotherapeutic preparations (Bubueanu et al., 2019).

ANTIOXIDANT ACTIVITY

Antioxidant property and free radical scavenging activity of plant materials are assumed to have a potential in the prevention of various diseases (Bubueanu et al., 2013). *Lamium* species, which have been subjected to many biological activity studies and traditional uses, are also searched for antioxidant activities.

The antioxidant activities of four *Lamium* species wildly grown in Turkey were comparatively tested. As a result, the *n*-butanol extracts of *Lamium* species showed the highest free radical scavenging effect (Yalçın et al., 2007). Several antioxidant assays were carried out regarding *L. album* that has a wide traditional use. In another study, butanolic extract of the flowers of *L. album* showed stronger antioxidant capability than *L. purpureum* (Matkowski & Piotrowska, 2006). The quantitative determination of phenolic compounds by Folin-Ciocalteu method was applied to the *Lamium* species possessing dose-dependent scavenger activity. The activity is most likely due to the presence of a variety of flavonoids and other phenolics (Bubueanu et al., 2013). The antioxidant activity studies were generally performed by DPPH (2,2-diphenyl-1-picrylhydrazyl) radical scavenging activity, chemiluminescence, lipid peroxidation measurement, FRAP (Ferric Reducing Antioxidant Power), and ABTS (2,2'-azinobis (3-ethylbenzothiazoline-6-sulfonic acid)) for total antioxidant capacity as *in vitro* assays (Uwineza et al., 2021). Until now, free radical scavenging activity or antioxidant capacity were tested on *L. album*, *L. purpureum*, *L. amplexicaule*, *L. galactophyllum*, *L. macrodon*, and *L. maculatum* (Bubueanu et al., 2013; Danila et al., 2015; Erbil et al., 2014; Matkowski & Piotrowska, 2006; Uwineza et al., 2021; Yumrutas & Saygideger, 2010). It is obvious that the extraction methods of the aerial parts or flowers with methanol or butanol withdraw more phenolic compounds than water or non-polar extractions (Salehi et al., 2019).

Anti-inflammatory Activity

Several *Lamium* species have been used to relieve pain in arthritic ailments in Turkish folk medicine. To evaluate the anti-inflammatory and antinociceptive activities, some *Lamium* species of Turkish origin have been tested on carrageenan-induced hind paw edema model, PGE2-induced hind paw edema model, and 12-O-tetradecanoyl-13-acetate (TPA)-induced mouse ear edema model and for the antinociceptive activity *p*-benzoquinone (PBQ)-induced writhing test in mice. As a result *Lamium garganicum* subsp. *laevigatum* and *Lamium garganicum* subsp. *pulchrum* displayed remarkable anti-inflammatory and antinociceptive activities (Küpeli Akkol et al., 2008). As many anti-inflammatory studies also evaluated free radical scavenging action of the plant extracts, we observed that free oxygen radicals play an important role in inflammatory processes (Salehi et al., 2019). Ocular inflammation is associated with the high-level expression of pro-inflammatory cytokines (TNF-α, IL-6, IL-18), and anti-inflammatory cytokines such as IL-10 are not altered or reduced (Kelayeh et al., 2019). The different extracts of *L. album* were tested to evaluate the cytotoxic, anti-inflammatory, and antioxidant effects on cultivated human corneal epithelial cells *in vitro*. The ethanolic extract of *L. album* was considered to be non-toxic to human corneal epithelial cells at concentrations up to 125 µg/mL and found active against DPPH as free radical scavenging and in nitric oxide production assays as well. Especially flavonoid glycoside composition, phenolic acids, and iridoids in the ethanolic extract of *L. album* were considered to be responsible for the activities (Paduch & Wozniak, 2015). Both ethanol and ethyl acetate extracts decreased pro-inflammatory cytokine (IL-1β, IL-6, TNF-α) levels in human corneal epithelial cells in several concentrations but not affected so much on anti-inflammatory IL-10 production levels. According to the results of this study, *L. album* extracts need to be applied to *in vivo* tests and be introduced into practice as supplements or even medicine for the treatment of ocular diseases (Paduch & Wozniak, 2015).

Anti-inflammatory activity and cytokine secretion inhibition of prepared ethanolic extracts from both the aerial parts and flowers of *L. album* were performed in human neutrophils (Czerwinska et al., 2017). Both extracts inhibited reactive oxygen species production and decreased the levels of myeloperoxidase and cytokine (interleukin 8 and TNF-α) secretion in human neutrophils. Bioactivity-guided isolation of iridoids and phenylpropanoids has been conducted from the extracts for further comprehension of activities. Phenylpropanoids (verbascoside and phlinoside D), as well as iridoids (lamalbid and shanzhiside methyl ester), and flavonoids were revealed to be more significant inhibitors of IL-8 secretion than TNF-α. After further study on the anti-inflammatory activity of *L. album*, the authors agreed that this medicinal plant might be a valuable source of bioactive compounds and may provide constituents to limit noninfectious inflammation diseases (Czerwinska et al., 2018).

Antimicrobial, Antifungal, and Antiviral Activity

Antimicrobial activity of some *Lamium* plants have been performed and compared with each other (Chipeva et al., 2013; Erbil et al., 2014; Yalçın et al., 2007), but only *L. album* extracts were likely to possess a broad spectrum of antibacterial activity, especially towards gram-positive bacteria (Chipeva et al., 2013). Yet, the antibacterial activity results were moderate (Konarska et al., 2021). The antifungal activity of Turkish endemic *L. tenuiflorum* against *Candida* and *Cryptococcus* species was tested, and the need for further pharmacological evaluations was highlighted, since it has a greater antifungal effect against *Candida* sp. (Dulger, 2009).

Above-ground part of *L. album* (wild form) and *in vitro* propagated version of the same species are extracted with 18 different solvents and methods. It is reported that all the other extracts possessed some antibacterial activity, whereas *in vitro* propagated plant extracts are shown to possess a broader spectrum of antibacterial activity. The highest antifungal activity against *Candida albicans* NBIMCC 72 and *Candida glabrata* NBIMCC 8673 is seen with the water extracts from the leaves of the plant. The lowest MIC was demonstrated by the chloroform extracts (from *in vivo* flowers and leaves) towards *E. faecalis, S. aureus, P. hauseri*, and *P. aeruginosa* (Chipeva et al., 2013).

As a result of enzymatic hydrolysis during water maceration/extraction process of the flowering tops of *L. album*, lamiridosins, iridoid aglycone epimers of the glycoside "Lamalbid" (lamiridoside), were obtained. This epimeric mixture together with other aglycones of shanzhiside methyl ester, loganin, loganic acid, geniposide, verbenalin, eurostoside, and picroside II exhibited significant anti-HCV entry and anti-infectivity activities (Zhang et al., 2009).

Antidiabetic Activity

Among members of the genus *Lamium*, antidiabetic activity was focused solely on *L. album* until now (Kelayeh et al., 2019). A group of researchers tested the hydroalcoholic extract of *L. album* and *Urtica dioica* on glucose, lipids, and serum hepatic enzyme levels in streptozotocin-induced diabetic rats. They observed both extracts caused a significant reduction of serum glucose levels in diabetic rats. Also, compared to the diabetic control group, the extract of *L. album* significantly reduced cholesterol, alkaline phosphatase, aspartate transaminase, and alanine transaminase. The beneficial effects of two medicinal plants on serum liver enzymes can be due to the antioxidant and cytoprotective properties of these extracts (Mehran et al., 2015). Khanaki et al. (2017) have measured mitochondrial oxidative stress in diabetic rats and found that *L. album* at a dose of 100 mg/kg could not decrease mitochondrial ROS production from neutrophils in diabetic rats (Khanaki et al., 2017). There was no significant positive result about *L. album* extracts on diabetic rats in the other studies, so the authors agreed to the need for further studies about antidiabetic activity mechanisms (Abedinzade et al., 2019; Khanaki et al., 2019).

Other Biological Activities

Effect of some Lamiaceae plants against osteoarthritis was examined by matrix metalloproteinase-3 (MMP-3) inhibitory method, which showed that *L. purpureum* is one of the medicinal plants that has significant matrix metalloproteinase-3 (MMP-3) inhibitory effect (Tahravi, 2021).

Antiproliferative features of the essential oil of *L. purpureum* on the melanoma cancer cell line B16F10 was determined via MTT assay, in which the essential oil of the aerial parts of the plant showed a cytotoxic activity at a relatively high dose level that can be dedicated to especially the palmitic acid context of the essential oil (Akkoyunlu & Dulger, 2019).

The effects of the hydroalcoholic extracts of *L. album* and *Urtica dioica* extracts on the contraction of the trachea were evaluated *in vitro*. The polyphenol-rich plant *L. album* showed a strong relaxant effect, which might be induced by its phenolic compounds like rutin, chlorogenic acid, and caffeic acid derivatives (Arefani et al., 2018).

CONCLUSION

The Lamiaceae family has a large number of medicinal and aromatic plants known for their various uses ethnopharmacologically for centuries. *Lamium* species have many biologically active compounds to which the activities are dedicated. The iridoid glucosides were considered as chemotaxonomic markers of the genus *Lamium*. The data compiled herein provides a means to understand the recent developments in the pharmacology and phytochemistry of the genus. The literature information demonstrates the importance of *Lamium* species as medicinal plant regarding traditional use, chemical composition, biological and pharmacological activities, as well as industrial potential and economic use.

REFERENCES

Abedinzade, M., Rostampour, M., Mirzajani, E., Khalesi, Z. B., Pourmirzaee, T., & Khanaki, K. (2019). *Urtica dioica* and *Lamium album* decrease Glycogen Synthase Kinase-3 beta and increase K-Ras in diabetic rats. *Journal of Pharmacopuncture*, *22*(4), 248–252. https://doi.org/10.3831/KPI.2019.22.033

Akkoyunlu, A., & Dulger, G. (2019). Chemical composition of *Lamium purpureum* L. and determination of anticancer activity of its essential oil on melanoma. *Düzce Üniversitesi Bilim ve Teknoloji Dergisi*, *7*(3), 1755–1763.

Alipieva, K., Kokubun, T., Taskova, R., Evstatieva, L., & Handjieva, N. (2007). LC–ESI-MS analysis of iridoid glucosides in *Lamium* species. *Biochemical Systematics and Ecology*, *35*(1), 17–22.

Alipieva, K. I., Taskova, R. M., Jensen, S. R., & Handjieva, N. V. (2006). Iridoid glucosides from *Lamium album* and *Lamium maculatum* (Lamiaceae). *Biochemical Systematics and Ecology*, *34*(1), 88–91. https://doi.org/10.1016/j.bse.2005.04.002

Arefani, S., Mehran, S. M. M., Moladoust, H., Norasfard, M. R., Ghorbani, A., & Abedinzade, M. (2018). Effects of standardized extracts of *Lamium album* and *Urtica dioica* on rat tracheal smooth muscle contraction. *Journal of Pharmacopuncture*, *21*(2), 70–75. https://doi.org/10.3831/KPI.2018.21.008

Baran, P., & Ozdemir, C. (2013). Morphological, anatomical and cytological studies on endemic *Lamium pisidicum*. *Pakistan Journal of Botany*, *45*(1), 73–85.

Baytop, T. (1999). *Türkiye'de bitkiler ile tedavi: geçmişte ve bugün*. Nobel Tıp Kitabevleri.

Bubueanu, C., Campeanu, G., Pirvu, L., & Bubueanu, G. (2013). Antioxidant activity of butanolic extracts of Romanian native species–*Lamium album* and *Lamium purpureum*. *Romanian Biotechnological Letters*, *18*(6), 8855–8862.

Bubueanu, C., Iuksel, R., & Panteli, M. (2019). Haemostatic activity of butanolic extracts of *Lamium album* and *Lamium purpureum* aerial parts. *Acta Pharm*, *69*(3), 443–449. https://doi.org/10.2478/acph-2019-0026

Celep, F. (2017). *Lamium bilgilii* (Lamiaceae), a new species from South-western Turkey (Burdur-Mugla). *Phytotaxa*, *312*(2), 263–270. https://doi.org/10.11646/phytotaxa.312.2.9

Chipeva, V. A., Petrova, D. C., Geneva, M. E., Dimitrova, M. A., Moncheva, P. A., & Kapchina- Toteva, V. M. (2013). Antimicrobial activity of extracts from in vivo and in vitro propagated *Lamium album* L. plants. *African Journal of Traditional, Complementary and Alternative Medicines*, *10*(6), 559–562. https://doi.org/10.4314/ajtcam.v10i6.30

Czerwinska, M. E., Kalinowska, E., Popowski, D., & Bazylko, A. (2020). Lamalbid, Chlorogenic acid, and Verbascoside as tools for standardization of *Lamium album* flowers-development and validation of HPLC-DAD method. *Molecules*, *25*(7), 1721. https://doi.org/10.3390/molecules25071721

Czerwinska, M. E., Swierczewska, A., & Granica, S. (2018). Bioactive constituents of *Lamium album* L. as inhibitors of cytokine secretion in human neutrophils. *Molecules*, *23*(11), 2770. https://doi.org/10.3390/molecules23112770

Czerwinska, M. E., Swierczewska, A., Wozniak, M., & Kiss, A. K. (2017). Bioassay-guided isolation of Iridoids and Phenylpropanoids from aerial parts of *Lamium album* and their anti-inflammatory activity in human neutrophils. *Planta Medica*, *83*(12–13), 1011–1019. https://doi.org/10.1055/s-0043-107031

Danila, D., Adriana, T., Camelia, S. P., Valentin, G., & Anca, M. (2015). Antioxidant activity of methanolic extracts of *Lamium album* and *Lamium maculatum* species from wild populations in the Romanian Eastern Carpathians [Congress Abstract]. *Planta Medica*, *81*(16), PW_192. https://www.thieme- connect.com/products/ejournals/abstract/10.1055/s-0035-1565816

Dulger, B. (2009). Antifungal activity of *Lamium tenuiflorum* against some medical yeast *Candida* and *Cryptococcus* species. *Pharmaceutical Biology*, *47*(5), 467–470.

Erbil, N., Alan, Y., & Diğrak, M. (2014). Antimicrobial and antioxidant properties of *Lamium galactophyllum* Boiss & Reuter, *L. macrodon* Boiss & Huet and *L. amplexicaule* from Turkish Flora. *Asian Journal of Chemistry*, *26*(2), 549–554.

Ersöz, T., Kaya, D., Yalcin, F. N., Kazaz, C., Palaska, E., Gotfredsen, C. H., Jensen, S. R., & Çaliş, İ. (2007). Iridoid glucosides from *Lamium garganicum* subsp. *laevigatum*. *Turkish Journal of Chemistry*, *31*(2), 155–162.

Flamini, G., Cioni, P. L., & Morelli, I. (2005). Composition of the essential oils and in vivo emission of volatiles of four Lamium species from Italy: *L. purpureum*, *L. hybridum*, *L. bifidum* and *L. amplexicaule*. *Food Chemistry*, *91*(1), 63–68. https://doi.org/10.1016/j.foodchem.2004.05.047

Frezza, C., Venditti, A., Serafini, M., & Bianco, A. (2019). Chapter 4 – Phytochemistry, chemotaxonomy, ethnopharmacology, and Nutraceutics of Lamiaceae. In A. U. Rahman (Ed.), *Studies in Natural Products Chemistry* (Vol. 62, pp. 125–178). Elsevier. https://doi.org/10.1016/B978-0-444-64185-4.00004-6

Grujić, S. M., Savković, Ž. D., Ristić, M. S., Džamić, A. M., Ljaljević-Grbić, M. V., Vukojević, J. B., & Marin, P. D. (2020). Glandular trichomes, essential oil composition, anti- Aspergillus and antioxidative activities of *Lamium purpureum* L. ethanolic extracts. *Archives of Biological Sciences*, *72*(2), 253–263.

Güler, B., Manav, E., & Uğurlu, E. (2015). Medicinal plants used by traditional healers in Bozüyük (Bilecik–Turkey). *Journal of Ethnopharmacology*, *173*, 39–47. https://doi.org/10.1016/j.jep.2015.07.007

Ito, N., Nihei, T., Kakuda, R., Yaoita, Y., & Kikuchi, M. (2006). Five new Phenylethanoid glycosides from the whole plants of *Lamium purpureum* L. *Chemical and Pharmaceutical Bulletin*, *54*(12), 1705–1708. https://doi.org/10.1248/cpb.54.1705

Jones, C. (2012). A chemical ecological investigation of the allelopathic potential of *Lamium amplexicaule* and *Lamium purpureum*. *Open Journal of Ecology*, *2*, 167–177. https://doi.org/10.4236/oje.2012.24020

Kargıoğlu, M., Cenkci, S., Serteser, A., Konuk, M., & Vural, G. (2010). Traditional uses of wild plants in the middle Aegean region of Turkey. *Human Ecology*, *38*(3), 429–450.

Kelayeh, T. P. S., Abedinzade, M., & Ghorbani, A. (2019). A review on biological effects of *Lamium album* (white dead nettle) and its components. *Journal of Herbmed Pharmacology*, *8*(3), 185–193. https://doi.org/10.15171/jhp.2019.28

Khanaki, K., Abedinzade, M., Gazor, R., Norasfard, M., & Jafari-Shakib, R. (2017). Effect of *Lamium album* on mitochondrial oxidative stress in diabetic rats. *Research in Molecular Medicine*, *5*(2), 9–13.

Khanaki, K., Abedinzade, M., & Hamidi, M. (2019). The effects of *Urtica dioica* and *Lamium album* extracts on the expression level of cyclooxygenase-2 and caspase-3 in the liver and kidney of streptozotocin-induced diabetic rats. *Pharmaceutical Sciences*, *25*(1), 37–43. https://doi.org/10.15171/ps.2019.6

Kikuchi, M., Onoguchi, R., & Yaoita, Y. (2009). Three new monoterpene glucosides from *Lamium amplexicaule*. *Helvetica Chimica Acta*, *92*(10), 2063–2070. https://doi.org/10.1002/hlca.200900158

Konarska, A., Weryszko-Chmielewska, E., Matysik-Wozniak, A., Sulborska, A., Polak, B., Dmitruk, M., Piotrowska-Weryszko, K., Stefanczyk, B., & Rejdak, R. (2021). Histochemical and phytochemical analysis of *Lamium album* subsp. *album* L. corolla: Essential oil, Triterpenes, and Iridoids. *Molecules*, *26*(14), 4166. https://doi.org/10.3390/molecules26144166

Küpeli Akkol, E., Yalçın, F. N., Kaya, D., Çalış, İ., Yesilada, E., & Ersöz, T. (2008). In vivo anti-inflammatory and antinociceptive actions of some *Lamium* species. *Journal of Ethnopharmacology*, *118*(1), 166–172. https://doi.org/10.1016/j.jep.2008.04.001

Matkowski, A., & Piotrowska, M. (2006). Antioxidant and free radical scavenging activities of some medicinal plants from the Lamiaceae. *Fitoterapia*, *77*(0367-326X (Print)), 346–353.

Mehran, M. M., Norasfard, M. R., Abedinzade, M., & Khanaki, K. (2015). *Lamium album* or *Urtica dioica*? Which is more effective in decreasing serum glucose, lipid and hepatic enzymes in streptozotocin induced diabetic rats: A comparative study. *African Journal of Traditional, Complementary and Alternative Medicines*, *12*(5), 84–88.

Mennema, J. (1989). *A taxonomic revision of Lamium* (Lamiaceae). In E. J. Brill Archive, (Vol 11). Leiden Botanical Series.

Mill, R. R. (1982). *Lamium* L. Flora of Turkey and the East Aegean Islands. In P. H. Davis (Ed.) Vol. 7, 126–148). University Press.

Morteza-Semnani, K., Saeedi, M., & Akbarzadeh, M. (2016). Chemical composition of the essential oil of the flowering aerial parts of *Lamium album* L. *Journal of Essential Oil Bearing Plants*, *19*(3), 773–777. https://doi.org/10.1080/0972060X.2016.1168321

Mulas, M. (2006). Traditional uses of Labiatae in the Mediterranean area. *I. International Symposium on the Labiatae: Advances in Production, Biotechnology and Utilisation 723*, 25–32. https://doi.org/10.17660/ActaHortic.2006.723.1

Nickavar, B., Mojab, F., & Bamasian, S. (2008). Volatile components from aerial parts of *Lamium amplexicaule* from Iran. *Journal of Essential Oil Bearing Plants*, *11*(1), 36–40. https://doi.org/10.1080/0972060X.2008.10643594

Paduch, R., & Wozniak, A. (2015). The effect of *Lamium album* extract on cultivated human corneal epithelial cells (10.014 pRSV-T). *Journal of Ophthalmic and Vision Research*, *10*(3), 229–237. https://doi.org/10.4103/2008-322X.170349

Pereira, O., Domingues, M. R., Silva, A., & Cardoso, S. (2012). Phenolic constituents of *Lamium album*: Focus on isoscutellarein derivatives. *Food Research International*, *48*, 330–335.

Sajjadi, S. E., & Ghannadi, A. (2012). Analysis of the essential oil of *Lamium amplexicaule* L. from Northeastern Iran. *Journal of Essential Oil Bearing Plants*, *15*(4), 577–581. https://doi.org/10.1080/0972060X.2012.10644091

Salehi, B., Armstrong, L., Rescigno, A., Yeskaliyeva, B., Seitimova, G., Beyatli, A., Sharmeen, J., Mahomoodally, M. F., Sharopov, F., & Durazzo, A. (2019). *Lamium* plants—A comprehensive review on health benefits and biological activities. *Molecules*, *24*(10), 1913.

Sulborska, A., Konarska, A., Matysik-Wozniak, A., Dmitruk, M., Weryszko-Chmielewska, E., Skalska-Kaminska, A., & Rejdak, R. (2020). Phenolic constituents of *Lamium album* L. subsp. *album* flowers: Anatomical, histochemical, and phytochemical study. *Molecules*, *25*(24). https://doi.org/10.3390/molecules25246025

Tahravi, L. (2021). *Characterization of the effects of some Lamiaceae family plants against osteoarthritis by Matrix Metalloproteinase-3 (MMP-3) inhibitory effect determination* [Master thesis, Hacettepe University Graduate School of Health Sciences, Pharmacognosy Program].

Tetik, F., Civelek, S., & Cakilcioglu, U. (2013). Traditional uses of some medicinal plants in Malatya (Turkey). *Journal of Ethnopharmacology*, *146*(1), 331–346. https://doi.org/10.1016/j.jep.2012.12.054

Uwineza, P. A., Gramza-Michałowska, A., Bryła, M., & Waśkiewicz, A. (2021). Antioxidant activity and bioactive compounds of *Lamium album* flower extracts obtained by supercritical fluid extraction. *Applied Sciences*, *11*(16), 7419. https://www.mdpi.com/2076-3417/11/16/7419

Valyova, MS, Dimitrova, MA, Ganeva, YA, Kapchina-Toteva, VM, & Yordanova, ZP (2011). Evaluation of antioxidant and free radical scavenging potential of *Lamium album* L. growing in Bulgaria. *J Pharm Res*, 4(4), 945–7.

Yalçın, F. N., Duygu, K., Kılıç, E., Özalp, M., Ersöz, T., & Çalış, İ. (2007). Antimicrobial and free radical scavenging activities of some *Lamium* species from Turkey. *Hacettepe University Journal of the Faculty of Pharmacy*, (1), 11–22.

Yalçın, F. N., & Kaya, D. (2006). Ethnobotany, pharmacology and phytochemistry of the genus *Lamium* (Lamiaceae). *FABAD Journal of Pharmaceutical Sciences*, *31*(1), 43. http://dergi.fabad.org.tr/pdf/volum31/issue1/43-52.pdf

Yordanova, Z. P., Zhiponova, M. K., Iakimova, E. T., Dimitrova, M. A., & Kapchina-Toteva, V. M. (2014). Revealing the reviving secret of the white dead nettle (*Lamium album* L.). *Phytochemistry Reviews*, *13*(2), 375–389.

Yumrutas, O., & Saygideger, S. D. (2010). Determination of in vitro antioxidant activities of different extracts of *Marrubium parviflorum* Fish et Mey. and *Lamium amplexicaule* L. from South east of Turkey. *Journal of Medicinal Plants Research*, *4*(20), 2164–2172.

Zhang, H., Rothwangl, K., Mesecar, A. D., Sabahi, A., Rong, L., & Fong, H. H. (2009). Lamiridosins, hepatitis C virus entry inhibitors from Lamium album. *Journal of Natural Products*, *72*(12), 2158–2162. https://doi.org/10.1021/np900549e

12 *Laurus nobilis* L.

Ceylan Dönmez

INTRODUCTION

Turkey is located in a unique geographical position. It has a landmass straddling both Asia and Europe. It has borders with Georgia, Armenia, Azerbaijan, and Iran in the east, Iraq and Syria to the southeast, and Greece and Bulgaria to the west. All the four seasons are experienced in the geographical regions of Turkey. The climate is milder along the shores and in the Aegean region (Ergener, 2002). In accordance with this temperate Mediterranean climate, *Laurus nobilis* L. grows well in large quantities (Baytop, 1999). *L. nobilis*, belongs to Lauraceae family, grows naturally in Turkey, Italy, Greece, North West Africa, and western sections of Libya and Syria (Demiriz, 1982). *L. nobilis* is commonly found on the coast of Aegean, Mediterranean, and Black Sea in Turkey (Figure 12.1). *L. nobilis* is a valuable plant, which has industrial importance and is used in foods, drugs, and cosmetics. Turkey is one of the most important laurel producers and exporting countries in the world. In the past 14 years, approximately five times increase has been achieved in production. While Turkey's laurel export revenue was approximately 12 million dollars in 2005, its export revenue in 2019 has increased to 40 million dollars. Supporting production by preventing unconscious use has greatly contributed to this increase (TUIK, 2020; OGM, 2021). The history, traditional uses, phytochemical constituents, pharmacological activities, and points to be considered in the use of *L. nobilis* will be presented in detail in this chapter.

HISTORY AND BOTANICAL FEATURES

The name of "daphne" originates from Greek mythology. According to the myth, Daphne is the daughter of the River Peneus and Earth. Apollo had fallen in love with her. Daphne's mother Earth turned her into a tree to protect her from the attentions of Apollo. It was sacred to Apollo. Apollo cuts a branch from Daphne's new appearance and wears it on his head as a wreath to console himself. From then on, he became known as Daphnephoros (Margaris, 2000). Daphne (bay or laurel) is the only member of the Lauraceae that has persisted to the present in southern Eurasia, in spite of a significant number of genera (e.g., *Cinammomum*, *Persea*, *Lindera*, …etc.) recorded in the Mio-Pliocene (Rodríguez-Sánchez et al., 2009).

The general botanical features of the *Laurus nobilis* may be summarized as follows: It is a shrub or small tree (about 2–20 meters in height), with slender, glabrous twigs. Leaf is oblong-lanceolate, acute or acuminate, glabrous, and have 5–10 x 2–7.5 cm size. Its flowering time is from February to April. Flower color is white or inconspicuous (Figure 12.2). Male flowers have eight to 12 stamens, most of them with two glands at the base. Female flowers have two to four staminodes. Fruits are 1–1.5 cm, ovoid, black when ripe (Mifsud, 2002).

TRADITIONAL USES

The dried leaves and essential oils are used extensively in the food industry. Laurel leaves are often used as a folk medicine for asthma, cardiac disorders, digestive diseases, healing constipation, and rheumatism (Bruni et al., 1997; Loi et al., 2004). The aqueous plant extract is used in Turkish folk medicine as a diuretic, antirheumatic, and antihemorrhoidal, as an antidote for snakebites, and for

 DOI: 10.1201/9781003146971-12

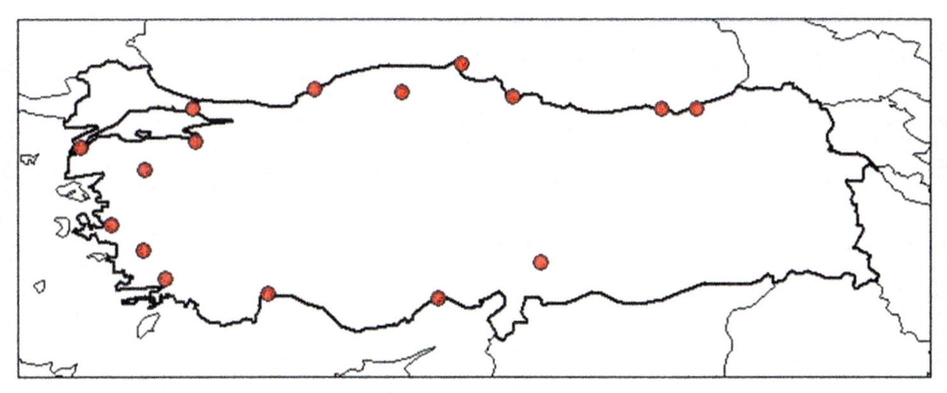

FIGURE 12.1 The distributions of laurel in Turkey (TUBIVES, 2021).

FIGURE 12.2 Laurel tree. (Photographed by Prof. Dr. Ayşegül Köroğlu.)

the treatment of stomachache (Baytop, 1999). Laurel leaves eliminate bad breath in the mouth and increase digestive secretions (Erden, 2005). It is used in rheumatism and dermatitis, for increasing appetite, facilitating digestion, and as deworming agent (Baytop, 1999; Kilic et al., 2004). Leaves have germicidal, mucolytic, antiviral, antifungal, antibacterial, muscle relaxant, and insecticidal properties. Mouthwash made from its leaves is effective in preventing tooth decay (Schnaubelt, 1999; Wyk and Wink, 2004). The laurel fruits have antiparasitic influence against parasites such as lice, fleas, ticks, etc., diuretic, relieving rheumatic pains (Zeybek and Zeybek, 1994; Baytop, 1999). The fruits contain both fixed and volatile oils, which are mainly used in soap making, while soap made from fruits has a curative effect on skin fungi and slows hair loss (Bozan ve Karakaplan, 2007). Also, when the bark of the root of the laurel is boiled and drunk, kidney stones and urinary disorders are cured because of diuretic and emmenagogue effects (Ali-Shtayeh et al., 2000).

PHARMACOLOGICAL ACTIVITIES

Acaricidal Effect

Acaricidal activity of laurel leaf oils was investigated against *Psoroptes cuniculi*. While the essential oil significantly reduced the activity of *P. cuniculi* to 51% at a concentration of 5%, it caused a mite's mortality rate of 73% at a concentration of 10% (Macchioni et al., 2006).

Analgesic and Anti-inflammatory Effects

Analgesic effect of the laurel leaf essential oil was evaluated in tail-flick (at 0.03 mL/kg dose) and formalin tests (at 0.25 mL/kg dose) on mice and rats. Anti-inflammatory activity of the laurel leaf essential oil was evaluated in the formalin-induced edema tests at 0.20 mL/kg dose. The essential oil showed significant, dose-dependent anti-inflammatory and sedative effects and analgesic activity (Sayyah et al., 2003). Anti-inflammatory activity of aqueous and ethanolic extracts obtained from leaves and seeds of laurel was also evaluated by using carrageenan-induced hind paw edema model in mice without inducing any gastric damage. It was determined that ethanolic seed extract showed a significant anti-inflammatory effect (Kupeli et al., 2007).

Anticholinergic Effect

Acetyl cholinesterase (AChE) enzyme inhibitory activities of essential oil, ethanolic extract, and decoction of laurel leaf were analyzed. It was shown that the essential oil (at 0.5 mg/mL), decoction (at 5 mg/mL), and ethanolic fraction (at 1 mg/mL) exhibited high AChE inhibition as about 51%, 56%, and 64%, respectively (Ferreira et al., 2006).

Anticonvulsant Effect

The essential oil of laurel was investigated on mice with tonic convulsions induced by maximal electroshock and especially by pentylenetetrazol. The essential oil showed anticonvulsant effect against experimental seizures. It was speculated that compounds responsible for the anticonvulsant activity might be methyl eugenol, eugenol, and pinene in the essential oil. While the essential oil produced sedation and motor impairment at anticonvulsant doses (1 mL/kg) (Sayyah et al., 2002).

Antifungal Effect

The laurel extract, which was obtained by supercritical carbon dioxide technique, was tested on seven strains of plant pathogenic fungi cultivated in Petri plates and at three different concentrations (50, 125, and 250 μg/mL). The results showed that laurel extract has the greatest antifungal effect on the fungus *Botrytis cinerea* Pers at 250 μg/mL concentration (Corato et al., 2007).

Antihyperlipidemic/Hypolipidemic Effects

The ethanolic laurel extract was tested for inhibitory activity against recombinant human lanosterol synthase, and it showed high activity. Another compound, eremanthine, obtained from methanolic laurel extract also showed potent activity (Tanaka et al., 2006). Aqueous laurel extract was investigated for its *in vitro* anti-atherosclerotic activity and *in vivo* hypolipidemic effects in a zebrafish model. The results showed that it suppressed the incidence of atherosclerosis with strong antioxidant potential, prevention of apolipoprotein A-I glycation and LDL phagocytosis, and inhibition of cholesteryl ester transfer protein (Jin et al., 2011).

Antimicrobial Effect

The antimicrobial activity of the essential oils of laurel was investigated at three different concentrations (1, 10, and 15%) on various microorganisms (*Salmonella typhimurium*, *Bacillus cereus*, *Staphylococcus aureus*, *Enterococcus faecalis*, *Escherichia coli*, *Yersinia enterocolitica*, *Saccharomyces cerevisiae*, *Candida rugosa*, *Rhizopus oryzae*, and *Aspergillus niger*). The results showed that the essential oil has antimicrobial effects against *B. cereus* and *E. faecalis* at three different concentrations (Ozcan ve Erkmen, 2001). The antimicrobial activity of 1, 8-cineole, which is a major compound of laurel essential oil, was evaluated against foodborne pathogens. The study showed that this compound has a high antibacterial effect (Dadalıoglu and Evrendilek, 2004). The extract of laurel leaves was evaluated against *S. aureus*, *B. subtilis*, *Pseudomonas aeruginosa*, *E. coli*, *C. albicans*, and *A. niger* and results showed that the extract exhibited antimicrobial effect for *S. aureus* (Santoyo et al., 2006).

In vitro antimycobacterial activity of whole laurel oil and its fractions and two sesquiterpene lactones (costunolide and dehydrocostuslactone) were determined against mycobacterial strains and clinical isolates by using the fluorometric Alamar Blue microassay. Antimycobacterial activity against drug-resistant *Mycobacterium tuberculosis* clinical isolates was better for the mixture than for individual compounds costunolide and dehydrocostuslactone (Luna-Herrera et al., 2007). Liu *et al.* reported that two kaempferol glycosides isolated from laurel, kaempferol-3-*O*-α -L-(2″,4″-di-*E*-*p*-coumaroyl)-rhamnoside and kaempferol-3-*O*-α -L-(2″-*E*-*p*-coumaroyl-4″-Z-p-coumaroyl)-rhamnoside, showed strong antibacterial activities against methicillin-resistant *Staphylococcus aureus* and vancomycin-resistant *Enterococci* (Liu et al., 2009). When antimicrobial and antioxidant activities of the essential oil, seed oil, and methanolic extract of seed oil obtained from laurel were compared *in vitro*, the methanolic extract of seed oil exhibited the most effective antibacterial activity (Ozcan et al., 2010). Laurel leaf extracts and infusions were evaluated in terms of the antioxidant activity such as scavenging activity, reducing power, and lipid peroxidation inhibition. Both wild and cultivated (mostly infusion) plant extracts showed high antioxidant effects (Dias et al., 2014a).

In a study which investigated the antibacterial effect of a mixture of laurel (fruits and leaves) oil and black cumin (seed) oil (1:1), it was proven that *Staphylococcus aureus* bacteria were sensitive to the blank discs impregnated with the oil mixture (Sarikaya et al., 2017).

Antimutagenic Effect

The compound 3-kaempferyl-p-coumarate isolated from the ethyl acetate extract of laurel was evaluated for mutagenicity. It was shown that this compound has antimutagenic activity against the dietary carcinogen 3-amino-1-methyl-5 H-pyrido [4, 3-b] indole (Trp-P-2) (Samejima et al., 1998).

Antioxidant Effect

It was determined that the methanolic laurel leaf extract and sesquiterpene lactones (especially costunolide, dehydrocostus lactone, eremanthine, zaluzanin C, magnolialide, santamarine, and spirafolide) isolated from the extract inhibited nitric oxide production in lipopolysaccharide (LPS)-activated mouse peritoneal macrophages (Matsuda et al., 2000). Antioxidant activities of essential oil, ethanolic extract, and decoction of laurel leaf were investigated using 2,2-diphenyl-1-picryl-hydrazyl (DPPH) and β-Carotene-linoleic acid assay. It was shown that all of them exhibited high antioxidant effects (Ferreira et al., 2006; Santoyo et al., 2006). Elmastas *et al.* demonstrated that aqueous and ethanolic laurel leaf extracts have high antioxidant activity. They evaluated this activity by reducing power, free radical scavenging, superoxide anion radical scavenging, hydrogen peroxide scavenging, and metal chelating activities (Elmastas et al., 2006). Besides, hydrodistilled extract from laurel was investigated in terms of antioxidant activity. Site- and non-site-specific techniques were used, and laurel shown the highest antioxidant activity (aside from iron chelation) in all of them. Iron(III) reduction, prevention of linoleic acid peroxidation, iron(II) chelation, DPPH, and hydroxyl radical-mediated 2-deoxy d-ribose degradation, were also investigated. (Hinneburg et al., 2006). Antioxidant activity of free volatile aglycones from laurel was compared to laurel essential oil. According to DPPH and ferric reducing/antioxidant power assay (FRAP) methods, volatile aglycones from laurel showed lower reducing power and free radical scavenging ability compared to essential oil as well as reference butylated hydroxytoluene (BHT) (Politeo et al., 2007).

Ozcan *et al.* showed that the 50% (IC_{50}) inhibition effect of the essential oil on the free radical DPPH was determined as 94.655 mg/mL. The inhibition value of the methanolic extract of seed oil was quite close to the synthetic antioxidant BHT, 92.46% inhibition (Ozcan et al., 2010). Laurel leaf extract and its major compounds exhibited inhibitory activities on LPS-induced microglial activation (Chen et al., 2014).

Anti-ulcerogenic Effect

Anti-ulcerogenic activity of *L. nobilis* seeds was tested using *in vivo* ethanol-induced gastric ulcer model. According to the results, 20 and 40% aqueous-ethanolic seed extract and oily fraction (obtained from ethanolic extract) showed anti-ulcerogenic activity (Afifi et al., 1997).

Antiviral Effect

The inhibitory activity of laurel oil was investigated against SARS-CoV and HSV-1 replication *in vitro* by visually scoring the virus-induced cytopathogenic effect post-infection. From the results, the oil showed antiviral activity against SARS-CoV with an IC_{50} value of 120 µg/mL (Loizzo et al., 2008).

Anti-ulcer Effect

Gastroprotective effects of the laurel fruit extracts (aqueous and methanolic) were evaluated on Sprague Dawley rats with ethanol-induced ulcerogenesis model. All stomach samples of animals treated with methanolic laurel extract were completely protected from severe damage (Gürbüz et al., 2002).

Cytotoxic Effect

When different *L. nobilis* leaves extracts (with n-hexane, ethanolic, and aqueous) were evaluated for cytotoxic properties using the Brine shrimp bioassay, the n-hexane extract only exhibited cytotoxic activity (Kivçak and Mert, 2002). Cytotoxicity of laurel essential oil and its constituents was investigated in amelanotic melanoma C32, renal cell adenocarcinoma ACHN, hormone-dependent prostate carcinoma LNCaP, and MCF-7 breast cancer cell lines using the sulforhodamine B (SRB) method. Laurel fruit oil showed the highest activity on C32 and ACHN, while α-humulene exhibited high effect on LNCaP cells (Loizzo et al., 2007). The potential antitumor effect of laurel leaf extract was tested on murine fibrosarcoma L929sA cells, human breast cancer cells MDA-MB231, and human breast cancer cells MCF7. The extract showed remarkable cytotoxic and immunomodulatory NFκB activities (Kaileh et al., 2007). The three extracts prepared from flowers, leaves, and fruits of *L. nobilis* were tested for both ovarian cytotoxic activity and DNA damaging properties. Among the extracts, the most cytotoxic active extract against ovarian cancer cell line was found to be the fruit extract with 98% inhibition. Among all tested extracts, only the fruit extract displayed marginal inhibition towards DNA repair-deficient yeast strain (pRAD52 Gal). Sesquiterpene lactones (costunolide, gazaniolide, reynosin, spirafolide santamarine, and 11,13-dehydrosantonin) obtained from laurel fruit were all determined to be highly cytotoxic against the A2780 ovarian cancer cell line except lauroxepine (Barla et al., 2007). Aqueous and methanolic laurel extracts were tested on MCF7 (breast adenocarcinoma), NCI-H460 (non-small cell lung cancer), HCT15 (colon carcinoma), HeLa (cervical carcinoma), and HepG2 (hepatocellular carcinoma). Both extracts displayed significant antitumoral activities (Dias et al., 2014b). In another study, ethanolic laurel leaf extract exhibited antiproliferative activity against MCF7 cells (Al-Kalaldeh et al., 2010). Cytotoxic effect of magnolialide, isolated from the laurel leaves, was tested on the rat basophilic leukemia mast cell line (RBL-2H3) *in vitro.* It inhibited the IL-4 mRNA expression with an IC_{50} value of 15.7 µM in IgE-sensitized RBL-2H3 cells (Lee et al., 2013).

Immunostimulant Effect

Immunostimulant effects of laurel powder were investigated on rainbow trout by dietary intake. After 14 days of adaptation, experimental diets containing 0.5 and 1% laurel leaf powder were administered to rainbow trout for 3 weeks. The fish were then switched back to the control diet. Non-specific immunity parameters as well as extracellular and intracellular respiratory burst activities, phagocytosis in blood leukocytes, lysozyme, and total plasma protein level were investigated at the end of 3 weeks and then later again on days 42 and 63. The nonspecific immune parameters were evaluated. However, this study showed that laurel has no effect on the immune system (Bilen and Bulut, 2010).

Insect Repellent Effect

The repellent activity of laurel leaf essential oil was investigated. In the study, the adult females of *Culex pipiens* were preferred by scientists as they are usually the most common pest mosquito in urban and suburban settings in the Antalya province in Turkey. According to the result, the essential oil showed repellent activity (Erler et al., 2006).

Neuroprotective Effect

Chloroform fraction of laurel leaf was investigated against cerebral ischemia neuronal damage (CIND). The chloroform fraction (at 4 mg/kg) significantly induced infarct dimensions by 79% of vehicle control in the middle cerebral artery occlusion *in vivo* model and inhibited death-associated protein kinase dephosphorylation (Cho et al., 2010). In another study, neuroprotective effect of *n*-hexane fraction of *L. nobilis* leaves was investigated on dopamine-induced intracellular reactive oxygen species production and apoptosis in human neuroblastoma SH-SY5Y cells. Compared with a positive control (apomorphine, IC_{50} = 18.1 M), the IC_{50} value of the fraction was found to be 3.0 g/mL for dopamine-induced apoptosis. Also, IC_{50} values of costunolide and dehydrocostus lactone, which were isolated from the fraction, were determined as 7.3 M and 3.6 M, respectively. It was shown that both hexane fraction and these major compounds significantly inhibited reactive oxygen species generation in dopamine-induced SH-SY5Y cells. In another experiment, *in vivo* 6-hydroxydopamine (6-OHDA) model was used in rats for evaluating Parkinson's disease. According to immunohistochemical analysis, results demonstrated that hexane fraction inhibited 6-hydroxydopamine-induced tyrosine hydroxylase-positive neurons in the substantia nigra and also decreased dopamine-induced α-synuclein formation in SH-SY5Y cells (Ham et al., 2011).

Wound Healing Effect

Wound healing effect of aqueous laurel leaf extract was evaluated by using excision and incision wound models on Sprague Dawley rats. The extract was administered to rats (200 mg/kg body weight/day) for two weeks (on excision model) and ten days (on incision model). In this study, which was compared with the *Allamanda cathartica* treated group of animals, laurel extract has lower therapeutic activity while both extracts significantly improved wound healing activity (Nayak et al., 2006).

PHYTOCHEMICAL CONSTITUENTS

Leaves of *L. nobilis* are known to contain essential oils, non-polar flavonoids, monoterpenes, sesquiterpenoid lactones, and isoquinoline alkaloids (Demo et al., 1998; Kang et al., 2002; Matsuda et al., 2002; Fang et al., 2005; Macchioni et al., 2006; Loizzo et al., 2008; Custodio and Veiga Junior, 2014). Fruits include essential oil components, flavonoids, anthocyanins, and lipids (Fiorini et al., 1998; PDR for Herbal Medicine, 2000; Longo and Vasapollo, 2005) (Figure 12.3).

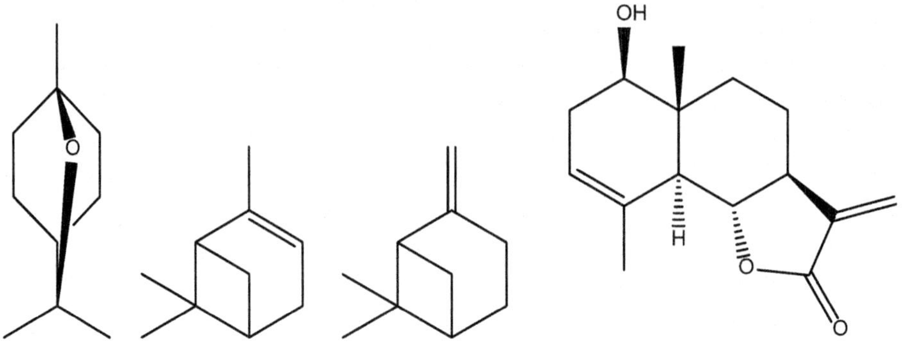

1,8-cineol α-pinene β-pinene santamarine

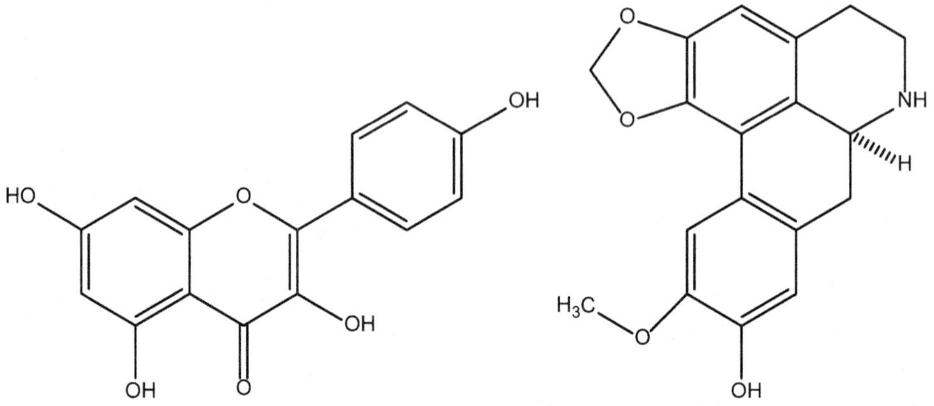

kaempferol actinodaphinine

FIGURE 12.3 Molecular structure of some phytochemical compounds of laurel.

Terpenoids and Essential Oil

According to most experimental results, 1,8-cineole is the major compound of laurel leaf (Ozcan and Chalchat, 2005). Laurel leaves contain *α* -eudesmol, *α* -humulene, *α* -pinene, *α* -phellandrene, *α* -terpinyl acetate, *α* -terpineol, *α* -terpinene, *β*-eudesmol, *β*-pinene, p-cymene, γ-terpinene, terpinene-4-ol, bornyl acetate, calamenene, camphene, limonene, linalool, myrcene, eugenol, methyl eugenol, sabinene, cis-/trans-sabinene hydrate, cis-/trans-2-thujene-ol, eudesmane lactones, and methyl esters (Kosar et al., 2005; Dall'Acqua et al., 2006; Sangun et al., 2007; Al-Kalaldeh et al., 2010; Julianti et al., 2012; Tomar et al., 2020).

Sesquiterpene lactones such as lauroxepine, costunolide, gazaniolide, reynosin, spirafolide santamarine, and 11,13-dehydrosantonin were isolated from methanolic fruit extract (Barla et al., 2007). *α* -pinene, *α* -terpinyl acetate, *β*-pinene, *β*-caryophyllene, and trans-*β*-*O*-simene were isolated from the leaves and fruits of laurel (Kovacevic et al., 2007; Sangun et al., 2007).

Flower essential oil's components are 1,8-cineole, *α* -pinene, α-farnesene, *α* -phellandrene, *α* -terpinyl acetate, *α* -thujene, *β*-pinene, *β*-caryophyllene, *β*-elemene, γ-muurolene, p-cymene, cis-α-bisbolene, trans-*β*-ocimene, (*E*)-ocimene, bornyl acetate, camphene, germacrene-A/-D, germacradienol, linalool, myrcene, sabinene, and viridiflorene (Fiorini et al., 1999; Kovacevic et al., 2007; Sangun et al., 2007).

Fatty acid components from laurel seed oil are caprinic acid, lauric acid, myristic acid, palmitic acid, linoleic acid, stearic acid, and eicosanoic acid (arachidic acid) (Ozcan et al., 2010). Methyleugenol, α -terpinyl acetate, 1,8-cineole, α -pinene, β-pinene, sabinene, and linalool are the bark oil constituents (Kovacevic et al., 2007). The twig essential oil contains 1,8-cineole, α -terpinyl acetate, methyl eugenol, β-linalool, β-pinene, sabinene, and terpinene-4-ol (Fidan et al., 2019).

Flavonoids and Anthocyanins

Flavone glycoside from laurel leaf and isoquercitrin and kaempferol from laurel fruit were isolated as flavonoids (Fiorini et al., 1998; Kang et al., 2002; De Marino et al., 2004). Anthocyanins such as cyanidin-3-*O*-glucoside, cyanidin-3-*O*-rutinoside, peonidine-3-*O*-rutinoside, and peonidine-3-*O*-glucoside were isolated from fruits of *L. nobilis* (Longo and Vasapollo, 2005).

Alkaloids

Actinodaphnine, boldine, reticuline, neolitsine, isodomesticine cryptodorine, launobine, *N*-methylactinodaphinine, nandigerine, and norisodomesticine were isolated from laurel leaves. Actinodaphnine, launobine, and reticuline from the twig; actinodaphnine, launobine, nandigerine, and reticuline from the roots (Pech and Bruneton, 1982; Custodio and Veiga Junior, 2014).

POINTS TO BE CONSIDERED IN THE USES OF *L. NOBILIS*

It has been reported that leaves and essential oils of *L. nobilis* cause allergic reactions such as contact dermatitis on the skin (PDR for Herbal Medicine, 2000; Adişen and Önder, 2007). It is a herb that can be used both internally and externally. For internal use, 5–10% infusion of leaves and 5–10% decoction of fruits are preferred for treatment. Essential oil and fruit oil are used in pomad and soap (Baytop, 1999; PDR for Herbal Medicine, 2000).

CONCLUSION

Turkey is one of the largest laurel exporters. Therefore, *L. nobilis* is a very important Mediterranean plant globally as well as in Turkey. 1,8-cineol, a major constituent in leaf, is responsible for the characteristic odor of laurel (Pino et al., 1993). In addition, measures should be taken to protect the national market product laurel that has an important place in the world market. When we look at the world trend, developed countries like the USA and EU countries increase their quality criteria day by day. As Turkey also has an increasing trend in medicinal and aromatic plants (MAPs) production, the sector should develop production processes and add more value by creating product varieties of MAPs in accordance with world criteria (Pakdemirli et al., 2021). This chapter will hopefully remind scientists and producers in terms of the protection and sustainability of this precious plant Laurel.

REFERENCES

Adişen, E., Önder, M. 2007. Allergic contact dermatitis from *Laurus nobilis* oil induced by massage. *Contact Dermatitis*, 56(6), 360–361.

Afifi, F. U., Khalil, E., Tamimi, S. O., Disi, A. 1997. Evaluation of the gastroprotective effect of *Laurus nobilis* seeds on ethanol induced gastric ulcer in rats. *Journal of Ethnopharmacology*, 58(1), 9–14.

Ali-Shtayeh, M. S., Yaniv, Z., Mahajna, J. 2000. Ethnobotanical survey in the Palestinian area: A classification of the healing potential of medicinal plants. *Journal of Ethnopharmacology*, 73(1–2), 221–232.

Al-Kalaldeh, J. Z., Abu-Dahab, R., Afifi, F. U. 2010. Volatile oil composition and antiproliferative activity of *Laurus nobilis*, *Origanum syriacum*, *Origanum vulgare*, and *Salvia triloba* against human breast adenocarcinoma cells. *Nutrition Research*, 30(4), 271–278.

Barla, A., Topçu, G., Öksüz, S., Tümen, G., Kingston, D. G. 2007. Identification of cytotoxic sesquiterpenes from *Laurus nobilis* L. *Food Chemistry*, 104(4), 1478–1484.

Baytop, T. 1999. *Therapy with Medicinal Plants in Turkey (Past and Present)*. Nobel Tip Press, Istanbul University.

Bilen, S., Bulut, M. 2010. Effects of laurel (*Laurus nobilis*) on the non-specific immune responses of rainbow trout (Oncorhynchus mykiss, Walbaum). *Journal of Animal and Veterinary Advances*, 9(8), 1275–1279.

Bozan, B., Karakaplan, U. 2007. Antioxidants from laurel (*Laurus nobilis* L.) berries: Influence of extraction procedure on yield and antioxidant activity of extracts. *Acta Alimentaria*, 36(3), 321–328.

Bruni, A., Ballero, M., Poli, F. 1997. Quantitative ethnopharmacological study of the Campidano Valley and Urzulei district, Sardinia, Italy. *Journal of Ethnopharmacology*, 57(2), 97–124.

Chen, H., Xie, C., Wang, H., Jin, D. Q., Li, S., Wang, M., Ren, Q., Xu, J., Ohizumi, Y., Guo, Y. 2014. Sesquiterpenes inhibiting the microglial activation from *Laurus nobilis*. *Journal of Agricultural and Food Chemistry*, 62(20), 4784–4788.

Cho, E. Y., Lee, S. J., Nam, K. W., Shin, J., Oh, K. B., Kim, K. H., Mar, W. 2010. Amelioration of oxygen and glucose deprivation-induced neuronal death by chloroform fraction of bay leaves (*Laurus nobilis*). *Bioscience, Biotechnology, and Biochemistry*, 74(10), 2029–2035.

Corato, D. U., Trupo, M., Leone, G. P., Sanzo, D. G., Zingarelli, G., Adami, M. 2007. Antifungal activity of the leaf extracts of laurel (*Laurus nobilis* L.), orange (Citrus sinensis Osbeck) and olive (Olea europaea L.) obtained by means of supercritical carbon dioxide technique. *Journal of Plant Pathology*, 89(3), 83–91.

Custodio, D. L., da Veiga Junior, V. F. 2014. Lauraceae alkaloids. *RSC Advances*, 4(42), 21864–21890.

Dadalıoglu, I., Evrendilek, G. A. 2004. Chemical compositions and antibacterial effects of essential oils of Turkish oregano (*Origanum minutiflorum*), bay laurel (*Laurus nobilis*), spanish lavender (*Lavandula stoechas* L.), and fennel (*Foeniculum vulgare*) on common foodborne pathogens. *Journal of Agricultural and Food Chemistry*, 52, 8255–8260.

Dall'Acqua, S., Viola, G., Giorgetti, M., Loi, M. C., Innocenti, G. 2006. Two new sesquiterpene lactones from the leaves of *Laurus nobilis*. *Chemical and Pharmaceutical Bulletin*, 54(8), 1187–1189.

De Marino, S., Borbone, N., Zollo, F., Ianaro, A., Di Meglio, P., Iorizzi, M. 2004. Megastigmane and phenolic components from *Laurus nobilis* L. leaves and their inhibitory effects on nitric oxide production. *Journal of Agricultural and Food Chemistry*, 52(25), 7525–7531.

Demiriz, H. 1982. Laurus L. In: *Flora of Turkey, and the East Aegean Islands*. Edinburgh University Press, Edinburgh, Vol. 7, 534–535.

Demo, A., Petrakis, C., Kefalas, P., Boskou, D. 1998. Nutrient antioxidants in some herbs and Mediterranean plant leaves. *Food Research International*, 31(5), 351–354.

Dias, M. I., Barros, L., Dueñas, M., Alves, R. C., Oliveira, M. B. P., Santos-Buelga, C., Ferreira, I. C. 2014a. Nutritional and antioxidant contributions of *Laurus nobilis* L. leaves: Would be more suitable a wild or a cultivated sample?. *Food Chemistry*, 156, 339–346.

Dias, M. I., Barreira, J., Calhelha, R. C., Queiroz, M. J. R., Oliveira, M. B. P. P., Soković, M., Ferreira, I. C. 2014b. Two-dimensional PCA highlights the differentiated antitumor and antimicrobial activity of methanolic and aqueous extracts of *Laurus nobilis* L. from different origins. *BioMed Research International*, 1–10.

Elmastas, M., Gülçin, I., Işildak, Ö., Küfrevioğlu, Ö. İ., İbaoğlu, K., Aboul-Enein, H. Y. 2006. Radical scavenging activity and antioxidant capacity of bay leaf extracts. *Journal of the Iranian Chemical Society*, 3(3), 258–266.

Erden, U. 2005. *Seasonal in Mediterranean Laurel (Laurus nobilis l.) Investigation of Variability and Optimal Drying Methods [Akdeniz Defnesi (Laurus nobilis l.)'nde Mevsimsel Varyabilite ve Optimal Kurutma Yöntemlerinin Araştırılması.]*. Çukurova University, Graduate School of Natural and Applied Sciences, Master Thesis.

Ergener, R. 2002. *About Turkey: Geography, Economy, Politics, Religion, and Culture*. Pilgrims Process, United States of America.

Erler, F., Ulug, I., Yalcinkaya, B. 2006. Repellent activity of five essential oils against Culex pipiens. *Fitoterapia*, 77(7–8), 491–494.

Fang, F., Sang, S., Chen, K. Y., Gosslau, A., Ho, C. T., Rosen, R. T. 2005. Isolation and identification of cytotoxic compounds from Bay leaf (*Laurus nobilis*). *Food Chemistry*, 93(3), 497–501.

Ferreira, A., Proença, C., Serralheiro, M. L. M., Araujo, M. E. M. 2006. The *in vitro* screening for acetylcholinesterase inhibition and antioxidant activity of medicinal plants from Portugal. *Journal of Ethnopharmacology*, 108(1), 31–37.

Fidan, H., Stefanova, G., Kostova, I., Stankov, S., Damyanova, S., Stoyanova, A., Zheljazkov, V. D. 2019. Chemical composition and antimicrobial activity of *Laurus nobilis* L. essential oils from Bulgaria. *Molecules*, 24(4), 804.

Fiorini, C., David, B., Fourasté, I., Bessière, J. M. 1999. Composition of the flower, leaf and stem essential oils from *Laurus nobilis*. *Flavour and Fragrance Journal*, 12(2), 91–93.

Fiorini, C., David, B., Fourasté, I., Vercauteren, J. 1998. Acylated kaempferol glycosides from *Laurus nobilis* leaves. *Phytochemistry*, 47(5), 821–824.

Gürbüz, İ., Üstün, O., Yeşilada, E., Sezik, E., Akyürek, N. 2002. *In vivo* gastroprotective effects of five Turkish folk remedies against ethanol-induced lesions. *Journal of Ethnopharmacology*, 83(3), 241–244.

Ham, A., Shin, J., Oh, K. B., Lee, S. J., Nam, K. W., Uk Koo, K. H. K., Mar, W. 2011. Neuroprotective effect of the n-Hexane extracts of Laurus nobilis L. in models of Parkinson' disease. *The Korean Society of Applied Pharmacology*, 19(1), 118–125.

Hinneburg, I., Dorman, H. D., Hiltunen, R. 2006. Antioxidant activities of extracts from selected culinary herbs and spices. *Food Chemistry*, 97(1), 122–129.

Jin, S., Hong, J. H., Jung, S. H., Cho, K. H. 2011. Turmeric and laurel aqueous extracts exhibit *in vitro* anti-atherosclerotic activity and *in vivo* hypolipidemic effects in a zebrafish model. *Journal of Medicinal Food*, 14(3), 247–256.

Julianti, E., Jang, K. H., Lee, S., Lee, D., Mar, W., Oh, K. B., Shin, J. 2012. Sesquiterpenes from the leaves of *Laurus nobilis* L. *Phytochemistry*, 80, 70–76.

Kaileh, M., Berghe, W. V., Boone, E., Essawi, T., Haegeman, G. 2007. Screening of indigenous Palestinian medicinal plants for potential anti-inflammatory and cytotoxic activity. *Journal of Ethnopharmacology*, 113(3), 510–516.

Kang, H. W., Yu, K. W., Jun, W. J., Chang, I. S., Han, S. B., Kim, H. Y., Cho, H. Y. 2002. Isolation and characterization of alkyl peroxy radical scavenging compound from leaves of *Laurus nobilis*. *Biological and Pharmaceutical Bulletin*, 25(1), 102–108.

Kilic, A., Hafizoglu, H., Kollmannsberger, H., Nitz, S. 2004. Volatile constituents and key odorants in leaves, buds, flowers, and fruits of *Laurus nobilis* L. *Journal of Agricultural and Food Chemistry*, 52(6), 1601–1606.

Kivçak, B., Mert, T. 2002. Preliminary evaluation of cytotoxic properties of *Laurus nobilis* leaf extracts. *Fitoterapia*, 73(3), 242–243.

Kosar, M., Tunalier, Z., Özek, T., Kürkcüoglu, M., Baser, K. H. C. 2005. A simple method to obtain essential oils from *Salvia triloba* L. and *Laurus nobilis* L. by using microwave-assisted hydrodistillation. *Zeitschrift für Naturforschung C*, 60(5–6), 501–504.

Kovacevic, N. N., Simic, M. D., Ristic, M. S. 2007. Essential oil of *Laurus nobilis* from Montenegro. *Chemistry of Natural Compounds*, 43(4), 408–411.

Kupeli, E., Orhan, I., Yesilada, E. 2007. Evaluation of some plants used in Turkish folk medicine for their anti-inflammatory and antinociceptive activities. *Pharmaceutical Biology*, 45(7), 547–555.

Lee, T., Lee, S., Kim, K. H., Oh, K. B., Shin, J., Mar, W. 2013. Effects of magnolialide isolated from the leaves of *Laurus nobilis* L. (Lauraceae) on immunoglobulin E-mediated type I hypersensitivity *in vitro*. *Journal of Ethnopharmacology*, 149(2), 550–556.

Liu, M. H., Otsuka, N., Noyori, K., Shiota, S., Ogawa, W., Kuroda, T., Tsutomu, H., Tsuchiya, T. 2009. Synergistic effect of kaempferol glycosides purified from *Laurus nobilis* and fluoroquinolones on methicillin-resistant *Staphylococcus aureus*. *Biological and Pharmaceutical Bulletin*, 32(3), 489–492.

Loi, M. C., Poli, F., Sacchetti, G., Selenu, M. B., Ballero, M. 2004. Ethnopharmacology of ogliastra (villagrande strisaili, sardinia, Italy). *Fitoterapia*, 75(3–4), 277–295.

Loizzo, M. R., Saab, A. M., Tundis, R., Statti, G. A., Menichini, F., Lampronti, I., Gambari, R., Cinatl, J., Doerr, H. W. 2008. Phytochemical analysis and *in vitro* antiviral activities of the essential oils of seven Lebanon species. *Chemistry & Biodiversity*, 5(3), 461–470.

Loizzo, M. R., Tundis, R., Menichini, F., Saab, A. M., Statti, G. A., Menichini, F. 2007. Cytotoxic activity of essential oils from Labiatae and Lauraceae families against *in vitro* human tumor models. *Anticancer Research*, 27(5A), 3293–3299.

Longo, L., Vasapollo, G. 2005. Anthocyanins from bay (*Laurus nobilis* L.) berries. *Journal of Agricultural and Food Chemistry*, 53(20), 8063–8067.

Luna-Herrera, J, Costa, MC, Gonzalez, HG, Rodrigues, AI, & Castilho, PC (2007). Synergistic antimycobacterial activities of sesquiterpene lactones from Laurus spp. *J. Antimicrob. Chemother*, 59(3), 548–552.

Macchioni, F., Perrucci, S., Cioni, P., Morelli, I., Castilho, P., Cecchi, F. 2006. Composition and acaricidal activity of Laurus novocanariensis and *Laurus nobilis* essential oils against Psoroptes cuniculi. *Journal of Essential Oil Research*, 18(1), 111–114.

Margaris, N. S. 2000. Flowers in Greek mythology. *Acta Horticulturae*, 541, 23–29. https://doi.org/10.17660/actahortic.2000.541.1.

Matsuda, H., Kagerura, T., Toguchida, I., Ueda, H., Morikawa, T., Yoshikawa, M. 2000. Inhibitory effects of sesquiterpenes from bay leaf on nitric oxide production in lipopolysaccharide-activated macrophages: Structure requirement and role of heat shock protein induction. *Life Sciences*, 66(22), 2151–2157.

Matsuda, H., Shimoda, H., Ninomiya, K., Yoshikawa, M. 2002. Inhibitory mechanism of costunolide, a sesquiterpene lactone isolated from *Laurus nobilis*, on blood-ethanol elevation in rats: Involvement of inhibition of gastric emptying and increase in gastric juice secretion. *Alcohol and Alcoholism*, 37(2), 121–127.

Mifsud, S. 2002. *Laurus nobilis* (Bay Laurel). MaltaWildPlants. com-the online Flora of the Maltese Islands [http://www.maltawildplants.com/LAUR/Laurus_nobilis.php, Access date: 29/06/2021].

Nayak, S., Nalabothu, P., Sandiford, S., Bhogadi, V., Adogwa, A. 2006. Evaluation of wound healing activity of Allamanda cathartica. L. and *Laurus nobilis*. L. extracts on rats. *BMC Complementary and Alternative Medicine*, 6(1), 1–6.

OGM. 2021 [https://www.tarimorman.gov.tr/Haber/4560/Tarim-Ve-Orman-Bakanligi-2019da-32-Bin-600-Ton-Defne-Uretimi-Gerceklestirdi%E2%80%A6, Access date: 30/06/2021].

Ozcan, B., Esen, M., Sangun, M. K., Coleri, A., Caliskan, M. 2010. Effective antibacterial and antioxidant properties of methanolic extract of *Laurus nobilis* seed oil. *Journal of Environmental Biology*, 31(5), 637–641.

Ozcan, M., Chalchat, J. C. 2005. Effect of different locations on the chemical composition of essential oils of laurel (*Laurus nobilis* L.) leaves growing wild in Turkey. *Journal of Medicinal Food*, 8(3), 408–411.

Ozcan, M., Erkmen, O. 2001. Antimicrobial activity of the essential oils of Turkish plant spices. *European Food Research and Technology*, 212(6), 658–660.

Pakdemirli, B., Birişik, N., Akay, M. 2021. General overview of medicinal and aromatic plants in Turkey. *Anatolian Journal of Aegean Agricultural Research Institute* [Anadolu Ege Tarımsal Araştırma Enstitüsü Dergisi], 31(1), 126–135.

PDR for Herbal Medicine. 2000. 2nd edition, Thomson Medical Economics, Montvale, NJ.

Pech, B., Bruneton, J. 1982. Alcaloïdes du laurier noble, *Laurus nobilis*. *Journal of Natural Products*, 45(5), 560–563.

Pino, J., Borges, P., Roncal, E. 1993. The chemical composition of laurel leaf oil from various origins. *Food/Nahrung*, 37(6), 592–595.

Politeo, O., Jukić, M., Miloš, M. 2007. Chemical composition and antioxidant activity of free volatile aglycones from laurel (*Laurus nobilis* L.) compared to its essential oil. *Croatica Chemica Acta*, 80(1), 121–126.

Rodríguez-Sánchez, F., Guzmán, B., Valido, A., Vargas, P., Arroyo, J. 2009. Late Neogene history of the laurel tree (Laurus L., Lauraceae) based on phylogeographical analyses of Mediterranean and Macaronesian populations. *Journal of Biogeography*, 36(7), 1270–1281.

Samejima, K., Kanazawa, K., Ashida, H., Danno, G. I. 1998. Bay laurel contains antimutagenic kaempferyl coumarate acting against the dietary carcinogen 3-amino-1-methyl-5 H-pyrido [4, 3-b] indole (Trp-P-2). *Journal of Agricultural and Food Chemistry*, 46(12), 4864–4868.

Sangun, M. K., Aydin, E., Timur, M., Karadeniz, H., Caliskan, M., Ozkan, A. 2007. Comparison of chemical composition of the essential oil of *Laurus nobilis* L. leaves and fruits from different regions of Hatay, Turkey. *Journal of Environmental Biology*, 28(4), 731–733.

Santoyo, S., Lloria, R., Jaime, L., Ibanez, E., Senorans, F. J., Reglero, G. 2006. Supercritical fluid extraction of antioxidant and antimicrobial compounds from *Laurus nobilis* L. Chemical and functional characterization. *European Food Research and Technology*, 222(5), 565–571.

Sarikaya, B., Alperen, H. H., Akgün, S. 2017. Antibacterial effect of pressed oil mixtures of Laurel (*Laurus nobilis*) and Black Cumin (*Nigella sativa*) plants [Defne (*Laurus nobilis*) ve Çörekotu (*Nigella sativa*) Bitkilerinin Pres Yağ karışımlarının Antibakteriyal Etkisi]. *Commagene Journal of Biology* [Kommagene Biyoloji Dergisi], 1(1), 57–59.

Sayyah, M., Saroukhani, G., Peirovi, A., Kamalinejad, M. 2003. Analgesic and anti-inflammatory activity of the leaf essential oil of *Laurus nobilis* Linn. *Phytotherapy Research*, 17(7), 733–736.

Sayyah, M., Valizadeh, J., Kamalinejad, M. 2002. Anticonvulsant activity of the leaf essential oil of *Laurus nobilis* against pentylenetetrazole-and maximal electroshock-induced seizures. *Phytomedicine*, 9(3), 212–216.

Schnaubelt, K. 1999. *Medical Aromateraphy. Healing with Essential Oils*. Frog Ltd. Berkeley, California, pp. 213–214.

Tanaka, R., Sakano, Y., Shimizu, K., Shibuya, M., Ebizuka, Y., Goda, Y. 2006. Constituents of *Laurus nobilis* L. inhibit recombinant human lanosterol synthase. *Journal of Natural Medicines*, 60(1), 78–81.

Tomar, O., Akarca, G., Gok, V., Ramadan, M. F. 2020. Composition and antibacterial effects of laurel (*Laurus nobilis* L.) leaves essential oil. *Journal of Essential Oil Bearing Plants*, 23(2), 414–421.

TUBIVES. 2021 [http://194.27.225.161/yasin/tubives/index.php?sayfa=1&tax_id=8237, Access date: 29/06/2021].

TUIK. 2020 [https://data.tuik.gov.tr/Kategori/GetKategori?p=DisTicaret-104, Access date: 30/06/2021].

Wyk, B. E., Wink, M. 2004. *Medicinal Plants of the World*. Timber Press, Portland, Oregon, p. 188.

Zeybek, N., Zeybek, U. 1994. *Pharmaceutical botanical Angiospermae systematics and important substances [Farmasotik Botanik Kapali Tohumlu Bitkiler (Angiospermae) Sistematigi ve Onemli Maddeleri]*. Ege University Faculty of Pharmacy Publication [Ege Universitesi Eczacilik Fakultesi Yayini], Izmir, Turkey.

13 *Lavandula* sp.

Gökşen Dilşat Durbilmez Üstün and Dudu Altıntaş

INTRODUCTION

Medicinal and aromatic plants have been used for thousands of years to treat diseases (Bousta, 2020). In recent years, the importance of aromatic plants has increased with the spread of aromatherapy science (Esposito, 2014). There are approximately 12,000 plant taxa in Türkiye (Şenkul, 2017).

Lavandula species are naturally found especially in Mediterranean countries. It is cultivated in Southern France, Spain, England, Italy, Hungary, Yugoslavia, Crimea, Moldavia the Caucasus, India, North Africa, South Africa, the USA, and Argentina (Kızılay, 2022).

Lavandula species belonging to the Lamiaceae family have been reported in Flora of Türkiye to classify as basically three taxa in Türkiye. These taxa are *Lavandula stoechas* L. subsp. *stoechas*, *Lavandula pedunculata* subsp. *cariensis* (Boiss.) Upson & S. Andrews, and *L. angustifolia* Mill. subsp. *angustifolia* (Küçük, 2019). *Lavandula stoechas* L. shows natural distribution in the country (Oraloğlu, 2019).

Lavandula angustifolia is, cultivated in the country, locally called "Lavender". *L. stoechas* L. is called as "Karabaş otu" and *L. pedunculata* is called as "Karan" (Küçük, 2019; Salih, 2019).

Traditionally, in Kuyucak village of Keçiborlu district and its surroundings, lavender (Lavandula x intermedia var. Super A), which has a hybrid structure, is cultivated in the province of Isparta. Lavenders are grown in the Kuyucak region contain high levels of volatile oil; however, the oil has a high level of camphor when compared to the European Pharmacopoeia standards (Kara, 2011). The distribution of lavender species in Türkiye is shown in Figure 13.1, Figure 13.2, and Figure 13.3.

Lavandula species are applied in cosmetics, perfumery, food industry, and agricultural control due to their odor properties (Kızılay, 2022).

In this chapter, we aimed to examine the lavender species grown in Türkiye in terms of ethnobotanical, phytochemical content, and biological activities.

BOTANICAL FEATURES/ETHNOBOTANICAL USAGE

In *Lavandula* species, the calyx and corolla are tubular, and the corolla has 5 lobes at the apex. The number of stamens is 4, the filaments are short (Tanker, 2007)

Lavandula species are generally used in brain diseases, bronchitis, and flu and as antispasmodic, antiseptic, and expectorant among the people in Türkiye. The drug prepared in the form of infusion of the plant is used in brain diseases in Malatya region (Tetik, 2011; Bozyel, 2020).

L. angustifolia **L.** is a plant in the form of a subshrub that grows in the Mediterranean region known as medicinal lavender, and is widely cultivated worldwide (Costea, 2019). The plant is in the form of a shrub, which does not naturally grow in Türkiye, has narrow leaves, and the upper lip of the corolla is 2-lobed and straight (Tanker, 2007). This plant is used in ethnomedical studies, insomnia, stress, diuretic, hair and skin care, as a gas reliever, and in diseases such as stomach disorders (Korkmaz, 2014; Güler, 2015; Ugulu, 2009).

L. stoechas L. is a 40–50 cm tall shrub that grows in Western and Southern Anatolia as well as in Thrace (Tanker, 2007). It is popularly used as an antiseptic, pain reliever, dizziness, gastrointestinal diseases, antiurolytic, wound healing, stimulant, antiepileptic, anti-asthmatic, expectorant, sedative,

DOI: 10.1201/9781003146971-13

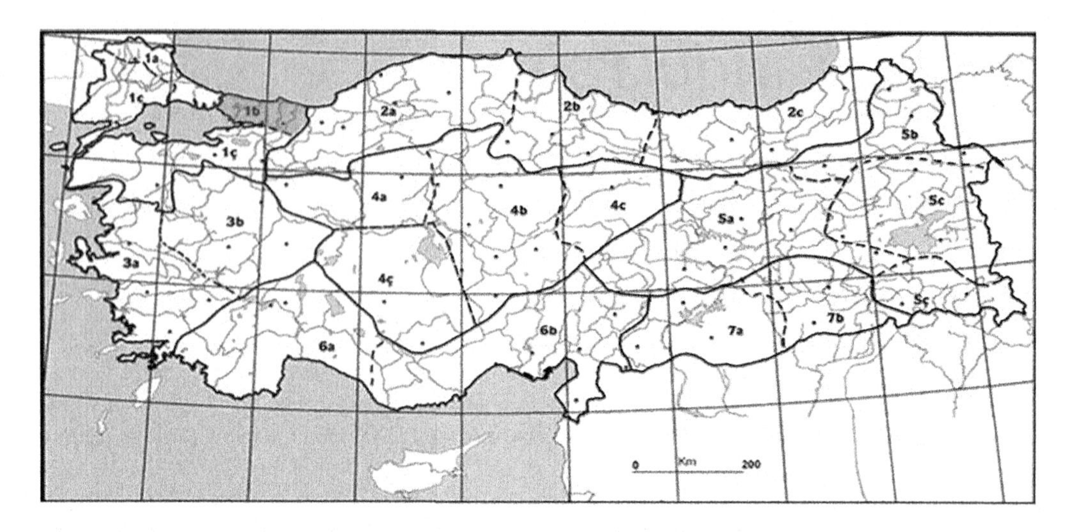

FIGURE 13.1 *Lavandula angustifolia* Mill. distribution of the species on the map of Türkiye. (1b) Çatalca-Kocaeli region (Dirmenci, 2012).

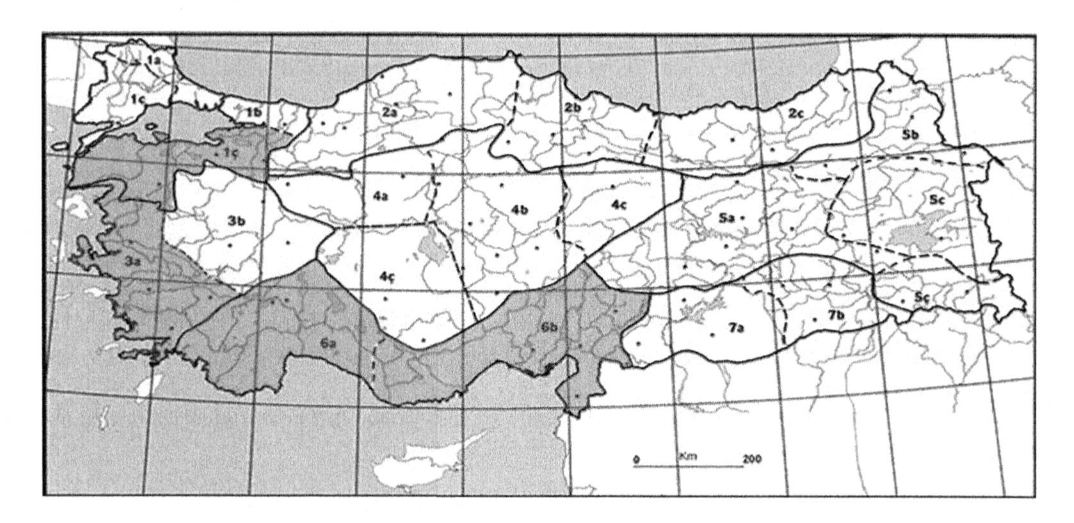

FIGURE 13.2 *Lavandula stoechas* L. distribution. (1ç) Southern Marmara region, (3a) Main Aegean region, (6a) Antalya region, (6b) Adana region (Dirmenci, 2012).

and as a heart strengthener (Sargın, 2013; Akbulut, 2013; Fakir, 2009; Bozyel, 2018). The aerial parts of *L. stoechas* are used for external pain. *Lavandula stoechas* L. subsp. *stoechas* subspecies is used ethnobotanically by preparing the infusion of the spike part in case of headache and vascular occlusion in and around Balıkesir. *L. stoechas* subsp. *stoechas* L. is popularly applied in cardiovascular disorders, high cholesterol, menstrual irregularity, stomach diseases, insomnia, headache; moreover, it has scientifically proven effects as analgesic, sedative, hypotensive, expectorant, and antihypertensive (Gürdal, 2013; Güzel, 2015; Selvi, 2013; Güner, 2016; Polat, 2012).

***L. pedunculata* subsp. *cariensis* (Boiss.) Upson & S. Andrews** L. grows in the arid hills of Southwest Anatolia. It is very similar to *L. stoechas*, but its inflorescence is long-stalked, making it easy to distinguish it from other species. The flower states of these two species are used interchangeably for the same purposes, especially in the form of tea (Oraloğlu, 2019). This subspecies, which is generally obtained in 2–5% infusions or by distillation method, has effects such as pain reliever,

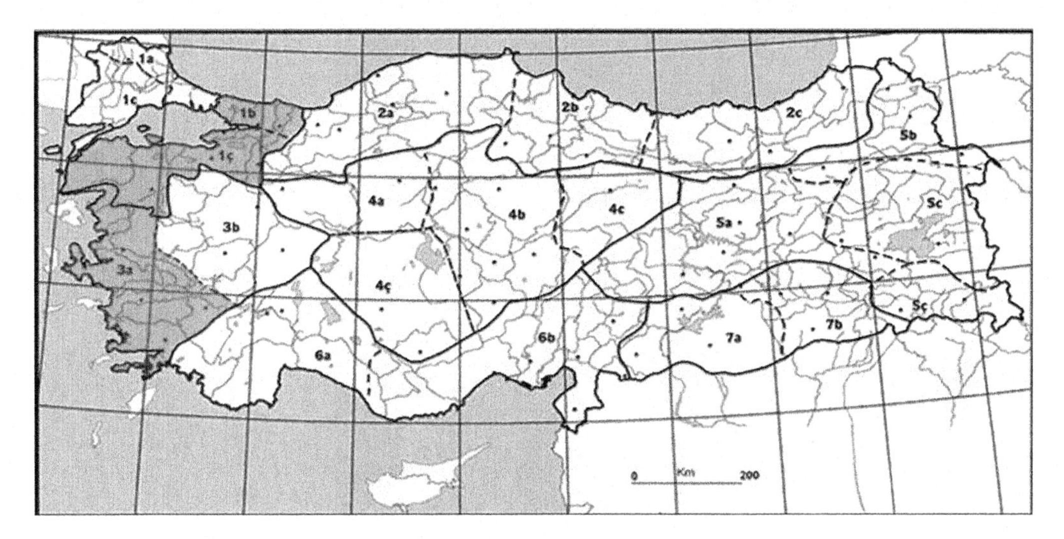

FIGURE 13.3 *Lavandula pedunculata* distribution. (1b) Çatalca-Kocaeli region, (1ç) Southern Marmara region, (3a) Main Aegean region (Dirmenci, 2012).

antiseptic, wound healing, soothing expectorant, relieving urinary tract inflammation, healing eczema wounds, and strengthening nerve and heart and has found a wide range of uses (Oraloğlu, 2018).

PHYTOCHEMICALS

In a study, *Lavandula* species contain 548 chemical components. The non-volatile compounds of *Lavandula* species consist of flavonoids, hydroxycinnamic acid, and triterpenoids. Its volatile components are hydrocarbons and monoterpenes (Ayaz, 2020). When the volatile oil of *L. angustifolia* L. plant was analyzed by GC-MS, 27 compounds were determined. Major compounds in the volatile oil are linalool, 1,8-cineol, camphor, linalyl acetate, borneol, α-terpineol, γ-muurolene, and α-bisabolol, respectively (Moussi, 2017).

HPLC-DAD-MS analysis of *L. stoechas* extract has demonstrated flavones, including di-glucuronide, which are determined as apigenin C- and di-C-hexoside, apigenin 7-O-glucoside, genkwanin (apigenin 7-methyl ether), luteolin 7-O-glucoside, luteolin 7-O-glucuronide, and luteolin. (Héral, 2021).

Another HPLC analysis of the methanolic extract of *Lavandula stoechas* has shown that isolated compounds such as rutin, caffeic acid, and rosmarinic acid may be partially responsible for the antioxidant activity of the methanolic extract of the plant (Ceylan, 2015). *Lavandula stoechas* L. subsp. *stoechas* in a study with stoechas, some compounds were seen as a result of GC/MS analysis of the hydrodistilled oil of the plant. The main components of the volatile oil were camphor (45.8%), α-fencon (31.8%), and bornyl acetate (4.2%) (Küçük, 2019).

L. pedunculata volatile oil was analyzed by GC and GC/MS. As a result of the analysis, it was shown that it contains highly oxygenated monoterpenes. It contains 1,8-cineol (2.4–55.5%), fencon (1.3–59.7%), and camphor (3.6–48.0%) as main compounds (Zuzarte, 2009).

BIOLOGICAL ACTIVITIES/EFFECTS

Studies have shown that *L. angustifolia* L. volatile oil may have antioxidant, antimicrobial, and antispasmodic effects on the central nervous system, anxiety, anxiolytic activity, and sleep (Prusinowska, 2014). The pharmacological effect of *L. stoechas* essential oils and extracts has been investigated in many studies. The antibacterial, antifungal, insecticidal, anti-leishmanial, antioxidant, and anti-inflammatory properties of these essential oils were evaluated (Bousta, 2020).

FIGURE 13.4 *Lavandula stoechas* by Ahmet Karataş.

A study with a hydroalcoholic extract of *L. stoechas* demonstrated an anti-inflammatory effect in the intestines, confirming their potential use as herbal remedies for gastrointestinal disorders (Algieri, 2016).

When the effect of the ethanolic extract *of Lavandula stoechas* L. on *Staphylococcus aureus* was examined in the study conducted by Sales *et al.*, in 2014, it was observed that the MIC and MBC values were 6.5 and 13 mg/mL, respectively. Thus, it was concluded that *Lavandula stoechas* L. may be antibacterial (Sales, 2014).

The inhibitory effect of the volatile oil obtained from the flower part of *Lavandula stoechas* plant against fungi was analyzed. According to the results obtained from the oil of lavender flower, *Botrytis cinerea* was the mostly affected fungi, followed by *Alternaria alternata* and *Fusarium oxysporum*. As a result, it was concluded that the dose of the volatile oil at 40 ppm completely inhibited mycelial growth of *B. cinerea*. In the study, it was observed that lavender oil showed fungistatic activity at different doses (Özcan, 2018).

The comparative activities of plant extracts of volatile oils at different concentrations against the fourth instar larva of *Culex pipiens molestus* were investigated. The results obtained showed that *Myrtus communis* extracts had the highest effect, while *L. stoechas* was the least effective (Traboulsi, 2002).

The antileishmanial effects of *L. stoechas* volatile oil on *Leishmania major*, *Leishmania tropica*, and *Leishmania infantum* were investigated. The dose that inhibited 50% of the parasite cells, namely the concentration of *L. stoechas* volatile oil, was determined. It was observed that the volatile oil of *L. stoechas* was effective against *Leishmania* species when compared to the positive control. *Leishmania major* was determined as the most sensitive strain, with $IC_{50} = 0.9 \pm 0.45$ μg/mL (Bouyahya, 2017).

The antioxidant activity of the ethanol extract of *L. stoechas* was determined by comparing it with the DPPH method against the BHT standard (Ezzoubi, 2014).

In addition, the results of study showed that the extract of *L. stoechas* prepared in water has cytotoxic and genotoxic activity (Çelik and Aslantürk, 2007).

A significant antifungal effect against dermatophyte strains was observed in a study among oils of *Lavandula pedunculata* species. The volatile oil with the highest camphor content is the most active with MIC and MLC values varying between 0.32 and 0.64 μL/mL (Zuzarte, 2009).

TABLE 13.1
Comparison of Major Components of Volatile Oils of *Lavandula* Species

Species	Major Components of Volatile Oils	References
L. angustifolia L.	1,8-Cineole *cis*-Linalool oxide *trans*-Linalool oxide Linalool Camphor Lavandulol Alpha terpineol Linalyl acetate Lavandulyl acetate 3-Octanone 1-Octen-3-ylacetate	Rai, 2020
L. stoechas L.	Fenchone Camphor p-Cymene Lavandulyl acetate Alpha pinene	Dob, 2006
L. pedunculata L.	Fenchone 1,8-Cineole Camphor	Zuzarte, 2009

TABLE 13.2
Biological Activity *Lavandula* Species

Species	Extract	Biological Activity	Experimental Method	References
***L. angustifolia* L.**	Volatile oil	Antioxidant	*In vitro*	Nikšić, 2017
	Volatile oil	Antibacterial	*In vitro*	Nikšić, 2017
	Volatile oil	Antispasmodic	Clinical experiment	Nasiri, 2016
	Volatile oil	Anxiolytic	Clinical experiment	Wotman, 2017
	Water extract	Anti-inflammatory	*In vivo* mice	Georgiev, 2017
	Ethanol extract	Antiproliferative for breast cancer	*In vitro*	Aboalhaija, 2022
***L. stoechas* L.**	Volatile oil	Antioxidant	*In vitro*	Benabdelkader, 2011
	Volatile oil	Antifungal	*In vitro*	Angioni, 2006
	Volatile oil	Antimicrobial	*In vitro*	Bouzouita, 2005
	Hydroalcoholic extract	Anti-inflammatory	*In vitro/in vivo* rat	Algieri, 2016
***L. pedunculata* L.**	Volatile oil	Antifungal	*In vitro*	Zuzarte, 2009
	Volatile oil	Antibacterial	*In vitro*	Chroho, 2022
	Volatile oil	Antioxidant	*In vitro*	Chroho, 2022

In another study, it was determined that *L. pedunculata* extract has antioxidant potential. In addition, it was concluded that the aqueous-ethanolic extract of the plant has anti-inflammatory activity and antiproliferative potential (Lopes, 2018).

The biological activities of *L. angustifolia* L., *L. stoechas*, and *L. pedunculata* species are given in Table 13.2 by considering the extracts and experimental methods used.

CONCLUSION

As a result, different taxa of Lavender species grown in Türkiye have wide usage areas both in terms of ethnobotany and biological activity with their rich chemical content. *L. angustifolia* cultivated in Türkiye and *L. stoechas* subsp. *stoechas* L. and *L. pedunculata* subsp. *cariensis* are widely used in various diseases among the people.

In this chapter, we concluded that *Lavender* sp. has especially microbiological, cosmetic, and biotechnological importance. Further studies are needed for *Lavandula* species.

REFERENCES

Aboalhaija, N. H., Syaj, H., Afifi, F., Sunoqrot, S., Al-Shalabi, E., & Talib, W. (2022). Chemical evaluation, *in vitro* and *in vivo* anticancer activity of *Lavandula angustifolia* grown in Jordan. *Molecules*, 27(18), 5910.

Akbulut, S., & Bayramoglu, M. M. (2013). The trade and use of some medical and aromatic herbs in Türkiye. *Studies on Ethno-Medicine*, 7(2), 67–77.

Algieri, F., Rodriguez-Nogales, A., Vezza, T., Garrido-Mesa, J., Garrido-Mesa, N., Utrilla, M. P., González-Tejero. M. R., Casares-Porcel M., Molero-Mesa J., Mar M., Segura-Carretero A., Pérez-Palacio J., Diaz C., Vergara N., Vicente F., Rodriguez-Cabezas E., & Galvez, J. (2016). Anti-inflammatory activity of hydroalcoholic extracts of *Lavandula dentata* L. and *Lavandula stoechas* L. *Journal of Ethnopharmacology*, 190, 142–158.

Angioni, A., Barra, A., Coroneo, V., Dessi, S., & Cabras, P. (2006). Chemical composition, seasonal variability, and antifungal activity of *Lavandula stoechas* L. ssp. *stoechas* volatile oils from stem/leaves and flowers. *Journal of Agricultural and Food Chemistry*, 54(12), 4364–4370.

Ayaz, A. P. D. F. (2020). *Lavandula stoechas* ssp. stoechas: *In vitro* antioxidant, and enzyme inhibitory activities. *Research In Medicinal and Aromatic Plants*, 25–42.

Benabdelkader, T., Zitouni, A., Guitton, Y., Jullien, F., Maitre, D., Casabianca, H., Legendre L., & Kameli, A. (2011). Volatile oils from wild populations of Algerian *Lavandula stoechas* L.: Composition, chemical variability, and *in vitro* biological properties. *Chemistry & Biodiversity*, 8(5), 937–953.

Bousta, D., & Farah, A. (2020). A Phytopharmacological review of a Mediterranean plant: *Lavandula stoechas* L. *Clinical Phytoscience*, 6(1), 1–9.

Bouyahya, A., Et-Touys, A., Abrini, J., Talbaoui, A., Fellah, H., Bakri, Y., & Dakka, N. (2017). Lavandula stoechas volatile oil from Morocco as novel source of antileishmanial, antibacterial and antioxidant activities. *Biocatalysis and Agricultural Biotechnology*, 12, 179–184.

Bouzouita, N., Kachouri, F., Hamdi, M., Chaabouni, M. M., Aissa, R. B., Zgoulli, S., Thonart, P., Marlier, M., & Lognay, G. C. (2005). Volatile constituents and antimicrobial activity of *Lavandula stoechas* L. oil from Tunisia. *Journal of Volatile Oil Research*, 17(5), 584–586.

Bozyel, M. E., & Merdamert, E. (2018). Antiurolithiatic activity of medicinal plants in Turkey. *Science, Ecology and Engineering Research in the Globalizing World*; St. Kliment Ohridski University Press: Sofia, Bulgaria, 152–167.

Bozyel, M. E., & Merdamert-Bozyel, E. (2020). Ethnomedicinal uses of genus *Lavandula* (Lamiaceae) in Turkish traditional medicine. *Rheumatism*, 13, 79.

Çelik T., & Aslantürk, Ö. (2007). Cytotoxic and genotoxic effects of *Lavandula stoechas* aqueous extracts. *Biologia*, 62(3), 292–296.

Ceylan, Y., Usta, K., Usta, A., Maltas, E., & Yildiz, S. (2015). Evaluation of antioxidant activity, phytochemicals and ESR analysis of *Lavandula stoechas*. *Acta Physica Polonica A*, 128, 483–487.

Chroho, M., El Karkouri, J., Hadi, N., Elmoumen, B., Zair, T., & Bouissane, L. (2022). Chemical composition, antibacterial and antioxidant activities of the volatile oil of *Lavandula pedunculata* from Khenifra Morocco. In *IOP Conference Series: Earth and Environmental Science* (Vol. 1090, No. 1, p. 012022). IOP Publishing.

Costea, T., Străinu, A. M., & Gîrd, C. E. (2019). Botanical characterization, chemical composition and antioxidant activity of romanian Lavender (*Lavandula angustifolia* Mill.) flowers. *Studia Universitatis Vasile Goldis Seria Stiintele Vietii (Life Sciences Series)*, 29(4), 159–167.

Dirmenci, T. (2012). *Lavandula. Şu sitede: Bizimbitkiler (2013).* <http://www.bizimbitkiler.org.tr>, [er. tar.: 07 02 2023].

Dob, T., Dahmane, D., Agli, M., & Chelghoum, C. (2006). Essential oil composition of Lavandula stoechas. from Algeria. *Pharmaceutical Biology*, 44(1), 60–64.

Esposito, E. R., Bystrek, M. V., & Klein, J. S. (2014). An elective course in aromatherapy science. *American Journal of Pharmaceutical Education*, ;78(4), 79.

Ezzoubi, Y., Bousta, D., Lachkar, M., & Farah, A. (2014). Antioxidant and anti-inflammatory properties of ethanolic extract of *Lavandula stoechas* L. from Taounate region in Morocco. *International Journal of Phytopharmacology*, 5(1), 21–26.

Fakir, H., Korkmaz, M., & Güller, B. (2009). Medicinal plant diversity of western Mediterranean region in Türkiye. *Journal of Applied Biological Sciences*, 3(2), 33–43.

Georgiev, Y. N., Paulsen, B. S., Kiyohara, H., Ciz, M., Ognyanov, M. H., Vasicek, O., Rice, F., Denev, P. N., Yamada, H., Lojek, A., Kussovski, V., Barsett, H., Krastanov, A. I., Yanakieva, I. Z., & Kratchanova, M. G. (2017). The common Lavender (*Lavandula angustifolia* Mill.) pectic polysaccharides modulate phagocytic leukocytes and intestinal Peyer's patch cells. *Carbohydrate Polymers*, 174, 948–959.

Güler, B., Manav, E., & Uğurlu, E. (2015). Medicinal plants used by traditional healers in Bozüyük (Bilecik–Türkiye). *Journal of Ethnopharmacology*, 173, 39–47.

Güner, Ö., & Selvi, S. (2016). Wild medicinal plants sold in Balıkesir/Türkiye herbal markets and their using properties. *Biological Diversity and Conservation*, 9(2), 96–101.

Gürdal, B., & Kültür, Ş. (2013). An ethnobotanical study of medicinal plants in Marmaris (Muğla, Türkiye). *Journal of Ethnopharmacology*, 146(1), 113–126.

Güzel, Y., Güzelşemme, M., & Miski, M. (2015). Ethnobotany of medicinal plants used in Antakya: A multicultural district in Hatay province of Türkiye. *Journal of Ethnopharmacology*, 174, 118–152.

Héral, B., Stierlin, É., Fernandez, X., & Michel, T. (2021). Phytochemicals from the genus Lavandula: A review. *Phytochemistry Reviews*, 20, 751–771.

Kara, N., & Baydar, H. (2011). Türkiye'de lavanta üretim merkezi olan Isparta ili Kuyucak yöresi lavantalarının (Lavandula x intermedia Emeric ex Loisel) uçucu yağ özellikleri. *Selçuk Tarım ve Gıda Bilimleri Dergisi*, 25(4), 42–46.

Kızılay, H., & Şarer, E. (2022). *Lavandula angustifolia* MILLER uçucu yağı üzerinde farmakognozik araştırmalar. *İnönü Üniversitesi Sağlık Hizmetleri Meslek Yüksek Okulu Dergisi*, 10(2), 554–562.

Korkmaz, M., & Karakurt, E. (2014). Kelkit (Gümüşhane) aktarlarında satılan tıbbi bitkiler. *Journal of Natural and Applied Science*, 18(3), 60–80.

Küçük, S., Altintaş, A., Demirci, B., Fehmiye, K. O. C. A., & Başer, K. H. C. (2019). Morphological, anatomical and phytochemical characterizations of *Lavandula stoechas* L. subsp. *stoechas* Growing in Türkiye. *Natural Volatiles and Volatile Oils*, 6(2), 9–19.

Lopes, C. L., Pereira, E., Soković, M., Carvalho, A. M., Barata, A. M., Lopes, V., ... Ferreira, I. C. (2018). Phenolic composition and bioactivity of Lavandula pedunculata (Mill.) Cav. samples from different geographical origin. *Molecules*, 23(5), 1037.

Moussi Imane, M., Houda, F., Said Amal, A. H., Kaotar, N., Mohammed, T., Imane, R., & Farid, H. (2017). Phytochemical composition and antibacterial activity of Moroccan *Lavandula angustifolia* Mill. *Journal of Volatile Oil Bearing Plants*, 20(4), 1074–1082.

Nasiri, A., Mahmodi, M. A., & Nobakht, Z. (2016). Effect of aromatherapy massage with lavender volatile oil on pain in patients with osteoarthritis of the knee: A randomized controlled clinical trial. *Complementary Therapies in Clinical Practice*, 25, 75–80.

Nikšić, H., Kovač-Bešović, E., Makarević, E., Durić, K., Kusturica, J., & Muratovic, S. (2017). Antiproliferative, antimicrobial, and antioxidant activity of *Lavandula angustifolia* Mill. essential oil. *Journal of Health Sciences*, 7(1).

Oraloğlu, Z. (2018). *Bitkisel drog monografı hazırlanması: Lavandula stoechas L* (Master's thesis, Sağlık Bilimleri Enstitüsü).

Oraloğlu, Z., & İşcan, G. (2019). Karabaşotu çaylarında kâfur miktar tayini. *The Journal of Food*, 44(5).

Özcan, M. M., Starovic, M., Aleksic, G., Figueredo, G., Juhaimi, F. A., & Chalchat, J. C. (2018). Chemical composition and antifungal activity of lavender (*Lavandula stoechas*) oil. *Natural Product Communications*, 13(7), 1934578X1801300728.

Polat, R., & Satıl, F. (2012). An ethnobotanical survey of medicinal plants in Edremit gulf (Balıkesir–Türkiye). *Journal of Ethnopharmacology*, 139(2), 626–641.

Prusinowska, R., & Smigielski, K. B. (2014). Composition, biological properties and therapeutic effects of lavender (*Lavandula angustifolia* L.). a review. *Herba Polonica*, 60(2).

Rai, V. K., Sinha, P., Yadav, K. S., Shukla, A., Saxena, A., Bawankule, D. U., Tandon S., Khan F., Chanotiya C. S., & Yadav, N. P. (2020). Anti-psoriatic effect of *Lavandula angustifolia* essential oil and its major components linalool and linalyl acetate. *Journal of Ethnopharmacology*, 261, 113127.

Sales, A. J. (2014). Evaluation of antibacterial activity of ethanol extract of *Lavandula stoechas* L. plant on antibiotic-resistant strains of *Staphylococcus aureus*. *Journal of Current Research in Science*, 2(6), 641–645.

Salih, P. A. Ş. A., Kaplan, E. N., & Çakı, M. N. (2019). *Lavandula stoechas* (Karabaş Otu) bitki özütünden kozmetik ürün eldesi. *Uluslararası Batı Karadeniz Mühendislik ve Fen Bilimleri Dergisi*, 1(1), 24–37.

Sargın, S. A., Ak icek, E., & Selvi, S. (2013). An ethnobotanical study of medicinal plants used by the local people of Alaşehir (Manisa) in Türkiye. *Journal of Ethnopharmacology*, 150(3), 860–874.

Selvi, S., Dağdelen, A., & Kara, S. (2013). Kazdağlarından (Balıkesir-Edremit) toplanan ve çay olarak tüketilen tıbbi ve aromatik bitkiler. *Journal of Tekirdağ Agricultural Faculty*, 10(2), 26–33.

Şenkul, Ç., & Seda, K. (2017). Türkiye endemik bitkilerinin coğrafi dağılışı. *Türk Coğrafya Dergisi*, (69), 109–120.

Tanker, N., Koyuncu, M., & Coşkun, M. (2007). *Farmasötik botanik. Ankara Üniversitesi Eczacılık Fakültesi Yayınları*, No. 93, 303.

Tetik, F. (2011). *Malatya ilinin etnobotanik değeri olan bitkileri üzerine bir araştırma*. Unpublished MSc Thesis. Fırat University, Türkiye.

Traboulsi, A. F., Taoubi, K., El-Haj, S., Bessiere, J. M., & Rammal, S. (2002). Insecticidal properties of essential plant oils against the mosquito Culex pipiens molestus (Diptera: Culicidae). *Pest Management Science*, 58(5), 491–495.

Ugulu, I., Baslar, S., Yorek, N., & Dogan, Y. (2009). The investigation and quantitative ethnobotanical evaluation of medicinal plants used around Izmir province, Türkiye. *Journal of Medicinal Plants Research*, 3(5), 345–367.

Wotman, M., Levinger, J., Leung, L., Kallush, A., Mauer, E., & Kacker, A. (2017). The efficacy of Lavender aromatherapy in reducing preoperative anxiety in ambulatory surgery patients undergoing procedures in general otolaryngology. *Laryngoscope Investigative Otolaryngology*, 2, 437–441.

Zuzarte, M., Gonçalves, M. J., Cavaleiro, C., Dinis, A. M., Canhoto, J. M., & Salgueiro, L. R. (2009). Chemical composition and antifungal activity of the volatile oils of *Lavandula pedunculata* (MILLER) CAV. *Chemistry & Biodiversity*, 6(8), 1283–1292.

14 *Malva sylvestris* L.

Golshan Zare and I. Irem Tatli Cankaya

MORPHOLOGICAL DESCRIPTION

Malva sylvestris L. (Figure 14.1) is native to Europe, Asia, and North Africa; it is a member of Malvaceae family (Barros et al. 2010). This species is an invasive plant in food crops and is known as a weed by some authors (Mas et al. 2007).

These plants are perennial to biennial herbs, with erect or ascending stems with pilose tuberculate hairs. Leaves are long petioles, simple 7–15 cm in diameter, membranous, more or less orbicular, 5–7-lobed, rounded or acute apexes, crenate, sparsely pilose to glabrous. The flowers are arranged in fascicles in the leaf axils. They are 3–5 cm wide and have an epicalyx, epicalyx segments are three linear to oblong or narrowly ovate situated immediately below it. The calyx has five pubescent triangular lobes 3–7 mm, gamosepalous at the bases, and corolla 18–25 mm, 4 or more times longer than the sepals, the calyx with five wedge-shaped, notched petals is fused to the stamen tube at their base, mauve to pink, emarginate or shallowly 3–5-fid. Numerous stamens, the filaments of which fuse into a stamina tube covered by small star-shaped trichomes, and numerous wrinkled carpels, glabrous or sometimes pubescent, enclosed in the stamen tube are arranged into a circle around a central style that ends with numerous filiform stigmas. Fruit are glabrous or hairy mericarps (Cullen 1967). In cultivated varieties, the epicalyx is three to seven partite, the calyx is five to eight partite, and the corolla is five to ten partite (Figure 14.1).

TRADITIONAL USES, BIOLOGICAL ACTIVITIES, AND PHYTOCHEMICAL PROPERTIES OF MALVA SYLVESTRIS L.

Malva sylvestris L. (Malvaceae) is among the numerous species used in folk medicine due to its important therapeutic properties since ancient times in Asia and Europe (Gasparettoa et al. 2011; Ghédira and Goetz 2016). This species is well known as an edible plant for its possible medicinal properties (Henry and Piperno 2008). The whole plant has some medicinal properties, but generally, the pharmacological properties of *Malva sylvestris* are assigned to the leaves and flowers of this plant (Pourrat et al. 1990; Classen and Blaschek 2002; Gasparetto et al. 2008).

The medicinal use and phytochemical properties of *Malva sylvestris* leaves and flowers have been documented in detail in several pharmacopeias including WHO, ESCOP, European Medicines Agency (Widy-Tyszkiewicz 2018), European Pharmacopoeia 5.0th, the Czechoslovak Pharmacopoeia (1947), Austrian Pharmacopoeia (Österreichisches Arzneibuch, 1960), Pharmacopoea Hungarica (1986), Pharmacopoea Helvetica (1997), Pharmacopoea Polonica III (1954), and Turkish Pharmacopoeia (2017).

Traditional Medicinal Use

Since ancient times, *Malva* was used by the Greeks and Romans for its emollient and laxative properties. In the Mediterranean area, as a medicinal food, raw fresh herbs are cooked and consumed as food in soups and salads (Barros et al. 2010; Guarrera 2003; Ishtiaq 2007).

M. sylvestris is traditionally used to treat several infections and diseases, such as cold, burn, cough, tonsillitis, bronchitis, eczema, gastrointestinal disorders as a mild laxative, abdominal pain, and diarrhea (Guarrera 2003; Ishtiaq 2007; Idolo et al. 2010, Pirbalouti and Azizi 2010).

DOI: 10.1201/9781003146971-14

FIGURE 14.1 *Malva sylvestris* L. (A) Habitus; (B) flowers; (C) fruits; and (D) male and female reproductive parts of a flower.

This plant generally is applied in the form of infusions, decoctions such as herbal tea or a gargle, and cutaneous use for the treatment of cuts, infected wounds, and skin burns in the form of poultices and lotions (Barros et al. 2010, Ishtiaq 2007).

The leaves and flowers of *M. sylvestris* are known due to their anti-inflammatory properties, abscesses and tooth pain, also use in the treatment of urological problems, insect bites, burns, furuncles, and ulcerous wounds (Idolo et al. 2010, Marc 2008, Pollio A et al. 2008; Leond et al. 2009). In Italy, decoctions of aerial parts are used as a laxative, against flu, cold, and stomachache (Quave et al. 2008), topically applied to the cheek or directly into aching teeth as an analgesic and as a mouthwash (Idolo et al. 2010), as well as for throat inflammation, bronchitis, and cough (Menale et al. 2016). In the Central Apennine area *M. sylvestris* leaves, flowers, and whole aerial parts are used in bronchitis, sore throat, and cough. In Spain, *M. sylvestris* flowers, leaves, and whole aerial parts are applied directly topically, often without pharmaceutical forms as a poultice or to prepare baths as antipruritic and external antiseptic treatment (Rigat et al. 2015). Mallow is traditionally used due to its high mucilage content for treatment of the symptoms of oral or pharyngeal irritations, associated dry cough, and gastrointestinal discomfort (Gasparetto et al. 2011).

The plant can cure wounds and inflammations and is used as an analgesic, but also as a herbal tea for respiratory problems and stomachaches (Calvo et al. 2011, 2015; González et al. 2011). In Portugal, both *Malva sylvestris* flowers and leaves are used as herbal tea as an anti-inflammatory, antiseptic, and antiodontalgic treatment (Novais et al. 2004). In Bosnia-Hercegovina, *M. sylvestris* herbal tea of leaves and flowers is used orally for the common cold, dry cough and crankiness, cough, pulmonary ailments, gastrointestinal spasms, and painful urination (Šarić-Kundalić et al. 2010). In Serbia, mallow leaves and flowers are topically applied as poultices on wounds and ulcers difficult to heal (Jarić et al. 2015).

Traditional Use in Turkey

In Turkey, *M. sylvestris* is called "Ebegümeci, ebegömeci, gümülcün, tolikotu, kömeç, or moloşa" and it has been used as a traditional medicinal plant for different purposes in Anatolia. Leaves and flowers of *M. sylvestris* are used both as food (Simsek et al. 2004; Kızılarslan and Özhatay 2012;

Yeşilyurt et al. 2017) and for medicinal purposes such as the treatment of cough, colds, and flu (Sezik et al. 2001; Kültür 2007; Fakir et al. 2009), shortness of breath, rheumatism (Dalar et al. 2018), stomach and intestinal diseases, as anticancer drug (Akbulut and Ozkan 2014). In addition, it is used for cancer, tonsillitis (Kültür 2007; Kültür and Sami 2009; Rexhepi, Akalın and Alpınar 1994; Bulut 2011; Genç G, Özhatay 2006; Tuzlacı and Alpaslan 2007; Tuzlacı E, İşbilen 2010), abdominal pain, as anti-inflammatory, antitussive, for burns, infertility, urinary inflammations, hemorrhoids, goiter, and wound (Bulut and Tuzlacı 2013; Cakilcıoglu and Turkoglu 2008; Miraldi et al. 2001, Azizi and Keshavarzi 2015).

Veterinary Uses

There are some records according to the use of *M. sylvestris* for veterinary purposes. Decoctions of whole plants can be administered to treat colic and to unblock rumens. Infusions and decoctions of aerial parts have been used to treat colic and to unblock rumens, and respiratory problems, as laxatives in horses, and also demonstrated activity against inflammation, wound infection, diarrhea in young calves, and intestinal inflammation in cows and sows (Angels 2007; Akerreta et al. 2010, Gasparetto et al. 2012).

Herbal Preparation(s)

Malva sylvestris flower and leaves are used as comminuted herbal substances, and also as an herbal tea for oral use. These substances are prepared by infusion or decoction for oromucosal use (Gasparetto et al. 2012).

PHYTOCHEMICAL CONSTITUENTS

Amino Acids/Protein Derivatives

Using high-performance thin-layer chromatography (HPTLC), the presence of the amino acids alanine, threonine, hydroxyproline, serine, glutamine, asparagine, and arginine have been detected in *M. sylvestris* (Classen and Blaschek 2002). Also, the most abundant components of arabinogalactan protein (AGP) were determined as galactose, arabinose, mannose, glucuronic acid, glucose, xylose, and rhamnose in this taxon. Blunden et al. (2001) reported the presence of trigonelline and glycine betaine in the leaves, flowers, and roots, respectively.

Phenolic Derivatives

Many derivatives of phenolics have been found in extracts from different parts of *M. sylvestris*. The total phenolic compounds were found to be 386.5 mg/g in the leaves, 317.0 mg/g in flowered stems, 258.7 mg/g in flowers, and 56.8 mg/g in immature fruits (Barros et al. 2010). Cutillo et al. (2006) introduced the presence of 4-hydroxybenzoic acid, 4-methoxybenzoic acid, 4-hydroxy-3-methoxybenzoic acid, 2-hydroxybenzoic acid, 4-hydroxy-2-methoxybenzoic acid, 4-hydroxybenzyl alcohol, 4-hydroxydihydrocinnamic acid, 4-hydroxy-3-methoxy dihydrocinnamic acid, 4-hydroxycinnamic acid, ferulic acid, and tyrosol in *M. sylvestris* (Cutillo et al. 2006).

Flavonoids

There are many types of research that indicated the presence of flavonoids in the *Malva*, especially in flowers including luteolin, apigenin, rutin, epicatechin, quercetin, kaempferol, malvidin, genistein myricetin, and their derivatives (Terninko et al. 2014; Alesiani et al. 2007; Pourrat et al. 1990;

Sikorska et al. 2004; Farina et al. 1995; Alesiani et al. 2007; Pourrat et al. 1990; Brouillard 1983; Schulz 2007).

Sikorska et al. (2004) calmed flavonols and flavones with additional OH groups at the C-8 A ring and/or the C-5′ B ring positions are characteristic properties of the Malvaceae family. In the leaves, gossypetin 3-sulphate-8-*O*-*β*-D-glucoside (gossypin) and hypolaetin 3′-sulphate were identified as the major constituents, followed by 3-*O*-*β*-D-glucopyranosyl-8-*O*-*β*-D glucuronopyranoside, hypolaetin 4′-methyl ether 8-O-*β*-D-glucuronopyranoside, hypolaetin 8-*O*-*β*-D-glucuronopyranoside, and isoscutellarein 8-O-*β*-D-glucuronopyranoside (Billeter et al. 1991; Nawwar and Buddrus 1981).

Additional phytochemicals such as leucoanthocyanins, cyanidin, and petunidin have been found, but in very low concentrations (Pourrat et al. 1990).

Coumarins

The presence of methoxycoumarin dimethoxycoumarin and 7-hydroxy-6-methoxycoumarin (scopoletin) has been reported in the leaves of *M. sylvestris* (Tosi et al. 1995; Terninko et al. 2014; Alesiani et al. 2007). Alesiani et al. mentioned phototoxic coumarin can be responsible for anticancer activity (2007).

Mucilage Polysaccharides

The Malvaceae family members generally possess the most abundant deposits of mucilages (Franz 1966). According to Wichtl (2004) and PDR (2000), the leaves of *Malva sylvestris* contain about 6–8% of mucilage. Mucilages are one of the important phytochemicals responsible for the therapeutic effects of *M. sylvestris*, mainly due to their anti-complementary and cough suppression activities (Tomoda et al. 1989, Franova et al. 2006).

The mucilages consist mainly of glucuronic acid, galacturonic acid, rhamnose, galactose, fructose, glucose, sucrose, and trehalose, but uronic acid, arabinose, mannose, xylose, fucose, raffinose, and 2′-*O*-*β*-(4-*O*-methyl-*β*-D-glucuronosyl)-xylotriose have also been found (Barros et al. 2010, Classen and Blaschek 1998).

The mucilage is found in idioblasts, mucilage ducts, cavities, and specialized epidermal cells (Classen 1998) and their contents can vary according to the plant organs they are produced. However, in general, high percentages of crude mucilages can be found in the leaves, flowers, and roots (Classen and Blaschek 1998; Hicsonmez et al. 2009). The main structural unit of polysaccharides consists of glucose, galactose, and rhamnose which are found in different concentrations in both flowers and leaves of *M. sylvestris*. Classen and Blaschek (1998) showed around 10% of mucilage presented in flowers.

Different authors isolated four polysaccharides from the flowers of *M. sylvestris* including a linear neutral polysaccharide, a branched neutral polysaccharide, high molecular weight acid polysaccharides, and a very branched acidic polysaccharide (Čapek and Kardošova 1995 and Čapek et al. 1999).

Terpenoids

Terpenoids, including monoterpenes, diterpenes, sesquiterpenes, and norterpenes, have been found in *M. sylvestris*. Aqueous extracts from fresh leaves of the plant have revealed the presence of linalool, linalool-1-oic acid, (6*R*,7*E*, 9*S*)-9-hydroxy-4,7-megastigmadien-3-one, (3*S*,5*R*,6*S*,7*E*, 9*R*)-5,6-epoxy-3,9-dihydroxy-7-megastigmene, blumenol A, (3*R*,7*E*)-3-hydroxy-5,7-megastigmadien-9-one, (+)- dehydrovomifoliol, (3*S*,5*R*,6*R*,7*E*,9*R*)-3,5,6,9-tetrahydroxy-7-megastigmene, and (6*E*,8*S*,10*E*,14*R*)-3,7,11,15-tetramethylhexadeca-1,6,10-trien-3,8,14,15-tetraol (Cutillo et al. 2006; DellaGreca et al. 2009, Gasparetto et al. 2012).

In seed oil, the main terpene presented was terpineol, and in the leaves, flowers, and immature fruits, carotenoids, which are tetraterpenoids, are present (Emets et al. 1994). Among these substances, malvone A (2-methyl-3-methoxy-5,6-dihydroxy-1,4-naphthoquinone) stands out due to its resistance against the pathogen *Verticillium dahliae*; it is therefore considered an important antimicrobial agent (Veshkurova et al. 2006).

Essential Oils

The essential oil was obtained from dried flowers containing 143 volatile compounds. The essential oil consisted mainly of hydrocarbons followed by, alcohols, acids, ethers, ketones, esters, aldehydes, and others, respectively. The main components found were hexadecanoic acid, 2-methoxy-4-vinylphenol, and pentacosane. β-Damascenone, phenylacetaldehyde, and (*E*)-β-ocimene were the most intense aroma active compounds (Usami et al. 2013).

Fatty Acids/Sterols

M. sylvestris, due to its rich fatty acids profile and the presence of essential fatty acids such as omega-3 and omega-6, has an important role as a nutraceutical food (Barros et al. 2010).

Lipids are found in the leaves, flowers, immature fruits, and flowering stems and mainly in seeds of this taxon (Barros et al. 2010). Conforti et al. (2008) indicated the presence of the steroids; campesterol, stigmasterol, and sitosterol in the leaves of *M. sylvestris*. Palmitic acid, pelargonic acid, stearic acid, and α -linolenic acid are the fatty acids found in *M. sylvestris* (Loizzo et al. 2016). The seed oil consists mainly of palmitic acid, oleic acid, malvalic acid, lauric acid, myristic acid, sterculic acid, palmitoleic acid, linoleic acid, vernolic acid, and traces of stearic acid (Emets et al. 1994; Mukarram et al. 1984).

Pigments

Carotenoids (Loizzo 2016, Alesiani et al. 2007, Redžić et al. 2005), chlorophyll A and B (Redžić et al. 2005), and tocopherols (Barros et al. 2010) have been determined.

Organic Acids

The presence of ascorbic acid, dehydroascorbic acid, and oxalic acid is referred to by Guil et al. (1996); citric acid, oxalic acid, and malic acid by Terninko et al. (2014) and Terninko and Onishcenko (2013) and (10*E*,15*Z*)-9,12,13-trihydroxyoctadeca-10,15-dienoic acid by DellaGreca et al. (2009).

In addition, some organic acids found by Pereira et al. (2013) are oxalic acid, malic acid, ascorbic acid, citric acid, and fumaric acid.

Vitamins

These species contain a considerable amount of tocopherols (vitamin E) and ascorbic acid (vitamin C) and their antioxidant activity is attributed to this substance. There are four forms of tocopherols (β, β, β, and β) in this taxon and the β-tocopherol type is the major form present in green plant tissues (Barros et al. 2010).

Quantitative analyses have demonstrated high concentrations of these substances in the leaves (106.5 mg%), as well as large quantities in flowered stems (34.9 mg%), flowers (17.4 mg%), and immature fruits (2.6 mg%) (Barros et al. 2010). In the same plant parts, ascorbic acid was also detected, at levels of 1.11 mg/g in flowers, 0.27 mg/g in immature fruits, 0.20 mg/g in flowered stems, and 0.17 mg/g in leaves. These results emphasized the importance of *M. sylvestris* as an antioxidant agent against reactive oxygen species (Barros et al. 2010).

Chemical Elements

Major and minor elements found in *M. sylvestris* are Ag, Al, B, Ba, Bi, Ca, Cd, Co, Cr, Cu, Fe, K, La, Mg, Mn, Na, Ni, Pb, Sn, Sr, Sb, Si, Ti, U, Zn, and Zr (Hiçsönmez et al. 2009). Also, *Malva* has a high potential to accumulate heavy metals such as Cd, Cu, Ni, Pb, and Zn. Therefore, there is a need to raise awareness about the health hazards of intake of these substances in populations that live in risk areas (Khan et al. 2010)

In conclusion, the quantitative and qualitative composition of all phytochemicals found in this plant directly depends on the soil and plant growth conditions (Emets et al. 1994).

PHARMACOLOGICAL ACTIVITIES

Several studies have been conducted to evaluate the pharmacological activity of *M. sylvestris*. *M. sylvestris* has classically been indicated for the treatment of oral diseases and thus its antimicrobial and anti-inflammatory properties have been assessed using different extracts and preparations.

Antimicrobial and Antibiofilm Activities

The results of the antibacterial assay indicated that both flowers and leaves of *M. sylvestris* methanol extracts exhibited high bactericidal activity against various bacterial and fungal species. The extracts were found to have strong antibacterial effects against *Erwinia carotovora*, a common plant pathogen bacteria, with MIC values of 128 and 256 μg/mL for the flowers and leaves extracts, respectively. The methanol extract of the plant flowers showed high antibacterial effects against some human pathogen or bacteria strains such as *Staph. aureus*, *Strep. agalactiae*, *Entro. faecalis*, with MIC values of 192, 200, and 256 μg/mL, respectively. Both flower and leave methanol extracts of the plant possessed modest antibacterial activity against another tested microorganism with MIC values in the range of 320–800 μg/mL (Razavi et al. 2011).

On the other hand, the antifungal assay showed that the methanol extract of the plant flowers indicated modest activity. The extract inhibited the growth of *Sclerotinia sclerotiorum*, a plant pathogen fungus, and some human pathogen fungi like *C. kefyr* and *C. albicans* with MIC values in the range of 640–800 μg/mL (Razavi et al. 2011).

An aqueous extract of leaves shows significant antimicrobial properties (concentration 0.60 g/m) against *Aspergillus candidus*, *Aspergillus niger*, *Penicillium* spp., and *Fusarium culmorum* and fungus (Sleiman and Daher 2009).

Ethanolic stem extracts show moderate activity, in biofilm formation assay against *S. aureus* (IC50 32 mg/mL) with limited bacteriostatic effects in planktonic growth tests (Quave et al. 2008). In addition ethanolic extracts obtained from the leaves and flowers possess a moderate to low antimicrobial activity against different strains of *Helicobacter pylori*.

Cogo et al. (2010) evaluated the antibacterial activity of the hydroalcoholic extract obtained from leaves and flowers against 11 clinical isolates of *Helicobacter pylori* and two reference strains. All the strains were previously evaluated against clarithromycin, amoxicillin, furazolidone, tetracycline, and metronidazole. Based on disk-diffusion test, the inhibition zone ranged from 8 to 10 mm for *H. pylori* 26695 and *H. pylori* J99 strains, respectively. Results indicated that the extracts were capable of inhibiting the *in vitro* growth of *H. pylori*.

In other studies using diffusion tests, lyophilized and crude methanolic extracts did not substantially inhibit strains of *Escherichia coli*, *Pseudomonas aeruginosa*, *Pseudomonas fluorescens*, *Klebsiella pneumoniae*, *Serratia marcescens*, *S. aureus*, *Staphylococcus epidermidis*, *Micrococcus luteus*, *Bacillus subtilis*, *Bacillus cereus*, *Bacillus pumilu*, *Bordetella bronchiseptica*, *Candida albicans*, and *Saccharomyces cerevisiae* (Bonjar et al. 2004; Souza et al. 2004).

Anti-Inflammatory Activity

Several researches have indicated the anti-inflammatory activity of *M. sylvestris*. The anti-inflammatory activity of *Malva* was evaluated in mice taking a 100 mg/kg oral dose of the aqueous extract. Edema and formalin were reduced by 60% in both the acute and chronic inflammation models. Sleiman and Daher (2009) have concluded leaves possess significant anti-inflammatory properties and malvidin 3-glucoside is responsible for these properties.

In other *in vivo* anti-inflammatory assays, creams containing different concentrations of *M. sylvestris* extract were evaluated in carrageenan-induced edema model and Carrageenan-induced edema was significantly reduced by *Malva* cream. This result was higher than that of cream with 2% indomethacin, which is a potent nonselective inhibitor of cyclooxygenase-2 (COX-2) (Chiclana et al. 2009).

According to Conforti et al. (2008), the hydroalcoholic extract of leaves possesses anti-inflammatory activity on croton oil-induced inflammation in the ears of mice (surface of 1 cm^2; application of 80 mg of aqueous ethanol). The local application of the extract in 300 mg/cm^2, concentration reduced edema by 21%. All these results have demonstrated the topical anti-inflammatory activity of *M. sylvestris* in inflammation.

Analgesic Activity

In vivo investigation on lyophilized aqueous extract of *M. sylvestris* leaves has indicated that this extract has significant analgesic effects. In addition, these results suggested that the inhibition of the prostaglandin synthesis pathway cyclooxygenase is related to the antinociceptive effects of extract (Esteves et al. 2009).

Antiproliferative Activity

The antiproliferative activity of ethanolic and hydroalcoholic leaf extracts of the *M. sylvestris* has been evaluated by MTT (3-(4,5-dimethylthiazol-2-yl)-2,5-diphenyl tetrazolium bromide) assay. Significant reductions in the proliferation of B16 murine melanoma and A375 human melanoma cells were observed (Daniela et al. 2007). In contrast, in Conforti et al (2008) research this plant did not show antiproliferative activity in the MCF-7 cancer cell line, the LNCaP prostate cancer cell line, C32 amelanotic melanoma, or renal adenocarcinoma in the sulphorhodamine B assay.

Antioxidant Activity

The antioxidant capacity of *M. sylvestris* in different extracts has been evaluated using several assays and methods.

The methanolic extracts from the leaves, flowers, immature fruits, and flowering stems of the *M. sylvestris* were evaluated using 1,1-diphenyl-2-picrylhydrazyl (DPPH) radical scavenging, neutralization of linoleate free radicals, malondialdehydethiobarbituric acid (MDA-TBARS) complex, and β-carotene models. Methanolic extracts obtained from the leaves show significant antioxidant properties including radical scavenging activity reducing power (EC_{50} = 0.07 mg/mL), and also lipid peroxidation inhibition in liposomes (EC_{50} = 0.04 mg/mL) and brain cell homogenates. The significant antioxidant activity of the leaf is attributed to the presence of phytochemicals including phenols, flavonoids, carotenoids, and tocopherols in this part of the plant. However, all parts of the plant have antioxidant activity, and the immature fruits were shown to have low activity in contrast (Barros et al. 2010).

The aqueous extract (20 and 100 mg/mL concentrations) shows a reduction in scavenging activity in a DPPH assay by 24% and 30%, respectively. In DellaGreca researches, 87% antioxidant activity was observed in 0.1 mg/mL aqueous extract when measured via the β-carotene-linoleic acid assay.

M. sylvestris essential oil indicated a considerable effect (77% antioxidant activity) at the same concentration (DellaGreca et al 2009; Ferreira et al. 2006). In conclusion, the ethanolic extracts of the leaves and aerial parts as well as the *n*-hexane, dichloromethane, and methanolic extracts obtained from the seeds of *M. sylvestris* possess moderate to low antioxidant activity (Conforti et al. 2008; Kumarasamy et al. 2007; Gasparetto et al. 2012). *M. sylvestris* has high antioxidant properties due to high phenolic contents such as flavenoid compounds.

M. sylvestris is commonly used for the maintenance of dermatological and mucous membrane integrity. Because of these properties, different products such as cosmetics, topical compounds, and moisturizers for the prevention of skin aging and damage have been developed from these plants (Gasparetto et al. 2012). These products have shown high efficiencies in relieving skin irritation, enhancing mucous production, and antioxidant properties (Seiberg et al. 2006). Aqueous extracts and ingestible compounds have also been shown to enhance the structural integrity of skin and other human tissues (Stone and Zhao 2010; Seiberg et al. 2006). In addition *Malva* solutions and lotions had successful experience in preventing alopecia and other capillary disorders (Ahmad 1994 and 1995, Gasparetto et al. 2012).

Anti-ulcerogenic Activity

The anti-ulcerogenic activity of *M. sylvestris* was demonstrated by Daher (2009). In this research aqueous extracts were evaluated in rats with induced gastric ulcers. After a month of treatment at a dosage of 500 mg/kg body weight, maximum protection of 37% was achieved. This level of anti-ulcerogenic activity is comparable with cimetidine, a reference drug that showed 30% maximum protection.

OTHER BIOLOGICAL ACTIVITIES

Sleiman and Daher (2009) evaluated the biochemical component profiles of rats which were fed with aqueous extracts of *M. sylvestris*. When rats were administered with doses of 400 and 800 mg/kg body weight, only serum triglycerides increased significantly, while the glycemic, liver enzymatic, and lipid profiles did not show any difference. But these results are controversial with those other studies that demonstrated reductions of triglyceride and uric acid levels after the intake of the aqueous extract (Seiberget al. 2006; Stone and Zhao 2010)

The pharmacological activities of *M. sylvestris* are summarized in Table 14.1.

TOXICITY

Despite the widespread use of *M. sylvestris* as a food or herbal medicine, there are a few researches on the toxicity of this plant. Seiberg M et al. (2006) used the Microtox Acute Toxicity assay and showed that the hydroalcoholic extract had toxicity very close to the established maximum (20% of bioluminescence inhibition).

USE IN CHILDREN AND ADOLESCENTS

Particular use in children has not been reported (Widy-Tyszkiewicz 2018).

SPECIAL WARNINGS AND PRECAUTIONS FOR USE

According to Widy-Tyszkiewicz (2018), if dyspnea, fever, or purulent sputum occurred during the use of these plants, a qualified healthcare practitioner or pharmacist should be consulted. There are no adequate data about the use in children and adolescent under 12 years of age.

TABLE 14.1
Pharmacological Activities of *M. sylvestris*

Activities	Extract/Pharmaceutical Preparations	Results
Acetylcholinesterase (AChE) inhibition	Ethanolic extract, decoction, and essential oil fraction (aerial parts)	Ferreira A et al. 2006
Anticancer	Ethanolic extract (leaves)	No antiproliferative activity was observed using four human cancer cell lines (breast cancer MCF-7, prostate cancer LNCaP, amelanotic melanoma C32, and renal adenocarcinoma) (Conforti F et al. 2008)
		Significant proliferative reduction of murine and human cancer cell lines (Daniela et al. 2007)
	Hydroalcoholic extract (leaves)	Significant proliferative reduction of B16 and A375 cancer cell lines (Daniela et al. 2007)
Anti-inflammatory	Hydroalcoholic extract (leaves)	Edema reduction 21% (Conforti et al. 2008)
	Malva extract cream	Significant inhibition of edema (Chiclana et al. 2009)
	Aqueous extract (aerial parts)	Significant anti-inflammatory activity (Sleiman and Daher 2009)
Antimicrobial	Commercial mouthwashes etylpyridinium chloride (CPC) and mallow combination	CPC+ mallow association inhibited the growth of all 28 *S. aureus* strains (Watanabe et al. 2008)
	Ethanolic extract (stems)	
	Ethanolic extract (aerial part)	Moderate activity by agar-diffusion test and low activity using minimal inhibitory concentration assay (Cogo et al. 2010)
	Methanolic extract (flower)	No activity against *Escherichia coli*, *Pseudomonas aeruginosa*, *Pseudomonas fluorescens*, *Klebsiella pneumoniae*, *Serratia marcescens*, *Staphylococcus aureus Staphylococcus epidermidis*, *Micrococcus luteus*, *Bacillus cereus* and *Bacillus pumilus*
		Moderate activity against *Bordetella bronchiseptica* (Bonjar 2004)
	Methanolic extract (aerial parts)	Low antimicrobial activity against *Saccharomyces cerevisiae*
		No activity for *Staphylococcus aureus*, *Staphylococcus epidermidis*, *Escherichia coli*, *Bacillus subtilis*, *Micrococcus luteus* and *Candida albicans* (Souza et al. 2004)
	Aqueous extract (leaves)	Growth colonies inhibition of *Aspergillus candidus*, *Aspergillus niger*, *Penicillium* spp., and *Fusarium culmorum* by use of 0.60 g/mL of the extract (Magro et al. 2006)
	Aqueous extract (aerial parts)	No potential against *Candida albicans* or against 11 hospital bacterial isolates (Sleiman and Daher 2009)

(*Continued*)

TABLE 14.1 (CONTINUED)
Pharmacological Activities of *M. sylvestris*

Activities	Extract/Pharmaceutical Preparations	Results
Antioxidant	Aqueous extract (aerial parts)	DPPH scavenging activity 24% at 20 mg/mL of the extract; antioxidant capacity observed using compounds isolated from the extracts (DellaGreca et al. 2009)
		DPPH assay 30% antioxidant activity; β-carotene-linoleic assay 87% of activity, both at 0.1 mg of dry plant/mL water (Ferreira et al. 2006)
	Essential oil hydroalcoholic extract (leaves and aerial parts)	DPPH assay, low radical scavenging activity (Ferreira et al. 2006)
		Low antioxidant activity using bovine brain peroxidation assay; moderate activity by β-carotene bleaching test (El and Karakaya 2004)
		No antioxidant effects on DPPH assay (El and Karakaya 2004)
	Methanolic extracts (leaves, flowers, immature fruits, and leafy flowered stems)	Antioxidant activity, particularly leaves (EC50 values < 0.4 mg/mL for all used models) (Barros et al. 2010)
	n-Hexane, dichloromethane, and methanolic extracts (seeds)	Moderate to low antioxidant activity (Kumarasamy et al. 2007)
	Ethanolic extracts (leaves and aerial parts)	Moderate to low antioxidant activity (Conforti et al. 2008; El and Karakaya 2004; Kumarasamy 2007)
Anti-aging	Seed extract	Increase of antioxidant gene expression (anti-aging properties) (Talbourdet et al. 2007)
	Lyophilized aqueous extract (leaves)	Significant antinociceptive effect against induced acetic acid abdominal constrictions (76.4%) and capsaicin-induced pain model (62.9%); inhibition of neurogenic (61.8%) and inflammatory (46.6%) phases of pain by formalin-induced pain test; no significant activity observed in central analgesic effect by hot-plate test (Esteves et al. 2009)
Anti-ulcerogenic	Aqueous extract (aerial parts)	Maximum protection (37%) at 500 mg/kg body weight (value higher than that observed with cimetidine (30%), a reference drug) (Sleiman and Daher 2009)
Immunomodulatory properties	Aqueous extract (leaves)	No adjuvant effect on anti-EA antibody production (El Ghaoui et al. 2008)
Hepatoprotective activity		Strong hepatoprotective effects against paracetamol-induced liver injury (Daniela et al. 2007)
		Hussain et al. (2014)
Antinociceptive activity	Aqueous extract	Significant antinociceptive activity in writhing test (76.4% of inhibition) and also inhibited the neurogenic (61.8%) and inflammatory (46.6%) phases of the formalin model (Esteves et al. 2009)
Wound healing activity	Diethyl ether extract (flowers)	Significant reduction in the wound (mice) (Pirbalouti et al. 2009; 2010)
	Chloroform extract (flowers)	Significant reduction in the wound (mice) (De Natale and Pollio 2007)
	Aqueous extract	Significant reduction in the wound (mice) (Cameio-Rodrigues 2003)
		Afshar et al.

PREGNANCY AND LACTATION

Due to the absence of adequate data, safety of using this plant during pregnancy and lactation has not been established and is not recommended (Widy-Tyszkiewicz 2018).

REFERENCES

Ahmad K. Alcoholic composition for preventing hair loss and promoting hair growth – Containing fruit, flowers, seeds, juice and plant parts. Patent number (s): DE4312109-A; DE4312109-A1, 1994.

Ahmad K. Hair growth promoting and loss preventing lotion also used as homeopathic compsn. or hebal tea – Comprises extract of plant components, e.g. apple, sandalwood, coriander, pimento, radish, beetroot, ets., aq. alcohol. Patent Number(s): DE4330597-A; DE4330597-A1, 1995.

Akalın E, Alpınar K. Research on the medicinal and food plants of Tekirdağ (Turkey). *Ege University, Journal of Faculty Pharmacy* 1994; 2: 1–11.

Akbulut S, Ozkan ZC. An ethnobotanical study of medicinal plants in Marmaris (Muğla, Turkey). *Kastamonu University Journal of Forestry Faculty* 2014; 14: 135–45.

Akerreta S, Calvo MI, Cavero RY. Ethnoveterinary knowledge in Navarra (Iberian Peninsula). *Journal of Ethnopharmacology* 2010; 130: 369–378.

Alesiani D, Pichichero E, Canuti L, Cicconi R, Karou D, D'Arcangelo G, Canini A. Identification of phenolic compounds from medicinal and melliferous plants and their cytotoxic activity in cancer cells. *Caryologia* 2007; 60: 90–95.

Angels BM, Valle's J. Ethnobotany of Montseny biosphere reserve (Catalonia, Iberian Peninsula): Plants used in veterinary medicine. *Journal of Ethnopharmacology- col* 2007; 110: 130–147.

Azizi H, Keshavarzi M. Ethnobotanical study of medicinal plants of Sardasht, Western Azerbaijan, Iran. *Journal of Medicinal Herbs* 2015; 6(2): 113–119.

Barros L, Carvalho AM, Ferreira ICFR. Leaves, flowers, immature fruits and leafy flowered stems of Malva sylvestris: A comparative study of the nutraceutical potential and composition. *Food and Chemical Toxicology* 2010; 48: 1466–1472.

Billeter M, Meier B, Sticher O. 8-Hydroxyflavonoid glucuronides from Malva sylvestris. *Phytochemistry* 1991; 30(3): 987–90.

Blunden G, Patel AV, Armstrong NJ, Gorham J. Betaine distribution in the Malvaceae. *Phytochemistry* 2001; 58(3): 451–454.

Bonjar S. Evaluation of antibacterial properties of some medicinal plants used in Iran. *Journal of Ethnopharmacology* 2004; 94: 301–305.

Brouillard R. The *in vivo* expression of anthocyanin colour in plants. *Phytochemistry* 1983; 22: 1311–1323.

Bulut G. Folk medicinal plants of Silivri (Istanbul Turkey). *Marmara Pharmaceutical Journal* 2011; 15(1): 25–29.

Bulut G, Tuzlacı E. Ethnobotanical study on medicinal plants in Turgutlu (Manisa- Turkey). *Journal of Ethnopharmacology* 2013; 149: 633–647.

Cakilcioglu U, Turkoglu I. Plants used for pass kidney stones by the folk in Elazığ. *Herbal Journal of Experimental Botany* 2008; 14: 133–144.

Capek P, Kardašová A. Polysaccharides from the flowers of Malva mauritiana L.: Structure of an arabinogalactan. *Collection of Czechoslovak Chemical Communications* 1995; 60(12): 2112–2118.

Calvo MI, Akerreta S, Cavero RY. Pharmaceutical ethnobotany in the riverside of Navarra (Iberian Peninsula). *Journal of Ethnopharmacology* 2011; 135(1): 22–33.

Calvo MI, Cavero RY. Medicinal plants used for neurological and mental disorders in Navarra and their validation from official sources. *Journal of Ethnopharmacology* 2015; 169: 263–268.

Čapek P, Kardošová A, Lath D. Heteropolysaccharide from the Flowers of Malva mauritiana L. *Chemical Papers* 1999; 53: 131–136.

Cameio-Rodrigues J, Ascensão L, Bonet MA, Vallès J. An ethnobotanical study of medicinal and aromatic plants in the Natural Park of 'Serra de São Mamede'(Portugal). *Journal of Ethnopharmacology* 2003; 89: 199–209.

Chiclana CF, Enrique A, Consolini AE. Topical antiinflam- matory activity of Malva sylvestris L. (Malvaceae) on carragenin-induced edema in rats. *Latin American Journal of Pharmacy* 2009; 28: 275–278.

Classen B, Blaschek W. An arabinogalactan-protein from cell culture of *Malva sylvestris. Planta Medica* 2002; 68: 232–236.

Classen B, Blaschek W. High molecular weight acidic polysaccharides from Malva sylvestris and Alcea rosea. *Planta Medica* 1998; 64: 640–644.

Cogo LL, Monteiro CL, Miguel MD, Miguel OG, Cunico MM, Ribeiro ML, et al. Anti-Helicobacter pylori activity of plant extracts traditionally used for the treatment of gastrointestinal disorders. *Brazilian Journal of Microbiology* 2010; 41(2): 304–309.

Conforti F, Ioele G, Statti GA, Marrelli M, Ragno G, Menichini F. Antiproliferative activity against human tumor cell lines and toxicity test on Mediterranean dietary plants. *Food and Chemical Toxicology* 2008; 46: 3325–3332.

Cullen J. Malva. In: Davis PH (ed.), *Flora of Turkey and the East Aegean Islands*, vol. 2, 1967; pp. 401–408, Edinburgh: Edinburgh University Press.

Cutillo F, D'Abrosca B, DellaGreca M, Fiorentino A, Zarrelli A. Terpenoids and phenol derivatives from Malva sylvestris. *Phytochemistry* 2006; 67: 481–485.

Dalara A, Mukemreb M, Unalc M, Ozgokceb F. Traditional medicinal plants of Ağrı Province, Turkey. *Journal of Ethnopharmacology* 2018; 226: 56–72. https://doi.org/10.1016/j.jep.2018.08.004.

Daniela A, Pichichero E, Canuti L, Cicconi R, Karou D, D'Arcangelo G, et al. Identification of phenolic compounds from medicinal and melliferous plants and their cytotoxic activity in cancer cells. *Caryologia* 2007; 60(1–2): 90–95.

De Natale A, Pollio A. Plants species in the folk medicine of Montecorvino Rovella (Inland Campania, Italy). *Journal of Ethnopharmacology* 2007; 109: 295–303.

de Souza GC, Haas AP, von Poser GL, Schapoval EE, Elisabetsky E. Ethnopharmacological studies of antimicrobial remedies in the south of Brazil. *Journal of Ethnopharmacology* 2004; 90(1): 135–43.

DellaGreca M, Cutillo F, D'Abrosca B, Fiorentino A, Pacifico S, Zarrelli A. Antioxidant and radical scavenging properties of Malva sylvestris. *Natural Product Communications* 2009; 4: 893–896.

El Ghaoui WB, Ghanem EB, Chedid LA, Abdelnoor AM, et al. The effects of *Alcea rosea* L., Malva sylvestris L. and Salvia libanotica L. Water extracts on the production of anti-egg albumin antibodies, interleukin-4, gamma interferon and interleukin-12 in BALB/c mice. *Phytotherapy Research* 2008; 22: 1599–1604.

El SN, Karakaya S. Radical scavenging and iron-chelating activities of some greens used as traditional dishes in Mediterranean diet. *International Journal of Food Sciences and Nutrition* 2004; 55: 67–74.

Emets TI, Steblyuk MV, Klyuev NA, Petrenko PP. Some components of the seed oil of *Malva sylvestris*. *Chemistry of Natural Compounds* 1994; 30: 322–325.

Esteves PF, Sato A, Esquibel MA, de Campos-Buzzi F, Meira AV, Cechinel-Filho V. Antinociceptive activity of *Malva sylvestris* L. *Latin American Journal of Pharmacy* 2009; 28(3): 454–456.

Fakir H, Korkmaz M, Güller B. Medicinal plant diversity of Western Mediterranean region in Turkey. *Journal of Applied Biological Sciences* 2009; 3: 33–43.

Farina A, Doldo A, Cotichini V, Rajevic M, Quaglia MG, Mulinacci N, et al. HPTLC and reflectance mode densitometry of anthocyanins in Malva silvestris L.: A comparison with gradient-elution reversed-phase HPLC. *Journal of Pharmaceutical and Biomedical Analysis* 1995; 14(1–2): 203–11.

Ferreira A, Proença C, Serralheiro MLM, Araújo MEM. The in vitro screening for acetylcholinesterase inhibition and antioxidant activity of medicinal plants from Portugal. *Journal of Ethnopharmacology- col* 2006; 108: 31–37.

Franova S, Nosalova G, Mokry J. Phytotherapy of cough. *Advances in Phytomedicine* 2006; 2: 111–131.

Franz G. Die Schleimpolysaccharide von *Althaea officinalis* and *Malva* sylvestris. *Planta Medica* 1966; 14: 90–110.

Gasparetto JC, Ferreira Martins CA, Sayomi Hayashi S, Otuky MF, Pontarolo P. Ethnobotanical and scientific aspects of *Malva sylvestris* L.: A millennial herbal medicine. *Journal of Pharmacy and Pharmacology* 2011; 64: 172–189. https://doi.org/10.1111/j.2042-7158.2011.01383.x.

Gasparetto JC, Martins CAF, Hayashi SS, Otuky MF, Pontarolo R. Ethnobotanical and scientific aspects of Malva sylvestris L.: A millennial herbal medicine. *Journal of Pharmacy and Pharmacology* 2012; 64(2): 172–189.

Guarrera PM. Food medicine and minor nourishment in the folk traditions of central Italy (Marche, Abruzzo and Latium). *Fitoterapia* 2003; 74: 515–544.

Ghédira K., Goetz P, Malva sylvestris L. (Malvaceae): Mauve. *Phytothérapie*, 2016; 14: 68–72.

Guil JL, Torija ME, Gimenez JJ, Rodriguez I. Identification of fatty acids in edible wild plants by gas chromatography. *Journal of Chromatography A* 1996; 719: 229–235.

Henry AG, Piperno DR. Using plant microfossils from dental calculus to recover human diet: A case study from Tell al-Raqā'i, Syria. *Journal of Archaeological Science* 2008; 35: 1943–1950.

Hicsonmez U, Ereeş FS, Ozdemir C, Ozdemir A, Cam S. Determination of major and minor elements in the *Malva sylvestris* L. from Turkey using ICP-OES techniques. *Biological Trace Element Research* 2009; 128: 248–257.

Hussain L, Ikram J, Rehman K, Tariq M, Ibrahim M, Akash MS. Hepatoprotective effects of *Malva sylvestris* L. against paracetamol-induced hepatotoxicity. *Turkish Journal of Biology* 2014; 38(3): 396–402.

Idolo M, Motti R, Mazzoleni S. Ethnobotanical and phytomedicinal knowledge in a long-history protected area, the Abruzzo, Lazio and Molise National Park (Italian Apennines). *Journal of Ethnopharmacology* 2010; 127: 379–395.

Ishtiaq M, Hanif W, Khan MA, Ashraf M, Butt AM. An ethnomedicinal survey and documentation of important medicinal folklore food phytonims of flora of samahni valley, (Azad kashmir) Pakistan. *Pakistan Journal of Biological Sciences* 2007; 10(13): 2241–2256.

Jarić S, Mačukanović-Jocić M, Djurdjević L, Mitrović M, Kostić O, Karadžić B, Pavlović P. An ethnobotanical survey of traditionally used plants on Suva planina mountain (south-eastern Serbia). *Journal of Ethnopharmacology* 2015; 175: 93–108.

Khan S, Rehmana S, Zeb Khana A, Khana MA, Shahb MT. Soil and vegetable enrichment with heavy metals from geological sources in Gilgit, northern Pakistan. *Ecotoxicology and Environmental Safety* 2010; 73: 1820–1827.

Kızılarslan Ç, Özhatay N. An ethnobotanical study of the useful and edible plants of İzmit. *Marmara Pharmaceutical Journal* 2012; 16: 134–140.

Kültür Ş. Medicinal plants used in Kırklareli Province (Turkey). *Journal of Ethnopharmacology* 2007; 111: 341–364.

Kültür Ş, Sami SN. Medicinal plants used in Isperih (Razgrad – Bulgaria) district. *Turkish Journal of Pharmaceutical Sciences* 2009; 6(2): 107–124.

Kültür Ş. Medicinal plants used in Kırklareli Province (Turkey). *Journal of Ethnopharmacology* 2007; 111: 341–364. https://doi.org/10.1016/j.jep.2006.11.035.

Kumarasamy Y, Byres M, Cox PJ, Jaspars M, Nahar L, Sarker SD. Screening seeds of some Scottish plants for free radical scavenging activity. *Phytotherapy Research* 2007; 21: 615–621.

Leond M, Casu L, Sanna F, Bonsignore L. A comparison of medicinal plant use in Sardinia and Sicily-de materia medica revisited? *Journal of Ethnopharmacology* 2009; 121: 55–67.

Loizzo MR, Pugliese A, Bonesi M, Tenuta MC, Menichini F, Xiao J, Tundis R. Edible flowers: A rich source of phytochemicals with antioxidant and hypoglycemic properties. *Journal of Agricultural and Food Chemistry* 2016; 64(12): 2467–2474.

Magro A, Carolino M, Bastos M, Mexia A. Efficacy of plant extracts against stored products fungi. *Revista Iber- oamericana De Micologia* 2007; 23(3): 176–178, 2006.

Marc EB, Arnold N, Delelis-Dusollier A, Dupont F. Plants are used as remedies for antirheumatic and antineuralgic in the traditional medicine of Lebanon. *Journal of Ethnopharmacology* 2008; 120: 315–334.

Mas TM, Poggiob SL, Verdúa AMC. Weed community structure of mandarin orchards under conventional and integrated management in northern Spain. *Agriculture, Ecosystems & Environment* 2007; 119: 305–310.

Menale B, Castro O, Cascone C, Muoio R. Ethnobotanical investigation on medicinal plants in the Vesuvio National Park (Campania, southern Italy). *Journal of Ethnopharmacology* 2016; 192: 320–349.

Miraldi E, Ferri S, Mostaghimi V. Botanical drugs and preparations in the traditional medicine of West Azerbaijan (Iran). *Journal of Ethnopharmacology* 2001; 75: 77–87.

Mukarram M, Ahmad I, Ahmad M. Hbr-reactive acids of Malva sylvestris seed oil. *Journal of the American Oil Chemists' Society* 1984; 61: 1060–1060.

Nawwar MA, Buddrus J. A gossypetin glucuronide sulphate from the leaves of Malva sylvestris. *Phytochemistry* 1981; 20(10): 2446–2448.

Pereira C, Barros L, Carvalho AM, Ferreira IC. Use of UFLC-PDA for the analysis of organic acids in thirty-five species of food and medicinal plants. *Food Analytical Methods* 2013; 6: 1337–1344.

Pirbalouti AG, Azizi S, Koohpayeh A, Hamedi B. Wound healing activity of Malva sylvestris and Punica granatum in alloxan-induced diabetic rats. *Acta Poloniae Pharmaceutica* 2010; 67(5): 511–516.

Pirbalouti AG, Yousefi M, Nazari H, Karimi I, Koohpayeh A. Evaluation of burn healing properties of Arnebia euchroma and Malva sylvestris. *Electronic Journal of Biology* 2009; 5(3): 62–66.

Pollio A, De Natale A, Appetiti E, Aliotta G, Touwaide A. Continuity and change in the Mediterranean medical tradition: *Ruta* spp. (Rutaceae) in Hippocratic medicine and present practices. *Journal of Ethnopharmacology* 2008; 116: 469–482.

Pourrat H, Texier O, Barthomeuf C. Identification and assay of anthocyanin pigments in Malva sylvestris L. *Pharmaceutica Acta Helvetiae* 1990; 65: 93–96.

Quave CL Plano LR, Pantuso T, Bennett BC. Effects of extracts from Italian medicinal plants on planktonic growth, biofilm formation and adherence of methicillin-resistant Staphylococcus aureus. *Journal of Ethnopharmacology* 2008; 118: 418–428.

Razavi SM, Zarrini G, Molavi G, Ghasemi G. Bioactivity of *Malva sylvestris* L., a medicinal plant from Iran. *Iranian Journal of Basic Medical Sciences* 2011; 14(6), 574–579.
Redžić, S, Tuka, M. Plant pigments (antioxidants) of medicinal plants Malva silvestris L. and Malva moschata L.(Malvaceae). *Bosnian Journal of Basic Medical Sciences* 2005; 5(2): 53.
Rigat M, Valles J, Gras A, Iglésias J, Garnatje T. Plants with topical uses in the Ripollès district (Pyrenees, Catalonia, Iberian Peninsula): Ethnobotanical survey and pharmacological validation in the literature. *Journal of Ethnopharmacology* 2015; 164: 162–179.
Šarić-Kundalić B, Dobeš C, Klatte-Asselmeyer V, Saukel J. Ethnobotanical study on medicinal use of wild and cultivated plants in middle, south and west Bosnia and Herzegovina. *Journal of Ethnopharmacology* 2010; 131(1): 33–55.
Schulz H, Baranska M. Identification and quantification of valuable plant substances by IR and Raman spectroscopy. *Vibrational Spectroscopy* 2007; 43: 13–25.
Seiberg Stone VI, Zhao R. Use of Malva species extract for enhancing the elasticity or structural integrity of skin or mucosal tissue, e.g. vaginal mucosal tissue. Patent number (s): US2006088615- A1, 2006.
Sezik E, Yesilada E, Honda G, Takaishi Y, Takeda Y, Tanaka T. Traditional medicine in Turkey X: folk medicine in Central Anatolia. *Journal of Ethnopharmacology* 2001; 75: 95–115.
Sikorska M, Matławska R, Fra ski R. 8-Hydroxyflavonoid glucuronides of *Malope trifida. Acta Physiologiae Plantarum* 2004; 26: 291–297.
Simsek I, Aytekin F, Yesilada E, Yildirimli Ş. An ethnobotanical survey of the Beypazarı, Ayas, and Güdül District towns of Ankara Province (Turkey). *Economic Botany* 2004; 58: 705–720.
Sleiman NH, Daher CF. *Malva sylvestris* water extract: A potential anti-inflammatory and anti-ulcerogenic remedy. *Planta Medica* 2009; 75: 1010–1010.
Stone VI, Zhao R. Ingestible composition for treating, e.g. sagging, lax and loose tissues and tightening and strengthening tissues such as skin, mucosal tissues comprises and aqueous Cotinus coggygria extract and Malva sylvestris extract. Patent Number(s): US2010040707-A1, 2010.
Takeda K, Enoki S, Harborne JB, Eagles J. Malonated anthocya- nins in Malvaceae – Malonylmalvin from *Malva sylvestris. Phytochemistry* 1989; 28: 499–500.
Talbourdet S, Sadick NS, Lazou K, Bonnet-Duquennoy M, Kurfurst R, Neveu M, Heusèle C, André P, Schnebert S, Draelos ZD, Perrier E. Modulation of gene expression as a new skin anti-aging strategy. *Journal of Drugs in Dermatology* 2007; 6 Supplement: 25–33.
Terninko II, Onishchenko UE, Frolova A. Research phenolic compounds Malva sylvestris by high performance liquid chromatography. *The Pharma Innovation* 2014, 3(4).
Terninko II, Onishchenko UE. Component composition of organic acids in leaves of Malva sylvestris. *Chemistry of Natural Compounds* 2013; 49: 332–333.
Tomoda M, Gonda R, Shimizu N, Yamada H. Plant mucilages 42. An anti-complementary mucilage from the leaves of *Malva sylvestris* var *mauritiana. Chemical and Pharmaceutical Bulletin* 1989; 37: 3029–3032.
Tuzlacı E, Alpaslan DF. Turkish folk medicinal plants. Part V: Babaeski (Kırklareli). *Journal of Faculty Pharmacy, Istanbul* 2007; 39: 11–23.
Tosi B, Tirillini B, Donini A, Bruni A. Presence of scopoletin in Malva sylvestris. *International Journal of Pharmacognosy* 1995; 33(4): 353–355.
Tuzlacı E, İşbilen DFA, Bulut G. Turkish folk medicinal plants, VIII: Lalapaşa (Edirne). *Marmara Pharmaceutical Journal* 2010; 14: 47–52.
Usami A, Kashima Y, Marumoto S, Miyazawa M, Characterization of aroma-active compounds in dry flower of Malva sylvestris L. by GC-MS-O analysis and OAV calculations. *Journal of Oleo Science* 2013; 62(8): 563–570.
Veshkurova O, Golubenko Z, Pshenichnov E, Arzanova I, Uzbekov V, Sultanova E, et al. Malvone A, a phytoalexin found in Malva sylvestris (family Malvaceae). *Phytochemistry* 2006; 67(21): 2376–2379.
Watanabe E, Guerreiro Tanomaru JM, Piacezzi Nascimento A, Júnior FM, Filho MT, Yoko Ito I. Determination of the maximum inhibitory dilution of cetylpyridinium chloride-based mouthwashes against *Staphylococcus aureus*: An *in vitro* study. *Journal of Applied Oral Science* 2008; 16: 275–279.
Wichtl, M. (Ed.). *Herbal drugs and phytopharmaceuticals: A handbook for practice on a scientific basis.* CRC Press, 2004.
Widy-Tyszkiewicz E. Assessment report on Malva sylvestris L. and/or Malva neglecta Wallr., folium and Malva sylvestris L., flos. EMA/HMPC/749518 Committee on Herbal Medicinal Products (HMPC), 2018.
Yeşilyurt EM, Şimşek I, Akaydin G, Yeşilada E. An ethnobotanical survey in selected districts of the Black Sea region (Turkey). *Turkish Journal of Botany* 2017; 41: 47–62. https://doi.org/10.3906/bot-1606-12.

15 *Origanum* sp.

Methiye Mancak and Ufuk Koca Çalışkan

INTRODUCTION

The *Origanum* genus, which belongs to the Lamiaceae family, is a medicinal and aromatic plant that has been utilized by humans for many years. The name Origanum is formed by the combination of Greek words "*oros*" (meaning mountain) and "*gano*" (meaning joy) (Fonnegra, 2007). *Origanum* is classified into ten sections and includes 38 species, six subspecies, and 17 hybrids. Most of the species (about 70%) grow in the Eastern Mediterranean region and are usually endemic to one region (Ietswaart, 1980). *O. vulgare* (oregano) and *O. majorana* (marjoram) are frequently used all over the world. *Origanum* grows 22 species or 32 taxa in Turkiye and 21 of them are endemic (Figure 15.1). *O. acutidens* is an endemic species widely distributed in Eastern Anatolia. The distribution of endemic species in Turkiye is given in Figure 15.2. Since 60% of the *Origanum* taxa in the world are grown in Turkiye, the *Origanum* gene center is thought to be in Turkiye. The country has an important place in the *Origanum* trade. In Turkiye, *O. onites* (Turkish oregano, Izmir kekigi) and *O. majorana* (marjoram, white oregano) are usually utilized as a spice and for medicinal purposes and exported (Baser, 2002).

Origanum species are rich in volatile oil containing carvacrol and thymol. *Origanum* sp. and their volatile oil are frequently used in the cosmetics, food, and pharmaceutical industries due to their antibacterial, antifungal, antiviral, antiproliferative, antidiabetic, antioxidant, and anti-inflammatory effects (Sharifi-Rad et al., 2021).

BOTANICAL FEATURES/ETHNOBOTANICAL USAGE

Origanum botanical features in brief: Subshrubby Lamiaceae with several erect, glandular punctate, ± ovate leaves, medium-sized stems. The length of the leaves is 2–40 mm, the width is 2–30 mm, and the length/width ratio is 1.3 on average. Leaves are usually petiolate at the lower nodes but in some species are sessile. Inflorescences are paniculate. Few (sub) sessile flowers in a verticillaster. Calyxes are highly variable in the *Origanum* genus while being straight in their outlines is common. Calyces the throats are pilose. Corollas, stamens, and styles can be of different shapes and/or sizes as adaptation to a variety of pollinating insects. The color of the corollas is pink, purple, or white. The most common type of corolla found in the *Origanum* has a straight tube, is 2-lipped for c. 1/3 (or 1/4), is 3–14 mm long, and has the lips at wide angles to the tube. Staminal filaments range in length from 0.5 to 14 mm and are usually glabrous. The styles can often be up to 22 mm long (rarely up to 40 mm). The nutlets appear in some species and are brown, ovoid, 1–1.5 mm long, and c. 0.5 mm wide. The underground parts of the plants are woody, especially those grown in the Eastern Mediterranean region have fairly thick, woody roots. Near all species, the stems are erect or ascending, but *O. vulgare* forms long superficial runners. Plants often have more than one stem and are usually 30–60 cm in length. In cross section, the stems are round or slightly quadrant and hairy but for a few species, they are glabrous (Ietswaart, 1980).

Origanum sp. are used as a spice, which is often added to pizzas, sauces, meats, and a variety of dishes in various countries for many years. The aerial parts of the plant are used to obtain paint. The species is traditionally used for the treatment of respiratory (such as cough and bronchitis), gastrointestinal (such as abdominal pain, nausea), kidney, and menstrual disorders, and additionally in

DOI: 10.1201/9781003146971-15

FIGURE 15.1 Some endemic *Origanum* species in Turkiye.

diabetes. Furthermore, oregano leaves are used as an antiseptic, to treat skin wounds and to relieve aching muscles (Singletary, 2010).

In Turkiye, various studies have been reported on the traditional use of *Origanum* sp. The infusion of *O. majorana* leaves is used for respiratory disorders such as influenza and asthma (Sargın et al., 2013). The maceration of the whole plant is used as a sedative, diaphoretic, and against stomachaches (Altundag and Ozturk, 2011; Özgökçe and Özçelik, 2004). The leaves of the plant mixed with seeds are used for stomach pain and against atherosclerosis (Uysal et al., 2010). The whole plant is used for circulation system and digestive disorders, sedative and diaphoretic effect (Everest ve Öztürk, 2005; Altundag and Ozturk, 2011). *O. vulgare* is used to treat asthma, atherosclerosis, cold flu, epilepsy, headache, hypertension, stomachache, toothache, and wound (Altundag and Ozturk, 2011; Hayta, Polat, and Selvi, 2014; Polat and Satıl, 2012; Tabata et al., 1994; Yesilada et al., 1993). *O. onites* is used to treat abdominal pain, cold, headache, toothache, and diabetes (Polat and Satıl, 2012; Yesilada et al., 1993). *O. acutidens* is used for abdominal pain, while *O.* x *majoricum* is used for cold, flu, nausea, stomachache, and urinary diseases (Altundag and Ozturk, 2011; Polat and Satıl, 2012; Yesilada et al., 1993). Additionally, *O. minutiflorum* is used for rheumatism, toothache, and stomachache (Yesilada et al., 1993).

PHYTOCHEMICALS

Origanum sp. contain volatile compounds, flavonoids, phenolic acids (caffeic acid, p-coumaric acid), rosmarinic acid, caffeoyl derivatives, carnosic acid, and ursolic acid. Rosmarinic acid is one of the most abundant phenolic compounds in aqueous extracts of oregano leaves (Singletary, 2010). The main constituents of volatile oils of *Origanum* sp. are terpenes, especially mono- and sesquiterpenes. The terpenes contained in different types of *Origanum* are carvacrol, thymol, linalool,

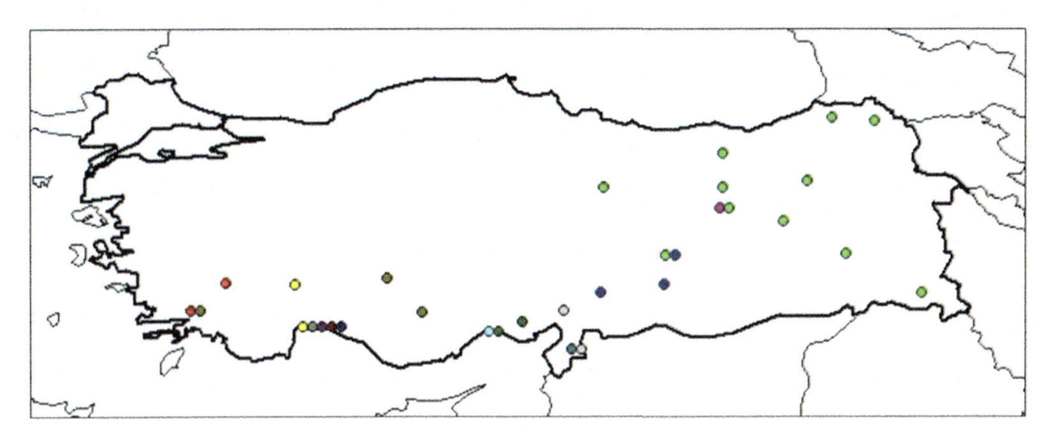

○ *O. boissieri*, ● *O. munzurense*, ○ *O. saccatum*, ● *O. solymicum*, ● *O. hypericifolium*,

● *O. acutidens*, ● *O. haussknechtii*, ● *O. brevidens*, ● *O. husnucan-baseri*,

● *O. leptocladum*, ○ *O. amanum*, ● *O. bilgeri*, ● *O. micranthum*, ● *O. minutiflorum*

FIGURE 15.2 Distribution of endemic species in Turkiye (Turkish Plants Data Service-TÜBİVES) *O. boissieri, O. munzurense, O. saccatum, O. solymicum, O. hypericifolium, O. acutidens, O. haussknechtii, O. brevidens, O. husnucan-baseri, O. leptocladum, O. amanum, O. bilgeri, O. micranthum, O. minutiflorum.*

β-caryophyllene, γ-terpinene, p-cymene, β-myrcene, terpinen-4-ol, and trans-sabinene hydrate. Various studies describe the phytochemical contents of *Origanum* sp. grown in Turkiye. *O. onites* consists of volatile oil, total 134 terpenoids, ten phenolics, six phenylpropanoids, four sterols, three triterpene acids, one quinone, 14 vonoids and related compounds, 14 fatty acids, five porphyrins, seven hydrocarbons, four tocopherols, inorganic compounds and/or minerals. In these studies performed with the aerial parts of *O. onites*, it was determined that it mostly contained terpenoids (especially monoterpene hydrocarbons and oxygenated monoterpenes) (Tepe et al., 2016). One study reported carvacrol (83.97–88.65%), thymol (0–7.48%), γ-terpinene (2.63–6.15%), p-cymene (1.52–3.16%), α-terpinene (0–1.05%), and α-pinene (0–1.88%) in *O. onites* volatile oil (Ozkan et al., 2010). Other similar study reported carvacrol (85.86%), γ-terpinene (4.43%), β-phellandrene (3.20%), and p-cymene (1.83%) in *O. onites* volatile oil as well (Arslan et al., 2012). The main components of *O. majorana* volatile oil were also carvacrol (52.5%) and linalool (45.4%) (Erdogan and Ozkan, 2017). Major compositions of essential oil of *Origanum* species are given in Table 15.1.

BIOLOGICAL ACTIVITIES/EFFECTS

Analgesic, Antinociceptive, and Spasmolytic Activity

Carvacrol was effective for orofacial pain management. Mice were administered carvacrol (CARV) (10 and 20 mg/kg, p.o.), β-cyclodextrin complex containing carvacrol (CARV-βCD) (10 and 20 mg/kg, p.o.), vehicle or morphine (10 mg/kg, i.p.) prior to induced nociceptive behavior. CARV-βCD reduces nociceptiveness during two phases of formalin testing (20 mg/kg, p.o.). While CARV did not reduce the face friction behavior in the first stage, it (20 mg/kg, p.o.) produced 28.7% analgesic inhibition in the second stage. The effect of CARV and CARV-βCD on opioid, vanilloid, and glutamate systems has been suggested as a possible mechanism of its analgesic activity (Silva et al., 2016).

The spasmolytic effect of thymol was investigated in circular smooth muscle strips from guinea pig stomach and vena portals. Thymol showed spasmolytic effect on α_1-, α_2-, and β-adrenergic

TABLE 15.1
Major Composition of Essential Oil of *Origanum* Species

Species name	Composition	Reference
Origanum acutidens	Carvacrol (61.8%); *p*-cymene (15.5%); thymol (12.7%)	Gulec et al., 2014
Origanum bilgeri	Carvacrol (90.20–84.30%); *p*-cymene (3.40–5.85%); γ-terpinene (0.47–1.20%); thymol (0.69–1.08%)	Sozmen et al., 2012
Origanum boissieri	Cymene (42.8%); carvacrol (17.57%)	Baser and Duman, 1998
Origanum haussknechtii	*p*-Cymene (15.56%); borneol (14.24%)	Baser et al., 1998
Origanum hypericifolium	*p*-Cymene (34.33%); carvacrol (21.76%); thymol (19.54%); γ-terpinene (13.91%)	Ocak et al., 2012
Origanum majorana	Carvacrol (52.5%); linalool (45.4%)	Erdogan and Ozkan, 2017
Origanum minutiflorum	Carvacrol (68.23%); *p*-cymene (11.84%)	Dadalioğlu and Evrendilek, 2004
Origanum munzurense	4-hydroxy-3-methylbenzaldehyde (44.84%); thymol (14.59%)	Yabalak et al., 2020
Origanum onites	Carvacrol (85.86%); γ-terpinene (4.43%); *β*-phellandrene (3.20%); *p*-cymene (1.83%)	Arslan et al., 2012
Origanum saccatum	*p*-Cymene (72.5–70.6%); thymol (9.32–8.11%); carvacrol (7.18–6.36%)	Sozmen et al., 2011

receptors at doses higher than 10^{-6} M. In the same study, it was determined that thymol affects α_2 adrenergic receptors and has an analgesic effect (Beer et al., 2007).

Antibacterial Activity

The antibacterial effect of *Origanum* species against both Gram-positive and Gram-negative bacteria has been determined. *Origanum* volatile oils, with their hydrophobic character, increase the permeability of the bacterial membrane, and then cause the cell contents to leak (Bouyahya et al., 2019). The volatile oil of *O. acutidens* (Hand.-Mazz.) Ietswaart was more effective against Gram-positives (*Enterobacter cloacae, Escherichia coli, Klebsiella pneumonia, Proteus vulgaris, Pseudomonas aeruginosa, Salmonella typhimurium, Serratia marcescens*) (Cosge et al., 2009). According MIC (minimum inhibitory concentration) values, *O. majorana* volatile oil (2000 µg/mL) was more effective on *E. coli* and *Pseudomonas aeruginosa* than *S. aureus* (31.25 µg. mL^{-1}) (Chaves et al., 2019). *O. marjorana* and *O. vulgare* volatile oils showed antibacterial effects on various bacteria (*Acinetobacter* spp., *Enterobacter* spp., *Klebsiella* spp., *Proteus* spp., *Staphylococcus aureus*, and *S. coagulase negative*) isolated from patients with conjunctivitis. The minimum inhibitory concentration was 2.5–10 µL/mL for *O. majorana* volatile oil and 5–20 µL/mL for *O. vulgare* volatile oil (Oliveira et al., 2009). Volatile oils of *O. marjorana*, *O. onites*, and *O. vulgare* showed antibacterial effect on *E. coli*, *Listonella anguillarum*, *Saccharomyces cerevisiae*, and *Vibrio* sp. Pure carvacrol had the greatest antibacterial activity on bacterial strains (Stefanakis et al., 2013). *O. majorana* volatile oil was more effective than streptomycin and chloramphenicol against foodborne bacteria (*Listeria monocytogenes, Salmonella typhimurium, Escherichia coli*, and *Staphylococcus aureus*) (Partovi et al., 2018). *O. majorana* methanol extract was effective against *H. pylori* associated with gastrointestinal diseases (MIC: 50 µg/mL) (Mahady et al., 2005). *O. vulgare* phenolics also inhibited the growth of *H. pylori* (Chun et al., 2005). The antibacterial activities of volatile oils of 12 plants (50–400 µg/mL), including *O. vulgare*, were investigated. The oils of *O. vulgare* showed the highest antibacterial activities with MIC values ranging between 0.11 and 0.76 mg/mL against examined bacteria with MBC (minimum bactericidal concentration) values ranging between 0.21 and 1.59. Moreover, it showed higher antibacterial activity than examined antibiotics in some cases (Elansary et al., 2018).

Antifungal Activity

O. acutidens volatile oil showed significant antifungal activity on 12 (67%) of the 18 fungi included in the study (Sokmen et al., 2004). *O. dictamnus*, *O. majorana*, and *O. vulgare* volatile oils inhibited *Penicillium digitatum* at low concentrations (250–400 µg/mL) (Daferera et al., 2000). *O. majorana* methanolic extract inhibited the growth of *Aspergillus niger* (Leeja et al., 2007). *O. onites* volatile oils showed antifungal activity against all tested fungi. They were most effective against *Aspergillus alternata* and later against *Penicillium roqueforti* (Korukluoglu et al., 2009). In another study, the antifungal activity of *O. onites* was associated with carvacrol (Soković et al., 2002). *O. vulgare* volatile oil was effective against *Candida* sp. and *Aspergillus niger* (Santoyo et al., 2006; Sauza et al., 2006). The efficacy of thymol and carvacrol on *Candida albicans* has been proven (Jirovetz et al., 2007; Tampieri et al., 2005). The combination of *O. vulgare* volatile oil with nystatin had a synergistic effect on *Candida* species (Rosato et al., 2009). *O. vulgare* subsp. *hirtum* and *O. minutiflorum* methanol extracts were effective against *Aspergillus flavus*, *A. niger*, *A. ochraceus*, and *Fusarium proliferatum* (Askun et al., 2008).

Anticancer and Antiproliferative Activity

In a study, extracts were prepared from aerial parts of *O. majorana*. Compounds (5,6,3′-Trihydroxy-7,8,4′-trimethoxyflavone, arbutin, hesperetin, hydroquinone, and rosmarinic acid) were then obtained from its water-soluble ethyl acetate extract. The antiproliferative effects of the extract and the isolated compounds were evaluated by BrdU cell proliferation enzyme-linked immunosorbent assay and xCELLigence assay. Hesperetin and hydroquinone showed stronger antiproliferative effect than 5-fluorouracil in C6 (rat brain tumor) and HeLa (human cervical cancer) cell lines (Erenler et al., 2016). The ethanol extract of *O. majorana* showed potent antiproliferative effect on two human colorectal cancer cell lines (HT-29 and Caco-2) in a concentration- and time-dependent manner (HT-29 IC_{50}: 498 and 342 µg/mL; Caco-2 cells IC_{50}: 506 and 296 µg/mL, at 24 and 48 h). Extracts (300, 450, and 600 µg/mL) inhibited cell cycle progression in G2/M (extract: 51%, control: 30%). Moreover, it induced autophagy and apoptosis (Benhalilou et al., 2019).

The antiproliferative effects of volatile oil of *O. onites* on breast cancer cells (MCF-7), colon cancer cells (HT-29, CT26), hepatocellular carcinoma cells (HepG2), and melanoma cells (A375) were investigated with sulforhodamine B test. Antiproliferative activity was mostly observed in HT-29 cells. It showed antiproliferative, and pro-apoptotic potential, especially on colorectal cancer. Oral administration of its (0.370 g/kg) for 13 days inhibited CT26-tumor growth in mice. Treated mice had a 52% lower tumor value compared to the control group (Spyridopoulou et al., 2019).

The anticancer activities of volatile oils of 12 plants (50–400 µg/mL), including *O. vulgare*, were investigated. For antiproliferative effects using HT-29, MCF-7, cervical adenocarcinoma (HeLa), T-cell lymphoblast (Jurkat), urinary bladder carcinoma (T24), normal human cell line (HEK-293), the highest anticancer activity belonged to *O. vulgare*, followed by *Citrus* sp. and *Artemisia monosperma*. No oils, including *O. vulgare*, showed inhibitory activity on HEK-293 (kidney epithelium) (Elansary et al., 2018). Volatile oil from *O. vulgare* (50 mg/mL) showed antiproliferative effect *in vitro* against human colon adenocarcinoma (HT-29) (60.8%) and human breast adenocarcinoma (MCF-7) (48.9%). Cytotoxic effect was higher in HT-29 (Begnini et al., 2014). In another study, *O. vulgare* volatile oil (0.011, 0.0037, 0.0012, and 0.00041%, v/v, concentration) had antiproliferative effect in human dermal fibroblasts. But, 0.011% concentration was cytotoxic (Han and Parker, 2017). *O. vulgare* volatile oil also showed antiproliferative effect in gastric cancer. It even inhibited the migration of cancer cells *in vitro*. It changed the colony-forming properties of the cancer cell and prevented them from multiplying (Balusamy et al., 2018). Crude *O. vulgare* extract had anticancer activity in adrenocortical tumor cell lines. The extract showed its effects by suppressing the MAPK and PI3 K/Akt pathways (Rubin et al., 2019).

The anticancer effect of carvacrol, which is largely found in *Origanum* species, on choriocarcinoma cell lines (JAR and JEG3) was investigated. The results showed that carvacrol suppressed PI3K-AKT and MAPK signaling, produced oxidative stress, and altered calcium homeostasis mediated by mitochondrial dysfunction in human choriocarcinoma JAR and JEG3 cells. It reduced the viability of JAR and JEG3 cells by approximately 76% and 49%, respectively (Lim et al., 2019). In another study, carvacrol showed antiproliferative effect in human prostate cancer cells. Carvacrol induced apoptosis, caspase activation, cell death in cancer cells, and ROS-mediated mitochondrial membrane depolarization. Additionally, it has cell cycle arrest in prostate cancer cells (Khan et al., 2017).

Antidiabetic and Metabolic Activity

Methanol extract of *O. majorana* leaves strongly inhibited rat intestinal α-glucosidase (Kawabata et al., 2003). Aqueous extract of *O. majorana* leaves (2 weeks, 20 and 40 mg/kg body weight) reduced amylase, glucose, insulin, AST, ALT, cholesterol, testosterone, urea, and creatinine levels in diabetic rats compared to the untreated diabetic group (Oaman and Abbas, 2010). In another study, volatile oil and methanolic extract of *O. majorana* leaves had greater antidiabetic activity than the aqueous extract (Pimple et al., 2012). Ethanol extract of *O. majorana* leaves (100, 200, 400 mg/kg body weight, 21 days) showed antidiabetic and antihyperlipidemic activity in diabetic rats. The extract significantly lowered blood sugar and increased serum insulin levels. The effects of the extract on diabetes-induced hyperlipidemia were also quite positive, and it was suggested that it could prevent complications caused by diabetes. It lowered serum total cholesterol, triglyceride, LDL, and VLDL levels and increased HDL levels. Effects were similar to standard drug glibenclamide (4 mg/kg) treatment (Tripathy et al., 2018). *O. onites* decreased LDL-C, Apo B, and lipoprotein(a) while lipid profiles increased HDL in patients with mild hyperlipidemia (Ozdemir et al., 2008). Methanolic extracts of *O. majorana* leaves inhibited AGE (advanced glycation end-products) formation. In addition, it reduced oxidative stress under diabetic conditions (Perez Gutierrez, 2012).

In the study examining the antidiabetic and antihypertensive activities of Lamiaceae species, *O. vulgare* aqueous extracts had the highest α-glucosidase inhibition activity (93.7%). It also had some ACE inhibitory activity (18.5%) (Kwon et al., 2006). Volatile oils from *O. vulgare* subsp. *vulgare* and *O. vulgare* subsp. *hirtum* showed α-amylase and α-glucosidase inhibitory activity (Sarikurkcu et al., 2015). *O. vulgare* leaf extract showed antihyperglycemic, antihyperlipidemic, and renoprotective effects in diabetic rats. Oregano extract is applied alone (150 or 300 mg/kg body) or in combination with chamomile extracts as an aqueous suspension (150 mg/kg of both extracts), for 6 weeks. Oregano application alone decreased blood sugar, amylase activity, and HbA1C while combination application increased serum insulin levels in addition to these effects (Prasanna et al., 2017). In another study, methanol extract of *O. vulgare* leaves (MOE) protected against Type 1 diabetes by showing antioxidant, immunomodulatory, and antiapoptotic effects in mice. Prophylactic treatment with MOE (5 mg/kg, 10 days) preserved normal insulin secretion and β-cells and reduced diabetes incidence. Prophylactic aqueous oregano extract treatment administered in the same way did not affect diabetes (Vujicic et al., 2015).

Anti-inflammatory Activity

Its volatile oil obtained by supercritical fluid extraction from *O. majorana* (10 μg/mL) showed anti-inflammatory activity *in vitro*. Volatile oil suppressed the production of pro-inflammatory cytokines and gene expression (Arranz et al., 2015). The ethanolic extract of *O. majorana* significantly reduced the rate of paw edema and showed anti-inflammatory effect in a model of carrageenan-induced paw edema in rats (Seoudi et al., 2009).

The combination of *T. vulgaris* and *O. vulgare* volatile oils at appropriate concentrations reduced the production of pro-inflammatory cytokines and showed anti-inflammatory effect in mice with colitis (Bukovská et al., 2007). *O. vulgare* volatile oil suppressed inflammatory biomarkers such as intracellular cell adhesion molecule 1 (ICAM-1), interferon gamma-induced protein 10 (IP-10), interferon-inducible T-cell alpha chemoattractant (I-TAC), monocyte chemoattractant protein 1 (MCP-1), monokine induced by gamma interferon (MIG), and vascular cell adhesion molecule 1 (VCAM-1) (Han and Parker, 2017). The ethanol *O. vulgare* extracts suppressed the production of interleukins and TNF-α and reduced skin inflammation from *Propionibacterium acnes* (Chuang et al., 2018). *O. vulgare* volatile oil showed an *in vitro* anti-inflammatory effect on lipopolysaccharide (LPS)-treated murine macrophage cells (RAW264.7). It inhibited the expression and secretion of TNF-α, IL-1β, and IL-6 (Cheng et al., 2018).

Carvacrol exhibited anti-inflammatory effect in human chondrocytes stimulated with interleukin 1β (IL-1β). It suppressed the activation of the NF-κB signaling pathway. It decreased nitric oxide (NO), prostaglandin E2 (PGE2) production, and expression of inducible NO synthase (iNOS) and cyclooxygenase (COX-2) (Xiao et al., 2018).

Antioxidant Activity

The antioxidant activities of volatile oils of 12 plants (50–400 μg/mL), including *O. vulgare*, were investigated. The volatile oil of *O. vulgare* showed significantly the highest antioxidant activities compared to other volatile oils (IC_{50} values of 2.8 and 1.1 mg/L in the DPPH and linoleic acid assays, respectively) (Elansary et al., 2018). *O. onites* showed antioxidant effects in patients with mild hyperlipidemia (Ozdemir et al., 2008). In addition, *O. onites* extracts exhibited free radical scavenging (EC_{50}: 0.489) and ferric-reducing antioxidant (EC50: 0.04) power assay antioxidant properties (Lagouri and Nisteropoulou, 2009).

Non-mutagenic, antioxidant properties of carvacrol were detected in human blood cells *in vitro*. Blood samples from individuals who had not been exposed to any genotoxic agents were treated with different concentrations of carvacrol (0–200 mg/L) for 24 and 48 hours. Carvacrol increased TAC (total antioxidant capacity) levels at 50, 75, and 100 mg/L concentrations, and TOS (total oxidant status) levels at 150 and 200 mg/L concentrations (Türkez and Aydın, 2016).

O. acutidents volatile oil showed DPPH radical scavenging activity due to its high carvacrol content (72 %, v/v). The antioxidant effect of the methanol extract was about as much as the synthetic antioxidant butylhydroxytoluene. The predominance of the methanol extract was attributed to the presence of polar phenolics (Sokmen et al., 2004).

Other Activities

The methanol extract of *O. majorana* inhibited platelet adhesion to laminin-coated wells by 40% and showed antiplatelet activity (Yazdanparast et al., 2008). The *in vitro* anticoagulant activity of *O. vulgare* ethanol extract has been determined and it has been nominated for the prevention and treatment of cardiovascular diseases (Alikhani et al., 2017).

O. onites volatile oil (0.1 or 1 mg/kg/day) administered intraperitoneally and rectally had a significant protective effect on colon damage. There was significant difference in the volatile oil groups in terms of inflammatory cell infiltration, mucosal atrophy, mucus cell depletion, ulceration, vascular dilatation compared to the control (Dundar et al., 2008). *O. majorana* ethanol extract (250 and 500 mg/kg) revealed anti-ulcerogenic activity. The extract decreased acid output, basal gastric secretion, and incidence of ulcers (Al-Howiriny et al., 2009).

According to one study, *O. majorana*, *O. minutiflorum*, *O. onites*, and *O. vulgare* volatile oils presented acetylcholinesterase and butyrylcholinesterase inhibitory activity *in vitro* (Orhan et al., 2008).

O. majorana ethanol extract reduced cisplatin-induced nephrotoxicity by modulating the oxidative stress (Soliman et al., 2016).

O. majorana volatile oil diminished prallethrin-induced oxidative damage and liver damage in rats (Mossa et al., 2013). Furthermore, carvacrol increased the liver regeneration rate in rats (Uyanoglu et al., 2008).

CONCLUSION

Origanum species are rich in carvacrol and thymol, which have predominately antimicrobial properties. At the same time, the complex nature of volatile oils makes it difficult to develop resistance against bacterial strains. These properties make them usable as food preservation additives. The rich phenolic content of the species makes it a potential option for the treatment of diseases such as diabetes and cancers associated with oxidative stress and inflammation. Their spasmolytic effect provides advantages for the treatment of disorders related to the gastrointestinal tract. On the other hand, these effects have been demonstrated by *in vitro* and *in vivo* studies. Considering its economic importance, medicinal benefits, and consumption as food ingredient, Turkiye's rich *Origanum* species should be protected. Amplification on the quality cultivation of *Origanum* and volatile oil production should be supported, and furthermore, with the intention of application *Origanum* derivatives in health treatments should be maintained by clinical studies.

REFERENCES

Al-Howiriny, T., Alsheikh, A., Alqasoumi, S., Al-Yahya, M., ElTahir, K., & Rafatullah, S. (2009). Protective effect of *Origanum majorana* L.'Marjoram'on various models of gastric mucosal injury in rats. *The American Journal of Chinese Medicine*, 37(3), 531–545.

Alikhani Pour, M., Sardari, S., Eslamifar, A., Azhar, A., Rezvani, M., & Nazari, M. (2017). Cheminformatics-based anticoagulant study of traditionally used medicinal plants. *Iranian Biomedical Journal*, 21(6), 400–405. doi: 10.18869/acadpub.ibj.21.6.400. Epub 2017 Apr 29. PMID: 28454485; PMCID: PMC5572436.

Altundag, E., & Ozturk, M. (2011). Ethnomedicinal studies on the plant resources of east Anatolia, Turkiye. *Procedia-Social and Behavioral Sciences*, 19, 756–777.

Arranz, E., Jaime, L., de las Hazas, M. L., Reglero, G., & Santoyo, S. (2015). Supercritical fluid extraction as an alternative process to obtain volatile oils with anti-inflammatory properties from marjoram and sweet basil. *Industrial Crops and Products*, 67, 121–129.

Arslan, M., Uremis, I., & Demirel, N. (2012). Effects of sage leafhopper feeding on herbage colour, volatile oil content and compositions of Turkish and Greek oregano. *Experimental Agriculture*, 48, 428–437.

Askun, T., Tumen, G., Satil, F., & Kilic, T. (2008). Effects of some Lamiaceae species methanol extracts on potential mycotoxin producer fungi. *Pharmaceutical Biology*, 46, 688–694.

Balusamy, S. R., Perumalsamy, H., Huq, M. A., & Balasubramanian, B. (2018). Anti-proliferative activity of *Origanum vulgare* inhibited lipogenesis and induced mitochondrial mediated apoptosis in human stomach cancer cell lines. *Biomedicine & Pharmacotherapy*, 108, 1835–1844. doi: 10.1016/j.biopha.2018.10.028. Epub 2018 Oct 19. PMID: 30372889.

Baser, K. H. C., & Duman, H. (1998). Composition of the essential oils of *Origanum boissieri* letswaart and *O. bargyli* Mouterde. *Journal of Essential Oil Research*, 10(1), 71–72.

Baser, K. H. C., Kürkçüoglu, M., & Tuman, G. (1998). Composition of the essential oil of *Origanum haussknechtii* Boiss. *Journal of Essential Oil Research*, 10(2), 227–228.

Baser, K. H. C. (2002). *The Turkish origanum species* (Vol. 109). London: Taylor & Francis.

Beer, A. M., Lukanov, J., & Sagorchev, P. (2007). Effect of Thymol on the spontaneous contractile activity of the smooth muscles. *Phytomedicine*, 14(1), 65–69.

Begnini, K. R., Nedel, F., Lund, R. G., Carvalho, P. H. D. A., Rodrigues, M. R. A., Beira, F. T. A., & Del-Pino, F. A. B. (2014). Composition and antiproliferative effect of volatile oil of *Origanum vulgare* against tumor cell lines. *Journal of Medicinal Food*, 17(10), 1129–1133.

Benhalilou, N., Alsamri, H., Alneyadi, A., Athamneh, K., Alrashedi, A., Altamimi, N., Al Dhaheri, Y., Eid, A. H., & Iratni, R. (2019). *Origanum majorana* ethanolic extract promotes colorectal cancer cell death by triggering abortive autophagy and activation of the extrinsic apoptotic pathway. *Frontiers in Oncology*, 9, 795. doi: 10.3389/fonc.2019.00795. PMID: 31497536; PMCID: PMC6712482.

Bouyahya, A., Abrini, J., Dakka, N., & Bakri, Y. (2019). Volatile oils of *Origanum compactum* increase membrane permeability, disturb cell membrane integrity, and suppress quorum-sensing phenotype in bacteria. *Journal of Pharmaceutical Analysis*, 9(5), 301–311.

Bukovská, A., Cikos, S., Juhás, S., Il'ková, G., Rehák, P., & Koppel, J. (2007). Effects of a combination of thyme and oregano volatile oils on TNBS-induced colitis in mice. *Mediators of Inflammation*, 2007, 23296. doi: 10.1155/2007/23296. PMID: 18288268; PMCID: PMC2233768.

Chaves, R. D. S. B., Martins, R. L., Rodrigues, A. B. L., de Menezes Rabelo, É., Farias, A. L. F., Araújo, C. M. D. C. V., ... Galardo, A. K. R. (2019). Larvicidal evaluation of the *Origanum majorana* L. Volatile oil against the larvae of the Aedes aegypti mosquito. *bioRxiv*, 595900.

Cheng, C., Zou, Y., & Peng, J. (2018). Oregano volatile oil attenuates RAW264.7 cells from lipopolysaccharide-induced inflammatory response through regulating NADPH oxidase activation-driven oxidative stress. *Molecules*, 23(8), 1857. doi: 10.3390/molecules23081857. PMID: 30049950; PMCID: PMC6222776.

Chuang, L. T., Tsai, T. H., Lien, T. J., Huang, W. C., Liu, J. J., Chang, H., Chang, M. L., & Tsai, P. J. (2018). Ethanolic extract of *Origanum vulgare* suppresses propionibacterium acnes-induced inflammatory responses in human monocyte and mouse ear edema models. *Molecules*, 23(8), 1987. doi: 10.3390/molecules23081987. PMID: 30096960; PMCID: PMC6222868.

Chun, S. S., Vattem, D. A., Lin, Y. T., & Shetty, K. (2005). Phenolic antioxidants from clonal oregano (*Origanum vulgare*) with antimicrobial activity against Helicobacter pylori. *Process Biochemistry*, 40(2), 809–816.

Cosge, B., Turker, A., Ipek, A., & Gurbuz, B. (2009). Chemical compositions and antibacterial activities of the volatile oils from aerial parts and corollas of *Origanum acutidens* (Hand.-Mazz.) Ietswaart, an endemic species to Turkiye. *Molecules*, 14(5), 1702–1712. doi: 10.3390/molecules14051702. PMID: 19471191; PMCID: PMC6254327.

Dadalioğlu, I., & Evrendilek, G. A. (2004). Chemical compositions and antibacterial effects of essential oils of Turkish oregano (*Origanum minutiflorum*), bay laurel (*Laurus nobilis*), Spanish lavender (*Lavandula stoechas* L.), and fennel (*Foeniculum vulgare*) on common foodborne pathogens. *Journal of Agricultural and Food Chemistry*, 52(26), 8255–8260.

Daferera, D. J., Ziogas, B. N., & Polissiou, M. G. (2000). GC-MS analysis of volatile oils from some Greek aromatic plants and their fungitoxicity on *Penicillium digitatum*. *Journal of Agricultural and Food Chemistry*, 48(6), 2576–2581.

Dundar, E., Olgun, E. G., Isiksoy, S., Kurkcuoglu, M., Baser, K. H. C., & Bal, C. (2008). The effects of intra-rectal and intra-peritoneal application of *Origanum onites* L. volatile oil on 2, 4, 6-trinitrobenzenesulfonic acid-induced colitis in the rat. *Experimental and Toxicologic Pathology*, 59(6), 399–408.

Elansary, H. O., Abdelgaleil, S. A. M., Mahmoud, E. A., Yessoufou, K., Elhindi, K., El-Hendawy, S. (2018). Effective antioxidant, antimicrobial and anticancer activities of volatile oils of horticultural aromatic crops in northern Egypt. *BMC Complementary and Alternative Medicine*, 18(1), 214. doi: 10.1186/s12906-018-2262-1. PMID: 30005652; PMCID: PMC6044011.

Erdogan, A., & Ozkan, A. (2017). Investigatıon of antioxıdative, cytotoxic, membranedamaging and membrane-protective effects of the volatile oil of *Origanum majorana* and its oxygenated monoterpene component linalool in human-derived Hep G2 cell line. *Iranian Journal of Pharmaceutical Research (IJPR)*, 16, 24.

Erenler, R., Sen, O., Aksit, H., Demirtas, I., Yaglioglu, A. S., Elmastas, M., & Telci, İ. (2016). Isolation and identification of chemical constituents from *Origanum majorana* and investigation of antiproliferative and antioxidant activities. *Journal of the Science of Food and Agriculture*, 96(3), 822–836. doi: 10.1002/jsfa.7155. Epub 2015 Apr 8. PMID: 25721137.

Everest, A., & Ozturk, E. (2005). Focusing on the ethnobotanical uses of plants in Mersin and Adana provinces (Turkiye). *Journal of Ethnobiology and Ethnomedicine*, 1, 6. doi: 10.1186/1746-4269-1-6. PMID: 16270936; PMCID: PMC1277086.

Fonnegra, F. G. (2007). *Plantas medicinales aprobadas en Colombia*. Universidad de Antioquia.

Gulec, A. K., Erecevit, P., Yuce, E., Arslan, A., Bagci, E., & Kirbag, S. (2014). Antimicrobial activity of the methanol extracts and essential oil with the composition of endemic *Origanum acutidens* (lamiaceae). *Journal of Essential Oil Bearing Plants*, 17(2), 353–358.

Han, X., & Parker, T. L. (2017). Anti-inflammatory, tissue remodeling, immunomodulatory, and anticancer activities of oregano (*Origanum vulgare*) volatile oil in a human skin disease model. *Biochimie Open*, 4, 73–77. doi: 10.1016/j.biopen.2017.02.005. PMID: 29450144; PMCID: PMC5801825.

Hayta, S., Polat, R., & Selvi, S. (2014). Traditional uses of medicinal plants in Elazıg (Turkiye). *Journal of Ethnopharmacology*, 154(3), 613–623.

http://194.27.225.161/yasin/tubives/index.php '*Origanum*'.

Ietswaart, J. H., & Ietswaart, J. H. (1980). *A taxonomic revision of the genus Origanum (Labiatae)* (Vol. 4, p. 158). The Hague: Leiden University Press.

Jirovetz, L., Wlcek, K., Buchbauer, G., Gochev, V., Girova, T., Stoyanova, A., & Schmidt, E. (2007). Antifungal activity of various Lamiaceae volatile oils rich in thymol and carvacrol against clinical isolates of pathogenic *Candida* species. *International Journal of Volatile oil Therapeutic*, 1, 153–157.

Kawabata, J., Mizuhata, K., Sato, E., Nishioka, T., Aoyama, Y., & Kasai, T. (2003). 6-Hydroxyflavonoids as α-glucosidase inhibitors from marjoram (*Origanum majorana*) leaves. *Bioscience, Biotechnology, and Biochemistry*, 67(2), 445–447.

Khan, F., Khan, I., Farooqui, A., & Ansari, I. A. (2017). Carvacrol induces reactive oxygen species (ROS)-mediated apoptosis along with cell cycle arrest at G0/G1 in human prostate cancer cells. *Nutrition and Cancer*, 69(7), 1075–1087. doi: 10.1080/01635581.2017.1359321. Epub 2017 Sep 5. PMID: 28872904.

Korukluoglu, M., Gurbuz, O., Sahan, Y., Yigit, A., Kacar, O. Y. A., & Rouseff, R. (2009). Chemical characterization and antifungal activity of *Origanum onites* L. volatile oils and extracts. *Journal of Food Safety*, 29(1), 144–161.

Kwon, Y. I., Vattem, D. A., & Shetty, K. (2006). Evaluation of clonal herbs of Lamiaceae species for management of diabetes and hypertension. *Asia Pacific Journal of Clinical Nutrition*, 15(1), 107–118. PMID: 16500886.

Lagouri, V., & Nisteropoulou, E. (2009). Antioxidant properties of *O. onites*, *T. vulgaris* and *O. basilicum* species grown in Greece and their total phenol and rosmarinic acid content. *Journal of Food Lipids*, 16(4), 484–498.

Leeja, L., & Thoppil, J. E. (2007). Antimicrobial activity of methanol extract of *Origanum majorana* L.(Sweet marjoram). *Journal of Environmental Biology*, 28(1), 145.

Lim, W., Ham, J., Bazer, F. W., & Song, G. (2019). Carvacrol induces mitochondria-mediated apoptosis via disruption of calcium homeostasis in human choriocarcinoma cells. *Journal of Cellular Physiology*, 234(2), 1803–1815. doi: 10.1002/jcp.27054. Epub 2018 Aug 2. PMID: 30070691.

Mahady, G. B., Pendland, S. L., Stoia, A., Hamill, F. A., Fabricant, D., Dietz, B. M., & Chadwick, L. R. (2005). In vitro susceptibility of Helicobacter pylori to botanical extracts used traditionally for the treatment of gastrointestinal disorders. *Phytotherapy Research: An International Journal Devoted to Pharmacological and Toxicological Evaluation of Natural Product Derivatives*, 19(11), 988–991.

Mossa, A. T. H., Refaie, A. A., Ramadan, A., & Bouajila, J. (2013). Amelioration of prallethrin-induced oxidative stress and hepatotoxicity in rat by the administration of *Origanum majorana* volatile oil. *BioMed Research International*, 2013.

Oaman, H. F., & Abbas, O. A. (2010). Beneficial effects of origanum majorana on some biochemical and histological changes in alloxan-induced diabetic rat. *Arab Journal of Nuclear Sciences and Applications*, 43(1), 269–280.

Ocak, I., Çelik, A., Özel, M. Z., Korcan, E., & Konuk, M. (2012). Antifungal activity and chemical composition of essential oil of *Origanum hypericifolium*. *International Journal of Food Properties*, 15(1), 38–48.

Oliveira, J. L. T. M. D., Diniz, M. D. F. M., Lima, E. D. O., Souza, E. L. D., Trajano, V. N., & Santos, B. H. C. (2009). Effectiveness of *Origanum vulgare* L. and *Origanum majorana* L. volatile oils in inhibiting the growth of bacterial strains isolated from the patients with conjunctivitis. *Brazilian Archives of Biology and Technology*, 52, 45–50.

Orhan, I., Kartal, M., Kan, Y., & Sener, B. (2008). Activity of volatile oils and individual components against acetyl- and butyrylcholinesterase. *Journal of Biosciences*, 63, 547–553.

Ozdemir, B., Ekbul, A., Topal, N. B., Sarandöl, E., Sağ, S., Başer, K. H., Cordan, J., Güllülü, S., Tuncel, E., Baran, I., & Aydinlar, A. (2008). Effects of *Origanum onites* on endothelial function and serum biochemical markers in hyperlipidaemic patients. *Journal of International Medical Research*, 36(6), 1326–1334. doi: 10.1177/147323000803600621. PMID: 19094443.

Ozkan, G., Baydar, H., & Erbas, S. (2010). The influence of harvest time on volatile oil composition, phenolic constituents and antioxidant properties of Turkish oregano (*Origanum onites* L.). *Journal of the Science of Food and Agriculture*, 90(2), 205–209. doi: 10.1002/jsfa.3788. PMID: 20355032.

Özgökçe, F., & Özçelik, H. (2004). Ethnobotanical aspects of some taxa in East Anatolia, Turkiye. *Economic Botany*, 58(4), 697–704.

Partovi, R, Talebi, F, & Sharifzadeh, A (2018). Antimicrobial efficacy and chemical properties of Caryophyllus aromaticus and Origanum majorana essential oils against foodborne bacteria alone and in combination. *Int. J. Enteric Pathog*, 6(4), 95–103.

Perez Gutierrez, R. M. (2012). Inhibition of advanced glycation end-product formation by *Origanum majorana* L. in vitro and in streptozotocin-induced diabetic rats. *Evidence-Based Complementary and Alternative Medicine*, 2012, 598638. doi: 10.1155/2012/598638. Epub 2012 Sep 12. PMID: 23008741; PMCID: PMC3447365.

Pimple, B. P., Kadam, P. V., & Patil, M. J. (2012). Ulcer healing properties of different extracts of *Origanum majorana* in streptozotocin-nicotinamide induced diabetic rats. *Asian Pacific Journal of Tropical Disease*, 2(4), 312–318.

Polat, R., & Satıl, F. (2012). An ethnobotanical survey of medicinal plants in Edremit gulf (Balıkesir–Turkiye). *Journal of Ethnopharmacology*, 139(2), 626–641.

Prasanna, R., Ashraf, E. A., & Essam, M. A. (2017). Chamomile and oregano extracts synergistically exhibit antihyperglycemic, antihyperlipidemic, and renal protective effects in alloxan-induced diabetic rats. *Canadian Journal of Physiology and Pharmacology*, 95(1), 84–92. doi: 10.1139/cjpp-2016-0189. Epub 2016 Sep 2. PMID: 27875075.

Rosato, A., Vitali, C., Piarulli, M., Mazzotta, M., Argentieri, M. P., & Mallamaci, R. (2009). In vitro synergic efficacy of the combination of Nystatin with the volatile oils of *Origanum vulgare* and *Pelargonium graveolens* against some *Candida* species. *Phytomedicine*, 16, 972–975.

Rubin, B., Manso, J., Monticelli, H., Bertazza, L., Redaelli, M., Sensi, F., Zorzan, M., Scaroni, C., Mian, C., Iacobone, M., Armanini, D., Bertolini, C., Barollo, S., Boscaro, M., Pezzani, R. (2019). Crude extract of *Origanum vulgare* L. induced cell death and suppressed MAPK and PI3/Akt signaling pathways in SW13 and H295R cell lines. *Natural Product Research*, 33(11), 1646–1649. doi: 10.1080/14786419.2018.1425846. Epub 2018 Jan 15. Erratum in: *Nat Prod Res.* 2019 Jun;33(12):1834. PMID: 29334260.

Santoyo, S., Cavero, S., Jaime, L., Ibanez, E., Senorans, F. J., & Reglero, G. (2006). Superitical carbon dioxide extraction of compounds with antimicrobial activity from *Origanum vulgare* L.: Determination of optimal extraction parameters. *Journal of Food Protection*, 69, 369–375.

Sargın, S. A., Akçicek, E., & Selvi, S. (2013). An ethnobotanical study of medicinal plants used by the local people of Alaşehir (Manisa) in Turkiye. *Journal of Ethnopharmacology*, 150(3), 860–874. doi: 10.1016/j.jep.2013.09.040. Epub 2013 Oct 11. PMID: 24126062.

Sarikurkcu, C., Zengin, G., Oskay, M., Uysal, S., Ceylan, R., & Aktumsek, A. (2015). Composition, antioxidant, antimicrobial and enzyme inhibition activities of two *Origanum vulgare* subspecies (subsp. vulgare and subsp. hirtum) volatile oils. *Industrial Crops and Products*, 70, 178–184.

Sauza, E. L., Stamford, T. L. M., Lima, E. O., & Triyano, V. N. (2006). Effectiveness of *Origanum vulgare* L. volatile oil to inhibit the growth of food spoiling yeast. *Food Control*, 18, 409–413.

Seoudi, D. M., Hewedi, I. H., Medhat, A. M., Osman, S. A., Arbid, M. S., & Mohamed, M. K. (2009). Evaluation of the antiinflammatory, analgesic, and antipyretic effects of *Origanum majorana* ethanolic extract in experimental animals. *Journal of Radiation Research and Applied Sciences*, 2(3), 513–534.

Sharifi-Rad, M., Berkay Yılmaz, Y., Antika, G., Salehi, B., Tumer, T. B., Kulandaisamy Venil, C., ... Sharifi-Rad, J. (2021). Phytochemical constituents, biological activities, and health-promoting effects of the genus Origanum. *Phytotherapy Research*, 35(1), 95–121.

Silva, J. C., Almeida, J. R., Quintans, J. S., Gopalsamy, R. G., Shanmugam, S., Serafini, M. R., ... Quintans-Júnior, L. J. (2016). Enhancement of orofacial antinociceptive effect of carvacrol, a monoterpene present in oregano and thyme oils, by β-cyclodextrin inclusion complex in mice. *Biomedicine & Pharmacotherapy*, 84, 454–461.

Singletary, K. (2010). Oregano: Overview of the literature on health benefits. *Nutrition Today*, 45(3), 129–138.

Sokmen, M., Serkedjieva, J., Daferera, D., Gulluce, M., Polissiou, M., Tepe, B., ... Sokmen, A. (2004). In vitro antioxidant, antimicrobial, and antiviral activities of the volatile oil and various extracts from herbal parts and callus cultures of *Origanum acutidens*. *Journal of agricultural and food chemistry*, 52(11), 3309–3312.

Soković, M., Tzakou, O., Pitarokili, D., & Couladis, M. (2002). Antifungal activities of selected aromatic plants growing wild in Greece. *Food/Nahrung*, 46(5), 317–320.

Soliman, A. M., Desouky, S., Marzouk, M., & Sayed, A. A. (2016). *Origanum majorana* attenuates nephrotoxicity of cisplatin anticancer drug through ameliorating oxidative stress. *Nutrients*, 8(5), 264.

Sozmen, F., Uysal, B., Oksal, B. S., Kose, E. O., & Deniz, I. G. (2011). Chemical composition and antibacterial activity of *Origanum saccatum* PH Davis essential oil obtained by solvent-free microwave extraction: Comparison with hydrodistillation. *Journal of AOAC International*, 94(1), 243–250.

Sozmen, F., Uysal, B., Köse, E. O., Aktaş, Ö., Cinbilgel, I., & Oksal, B. S. (2012). Extraction of the essential oil from endemic *Origanum bilgeri* PH Davis with two different methods: Comparison of the oil composition and antibacterial activity. *Chemistry & biodiversity*, 9(7), 1356–1363.

Spyridopoulou, K., Fitsiou, E., Bouloukosta, E., Tiptiri-Kourpeti, A., Vamvakias, M., Oreopoulou, A., Papavassilopoulou, E., Pappa, A., & Chlichlia, K. (2019). Extraction, chemical composition, and anticancer potential of *Origanum onites* L. volatile oil. *Molecules*, 24(14), 2612. doi: 10.3390/molecules24142612. PMID: 31323754; PMCID: PMC6680447.

Stefanakis, M. K., Touloupakis, E., Anastasopoulos, E., Ghanotakis, D., Katerinopoulos, H. E., & Makridis, P. (2013). Antibacterial activity of volatile oils from plants of the genus *Origanum*. *Food Control*, 34(2), 539–546.

Tabata, M., Sezik, E., Honda, G., Yesilada, E., Fukui, H., Goto, K., & Ikeshiro, Y. (1994). Traditional medicine in Turkiye III. Folk medicine in East Anatolia, van and Bitlis provinces. *International Journal of Pharmacognosy*, 32(1), 3–12.

Tampieri, M. P., Galuppi, R., Macchioni, F., Carelle, M. S., Falcioni, L., Cioni, P. L., Morelli, I. (2005). The inhibition of *Candida albicans* by selected volatile oils and their major components. *Mycopathologia*, 159, 339–345.

Tepe, B., Cakir, A., & Sihoglu Tepe, A. (2016). Medicinal uses, phytochemistry, and pharmacology of *Origanum onites* (L.): A review. *Chemistry & Biodiversity*, 13(5), 504–520. doi: 10.1002/cbdv.201500069. PMID: 27062715.

Tripathy, B., Satyanarayana, S., Khan, K. A., & Raja, K. (2018). Evaluation of anti-diabetic and anti-hyperlipidemic activities of ethanolic leaf extract of *Origanum majorana* in streptozotocin induced diabetic rats. *International Journal of Pharmaceutical Sciences*, 9, 1529–1536.

Türkez, H., & Aydın, E. (2016). Investigation of cytotoxic, genotoxic and oxidative properties of carvacrol in human blood cells. *Toxicology and Industrial Health*, 32(4), 625–633.

Uyanoglu, M., Canbek, M., Aral, E., & Baser, K. H. C. (2008). Effects of carvacrol upon the liver of rats undergoing partial hepatectomy. *Phytomedicine*, 153, 226–229.

Uysal, İ., Onar, S., Karabacak, E., & Çelik, S. (2010). Ethnobotanical aspects of Kapıdağ peninsula (Turkiye). *Biological Diversity and Conservation*, 3(3), 15–22.

Vujicic, M., Nikolic, I., Kontogianni, V. G., Saksida, T., Charisiadis, P., Orescanin-Dusic, Z., Blagojevic, D., Stosic-Grujicic, S., Tzakos, A. G., & Stojanovic, I. (2015). Methanolic extract of *Origanum vulgare* ameliorates type 1 diabetes through antioxidant, anti-inflammatory and anti-apoptotic activity. *British Journal of Nutrition*, 113(5), 770–782. doi: 10.1017/S0007114514004048. Epub 2015 Feb 11. PMID: 25671817.

Xiao, Y., Li, B., Liu, J., & Ma, X. (2018). Carvacrol ameliorates inflammatory response in interleukin 1β-stimulated human chondrocytes. *Molecular Medicine Reports*, 17(3), 3987–3992. doi: 10.3892/mmr.2017.8308. Epub 2017 Dec 19. PMID: 29257341.

Yabalak, E, Emire, Z, Adıgüzel, AO, Könen Adıgüzel, S, & Gizir, AM (2020). Wide-scale evaluation of *Origanum munzurense* Kit Tan & Sorger using different extraction techniques: Antioxidant capacity, chemical compounds, trace element content, total phenolic content, antibacterial activity and genotoxic effect. *Flavour Fragr.* J, 35(4), 394–410.

Yazdanparast, R., & Shahriyary, L. (2008). Comparative effects of *Artemisia dracunculus*, *Satureja hortensis* and *Origanum majorana* on inhibition of blood platelet adhesion, aggregation and secretion. *Vascular Pharmacology*, 48(1), 32–37.

Yesilada, E., Honda, G., Sezik, E., Tabata, M., Goto, K., & Ikeshiro, Y. (1993). Traditional medicine in Turkiye IV. Folk medicine in the Mediterranean subdivision. *Journal of Ethnopharmacology*, 39(1), 31–38.

https://commons.wikimedia.org/wiki/File:Origanum_acutidens_1.jpg

https://turkiyebitkileri.com/tr/foto%C4%9Fraf-galerisi/lamiaceae-ball%C4%B1babagiller/origanum-mercanko%C5%9Fk/origanum-boissieri.html

https://turkiyebitkileri.com/tr/foto%C4%9Fraf-galerisi/lamiaceae-ball%C4%B1babagiller/origanum-mercanko%C5%9Fk/origanum-munzurense.html

https://turkiyebitkileri.com/en/photo-gallery/lamiaceae-ball%C4%B1babagiller/origanum-mercanko%C5%9Fk/origanum-hypericifolium.html?start=0

https://turkiyebitkileri.com/tr/foto%C4%9Fraf-galerisi/lamiaceae-ball%C4%B1babagiller/origanum-mercanko%C5%9Fk/origanum-amanum.html

https://turkiyebitkileri.com/tr/foto%C4%9Fraf-galerisi/lamiaceae-ball%C4%B1babagiller/origanum-mercanko%C5%9Fk/origanum-husnucanbaseri.html

https://turkiyebitkileri.com/tr/foto%C4%9Fraf-galerisi/lamiaceae-ball%C4%B1babagiller/origanum-mercanko%C5%9Fk/origanum-bilgeri/45932-akseki-antalya.html

https://turkiyebitkileri.com/tr/foto%C4%9Fraf-galerisi/lamiaceae-ball%C4%B1babagiller/origanum-mercanko%C5%9Fk/origanum-minutiflorum/46220-i%CC%87brad%C4%B1-antalya.html

http://yabanicicekler.com/flower/origanum-saccatum-530

16 *Papaver* sp.

Günay Sariyar and Gizem Gülsoy Toplan

INTRODUCTION

Papaver somniferum (Opium poppy) (Figure 16.1) is one of the oldest medicinal plants which has been cultivated since the time of the Hittites in Anatolia (2000 BC). The cultivation of this plant for the production of opium, a product obtained from the latex by incising the immature capsules, continued until 1971 in Turkey. Due to the strong belief that Turkey is one of the largest sources of illicit opium which contained high amounts of the narcotic-analgesic alkaloid, morphine (Figure 16.1), the government decided to ban the production of opium in 1971. When opium production was prohibited, opium poppy cultivation was also banned. However, in 1974, controlled opium poppy cultivation resumed to obtain its major alkaloid, morphine, by extraction from the poppy straws and to prepare morphine derivatives in the "Opium Alkaloid Plant", which was carried out in the district of Bolvadin, Afyonkarahisar. Since 1981, this plant operates to meet the legal opium poppy alkaloids' requirement for domestic and international markets **(1)**.

Increased morphine abuse has made it necessary to find alternative sources to *P. somniferum* as a raw material for the pharmaceutical industry. About 90% of the obtained morphine is converted into other bases, mainly codeine (Figure 16.2) which is the medicinally most important opiate with a very low abuse potential. Discovery of thebaine (Figure 16.3) as a major alkaloid in another *Papaver* species, *P. bracteatum*, in the 1970s drew the attention of scientists to use this alkaloid as a starting material for the preparation of other opioids, mainly codeine (Nyrnan U & Bruhn J.G. 1979) **(2)**. In fact, today thebaine is one of the most important alkaloids that is used as a starting material for the manufacture not only of codeine but also of a number of other opioids such as oxycodone, dihydrocodeine, etorphine, hydrocodone, and oxymorphone. In addition to thebaine, another morphinane alkaloid, oripavineIV, which can be converted to thebaine, is also attractive as a starting material for the production of selected semisynthetic opioids. Today, high thebaine and oripavine containing strains of *P. somniferum* are produced in countries, such as the USA, Australia, Spain, and France (Narcotic Drugs, Reports published by the International Narcotic Control Board for 2020) **(3)**.

Information presented above displays the significance of *P. somniferum* as a medicinal plant. Similar to *P. somniferum*, many other *Papaver* species including *P. bracteatum* are wildly grown in Turkey. Studies on the alkaloids of these species have been undertaken to reveal their potential as alternative sources for the manufacture of medicinally important alkaloids. The distribution of *Papaver* species varies according to Flora of Turkey **(4)** and recent modern taxonomic revisions are as follows:

In the Flora of Turkey, the genus *Papaver* is represented by 35 species belonging to seven sections **(4)**. Then, in his revised sectional nomenclature in *Papaver*, Kiger recognized nine sections, seven of which were identical to those present in Turkey, with two perennials (Macrantha and Pilosa), one biennial (Meconidium), and four annuals (Argemonidium, Carinata, Papaver, and Rhoeadium) **(5)**. In the same work, some sections have also been renamed. In countries covered by Med-Checklist, the genus *Papaver* is represented by 32 species **(6)** in the Mediterranean area. Based on Med-checklist, 23 species with 19 subspecies belonging to eight sections (with the addition of a new

DOI: 10.1201/9781003146971-16

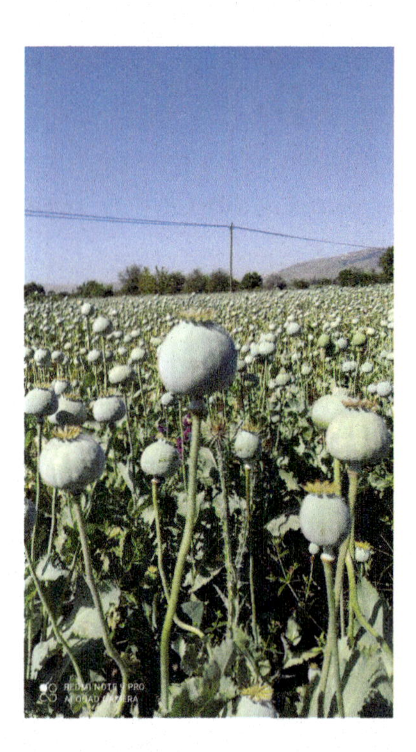

FIGURE 16.1 *Papaver somniferum* L.

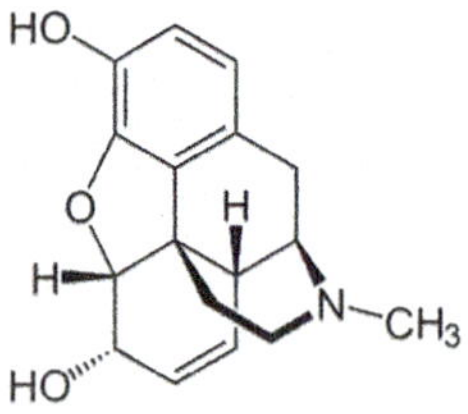

FIGURE 16.2 Morphine.

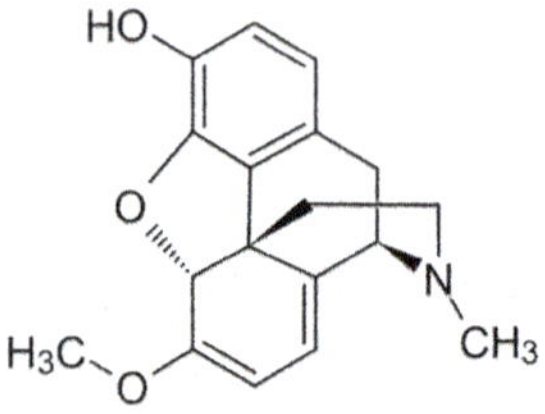

FIGURE 16.3 Codeine.

section, Pseudopilosa) in Turkey have been reported by Kadereit (**7**). The classification of sections proposed by Cullen, Kiger, and Kadereit is summarized in Table 16.1.

The results of our studies on the alkaloids of Turkish *Papaver* species have been summarized in separate sections based on their classification. With the exception of some species from section Argemonidium and Rhoeadium, most of our samples have been identified as described by Cullen in the Flora of Turkey (**4**).

TABLE 16.1
Sections of Genus *Papaver*

Cullen 1965	Kieger 1985, Kadereit 1997
Argemonorhoeades Fedde	Argemonidium Spach
Papaver (Syn. Sect. Orthorhoeades Fedde)	Rhoeadium Spach
Carinata Fedde	Carinata Fedde
Miltahtha Bernh.	Meconidium Spach
Pilosa Prantl.	Pilosa Prantl
Macrantha Elk.	Macrantha Elkan
Mecones Bernh.	Papaver
	Pseudopilosa M. Popov ex Günther Per.

TABLE 16.2
Species from Section Argemonidium SPACH

Argemonorhoeades (Cullen 1965)	Argemonidium SPACH (Kadereit 1997)
P. argemone L.	*P. argemone* L.
	subsp. *nigrotinctum* (Fedde) Kadereit
	subsp. *davisii* Kadereit
	subsp. *belangeri* Taht
P. hybridum L.	*P. hybridum* L.
P. virchowii Aschers	
P. apulum Ten	
P. ramossisimum Fedde	
P. shepardii Post	
P. pinnatifidum Moris	

DISTRIBUTION AND CHEMISTRY OF *PAPAVER* SPECIES

Section Argemonidium SPACH (syn. Argemonorhoeades FEDDE)

Table 16.2 shows the number and the names of the species from the annual Argemonidium SPACH section reported by Cullen and Kadereit (**4, 7**).

Papaver argemone subsp. *davisii*, which is widespread in Turkey, is the only one that has been studied for its alkaloids. The alkaloid content of this sample is quite low, both qualitatively and quantitatively. Table 16.3 summarizes the alkaloids isolated from this species. The presence of spirobenzylisoquinoline types, fumariline and fumarophycine, has been shown for the first time in genus *Papaver* (**8**).

Section Carinata FEDDE

This section is represented by only one species named as *P. macrostomum* Boiss. & Huet ex Boiss (**4,7**). Two samples from this species have been studied for both their alkaloid and flavonoid contents (**9**), while one sample has been investigated for its alkaloids only (**8,9**). All these three samples have been shown to contain different types of alkaloids, which suggests the presence of chemotypes. Table 16.4 summarizes the alkaloid and flavonoid contents of this species.

TABLE 16.3
Alkaloids Present in *Papaver* Species from Section Argemonidium SPACH

Alkaloids from *Papaver argemone* ssp. Davisii
Protopine
Protopine
Spirobenzylisoquinoline
Fumariline
Fumarophycine

TABLE 16.4
Alkaloids and Flavonoids Isolated from *P. macrostomum*

Alkaloids
Isopavine
Amurensine
Amurensinine
Proaporphine
Mecambrine
Aporphine
Isocorydine
Laudonosine
Protoberberine
Cheilantifoline
Flavonoids
Luteoline
Tricine

SECTION PAPAVER (SYN. MECONES)

Distribution of *Papaver* species in section Papaver by Kadereit (**7**) (Syn. Mecones, CULLEN) (**4**) is given in Table 16.5.

P. somniferum is the most important species among all species of the genus *Papaver*. Medicinal importance, history, and production of this species in Turkey have been outlined above (see Introduction). Investigation of the alkaloids of *P. gracile* (**8**) has revealed the presence of the alkaloids as summarized in Table 16.**6**. Although *P. gracile* and *P. somniferum* are two species of the same section, they differ in their alkaloid contents; the former does not contain any medicinally important alkaloids.

SECTION RHOEADIUM SPACH (SYN. ORTHORHOEADES)

This annual section is the richest one for the number of species. Cullen (**4**) reported the existence of ten species whereas taxonomic revision by Kaderiet (**7, 10**) included nine species with four subspecies in this section (Table 16.7).

Two widespread species in this section are *P. rhoeas* and *P. dubium*. Studies on the alkaloids of samples of *P. rhoeas* collected from different locations revealed the presence of several chemotypes (**8,11,12**) (Table 16.7). This species, which is known as "Gelincik" in Turkish, has been used

TABLE 16.5
Species from Section Papaver

Section Papaver
P. somniferum L.
P. glaucum Boiss. & Hausskn
P. gracile Boiss.

TABLE 16.6
Alkaloids from *P. gracile* of Section *Papaver* (syn. Mecones)

Alkaloids
Proaporphine
Mecambrine
Aporphine
Roemerine
Dehydroroemerine
Roemerine N-oxide

TABLE 16.7
***Papaver* Species in Section Rhoeadium (syn. Papaver)**

Papaver (syn. Orthorhoeades) (Cullen 1965)	Rhoeadium (Kadereit 1997)
-	*P. arachnoideum* Kadereit
P. arenarium Bieb.	*P. arenarium* M. Bieb.
P. clavatum Boiss. & Hausskn. Ex Boiss.	*P. clavatum* Boiss. & Hausskn. Ex Boiss.
P. commutatum Fisch. & Mey.	*P. commutatum* Fischer & C. Meyer subsp. *euxinum* Kadereit
P. dubium L.	P. *dubium* L. subsp. *dubium* Subsp. *lecoqii* (Lamotte Syme) Subsp. *laevigatum* (M. Bieb) Kadereit
P. lacerum Popov	*P. dubium* subsp. *laevigatum*
P. postii Fedde	-
-	*P. purpureamarginatum* Kadereit
P. rhoeas L.	*P. rhoeas* L.
P. rhopalothece Staph	Syn. *P. guerlekense* Staph
P. stylatum Boiss. & Bal.	*P.stylatum* Boiss. & Balansa ex Boiss
P. syriacum Boiss.& Blanche	-

extensively for the treatment of various diseases as part of traditional medicine. One of the subspecies of *P. dubium*, *P. dubium* subsp. *lecoqii*, contains the medicinally important alkaloid, thebaine (Figure 16.4) (**13**). The presence of this alkaloid in an annual species is quite significant since this chemotype can be used as a potential source for this alkaloid. Two other subspecies of *P. dubium*, subsp. *leavigatum* and subsp. *dubium*, differ in their alkaloid content (**8,14**). Another interesting species is *P. rhopalothece* which has now been accepted as a synonym for *P. guerlekense* (**7,10**). During the field studies, it has been observed that this species, together with *P. rhoeas*, has been

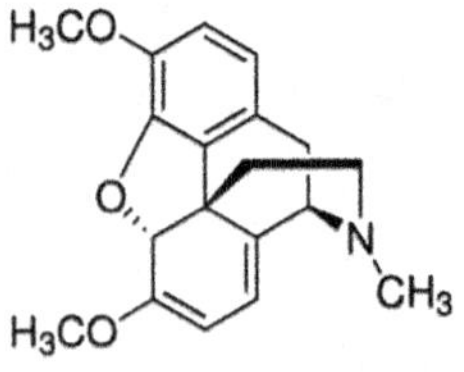

FIGURE 16.4 Thebaine.

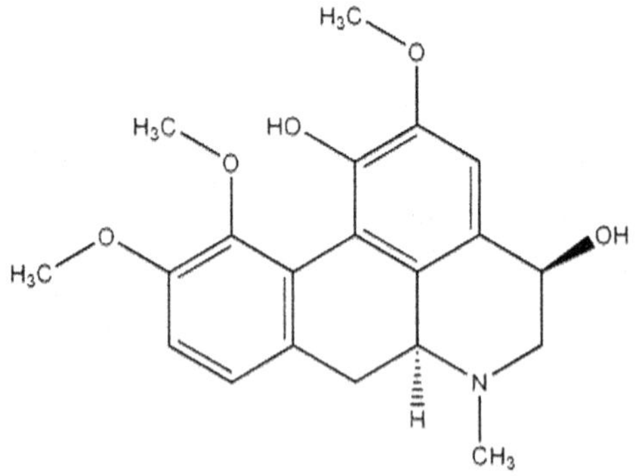

FIGURE 16.5 Rhopalotine.

used for the treatment of cough in the southwest of Turkey as a folk medicine. Isolation of noscapine as a major alkaloid from this species has justified its use. A new aporphine type alkaloid named as rhopalotine (Figure 16.5) has also been obtained from this interesting species **(15)**. Isolation of cularine-type molecules has been reported from the genus *Papaver.* Investigation of the alkaloids of two samples of *P. commutatum* subsp. *euxinum*, which are endemic to Northern Turkey (northeast and northwest) both revealed the presence of isocorydine as their major alkaloids. The minor alkaloids isolated from these two samples were different **(16)**. Similar to *P. rhopalothece*, this species is also used for the treatment of cough in folk medicine. Another endemic species from this section *P. aracnoideum* proved to contain an aporphine-type alkaloid, roemerine, an oxoaporhine-type alkaloid, liriodenine, and three proaporhines, namely, mecambrine, roemeronine, and dihydromecambrinol **(17)**. Other species of the section, *P. clavatum*, *P. lacerum*, and *P. stylatum*, have also been studied for their alkaloid content **(18,19).** Table 16.8 summarizes the alkaloids obtained from the species of section Rhoeadium.

Section Meconidium SPACH (Syn. Miltantha ELKAN)

Biennial section Miltantha of the genus *Papaver* is represented by eight species in the Flora of Turkey **(4)**. This section, which was renamed as Meconidium by Kieger **(5)** and in his revised study on this section Kadereit **(7)** indicated the presence of four species with five subspecies, as summarized in Table 16.9.

All the species investigated for their alkaloid content in this section were identified as described by **Cullen (4)**. *P. fugax* was the first species of Turkish origin studied for its alkaloids **(20).** Isolation of thebaine, as a major alkaloid from this species was not only surprising but was also quite exciting since this alkaloid was found to be an important source for the production of several opioids by semisynthesis, as explained in the introduction. Distribution of this major alkaloid was previously shown to be limited in sections Macrantha **(21,22)** and in *P. somniferum* and *P. setigerum.* This exciting result has led to continue work on the alkaloids of other species from this section. With

TABLE 16.8A
Alkaloids of *Papaver* Species from Section Rhoeadium (syn. Orthorhoeades)

Alkaloids	*P. arachnoideum*	*P. clavatum*	*P. commutatum* ssp. *euxinum*	*P. dubium* ssp. *dubium*	*P. dubium* ssp. *Leavigatum*	*P. dubium* ssp. *lecoqii*
Isopavine						
Amurensinine			+			
Cularine						
Cularine						
Proaporphine						
Dihydromecambrinol	+					
Mecambrine	+				+	+
Pronuciferine						
Roemeronine	+					
Aporphine						
Corydine						
Isocorydine			+	+		
Liriodenine	+					
N-methylasimilobine						
Roemerine	+				+	+
Rhopalotine*						
Protoberberine & teterahydroprotoberberine						
Berberine		+		+		+
Cheilanthifoline			+			
Coptisine						
Coulteropine						
Sinactine						
Stylopine				+		
Tetrahydropseudocoptisine				+		
Thalifendine				+		
Protopine						
Allocryptopine					+	+
Coulteropine						
Cryptopine						+
Protopine					+	+
Phthalideisoquinoline						
Narceine						
Narcotine						
Rhoeadine and Papaverrubines						
Epiglaucamine						
Glaucamine						
Glaudine						
Isorhoeagenine						
Papaverrubine A			+			
Rhoeagenine			+			
Rhoeadine						
Rhoeadine						
Morphinanedienone						
Amurine						
Meconoquintupline						
Salutaridine						
Morphinane						
Thebaine						+
Thebaine N oxide						+
Cularine						
Cularine						

TABLE 16.8B
Alkaloids of *Papaver* Species from Section Rhoeadium (Cont.)

Alkaloids	*P. lacerum*	*P. rhoeas*	*P. rhopalothece**	*P. stylatum*
Isopavine				
Amurensinine				
Cularine				
Cularine			+	
Proapaorphine				
Mecambrine	+	+		
Pronuciferine	+	+		
Aporphine				
Corydine				
Isocorydine		+	+	+
N-methylasimilobine	+	+		
Roemerine	+	+	+	
Rhopalotine* **5**			+	
Protoberberine and teterahydroprotoberbererine				
Berberine		+	+	+
Cheilanthifoline		+		
Coptisine		+	+	
Sinactine		+		
Stylopine				
Tetrahydropseudocoptisine				
Thalifendine				
Protopine				
Allocryptopine		+		
Coulteropine		+	+	
Cryptopine			+	
Protopine		+	+	
Phytalideisoquinoline				
Narceine			+	
Narcotine			+++	
Rhoeadine and Papaverrubines				
Epiglaucamine		+		
Glaucamine		+		
Glaudine		+		
Isorhoeagenine		+		
Papaverrubine A		+		
Rhoeagenine		+		
Rhoeadine		+		
Morphinanedienone				
Salutaridine		+		
Meconoquintupline				+
Morphinane				
Thebaine				
Thebaine N-Oxide				

*This species has been renamed as *P. guerlekense* by Kadereit (1997)

TABLE 16.9
Species from Section Meconidium SPACH (syn. Miltantha ELKAN)

Miltantha ELKAN (Cullen 1965)	Meconidium SPACH (Kadereit 1997)
P. acrochaeum Bornm.	-
P. armeniacum (L.) DC.	*P. armeniacum* (L.) DC.
	subsp. *armeniacum*
P. curviscapum Nábělek	*P. curviscapum* Nab
P. cylindricum Cullen	Syn. *P. armeniacum*
P. fugax Poiret	Syn. *P. armeniacum*
P. polychaetum Schott & Kotschy	*P. libanotecum*
	subsp. *polychaetum* (Schott & Kotschy ex Boiss.) Kadereit
P. tauricola Boiss.	Syn. *P. persicum* Lindl.
	subsp. *persicum*
	subsp. *tauricolum* (Boiss.) Kadereit
	subsp. *microcarpum* (Boiss.) Kadereit
P. triniifolium Boiss.	Syn. *P. armeniacum*

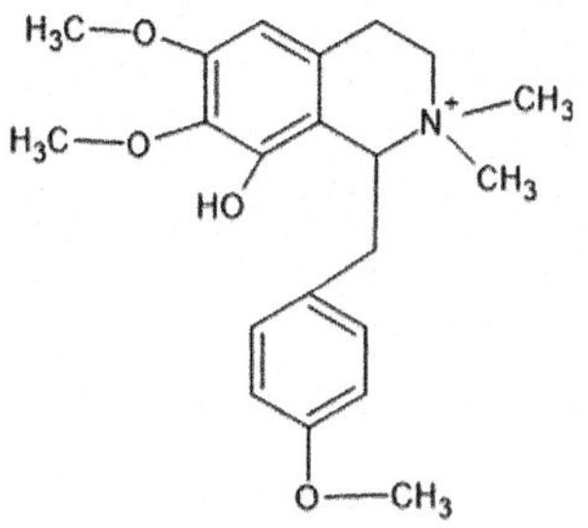

FIGURE 16.6 Miltanthaline.

the exception of *P. acrochaetum*, the alkaloid contents of all the species of *P. armeniacum*, *P. curviscapum*, *P. cylindricum*, *P. fugax*, *P. polychaetum*, *P. tauricola*, and *P. triniifolium* were investigated. Between 1973 and 2006, 30 samples were screened for their alkaloid contents **(20, 23-34).** *P. curviscapum* and *P. polychaetum*, which have limited distribution, differ chemically, both containing berberine as their major alkaloids **(28,31).** Similarities among five other species for their major alkaloids, some of which containing medicinally important alkaloids mainly noscapine, oripavine, papaverine, and thebaine, have been shown. Only one sample of *P. fugax*, collected from Hakkari, which yielded secoberbine-type papaveroxine as its major alkaloid is different. Nine new alkaloids, namely miltantaline (Figure 16.6), miltantoridine (Figure 16.7), miltantoridinone (Figure 16.8) **(30,32)**, narcotinehemiacetale (Figure 16.9), papaveroxine (Figure 16.10a) **(27)**, N-methyl-narcotine (Figure 16.11) **(29)**, 1-methoxyallocryptopine (Figure 16.12) **(28),** triniifoline (Figure 16.13) **(30),** and (–)-6a,7-dehydrofloripavidine (Figure 16.14) **(34)**, have been isolated as novel alkaloids. The following chemotypes can be distinguished in the species in this section for their major alkaloid contents:

1. Rhoeadine
2. Benzylisoquinoline-proapaorphine-aporphine
3. Morphinane-phtalideisoquinoline-rhoeadine
4. Morphinane type

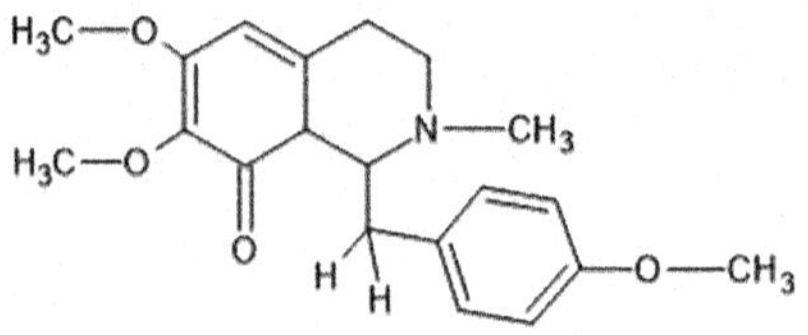

FIGURE 16.7 Miltantoridine.

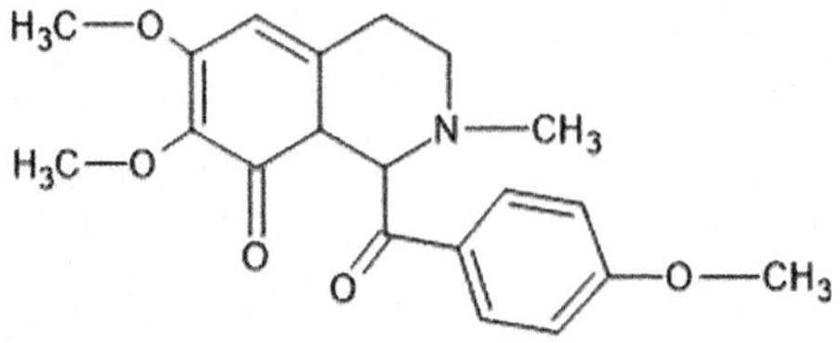

FIGURE 16.8 Miltanthoridinone.

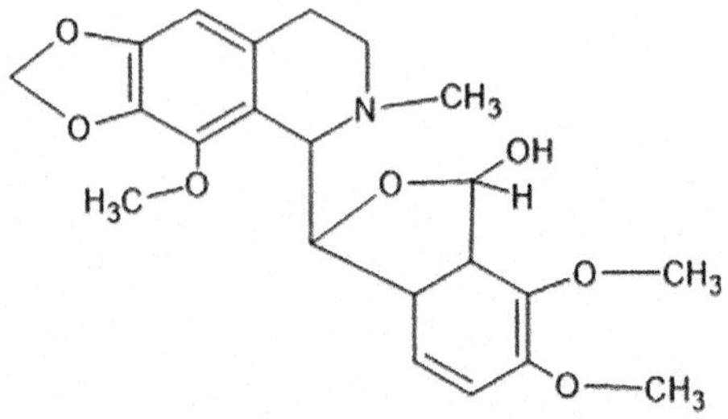

FIGURE 16.9 Narcotinhemiacetale.

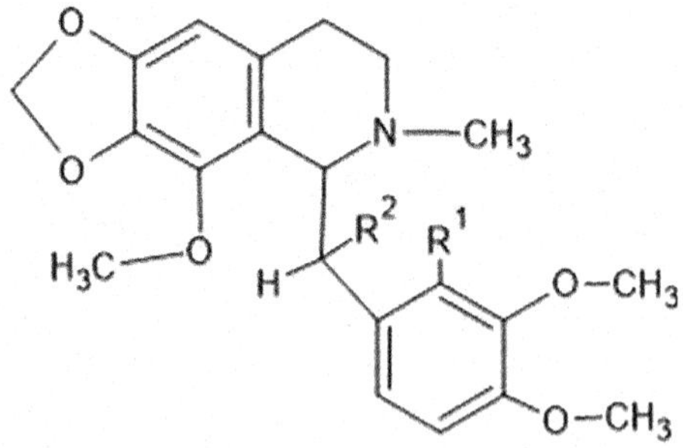

FIGURE 16.10 (a) Papaveroxine CHO COCH3, (b) Papaveroxinoline CH2OH COCH3, (c) Papaveroxidine COOH COCH3, (d) Macrantaline CH2OH H, (e) Macrantaldehyde CHO H, (f) Macrantoridine COOH H, and (g) Narcotindiol CH2OH OH.

5. Proaporphine-aporphine types
6. Secoberbine
7. Benzylisoquinoline-rhoeadine types
8. Proaporphine-promorphinane types
9. Aporphine-promorphinane
10. Benzylisoquinoline-aporphine

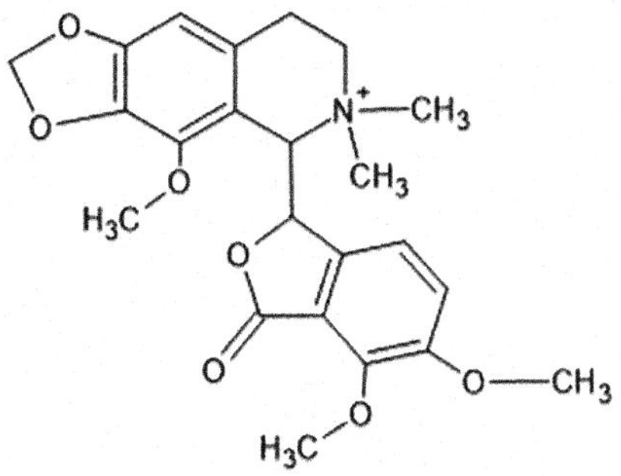

FIGURE 16.11 N-methylnarcotine.

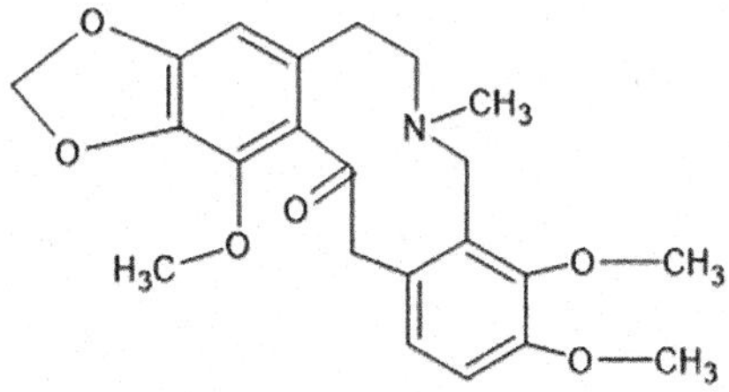

FIGURE 16.12 1-Methoxyallocryptopine.

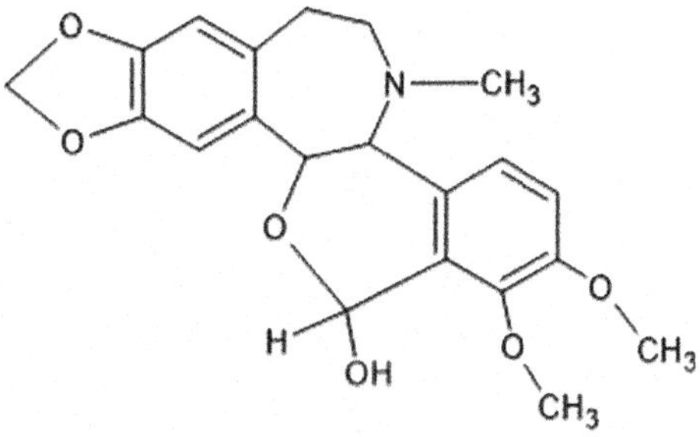

FIGURE 16.13 Triniifoline.

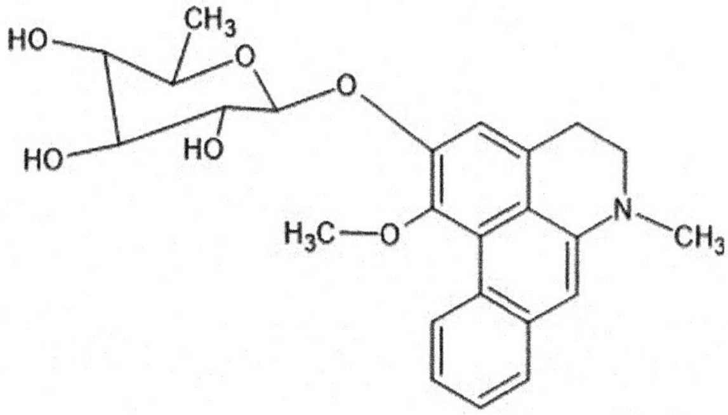

FIGURE 16.14 (–)-6a,7-dehydrofloripavidine.

11. Aporphine-rhoeadine
12. Rhoeadine-morphinane
13. Proaporphine-benzylisoquinoline-phtalideisoquinoline
14. Rhoeadine poaporphine types

Table 16.10 outlines the alkaloid contents of the species from this section.

Section Macrantha Elkan (syn. Oxytona)

In the Flora of Turkey, Cullen reported the presence of five species in the perennial section Macrantha **(4)**. Revision of this section by Goldblatt under the name Oxtona recognized three species, viz. *P. bracteatum*, *P. pseudo-orientale*, and *P. orientale*, excluding the distribution of *P. bracteatum* in Turkey **(35)** (Table 16.11).

Goldblatt distinguished these species in his work by following their cytological and chemical characteristics: *P. bracteatum* (diploid, 2n = 14) with the major alkaloid thebaine, *P. orientale* (tetraploid 2n = 28) with the major alkaloid oripavine, *P. pseudo-orientale* (hexaploid 2n = 42) with the major alkaloid being isothebaine **(35).** The isolation of high amounts of thebaine from an Iranian *P. bracteatum* sample as a major alkaloid **(22)** has led to the investigation of the alkaloids of all the species of this section growing in Turkey. During the extensive excursions to different parts of Turkey between 1973 and 2000, more than 20 samples from different locations were collected. Chromosome numbers of some samples were counted and the major alkaloids were determined. The result was published based on Goldblatt's report **(35)** indicating the distribution of two samples in Turkey, although there were several conflicting results regarding the alkaloid contents and chromosome numbers, mainly in two samples from Niğde and Tunceli, which were identified as *P. pseudo-orientale* **(36)**. It was decided to repeat the investigation on the samples from these two locations with careful field observation on their morphological characters (petal colors and marking shapes), chromosome counts, and major alkaloid contents. These two samples had diploid chromosome numbers of 2n = 14 and thebaine as their major alkaloid. The only difference between *P. bracteatum* of Iranian origin and these two samples were the presence of salutaridine as the second major alkaloid together with thebaine in Turkish samples. Another species which was recognized as *P. lasiothrix* by Cullen and distributed in two locations, Sivas and Gümüşhane, were reinvestigated. These were previously treated as *P. pseudo-orientale* by Goldblatt **(35)**. The sample from Sivas was found to contain salutaridine together with a novel alkaloid macrantaline (Figure 16.10d) **(37)**, as the major alkaloids and chromosome number was diploid as 2n = 14. It is now suggested that this sample could better be treated as a subspecies of *P. bracteatum* rather than *P. pseudo-orientale*, as recognized by Goldblatt **(35)**. The sample from Gümüşhane proved to have salutaridine as its major alkaloid and diploid chromosome numbers of 2n = 28. When chromosome numbers and major alkaloid contents have been taken into consideration, this sample could be accepted as *P. lasiothrix* or subspecies of *P. orientale* rather than *P. pseudo-orientale* **(37,38).** Together with the medicinally important alkaloid thebaine, several novel alkaloids, namely macrantaline (Figure 16.10d), macrantoridine (Figure 16.10f) **(37),** macrantaldehyde (Figure 16.10e), narcotinhemiacetale (Figure 16.9), narcotolinol (Figure 16.15), papaveroxine (Figure 16.10a), papaveroxinoline (Figure 16.10b)**,** papaveroxidine (Figure 16.10c) **(27**, **39)**, a dimeric morphinanedienone alkaloid salutadimerine (Figure 16.16) **(27,38, 40),** quaternary alkaloids cotarnine (Figure 16.17), cotarnoline (Figure 16.18) and papaverberbine (Figure 16.19) **(29)**, and salutaridine N-oxide (Figure 16.20) **(41)**, have been identified in Turkish samples from Section Macrantha **(40, 27, 29, 38, 41)**. High content of thebaine in two samples, which have now been recognized as *P. bracteatum*, was promising since they could be used for the production of thebaine **(42)**. All the alkaloids isolated from the samples of section Macrantha are summarized in Table 16.12. More recently, a new species has been described as *Papaver yildirimli* collected from Siirt in this section **(43)**.

TABLE 16.10
Alkaloids of *Papaver* Species from Section Meconidium (Syn: Miltantha Cullen 1965)

Alkaloids	*P. armeniacum*	*P. curviscapum*	*P. cylindricum*	*P. fugax*	*P. presicum*	*P. polychaetum*	*P. triniifolium*
Benzylisoquinoline (with benzyltetra-hydroisoquinoline)							
Armepavine			+	+	+		
Cryconisine							+
Miltanthaline* **6**							+
Miltanthoridine* **7**							+
Miltanthoridinone* **8**							+
Papaverine			+				+
Isopavine							
Amurine							+
Amurensinine					+		
Proaporphine							
Mecambrine				+	+		+
N-methylcrotonosine							+
Aporphine							
Floripavidine			+	+	+		+
Lirinidine	+						
(-)-6a,7-Dehydrofloripavidine***14**							
N-Methylasimilobine			+	+	+		
Secoberbine							
Papaveroxine* **10a**				+			
Protoberberine (with tetrahydroprotoberberine)							
Berberine		+				+	
Cheilanthifoline			+				+
Isocorypalmine							+
N-Methylsinactine							+
Sinactine					+		+
Scoulerine			+		+		
Protopine							
Allocryptopine		+					
Cryptopine	+						
1-methoxyallocryptopine***12**		+					
1-methoxy-13-oxo-allocryptopine		+					
Protopine		+					
Phthalideisoquinoline							
Narceine							+
Narcotine			+	+			+
Narcotinhemiacetale* **9**				+			
N-methylnarcotine***11**			+				
Rhoeadine							
Epiglaudine					+		
Glaucamine					+		
Glaudine	+				+		
Oreodine				+	+		+
Oreogenine					+		+
O-Ethylrhoeagenine							+
O-Ethyloreogenine							+
O-Ethyltriniifoline							+
Rhoeadine	+		+	+			+
Rhoeagenine	+						+
Triniifoline* **13**							+
Promorphinane							
Salutaridine			+	+	+		+
Morphinane							
Oripavine			+				
Thebaine			+	+			+
Thebaine N-methyl			+				

*Novel Alkaloids

TABLE 16.11
Species from Section Macrantha Elk. (Syn. Oxytona)

Macrantha, Elk. Cullen 1965	Oytona, Goldblatt 1974
P. bracteatum Lindl.	.
P. lasiothrix Fedde	*Syn. P. pseudo-orientale* (Fedde) Medw.
P. orientale L.	*P. orientale* L.
P. paucifoliatum (Trautv.)Fedde	*Syn. P. orientale*
	P. pseudo-orientale (Fedde) Medw.

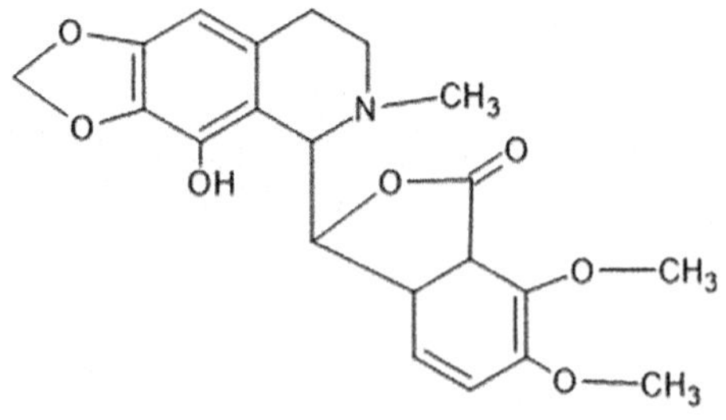

FIGURE 16.15 Narcotolinol.

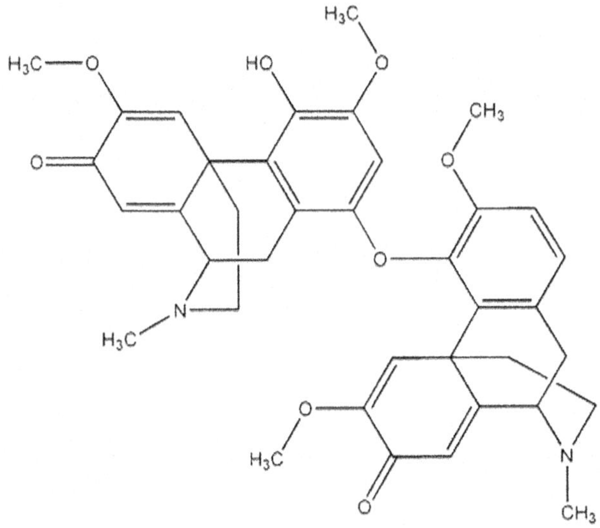

FIGURE 16.16 Salutadimerine.

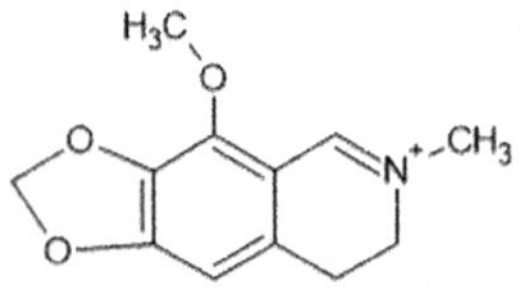

FIGURE 16.17 Cotarnine.

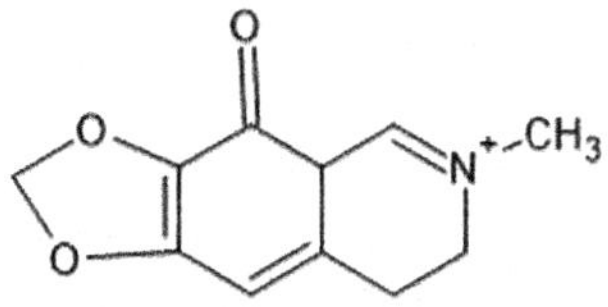

FIGURE 16.18 Cotarnoline.

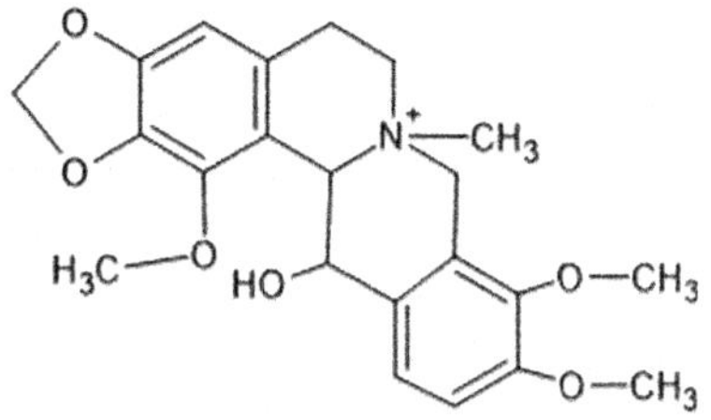

FIGURE 16.19 N-metho-Papaverberbine.

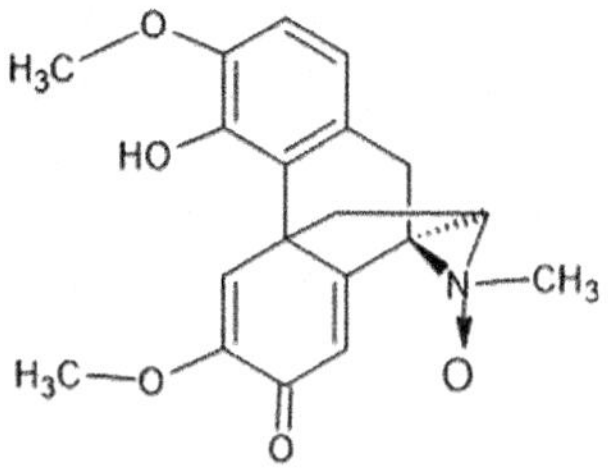

FIGURE 16.20 Salutaridine N-oxide.

Sections Pilosa PRANTL and Pseudopilosa M.POPOV ex GÜNTHER

The first report by Cullen recognized five species in section Pilosa **(4).** The revision of section Pilosa by Kadereit excluded *P. lateritium* from this section and included it in a separate section named as Pseudopilosa **(7).** These two sections, Pilosa and Pseudopilosa, are both represented by only one species as *P. pilosum* with five subspecies and *P. lateritium* with two subspecies, respectively, as shown in Tables **16.**13 and **16.**14. His classification was based on morphological characteristics of these two species as well as their major alkaloid content, which is based on the result of the studies on the alkaloids of these species collected from Turkey **(44)**. All the samples in this section were investigated in detail for their alkaloid content and identified as described by Cullen **(4)**. In studies on the alkaloids of five species, the presence of the same type of major alkaloids has been shown in *P. apokrinomenon*, *P. pilosum*, *P. spicatum*, and *P. stylatum* as aporphine and promorphinane types, whereas the major alkaloids isolated from two samples of *P. lateritium* were found to be completely different as protopine and rhoeadine types **(44-46)**. These differences were taken into consideration by Kadereit to include *P. lateritium* in a separate section as pseudopilosa (Table 16.14). Two new alkaloids of promorphinane type from *P. pilosum* named as amurinine and epiamurinine (Figure 16.21) have been isolated **(47)**. All the alkaloids isolated from the species of these two sections are summarized in Table **16.**15.

USES OF *PAPAVER* SPECIES IN FOLK MEDICINE

Among all the *Papaver* species, *P. rhoeas* is one of the most widespread one and is used as a folk medicine for several disorders. The uses of this species in folk medicine have been summarized

TABLE 16.12
Alkaloids of *Papaver* Species from Section Macrantha

Alkaloids	*P. bracteatum*	*P. lasiothrix*	*P. pseudo-orientale*	*P. orientale*
Simple isoquinoline				
Cotarnine* **17**			+	
Cotarnoline* **18**			+	
Proaporphine				
Orientalinone			+	
Aporphine				
Bracteoline			+	
Isothebaine			+++	+
Secoberbine				
Macranthaline* **10d**			+	
Macrantaldehyde* **10e**			+	
Macrantoridine* **10f**			+	
Narcotindiol 10g				
Papaveroxine* **10a**			+	
Papaveroxidine* **10c**			+	
Papaveroxinoline* **10b**			+	
Protoberberine				
Mecambridine		+	+	+
N-methopapaverberbine* **19**			+	
Orientalidine		+	+	+
Phtalideisoquinoline				
Narcotine			+	
Narcotinhemiacetale* **9**			+	
Narcotolinol* **15**			+	
Rhoeadine				
Alpinigenine			+	+
Alpinine			+	
Promorphinane				
Norsalutaridine		+		
Salutaridimerine* **16**		+		
Salutaridine	+	+	+	+
Salutaridine-N-oxide* **20**	+	+		
Morphinane				
Neopine	+			
Oripavine	+			+
Thebaine	+	+		+
Thebaine-N-oxide	+	+		

*Novel alkaloids

in Table 16.16. Two other species of section Rhoeadium, *P. rhopalothece* and *P. euxinum* subsp. *Euxinum*, have also been reported to be used as an antitussive **(15,16).**

BIOLOGICAL ACTIVITIES OF *PAPAVER* SPECIES

Antimicrobial Activities of *Papaver* Species

Five annual *Papaver* species, *P. argemone* subsp. *davisii*, *P. clavatum*, *P. dubium* subsp. *lecoqii* var. *lecoqii*, *P. macrostomum,* and *P. rhoeas* have been screened for their antimicrobial activities (**9, 12, 58**).

TABLE 16.13
Species from Sections Pilosa PRANTL

Cullen 1965	Kadereit 1997
P. apokrinomenon Fedde	*Syn.P. pilosum* Sibth.& Sm. subsp. *pilosum**
P. lateritium Koch	
P. pilosum Sibth. & Sm	*P. pilosum* Sibth.& Sm subsp. *glabrisepalum* Kadereit Subsp. *sparsipilosum* (Boiss.) Kadereit
P. spicatum Boiss.& Bal	*Syn P. pilosum* subsp. *spicatum* (Boiss. & Balansa) Wendt ex Kadereit
P. strictum Boiss. & Bal.	*Syn P. pilosum* subsp. *strictum* (Boiss. & Balansa) Wendt ex Kadereit

*This species has now been included in a new section by Kadereit named Pseudopilosa

TABLE 16.14
Species from Section Pseudopilosa M. POPOV ex GÜNTHER

KADEREIT 1997
P. lateritium K.Koch
subsp. *lateritium*
subsp. *monanthum* (Tautv.) Kadereit

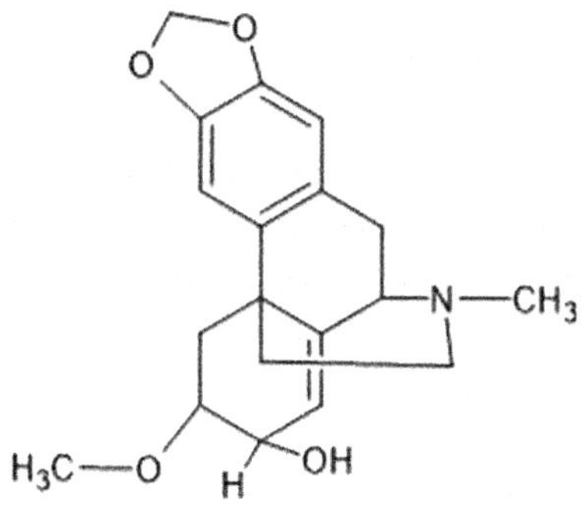

FIGURE 16.21 Amurinine.

Two samples of *P. macrostomum* which have been shown to exist in two chemotypes have been screened for their antimicrobial activities. The tests were performed on petroleum, diethylether, chloroform, acetone, and ethanol extracts obtained from the aerial parts. Extracts were tested against six bacterial and six fungal strains. Although two samples differed chemically, they displayed similarities in their antimicrobial activities. It has been found that diethyl ether and acetone extracts of two samples of *P. macrostomum* had antimicrobial activity against almost all microorganisms tested **(9).**

Another activity test was performed on the petroleum, diethylether, chloroform, acetone, and ethanol extracts obtained from the aerial parts of four annual *Papaver* species from sections Argemonidium (*P. argemone* subsp. *davisii*) and Rhoeadium (*P. clavatum*, *P. dubium* subsp. *lecoqii* var. *lecoqii*, and *P. rhoeas*) by using the microbroth dilution technique. The petroleum

TABLE 16.15
Alkaloids of *Papaver* Species from Section Pilosa and Pseudopilosa

Alkaloids	*P. apokrinomenon*	*P. lateritum**	*P. pilosum*	*P. spicatum* var. *luschanii*	*P. spicatum* var. *spicatum*	*P. strictum*
Secoberbine						
Macranthaline		+				
Aporphine						
Dehydroglaucine	+		+	+	+	+
Dehydroroemerine	+					
Glaucine	+		+	+	+	
N-Methylglaucine						+
N-Methyllaurotetanine	+					+
N-Methylroemerine						+
Roemerine	+		+	+	+	+
Protoberberine (with tetrahydro-protoberberine)						
Mecambridine		+				
N-methyl-tetrahydropalmatine		+				
Protopine						
Cryptopine		+				
Protopine		+				
Rhoeadine						
Rhoeagenine		+				
Rhoeadine		+				
Promorphinane						
Amurine			+	+	+	+
Amurinine* **21**			+	+	+	+
Dihydronudaurine			+	+	+	+
Epiamurinine*			+	+	+	+

*Novel alkaloids

ether and diethyl ether extracts of *P. dubium* subsp. *lecoqii* showed antibacterial activity against *Staphylococcus aureus* (MIC: 9.76 and 19.52 μg/mL, respectively). The diethyl ether and chloroform extracts of *P. argemone* subsp. *davisii* and diethyl ether, chloroform, and acetone extracts of *P. rhoeas* also had activities at the same range against *S. aureus* with a MIC value of 39.06 μg/mL **(58).**

In a recent study, the antimicrobial activities of tertiary alkaloids and methanol extracts of eight samples of *P. rhoeas* collected from different locations have been screened by the microbroth dilution method. The alkaloid extract of one sample, which contained roemerine as the major alkaloid showed potent antimicrobial activity against *S. aureus* with a MIC value of 1.22 μg/mL, together with a remarkable activity against *Candida albicans* with a MIC value of 2.4 μg/mL. This alkaloid extract also displayed moderate activity against *Klebsiella pneumoniae* with a MIC value of 39 μg/mL. This significant activity has been attributed to the presence of roemerine content in this sample. Another alkaloid extract, which has been shown to contain roemerine, also exhibited good activity against *S. aureus* and *S. epidermidis* with MIC values of 9.7 μg/mL and 19 μg/mL, respectively. Among the methanol extracts tested, one sample showed moderate antimicrobial activity against *S. aureus* with a MIC value of 39 μg/mL **(12).**

TABLE 16.16
Uses of *Papaver rhoeas* in Folk Medicine

Plant Parts Used	Uses	References
Petals	Sedative in children, blood sugar lowering	**48**
Young sprouts	Mild sedative	**49**
Flowers	Against hemorrhoids	**50**
Roots	Anthelminthic	**51, 52**
Petals	Against kidney stones	**53**
Flowers	Antipyretic	**51, 54**
Flowers	Against asthma	**51, 54**
Flowers	Insomnia	**51, 54**
Petals	Antidiarrhoeal	**51, 54**
Flowers	Antitussive	**51, 55**
Flowers	Antispasmodic	**51, 55**
Flowers	Menstrual disorders	**53, 55**
Aerial parts	Antirheumatism	**56**
Petals	Sore throat	**56**
Petals	Demulcent	**56**
Petals	Immunostimulant	**56**
Petals	Against nose bleeding	**56**
Petals	Mouth wounds in children	**51**
Leaves	Tonic	**57**
Leaves	Against jaundice	**51**
Petals	Galactagogue	**56**

Cytotoxicity of *Papaver* Species

In vitro anticancer activity and cytotoxicity of some alkaloids isolated from *Papaver* species of Turkish origin have also been studied. Two alkaloids, berberine and macranthaline, showed significant SI (SI = IC_{50} for normal cells/IC_{50} for cancer cell) values, which were 12.42 for macrantaline and 5.89 berberine **(59).**

From *P. lacerum* (Syn: *P. dubium* subsp. *leavigatum*), two compounds, Tyrosol-1-*O*-β-xylopyranosyl-(1→6)-*O*-β-glucopyranoside) and a new one, 5-*O*-(6-*O*-α -rhamnopyronosyl-β-glucopyronosyl) mevalonic acid, were isolated. These compounds possess moderate cytotoxic effects with an IC_{50} of 66.4 μM ($p < 0.0001$) and 54 μM ($p < 0.0001$), respectively **(60).**

CONCLUSION

In Turkey, the genus *Papaver* is represented by 23 species with 19 subspecies belonging to eight sections based on med-checklist area (Greuter 1989) and Kadereit's work on the genus *Papaver* L. in the Mediterranean area **(6, 7)** while earlier reports by Cullen **(4)** included 40 species in the Flora of Turkey. With the exception of three species and four subspecies from section Rhoeadium revised by Kadereit (1988) **(10),** all other 23 species were identified as described by Cullen in the Flora of Turkey **(4).** The alkaloids of 26 species were studied and 100 alkaloids belonging to 16 different types (simple isoquinoline and benzyisoquinoline, isopavine, proaporphine, aporphine, phenanthrene, cularine, morphinanedienone, morphinane, protobeberine, secoberbine, protopine, narceine, phtalideisoquinoline, roeadine-papaverrubine, spirobenzylisoquinoline) were isolated. The presence of cularine and spirobenzylisoquinoline types has been shown for the first time in the genus *Papaver.* Including a dimeric morphinanedienone, 24 other compounds were identified as novel alkaloids. The existing chemotypes were found in the species of six sections. Species from the section Meconidium possesses the highest number of chemotypes. The presence of medicinally important alkaloids, mainly thebaine, oripavine, noscapine, and papaverine, has been shown in the species from sections Rhoeadium, Macrantha, and Meconidium. Two species, *P. macrostomum* and *P. lacerum* (syn. *P. dubium* subsp. *leavigatum*), have also been studied for flavonoid and glycoside content, respectively. Three annual species from the section Rhoeadium, *P. rhoeas*, *P. rhopalothece*, and *P. commutatum*, have been shown to be used as folk medicines in different parts of Turkey. Several annual *Papaver* species and their alkaloids have also been found to possess antimicrobial activities. Cytotoxicity of the extracts of some species and their alkaloids has recently been reported.

BIBLIOGRAPHY

1. Turkish Grain Board. *Poppy and Alkaloid Affairs.* Ankara, 2013.
2. Nyman U, Bruhn J G. Papaver bracteatum - A summary of current knowledge. *Planta Med.* 1979; 35: 96–117.
3. *Narcotic Drugs, Reports Published by the International Narcotic Control Board for 2020.*
4. Cullen P H, Papaver L. *Flora of Turkey and the East Aegean Islands.* Edinburgh: Edinburgh University Press, 1965; 219–236.
5. Kiger R W. Revised sectional nomenclature in Papaver L. *Taxon.* 1985; 34: 150–152.
6. Greuter W, Burdet H M, Long G. (eds.). *Med-Checklist*, 4. Genève, 1989.
7. Kadereit J W. The genus Papaver L. in the Mediterranean area. *Lagascalia* 1997; 19: 83–92.
8. Sarıyar G, Mat A, Ünsal Ç. et al. Biodiversity in the alkaloids of annual Papaver species of Turkish origin. *Acta Pharmaceutica Turcica.* 2002; 44: 159–168.
9. Ünsal Ç, Sarıyar G, Gürbüz-Akarsu B, et al. Antibacterial and phytochemical studies on Turkish samples of Papaver macrostomum. *Pharmaceutical Biology* 2007; 45: 626–630.
10. Kadereit J W. A revision of Papaver L section Rhoeadium Spach. *Notes Roy.Bot.Gard.* 1988; 45: 225–286.
11. KalavY N, Sariyar G. Alkaloids from Turkish Papaver rhoeas. *Planta Med.* 1989; 55: 488.
12. Çoban İ, Toplan-Gülsoy G, Özbek B, et al. Variation of alkaloid contents and antimicrobial activities of Papaver rhoeas l. growing in Turkey and Northern Cyprus. *Pharmaceutical Biology* 2017; 55: 1894–1898.
13. Ünsal Ç, Sariyar G, Mat A, et al. Distribution of alkaloids in the samples of *Papaver dubium* subsp. *lecoqii* var. *lecoqii* from Turkey: A potential source for thebaine. *Biochemical Systematics and Ecology* 2006; 34: 170–173.
14. Mat A, Sariyar G, Ünsal Ç, et al. Alkaloids and bioactivity of Papaver dubium subsp. dubium and P. dubium subsp.laevigatum. *Natural Product Letters* 2000; 14: 205–210.
15. Sariyar G, Kalav Y N. Alkaloids from Papaver rhopalothece used as folk medicine in Anatolia. *Planta Med.* 1990; 56: 232.
16. Atay M, Sariyar G, Özhatay N. Alkaloids from *Papaver commutatum* subsp. *euxinum. Planta Med.* 1996; 62: 483–484.
17. Şerbetçi T, Sarıyar G, Michel S, et al. Isolation and chemistry of alkaloids from *Papaver arachnoideum* Kadereit. *Biochemical Systematics and Ecolgy* 2009; 37: 501–503.
18. Ünsal Ç, Eroğlu E, Şerbetçi T, et al. Alkaloids of Papaver clavatum and P. stylatum. *Biochemical Systematics and Ecology* 2008; 36: 497–499.
19. Sariyar G, Phillipson J D. Alkaloids of papaver lacerum. *J. Nat. Prod.* 1981; 44: 239–240.
20. Phillipson J D, Sariyar G, Baytop T. Alkaloids from Papaver fugax of Turkish origin. *Phytochemistry* 1973; 12: 2431–2434.
21. Lalezari I, Shafii A, Asgharian R. Isolation of Alpinigenine from Papaver bracteatum. *J. Pharm. Sci.* 1973; 62: 1718.
22. Lalezari I, Nasseri-Nouri P, Asgharian R. Papaver bracteatum Lindl.: Population Arya II. *J. Pharm. Sci.* 1974; 63: 1331.
23. Sariyar G, Phillipson J D. Alkaloids of Turkish Papaver tauricola. *Phytochemistry* 1980; 19: 2189–2192.
24. Phillipson J D, Thomas O O, Gray A I et al. Alkaloids from Papaver armeniacum, P. fugax, P. tauricola. *Planta Med.* 1981; 41: 105–118.

25. Sariyar G. Alkaloids from Papaver cylindricum. *Planta Med.* 1982; 46: 175–178.
26. Sarıyar G. Alkaloids from Papaver triniifolium of Turkish origin. *Planta Med.* 1983; 49: 43–45.
27. Sariyar G, Shamma M. Six alkaloids from Papaver species. *Phytochemistry* 1986; 25: 2403–2406.
28. Sariyar G, Baytop T, Phillipson J D. A new protopine alkaloid from Turkish Papaver curviscapum: 1-methoxyallocryptopine. *Planta Med.* 1989; 55: 89–90.
29. Sariyar G, Sarı A, Freyer A J et al. Quaternary Isoquinoline alkaloids from Papaver species. *J. Nat. Prod.* 1990; 53: 1302–1306.
30. Sarı A, Sariyar G. Isolation of triniifoline, miltanthaline and some medicinally important alkaloids from Papaver triniifolium. *Planta Med.* 1997; 63: 575–576.
31. Sariyar G, Özhatay N, Atay M, et al. Alkaloids and chromosome numbers of P. polychaetum and P. triniifolium. *Istanbul Ecz.Fak. Dergi.* 1998; 32: 59–68.
32. Sari A, Gray A I, Sariyar G. Two new benzylisoquinoline alkaloids from Papaver triniifolium. *Natural Product Research* 2006; 20: 493–496.
33. Sari A, Sariyar G. Alkaloids and bioactivity of Papaver triniifolium. *Pharmazie* 2000; 55: 471–472.
34. Sari A, Gray A I, Sariyar G. A new dehydroaporphine alkaloid from Papaver fugax. *Natural Product Research* 2004; 18: 265–268.
35. Goldblatt P. Biosystematic studies on Papaver section Oxytona. *Ann.Missouri Bot. Gard.* 1974; 61: 264–296.
36. Phillipson, J D, Scutt A, Baytop A, et al. Alkaloids from Turkish samples of Papaver orientale and P. pseudo-orientale. *Planta Med.* 1981; 43: 261–271.
37. Sariyar G, Phillipson J D. Macranthaline and macrantoridine, new alkaloids from a Turkish sample of Papaver pseudo-orientale. *Phytochemistry* 1977; 16: 2009–2013.
38. Sariyar G, Baytop T. Alkaloids from Papaver pseudo-orientale (P. lasiothrix) of Turkish origin. *Planta Med.* 1980; 38: 378–380.
39. Sariyar G, Shamma M. (-)-Papaveroxidine, a modified phtalideisoquinoline alkaloid from Papaver pseudo-orientale. *J. Nat. Prod.* 1988; 51: 802–803.
40. Sariyar G, Freyer A J, Guinaudeau H, et al. (+)-Salutadimerine: A dimeric morphinonedienone alkaloid. *J. Nat.Prod.* 1990; 53: 1383–1386.
41. Sariyar G, Gülgeze H B, Gözler B. Salutaridine N-oxide from the capsules of Papaver bracteatum. *Planta Med.* 1992; 58: 368–369.
42. Sarıyar G. Türkiye'de yetişen *Papaver bracteatum* Lindl. ve *P.pseudo-orientale* (Fedde) Medw.in alkaloitleri. *Istanbul Ecz. Fak. Mec.* 1977; 13: 171–177.
43. Yildirimli Ş, Ertekin S. Two new species Papaver yildirimli ertekin (Papaveraceae) and salvia ertekinl*i* yildirimli (Lamaiceae) from Siirt, Turkey. *Ot. Sistemeatik Botanik Dergisi.* 2008; 15: 1–8.
44. Öztekin A, Baytop T, Hutin M, et al. Comparison of chemical and botanical studies of Turkish Papaver belonging to section Pilosa. *Planta Med.* 1985; 51: 431–434.
45. Sariyar G, Sari A, Mat A. Quaternary alkaloids of Papaver spicatum. *Planta Med.* 1994; 60: 293.
46. Sari A, Sariyar G, Mat A, et al. Alkaloids and bioactivity of Papaver lateritium occuring in Turkey. *Planta Med.* 1998; 64: 582.
47. Hocquemiller R, Öztekin A, Roblot F, et al. Alkaloides de Papaver pilosum. *J. Nat. Prod.* 1984; 47: 342–346.
48. Yucel E, Tülükoğlu A. 2000. Plants used as folk medicine in Gediz (Kutahya) province. *Ekoloji* 2000; 36: 12–14.
49. Tuzlacı E, Emre Bulut G. Turkish folk medicinal plants, Part VII: Ezine (Çanakkale). *J Fac Pharm. Istanbul.* 2007; 39: 39–51.
50. Ecevit Genc G, Ozhatay N. 2006. An ethnobotanical study in Çatalca (European part of Istanbul) II. *Turk J Pharm Sci.* 2006; 3: 73–89.
51. Tuzlacı E. *Şifa Niyetine Turkiye 'nin Bitkisel Halk İlaçları.* Istanbul, Turkey: Alfa Yayınları, 2006.
52. Tuzlacı E, Erol M K. Turkish folk medicinal plants, Part II: Eğirdir (Isparta). *Fitoterapia* 1999; 70: 593–610.
53. Aslan A, Mat A, Ozhatay N, Sarıyar G. A contribution to traditional medicine in West Anatolia. *J Fac Pharm. Istanbul.* 2007; 39: 73–83.
54. Tuzlacı E, Sadıkoglu E. Turkish folk medicinal plants, Part VI: Koc¸arlı (Aydın). *J Fac Pharm. Istanbul.* 2007; 39: 25–37.
55. Tuzlacı E, Alparslan D F. Turkish folk medicinal plants, Part V: Babaeski (Kırklareli). *J Fac Pharm. Istanbul.* 2007; 39: 11–23.
56. Kultür Ş. Medicinal plants used in Kırklareli province (Turkey). *J Ethnopharmacol.* 2007; 111: 341–364.

57. Tuzlacı E, Eryaşar Aymaz P. Turkish folk medicinal plants, Part IV: Gonen (Balikesir). *Fitoterapia* 2001; 72: 323–343.
58. Ünsal Ç, Özbek B, Sarıyar G, et al. Antimicrobial activity of four annual Papaver species growing in Turkey. *Pharmaceutical Biology* 2009; 47(1): 4–6.
59. Demirgan R, Karagöz A, Pekmez M, et al. In vıtro antıcancer actıvıty and cytotoxıcıty of some Papaver alkaloıds on cancer and normal cell lınes. *Afr J. Tradit Complement Altern Med.* 2016; 13: 22–26.
60. Bayazeid O, Bedir E, Yalçin F. Ligand-based virtual screening and molecular docking of two cytotoxic compounds isolated from Papaver lacerum. *Phytochemistry Letters* 2019; 30: 26–30.

17 *Pistacia* sp.

Seren Gündoğdu and Ayşe Uz

INTRODUCTION

Pistacia L. (pistache, pistachio) species belongs to the Anacardiaceae family, a wide-range family that includes approximately 81 genera and 800 species. The species of the genus *Pistacia* are deciduous or evergreen shrubs and small trees ranging from 5 to 15 m in height (Pell et al., 2010; Rauf et al., 2017). *Pistacia* species are distributed in different countries like Iran, Pakistan, Greece, Turkey, and North Africa (Mahjoub et al., 2018). *Pistacia* species are economically important commercial crops in Turkey. Turkey is among the countries that are the main pistachio producers in the world, contributing over 90% of production (Faostat, 2014). Dry and fresh, toasted or raw pistachios are commonly applied as food, flavoring agent, and food additive. Many species belonging to the genus *Pistacia* are known and used as rootstocks for the only commercially cultivated species *P. vera* (Bozorgi et al., 2013). Turkey is among the countries where pistachio (*Pistacia vera* L.) can be produced efficiently due to suitable ecological conditions. *P. vera* grafted on other *Pistacia* species was also cultured in other cities in Turkey.

The genus *Pistacia* comprises plants known for their medicinal properties (*P. lentiscus*) and as food (*P. vera* as food additive, *P. terebinthus* as snack food) (Rauf et al., 2017). *P. lentiscus* var. *Chia* which grows in the southern region of Chios Island, Greece, Ege region in Turkey, and Mediterranean region in Africa produces a resin known as Chios mastic gum. It has been used in traditional Greek and Anatolian medicine for diverse gastrointestinal disorders including dyspepsia or peptic ulcer disease for more than 2500 years. Pistachios, compared with nuts such as walnuts, hazelnuts, almonds, and pine nuts, it has a lower fat and energy content and higher fiber (both soluble and insoluble) content, vitamin K, phytosterols, gamma-tocopherol, and xanthophyll carotenoids. Also when compared to oil seeds such as almonds, walnuts, and hazelnuts, it ranks first in terms of carbohydrate, protein, vitamins B1, B6, E and A, iron, potassium, beta-carotene, total phytosterol, and lutein values (Çağlar et al., 2017). Over 100 chemical compounds such as phenolic compounds, terpenoids, monoterpenes, flavonoids, alkaloids, saponins, alkyl salicylic acids, fatty acids, and sterols have been identified in different parts of the *Pistacia* species (Rauf et al., 2017; Ben Ahmed et al., 2021; Benalia et al., 2021). Traditionally they are used for the treatment of several common diseases such as asthma, rheumatism, hypertension, diabetes, diarrhea, and hemorrhoids (Sezik et al., 2001). The pistachio hull (outer skin/ fruit peels) which is removed during industrial processing is a significant by-product of pistachio. It has been known that bioactive molecules that could be obtained from pistachio hulls, which are agricultural food wastes, have the potential to be evaluated in the field of pharmacy (Hürkul, 2021). Ethnomedical uses, phytochemical profiles, and pharmacological significance of *Pistacia* species are summarized in this chapter.

BOTANICAL FEATURES AND ETHNOBOTANICAL USAGE

The pistachio tree (Anacardiaceae family) (Figure 17.1), which can be up to 10 meters tall, has a single-seeded fruit. Members of the genus are dioecious trees or shrubs with well-developed vertical resin canals. Leaves are alternate, deciduous or persistent, evergreen, pinnately compound, but sometimes trifoliolate or unifoliolate, pari- or im-paripinnate, membranous, or leathery. Inflorescences are determinate and terminal or axillary. Flowers of species are dioecious and

DOI: 10.1201/9781003146971-17

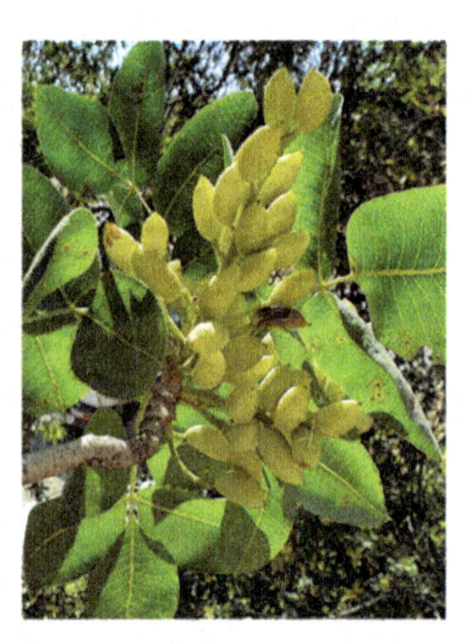

FIGURE 17.1 Tree and hulls of *P. vera*.

unisexual. Flowers are almost always unisexual, radial, small, with well-developed staminodes or carpellodes, and apetalous; sepals are usually 5, distinct to slightly connate. Stamens are 4–5, filaments are very short, usually glabrous, and usually distinct; pollen grains tricolporate or triporate. Carpels are usually 3 and variously connate; ovary superior, with only 1 fertile and well-developed carpel; gynoecium asymmetrical and unilocular with apical placentation, ovule 1; stigma capitate. Nectariferous disk present, usually intrastaminal. Fruit with flattened asymmetrical drupe, embryo curved; endosperm scanty or lacking. The seed consists of two fleshy cotyledons, surrounded by a thin testa, surrounded by the bony, hard endocarp, while the epicarp and mesocarp form the rind of the fruit (Pell et al., 2010; Ben Ahmed et al., 2021; Hürkul, 2021).

In the flora of Turkey, eight *Pistacia* species [(*P. atlantica*, *P. eurycarpa*, *P. khinjuk* (bıttım, menengiç), *P. lentiscus* (sakız ağacı), *P. palaestina*, *P. terebinthus* (menengiç), *P. vera* (antep fıstığı), and *P. saportae* (çetem)] grow and called with their Turkish name. *Pistacia* species have been commonly used as Turkish traditional medicine for various diseases such as eczema, peptic ulcer, diabetes, respiratory tract infections, renal stones, asthma, and stomachache and as astringent, anti-inflammatory, and antibacterial agents (Kordali et al., 2003). In traditional Iranian medicine, pistachio fruit kernels are known as an effective tonic for heart, brain, liver, and stomach and used for the treatment of coughs, while pistachio hulls are used as tonic, sedative, and antidiarrheal (Bozorgi et al., 2013; Sarkhail et al., 2017; Mahjoub et al., 2018). It has been reported that roasted *P. khinjuk* fruit and *P. terebinthus* decoction were used for their healing effects on stomachaches in Turkey. Also essential oils of some *Pistacia* species are used as antimicrobial, anti-inflammatory, and anti-ulcer medicine by region folk (Hayta et al., 2014). The essential oil and mastic gum from *P. lentiscus* var. *Chia* (lentisk, mastic tree) are currently broadly used in medicine owing to their antimicrobial, antioxidant, and hepatoprotective properties (Kottakis et al., 2009). The aerial part was used traditionally in the treatment of arterial hypertension due to its diuretic properties. The leaves are known to have anti-inflammatory, antibacterial, antifungal, antipyretic, astringent, hepatoprotective, expectorant, and stimulant activities. They are also used in the treatment of eczema, oral infections, diarrhea, kidney stones, jaundice, headaches, asthma, and respiratory problems. The essential oil of lentisk is known for its therapeutic virtues with regard to lymphatic and circulatory problems (Hafsé et al., 2015). The fruit oil of *P. lentiscus* is used in traditional medicine to treat burns, skin impairments as well as inflammatory diseases as soothing massage or internal use (Benalia et al., 2021). This genus is also considered as one of the valuable natural resources for neuroprotection research based on experiences in traditional medicine. The oleoresin is used in folkloric medicine as chewing gum in Europe and the Middle East (Moeini et al., 2019).

Pistacia coffee is made from the fruits of *P. terebinthus*. Another product in Southern Turkey is the *Pistacia* soap (bıttım sabunu as called in Turkish). The *Pistacia* oil is added to olive oil to make the soap in a traditional way. It is known that such soap is good for hair and skin traditionally. It is claimed to be a good treatment for hair loss and dandruff with healing effects.

PHYTOCHEMICALS

A series of bioactive phytochemicals such as phenolic compounds, essential oil, and gum have been isolated and identified from different *Pistacia* species. The list of key phytochemicals in this genus and their molecular formulae are presented in Table 17.1 (Bozorgi et al., 2013).

Apart from the above, flavonoid glycosides have been reported in some *Pistacia* species. For instance, quercetin-3-*O*-rutinoside was found as the main constituent of the seeds from *P. vera* (Tomaino et al., 2010). On the other hand, major flavonoid glycosides from *P. khinjuk* are myricetin-3-*O*-glucoside, myricetin-3-*O*-galactoside, and myricetin-3-*O*-rutinoside (Kawashty et al., 2000). Also, cyanidin-3-*O*-glucoside, cyanidin-3-*O*-galactoside, delphinidin-3-*O*-glucoside, and quercetin-3-*O*-rutinoside as main anthocyanins have been reported in *P. vera* fruits and *P. lentiscus* leaves (Romani et al., 2002; Tomaino et al., 2010). The essential oil profile of *Pistacia* species is rich in both monoterpenes and oxygenated sesquiterpenes. The common constituents of essential oil are myrcene, limonene, β-gurjunene, germacrene, α-pinene, β-pinene, muurolene, α-humulene, delta-3-carene, camphene, terpinen-4-ol, spathulenol, epi-bicyclosesquiphellandrene, Germacrene-D, β-caryophyllene, etc. (Duru et al., 2003; Bozorgi et al., 2013; Rand et al., 2014). The parts that are rich in essential oil are fresh unripe pistachio fruits. Nuts of *Pistacia* species are also rich in mono- and polyunsaturated fatty acids (Rauf et al., 2017). In *Pistacia* species with rich active phytochemical content, phenolic compounds are concentrated in the nut skin. The results of chromatographic (LC-MS/MS) analyses indicated that *P. vera* skin studied in Turkey is very rich in quercetin and gallic acid. Methanol extract of *P. vera* skin could be regarded as a strong antioxidant agent relative to phenol and flavonoid contents and moderate-to-strong antimicrobial agent against different bacteria strains (Şermet, 2015). A few previous studies showed that Iranian pistachio hull is rich in polyphenolic compounds and acts as an antioxidant and antimicrobial agent and to promote wound healing and antimutagenic activity, while the Italian pistachio hull was found to have better antioxidant activity as compared to the seed due to its higher phenolic content such as gallic acid, catechin, epicatechin, quercetin, luteolin, naringenin, and cyanidin-3-*O*-galactoside (Sarkhail et al., 2017).

BIOLOGICAL ACTIVITY

Antioxidant Activity

The high antioxidant effect is mostly attributed to phenolic compounds which are widely distributed in nature. In this context, different parts of *Pistacia* species show radical scavenging properties *in vitro*. The major components in this *Pistacia* responsible for the antioxidant activity were quercetin, rutin, kaempferol, apigenin, etc. (Rauf et al., 2017). *P. vera* which is a species of commercial importance, the aqueous extract with high phenolic content (32.0–34.0 mg/g dry weight of the sample) showed antioxidant activity as important as synthetic antioxidants (Goli et al., 2005). In the study by Kavak *et al.* (2010), *P. terebinthus* (from Karaburun region in Turkey) leaf extract (80% ethanol-water) showed antioxidant capacity of 85 TEAC (Trolox equivalent antioxidant capacity) value due to its flavonoid contents such as luteolin, luteolin-7-glucoside, and apigenin-7-glucoside. It was reported that this leaf extract had about 12-fold higher antioxidant capacity than those of butylate hydroxyanisole (BHA) and ascorbic acid and also exhibits 10-fold higher antioxidant activity than (+)-catechin. However, *P. terebinthus* fruits showed notable metal-chelation properties as compared to that of the standard (ethylenediamine tetraacetic acid, EDTA) and a remarkable scavenging effect against DPPH radical (Orhan et al., 2012). It was also observed that the hydroxyl radical scavenging

TABLE 17.1
List of Major Compounds Isolated from Various *Pistacia* L. Species

Name and Structure of Compounds	Molecular Formulae	Species	Plant Parts	References
Phenolic Compounds				
Gallic acid 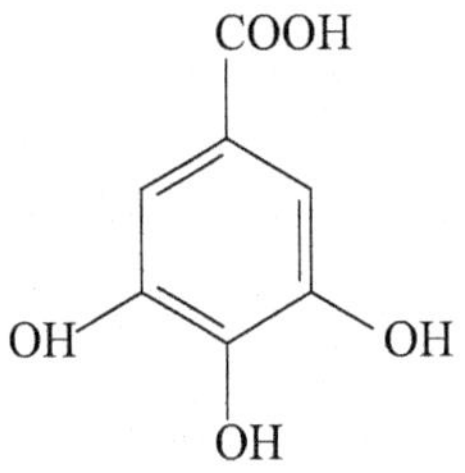	$C_6H_2(OH)_3CO_2H$	*P. atlantica* *P. lentiscus* *P. vera*	Seed and skin Leaf Leaf	Yousfi et al., 2009; Romani et al., 2002; Tomaino et al., 2010
Catechin 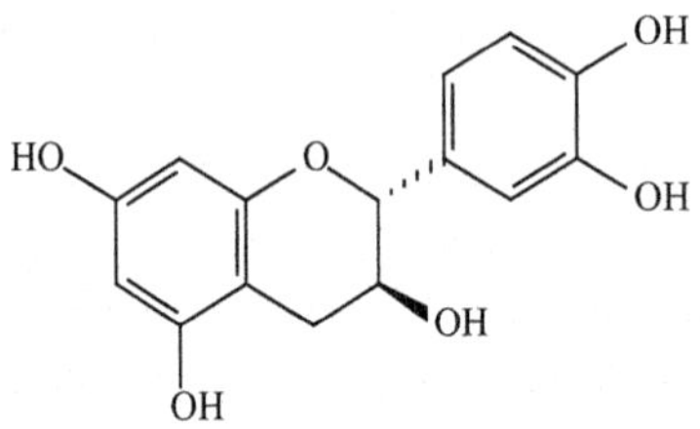	$C_{15}H_{14}O_6$	*P. lentiscus* *P. khinjuk* *P. vera*	Leaf Leaf Skin and kernel	Romani et al., 2002; Moussa et al., 2013; Grace et al., 2016
Epicatechin 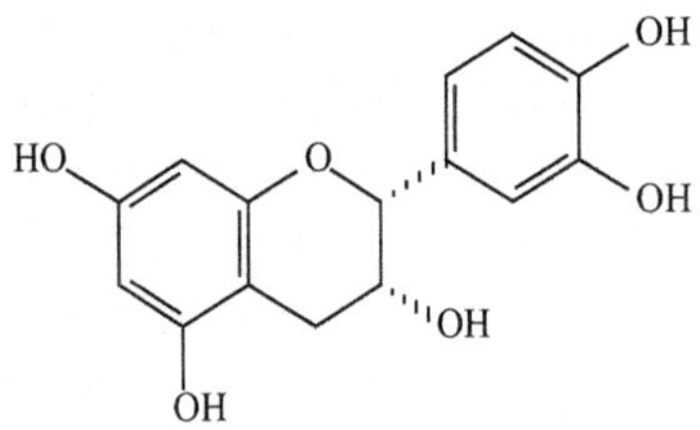	$C_{15}H_{14}O_6$	*P. vera*	Skin and kernel	Grace et al., 2016
Digallic acid 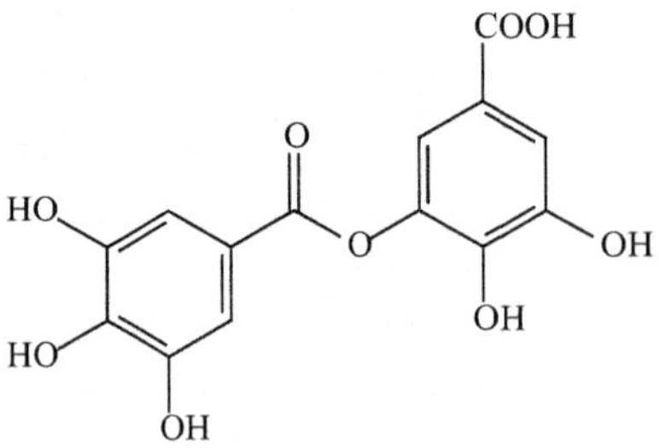	$C_{14}H_{10}O_9$	*P. lentiscus* *P. vera*	Fruit Hull	Bhouri et al., 2010; Erşan et al., 2018

(Continued)

TABLE 17.1 (CONTINUED)
List of Major Compounds Isolated from Various *Pistacia* L. Species

Name and Structure of Compounds	Molecular Formulae	Species	Plant Parts	References
***Trans*-resveratrol** 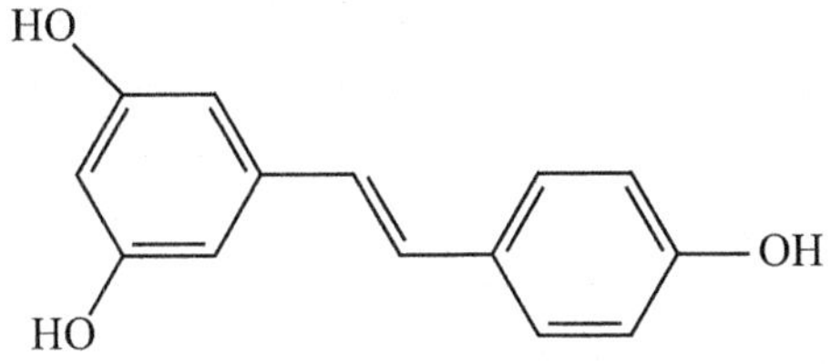	$C_{14}H_{12}O_3$	*P. vera*	Skin and kernel	Tokuşoğlu et al., 2005; Grippi et al., 2008
***Cis*-resveratrol** 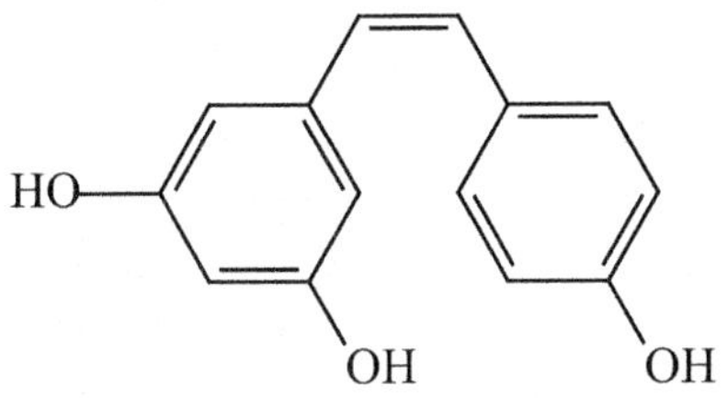	$C_{14}H_{12}O_3$	*P. vera*	Kernel	Tokuşoğlu et al., 2005
3-(8-Pentadecenyl)-phenol 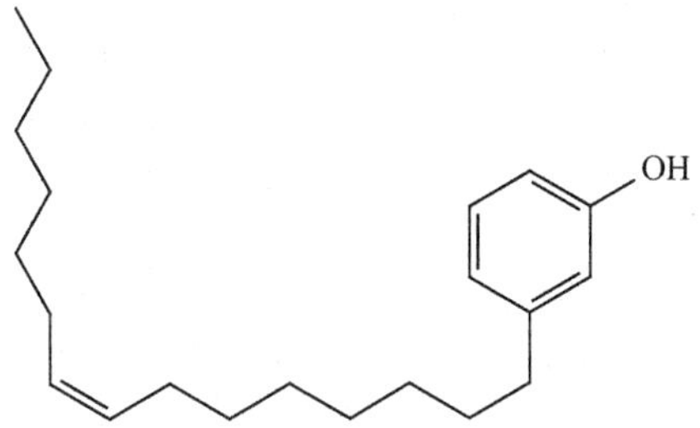	$C_{21}H_{34}O$	*P. vera*	Kernel	Saitta et al., 2009
Naringenin 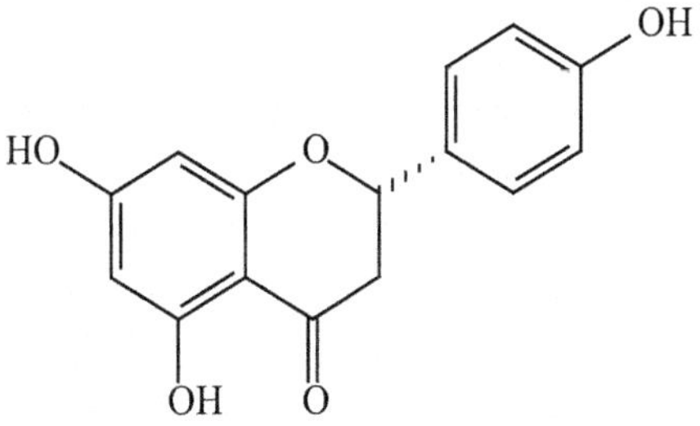	$C_{15}H_{12}O_5$	*P. lentiscus* *P. vera*	Leaf Seed and skin	Amel et al., 2016;Tomaino et al., 2010

(*Continued*)

TABLE 17.1 (CONTINUED)
List of Major Compounds Isolated from Various *Pistacia* L. Species

Name and Structure of Compounds	Molecular Formulae	Species	Plant Parts	References
Eriodictyol 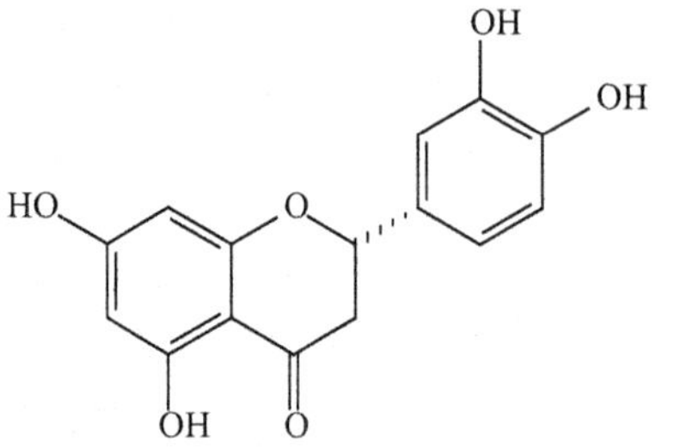	$C_{15}H_{12}O_6$	*P. atlantica* *P. vera*	Fruit Seed, skin and kernel	Khallouki et al., 2017; Ballistreri et al., 2009; Martorana et al., 2013
Daidzein 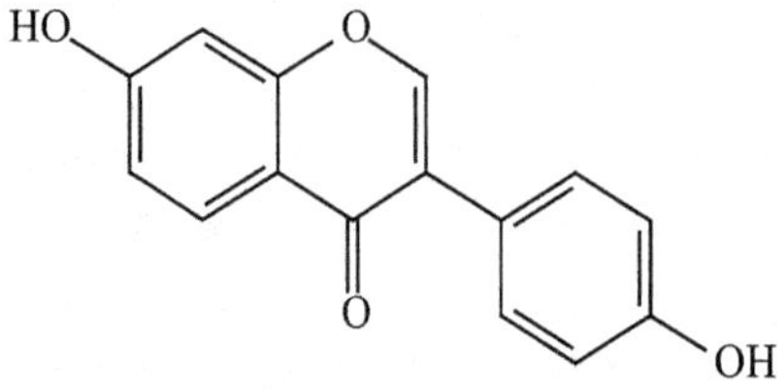	$C_{15}H_{10}O_4$	*P. vera*	Kernel	Ballistreri et al., 2009
Quercetin 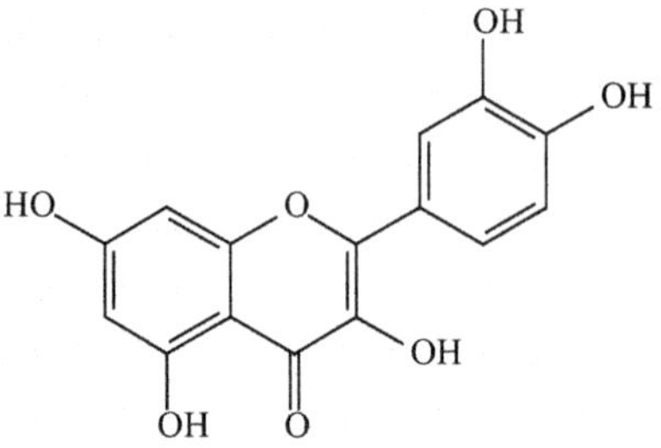	$C_{15}H_{10}O_7$	*P. khinjuk* *P. terebinthus* *P. vera*	Leaf Fruit Skin	Moussa et al., 2013; Topçu et al., 2007; Tomaino et al., 2010
Kaempferol 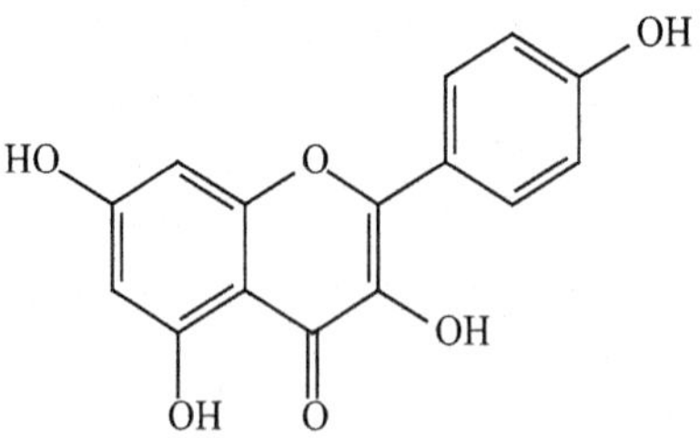	$C_{15}H_{10}O_6$	*P. atlantica* *P. khinjuk* *P. lentiscus* *P. vera*	Leaf Leaf Leaf Skin	Amri et al., 2018; Moussa et al., 2013; Vaya & Mahmood, 2006; Tomaino et al., 2010

(Continued)

TABLE 17.1 (CONTINUED)
List of Major Compounds Isolated from Various *Pistacia* L. Species

Name and Structure of Compounds	Molecular Formulae	Species	Plant Parts	References
Apigenin 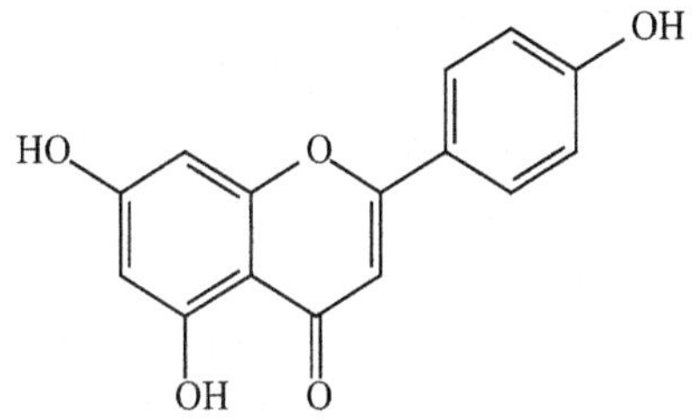	$C_{15}H_{10}O_5$	*P. atlantica* *P. lentiscus* *P. terebinthus* *P. vera*	Fruit Leaf Fruit Seed	Khallouki et al., 2017; Vaya & Mahmood, 2006; Topçu et al., 2007; Tomaino et al., 2010
Luteolin 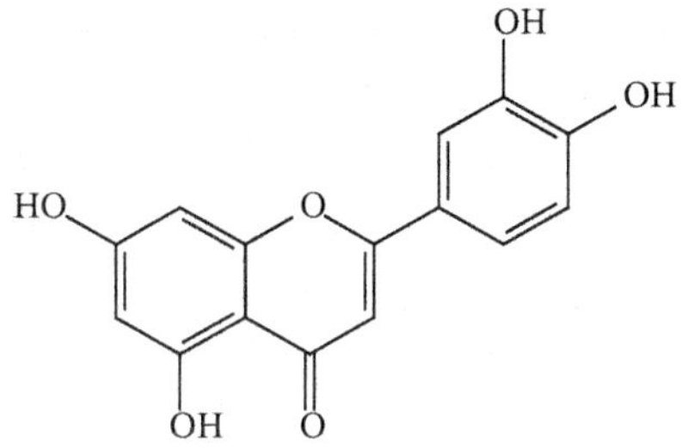	$C_{15}H_{10}O_6$	*P. khinjuk* *P. terebinthus* *P. vera*	Leaf Fruit Skin	Moussa et al., 2013; Topçu et al., 2007; Tomaino et al., 2010
Myricetin 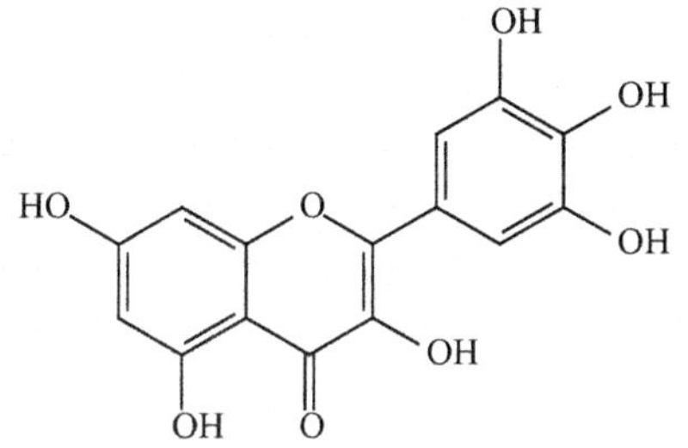	$C_{15}H_{10}O_8$	*P. khinjuk* *P. lentiscus*	Leaf Leaf	Moussa et al., 2013; Vaya & Mahmood, 2006
Alkylsalicylic Acids				
Anacardic acid 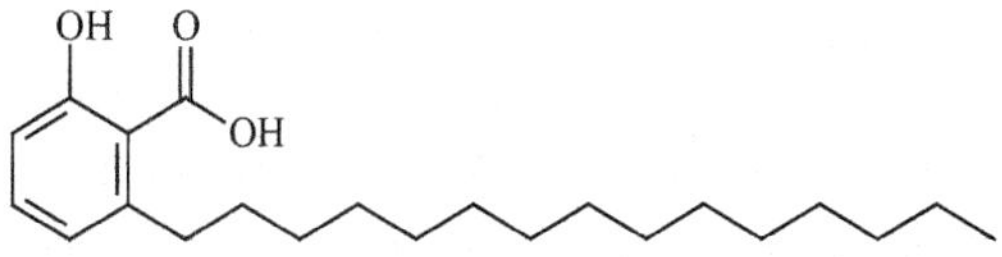	$C_{22}H_{36}O_3$	*P. lentiscus* *P. vera*	Hull Fruit oil	Erşan et al., 2018; Benalia et al., 2021

(*Continued*)

TABLE 17.1 (CONTINUED)
List of Major Compounds Isolated from Various *Pistacia* L. Species

Name and Structure of Compounds	Molecular Formulae	Species	Plant Parts	References
Ginkgolic acid 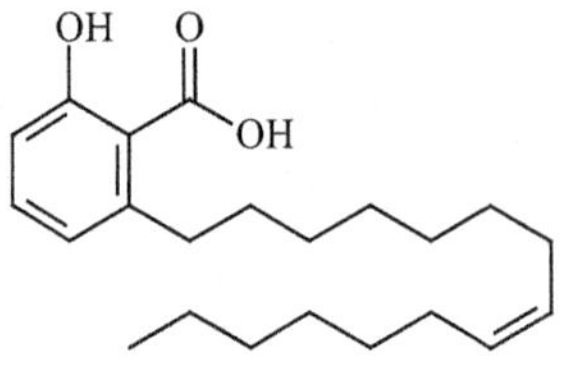 OH O OH	$C_{22}H_{34}O_3$	*P. lentiscus*	Fruit oil	Benalia et al., 2021
Monoterpenoids and Sesquiterpenoids				
α-pinene 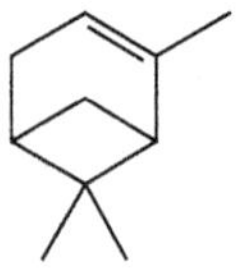	$C_{10}H_{16}$	*P. atlantica* *P. eurycarpa* *P. lentiscus* *P. palaestina* *P. terebinthus* *P. vera*	Leaf and fruit Oleoresin Leaf Leaf and gall Leaf and fruit Gum	Mecherara-Idjeri et al., 2008; Benhassaini et al., 2008; Kıvçak et al., 2004; Flamini et al., 2004; Özcan et al., 2009; Alma et al., 2004
Limonene 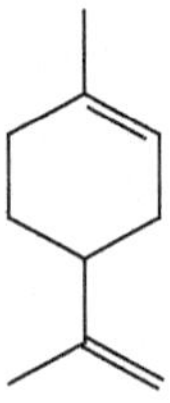	$C_{10}H_{16}$	*P. eurycarpa* *P. lentiscus* *P. palaestina* *P. terebinthus* *P. vera*	Oleoresin Leaf Gall Leaf and twig Gum	Benhassaini et al., 2008; Kıvçak et al., 2004; Flamini et sl., 2004; Alma et al., 2004
Terpinolene 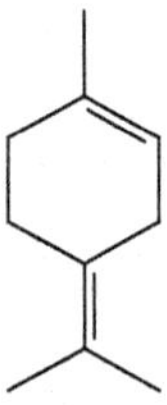	$C_{10}H_{16}$	*P. lentiscus* *P. terebinthus* *P. vera*	Twig Leaf Nut	Kıvçak et al., 2004; Kendirci & Onoğur, 2011
Sabinene 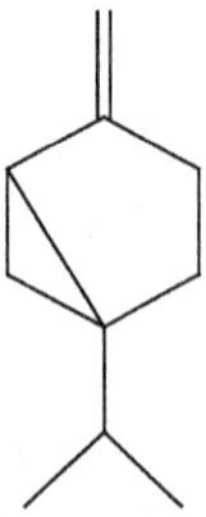	$C_{10}H_{16}$	*P. lentiscus* *P. palaestiana* *P. terebinthus*	Leaf and twig Gall, ripe fruit Fruit	Kıvçak et al., 2004; Flamini et al., 2004; Özcan et al., 2009

(*Continued*)

TABLE 17.1 (CONTINUED)
List of Major Compounds Isolated from Various *Pistacia* L. Species

Name and Structure of Compounds	Molecular Formulae	Species	Plant Parts	References
Myrcene 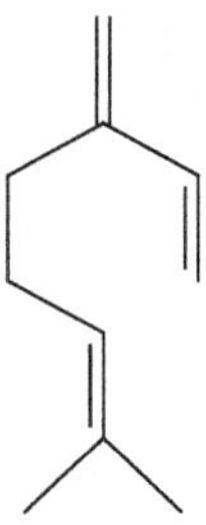	$C_{10}H_{16}$	*P. lentiscus* *P. palaestiana* *P. terebinthus* *P. vera*	Leaf and twig Leaf Fruit Nut	Kıvçak et al., 2004; Flamini et al., 2004; Özcan et al., 2009; Kendirci & Onoğur, 2011
Spathulenol 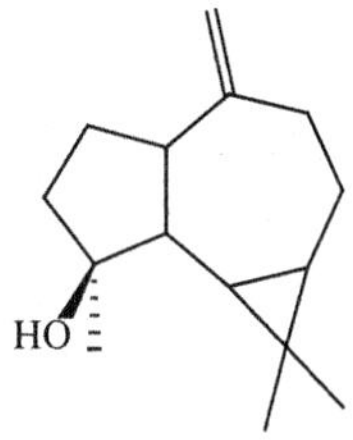	$C_{15}H_{24}O$	*P. atlantica* *P. khinjuk* *P. lentiscus* *P. terebinthus* *P. vera*	Leaf Leaf Leaf Leaf and fruit Leaf	Mahjoub et al., 2018; Ravan et al., 2019; Kıvçak et al., 2004; Özcan et al., 2009; Duru et al., 2003
Diterpenoids				
Abietadiene 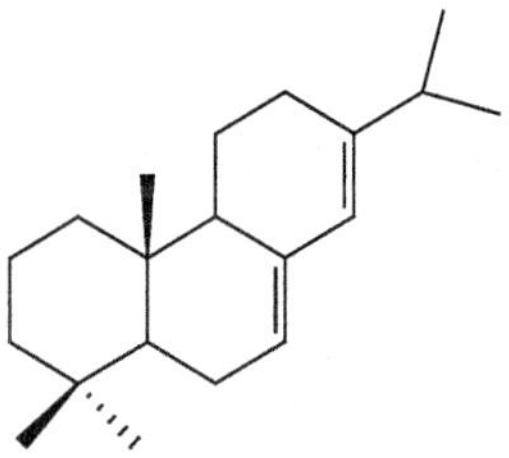	$C_{20}H_{32}$	*P. vera*	Gum	Alma et al., 2004
Triterpenoids				
Oleanolic acid 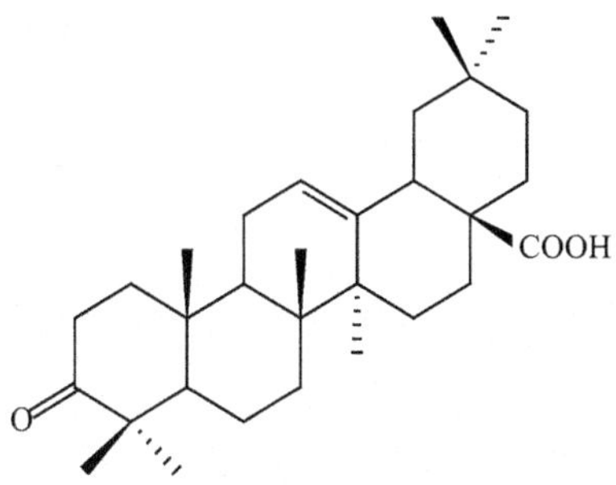	$C_{30}H_{48}O_3$	*P. atlantica* *P. terebinthus*	Resin Gall	Mahjoub et al., 2018; Giner-Larza et al., 2001

(*Continued*)

TABLE 17.1 (CONTINUED)
List of Major Compounds Isolated from Various *Pistacia* L. Species

Name and Structure of Compounds	Molecular Formulae	Species	Plant Parts	References
Ursonic acid 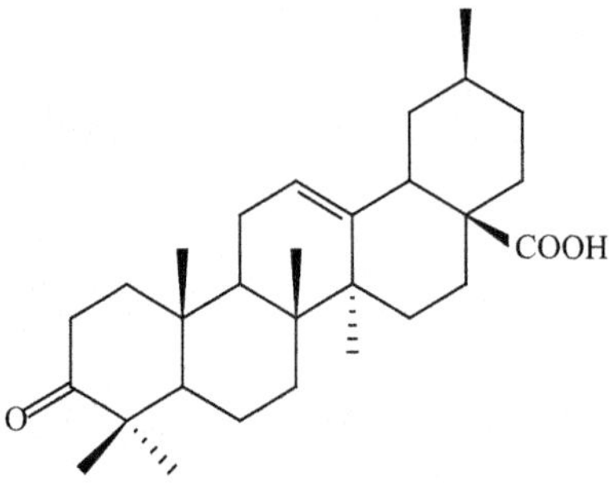	$C_{30}H_{48}O_3$	*P. atlantica*	Resin	Mahjoub et al., 2018
Masticadienonic acid 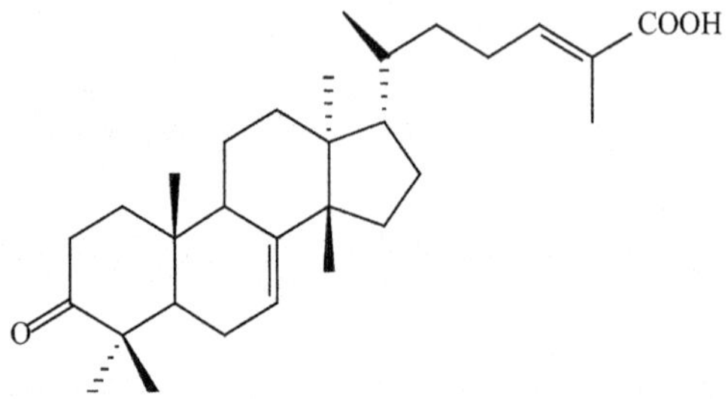	$C_{30}H_{46}O_3$	*P. atlantica* *P. terebinthus*	Resin Leaf	Mahjoub et al., 2018; Rauf et al., 2017
Morolic acid 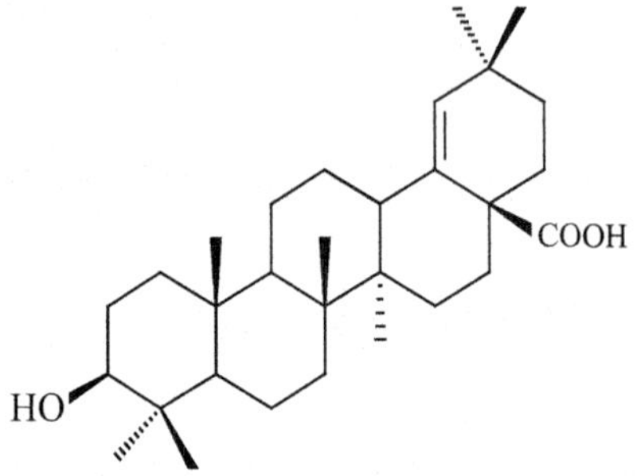	$C_{30}H_{48}O_3$	*P. atlantica*	Resin	Mahjoub et al., 2018
Moronic acid 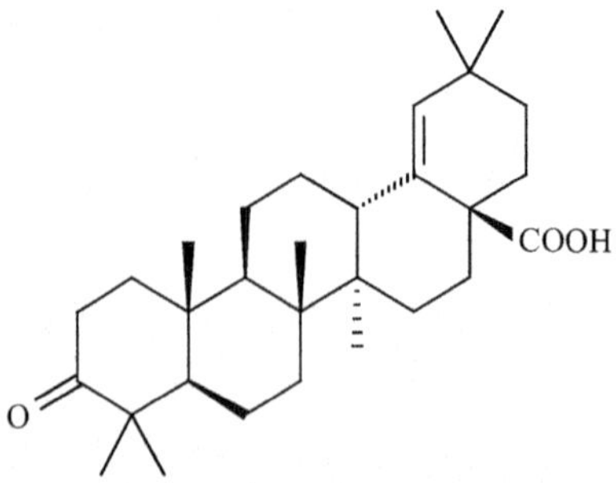	$C_{30}H_{46}O_3$	*P. terebinthus*	Resin	Assimopoulou & Papageorgiou, 2005

activity of *P. atlantica* (from Istanbul region in Turkey) aqueous extract, which contains significant amounts of flavonoids (33.52 ± 2.04 µg catechin equivalents/mg extract), was remarkably better than ascorbic acid and BHA (Peksel et al., 2013). The results obtained from different assays indicated that the hull of *P. vera* had a stronger antioxidant activity than its kernel. That is because the hull has higher amounts of phenolic compounds acting as antioxidants. Besides, various *in vitro* antioxidant assays revealed that the leaves and fruits of *P. atlantica* showed antioxidant activities similar to or considerably higher than those of standard antioxidant compounds (Mohamadi & Karimabad, 2020).

Anti-inflammatory Activity and Wound Healing

P. terebinthus gall was found to be effective against chronic and acute inflammation in different *in vivo* models of inflammation induced by TPA, EPP, PLA_2, and this anti-inflammatory activity was attributed to masticadienonic acid, masticadienolic acid, and morolic acid isolated from *P. terebinthus* (Giner-Lazza et al., 2002). The anti-inflammatory activity of aqueous and ethanolic extracts of *P. vera* leaves was investigated in mice and the effects of these extracts against acute and chronic inflammation were determined using xylene-induced ear edema and cotton pellet test, respectively. The ethanolic extract was found to be as effective as diclofenac, a nonsteroidal anti-inflammatory drug (Hosseinzadeh et al., 2011). Early wound healing is highly important to prevent pathogenic infections. *Pistacia* is used for the treatment of wound inflammation in folk medicine. Based on this knowledge; the wound-healing effect of *P. atlantica* bark ointment on wounds created by excision and incision in rats has been evaluated *in vivo*. The hydroethanolic extract of *P. atlantica* bark ointment when applied for three weeks had significantly elevated wound contraction percentage and upregulated hydroxyproline content. And by the end of the 7th day, neovascularization has increased in a dose-dependent manner. The data in this study showed that the wound healing effect of *P. atlantica* bark ointment was due to shortening the inflammation phase by provoking fibroblast proliferation, increasing the infiltration of mast cells, and subsequently promoting neovascularization. In conclusion, promoted angiogenesis has also benefited by increasing RNA stability (Farahpour et al., 2015). The results suggest that topical application of *P. atlantica* oil gel has improved the morphological, biochemical, and biomechanical properties of experimentally induced wound defects in rats (Hamidi et al., 2017).

Antimicrobial and Antiviral Activity

Lipophylic extracts from different parts (leaf, branch, stem, kernel, etc.) of *P. vera* showed low antibacterial activity between the range of 128 and 256 µg/mL concentrations whereas at the same concentrations they had noticeable antifungal activity against pathogen fungi (*C. albicans* and *C. parapsilosis*). Besides, *P. vera* kernel and seed extracts showed remarkable antiviral activity as well as control (acyclovir and oseltamivir) (Özçelik et al., 2005). Antibacterial activities of different extracts of *P. atlantica* and *P. khinjuk* species have been measured on three bacterial strains (*Escherichia coli*, *Staphylococcus aureus*, and *Staphylococcus epidermidis*) using the disk diffusion method. Among these extracts, the hydroalcoholic extract of the fruit skin of *P. khinjuk* (75 mg/mL) has been shown to have higher antibacterial activity than Tobramycin and identical to Gentamicin and Kanamycin on *E. coli* (Tohidi et al., 2011). In a similar study, the antimicrobial activity of essential oils of gum obtained from *P. vera* (from Gaziantep province in southeast Turkey) has been evaluated on 13 bacterial growth and compared with those of ampicillin, streptomycin, and nystatin used as positive controls. Antimicrobial activity results showed that *E. coli*, one of the most common pathogenic bacteria, was significantly inhibited by the essential oil at a concentration of 4 µL/disk. Besides, carvacrol was found to be the most effective constituent (Alma et al., 2004). The crude extracts (petroleum ether, chloroform, ethyl acetate, and ethyl alcohol) which were obtained from the leaves of *P. vera*, *P. terebinthus*, and *P. lentiscus* were examined for antifungal activities against three pathogenic fungi for plants, *Phythium ultimum*, *Rhizoctania solani*, and *Fusarium sambucinum*. The extracts considerably inhibited the growth of *P. ultimum* and *R. solani* whereas the antifungal activity has not been observed against *F. sambucinum*. In particular, the highest inhibition effects were found for *P. ultimum* in the ethyl alcohol extract (2500 ppm, 34.3 %) and *R. solani* in the petroleum ether extract (5000 ppm, 48.4 %) of *P. vera*. (Kordali et al., 2003).

Antidiabetic Activity

Water extracts of *P. atlantica* showed an important inhibitory effect on both α -amylase and α -glucosidase in a dose-dependent manner compared with acarbose. The extracts of *P. atlantica* proved to be dual inhibitors of the two enzymes tested *in vitro* and *in vivo*. It also improved glucose intolerance (Kasabri et al., 2011). The antidiabetic activity of *P. atlantica* leaves was evaluated using normoglycemic and streptozocin-induced diabetic rats. *P. atlantica* leaf extract did not reduce the blood glucose level of streptozocin-induced diabetic or normoglycemic rats, thus it was indicated that it had no hypo- or hyperglycemic effects. Including pistachios in snacks is beneficial in terms of providing glycemic control, especially in diabetic patients who and improved glucose intolerance (Hamdan & Afifi, 2004). Studies have shown that peanut oil wild pistachio (*Pistacia atlantica mutica*) consumption in the prevention of hyperglycemia, hypertriglyceridemia, hypercholesterolemia, inflammation, and pancreatic secretion disorders have demonstrated to have beneficial effects (Jamshidi et al., 2018). In the study to evaluate the role of *P. atlantica* extract in the protection against ovary damage by streptozotocin (STZ)-induced DM in rats, it was observed that it has anti-hyperglycemic and antioxidative properties, and it decreased ovarian complications (Behmanesh et al., 2021).

Anticholinesterase and Neuroprotective Activity

There is a lot of pharmacological evidence related to neuroprotection in the *Pistacia* genus (Moeini et al., 2019). For instance, aqueous extract of *P. atlantica* and *P. lentiscus* exhibited strong AChE inhibition (Benamar et al., 2010). However, the anticholinesterase effects of the ethyl acetate and methanol extracts of fruits of *P. terebinthus* were investigated through an enzyme inhibition test against acetylcholinesterase (AChE) and butyrylcholinesterase (BChE), which are linked to Alzheimer's disease as well as tyrosinase related to Parkinson's disease. Fruits of *P. terebinthus* did not have inhibitory activity against AChE and tyrosinase, while they selectively inhibited butyrylcholinesterase (BChE) at moderate levels (Orhan et al., 2012). Recently, polyunsaturated fatty acids, terpenes, triterpenoid derivatives, and phenolic compounds were found in *Pistacia* that have been identified as potential neuroprotective agents with ability to protect the brain against neurodegenerative and neuroinflammatory processes (Moeini et al., 2019).

Antihyperlipidemic Activity

P. atlantica fruit oil decreased low-density lipoprotein (LDL) cholesterol, very-low-density lipoprotein (VLDL) cholesterol, triglycerides, and increased high-density lipoprotein (HDL) cholesterol (Saeb et al., 2004).

Cytotoxic and Anticancer Activity

The results of the studies highlight the intense cytoprotective and anticancer activities of the components of pistachios. Evidence shows that the anticancer effects of pistachios result from their influence on numerous apoptosis-related pathways in tumor cells. Mastic oil of *P. lentiscus* has significantly prevented the proliferation of cancer cells in immune-competent mice with no signs of toxicity. Research has shown antiproliferative and pro-apoptotic effects of mastic oil on leukemia cell lines via inhibiting the release of the vascular endothelial growth factor from these cells. The extract of the *P. atlantica* fruit and the gum of *P. lentiscus* var. *chia* have exerted inhibitory effects on the growth of colon cancer cells and prevented the growth of colorectal cancer cell lines, respectively. Oleoresin extracted from *P. vera* has shown moderate cytotoxic effects on hepatocellular carcinoma, cervical tumor, breast tumor cells, and normal melanocytes. The ethyl acetate extract of pistachio hull had no significant cytotoxic effects on normal fibroblast cells; however, it showed significant cytotoxicity for all tested human cancer cells (HT-29, HCT-116, MCF-7, H23, and HepG2), suggesting its selective cytotoxicity (Mohamadi & Karimabad, 2020).

CONCLUSION

The use of pistachios seeds as food and snack is very common in Turkey and all over the world. *Pistacia* L. species have an increasing significance due to their rich phytochemical content, active potential drug candidate compounds, functional foods, and commercial value. It has been demonstrated by scientific studies that pistachios play an important role in the healthy diet of individuals and in reducing the risk of nutrition-related diseases, due to their fiber, unsaturated fat, phenolic compounds, vitamins, and minerals. Current research on this plant highlighted that these species are valuable sources of medicinal compounds. Many traditional uses of *Pistacia* species have now been confirmed by some biological research. It has been shown in many studies that their antioxidant, antimicrobial, and anti-inflammatory effects on various chronic diseases, especially cardiovascular diseases and Alzheimer's disease. It is also recommended to be consumed in order to maintain a healthy life. In addition, scientific studies would be important to evaluate the possible biological effects and the functional properties of the other parts (especially its nut hulls) of this plant except for seeds. Usage of *Pistacia* species in traditional medicine should be supported by detailed and extensive studies and clinical evaluations should be carried out for safety and risk approval of their therapeutic applications.

REFERENCES

Alma, M. H., Nitz, S., Kollmannsberger, H., Digrak, M., Efe, F. T. & Yilmaz, N. (2004). Chemical composition and antimicrobial activity of the essential oils from the gum of Turkish pistachio (*Pistacia vera* L.). *Journal of Agricultural and Food Chemistry*, 52(12), 3911–3914.

Amel, Z., Nabila, B. B., Nacéra, G., Fethi, T. & Fawzia, A. B. (2016). Assessment of phytochemical composition and antioxidant properties of extracts from the leaf, stem, fruit and root of *Pistacia lentiscus* L. *International Journal of Pharmacy and Pharmaceutical Research*, 8, 627–633.

Amri, O., Zekhnini, A., Bouhaimi, A., Tahrouch, S. & Hatimi, A. (2018). Anti-inflammatory activity of methanolic extract from *Pistacia atlantica* desf. leaves. *Pharmacognosy Journal*, 10(1).

Assimopoulou, A. N. & Papageorgiou, V. P. (2005). GC-MS analysis of penta-and tetra-cyclic triterpenes from resins of *Pistacia* species. Part I. *Pistacia* lentiscus var. Chia. *Biomedical Chromatography*, 19(4), 285–311.

Ballistreri, G., Arena, E. & Fallico, B. (2009). Influence of ripeness and drying process on the polyphenols and tocopherols of *Pistacia vera* L. *Molecules*, 14(11), 4358–4369.

Behmanesh, M. A., Poormoosavi, S. M., Pareidar, Y., Ghorbanzadeh, B., Mahmoodi-Kouhi, A. & Najafzadehvarzi, H. (2021). *Pistacia atlantica*'s effect on ovary damage and oxidative stress in streptozotocin-induced diabetic rats. *JBRA Assisted Reproduction*, 25(1), 28–33.

Ben Ahmed, Z. B., Yousfi, M., Viaene, J., Dejaegher, B., Demeyer, K. & Vander Heyden, Y. (2021). Four Pistacia atlantica subspecies (atlantica, cabulica, kurdica and mutica): A review of their botany, ethnobotany, phytochemistry and pharmacology. *Journal of Ethnopharmacology*, 113329.

Benalia, N., Boumechhour, A., Ortiz, S., Echague, C. A., Rose, T., Fiebich, B. L, Chemat, S., Michel, S., Deguin, B., Dahamna, S. & Boutefnouchet, S. (2021). Identification of alkylsalicylic acids in Lentisk oil (*Pistacia lentiscus* L.) and viability assay on Human Normal Dermal Fibroblasts. *Oil Seeds and Fats Crops and Lipids*, 28, 22.

Benamar, H., Rached, W., Derdour, A. & Marouf, A. (2010). Screening of Algerian medicinal plants for acetylcholinesterase inhibitory activity. *Journal of Biological Sciences*, 10(1), 1–9.

Benhassaini, H., Bendeddouche, F. Z., Mehdadi, Z. & Romane, A. (2008). GC/MS analysis of the essential oil from the oleoresin of *Pistacia atlantica* Desf. subsp. *atlantica* from Algeria. *Natural Product Communications*, 3(6), 1934578X0800300621.

Bhouri, W., Derbel, S., Skandrani, I., Boubaker, J., Bouhlel, I., Sghaier, M. B., Kilania S., Mariotte, A. M., Dijoux-Franca M. G., Ghedira K. & Chekir-Ghedira, L. (2010). Study of genotoxic, antigenotoxic and antioxidant activities of the digallic acid isolated from *Pistacia lentiscus* fruits. *Toxicology in Vitro*, 24(2), 509–515.

Bozorgi, M., Memariani, Z., Mobli, M., Salehi Surmaghi, M. H., Shams-Ardekani, M. R. & Rahimi, R. (2013). Five *Pistacia* species (*P. vera, P. atlantica, P. terebinthus, P. khinjuk*, and *P. lentiscus*): A review of their traditional uses, phytochemistry, and pharmacology. *The Scientific World Journal*, 219815.

Çağlar, A., Tomar, O., Vatansever, H. & Ekmekçi, E. (2017). Antepfıstığı (*Pistacia vera* L.) ve insan sağlığı üzerine etkileri. *Akademik Gıda*, 15(4), 436–447.

Duru, M. E., Cakir, A., Kordali, S., Zengin, H., Harmandar, M., Izumi, S. & Hirata, T. (2003). Chemical composition and antifungal properties of essential oils of three *Pistacia* species. *Fitoterapia*, 74(1), 170–176.

Erşan, S., Üstündağ, Ö. G., Carle, R. & Schweiggert, R. M. (2018). Subcritical water extraction of phenolic and antioxidant constituents from pistachio (*Pistacia vera* L.) hulls. *Food Chemistry*, 253, 46–54.

FAOSTAT, 2014: FAO web page (http://faostat.fao.org).

Farahpour, M. R., Mirzakhani, N., Doostmohammadi, J. & Ebrahimzadeh, M. (2015). Hydroethanolic *Pistacia atlantica* hulls extract improved wound healing process; evidence for mast cells infiltration, angiogenesis and RNA stability. *International Journal of Surgery*, 17, 88–98.

Flamini, G., Bader, A., Cioni, P. L., Katbeh-Bader, A. & Morelli, I. (2004). Composition of the essential oil of leaves, galls, and ripe and unripe fruits of Jordanian *Pistacia palaestina* Boiss. *Journal of Agricultural and Food Chemistry*, 52(3), 572–576.

Giner-Larza, E. M., Máñez, S., Recio, M. C., Giner, R. M., Prieto, J. M., Cerdá-Nicolás, M. & Ríos, J. L. (2001). Oleanonic acid, a 3-oxotriterpene from *Pistacia*, inhibits leukotriene synthesis and has anti-inflammatory activity. *European Journal of Pharmacology*, 428(1), 137–143.

Giner-Larza, E. M., Máñez, S., Giner, R. M., Recio, M. C., Prieto, J. M., Cerdá-Nicolás, M. & Ríos, J. (2002). Anti-inflammatory triterpenes from *Pistacia terebinthus* galls. *Planta Medica*, 68(4), 311–315.

Goli, A. H., Barzegar, M. & Sahari, M. A. (2005). Antioxidant activity and total phenolic compounds of pistachio (*Pistacia vera*) hull extracts. *Food Chemistry*, 92(3), 521–525.

Grace, M. H., Esposito, D., Timmers, M. A., Xiong, J., Yousef, G., Komarnytsky, S. & Lila, M. A. (2016). *In vitro* lipolytic, antioxidant and anti-inflammatory activities of roasted pistachio kernel and skin constituents. *Food & Function*, 7(10), 4285–4298.

Grippi, F., Crosta, L., Aiello, G., Tolomeo, M., Oliveri, F., Gebbia, N. & Curione, A. (2008). Determination of stilbenes in Sicilian pistachio by high-performance liquid chromatographic diode array (HPLC-DAD/FLD) and evaluation of eventually mycotoxin contamination. *Food Chemistry*, 107(1), 483–488.

Hafsé, M., Fikri, B. K. & Farah, A. (2015). Enquête ethnobotanique sur l'utilisation de *Pistacia lentiscus* au Nord du Maroc (Taounate). *International Journal Innovative Appl Studies*, 13(4), 864–872.

Hamdan, I. I. & Afifi, F. U. (2004). Studies on the *in vitro* and *in vivo* hypoglycemic activities of some medicinal plants used in treatment of diabetes in Jordanian traditional medicine. *Journal of Ethnopharmacology*, 93(1), 117–121.

Hamidi, S. A., Naeini, A. T., Oryan, A., Tabandeh, M. R., Tanideh, N. & Nazifi, S. (2017). Cutaneous wound healing after topical application of *Pistacia atlantica* gel formulation in rats. *Turkish Journal of Pharmaceutical Sciences*, 14(1), 65–74.

Hayta, S., Polat, R. & Selvi, S. (2014). Traditional uses of medicinal plants in Elazığ (Turkey). *Journal of Ethnopharmacology*, 154(3), 613–623.

Hosseinzadeh, H., Behravan, E. & Soleimani, M. M. (2011). Antinociceptive and anti-inflammatory effects of *Pistacia vera* leaf extract in mice. *Iranian Journal of Pharmaceutical Research: IJPR*, 10(4), 821.

Hürkul, M. M. (2021). Antep fıstığı (*Pistacia vera* L.) meyve kabuğu: Biyoaktif bileşikler için potansiyel kaynak. *Journal of Faculty of Pharmacy of Ankara University*, 45(3), 586–597.

Jamshidi, S., Hejazi, N., Golmakani, M. T. & Tanideh, N. (2018). Wild pistachio (*Pistacia atlantica mutica*) oil improve metabolic syndrome features in rats with high fructose ingestion. *Iranian Journal of Basic Medical Sciences*, 21, 1255–1261.

Kasabri, V., Afifi, F. U. & Hamdan, I. (2011). *In vitro* and *in vivo* acute antihyperglycemic effects of five selected indigenous plants from Jordan used in traditional medicine. *Journal of Ethnopharmacology*, 133(2), 888–896.

Kavak, D. D., Altıok, E., Bayraktar, O. & Ülkü, S. (2010). *Pistacia terebinthus* extract: As a potential antioxidant, antimicrobial and possible β-glucuronidase inhibitor. *Journal of Molecular Catalysis B: Enzymatic*, 64(3–4), 167–171.
Kawashty, S. A., Mosharrafa, S. A. M., El-Gibali, M. & Saleh, N. A. M. (2000). The flavonoids of four *Pistacia* species in Egypt. *Biochemical Systematics and Ecology*, 28(9), 915–917.
Kendirci, P. & Onoğur, T. A. (2011). Investigation of volatile compounds and characterization of flavor profiles of fresh pistachio nuts (*Pistacia vera* L.). *International Journal of Food Properties*, 14(2), 319–330.
Khallouki, F., Breuer, A., Merieme, E., Ulrich, C. M. & Owen, R. W. (2017). Characterization and quantitation of the polyphenolic compounds detected in methanol extracts of *Pistacia atlantica* Desf. fruits from the Guelmim region of Morocco. *Journal of Pharmaceutical and Biomedical Analysis*, 134, 310–318.
Kıvçak, B., Akay, S., Demirci, B. & Başer, K. (2004). Chemical composition of essential oils from leaves and twigs of *Pistacia lentiscus*, *Pistacia lentiscus* var. *chia*, and *Pistacia terebinthus* from Turkey. *Pharmaceutical Biology*, 42(4–5), 360–366.
Kordali, S., Cakir, A., Zengin, H. & Duru, M. E. (2003). Antifungal activities of the leaves of three *Pistacia* species grown in Turkey. *Fitoterapia*, 74(1–2), 164–167.
Kottakis, F., Kouzi-Koliakou, K., Pendas, S., Kountouras, J. Ü. & Choli-Papadopoulou, T. (2009). Effects of mastic gum *Pistacia lentiscus* var. Chia on innate cellular immune effectors. *European Journal of Gastroenterology & Hepatology*, 21(2), 143–149.
Mahjoub, F., Rezayat, K. A., Yousefi, M., Mohebbi, M. Ü. & Salari, R. (2018). *Pistacia atlantica* Desf. A review of its traditional uses, phytochemicals and pharmacology. *Journal of Medicine and Life*, 11(3), 180–186.
Martorana, M., Arcoraci, T., Rizza, L., Cristani, M., Bonina, F. P., Saija, A. & Tomaino, A. (2013). *In vitro* antioxidant and *in vivo* photoprotective effect of pistachio (*Pistacia vera* L., variety Bronte) seed and skin extracts. *Fitoterapia*, 85, 41–48.
Mecherara-Idjeri, S., Hassani, A., Castola, V. & Casanova, J. (2008). Composition of leaf, fruit and gall essential oils of Algerian *Pistacia atlantica* Desf. *Journal of Essential Oil Research*, 20(3), 215–219.
Mohamadi, M. & Noroozi Karimabad, M. (2020). Potential anticancer activity of the genus *Pistacia* through apoptosis induction in cancer cells. *Pistachio and Health Journal*, 3(3), 18–32.
Moeini, R., Memariani, Z., Asadi, F., Bozorgi, M. & Gorji, N. (2019). *Pistacia* genus as a potential source of neuroprotective natural products. *Planta Medica*, 85(17), 1326–1350.
Moussa, A. Y., Labib, R. M. & Ayoub, N. A. (2013). Isolation of chemical constituents and protective effect of *Pistacia khinjuk* against CCl4–induced damage on HepG2 cells. *Phytopharmacology*, 4(4), 1–9.
Orhan, I. E., Senol, F. S., Gulpinar, A. R., Sekeroglu, N., Kartal, M. & Sener, B. (2012). Neuroprotective potential of some terebinth coffee brands and the unprocessed fruits of *Pistacia terebinthus* L. and their fatty and essential oil analyses. *Food Chemistry*, 130(4), 882–888.
Özcan, M. M., Tzakou, O. & Couladis, M. (2009). Essential oil composition of the turpentine tree (*Pistacia terebinthus* L.) fruits growing wild in Turkey. *Food Chemistry*, 114(1), 282–285.
Özçelik, B., Aslan, M., Orhan, I. & Karaoglu, T. (2005). Antibacterial, antifungal, and antiviral activities of the lipophylic extracts of *Pistacia vera*. *Microbiological Research*, 160(2), 159–164.
Peksel, A., Arisan, I. & Yanardag, R. (2013). Radical scavenging and anti-acetylcholinesterase activities of aqueous extract of wild pistachio (*Pistacia atlantica* Desf.) leaves. *Food Science and Biotechnology*, 22(2), 515–522.
Pell, S. K., Mitchell, J. D., Miller, A. J. & Lobova, T. A. (2010). Anacardiaceae. In *Flowering Plants. Eudicots* (pp. 7–50). Springer, Berlin, Heidelberg.
Rand, K., Bar, E., Ben-Ari, M., Lewinsohn, E. & Inbar, M. (2014). The mono-and sesquiterpene content of aphid-induced galls on *Pistacia palaestina* is not a simple reflection of their composition in intact leaves. *Journal of Chemical Ecology*, 40(6), 632–642.
Ravan, S., Khani, A. & Veysi, N. (2019). GC-MS analysis and insecticidal effect of methanol extract of *Pistacia khinjuk* Stocks leaves. *Acta Agriculturae Slovenica*, 113(2), 231–237.
Rauf, A., Patel, S., Uddin, G., Siddiqui, B. S., Ahmad, B., Muhammad, N., Mabkhot, Y. N. & Hadda, T. B. (2017). Phytochemical, ethnomedicinal uses and pharmacological profile of genus *Pistacia*. *Biomedicine & Pharmacotherapy*, 86, 393–404.
Romani, A., Pinelli, P., Galardi, C., Mulinacci, N. & Tattini, M. (2002). Identification and quantification of galloyl derivatives, flavonoid glycosides and anthocyanins in leaves of *Pistacia lentiscus* L. *Phytochemical Analysis: An International Journal of Plant Chemical and Biochemical Techniques*, 13(2), 79–86.
Saitta, M., Giuffrida, D., La Torre, G. L., Potortì, A. G. & Dugo, G. (2009). Characterisation of alkylphenols in pistachio (*Pistacia vera* L.) kernels. *Food Chemistry*, 117(3), 451–455.

Saeb, M., Nazifi, S. & Mirzaei, A. (2004). The effects of wild pistacio fruit oil on the serum concentration of lipids and lipoproteins of female rabbits. *Journal of Shahid Sadoughi University of Medical Sciences and Health Services*. URL: http://jssu.ssu.ac.ir/article-1-1248-en.html.

Sezik, E., Yeşilada, E., Honda, G., Takaishi, Y., Takeda, Y. & Tanaka, T. (2001). Traditional medicine in Turkey X. Folk medicine in central Anatolia. *Journal of Ethnopharmacology*, 75(2–3), 95–115.

Şermet, M. Ö. (2015). Antioxidant, antimicrobial activity and phytochemical analysis of Pistacia vera L. skin (Master's thesis, Middle East Technical University). URL: http://etd.lib.metu.edu.tr/upload/12618811/index.pdf.

Tohidi, M., Khayami, M., Nejati, V. & Meftahizade, H. (2011). Evaluation of antibacterial activity and wound healing of *Pistacia atlantica* and *Pistacia khinjuk*. *Journal of Medicinal Plants Research*, 5(17), 4310–4314.

Tokuşoğlu, Ö., Ünal, M. K. & Yemiş, F. (2005). Determination of the phytoalexin resveratrol (3, 5, 4 '-trihydroxystilbene) in peanuts and pistachios by high-performance liquid chromatographic diode array (HPLC-DAD) and gas chromatography– mass spectrometry (GC-MS). *Journal of Agricultural and Food Chemistry*, 53(12), 5003–5009.

Tomaino, A., Martorana, M., Arcoraci, T., Monteleone, D., Giovinazzo, C. & Saija, A. (2010). Antioxidant activity and phenolic profile of pistachio (*Pistacia vera* L., variety Bronte) seeds and skins. *Biochimie*, 92(9), 1115–1122.

Topçu, G., Ay, M., Bilici, A., Sarıkürkcü, C., Öztürk, M. & Ulubelen, A. (2007). A new flavone from antioxidant extracts of *Pistacia terebinthus*. *Food Chemistry*, 103(3), 816–822.

Vaya, J. & Mahmood, S. (2006). Flavonoid content in leaf extracts of the fig (*Ficus carica* L.), carob (*Ceratonia siliqua* L.) and pistachio (*Pistacia lentiscus* L.). *Biofactors*, 28(3–4), 169–175.

Yousfi, M., Djeridane, A., Bombarda, I., Duhem, B. & Gaydou, E. M. (2009). Isolation and characterization of a new hispolone derivative from antioxidant extracts of *Pistacia atlantica*. *Phytotherapy Research: An International Journal Devoted to Pharmacological and Toxicological Evaluation of Natural Product Derivatives*, 23(9), 1237–1242.

18 *Rhus coriaria* L.

Ayşegül Köroğlu

INTRODUCTION

Rhus coriaria L. (Sumac, Sumak; Anacardiaceae) is a shrub or small tree that grows almost all over Anatolia. The name sumac comes from the Syriac word sumaga, meaning red (Shabbir, 2012). *R. coriaria* has been of economic importance in food technology, the leather industry (for its high tannin content), and traditional medicine; it is also grown as an ornamental tree (Giovanelli et al., 2017; Matloobi & Tahmasebi, 2019; Arena et al., 2022). Sumac is found in temperate and tropical regions worldwide, often growing in areas of marginal agricultural capacity (Golzadeh et al., 2012). The leaves of the plant contain gallo-tannins (Baytop, 1999; Alsamri et al., 2021), gallic acid (Arena et al., 2022), flavonoids (Mehrdad et al., 2009) together with biflavonoids (Van Loo et al., 1988), sugars, wax (Baytop, 1999), and essential oil (Güvenç & Koyuncu, 1994; Kurucu et al., 1993; Morshedloo et al., 2018; Arena et al., 2022). Since the leaves are rich in tannins, they are also utilized as a tanning agent in the leather industry and are exported for this purpose and also used in fabric dyeing. So the role of the plant leaf in the leather and textile industry is also significant (Baytop, 1999; Abu-Reidah et al., 2014). *R. coriaria* has been prescribed for the treatment of many ailments, including diarrhea, ulcer, hemorrhoids, hemorrhage, wound healing, hematemesis, and eye ailments like ophthalmia and conjunctivitis. The plant is also used as diuretic, antimicrobial, abortifacient, and stomach tonic (Elagbar et al., 2020). Sumac fruits have long been used as a flavoring spice, drink, appetizer, and as sour in food recipes, in addition to their use in traditional medicine. The fruits contain tannins (Baytop, 1999), various organic acids (Baytop, 1999), anthocyanins (Güvenç & Koyuncu, 1994; Abu- Abu-Reidah et al., 2014), essential oil (Güvenç & Koyuncu, 1994; Kurucu et al., 1993; Giovanelli et al., 2017), and fixed oil (Güvenç & Koyuncu, 1994; Abu-Reidah et al., 2014). Either the whole fruit or only the pericarp is used as a condiment in Turkey and is called "Sumak" (Baytop, 1999), and it is a very popular condiment in Turkey. Fruits of sumac are mixed with freshly cut onions; it is frequently eaten as an appetizer. A mixture of yogurt and these fruits is often served with kebabs in Turkey (Özcan, 2003).

Morphological Character

Rhus coriaria are shrubs or small trees that can grow up to 2–3 m. Leaves are imparipinnate with 9–15 folioles; folioles are elliptic to lanceolate, petiole very small or almost sessile, the margin is serrate to deeply serrate; leaf rachis sometimes winged. The inflorescence is 15–25 cm tall, a frequent paniculate; flowers are hermaphrodite, greenish, small, and modest; sepals are ovate, green, and pilosus; petals are greenish white, 3–4.5 mm. There are three bracts in the base of the flower. Stamens are five. The ovary is ovate with three carpels; style three pairs. Fruit is like compressed globose, 5–6 mm, when raw is green; ripe is dark red or maroon, mixed glandular-pubescent and pilose, sour and 1-seeded drupe (Figure 18.1). It is found growing naturally in Aegean, Mediterranean, Southeast, Central, and Northern regions of Turkey and Mediterranean bordering countries, South Europe, North Africa, Iran, and Afghanistan (Davis, 1967; Tutin et al., 1968; Koyuncu & Köroğlu, 1991; Abu-Reidah et al., 2014; Yılmaz & Eminağaoğlu, 2020; worldfloraonline, 2022).

DOI: 10.1201/9781003146971-18

FIGURE 18.1 *Rhus coriaria* to must be native at Kemer (Antalya). (Photo: A. Köroğlu.)

Used Part of the Plant

Rhus coriaria is an ancient spice that tempts in and on foods. *R. coriaria* bark powder is an effective tooth-cleaning agent. An infusion prepared from bark powder is useful in the treatment of viral eye infections (Abdul-Jalil, 2020). The aqueous extract prepared from the leaves and fruits is used as a wound healer (Sezik et al., 1991). Both the ripe fruits and leaves of the sumac are used (Figures 18.2–18.4).

Traditional Use

Leaf

Sumac leaves were used by Dioscorides and İbn-i Sina (Avicenna) in the treatment of diarrhea, hemorrhoids, mouth sores, eye diseases, and cracked hands and feet (Demirhan, 1975; Kâhya, 2019). Sumac leaves are used in folk medicine in the treatment of diarrhea and diabetes and are still used (Güvenç & Koyuncu, 1994). The leaves have been utilized in Turkey folk medicine for mouth and throat diseases (Baytop, 1999). The decoction prepared from leaves and branches is mixed with salt in Gözne (Mersin) and used externally for wound healing (Yeşilada et al., 1993). The decoction prepared from its leaves and bark is used externally to relieve animal bites in Beyşehir (Konya) and edema in the legs of animals in Bozkır (Konya). To accelerate the maturation of the abscess, sumac leaves are cooked in milk in Karaman and then used externally by adding barley flour. In Bilecik, a decoction prepared from sumac leaves is used internally for stomach ulcers and stomach pain (Fujita et al., 1995). In the past, sumac leaves are used to dye wool, silk, and leather yellow and khaki. Wool threads were boiled together with sumac leaves and dyed black (Harmancıoğlu, 1955; Baytop, 1999). It is used in the treatment of mouth sores, diarrhea, and foot and mouth diseases of animals in Anatolian villages (Dinçer, 1967). This plant is used in the traditional medicine of Jordan for sweating and cholesterol reduction and in the treatment of diarrhea (Giovanelli et al., 2017). Leaves of *R. coriaria* are used in the treatment of chronic diseases such as osteoarthritis, and it has been reported to be beneficial in the treatment of joint diseases (Abu-Reidah et al., 2014).

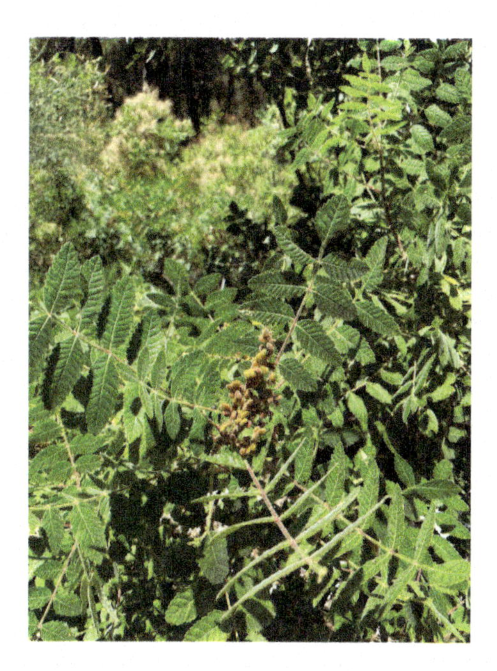

FIGURE 18.2 *Rhus coriaria* leaves and immature fruit. (Photo: A. Köroğlu.)

FIGURE 18.3 *Rhus coriaria* leaves. (Photo: A. Köroğlu.)

Fruit

Sumac is added to dishes as a spice and has been used for thousands of years (Shabbir, 2012; Abu-Reidah et al., 2014; Kâhya, 2019). Either the whole fruit (pericarp and seed) or the pericarp is used as a condiment in Turkey and is called "sumac" (Güvenç & Koyuncu, 1994). In Turkey, spices are prepared from sumac fruits in two ways:

1- The collected fruits are pounded in mortars, the pericarp and seed (semen) parts are separated, then the pericarp part is wetted with water, rubbed with salt, and dried. It is then mixed with olive oil and used as a spice in this way.
2- After the fruits are dried, they are ground in the mill without any separation process and offered for sale.

FIGURE 18.4 *Rhus coriaria* immature fruit. (Photo: A. Köroğlu.)

Pedanius Dioscorides (40–90 AD) wrote in his *De Materia Medica* that sumac fruits were added to salads as a spice for intestinal ailments. He also noted that the fruits were used as a powder for hemorrhoids and in the form of mush to relieve toothache (Kâhya, 2019). However, Dioscorides described sumac as a carminative, stomach tonic, and diuretic (Abdul-Jalil, 2020). İbn-i Sina described sumac in his work *El-Kanun Fi't-Tıbb* as it relieves mouth inflammation, prevents ear inflammation, relieves toothache, relieves bleeding, and stops diarrhea. He wrote that it is applied as an enema in uterine discharges and hemorrhoids. He stated that it would be used in diarrhea, intestinal wounds, and liver diseases by adding to food (Kâhya, 2017). It has been reported that fruit and seeds were used together in burns in Bitlis (Tabata et al., 1994), and also in Gözne (Mersin) decoction is prepared and drunk to reduce kidney stones (Yeşilada et al., 1993). In Dötryol (Hatay), sumac fruits are prepared by boiling with water and adding salt. It is used in the treatment of sour internally, dysentery, and diarrhea (Yeşilada et al., 1995). In Hisarcık (Kütahya), the water extract of sumac fruits is used internally against colds (Honda et al., 1996). In Develi (Kayseri), sumac fruits are used internally for diarrhea; powdered fruits are sprinkled on a boiled egg and swallowed (Sezik et al., 2001). It was determined that the decoction prepared from the fruits in Gönen (Balıkesir, Turkey) had been used in the treatment of ulcers (Tuzlacı & Aymaz, 2001). The decoction prepared from sumac fruits is used externally to disinfect wounds (Tuzlacı, 2016). In Turkey, the fruits that are kept in water due to their sour taste are drained and consumed as "sumac syrup" in meals (Ünder & Saltan, 2019).

The fruit has both nutritional qualities and medicinal values as it is used as a spice by crushing and mixing the dried fruits with salt and is commonly used as a medicinal herb in Persia, the Mediterranean, Palestinian, Israel, Jordan, and Cyprus (Shabbir, 2012). In Iran and Palestine, sumac represents pure ground fruit epicarps of the plant (Abu-Reidah et al., 2014). *R. coriaria* is a very important ingredient in Middle Eastern cuisine from Jordan to Egypt, where it is used as a spice to give a sour lemon taste to grilled meats and stews, but in rice and vegetable dishes too. These fruits are used in the traditional medicine of Jordan for sweating and cholesterol reduction and in the treatment of diarrhea (Giovanelli et al., 2017). Powdered fruits are sparsed on boiled eggs and eaten for the treatment of diarrhea. A fruit decoction is set and given orally (150 mL) twice daily for the treatment of hepatic diseases, urinary system disorders, and diarrhea till improvement occurs (Abdul-Jalil, 2020). In the past, sumac fruits are used to dye wool, silk, and leather red and khaki (Harmancıoğlu, 1955; Baytop, 1999; Arena et al., 2022).

Biological Activity

Various bioactivity studies on *Rhus coriaria* have shown that it has antioxidant, antiviral, antimicrobial (antifungal, antibacterial), antiseptic, fever-reducing, anti-inflammatory, DNA protective, antitumor, antimutagenic, antifibrogenic, anti-migratory, anti-ischemic, cardioprotective, hepatoprotective, hypouricemic, hypoglycemic, hypolipidemic properties, etc. (Shabbir, 2012; Abu-Reidah et al., 2014; Doğan & Çelik, 2016; Ünder & Saltan, 2019; Abdul-Jalil, 2020; Elagbar et al., 2020; Karadaş et al., 2020; Alsamri et al., 2021, Martinelli et al., 2022).

IN VITRO

Antioxidant Activities

Although leaves and fruits of *Rhus coriaria* antioxidant properties have been investigated more recently, there are many studies on this subject (Nasar-Abbas & Halkman, 2004; Kosar et al., 2007; Giovanelli et al., 2017; Abdul-Jalil, 2020; Alsamri et al., 2021). *R. coriaria* fruit extracts have been investigated by ORAC assay, one of the most common methods used to estimate the antioxidant capacity in food. The results showed that fruits have significant antioxidant activity by expressing their oxygen radical absorbance capacity as Trolox equivalents, in line with their polyphenolic content and composition (Arena et al., 2022). Extract prepared from sumac fruit peels (pericarp) with methanol has an antioxidant effect at high concentrations. The data obtained showed that using *R. coriaria* fruits as a natural antioxidant in fatty foods could be supported (Özcan, 2003).

Antiviral and Antimicrobial Activities

Many reports noted that sumac showed antiviral, antibacterial, and antifungal activities (Iauk et al., 1998; Nasar-Abbas and Halkman, 2004; Fazeli et al., 2007; Adwan et al., 2009; Onkar et al., 2011; Gharabolagh et al., 2018; Abdul-Jalil, 2020; Alsamri et al., 2021).

The antiviral effects of *Rhus coriaria* fruit extract were evaluated at different concentrations of the extract before, during, and after infection against acyclovir-resistant HSV-1 in the Hela cell line. The effective minimal cytotoxic concentration was evaluated at different times of virus replication after virus adsorption. Virus titer was determined by tissue culture infectious dose 50 methods. The antiviral effects of *R. coriaria* fruit extract on acyclovir-resistant HSV-1 after virus infection were more remarkable than treatment of the virus with the extract before virus adsorption (Parsania et al., 2017).

Some active polyphenolic constituents of *Rhus* spp. were studied for antiviral activity by molecular docking, drug-likeness, and synthetic accessibility score (SAS) as inhibitors against the SARS-CoV-2 M. Six phenolic compounds of *Rhus* spp. (methyl 3,4,5-trihydroxybenzoate, (Z)-1-(2,4-d ihydroxyphenyl)-3-(3,4-dihydroxyphenyl)-2-hydroxyprop-2-en-1-one, (Z)-2-(3,4-dihydroxybenzyli dene)-6-hydroxybenzofuran-3(2H)-one, 3,5,7-trihydroxy-2-(4-hydroxyphenyl)chroman-4-one, 2-(3, 4-dihydroxyphenyl)-3,5-dihydroxy-7-methoxy-4H-chroman-4-one, and 3,7-dihydroxy-2-(4-hydrox yphenyl)chroman-4-one) were proposed by drug-likeness, solubility in water, and SAS analysis as potential inhibitors of the main protease enzyme that may be used for the treatment of COVID-19. Based on the data obtained, these phenolic substances are recommended as potential inhibitors to the main protease enzyme for the treatment of COVID-19 (Sherif et al., 2021).

The inhibitory effect of *Rhus coriaria* (sumac) aqueous extract of fruit on hepatitis B virus replication and HBsAg secretion in the HuH-7 cell line after transfection by HBV has been evaluated. Huh-7 cell line was transfected by a plasmid (pCH-9/3091) containing the HBV genome using Lipofectamine 2000. The aqueous extract of sumac was prepared, and its cytotoxicity on transfected cells was assessed by MTT assay. The inhibitory effect of the aqueous extract of sumac on the release of HBsAg as a consequence of HBV replication was measured by ELISA. Data showed a significantly lower concentration of HBsAg after exposure to *R. coriaria* extract compared to the untreated control group and a positive control group treated with Lamivudine of HBV transfected Huh-7 cell line (Gharabolagh et al., 2018).

Kacergius et al. (2017) showed that methanolic extract of *Rhus coriaria* fruit significantly inhibits the growth of *Streptococcus mutans*. A greater inhibitory effect was obtained with methyl gallate, a major constituent of the methanol extract of sumac fruit (Kacergius et al., 2017).

In an antifungal effect study using fluconazole as a standard compound, coriarianthracenyl ester was found to be active against *Aspergillus flavus*, *Candida albicans*, and *Penicillium citrinium* strains at four different (25–200 μg/mL) concentrations (Onkar et al., 2011).

Dental Protection Activities

The effect of *Rhus coriaria* fruit peel (pericarp) water extract on five common oral bacteria and bacterial biofilm formation on the orthodontic wire was evaluated. The diameter of the zone of growth inhibition was proportionate to the tested concentrations of the extract. The lowest MIC (0.390 mg/mL) and MBC (1.5 mg/mL) of the sumac fruit peel were found to be against *Streptococcus sobrinus* ATCC 27607. *R. coriaria* water extract decreased bacterial biofilm formation on the orthodontic wire at MIC and 1/8 of MIC by *Streptococcus sanguinis*, *S. sobrinus*, *S. salivarius*, *S. mutans*, and *Enterococcus faecalis* (Vahid-Dastjerdi et al., 2014).

A study by Seseogullari-Dirihan et al. (2016) evaluated the matrix metalloproteinase (MMP) activity of demineralized dentin matrix following pretreatment by various collagen cross-linkers, including *R. coriaria* fruit extract. This extract significantly reduced total dentin MMP activity compared to the control. Moreover, zymogram analysis of dentin powder treated with *R. coriaria* extract confirmed a decrease in MMP-2 and MMP-9 activities. Thus, it was stated that *R. coriaria* might serve as a source for new MMP inhibitors that may be useful in the prevention or treatment of dental diseases (Seseogullari-Dirihan et al., 2016).

Anticancer Activities

The anticancer activity of hydro-alcoholic (70:20) extract of air-dried *Rhus coriaria* fruit against different human cancer MCF-7, PC-3, and SKOV3 cell lines was investigated. Gallic acid, isohyperoside, dihydroxy methyl xanthone, β-sitosterol-hexoside, α-tocopherol, linoleic acid, quercitrin, myricetin glucuronide, and myricetin rutinoside were identified in this extract of the sumac fruit. These compounds are proposed to be potentially involved in the sumac effect on cancer cells and inhibition activity of CA isoforms of the studied human cancer cell lines; MCF-7, PC-3, and SKOV3. Sumac at doses of 50 and 100 μM significantly inhibited the growth, proliferation, and viability of cancer cells by activating the apoptotic process via caspase-3 overexpression and the regulation of Bcl-2 anti-apoptotic protein (Gabr & Alghadir, 2021).

The cytotoxic and apoptotic effects of silver NPs (NPs are used to prevent tumor formation and progression because of their intrinsic anticancer properties) synthesized from sumac (*Rhus coriaria*) fruit aqueous extract on human breast cancer cells (MCF-7) were investigated. The antiproliferative effect of sumac extract was determined by MTT assay. The mechanism of apoptosis induction in treated cells was investigated using molecular analysis. Overall results of the morphological examination and cytotoxic assay revealed that sumac extract exerts a concentration-dependent inhibitory effect on the viability of MCF-7 cells (IC50 of ~10 μmol/48 h). Acridine orange/propidium iodide staining confirmed the occurrence of apoptosis in cells treated with sumac fruit aqueous extract. In addition, the molecular analysis demonstrated that the apoptosis in MCF-7 cells exposed to sumac fruit aqueous extract was induced via upregulation of Bax and downregulation of Bcl-2 (Ghorbani et al., 2018).

Cytotoxic and angiogenic effects of oleo gum resin extracts of *Rhus coriaria* against human umbilical vein endothelial normal cell and Y79 cell lines using 3-(4, 5-dimethylthiazol-2-yl)-2,5-diphenyl tetrazolium bromide assay were studied. *In vitro* tube formation assay also showed that extract of *R. coriaria* resin inhibited angiogenesis significantly tested extracts ($P< 0.05$) (Mirian et al., 2015).

Anti-inflammatory Activities

Martinelli et al. (2022) suggested ripe sumac (*Rhus coriaria*) fruits as a useful nutraceutical in *Helicobacter pylori*-associated gastritis. The ethanol-water (50:50 v/v) extract of these fruits maintains a similar MIC value after *in vitro* simulated gastric digestion and shows strong anti-inflammatory activity and inhibits IL-8 release induced by *H. pylori* or TNF F-α. The phenolic content and the bioactivity of ethanol-water extract were maintained after simulated gastric digestion and were associated with nuclear factor kappa B (NF-κB) impairment, considered the main putative anti-inflammatory mechanism. This extract has not affected urease activity under *in vitro* experimental conditions (Martinelli et al., 2022).

Hypoglycemic Activities

The hypoglycemic activity of ethyl acetate extracts of *Rhus coriaria* fruits was investigated through the inhibition of α-amylase, a glycoside hydrolase. According to the results obtained, it has been suggested that sumac ethyl acetate extract can be used in the treatment and prevention of hyperglycemia, diabetes, and obesity with an IC50 value of 28.7 mg mL^{-1} (Giancarlo et al., 2006). The effects of ethanol extract of *R. coriaria* fruits on blood sugar, lipids, and antioxidant enzymes were measured with commercial kits. In addition, their effects on intestinal α-glucosidases were measured using an in vitro method. In the study, it was determined that fruits have important antioxidant effects by increasing superoxide dismutase and catalase activities (Mohammadi et al., 2010).

IN VIVO

Antimicrobial Activities

The effects of antimicrobial activities of sumac (*Rhus coriaria*) and dried whey powder on growth performance, carcass characteristics, intestinal morphology, microbial population, and antibody titer of female broilers against Newcastle disease were investigated. The response of broiler consumption with symbiotic dried whey powder and sumac powder as prebiotic on nutritional performance, intestinal morphology, and microbial population of *Escherichia coli* and *Lactobacillus* sp. in the gastrointestinal tract was investigated. The results showed that sumac and dried whey had significant ($P < 0.05$) effects on the intestinal microbial population compared to the control group. The data showed that *Lactobacillus* population was significantly increased and *E. coli* population decreased in sumac and dried whey powder, respectively, compared with the control group ($P < 0.05$) (Kheiri et al., 2015).

Antisecretory and Antidiarrheal Activities

In mice, the crude methanolic extract of dried fruits of *R. coriaria* demonstrated antisecretory and antidiarrheal effects against castor oil-induced fluid accumulation and diarrhea at 100 and 300 mg/kg, respectively, and was found safe up to a tested dose of 5 g/kg. Sumac fruit extract reduced in mice the total number of wet feces to 4.33 ± 1.51 and 2.75 ± 2.64 (mean ± SEM, n=6), respectively, when compared with saline (8.75 ± 2.32, n=6), while loperamide administration (a positive control) reduced the number of wet feces to 1.17 ± 1.47. In isolated rabbit jejunum, the crude extract of *R. coriaria* exhibited concentration-dependent (0.01–3.0 mg/mL) relaxation of spontaneous and high K^+ (80 mM)-induced contractions, similar to verapamil (Janbaz et al., 2014).

Hypoglycemic Activities

The fruits of *Rhus coriaria* have antihyperglycemic and antioxidant activities. Moreover, the ethanol extract of its fruits can reverse diabetes-associated dyslipidemia by increasing HDL and reducing LDL levels. In the study by Mohammadi et al. (2010), two other properties of *R. coriaria* fruits that can improve the lifespan of patients with type 2 diabetes in diabetic rats are described: mild antihyperglycemic and potent antioxidant properties (Mohammadi et al., 2010).

The protective and therapeutic effect of *R. coriaria* lyophilized fruit extract on streptozotocin-induced diabetic rats was investigated. The extract decreased the levels of blood glucose in diabetic groups (an average of 31%). Triglyceride, total cholesterol, high-density lipoprotein, and low-density lipoprotein levels were balanced by plant extract (500 mg/kg) supplementation in the diabetic group. Decreased levels of aspartate aminotransferase, alanine aminotransferase, lactate dehydrogenase, alkaline phosphatase, creatinine, and urea were detected in the 500 mg/kg dose of plant extract supplemented diabetic group (Doğan & Çelik, 2016).

The effect of methanolic extract of the dried seed of *Rhus coriaria* on hyperinsulinemia, glucose intolerance, and insulin sensitivity in non-insulin-dependent diabetes mellitus rats was investigated. The extract (200 mg/kg and 400 mg/kg) was administered orally once daily for five weeks after the animals were confirmed to be diabetic (i.e., 90 days after STZ injection). These study data show that the methanolic extract of *R. coriaria* significantly delays the onset of hyperinsulinemia and glucose intolerance and ameliorates it (Anwer et al., 2013).

Seed powder of *R. coriaria* was soaked in distilled water-ethanol mixture (60:40) and maintained for 72 hours at room temperature. Diabetes was induced by intraperitoneal administration of streptozotocin 15 minutes after an injection of nicotinamide. Then, glibenclamide and the above-mentioned extract were administered orally for 28 consecutive days. Twenty-four hours after the last treatment, serum samples, the testes, and the cauda epididymis were removed immediately for hormonal, testis morphology, and sperm parameter assessments. The results indicated that the hydro-alcoholic extract of *R. coriaria* seeds has anti-infertility effects in Type-2 diabetes in male mice (Ahangarpour et al., 2014).

The antidiabetic and hypolipidemic properties of *R. coriaria* seed hydro-alcoholic extract on nicotinamide-streptozotocin-induced type 2 diabetic mice were investigated. Diabetic mice were treated with glibenclamide (0.25 mg/kg, as a standard antidiabetic drug) or *R. coriaria* seed extract in doses of 200 and 300 mg/kg, and control groups received these two doses of extract orally for 28 days. After induction of type 2 diabetes, the level of glucose, cholesterol, low-density lipoprotein, serum glutamic oxaloacetic transaminase, and serum glutamic pyruvic transaminase increased, and the level of insulin and high-density lipoprotein decreased remarkably. Administration of both doses of extract decreased the level of glucose and cholesterol significantly in diabetic mice. LDL level decreased in the treated group with a dose of 300 mg/kg of the extract. Although usage of the extract improved level of other lipid profiles, insulin, and hepatic enzymes, changes weren't significant (Ahangarpour et al., 2017).

Hypolipidemic Activity

The effects of symbiotic dried whey powder and prebiotic sumac powder on the nutritional performance, carcass characteristics, some biochemical parameters, and antibody titer of female broilers against Newcastle disease were investigated. A total of 360 1-day-old female broiler chicks with an average weight of 38±0.42 g were randomly divided into three treatment groups with six replicates each. The data revealed an improvement in the body weight gain, feed conversion ratio because of increased intestinal morphology, antibody level, and some useful microbial population in female broiler chicks receiving the sumac and dried whey powder. Results showed that the feed intake of chicks increased significantly in sumac powder and dried whey powder in comparison with the control group ($P < 0.05$). Body weight gain was also significantly higher in the treated groups. The serum concentration triglyceride and cholesterol of chicks decreased significantly with sumac powder and dried whey powder feeding. While low-density lipoprotein levels decreased significantly, high-density lipoprotein levels increased in the sumac powder and dried whey powder group. Antibody levels increased titer against Newcastle disease significantly by feeding treated diet compared to the control group. It is concluded that dietary supplementation of sumac (0.02 %)

and dried whey powder (0.02 %) can reduce triglyceride, cholesterol, and low-density lipoprotein in the plasma of female broiler chicks at 42 days of age (Kheiri et al., 2015). The effect of different levels of sumac fruit powder as an antibiotic growth promoter substitute on the growth performance, immune responses, and serum lipid profile of broilers was investigated. A total of 260 day-old broiler chicks (Ross 308) were used in this study with four treatments in a completely randomized design. Treatments consisted of the basal diet (control) and antibiotic growth enhancer (flavophosphosfolipol), a basal diet with 3 or 7 g/kg sumac powder. Feed consumption was significantly reduced ($P< 0.05$) in broilers fed 3 and 7 g/kg sumac powder, but weight gain was not affected by dietary treatments ($P> 0.05$). In addition, serum triglyceride, cholesterol, HDL, and LDL cholesterol were not affected by the treatments ($P> 0.05$). It was determined that 7 g/kg of sumac powder promoted growth and had no negative effect on immune responses to diet. It was indicated that it is appropriate to add chicken chicks to their diets as an alternative (Toghyani & Faghan, 2017).

Hepatoprotective Activities

The protective effects of the aqueous extract of *Rhus coriaria* fruit (75 and 100 lg/mL) and one of its main components, gallic acid (100 lM), at different concentrations were investigated against oxidative stress toxicity caused by cumene hydroperoxide in isolated plants. Aqueous extract of *R. coriaria* fruit at concentrations of 75 and 100 lg/mL and gallic acid (100 μM) significantly inhibited ROS formation and lipid peroxidation ($P < 0.05$), as well as cumene hydroperoxide-induced hepatocyte membrane lysis. In addition, the H_2O_2 scavenging effect of the aqueous extract of *R. coriaria* fruit and gallic acid at concentrations (75 and 100 lg/mL) was determined in hepatocytes and compared with gallic acid (100 lM). Gallic acid (100 μM) was found to have a more effective H_2O_2 scavenging activity ($P < 0.05$) than aqueous sumac extracts (75 and 100 lg/mL) (Pourahmad et al., 2010).

CLINICAL TRIALS

Anti-obesity Activities

To determine the efficacy and safety of *Rhus coriaria* fruit supplementation in overweight or obese patients in the affiliated endocrinology clinic, a parallel two-arm, double-blind, randomized, and placebo-controlled clinical trial was designed. The observed effect of *R. coriaria* against obesity in this study was not through the change in leptin-associated pathways. However, the aqueous extracts of *R. coriaria* fruit showed a significant anti-lipase effect. In conclusion, the study showed the anti-obesity and insulin-sensitizing effect of *R. coriaria* fruit supplementation in patients who are overweight and obese. Although this effect was clinically small, it has been found statistically significant (Heydari et al., 2019).

Anti-hypertensive Activity

The effects of sumac (*Rhus coriaria*) fruits in hypertensive patients were investigated. A randomized, double-blind, placebo-controlled clinical trial was conducted on 80 hypertensive patients receiving captopril (25 mg day^{-1}). The patients were randomly divided into two groups: the first group received *R. coriaria* fruit capsules (500 mg twice a day) and captopril (25 mg once a day), and the second one received placebo capsules (500 mg starch twice a day) and captopril (25 mg once a day), for eight weeks. Blood pressure and body weight index in all patients were determined every week. Data indicated that hypertension was decreased significantly in the sumac group compared to baseline and placebo groups after eight weeks. Still, body weight index did not demonstrate a marked change in comparison with baseline and placebo groups. Moreover, the most abundant phenolic compounds identified in *R. coriaria* fruits were luteolin, apigenin, and quercetin flavonoids (Ardalani et al., 2016).

TOXICOLOGY STUDIES

Rhus coriaria was shown to be safe to be consumed by both humans and animals (Alsamri et al., 2021). Toxicity tests of this fruit extract were conducted using three different dosages (250, 500, and 1000 mg/kg). Optimal tolerance and non-lethal per os (*p.o.*) intake of *R. coriaria* lyophilized fruit extract even at a very high dose (1000 mg/kg) and no signs of toxicity and death were observed after daily administration of the extract for three days (Doğan & Çelik, 2016). The hydroalcoholic sumac seed extract *p.o.* administration of 300 mg/kg has a favorable outcome in controlling some blood parameters in type 2 diabetic mice without causing undesirable side effects (Ahangarpour et al., 2017). Janbaz et al. (2014) found that the crude methanolic extract of the fruit of *R. coriaria* was safe up to the 5 g/kg dose tested. These findings strongly indicate that this plant and its extracts are safe, making them even more attractive for medicinal use or drug discovery approaches (Alsamri et al., 2021).

Acute Toxicity

Twenty-four (male/female) BALB/c mice (20–25 g) were divided into four groups, each containing six animals. The groups of first, second, and third mice served as test groups and were treated orally with the respective crude sumac extract doses of 1, 3, and 5 g/kg. The group 4 animals served as negative control and were treated with saline (10 mL/kg). The animals were deprived of food 12 hours before the test, while water remained freely available. The animals had been under strict observation for 8 hours for possible behavioral change and 24 hours of monitoring for possible mortality. The crude methanolic extract of *R. coriaria* was found safe and did not exert any harmful effect in mice up to oral doses of 1, 3, and 5 g/kg (Janbaz et al., 2014).

CHEMISTRY

As of today, over 260 phytochemicals have been isolated from *Rhus coriaria*, and these include organic acids such as malic acid derivatives; flavonoids (myricetin, quercetin, kaempferol, etc.), isoflavonoids (amenthoflavone, agathisflavone, hinokiflavone, sumaflavone, etc.); hydrolyzable tannins such as gallotannins, gallic acid, methyl gallate, digallic acid, tri-gallic acid, and ellagic acid, together with mono-, di-, tri-, tetra-, penta-, deca-, undeca-, and dodecagallolyl glycoside derivatives; anthocyanins, terpenoids, and other compounds such as butein, iridoid, and coumarin derivatives (Mehrdad et al., 2009; Shabbir, 2012; Abu-Reidah et al., 2014; Abdul-Jalil-2020; Alsamri et al., 2021; Gabr & Alghadir, 2021; Arena et al., 2022). Furthermore, the leaf and fruit parts of the plant also contain essential oil (Kurucu et al., 1993).

Leaf

Rhus coriaria leaves contain about 15–20% polyphenolic compounds. These are mostly hydrolyzable tannins with a central glucose unit, to which several gallic acid rests are bound depsidically (Van Loo et al., 1988). Leaves also contain volatile oil, wax, and sugars (Güvenç & Koyuncu, 1994). Five to ten percent of the total polyphenolic fraction consists of condensed tannins or flavonoids. Myricetin has been found to be the major flavonol in 80% methanol extracts of the leaves and fruits of *R. coriaria*, followed by minor quantities of quercetin and kaempferol. In the aqueous extract of sumac leaves, a group of gallotannins from mono- to deca-galloyl glycosides of the class of hydrolyzable tannins and a set of co-extracted flavonoid derivatives including myricetin, quercetin-3-*O*-rhamnoside, myricetin-3-*O*-glucoside, myricetin-3-*O*-glucuronide, and myricetin-3-*O*-rhamnoglucoside were identified (Van Loo et al., 1988; Mehrdad et al., 2009; Regazzoni et al., 2013). The presence of the dimeric flavonoids agathisflavone, amentoflavone, and hinokiflavone is proved. Van loo et al. (1988) isolated and identified a dimeric flavonoid, sumaflavone, from the leaves of *R. coriaria* (Van Loo et al., 1988).

Fruit

Dried sumac fruits are mainly composed of moisture, protein, fiber, ash, and water-soluble extract (Kızıl & Türk, 2010; Alsamri et al., 2021). They also contain tannins, flavones, flavonoids, anthocyanins, organic acids such as malic acid, essential oils, minerals, and vitamins (Kurucu et al., 1993; Güvenç & Koyuncu, 1994; Kossah et al., 2009; Shabbir, 2012; Abu-Reidah et al., 2015; Karadaş et al., 2020). Linoleic, oleic, palmitic, and stearic acids have been found as the major components of sumac fixed oil. Al, Cd, Cr, Cu, Fe, Mg, Na, Ni, Pb, and Zn were determined in ripe fruits, mainly K and Ca (Kossah et al., 2009; Kızıl & Türk, 2010; Abu-Reidah et al., 2015; Karadaş et al., 2020; Alsamri et al., 2021). It is worth mentioning that mineral contents were affected by environmental factors and the geographic locations where sumac fruits were collected (Kossah et al., 2009). As for the vitamin content, sumac fruit contained thiamine, riboflavin, pyridoxine, cyanocobalamin, nicotinamide, biotin, and ascorbic acid (Kossah et al., 2009; Karadaş et al., 2020; Alsamri et al., 2021). Specifically, headspace solid-phase microextraction gas chromatography coupled to mass spectrometry and comprehensive two-dimensional liquid chromatography coupled to photodiode array and mass spectrometry detection has been employed. A total of 263 volatile compounds, including terpene hydrocarbons, acids, aldehydes, and polyphenolic compounds, mainly gallic acid derivatives, in the fruits of *Rhus coriaria* have been identified (Arena et al., 2022). In the fruits of *R. coriaria* cyanidin, peonidin, pelargonidin, petunidin, and delphinidin and their glucosides have been identified (Abu-Reidah et al., 2014). Isohyperoside, quercitrin, myricetin glucuronide, and myricetin rutinoside were identified in the hydro-alcoholic extract of the sumac fruit (Gabr & Alghadir, 2021). Among the isolated compounds, isoquercitrin and myricetin-3-O-α-L-rhamnoside were reported from the fruits by Shabana et al. (2011). Sumac fruit mainly contains organic acids such as p-benzoic acid, citric acid, malic acid, protocatechic acid, tartaric acid, and vanillic acid. The sourness of sumac is mainly due to malic, citric, and tartaric acids (Abu-Reidah et al., 2014).

Aromatic compounds were isolated from the ethanolic extract of *R. coriaria* seeds and identified as 1-methoxy-4-hydroxy-methylene naphthalene (choriarianaphthyl ether), 7-methoxy-5-methyl benzene-4-al-1-oic acid (choriariaoic acid), and 1-dodecanoxy-2,8-dihydroxy-anthracene-15-oic acid (choriariantracenyl ester). And also, these phyto-components together with *n*-tetracosane, *n*-pentacosane, anise alcohol, *p*-hydroxybenzyl alcohol, methyl lawsone, and 2-hydroxymethylene naphthaquinone were obtained (Onkar et al., 2011).

CONCLUSION

Rhus coriaria L. (Sumak, Sumac) grows naturally in the Mediterranean, Southeast, Central, and Northern regions of Türkiye and other countries such as coast to the Mediterranean, Southern Europe, North Africa, Iran, and Afghanistan. Sumac has been used as food, spice, and medicine for centuries and is also used in the leather industry. Its leaves and fruits are used as medicine. Various bioactivity *(in vitro and in vivo)* studies and clinical trials on R. coriaria have shown that it has positive effects on health. To date, over 260 phytochemicals have been isolated from *R. coriaria* such as organic acids, flavonoids, isoflavonoids, hydrolyzable tannins, anthocyanins, terpenoids, iridoid, and coumarin derivatives. In light of these scientific data, it can be recommended to consume standard sumac products to lead a healthy life. For this reason, the cultivation of this plant, which grows naturally in our country, and the production of standard products should be encouraged.

REFERENCES

Abdul-Jalil, T.Z. 2020. *Rhus coriaria (Sumac): A Magical Spice.* IntechOpen. http://dx.doi.org/10.5772/intechopen.92676

Abu-Reidah, I.M., Jamous, R.M., Ali-Shtayeh, M.S. 2014. Phytochemistry, pharmacological properties and industrial applications of *Rhus coriaria* L. (Sumac). *Jordan Journal of Biological Sciences*, 7(4), 233–244.

Abu-Reidah, I.M., Ali-Shtayeh, M.S., Jamous, R.M., Arráez-Román, D., Segura-Carretero, A. 2015. HPLC-DAD-ESI-MS/MS screening of bioactive components from *Rhus coriaria* L. (sumac) fruits. *Food Chem.*, 166, 179–191.

Adwan, G.M., Abu-Shanab, B.A., Adwan, K.M. 2009. In vitro activity of certain drugs in combination with plant extracts against *Staphylococcus aureus* infections. *African Journal of Biotechnology*, 8(17), 4239–4241. https://doi.org/10.1016/s1995-7645(10)60064-8

Ahangarpour, A., Heidari, H., Junghani, M.S., Absari, R., Khoogar, M., Ghaedi, E. 2017. Effects of hydroalcoholic extract of *Rhus coriaria* seed on glucose and insulin related biomarkers, lipid profile, and hepatic enzymes in nicotinamide-streptozotocin-induced type II diabetic male mice. *Research in Pharmaceutical Sciences*, 12(5), 416–424.

Ahangarpour, A., Oroojan, A.A., Heidari, H., Ghaedi, E., Nooshabadi M.R.R. 2014. Effects of hydro-alcoholic extract of *Rhus coriaria* (sumac) seeds on reproductive complications of nicotinamide-streptozotocin induced type-2 diabetes in male mice. *World Journal of Men's Health*, 32(3), 151–158. http://dx.doi.org/10.5534/wjmh.2014.32.3.151

Alsamri, H., Athamneh, K., Pintus, G., Eid, A.H., Iratni, R. 2021. Pharmacological and antioxidant activities of *Rhus coriaria* L. (Sumac). *Antioxidants*, 10, 73. https://doi.org/10.3390/antiox10010073

Anwer, T., Sharma, M., Khan, G., Iqbal, M., Ali, M.S., Alam, M.S., Safhi, M.M., Gupta, N. 2013. *Rhus coriaria* ameliorates insulin resistance in non-insulin-dependent diabetes mellitus (niddm) rats. *Acta Poloniae Pharmaceutica - Drug Research*, 70(5), 861–867.

Ardalani, H., Moghadam, M.H., Rahimi, R., Soltani, J., Mozayanimonfared, A., Moradie, M., Azizi, A. 2016. Sumac as a novel adjunctive treatment in hypertension: A randomized, double-blind, placebo-controlled clinical trial. *RSC Advances*, 14.

Arena, K., Trovato, E., Cacciola, F., Spagnuolo, L., Pannucci, E., Guarnaccia, P., Santi, L., Dugo, P., Mondello, L., Dugo, L. 2022. Phytochemical characterization of *Rhus coriaria* L. extracts by headspace solid-phase micro extraction gas chromatography, comprehensive two-dimensional liquid chromatography, and antioxidant activity evaluation. *Molecules*, 27(5), 1727. https://doi.org/10.3390/molecules27051727

Baytop, T. 1999. Türkiye'de Bitkiler ile Tedavi-Geçmişte ve Bugün, İlaveli 2. Baskı. Nobel Tıp Kitabevleri, İstanbul. (ISBN: 975-420-021-1).

Bonnier, G. 1934. *Flore Compléte Illustrée en Couleurs de France*. Suisse et Belgique. Tome II, Paris.

Davis, P.H. 1967. *Flora of Turkey and the Aegean Islands*. Vol. 2, University Press, Edinburgh.

Demirhan, A. 1975. *Mısır Çarş ısı Drogları*. Sermet Matbaası, İstanbul.

Dinçer, F. 1967. Türk folklorunda veteriner hekimliği üzerine araştırmalar. Ankara Ü. Veteriner Fakültesi Yayınları, Yayın No: 214, A.Ü. Veteriner ve Ziraat Fakültesi Basımevi.

Doğan, A., Çelik, İ. 2016. Healing effects of sumac (*Rhus coriaria*) in streptozotocin-induced diabetic rats. *Pharmaceutical Biology*, 54(10), 2092–2102. http://dx.doi.org/10.3109/13880209.2016.1145702

Elagbar, Z.A., Shakya, A.K., Barhoumi, L.M., Al-Jaber, H.I. 2020. Phytochemical diversity and pharmacological properties of *Rhus coriaria*. *Chemistry & Biodiversity*, 17(4), e1900561. https://doi.org/10.1002/cbdv.201900561

Fazeli, M.R., Amin, G., Attari, M.M.A., Ashtiani, H., Jamalifar, H., Samadi, N. 2007. Antimicrobial activities of Iranian sumac and avishan-e shirazi (*Zataria multiflora*) against some food-borne bacteria. *Food Control*, 18(6), 646–649.

Fujita, T., Sezik, E., Tabata, M., Yeşilada, E., Honda, G., Takeda, Y., Tanaka, T., Takaishi, Y. 1995. Traditional medicine in turkey VII. folk medicine in Middle and West Black Sea regions. *Economic Botany*, 49(4), 406–422.

Gabr, S.A., Alghadir, A.H. 2021. Potential anticancer activities of *Rhus coriaria* (sumac) extract against human cancer cell lines. *Bioscience Reports*, 41, BSR20204384 https://doi.org/10.1042/BSR20204384

Gharabolagha, A.F., Sabahia, F., Karimi, M., Kamalinejad, M., Mirshahabi, H., Seyed Dawood Nasab, M., Ahmadi, N.A. 2018. Effects of *Rhus Coriaria* L. (Sumac) extract on hepatitis B virus replication and Hbs Ag secretion. *Journal of Reports in Pharmaceutical Sciences*, 7(1), 100–107.

Giancarlo, S., Rosa, L.M., Nadjafi, F., Francesco, M. 2006. Hypoglycaemic activity of two spices extracts: *Rhus coriaria* L. and *Bunium persicum* Boiss. *Natural Product Research*, 20(9), 882–886. https://doi.org/10.1080/14786410500520186

Giovanelli, S., Giusti, G., Cioni, P.L., Minissale, P., Ciccarelli, D., Pistelli L. 2017. Aroma profile and essential oil composition of *Rhus coriaria* fruitsfrom four Sicilian sites of collection. *Industrial Crops and Products*, 97, 166–174.

Ghorbani, P., Namvar, F., Homayouni-Tabrizi, M., Soltani, M., Karimi, E., Yaghmaei, P. 2018. Apoptotic efficacy and antiproliferative potential of silver nanoparticles synthesised from aqueous extract of sumac (*Rhus coriaria* L.). *IET. Nanobiotechnology*, 12, 600–603. https://doi.org/10.1049/iet-nbt.2017.0080

Golzadeh, M., Farhoomand, P., Daneshyar, M. 2012. Dietary *Rhus coriaria* L. powder reduces the blood cholesterol, VLDL-c and glucose, but increases abdominal fat in broilers. *South African Journal of Animal Science*, 42(4). http://dx.doi.org/10.4314/sajas.v42i4.8

Güvenç, A., Koyuncu, M. 1994.A study on the main active compounds of leaves and fruits of *Rhus coriaria* L. *Turkish Journal of Medical Sciences*, 20(1), 11–13.

Harmancıoğlu, M. 1955. Türkiye'de bulunan önemli bitki boyalarından elde olunan renklerin çeşitli müessirlere karşı yün üzerinde haslık dereceleri. A.Ü. Ziraat Fakültesi Yayınları, Yayın No: 77, A.Ü. Basımevi, Ankara.

Heydari, M., Nimruzi, M., Hajmohammadi, Z., Faridi, P., Omrani, G.R., Shams, M. 2019. *Rhus coriaria* L. (sumac) in patients who are overweight or have obesity: A placebo-controlled randomized clinical trial. *Shiraz E-Medical Journal*, 20(10), e87301. https://doi.org/10.5812/semj.87301

Honda, G., Yeşilada, E., Tabata, M., Sezik, E., Fujita, T., Takeda, Y., Takaishi, Y., Tanaka, T. 1996. Traditional medicine in Turkey VI. Folk medicine in West Anatolia: Afyon, Kütahya, Denizli, Muğla, Aydın provinces. *Journal of Ethnopharmacology*, 53, 75–87.

Iauk, L., Caccamo, F., Speciale, A.M., Tempera, G., Ragusa, S., Pante, G. 1998. Antimicrobial activity of *Rhus coriaria* L. leaf extract. *Phytotherapy Research: An International Journal Devoted to Pharmacological and Toxicological Evaluation of Natural Product Derivatives*, 12(S1), S152–S153.

Janbaz, K.H., Shabbir, A., Mehmood, M.H., Gilani, A.H. 2014. Pharmacological basis for the medicinal use of *Rhus coriaria* in hyperactive gut disorders. *Bangladesh Journal of Pharmacology*, 9, 636–644.

Kacergius, T., Abu-Lafi, S., Kirkliauskiene, A., Gabe, V., Adawi, A., Rayan, M., Qutob, M., Stukas, R., Utkus, A., Zeidan, M., Rayan, A. 2017. Inhibitory capacity of *Rhus coriaria* L. extract and its major component methyl gallate on *Streptococcus mutans* biofilm formation by optical profilometry: Potential Applications for Oral Health. *Molecular Medicine Reports*, 16, 949–956. https://doi.org/10.3892/mmr.2017.6674

Kâhya, E. 2017. İbn-I Sina, El-Kanun Fi't-Tıbb. İkinci Kitap, p. 446, T.C. Başbakanlık Atatürk Kültür, Dil ve Tarih Yüksek Kurumu, Atatürk Kültür Merkezi Yayını: 470, Araştırma İnceleme Dizisi: 123, Semih Ofset, Ankara.

Kâhya, E. 2019. Materia Medica/Dioscorides. Ankara Nobel Tıp Kitabevleri Ltd. Şti. Ankara.

Karadaş, Ö., Yılmaz, İ., Geçgel, Ü. 2020. Properties of Sumac plant and its importance in nutrition. *International Journal of Innovative Approaches in Agricultural Research*, 4(3), 377–383. https://doi.org/10.29329/ijiaar.2020.274.10

Kheiri, F., Rahimian, Y., Nasr, J. 2015. Application of sumac and dried whey in female broiler feed. *Archives Animal Breeding*, 58, 205–210. https://doi.org/10.5194/aab-58-205-2015

Kızıl, S., Türk, M. 2010. Microelement contents and fatty acid compositions of *Rhus coriaria* L. and *Pistacia terebinthus* L. fruits spread commonly in the south eastern Anatolia region of Turkey. *Natural Product* Research, 24, 92–98.

Kossah, R., Nsabimana, C., Zhao, J., Chen, H., Tian, F., Zhang H., Chen, W. 2009. Comparative study on the chemical composition of Syrian sumac (*Rhus coriaria* L.) and Chinese sumac (*Rhus typhina* L.) fruits. *Pakistan Journal of Nutrition*, 8(10), 1570–1574. https://doi.org/10.3923/pjn.2009.1570.1574

Koşar, M., Bozan, B., Temelli, F., Başer, K.H.C. 2007. Antioxidant activity and phenolic composition of Sumac (*Rhus coriaria* L.) extracts. *Food Chemistry*, 103, 952–959.

Koyuncu, M., Köroğlu, A. 1991. *Rhus coriaria* L. yaprak ve meyvalarının anatomik incelenmesi. *Doğa Türk Ecz. Derg.*, 1, 89–96.

Kurucu, S., Koyuncu, M., Güvenç (Köroğlu), A., Baser, K. H. C., Özek, T. 1993. The essential oils of Rhus coriaria L. (Sumac). *Journal of Essential Oil Research*, 5(5), 481–486.

Martinelli, G., Angarano, M., Piazza, S., Fumagalli, M., Magnavacca, A., Pozzoli, C., Khalilpour, S., Dell'Agli, M., Sangiovanni, E. 2022. The Nutraceutical properties of sumac (*Rhus coriaria* L.) against gastritis: Antibacterial and anti-Inflammatory activities in gastric epithelial cells infected with *H. pylori*. *Nutrients*, 14, 1757. https://doi.org/10.3390/nu14091757

Matloobia, M., Tahmasebi, S. 2019. Iranian sumac (*Rhus coriaria*) can spice up urban landscapes by its ornamental aspects. *Acta Horticulturae*, 1240, 77–82.

Mehrdad, M., Zebardast, M., Abedi, G., Koupaei, M.N., Rasouli, H., Talebi, M. 2009. Validated High-Throughput HPLC method for the analysis of flavonol aglycones myricetin, quercetin, and kaempferol in *Rhus coriaria* L. using a monolithic column. *Journal of AOAC International*, 92(4), 1035–1043.

Mirian, M., Behrooeian, M., Ghanadian, M., Dana, N. Sadeghi-Aliabadi, H. 2015. Cytotoxicity and antiangiogenic effects of *Rhus coriaria*, *Pistacia vera* and *Pistacia khinjuk* oleoresin methanol extracts. *Research in Pharmaceutical Sciences*, 10(3), 233–240.

Mohammadi, S., Kouhsari, S.M., Feshani, A.M. 2010. Antidiabetic properties of the ethanolic extract of *Rhus coriaria* fruits in rats. *Daru*, 18(4), 270–274.

Morshedloo, M.R., Maggi, F., Neko, H.T., Aghdam, M.S. 2018. Sumac (*Rhus coriaria* L.) fruit: Essential oil variability in Iranian populations. *Industrial Crops & Products*, 111, 1–7.

Nasar-Abbas, S.M., Halkman A.K. 2004. Antimicrobial effect of water extract of sumac (*Rhus coriaria* L.) on the growth of some food borne bacteria including pathogens. *International Journal of Food Microbiology*, 97, 63–69.

Onkar, S., Mohammed, A., Nida, A. 2011. New antifungal aromatic compounds from the seeds of *Rhus coriaria* L. *International Research Journal of Pharmacy*, 2(1), 188–194. http://www.irjponline.com

Özcan, M. 2003. Effect of sumach (*Rhus coriaria* L.) Extracts on the oxidative stability of peanut oil. *Journal of Medicinal Food*, 6, 63–66.

Parsania, M., Rezaee, M.B., Monavari, S.H., Jaimand, K., Jazayeri, S.M.M., Razazian, M., Nadjarha, M.H. 2017. Evaluation of antiviral effects of sumac (*Rhus coriaria* L.) fruit extract on acyclovir resistant Herpes simplex virus type 1. 27(1), 1–8. https://doi.org/20.1001.1.10235922.1396.27.1.2.2

Pourahmad, J., Eskandari, M.R., Shakibaei, R., Kamalinejad, M. 2010. A search for hepatoprotective activity of aqueous extract of *Rhus coriaria* L. against oxidative stress cytotoxicity. *Food and Chemical Toxicology*, 48, 854–858. https://doi.org/10.1016/j.fct.2009.12.021

Regazzoni, L., Arlandini, E., Garzon, D., Santagati, N.A., Berettaa, G., Facino, R.M. 2013. A rapid profiling of gallotannins and flavonoids of the aqueous extract of Sumac L. by flow injection analysis with high-resolution mass spectrometry assisted with database searching. *Journal of Pharmaceutical and Biomedical Analysis*, 72, 202–207. https://doi.org/10.1016/j.jpba.2012.08.017

Seseogullari-Dirihan, R., Apollonio, F., Mazzoni, A., Tjaderhane, L., Pashley, D., Breschi, L., Tezvergil-Mutluay, A. 2016. Use of crosslinkers to inactivate dentin MMPsR. *Dental Materials*, 32, 423–432.

Sezik, E., Yeşilada, E., Honda, G., Takaishi, Y., Takeda, Y., Tanaka, T. 2001. Traditional medicine in Turkey X. Folk medicine in Central Anatolia. *Journal of Ethnopharmacology*, 75, 95–115.

Sezik, E., Tabata, M., Yeşilada, E., Honda, G., Goto, K., Ikeshiro, Y. 1991. Traditional medicine in Turkey. I. Folk medicine in northeast Anatolia. *Journal of Ethnopharmacology*, 35(2), 191–196. https://doi.org/10.1016/0378-8741(91)90072-1

Shabana, M.M., El Sayed, A.M., Yousif, M.F., El Sayed, A.M., Sleem, A.A. 2011. Bioactive constituents from *Harpephyllum caffrum* Bernh. and *Rhus coriaria* L. *Pharmacognosy Magazine*, 7, 298–306.

Shabbir, A. 2012. *Rhus coriaria* Linn, a plant of medicinal, nutritional and industrial importance: A review. *Journal of Animal and Plant Sciences*, 22, 505–512.

Sherif, Y.E., Gabr, S.A., Hosny, N.M., Alghadir, A.H., Alansari, R. 2021. Phytochemicals of *Rhus* spp. as Potential inhibitors of the SARS-CoV-2 main protease: Molecular docking and drug-likeness study. *Evidence-Based Complementary and Alternative Medicine*, Article ID 8814890. https://doi.org/10.1155/2021/8814890

Tabata, M., Sezik, E., Honda, G., Yeşilada, E., Fukui, H., Goto, K., Ikeshiro, Y. 1994. Traditional medicine in Turkey. III. Folk medicine in East Anatolia, Van and Bitlis province. *International Journal of Pharmacognosy*, 32(1), 3–12.

Toghyani, M., Faghan, N. 2017. Effect of sumac (*Rhus coriaria* L.) fruit powder as an antibiotic growth promoter substitution on growth performance, immune responses and serum lipid profile of broiler chicks. *Indian Journal of Pharmaceutical Education and Research*, 51(3), (Special Issue). https://doi.org/10.5530/ijper.51.3s.33

Tutin, T.G., Heywood, V.H., Burges, N.A., Moore, D.M., Valentine, D.H., Walters, S.M., Webb, D.A. 1968. *Flora Europaea*. Vol: 3. Cambridge University Press.

Tuzlacı, E. 2016. Türkiye Bitkileri Geleneksel İlaç Rehberi. İstanbul Tıp Kitabevleri, İstanbul.

Tuzlacı, E., Aymaz, P.E. 2001. Turkish folk medicinal plants, Part IV: Gönen (Balıkesir). *Fitoterap*, 72, 323–343.

Ünder, D., Saltan, F.Z. 2019. Sumak ve Önemli Biyolojik Etkileri. *Çukurova Journal of Agricultural and Food Sciences*, 34(1), 51–60.

Vahid-Dastjerdi, E., Sarmast, Z., Abdolazimi, Z., Mahboubi, A., Amdjadi, P., Kamalinejad, M. 2014. Effect of *Rhus coriaria* L. water extract on five common oral bacteria and bacterial biofilm formation on orthodontic wire. *Iranian Journal of Microbiology*, 6, 269–275.

Van Loo, P., de Bruyn, A., Verzele, M. 1988. Liquid chromatography and identification of the flavonoids present in the "sumach tannic acid" extracted from *Rhus coriairia*. *Chromatographia*, 25(1), 15–20.

Yeşilada, E., Honda, G., Sezik, E., Tabata, M., Goto, K., Ikeshiro, Y. 1993. Traditional medicine in Turkey. IV. Folk medicine in Mediterranean subdivisio. *Journal of Ethnopharmacology*, 39, 31–38.

Yeşilada, E., Honda, G., Sezik, E., Tabata, M., Fujita, T., Tanaka, T., Takeda Y., Takaishi, Y. 1995. Traditional medicine in Turkey V. Folk medicine in the Inner Taurus Mountains. *Journal of Ethnopharmacology*, 46, 133–152.

Yılmaz, H., Eminağaoğlu, Ö. 2020. Anacardiaceae. In: Akkemik, Ü. (ed.). *Türkiye'nin Bütün Ağaçları ve Çalıları*. Türkiye İş Bankası Kültür Yayınları, p. 272.

http://www.worldfloraonline.org/.

19 *Rosa canina* L.

Ayşegül Köroğlu

INTRODUCTION

Rosa canina L. (Rose hip, Dog rose, Kuşburnu, Gül burnu; Rosaceae) is a shrub that grows almost all over Anatolia. Many apocarpous pistils mature separately in the fruit of rose hip to form achenes. These achene fruits are contained within and attached to a fleshy enlarged receptacle. These types of fruits are called pseudo fruits (Nilsson, 1972). It is estimated that roses have been used for beauty, fragrance, cosmetic, and medicinal purposes in human history for over 5000 years. Especially rose hip fruits are used traditionally, and they are essential because of their pharmaceutical, nutraceutical, and commercial value (Güvenç, 2011; Šindrak et al., 2012; Ahmad et al., 2016). The pseudo fruits of rose hips are usually consumed in dried form, but the ripe fresh fruits are used to make marmalades, jams, juices, teas, and syrups. Furthermore, these fruits are used in foodstuffs and drinks, including jellies and alcoholic beverages. In our country, it is used as a tea in the form of filtered teas, especially mixed with hibiscus flowers (Bisset, 1994; Coşkun et al., 1997; Šindrak et al., 2012; Czyzowska et al., 2015). It is known as "kuşburnu" among the people in Turkey and is widely used. Rose hips are rich in vitamins (A, B1, B2, C, K, P), especially vitamin C, as well as phenolic compounds anthocyanins, flavonoids, bioflavonoids, tannins, calchon, organic acids, carotenoids, tocopherol, essential oils, and pectins (PDR for Herbal Medicines, 2000; Hvattum, 2002; Valsta et al., 2003; Erciṣli, 2007; Güvenç, 2011, Tumbas et al., 2012; Selahvarzian et al., 2018). The lipid fraction of pseudo fruit consists of neutral lipids, glycolipids, and phospholipids. It has been shown that fruits contain fixed oil and the main components of this oil are lauric acid, palmitic acid, α -linolenic acid, and linoleic acid (Ul'chenko et al., 1995; Nowak, 2005a; Nowak, 2005b; Erciṣli, 2007; İlyasoğlu, 2014; Winther et al., 2016; Ahmad et al., 2016). Rose hip is known worldwide as a good source of vitamin C. *R. canina* pseudo fruits are used in kidney and urinary system diseases, in reducing kidney stones, in the treatment of edema, rheumatism, and gout, in cases of cold, fever, and infection, in bleeding, as a blood purifier, in scurvy, and malaria. It is used to increase body resistance in the treatment and prevention of vitamin C deficiency. It is also a diuretic and laxative due to the pectin and organic acids it contains. Adults: Infusion is prepared from 2 to 2.5 g of coarsely powdered dry drug with 150 mL of hot water, waiting for 10–15 minutes. Drink this infusion 1–2 times a day (Bisset, 1994; Commission E Monographs, 1998; PDR for Herbal Medicines, 2000; Czyzowska et al., 2015). The seed oil of *R. canina* is one of the richest sources of linoleic acid and linolenic acid, with content as high as 70–80%. Due to its content of these fatty acids, carotenoids, and vitamin A with skin rejuvenating properties, rose hip seed oil acts as a natural skin care remedy, moisturizer, and antiaging agent. Moreover, the oil has healing power to treat skin problems such as scars, dermatitis, acne, eczema, and burns (Ahmad et al., 2016).

It has antimicrobial (antibacterial, antifungal), antiviral, antioxidant, anticarcinogenic, and anti-inflammatory effects. It is used in the treatment of rheumatic diseases such as rheumatism, arthritis, and osteoarthritis. It was reported in a systematic review that the reduction of osteoarthritis signs through the consumption of rose hip was much more significant than those of glucosamine which is a widely used medicine. *In vivo* studies have shown antigenotoxic, antidiabetic, antiulcerogenic, antidiarrheal, immunomodulatory, antiproliferative, and antihepatotoxic. It has also been shown to be effective in kidney disorders, diarrhea, inflammatory disorders, arthritis, hyperlipidemia,

DOI: 10.1201/9781003146971-19

obesity, and cancer (Chrubasik et al., 2008; Orhan et al., 2007, Güvenc, 2011, Christensen et al., 2013; Mármol et al., 2017; Gruenwald et al., 2019). This plant grows wildly in the northern hemisphere in Europe, northern Africa, temperate Eurasia, Western Asia, and North and South America. It is widely found in Anatolia (Nilsson, 1972, Ahmad et al., 2016).

MORPHOLOGICAL CHARACTER

(0.5-)1.5-3.5(-7) m erect, sometimes a climbing shrub, arching, not rhizomatous; distal branches arching, bark green; infrastipular prickles paired, curved or appressed, 6–7 × 4–9 mm, lengths ± uniform, internodal prickles rare, single, rarely absent. Leaves deciduous, 4.5–11 cm; stipules 10–22 × 3–5 mm, erect, auricles, margins stipitate-glandular or eglandular, surfaces glabrous, eglandular or sparsely stipitate-glandular; petiole and rachis sometimes with prickles, glabrous, eglandular; leaflets 5–7, terminal: petiolule 5–11 mm, blade ovate, obovate, or elliptic, 15–40 × 12–20 mm, base obtuse to slightly cuneate, margins serrate, teeth 20–30 per side, rarely glandular, apex acute, sometimes acuminate, abaxial surfaces glabrous, rarely pubescent or tomentose on midveins, eglandular, adaxially dark green to green, glabrous, rarely tomentose. Inflorescences panicles, sometimes corymbs, solitary, sometimes 2-3(-15) flowered. Pedicels 8–20 mm, erect to reflexed as hips mature, eglandular or stipitate-glandular; bracts 2, ovate-lanceolate, 6–18 × 4–5 mm, margins glandular-serrate, abaxial surfaces puberulent or hispid, adaxial surfaces glabrous, eglandular. Flowers 3.5–5 cm diam.; hypanthium narrowly urceolate, 7–9 × 3–6 mm, eglandular, neck 2–3 × 1–2 mm; sepals ovate-lanceolate, 10–17 × 3–5 mm, appressed-reflexed, spreading or erect, margins deeply pinnatifid, tip 4–6 × 0.5 mm, abaxially eglandular, pubescent or rarely glandular, sepals deciduous as hips mature; petals rose, pink, or white, rarely dark pink, 18–25 × 15–18 mm; carpels 26–36, styles villous or sometimes eglandular, hypanthial disc (4–5 mm diam.), conically. Hips red to dark red, globose, ovoid, urceolate, or ellipsoid, 10-16(-25-30) × 6–16 mm, glabrous, sometimes eglandular. Achenes 14–23, tan, 5–6 × 3–3.5 mm (Figures 19.1 and 19.2) (Nilsson, 1972; worldfloraonline, 2022).

USED PART OF THE PLANT

Fruits of *Rosa canina* is an aggregate fruit (= pseudo fruit) type, which is the Rosaceae family's fruit character of the Rosoidae subfamily of the Rosaceae family and formed as a result of the rise or

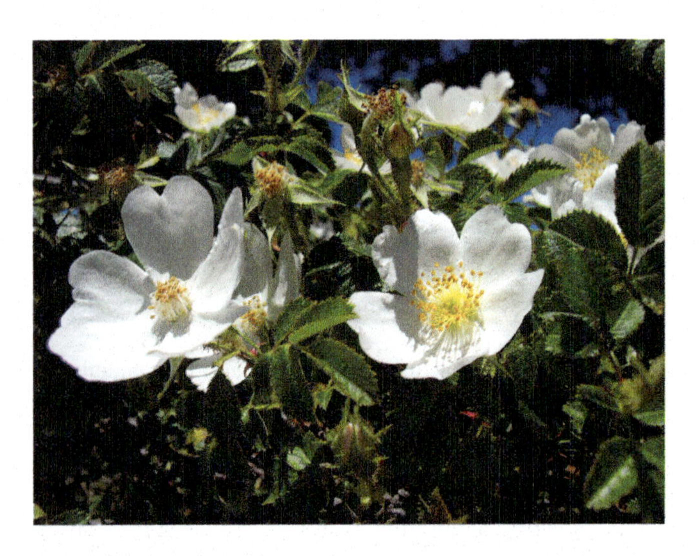

FIGURE 19.1 *Rosa canina* native to Elmadağ (Ankara). (Photo: Prof. Dr. Mecit Vural.)

FIGURE 19.2 *Rosa canina* hips native to Samsun. (Photo: Prof. Dr. Ayşegül Köroğlu.)

FIGURE 19.3 Rose hips sold in traditional bazar in Samsun. (Photo: Prof. Dr. Ayşegül Köroğlu.)

pitting of the receptaculum and its participation in fruit formation (Nilsson, 1972). Of these pseudo fruits, 30–35% are seeds, and 65–70% are receptacles (hypanthium or floral cup shells) (Ahmad et al., 2016; Winther et al., 2016). Therefore, rose hip fruits also contain many achene-type fruits in the receptaculum, which is fleshy and pitted at maturity. In fact, achenes, which are the primary fruit, are considered seeds in these fruits (Figure 19.3).

According to the European Pharmacopoeia, it is desirable that the Rosae pseudo fructus seed pod be mature, fresh, or dry, free from seeds (achenes) and attached trichomes, and have an ascorbic acid content of not less than 0.3% (European Pharmacopoeia, 2010; TF, 2017). According to the Commission E Monographs, all parts of the fruit were evaluated in three different ways: (1) rose hip (Rosae pseudofructus, the ripe, fresh or dried seed receptacle, freed from seed and attached trichomes), (2) rose hip and seed (Rosae pseudofructus cum fructibus, the ripe, fresh or pseudo dried fruits including the seed), and (3) rose hip seed (Rosae fructus, the ripe, dried seed) (Commission E Monographs, 1998; Chrubasik et al., 2008; Güvenç, 2011).

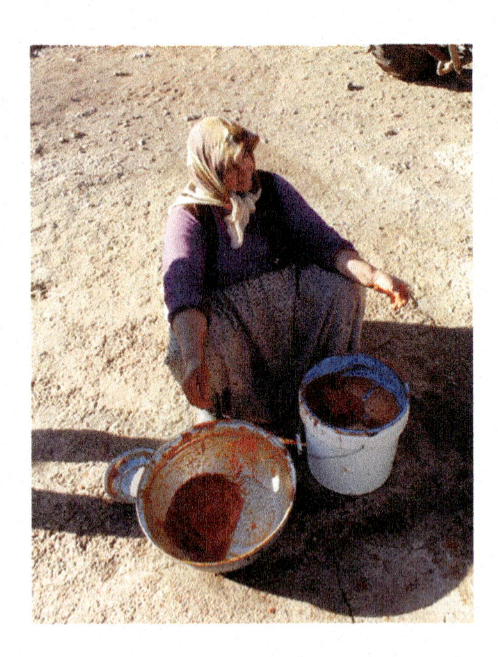

FIGURE 19.4 Preparation of marmalade of *Rosa canina* (Elmadağ, Ankara). (Photo: S.T. Körüklü.)

TRADITIONAL USE

Rose hips are known all over the world as a good source of vitamin C. These pseudo fruits are used in many cultures of the world, such as teas, jellies, jams, and alcoholic beverages. The use of rose hip fruits to prepare jam and marmalade is widespread in Turkey (Figure 19.4). Rose hips are traditionally employed to treat influenza, inflammation, and chronic pain. These pseudo fruits have also been used traditionally for the prevention and treatment of rheumatoid arthritis. Moreover, they are useful for the treatment of skin disorders and ulcers. *R. canina* fruits are popularly used in kidney and urinary system diseases, in reducing kidney stones, and in the treatment of edema, rheumatism, and gout. The fruits, boiled in water, can be used as a diuretic as well as a remedy for the common cold (Bisset, 1994; Coşkun et al., 1997; Commission E Monographs, 1998; PDR for Herbal Medicines, 2000; Ameye & Chee, 2006; Czyzowska et al., 2015; Ahmad et al., 2016; Winther et al., 2016; Mármol et al., 2017). It is used in Turkey against diabetes, colds, liver diseases, and hemorrhoids (Başer et al., 1986; Baytop, 1999; Fujita et al., 1995; Orhan et al., 2007). And also, there is a record of its external use in the treatment of hemorrhoids among the people in the Western Black Sea Region (Fujita et al., 1995).

BIOLOGICAL ACTIVITY

Various bioactivity studies on several parts of *Rosa canina* [rose hip (pseudo fruit), hypanthium (floral cup or shells) or seed (achenes) flowers, leaves, etc.] have demonstrated that it has antioxidant, antiviral, antimicrobial (antifungal, antibacterial), anti-inflammatory, antiulcer, hypoglycemic, hypolipidemic, antimutagenic, anticancerogenic, pain-relieving properties, etc. Especially the antioxidant, anti-inflammatory, immunomodulatory effects, and also antiobesity and antidiabetic activities of the rose hip have been well documented in numerous studies (Özcan, 2000; Choi et al., 2002; Campanella et al., 2003; Kumarasamy et al., 2003; Ameye & Chee, 2006; Chrubasik et al., 2008; Güvenç, 2011; Lattanzio et al., 2011; Kızılet et al., 2013; Fan et al., 2014; İlyasoğlu, 2014; Czyzowska et al., 2015; Jiménez et al., 2016; Winther et al., 2016; Jemaa et al., 2017; Mármol et al., 2017; Pehlivan et al., 2018; Selahvarzian et al., 2018; Gulbagca et al., 2019; Gruenwald et al., 2019; Paunović et al., 2019; Kilinc et al., 2020).

IN VITRO

Antioxidant Activities

Rose hip is a good source of phytonutrients, including vitamin C and lycopene (Daels-Rakotoarison et al., 2002; Fan et al., 2014). The antioxidant properties of these fruits are attributed to ascorbic acid. It was determined that the ethyl acetate extract did not have any effect on the polymorphonuclear neutrophil mechanism. It has been stated that the antioxidant activity will depend not only on the ascorbic acid contained in the fruits but also on the polyphenolic compounds in their composition (Daels-Rakotoarison et al., 2002). The antioxidant capacity of *R. canina* pseudo fruits was investigated by catalase, glutathione peroxidase, glutathione-S-transferase, and lipid peroxidation malondialdehyde methods. As a result of the study, the lipid peroxidation activity of the pseudo fruits was found to be high (Özmen et al., 2005). The polyphenol content and antioxidant activity of the aqueous extracts of 23 plants used in the treatment of various diseases among the people in Bulgaria were investigated with ABTS (2,2'-azinobis (3-ethylbenzothiazoline-6-sulfonic acid). The phenolic content and antioxidant capacity of *Rosa canina* were much higher than that of the other medicinal plants (Kiselova et al., 2006). Rose hip tea prepared from dried fruits of *R. canina* was separated by solid phase extraction (SPE) into vitamin C, flavonoids, and phenolic acids. This study screened these different fractions for antioxidant activity against stable 2,2-diphenyl-1-picrylhydrazyl radicals (DPPH•) using electron spin resonance spectroscopy. The data of this study confirm that vitamin C and flavonoids are responsible for the antioxidant activity of rose hip tea (Tumbas et al., 2012). The antioxidant effects of purees and jams prepared according to traditional Serbian recipes were investigated. These extracts were evaluated by DPPH, nitric oxide, superoxide anion, and hydroxyl radical, as well as the reducing power (FRAP assay) and inhibition of lipid peroxidation. The high antioxidant potential of *R. canina* in extracts was determined through several assays (Nadpal et al., 2016). Eight different tinctures were prepared from raw *R. canina* rose hip from Malopolska (Poland). And some parameters such as total phenolic content, phenylpropanoids, flavonols, anthocyanins, L-ascorbic acid, dehydroascorbic acid contents of these tinctures, antioxidant activity against ABTS radical, and ferric reducing antioxidant power were determined. A higher amount of total polyphenols containing vitamin C, including (+)-catechin, chlorogenic, and salicylic acids, were detected in tinctures prepared from seedless pseudo fruits; this tincture was also found to have higher antioxidant activity (Tabaszewska & Najgebauer-Lejko, 2020).

Antiviral and Antimicrobial Activities

Antiviral efficacies of 75 different plants used in the treatment of viral agents among the people in Morocco were screened. It was found that methanolic extracts of *R. canina* pseudo fruits in the screened plants were effective against *Sinbis virus* strains (Mouhajir et al., 2001).

It has been determined that the extracts obtained from *R. canina*, which are known to contain flavonoids and prepared with three different solvents – water, ethanol, and petroleum ether, show antifungal activity against *Candida albicans* (Trovato et al., 2000). Antibacterial (antibacterial and antifungal) properties of silver nanoparticles synthesized by *R. canina* pseudo fruits were performed by using microdilution methods in 96 wells of the microplate. *Bacillus cereus*, *Enterococcus hirae* (ATCC 10541), *Escherichia coli* (ATCC 10536), *Staphylococcus aureus* (ATCC 6538), *Legionella pneumophila* subsp. *pneumophila* (ATCC 33152), *Pseudomonas aeruginosa* (ATCC 9027), and *Candida albicans* were used as test microorganisms. According to this study's results, silver nanoparticles with the help of *R. canina* pseudo fruits have been showing the most effective antimicrobial activity against *L. pneumophila* subsp. *pneumophila* (Gulbagca et al., 2019).

Ethanolic and methanolic extracts of *Rosa canina* flowers were screened by agar-well diffusion method against two Gram-negative (*Escherichia coli* CCM 3988 and *Pseudomonas aeruginosa*

CCM 1960) bacteria and three microscopic filamentous fungal strains (*Aspergillus niger, Fusarium culmorum*, and *Alternaria alterna*). The best antimicrobial effect of *R. canina* flowers was found against *P. aeruginosa* (ethanolic extract) and *E. coli* (methanolic extract) than against microscopic fungi (Rovná et al., 2015).

It was determined that the methanolic extract prepared from *R. canina* seeds (achenes) showed strong inhibition, especially against *Escherichia coli*. Later in the study, kaempferol 3-*O*-(6''-*O*-*E*-*p*-coumaroyl)-β-D-glucopyranoside and kaempferol 3-*O*-(6''-*O*-*Z*-*p*-coumaroyl)-β-D-glucopyranoside were isolated from this extract. The antibacterial effects of these two compounds were investigated. The mixture of these two compounds showed strong inhibition against *Lactobacillus plantarum, Proteus mirabilis*, and *Staphylococcus epidermidis* at low doses (Kumarasamy et al., 2002, 2003).

Chemical constituents, as well as the antimicrobial and antibiofilm potential of *R. canina* methanolic leaf extract, were investigated. According to HPLC analysis, the leaves contained phenolic compounds quercetin and isorhamnetin derivatives (isoquercetin and isorhamnetin- 3-*O*-rutinoside). The antibacterial and antifungal activities were tested on the Gram-positive [*Staphylococcus aureus* (ATCC 6538), *Micrococcus flavus* (ATCC 10240), and *Listeria monocytogenes* (NCTC 7973)], the Gram-negative [*Pseudomonas aeruginosa* (ATCC 27853), *Salmonella typhimurium* (ATCC 13311), *Escherichia coli* (ATCC 35210), and *Enterobacter cloacae* (human isolate)], and six *Candida* strains [*Candida albicans* (ATCC 10231), two clinical isolates of *C. albicans, C. tropicalis* (ATCC 750), *C. glabrata* (clinical isolate), and *C. krusei* (clinical isolate)]. Among the tested bacteria, *P. aeruginosa* and *S. typhimurium* were found to be most sensitive to the antimicrobial activity of *R. canina* leaf methanol extract with both MIC and MBC values (Živković et al., 2015).

Anti-inflammatory Activities

Studies have shown that rose hip inhibits chemotaxis of peripheral blood polymorphonuclear leukocytes and monocytes in vitro (Kharazmi & Winther, 1999; Warholm et al., 2003; Winther et al., 1999). Considering these studies, an anti-inflammatory compound with an inhibitory effect on the chemotaxis of human peripheral blood neutrophils was isolated from *R. canina* pseudo fruits by the biological activity-directed fractionation method in vitro. This compound is (2*S*)-1,2-di-*O*-[(9Z,12Z,15Z)-octadeca-9,12,15-trienoyl]-3-*O*-β-D-galactopyranosyl glycerol, which is a galactolipid (Larsen et al., 2003).

Antidiabetic Activities

The α -amylase inhibitory assay of methanol extract of *Rosa canina* pseudo-fruits was determined by measuring diameter inhibition on Petri plates. The α -amylase inhibition assays of methanol extract of *R. canina* showed that it provided complete inhibition of α -amylase enzyme without starch hydrolysis. According to these results, it has been suggested that habitual drinking of *R. canina* infusion may be beneficial for diabetes control (Ben Jemaa et al., 2017).

An *in vitro* study was conducted to evaluate the action mechanism of water extract of the ripe pseudo fruits of *R. canina* in diabetes mellitus. The protective activity of the water extract of the rose hip on streptozotocin-induced death in bTC6 cells was investigated. In addition, the effect of *R. canina* on glucose metabolism in HepG2, a hepatocellular carcinoma cell line, was evaluated. A novel mechanism has been proposed for the observed antidiabetic effect of *R. canina* water extract that could act as a growth factor for the pancreatic b-cell line (Fattahi et al., 2017).

Activities on DNA

DNA separation activities of silver nanoparticles synthesized from *Rosa canina* pseudo fruits were investigated using agarose gel electrophoresis. The electrophoresis study clearly revealed that Rc-Ag

NPs act on plasmid DNA molecules. Synthesized Rc-Ag NPs were found to be highly effective in their DNA cleavage activities (Gulbagca et al., 2019).

Antiproliferative Activities

Natural antioxidants from dried pseudo fruits of *R. canina*, and rose hip tea, were separated by solid phase extraction into vitamin C, flavonoids, and phenolic acids. These three fractions were then screened for their antiproliferative activity on the growth of three human tumor cell lines, HeLa (epitheloid cervix carcinoma), MCF7 (breast adenocarcinoma, estrogen receptor-positive), and HT-29 (colon adenocarcinoma). The results of this study demonstrated that polyphenols of rose hip tea are responsible for antiproliferative activity (Tumbas et al., 2012). The effects of total extract, vitamin C, neutral polyphenols, and acidic polyphenols from rose hips on human colon cancer (Caco-2) cell lines were investigated. These four different extracts have been shown that high cytotoxicity after 72 h, both low and high concentrations. Vitamin C and polyphenols obtained from rose hips have been demonstrated antiproliferative effects against cancer cells (Caco-2). These effects are provided by vitamin C, polyphenols, and flavonoids, both together and separately (Jiménez et al., 2016). Both the phenolic content and cytotoxic effects on human lung (A549) and prostate (PC-3) cancer cells of the dimethyl sulfoxide (DMSO) extract of the ripe pseudo fruits of *R. canina* were determined and possible mechanisms were investigated. It was determined that the main phenolic compounds in the DMSO extract were ascorbic acid and p-coumaric acid. *R. canina* extract exhibited a selective cytotoxic effect on A549 and PC-3 cells compared to normal fibroblast cells. This extract induced cell cycle arrest at the G1 phase and apoptosis via reduced mitochondrial membrane potential and increased caspase activity in these cells (Kilinc et al., 2020).

Anticancer Activities

The antimutagenic effect of a group of plants used in the treatment of cancer by the Turkish people was screened with the Ames test, and the antimutagenic effect of rose hip was found to be significantly higher. It has been stated that this effect may be due to the flavonoids, tannins, and proanthocyanins carried by the plant (Karakaya & Kavas, 1999).

The cytotoxic effect of the dimethyl sulfoxide extract obtained from *R. canina* pseudo fruit powder was determined using an MTT assay. In this study, human colon adenocarcinoma (WiDr, ATCC-CCL-218) cancer and human colon normal epithelial cell lines (CCD841CoN, ATCC- CRL-1790) were used in the study. Dimethyl sulfoxide extract of *R. canina* has been determined to exert a selective cytotoxic effect on WiDr cells compared to normal colon cells (Turan et al., 2018).

IN VIVO

Anti-inflammatory Activities

In mice, the anti-inflammatory and antinociceptive effects of aqueous and ethanolic extracts prepared from *R. canina* pseudo fruits were investigated in various *in vivo* models. It has been determined that ethanol extract has a significant inhibitory effect against inflammation models. According to this result, the ethanol extract was fractionated with solvents of different polarities (hexane, chloroform, ethyl acetate, *n*-butanol, and the remaining water fractions). From these fractions obtained by bioassay-guided fractionation, it was determined that ethyl acetate and *n*-butanol fractions showed anti-inflammatory and antinociceptive effects without causing acute toxicity (Deliorman Orhan et al., 2007).

The anti-inflammatory effects of a hydroalcoholic crude extract of crushed *Rosa canina* pseudo fruits were tested in male Sprague-Dawley rats weighing 180–200 g on the carrageenin-induced rat paw edema assay. Data showed that the *Rosa canina* fruit extract inhibits the development of carrageenin-induced edema; the anti-inflammatory power is similar to indomethacin (Lattanzio et al., 2011).

Anti-ulcerogenic Activities

Five plants, including *Rosa canina* fresh fruits, used in Turkey against gastric diseases, were examined in terms of these activities in an ethanol-induced ulcer model created in rats. As a result of the histopathological examination of the stomachs of the animals (Sprague-Dawley rats), it was observed that the aqueous extract prepared from the fresh fruits of rose hip had as strong an effect as misoprostol used as a positive control (Gürbüz et al., 2003). The gastroprotective effects of a hydroalcoholic crude extract of crushed *R. canina* fruits were investigated in male Sprague-Dawley rats weighing 180–200 g on the ethanol-induced gastric damage model. The data in this study were found to be ineffective on the administered dose-dependent ulcerogenic damage that might be in the therapeutic regimen against ethanol-induced ulcerogenic damage to *R. canina*, although their quantitative statistical evaluation was not significant (Lattanzio et al., 2011).

Antidiabetic Activities

Diabetic rats were treated by oral gavage for eight weeks with crude extract (40% w/v) and purified oligosaccharide (2 mg/kg) from ripe fruits of *Rosa canina*. Post-treatment body weight, fasting blood glucose, serum insulin levels, and islet beta cell repair and proliferation were investigated. In the animal model of diabetes, insulin levels were observed to increase significantly due to the regeneration of beta cells in the isles of Langerhans by the purified oligosaccharide (Bahrami et al., 2020a). In another study, the oligosaccharide fraction isolated from *R. canina* revealed that DNA methylation played a causal role in the efficacy of isolated oligosaccharides. Moreover, it was concluded that the possible regenerative potential of the oligosaccharide in diabetic rats may have contributed to the modulation of DNA methylation (Bahrami et al., 2020b). The oligosaccharide isolates from *Rosa canina* pseudo fruits were characterized, and the antidiabetic mechanisms of the main component were evaluated. The structure of the oligosaccharide was identified by using a high-performance liquid chromatography diode array detector tandem mass spectroscopy, infra-red radiation, and nuclear magnetic resonance antidiabetic mechanisms of action of oligosaccharide was investigated in STZ-induced diabetic rats with oral glucose tolerance. Gluconeogenesis and α -glucosidase activity were inhibited by this oligosaccharide in diabetic rats. Significant improvement was observed by the oral glucose tolerance test in the oligosaccharide-treated group. Pathological examination of pancreatic β-cells and tissues revealed a considerable improvement post-treatment period (Rahimi et al., 2020).

Antiobesity Activities

Obesity is a health problem that has increased in recent years and is widespread all over the world. Obesity is also recognized as an important risk factor for type 2 diabetes. It is also thought to be the trigger of metabolic diseases such as hyperlipidemia, hypertension, and arteriosclerosis. An 80% aqueous acetone extract prepared from *R. canina* seeds (achenes) was administered to mice for two weeks. The extract significantly suppressed visceral fat weight and produced a significant reduction in plasma triglyceride and free fatty acid values on the 14th day of administration without any toxic side effects. In the studies carried out to determine the active compounds, *trans*-tiliroside, *cis*-tiliroside, buddlenoids A and B, dihydrodehydrodiconiferyl alcohol, and urolignosid were obtained from the *n*-butanol fraction prepared from the seed extract by column chromatography and HPLC. *trans*-Tiliroside, identified as the main component, strongly inhibited the gain of body weight, especially visceral fat weight, and increased plasma glucose levels after glucose loading in mice. A single oral administration of *trans*-tiliroside at a dose of 10 mg/kg increased the expression of peroxisome proliferator-activated receptor α mRNA in liver tissue in mice. The effect of *trans*-tiliroside's anti-obese compound was found to be stronger than orlistat (Ninomiya et al., 2007).

Effects on the Urinary System

The effects of the infusion prepared from *R. canina* pseudo fruit and magnesium-supplemented *R. canina* on the urinary risk factor caused by calcium oxalate stones were investigated in female Wistar rats. As a result of the study, it was observed that the rose hip herbal infusion did not cause any increase in urine (diuretic effect). Urine pH decreased when only magnesium chloride was applied; however, this effect was not observed in herbal infusions prepared with magnesium. A significant decrease in citraturia was observed when magnesium was combined with the infusion prepared from *R. canina* pseudo fruit (Grases et al., 1992).

Activities on DNA

In a study examining the *in vivo* protective effects of *Rosa canina* against doxorubicin-induced testicular toxicity in mice, water extracts of *R. canina* dried fruit pulp appeared to have protective effects against doxorubicin-induced testicular toxicity. Male NMRI mice randomly assigned to six treatment groups were used in the study. Mice were given 100 and 200 mg/kg *R. canina* pulp water extract by the intraperitoneal route for 28 days. This study's results indicate that *R. canina* extract is able to improve testicular and sperm parameters in mice treated with doxorubicin (Nowrouzi et al., 2019).

Hepatoprotective Activities

The hepatoprotective activity of hydro-ethanolic (50/50, v/v) pseudo fruit extract of *R. canina* against carbon tetrachloride (CCl4)-induced hepatotoxicity in male albino Wistar rats was investigated. Rats were randomly divided into six groups of eight animals each, and animals were administered *R. canina* (p.o., daily) and CCl4 (1 mL/kg twice a week, 50% v/v in olive oil, i.p.) for six weeks. Biochemical analyzes were performed to test the levels of aspartate aminotransferase (AST), alanine amino transaminase (ALT), alkaline phosphatase (ALP), albumin (ALB), total protein (TP), and malondialdehyde (MDA). These biochemical analysis results were also supplemented with a histopathological examination of the liver section. It was determined that treatment with hydro-alcoholic pseudo fruit extract of *R. canina* at doses of 500 and 750 mg/kg significantly reduced CCl4-high ALT, AST, ALP, and MDA levels ($p<0.01$), which was also supported by data obtained from histopathological studies. The results of this study showed that the hydro-alcoholic fruit extract of *R. canina* has a hepatoprotective effect on CCl4-induced liver injury in rats (Sadeghi et al., 2016).

CLINICAL TRIALS

There are many clinical trials done with rose hip and seed (Winther et al., 2005; Chrubasik et al., 2006; Chrubasik et al., 2008; Christensen et al., 2013; Mármol et al., 2017; Selahvarzian et al., 2018; Gruenwald et al., 2019).

Antiobesity Activities

A randomized, double-blind, single-center, placebo-controlled clinical trial was conducted examining the effect of rose hip extract (aqueous ethanol extract of rose hip containing its seeds, dextrin, and cyclodextrin) on human body fat in preobese subjects. Thirty-two subjects participated in the study and were divided into two equal groups to avoid bias regarding gender, body mass index, and body fat percentage. All subjects received one rose hip (n= 16) or placebo tablet (n= 16) once a day for 12 weeks by chewing. This clinical study showed that a daily intake of 100 mg/kg of rose hip extract was able to reduce abdominal visceral fat and significantly reduced both body weight and body mass index in pre-obese subjects. No abnormality, subjective symptoms, or signs that could indicate clinical problems were encountered during the study period (Nagatomo et al., 2015).

Anti-inflammatory Activities

Hyben Vital®, a standardized preparation in capsule form prepared from *R. canina* fruits, was studied on eight male volunteers for its anti-inflammatory effect. Four of the volunteers participating in the study had no muscle and joint complaints throughout their lives, and the other four were selected from people with long-term osteoarthritis complaints. Both groups of volunteers were given Hyben Vital® at a dose of 45 g/day for four weeks. At the end of the period, the effect of the capsules was examined in the blood samples taken from the volunteers. In the study, there was a significant decrease in blood serum keratin and C-reactive protein levels, but no impairment in liver functions was observed. As a result of the study, it was observed that the seed extract had more inhibition than the fruit extract. These results showed that the inhibition was not caused by the vitamin C carried by the fruits, and the compounds carried by the seeds were effective (Winther et al., 1999; Ameye & Chee, 2006).

Osteoarthritis, Rheumatoid Arthritis, and Back Pain Activities

In a double-blind, placebo-controlled, and parallel study, 18 years and over 89 patients with rheumatoid arthritis, according to the revised American Rheumatism Association criteria for the classification of rheumatoid arthritis, 5 g of rose hip powder capsules per day or the same amount of placebo applied for six months. The quality of physical life score was significantly improved in the rose hip group compared to placebo. In addition, the relieving effect of rheumatic pains of 5 g rose hip capsules was not different from the other groups taking pain medication (Willich et al., 2010).

One hundred patients (65 women, 53 men) with osteoarthritis of the hip and knee joints were divided into two groups: the first group (50 patients) received 5 g standardized rose hip powder (capsule) twice daily for four months, and the second group (50 patients) a placebo capsule for four months given. In the study, mobility in the knee and hip joints, improvement in quality of life, and reduction in osteoarthritis pain were followed. As a result of the study, it was noted that 64.6% of patients treated with standardized rose hip powder had a reduction in complaints (Warholm et al., 2003).

The effect of Hyben Vital® prepared from *R. canina* fruits on osteoarthritis symptoms was evaluated in 112 patients. The data obtained as a result of the study showed that Hyben Vital® reduces the symptoms of osteoarthritis. In addition, it has been determined that gastrointestinal side effects are less when compared to nonsteroidal anti-inflammatory drugs used in the treatment of osteoarthritis (Rein et al., 2004).

The effect of rose hip peels and achenes on 94 volunteers with osteoarthritis was studied in a double-blind, placebo-controlled, and randomized study. The rose hip preparation was given to 47 patients at a dose of 5 g/day for three months. The other patient group was administered a placebo in the same way. When the rose hip treatment was compared with the placebo treatment, a reduction in pain was observed in the patients. The data obtained showed that the pseudo fruits of *R. canina* can be recommended for alleviating the symptoms of osteoarthritis (Winther et al., 2005).

Skin Disorders and Antiaging Activities

A standardized rose hip powder (Hyben Vital®) made from seeds and shells on cell senescence, skin wrinkling, and aging activities was carried out in a randomized, double-blind controlled clinical trial on 34 healthy middle-aged male and female volunteers. The effect of rose hip powder on skin wrinkles and red blood cells longevity was studied. The effect of the rose hip preparation on cell longevity was measured in terms of the hemolytic index in blood samples kept in a blood bank for five weeks. After eight weeks of treatment with rose hip, statistically significant improvements in crow's feet, wrinkles, skin moisture, and elasticity were observed (Phetcharat et al., 2015).

Antihemorrhoidal Activities

The antioxidant capacity of a formulation containing *R. canina* fruits was tested by the photochemiluminescence method. In addition, its efficacy and safety were tested in double-blind and placebo-controlled clinical studies in 96 patients. The results showed that this formulation containing *Rosa canina* (70% titered in ascorbic acid) was effective in relieving the symptoms of hemorrhoids (Vertuani et al., 2004).

TOXICOLOGY STUDIES

Toxicological experiments were performed with aqueous and ethanol extracts of rose hip achenes in rodents, frogs, and mice. A decoction equivalent to 0.25–0.75 g crude rose hip or 0.5–1.5 mL of a decoction of rose hip seed was also well tolerated, as were subcutaneous administration of rose hip aqueous and ethanol extracts in rodents. For the rose hip ethanol fluid extract, a minimum lethal dose could not be determined in frogs. In mice, the dose was 0.9–1.1 mL, and in rats, 0.7–0.9 mL. For rose hip seed, a minimum lethal dose could not be determined in any of the species (Chrubasik et al., 2008)

The general toxicity of kaempferol 3-*O*-(6″-*O*-*E*-*p*-coumaroyl)-*β*-D-glucopyranoside and kaempferol 3-*O*-(6''-*O*-*Z*-*p*-coumaroyl)-*β*-D-glucopyranoside isolated from methanol extract of *R. canina* seeds was investigated with Brine shrimp (*Artemia salina*) lethality assay. In this toxicity test, the toxicity of the isomeric mixture (2:1) of compounds was found to be compared to (0.017 mg/mL) podophyllotoxin (0.0029 mg/mL; positive control) as a known cytotoxic agent (Kumarasamy et al., 2002; Kumarasamy et al., 2003).

The genotoxicity effect of ethanol extract of *R. canina* pseudo fruits was evaluated by the somatic mutation and recombination genotoxicity test of ethyl methanesulfonate on *Drosophila melanogaster*. According to the results of this study, it was determined that the statistically significant genotoxic effects of ethyl methanesulfonate ($P<0.05$) were abolished by the ethanol extract of *R. canina* (Kızılet et al., 2013).

Acute Toxicity

Animals (male Swiss albino mice) used in the Carrageenan-induced paw edema experiment were observed for 48 hours. No significant acute toxicity was observed in the models in which ethanol extract and its active fractions (ethylacetate, *n*-butanol) were administered during the 48-hour observation period (Deliorman Orhan et al., 2007).

CHEMISTRY

It was determined that the infusion prepared from the pseudo fruits of *Rosa canina* contains high levels of Ca, Mg, Fe, and Ag together with Cu, K, Mn, Na, P, S, Se, Tl, and Zn (Demir & Özcan, 2001; Başgel & Erdemoğlu, 2005; Ercişli, 2007; Özcan et al., 2008; Şekeroğlu et al., 2008; Kazaz et al., 2009; Fan et al., 2014, Paunović et al., 2019). It has been reported that the carbohydrate composition of *R. canina* pseudofruit consists of mono- and polysaccharides, and the pectin content of the fruits is high (Bisset, 1994; Khodzhaeva et al., 1998; PDR for Herbal Medicines, 2000). The lipid fraction of pseudofruit consists of neutral lipids (7.3%), glycolipids (83.6%), and phospholipids

(9.1%). It has been shown that fruits contain 1.78% fixed oil, and the main components of this oil are lauric acid, palmitic acid, α -linolenic acid, and linoleic acid. (2*S*)-1,2-di-O-[(9Z,12Z,15Z)-octadeca-9,12,15-trienoyl]-3-*O*-β-D-galactopyranosyl glycerol were isolated from the dried fruits of *R. canina* (Ul'chenko et al., 1995; Nowak, 2005a; Nowak, 2005b; Ercişli, 2007; Kazaz et al., 2009; Fan et al., 2014; Ahmad et al., 2016; Mármol et al., 2017). The achene (seeds) of the plant have been found to have an oil rich in unsaturated fatty acids (~94%), and the achene oil carries palmitic, stearic, oleic, linoleic, linolenic, and arachidic acids (Özcan, 2002; Szenthmihalyi et al., 2002; Deineka, 2003; İlyasoğlu, 2014; Ahmad et al., 2016; Winther et al., 2016). Pseudo fruits contain 0.3% essential oil, the main component of which is β-ionone (PDR for Herbal Medicines, 2000; Nowak, 2005a; Nowak, 2005b; Selahvarzian et al., 2018). *R. canina* hypanthium is rich in carotenoid pigments. The presence of xanthophyll (neoxanthin, *trans*-violaxanthin, *cis*-violaxanthin, 5,6-epoxylutein, lutein, rubixanthin, zeaxanthin, and β-cryptoxanthin) and carotene (lycopene, β-carotene, ε -carotene) was determined by studies performed with TLC and HPLC. It has been emphasized that rose hip fruits are a good source of lycopene (12.9–35.2 mg/100 g) (Razungles et al., 1989; Hodisan et al., 1997; Böhm et al., 2003; Fan et al., 2014). The presence of phenolic compounds has been determined in studies on fruits. These compounds are anthocyanosides (cyanidin-3-*O*-glucoside); flavonoids [quercetin, (+)-catechin, (-)-epicatechin, taxifolin, eriodicthiol, rutin, quercitrin]; tannin (methyl gallate); chalcone (fluridzin); and lignans (matairezinol, secoisolariciresinol) (PDR for Herbal Medicines, 2000; Hvattum, 2002; Valsta et al., 2003; Ercişli, 2007; Czyzowska et al., 2015). In dog rose pseudo fruits found flavonol glycosides such as quercetin galactoside, glucuronide, and arabinofuranoside, as well as kaempferol rutinoside and coumaroylglucoside. In seeds chaemferol 3-*O*-(6''-*O*-*E*-*p*-kumaroyl)-β-D-glucopyranoside and chemferol 3-O-(6''-*O*-*Z*-*p*-kumaroyl)-β-D-glucopyranoside were determined. It has been found that the pseudo fruits of the plant carry various organic acids and the main of these organic acids are gallic, ellagic, ascorbic, malic, and citric acids. In studies on the fruits of different species of the genus *Rosa*, it has been shown that they are very rich in terms of ascorbic acid (vitamin C), and the pseudo fruits of *R. canina* contain 0.2–2.4% ascorbic acid (PDR for Herbal Medicines, 2000; Bisset, 1994; Coşkun et al., 1997; Kurucu et al., 1997; Bozan et al., 1998; Baytop, 1999; Kumarasamy et al., 2003; Erentürk et al., 2005; Nowak, 2006a, 2006b; Ercişli, 2007; Czyzowska et al., 2015).

CONCLUSION

Rosa canina L. (Rose hip, Dog rose, Kuşburnu, Gül burnu) grows naturally in the northern hemisphere in Europe, northern Africa, temperate Eurasia, Western Asia, and North and South America. It is widely found in Türkiye. Rosae pseudo fructus (rose hip fruits) contain many achene-type fruits in the receptaculum. Rose hip has been used as food, tea and medicine for centuries. *In vitro, in vivo* and clinical studies on *R. canina* have shown that it has positive effects on health. Various inorganic and organic phytochemical compounds have been identified from *R. canina* to date. Rose hip, which is widely used in Türkiye and has products in the form of jam and herbal tea, can be evaluated more comprehensively in the field of health in line with the scientific data obtained. The fact that there are products used in the treatment of osteoarthritis and antiaging activities in different countries can set an example for Türkiye. It can be recommended for the pharmaceutical industry in Türkiye to design and manufacture new pharmaceutical products containing rose hip.

REFERENCES

Ahmad, N., Anwar, F., Gilani, A.H. 2016. Rose hip (Rosa canina L.) oils. Chapter – 76. In *Essential Oils in Food Preservation, Flavor and Safety*. http://dx.doi.org/10.1016/B978-0-12-416641-7.00076-6.

Ameye, L.G., Chee, W.S.S. 2006. Osteoarthritis and nutrition. From nutraceuticals to functional foods: A systematic review of the scientific evidence. *Arthritis Res. & Ther.*, 8, R127–149.

Bahrami, G., Miraghaee, S.S., Mohammadi, B., Bahrami, M.T., Taheripak, G., Keshavarzi, S., Babaei, A., Sajadimajd, S., Hatami, R. 2020a. Molecular mechanism of the anti-diabetic activity of an identified oligosaccharide from *Rosa canina*. *Research in Pharmaceutical Sciences*, 15(1), 36–47.

Bahrami, G., Sajadimajd, S., Mohammadi, B., Hatami, R., Miraghaee, S., Keshavarzi, S., Khazaei, M., Madani, S.H. 2020b. Anti-diabetic effect of a novel oligosaccharide isolated from *Rosa canina* via modulation of DNA methylation in Streptozotocin-diabetic rats. *DARU Journal of Pharmaceutical Sciences*, 28, 581–590.

Başer, K.H.C., Honda, G., Miki, W. 1986. *Herb Drugs and Herbalists in Turkey*. Tokyo: Institute for the Study of Languages and Cultures of Asia and Africa.

Başgel, S., Erdemoğlu, S.B. 2005. Determination of mineral and trace elements in some medicinal herbs and their infusions consumed in Turkey. *Sci. Total Envir.*, 359, 82–89.

Baytop, T. 1999. *Türkiye'de Bitkiler ile Tedavi (Geçmişte ve Bugün). İlaveli 2*. İstanbul: Baskı, Nobel Tıp Kitabevleri.

Bisset, N.G. 1994. *Max Wictl's Herbal Drugs and Phytopharmaceuticals*. Boca Raton FL: CRC Press.

Bozan, B., Sagdullaev, B.T., Kozar, M., Aripov, KhN., Başer, K.H.C. 1998. Comparison of ascorbic and citric acid contents in *Rosa canina* L. fruit growing in the central Asia region. *Chem. Nat. Comp.*, 34, 687–689.

Böhm, V., Fröhlich, K., Bitsch, R. 2003. Rosehip-a "new" source of lycopene? *Mol. Aspects Med.*, 24, 385–389.

Campanella, L., Bonanni, A., Tomassetti, M. 2003. Determination of the antioxidant capacity of samples of different types of tea, or of beverages based on tea or other herbal products, using a superoxide dismutase biosensor. *J. Pharm. Biomed. Anal.*, 32, 725–736.

Choi, H.R., Choi, J.S., Han, Y.N., Bae, S.J., Chung, H.Y. 2002. Peoxynitrite scavenging activity of herb. *Phytother. Res.*, 16, 364–367.

Christensen, R., Bartels, E.M., Bliddal, H. 2013. Superiority trials in osteoarthritis using glucosamine hydrochloride as comparator: Overview of reviews and indirect comparison with *Rosa canina* (a novel nutraceutical). *OA Arthritis*, 1(1), 1–5.

Chrubasik, C., Duke, R.K., Chrubasik, S. 2006. The evidence for clinical efficacy of rose hip and seed: A systematic review. *Phytother. Res.*, 20, 1–3. https://doi.org/10.13172/2052-9554-1-1-305.

Chrubasik, C., Roufogalis, B.D., Müller-Ladner, U., Chrubasik, S. 2008. A systematic review on the *Rosa canina* effect and efficacy profiles. *Phytother. Res.*, 22, 725–733. https://doi.org/10.1002/ptr.2400.

Commission E Monographs: The Complete German Commission E Monographs: Therapeutic Guide to Herbal Medicines. 1998. (eds. Blumenthal, M., Busse, W.R.), 1st ed. Austin, TX: American Botanical Council, Lippincott Williams & Wilkins.

Coşkun, M., Kurucu, S., Kartal, M. 1997. HPLC analysis of ascorbic acid in the fruits of some Turkish *Rosa* species. *Sci. Pharm.*, 65, 169–174.

Czyzowska, A., Klewicka, E., Pogorzelski, E., Nowak, A. 2015. Polyphenols, vitamin C and antioxidant activity in wines from *Rosa canina* L. and *Rosa rugosa* Thunb. *J. Food. Compost. Anal.*, 39, 62–68.

Daels-Rakotoarison, D.A., Gressier, B., Trotin, F., Brunet, C., Luyckx, M., Dine, T., Bailleul, F., Cazin, M., Cazin, J-C. 2002. Effects of *Rosa canina* fruit extract on neutrophil respiratory burst. *Phytother. Res.*, 16, 157–161.

Deineka, V.I. 2003. Triglyceride composition of seed oils from certain plants. *Chem. Nat. Comp.*, 39, 523–527.

Deliorman Orhan, D., Hartevioğlu, A., Küpeli, E., Yeşilada, E. 2007. *In vivo* anti-inflammatory and antinociceptive activity of the crude extract and fractions from *Rosa canina* L. fruits. *J. Ethnopharmacol.*, 112, 394–400.

Demir, F., Özcan, M. 2001. Chemical and technological properties of Rose (*Rosa canina* L.) fruits wild in Turkey. *J. Food Eng.*, 47, 333–336.

European Pharmacopoeia. 2010. 11th Edition (Books (11.0 - 11.1- 11.2), ISBN: 978-92-871-9105-2, Language: English) Council of Europe: Strabourg.

Ercişli, S. 2007. Chemical composition of fruits in some rose (*Rosa* ssp.) species. *Food Chem.*, 104, 1379–1384.

Erentürk, S., Gulaboğlu, M.Ş., Gültekin, S. 2005. The effects of cutting and drying medium on the vitamin C content of rosehip during drying. *J. Food Eng.*, 68, 513–518.

Fan, C., Pacier, C., Martirosyan, D.M. 2014. Rose hip (*Rosa canina* L): A functional food perspective. *Functional Foods in Health and Disease*, 4(11), 493–509.

Fattahi, A., Niyazi, F., Shahbazi, B., Farzaei, M.H., Bahrami, G. 2017. Antidiabetic mechanisms of *Rosa canina* fruits: An in vitro evaluation. *Journal of Evidence-Based Complementary & Alternative Medicine*, 22(1), 127–133.

Fujita, T., Sezik, E., Tabata, M., Yeşilada, E., Honda, G., Takeda, Y., Tanaka, T., Takaishi, Y. 1995. Traditional medicine in Turkey VII. Folk medicine in middle and west black sea regions. *Econ. Bot.*, 49, 406–422.

Gulbagca, F., Özdemir, S., Gulcan, M., Sen, F. 2019. Synthesis and characterization of *Rosa canina*-mediated biogenic silver nanoparticles for anti-oxidant, antibacterial, antifungal, and DNA cleavage activities. *Heliyon*, 5, e02980. https://doi.org/10.1016/j.heliyon.2019.e02980.

Grases, F., Masarova, L., Costa-Bauza, A., March, J.G., Prieto, R., Tur, J.A. 1992. Effect of "*Rosa canina*" infusion and magnesium on the urinary risk factors of calcium oxalate urolithiasis. *Planta Medica*, 58, 499–512. https://doi.org.10.1055/s-2006-961537.

Gruenwald, J., Uebelhack, R., Moré, M.I. 2019. *Rosa canina* – Rose hip pharmacological ingredients and molecular mechanics counteracting osteoarthritis – A systematic review. *Phytomedicine*, 60, 152958. https://doi.org/10.1016/j.phymed.2019.152958.

Gürbüz, İ., Üstün, O., Yeşilada, E., Sezik, E., Kutsal, O. 2003. Anti-ulcerogenic activity of some plants used as folk remedy in Turkey. *J. Ethnopharmacol.*, 88, 93–97.

Güvenç, A. 2011. *Rosa canina* (Kuşburnu). In *FFD Monografları "Tedavide Kullanılan Bitkiler"*. Eds. Demirezer, Ö., Ersöz, T., Saraçoğlu, İ., Şener, B. Ankara: MN Medikal & Nobel Tıp Kitap Sarayı.

Hodisan, T., Socaciu, C., Ropan, I., Neamtu, G. 1997. Carotenoid composition of *Rosa canina* fruits determined by thin-layer chromatography and high-performance liquid chromatography. *J. Pharm. Biomed. Anal.*, 16, 521–528.

Hvattum, E. 2002. Determination of phenolic compounds in rose hip (*Rosa canina*) using liquid chromatography coupled to electrospray ionisation tandem mass spectrometry and diode-array detection. *Rapid Commun. Mass Spect.*, 16, 655–662.

İlyasoğlu, H. 2014. Characterization of rosehip (*Rosa canina* L.) seed and seed oil. *Int. J. Food Prop.*, 17(7), 1591–1598. https://doi.org/10.1080/10942912.2013.777075.

Jemaa, H.B., Jemia, A.B., Khlifi, S., Ahmed, H.B, Slama, F.B. Benzarti, A. Elati, J. Aouidet, A. 2017. Antioxidant activity and α-amylase inhibitory potential of Rosa canina L. African journal of traditional, complementary and alternative medicines. 14(2), 1–8.

Jiménez, S., Gascón, S., Luquin, A., Laguna, M., Ancin-Azpilicueta, C., Rodríguez-Yoldi, M.J. 2016. *Rosa canina* extracts have antiproliferative and antioxidant effects on Caco-2 human colon cancer. *PLoS ONE*, 11(7), e0159136. https://doi.org/10.1371/journal.pone.0159136.

Kasimoğlu, C. Uysal, H.2016. Genoprotective Effects of Aqueous Extracts of Rosa Canina L. Fruits on Ethyl Methanesulfonate-Induced DNA Damage in Drosophila Melanogaster. *Cumhuriyet Üniversitesi Fen Edebiyat Fakültesi Fen Bilimleri Dergisi*, 37(3), 241–247.

Karakaya, S., Kavas, A. 1999. Antimutagenic activities of some foods. *J. Sci. Food Agric.*, 79, 237–242.

Kazaz, S., Baydar, H., Erbaş, S. 2009. Variations in chemical compositions of *Rosa damascena* Mill. and *Rosa canina* L. fruits. *Czech J. Food Sci.*, 27, 178–184.

Kharazmi, A., Winther, K. 1999. Rose hip inhibits chemotaxis and chemiluminescence of human peripheral blood neutrophils *in vitro* and reduced certain inflammatory parameters *in vivo. Inflammopharmacol.*, 7, 377–386.

Khodzhaeva, M.A., Sagdullaev, B.T., Turakhozaev, M.T., Aripov, Kh.N. 1998. Carbohydrates of the fruit of *Rosa canina. Chem. Nat. Comp.*, 34, 736–737.

Kızılet, H., Kasımoğlu, C., Uysal, H. 2013. Can the *Rosa canina* plant be used against alkylating agents as a radical scavenger? *Pol. J. Environ. Stud.*, 22(4), 1263–1267.

Kilinc, K., Demir, S., Turan, İ., Mentese, A., Orem, A., Sonmez, M., Aliyazicioglu, Y. 2020. *Rosa canina* extract has antiproliferative and proapoptotic effects on human lung and prostate cancer cells. *Nutrition and Cancer*, 72(2), 273–282. https://doi.org/10.1080/01635581.2019.1625936.

Kiselova, Y., Ivanova, D., Chervenkov, T., Gerova, D., Galunska, B., Yankova, T. 2006. Correlation between the *in vitro* antioxidant activity and poyphenol content of aqueous extracts from Bulgarian herbs. *Phytother. Res.*, 20, 961–965.

Kumarasamy, Y., Cox, P.J., Jaspars, M., Nahar, L., Sarker, S.D. 2002. Screening seeds of Scottish plants for antibacterial activity. *J. Ethnopharmacol.*, 83, 73–77.

Kumarasamy, Y., Cox, P.J., Jaspars, M., Rashid, M.A., Sarker, S.D. 2003. Bioactive flavonoid glycosides from the seeds of *Rosa canina. Pharm. Biol.*, 41, 237–242.

Kurucu, S., Coşkun, M., Kartal, M. 1997. HPLC analysis of ascorbic acid in the fruits of *Rosa* species growing at Işık Mountain. *FABAD-J. Pharm Sci.*, 22, 9–11.

Larsen, E., Kharazmi, A., Christensen, L.P., Christensen, S.B. 2003. An antiinflammatory galactolipid from rose hip (*Rosa canina*) that inhibits chemotaxis of human peripheral blood neutrophils *in vitro. J. Nat. Prod.*, 66, 994–995.

Lattanzio, F., Greco, E., Carretta, D., Cervellati, R., Govoni, P., Speroni, E. 2011. *In vivo* anti-inflammatory effect of *Rosa canina* L. extract. *J. Ethnopharmacol.*, 137, 880–885.

Mármol, I., Sánchez-de-Diego, C., Jiménez-Moreno N., Ancín-Azpilicueta, C., Rodríguez-Yoldi, M.J. 2017. Therapeutic applications of rose hips from different *Rosa* species. *Int. J. Mol. Sci.*, 18, 1137. https://doi.org/10.3390/ijms18061137.

Mouhajir, F., Hudson, J.B., Rejdali, M., Towers, G.H.N. 2001. Multiple antiviral activities of endemic medicinal plants used by berber peoples of Morocco. *Pharm. Biol.*, 39, 364–374.

Nadpal, J.D., Lesjak, M.M., Šibul, F.S., Anackov, G.T., Cetojevic-Simin, D.D., Mimica-Dukic, N.M., Beara, I.N. 2016. Comparative study of biological activities and phytochemical composition of two rose hips and their preserves: *Rosa canina* L. and *Rosa arvensis* Huds. *Food Chemistry*, 192, 907, 914.

Nagatomo, A., Nishida, N., Fukuhara, I., Noro, A., Kozai, Y., Sato, H., Matsuura, Y. 2015. Daily intake of rosehip extract decreases abdominal visceral fat in preobese subjects: A randomized, double-blind, placebo-controlled clinical trial. *Diabetes Metab. Syndr. Obes.: Targets Ther.*, 8, 147–156.

Nilsson, Ö. 1972. *Rosa* L. In *Flora of Turkey and the East Aegean Islands*. Ed. Davis, P.H. Vol. 4. Edinburgh: University Press.

Ninomiya, K., Matsuda, H., Kubo, M., Morikawa, T., Nishida, N., Yoshikawa, M. 2007. Potent anti-obese principle from Rosa canina: Structural requirements and mode of action of trans-tiliroside. *Bioorg. Med. Chem. Lett.*, 17, 3059–3064.

Nowrouzi, F., Azadbakht, M., Kalehoei, E., Modarresi, M. 2019. Protective effect of *Rosa canina* Extract against doxorubicin-induced testicular toxicity in mice. *Braz. Arch. Biol. Technol.*, 62, e19180017. http://dx.doi.org/10.1590/1678-4324-2019180017.

Nowak, R. 2005a. Fatty acids composition in fruits of wild rose species. *Acta Soc. Bot. Pol.*, 74, 229–235.

Nowak, R. 2005b. Chemical composition of hips essential oils of some *Rosa* L. species. *Z. Naturforsch. C*, 60, 369–378.

Nowak, R. 2006a. Comparative study of phenolic acids in pseudofruits of some species of roses. *Acta Pol. Pharm.*, 63, 281–288.

Nowak, R. 2006b. Determination of ellagic acid in pseudofruits of some species of roses. *Acta Pol. Pharm.*, 63, 289–292.

Orhan, D. D., Hartevioğlu, A., Küpeli, E., & Yesilada, D. 2007. In vivo anti-inflammatory and antinociceptive activity of the crude extract and fractions from *Rosa canina* L. fruits. *Journal of Ethnopharmacology*, 112(2), 394–400.

Özcan, M. 2000. Antioxidant activity of seafennel (*Crithmum maritimum* L.) essential oil and rose (*Rosa canina*) extract on natural olive oil. *Acta Alim.*, 29, 377–384.

Özcan, M. 2002. Nutrient composition of rose (*Rosa canina* L.) seed and oils. *J. Med. Food*, 5, 137–140.

Özcan, M.M., Ünver, A., Uçar, T., Arslan, D. 2008. Mineral content of some herbs and herbal teas by infusion and decoction. *Food Chem.*, 106, 1120–1127.

Özmen, İ., Erciş̧li, S., Hızarcı, Y., Orhan, E. 2005. Investigation of antioxidant enzyme activities and lipid peroxydation of *Rosa canina* and *Rosa dumalis* fruits, *Acta Horticult.*, 690(*Proceedings of the 1st International Rose hip Conference*, 2004), 245–248.

Paunović, D., Kalušević, A., Petrović, T., Urošević, T., Djinović, D., Nedović, V., Popović-Djordjević, J. 2019. Assessment of chemical and antioxidant properties of fresh and dried rosehip (*Rosa canina* L.). *Not Bot Horti Agrobo*, 47(1), 108–113. https://doi.org/47.15835/nbha47111221.

PDR for Herbal Medicines, 2nd ed. 2000. Montvale, NJ: Thomson Medical Economics.

Pehlivan, M., Mohammed, F.S., Sevindik, M., Akgül, H. 2018. Antioxidant and oxidant potential of *Rosa canina*. *Eurasian Journal of Forest Science*, 6(4), 22–25.

Rahimi, M., Sajadimajd, S., Mahdian, Z., Hemmati, M., Malekkhatabi, P., Bahrami, G., Mohammadi, B., Miraghaee, S., Hatami, R., Mansouri, K., Motlagh, H.R.M., Keshavarzi, S., Derakhshankhah, H. 2020. Characterization and anti-diabetic effects of the oligosaccharide fraction isolated from Rosa canina in STZ-Induced diabetic rats. *Carbohydr. Res.*, 489, 107927.

Rein, E., Winther, K., Kharazmi, A. 2004. A herbal remedy, Hyben Vital (stand. Powder of a subspecies of Rosa canina fruits), reduces pain and improves general wellbeing in patients with osteoarthritis: double-blind, placebo-controlled, randomized clinical trial. *Phytomedicine*, 11, 383–391.

Phetcharat, L., Wongsuphasawat, K., Winther, K. 2015. The effectiveness of a standardized rose hip powder, containing seeds and shells of *Rosa canina*, on cell longevity, skin wrinkles, moisture, and elasticity. *Clin. Interv. Aging*, 10, 1849.

Razungles, A., Oszmianski, J., Sapis, J.C. 1989. Determination of carotenoids in fruits of *Rosa* sp. (*Rosa canina* and *Rosa rugosa*) and of chokeberry (*Aronia melanocarpa*). *J.Food Sci.*, 54, 774–775.

Rovná, K., Petrová, J., Terentjeva, M., Černá, J., Kačániová, M. 2015. Antimicrobial activity of *Rosa canina* flowers against selected microorganisms. *J Microbiol Biotech Food Sci.*, 4(special issue 1), 62–64. https://doi.org/10.15414/jmbfs.2015.4.special1.62-64.

Sadeghi, H., Hosseinzadeh, A.S., Akbartabar Touri, M., Ghavamzadeh, M., Jafari Barmak, M., Sayahi, M., Sadeghi, H. 2016. Hepatoprotective effect of *Rosa canina* fruit extract against carbon tetrachloride induced hepatotoxicity in rat. *Avicenna J Phytomed*, 6(2), 181–188.

Selahvarzian, A., Alizadeh, A., Amanolahi Baharvand, P., Eldahshan, A.O., Rasoulian, B. 2018. Medicinal Properties of *Rosa canina* L.. *Herb. Med. J.*, 3(2), 77–84. https://doi.org/10.22087/hmj.v0i0.615.

Šindrak, Z., Jemrić, T., Baričević, L., Han Dovedan, I., Fruk, G. 2012. Fruit quality of dog rose seedlings (*Rosa canina* L.). *J. Cent. Eur. Agric*, 13(2), 321–330. https://doi.org/10.5513/JCEA01/13.2.1053.

Szenthmihalyi, K., Vinkler, P., Lakatos, B., Illes, V., Then, M. 2002. Rose hip (*Rosa canina* L.) oil obtained from waste hip seeds by different extraction methods. *Biores. Technol.*, 82, 195–201.

Şekeroğlu, N., Ozkutlu, F., Kara, S.M., Özgüven, M. 2008. Determination of cadmium and selected micronutrients in commonly used and traded medicinal plants in Turkey. *J Sci Food Agric.*, 88, 86–90.

Tabaszewska, M., Najgebauer-Lejko, D. 2020. The content of selected phytochemicals and in vitro antioxidant properties of rose hip (*Rosa canina* L.) tinctures. *NFS J.*, 21, 50–56.

TF 2017. Türk Farmakopesi 2017, Genel Monograflar, IV, H-N. T.C. Sağlık Bakanlığı Türkiye İlaç ve Tıbbi Cihaz Kurumu, T.C. Sağlık Bakanlığı Yayın No: 1088, TİTCK Yayın No: 21, Artı 6 Reklam Matbaa Ltd. Şti. Basım Yılı 2018, 3027–3028.

Trovato, A., Monforte, M.T., Forestieri, A.M., Pizzimenti, F. 2000. *In vitro* anti-mycotic activity of some medicinal plants containing flavonoids. *Bollet. Chim. Farm.*, 139, 225–227.

Tumbas, V.T., Canadanovic-Brunet, J.M., Cetojevic-Simin, D.D., Cetkovic, G.S., Dilas, S.M., Gille, L. 2012. Effect of rosehip (*Rosa canina* L.) phytochemicals on stable free radicals and human cancer cells. *J Sci Food Agric.*, 92, 1273–1281.

Turan, İ., Demir, S., Kilinc, K., Özer Yaman, S., Misir, S., Kara, H., Genc, B., Mentese, A., Aliyazicioglu, Y., Değer, O. 2018. Cytotoxic effect of *Rosa canina* extract on human colon cancer cells through repression of telomerase expression. *J. Pharm. Anal.*, 8, 394–399.

Ul'chenko, N.T., Mukhamedova, KhS., Glushenkova, A.I., Nabiev, A.A. 1995. Lipids of dog rose hips. *Khim. Prir. Soedin.*, 6, 799–800.

Valsta, L.M., Kilkkinen, A., Mazur, W., Nurmi, T., Lampi, A.M., Ovaskainen, M.L., Korhonen, T., Adlercreutz, H., Pietinen, P. 2003. Phyto-oestrogen database of foods and average intake in Finland, *British J. Nut.*, 89, 31–38.

Vertuani, S., Bosco, E., Testoni, M., Ascanelli, S., Azzena, G., Manfredini S. 2004. Antioxidant herbal supplements for hemorrhoids developing a new formula. *Nutrafoods*, 3, 19–26.

Warholm, O., Skaar, S., Hedman, E., Molmen, H.M., Eik, L. 2003. The effects of a standardized herbal remedy made from a subtype of *Rosa canina* in patients with osteoartritis: A double-blind, randomized, plasebo controlled clinical trial. *Curr. Ther.Res.*, 64, 21–31.

Willich, S.N., Rossnagel, K., Roll, S., Wagner, A., Mune, O., Erlendson, J., Kharazmi, A., Sörensen, H., Winther, K. 2010. Rose hip herbal remedy in patients with rheumatoid arthritis -A randomized controlled trial. *Phytomedicine*, 17, 87–93.

Winther, K., Rein, E., Kharazmi, A. 1999. The anti-inflammatory properties of rose-hip. *Inflammopharm.*, 7, 63–68.

Winther, K., Apel, K., Thamsborg, G. 2005. A powder made from seeds and shells of a rose-hip subspecies (*Rosa canina*) reduces symptoms of knee and hip osteoarthritis: A randomized, double-blind, placebo-controlled clinical trial. *Scan. J. Rheum.*, 34, 302–308.

Winther, K., Vinther Hansen, A.S., Campbell-Tofte, J. 2016. Bioactive ingredients of rose hips (*Rosa canina* L) with special reference to antioxidative and anti-inflammatory properties: In vitro studies. *Bot. Targets Ther.*, 6, 11–23.

Živković, J., Stojković, D., Petrović, J., Zdunić, G., Glamočlija, J., Soković, M. 2015. *Rosa canina* L. – New possibilities for an old medicinal herb. *Food Funct.*, 6, 3687–3692. http://www.worldfloraonline.org/; 2022.

Zlatanov, M.D 1999. Lipid composition of Bulgarian chokeberry, black currant and rose hip seed oils. *Journal of the Science of Food and Agriculture*, 79(12), 1620–1624.

20 *Rosa damascena Mill.* DAMASK ROSE

O. Tuncay Ağar and L. Ömür Demirezer

BOTANICAL INFORMATION OF *ROSA L.*

Rosaceae is a flowering plant family containing 91 genera and 2950 species [1]. It is divided into three subfamilies as Dryadoideae, Rosoideae, and Amygdaloideae [2]. The *Rosa* L. genus, which is widely found in the temperate and subtropical regions of the northern hemisphere, is represented by 150–200 species worldwide. The *Rosa* genus is located in the Rosaceae family and Rosoideae subfamily (Table 20.1) [2, 3]. The *Rosa* genus is represented by a total of 32 species (35 taxa) that grow in Turkey [4].

ROSA DAMASCENA MILL.

Rosa damascena Mill. is a natural hybrid of *R. gallica* L. and *R. phenicia* Boiss. species, and it is cultivated in the Turkish Lakes Region, which includes a part of Isparta, Burdur, Afyon, Denizli, Konya, and Antalya [5, 6]. The natural distribution area of *R. damascena*, whose homeland is Iran, is the Northern Hemisphere and it is cultured in France, Lebanon, India, Russia, China, Morocco, Mexico, Italy, and other European countries, especially Turkey and Bulgaria [6, 7]. *R. damascena*, which is formed by the hybridization of *R. gallica* and *R. phoenicia* plants, is a very thorny, less stratified, and pink-flowered plant [8].

The description of the plant is as follows: perennial shrubs with large, showy, and colorful flowers, reaching a height of 1–2 meters; leaves are compound, 5–7 leaflets, imparipinnate; the inflorescence is corymb which consists of several pinkish-yellow flowers with pedicel; sepals, petals, and stamens attached to pedicel with receptacle at the base; pedicel, 0.6–3.5 cm long, light green, thin, covered with numerous spines and hairs; receptacle 1.0–1.8 cm long, covered with numerous spines and hairs; sepals 5, 1.3–2.4 cm long, unequal, leaf-like, upper part cream-green and lower yellowish-green color, glandular hairy; petals numerous, pinkish yellow, 1.5–4.2 cm long, slightly obovate to subcordate; stamens numerous, free, unequal, anther dorsifixed, dark brown, filament 0.3–0.5 mm long, carpel numerous, free, ovary inferior; stylus lateral, hairy, free; stigma is terminal. Flowering is in June and July [8-11].

PHYTOCHEMISTRY

FLAVONOIDS

Flavonoids isolated from the *Rosa damascena* plant are as follows:

- **Flavonol aglycones:** Kaempferol, quercetin [12], myrcetin [13].
- **Flavonol glycosides:** Afzelin, juglanin [14], astragalin, quercetin 3-*O*-arabinoside [15], trifolin, tribulosite[16],kaempferol-3-*O*-(6-*O*-trans-*p*-coumaroyl)-*β*-D-glucopyranoside,populnin,quercimeritrin [12], roxyloside, quercetin gentiobioside [17], kemferol-3-*O*-galactosyl-arabinoside,

DOI: 10.1201/9781003146971-20

TABLE 20.1
The Taxonomy of *Rosa damascena* Mill

Kingdom	Plantae
Division	Tracheophyta
Phylum	Spermatophyta
Subphylum	Angiospermae
Class	Magnoliopsida
Superorder	Rosanae
Order	Rosales
Family	Rosaceae
Subfamily	Rosoideae
Genus	*Rosa* L.
Species	*Rosa damascena* Mill.

kempferol-3-*O*-glucosyl-arabinoside, kempferol-3-*O*-rhamnosyl-galactoside, kemferol-3-*O*-rhamnosyl-glucoside, kempferol-3-*O*-rutinosyl-7-*O*-arabinoside, isoquecitrin, quercetin-3-*O*-β-D-glucosyl-galactoside, foeniculin [18], quercitrin, rutin [13], quercetin 3-*O*-galactoside, quercetin 3-*O*-glucoside, quercetin 3-*O*-rutinoside, quercetin 3-*O*-rhamnoside, kaempferol 3-*O*-β-D-xylopyranoside [19].

- **Flavon aglycones:** Apigenin [13], luteolin [20].
- **Isoflavonoids:** 5,6,4′-Trihidroxy-7,8-dimetoxy flavone, 5,3′-Dihidroxy-7,8-dimetoxy flavone, 5-Hydroxy-6,7,8,3′,4′-pentametoxy flavone, formononetin [21].
- **Flavan 3-ols:** Catechin, epicatechin [13], liquiritin [15].
- **Aurones:** Maritimein, 6,7,3′,4′ tetrahydroxyiaurone, cernuoside, damauron B, 4′,6-dihydroxy-4-metoxyisoauron [22], rugauron B, (E)-3′-*O*-β-D- glucopyranosyl-4,5,6,4'-tetrahydroxy-7,2′-dimetoxyaurone, siamaurone A, damaurone C, damaurone D, damaurone E [23], (Z)-2-(4- methoxybenzylidene)-7,7-dimethyl-7,8-dihydro-2*H*-furo[2,3-f] chromene-3,9-dion [24].
- **Chalcone:** Davidioside [15].
- **Anthocyanins:** Delphinidin [22], cyanidin-3-*O*-β-glucoside (17), cyanidin-3,5-di-*O*-β-D-glucoside [18].

Coumarins

Coumarins isolated from the *Rosa damascena* plant are as follows: rosacoumarin [25].

Terpenic and Volatile Compounds

Terpenic and volatile compounds isolated from the *Rosa damascena* plant are as follows:

- **Cyclic Monoterpenes:** p-cymene, sabinene, limonene, terpinene, terpinolene, nerol oxide, 1,8-cineole [26], terpinene-4-ol, trans-rose oxide [27], carvone [28], α-pinene, β-pinene, cis- rose oxide [29], β-damascenone [30], camphene [31].
- **Acyclic Monoterpenes**: Geraniol, nerol, myrcene, geranyl acetate, linalyl acetate, geranial, linalool [26], geranyl acetone [32], citronellyl acetate [29], citronellol, citronellal, citronellyl formate, neral [33].
- **Sesquiterpenes:** α-eudesmol, α-cadinol, caryophyllene oxide, β-caryophyllene, α-guajane, germacrene D, α-humulene [26], γ-muurolene, γ-cadinene [27], δ- cadinene [29], β-eudesmol [34], 10-epi-γ- eudesmol [35], t-muurolol [36].

- **Hyrocarbons:** Eicosane, pentacosane, heptacosane, hexatriacontane, tricosane, heptadecane, octadecane, nonadecane [14], 1-heptanol, 1-heptadecene, (E)-9- eicosane, 1- heneicosane, tetracosane, nonanal, dodecanoic acid [27], undecanal, 1-undecanol, 1-nonadecene, 1-octadecene, heneicosane, docosane [33], hexadecanoic acid, tetradecanoic acid [37], hexadecane [29].
- **Benzenoids:** Phenyl ethyl benzoate [37], phenyl ethyl alcohol [14].
- **Diterpenes:** Geranyl-6-*O*-α-L-arabinofuranosyl-*β*-D-glucopyranoside, geranyl-6-*O*-*β*-L-arabinofuranosyl-*β*-D- glucopyranoside [38].
- **Triterpenes:** Ursolic acid [39], 2-hydroxy ursolic acid, Ursolic acid methyl ester, β-amyrin [40].
- **Steroidal compounds:** *β*-sitosterol, stigmasterol [39].

Phenylpropanoids

Phenylpropanoids isolated from the *Rosa damascena* plant are as follows: Eugenol [14], eugenol methyl ester, cinnamaldehyde [28], eugenol-*β*-D-xylopyranosyl-(1→6)-*β*-D-glucopyranoside, eugenol-α-L-arabinopyranosyl-(1→6)-*β*-D-glucopyranoside, eugenol-α-L-arabinofuranosyl-(1→6)-β-D-glucopyrano side [38].

Plant Acids

Plant acids isolated from the *Rosa damascena* plant are as follows: Quinic acid [12], gallic acid, and rosacedrenoic acid (cedr-6-en-12-ol-14-oic acid) [41].

TRADITIONAL USES *OF ROSA DAMASCENA*

Decoctions prepared from *R. damascena* flowers in Iran are traditionally used to relieve ailments such as chest and abdominal pain complaints, menstrual bleeding, digestive system ailments, constipation complaints, and to strengthen the heart. In addition to the traditional use of rose water as an antiseptic in eye washing and as a disinfectant for the mouth, it is also used as a muscle relaxant for the treatment of disorders such as abdominal pain and bronchial and chest congestion. Decoctions prepared from dried rose petals are used as diuretics. In addition to being prescribed as a blood purifier, rose seeds are also used by the public in bread making [6]. Flowers, petals, and fruits are traditionally used for insomnia treatment [9]. In Morocco, *R. damascena* flowers are used to relieve abdominal pain complaints [42]. It has been reported that in Syria, *R. damascena* flowers are traditionally used as a laxative, diuretic, and refresher, as well as for relieving ailments such as general weakness, cold, cough, flu, abdominal pain, and gallstones [43]. In Jordan, infusions prepared from *R. damascena* receptacles and seeds are traditionally used for antibacterial purposes and in the treatment of diseases such as intestinal cancer and heart diseases [44]. In North Africa and Near East Countries, *R. damascena* leaves and flowers are used in the treatment of the digestive system, urinary system, nervous system, respiratory system disorders, and skin diseases [45]. In ancient medical history, *R. damascena* was used in the treatment of ailments such as abdominal and chest pain, menstrual pain, and digestive problems. It is also known to be prescribed to strengthen the heart, especially to reduce inflammation in the neck. Decoctions prepared from *R. damascena* root were used in the indigenous peoples of North America to relieve cough in children. The herb is also known to be used as a mild laxative. Rose oil is used to heal depression, sadness, nervous stress, and blood pressure problems. It is used for quenching thirst, reducing prolonged cough and gynecological complaints, healing wounds, and treating skin diseases. Treatment using rose oil vapor has been reported to be beneficial for some allergies, headaches, and migraines [10]. Rose oil is also used in aromatherapy for the treatment of depression, stress-related conditions, and heart diseases. *R. damascena* is traditionally used in many cuisines as a flavor enhancer. Rose water is often used in many meat dishes. It has also been reported that it is added to foods such as ice cream, jam, pudding, cake, and yogurt in terms of flavoring [7].

IN VITRO BIOLOGICAL ACTIVITY TESTS OF *ROSA DAMASCENA*

ANTIMICROBIAL EFFECT

Antibacterial effects of ethanol and methanol extracts prepared from *R. damascena* petals on *Streptococcus mutans*, a dental pathogen at different concentrations, were investigated by the well diffusion method. While ethanol extract creates 23 ± 1.632 mm inhibition zone at 300 μL concentration the methanol extract showed an antibacterial effect by forming a 25 ± 1.632 mm inhibition zone at the same concentration [46]. The effects of aqueous and ethanolic extracts and chloroform, ethyl acetate, and butanol fractions by liquid partitioning of ethanol extract and essential oil of *R. damascena* flowers were studied on various Gram-negative bacteria by disk diffusion method. The essential oil had an antibacterial effect on *Streptococcus pyogenes* (MIC=0.25 mg/mL), *Bacillus subtilis* (MIC=0.25 mg/mL), and *Staphylococcus aureus* (MIC=0.25 mg/mL). Aqueous extract showed antibacterial effect on *B. subtilis* (MIC=0.25 mg/mL) and *S. aureus* (MIC=0.25 mg/mL). Ethanolic extract showed antibacterial effect on *S. pyogenes* (MIC=0.125 mg/mL), *B. subtilis* (MIC=0.5 mg/mL), and *S. aureus* (MIC=0.5 mg/mL). Chloroform fraction showed antibacterial effect on *B. subtilis* (MIC=0.125 mg/mL). Ethyl acetate fraction showed antibacterial effect on *S. pyogenes* (MIC=0.125 mg/mL), *B. subtilis* (MIC=0.125 mg/mL), *S. aureus* (MIC=0.125 mg/mL), and *Mycobacterium phlei* (MIC=0.25 mg/mL). Butanol fraction showed antibacterial effect on *B. subtilis* (MIC=0.25 mg/mL) and *S. aureus* (MIC=0.5 mg/mL) [47]. Antibacterial effects of ethanolic extract prepared from *R. damascena* petals on various bacteria were investigated by liquid microdilution method, and for this purpose, MIC, MBK, and inhibition zone diameter (IZD) values were determined. For ethanol extract, *Pseudomonas aeruginosa* MIC and MBK values were found to be 62.5 μg/mL and IZD was measured as 34 mm. MIC and MBK values of the extract in *Escherichia coli* were found to be 62.5 μg/mL, and IZD was measured as 30 mm [48]. The effects of 2% standardized *R. damascena* ethanol extract (Barij essence) on some endodontic microorganisms were investigated by using the liquid macrodilution method. As a result, it was reported that the extract had an antimicrobial effect on *Enterococcus faecalis* (MIC = 0.06%), *Actinomyces naeslundii* (MIC = 0.03%), *Porphyromona gingivalis* (MIC = 0.03%), and *Fusobacterium nucleatum* (MIC = 0.06%) [49]. The efficiency of the methanolic extract prepared from *R. damascena* fresh flowers and the remaining flowers (consumed flower) after the essential oil was obtained by distillation on various microorganisms by agar well diffusion method was investigated. The extract has been found effective on *Aeromonas hydrophila, Bacillus cereus, Enterobacter aerogenes, Enterococcus faecalis, Escherichia coli, Klebsiella pneumoniae, Mycobacterium smegmatis, Proteus vulgaris, Pseudomonas aeruginosa, Pseudomonas fluorescens, Salmonella enteritidis, Salmonella typhimurium, Staphylococcus aureus,* and *Yersinia enterocolitica* ($p<0.05$). For the methanol extract prepared from *R. damascena* fresh flowers, the highest effect was observed on *S. enteritidis* (inhibition zone = 21 mm) at a concentration of 10%. The extract obtained from the consumed flowers showed an inhibition zone of 16 mm at the same concentration. Fresh flower extract at 10% concentration has shown an antimicrobial effect by creating a 15 mm inhibition zone on *M. smegmatis.* The consumed flower extract showed antimicrobial effect by 21 mm inhibition zone at the same concentration [50].

ANTI-HIV EFFECT

Aqueous and methanol extracts prepared from *R. damascena* dried petals and nine pure compounds isolated from methanol extract was examined for anti-HIV effect on C8166 human T lymphoblastoid cell line infected with HIV-1_{MN} by XTT-formazan method and H9 human T-cell lymphoma cell chronically infected with HIV-1_{IIIB}. While aqueous and methanol extracts exhibited moderate anti-HIV effect on the infected C8166 cell line with EC_{50} value of 10 μg/mL, pure compounds kaempferol, kaempferol-3-*O*-*β*-D-glucopyranoside, and kaempferol-3-*O*-(6-*O*-trans-*p*-coumaroyl)-β-glucopyranoside showed high anti-HIV activity with the values of EC_{50} = 4 μg/mL, EC_{50} = 8 μg/mL, and EC_{50} = 8 μg/mL, respectively. At the same time, it has been reported that kaempferol and quercetin compounds reduce the HIV effect on the infected H9 cell line with EC_{50} = 0.8 μg/mL and EC_{50} = 10 μg/mL values, respectively, and this effect is due to the prevention of Gp120 binding to CD4 [12].

Antiparasitic Effect

Nine compounds isolated from *R. damascena* flowers were investigated in terms of anti-babesyial effect using the method given by Subeki et al. [51]. 2-phenylethyl 6-*O*-galloyl-*β* -D-glucopyranoside, gallic acid, 3,4-dihydroxy benzoic acid, and quercetin 3-*O*-*β*-D-(6-O-acetyl)-glucopyranosyl-(1→4)-α-L-rhamnopyranoside compounds showed positive results with IC_{50} values of 11.78 μg/mL, 15.74 μg/mL, 78.00 μg/mL, and 90.88 μg/mL, respectively [52].

Anti-inflammatory and Cytoprotective Effect

In order to determine the anti-inflammatory and cytoprotective effects of 70% ethanol extract prepared from *R. damascena* flowers on *Helicobacter pylori*-infected gastric epithelial cells, interleukin-8 (IL-8) secretion and formation of reactive oxygen species were investigated. Anti-*Helicobacter pylori* activity and cytotoxic effects of the extract were evaluated by serial dilution method and DNA fragmentation test, respectively. Enzyme-linked immunosorbent assay (ELISA) and flow cytometry methods were used to evaluate IL-8 release and formation of reactive oxygen species in cells infected with *Helicobacter pylori*. *R. damascena* extract showed moderate (1000–2000 pg/mL) and strong (<1000 pg/mL) inhibition against IL-8 secretion at the concentrations of 50 μg/mL and 100 μg/mL, respectively. *Helicobacter pylori* viability was measured as 90% CFU (colony forming unit) in the presence of 100 μg/mL extract, which is five times lower than the minimum bactericidal concentration value. DNA fragmentation was measured at a concentration of 100 μg/mL, and it was determined that the extract did not show a significant induction of DNA fragmentation (7.4 ± 0.9%) compared to the control (4.5 ± 0.8%). According to these results, it was determined that the effect of the extract on the Helicobacter pylori/cell co-culture system did not result from changes in *Helicobacter pylori* viability or its toxicity to cells [53].

Antiproliferative and Cytotoxic Effect

Cytotoxic effect of methanol extract prepared from *R. damascena* flowers on human HeLa cervical cancer cell line was evaluated by MTT colorimetric method and the results were compared with the normal African green monkey kidney epithelial cell line (Vero). The extract was noted to have an antiproliferative effect on HeLa cell line and the extract's IC_{50} = 265 μg/mL for the HeLa cell line. The IC_{50} value in the extract, vero cell line was found to be >1000 μg/mL, and it was reported to have a cytotoxic effect against HeLa cervical cancer cell line [54]. Potential cytotoxicity of an auron (Z)-2-(4-methoxybenzylidene)-7,7-dimethyl-7,8-dihydro-2H-furo [2,3-f] chromene-3,9-dione isolated from dried flowers of *R. damascena* was tested against cell lines NB4, A549, SHSY5Y, PC3, and MCF7, according to the method developed by Gao et al. [55]. It has been reported that the compound has a cytotoxic effect against NB4 and SHSY5Y cell lines with IC_{50} = 4.8 μM and IC_{50} = 3.4 μM, respectively [24].

Antioxidant Activity and Radical Scavenging Capacity

The radical scavenging effect of methanolic extract prepared from *R. damascena* flowers was investigated by 1,1-diphenyl-2-picrylhydrazyl (DPPH) method and its effect on lipid peroxidation by ferric thiocyanate (FTC) and thiobarbituric acid (TBA) method. When compared with the standards (BHA = 89.49 ± 4.63%, Vitamin E = 72.01 ± 5.85%, Ascorbic acid = 96.78 ± 0.24%), it was determined that the extract showed a high level of DPPH radical scavenging effect with 95.67 ± 0.6%. The extract has been reported to have a strong antioxidant capacity with 84.27% inhibition in the FTC method and 92.93% in the TBA method when compared with the standard vitamin E and BHA methods [56]. The *n*-butanol fraction obtained from 50% methanol extract prepared from *R. damascena* petals was investigated in terms of antiradical efficacy by using DPPH, ABTS, FRAP, OH, O_2^-, $ONOO^-$ and NO methods. Fraction showed radical scavenging effect according to the following results: IC_{50} = 8.231 ± 0.056 µg/mL (Std-Quercetin: IC_{50} = 3.21 ± 0.11 µg/mL) in DPPH method; IC_{50} = 3.315 ± 0.55 µg/mL (Std- Quercetin: IC_{50} = 1.34 ± 0.08 µg/mL) in ABTS method; IC_{50} = 8.24 ± 0.77 µg/mL (Std-Quercetin: IC_{50} = 7.42 ± 0.32 µg/mL) in OH method; IC_{50} = 39.56 ± 0.85 µg/mL in O_2^- method (Std-Quercetin: IC_{50} = 41.98 ± 0.95 µg/mL); IC_{50} = 590.57 ± 2.78 µg/mL (Std-Gallic acid: IC_{50} = 820.12 ± 27.34 µg/mL) in $ONOO^-$ method; and IC_{50} = 209.4 ± 0.69 µg/mL (Std-Quercetin: IC_{50} = 19.23 ± 0.42 µg/mL) in NO method. It also showed antioxidant activity in the FRAP method with a value of 4.23 ± 0.06 mmol Fe_2^+/g [57]. Methanol extract prepared from *R. damascena* petals showed a radical scavenging effect close to gallic acid with an IC_{50} value of 1698.3 µg/mL in the DPPH method [8]. The antiradical effect of the methanolic-aqueous (80:20) extract prepared from fresh petals of *R. damascena* was investigated by DPPH method and it was reported to have a strong radical scavenging effect with IC_{50} = 2.24 µg/mL when compared to standards (BHT = 110.98 ± 8.25 µg/mL, vitamin E = 22.72 ± 1.91 µg/mL) [14]. It has been reported that the methanolic extract (IC_{50} = 21.4 ± 0.52 µg/mL) prepared from *R. damascena* fresh flowers showed a strong antiradical effect with the DPPH method when compared with the standard L-ascorbic acid (IC_{50} = 30.0 ± 0.08 µg/mL) [58]. Methanol extract prepared from *R. damascena* fresh flowers showed a radical scavenging effect of 74.51 ± 1.65% at 100 ppm by DPPH method and showed antioxidant effect with a value of 372.26 ± 0.96 mg/g with phosphomolybdenum method [50]. The effects of the acetone fraction obtained from *R. damascena* fresh petals on superoxide radical formation, hydroxyl radical formation, and lipid peroxide formation were investigated, and the results were compared with curcumin. The fresh petals of *R. damascena* were crushed in a mortar to dehydrated, filtered, and the filtrate was diluted with distilled water and then made into a lyophilized powder. The acetone fraction obtained from this lyophilized powder by column chromatography was found to be antiradical and antioxidant effective as follows: IC_{50} = 13.75 ± 0.86 µg/mL (std-curcumin: IC_{50} = 6.25 ± 0.27 µg/mL) in the superoxide radical formation test; IC_{50}=135±0.75 µg/mL (std-curcumin: 2.3±0.05 µg/mL) in hydroxyl radical formation test and IC_{50} = 410±0.74 µg/mL (std-curcumin: IC_{50}=20±0.58 µg/mL) in lipid peroxide formation test [59]. The antioxidant effect of the extract with methanol prepared from *R. damascena* dried flowers was investigated by the trolox equivalent antioxidant capacity (TEAC) method. The extract has been reported to have antioxidant capacity with its IC_{50}=25.10 µg/50µL (TEAC=0.38) when compared with trolox (TEAC=1) [60]. Radical scavenging effects of hot and cold methanol extracts prepared from *R. damascena* green leaves were investigated by the DPPH method and antioxidant capacity by the FRAP method. The extract has been reported to have a potent radical scavenging effect at a concentration of 50 g/mL (hot: 84.56±0.27%; cold: 88.58±0.80%) when compared to synthetic antioxidants trolox (82.96 ± 0.30%), BHA (65.23 ± 0.57%), and BHT (35.55 ± 0.04%). It has been reported that the cold extract of leaves showed the closest antioxidant effect to the standard (BHA=1.79 and 2.37 µg/mL) at concentrations of 100 and 150 µg/mL (1.04 ± 0.06 and 1.44 ± 0.14 µg/mL, respectively) with the FRAP method [61].

Sun-Protection Effect

Whether aqueous-ethanol (50:50), ethylacetate-ethanol (80:20), and ether extracts obtained from *R. damascena* dried flowers can be used as an antisolar agent was investigated spectrophotometrically by measuring their ultraviolet absorption capability. It was observed that all three extracts effectively absorb UV radiation between the 200 and 400 nm band. The maximum absorption range for the three extracts was measured at 200–320 nm, 250–360 nm, and 230–370 nm, respectively. The method used by Diffey and Robson [62] for SPF measurements of creams was modified, and the FDA standard (8% homosalate) was used to standardize the method. SPF measurements were made by adding the extracts to water-based creams prepared in two concentrations of 5% and 8% (w/w). According to the measurements made in 5% and 8% creams, aqueous ethanol extract (5.73 ± 0.76 $\mu L/cm^2$; 7.35 ± 1.09 $\mu L/cm^2$, respectively) showed the highest SPF value. SPF values were measured as 2.39 ± 0.11 $\mu L/cm^2$ in 5% cream and 3.64 ± 0.28 $\mu L/cm^2$ in 8% cream. Cream containing 8% ether extract showed an SPF value of 5.22±0.85 $\mu L/cm^2$. According to the results of physicochemical studies, it was stated that the cream containing 5% ether extract has the most appropriate appearance and stability with a value of 4.24 ± 0.24 $\mu L/cm^2$ [63].

Cardiovascular Effect

Anti-3-hydroxy-3-methylglutaryl coenzyme A (anti-HMG-CoA) effect of methanol extract prepared from *R. damascena* flowers was determined spectrophotometrically by measuring NADPH oxidation. In the method where HMG-CoA is used as a substrate, it has been determined that the extract has an inhibitory effect of 70.0% on enzyme activity. Kinetic properties of HMG-CoA reductase were determined by Lineweaver Burk plot analysis of different concentrations of HMG-CoA (0.3, 0.6, 0.9, and 1.2 mmol/L) and *R. damascena* extract (0.05 and 0.15 mg/mL) at 340 nanometers. The extracts have been reported to show non-competitive inhibition. When the extract was added to the enzyme mixture at a concentration of 0.15 mg/mL, the Vmax value was determined as 0.0031 mM/min [64]. HMG-CoA reductase effect of the kaempferol-3-*O*-*β*-D-glucopyranosyl-(1→4)-*β*-D-xylopyranoside (Roxyloside A), isococitrin, apheline, cyanidin-3-*O*-*β*-D-glucoside, and quercetin gentiobioside compounds isolated from the buds of *R. damascena* was investigated by making minor changes in the method specified by Kleinsek et al. [65]. Angiotensin I-converting enzyme effect was tested by making minor changes in the method specified by Cushman and Cheung [66]. Except for cyanidin-3-*O*-*β*-D-glucoside, all compounds showed a high level of HMG-CoA reductase effect with IC50 values of 47.1, 80.6, 80.1, and 50.6 μM, respectively. Cyanidin-3-*O*-*β*-D-glucoside significantly stopped angiotensin I-converting enzyme activity with an IC_{50} value of 138.8 μM [17].

Other Effects

Rosa damascena aqueous-methanolic extract (1:1;v/v) was evaluated for anticholinesterase effect by Ellman method and the extract was found to be effective with IC50 = 93.10 μg/mL [67].

The effect of methanol extract prepared from *R. damascena* flowers on pancreatic lipase enzyme inhibition by turbidimetry method was investigated. It has been reported that the extract inhibits the pancreatic lipase enzyme by 57.0% when used in the concentration range of 0.05 and 0.15 mg/mL [68].

IN VIVO BIOLOGICAL EFFECTS OF *ROSA DAMASCENA*

Antidiabetic Effect

The antidiabetic effect of the methanolic extract prepared from *R. damascena* flowers was compared with acarbose, an α-glucosidase inhibitor, on normal and diabetic rats. To examine the hypoglycemic effect of *R. damascena* extract, 18 normal male rats were divided into three equal groups. The rats in group 1 were used as a control group and were treated only with maltose (2 g/kg body weight). Maltose (2 g/kg body weight) and acarbose (50 mg/kg body weight) were administered to the rats in Group 2. A single dose (1000 mg/kg body weight) of *R. damascena* extract was given to the rats in Group 3. The glucose level in the blood was initially defined by glucometry at the 30th,

60th, and 120th minute. The same procedure was also applied to 18 diabetic rats, which were also divided into three equal groups. To evaluate hypoglycemic activity in diabetic rats, 36 rats were divided into six groups, each containing six rats. The first two groups were treated in the same way as the procedures given above. Groups 3–6 were treated with maltose (2 g/kg) and increasing doses of *R. damascena* extract (100, 250, 500, and 1000 mg/kg, respectively). The glucose level in the blood was likewise determined by glucometry at the beginning, 30th, 60th, and 120th minute. When the postprandial blood glucose level in normal rats was examined, after 30 minutes of dieting it was observed that the blood glucose level in the control group increased from 83 mg/dL to 148 mg/dL and then decreased. However, the rise in blood glucose level after high maltose diet was significantly ($p<0.0001$) suppressed by the application of *R. damascena* extract. Although acarbose, used as a positive control, showed a significant suppression in postprandial blood glucose level, there was no significant difference between it and *R. damascena* extract. When the postprandial blood glucose level in diabetic rats was examined, the postprandial blood glucose level was found to be lower in rats treated with *R. damascena* extract and acarbose compared to the control group. In the control group, 30 minutes after the treatment, it was observed that the glucose concentration in the blood increased from 382 mg/dL to 583 mg/dL. Postprandial blood glucose was significantly reduced in rats treated with *R. damascena* extract. The glucose level in the blood changed from 398 mg/dL to 432 mg/dL 30 minutes after the treatment, and there was a significant difference between the control group ($p<0.0001$). It was observed that postprandial blood glucose level in diabetic rats decreased depending on the dose with the administration of different doses of *R. damascena* extract. In the control group, the blood glucose level increased from 348 mg/dL to 580 mg/dL after 30 minutes of the maltose treatment (2.0 g/kg) of the rats. However, in rats treated with different doses of *R. damascena* extract (100, 250, 500, and 1000 mg/kg), the blood glucose level was significantly reduced depending on the dose [69].

Anti-inflammatory Effect

The effect of 70% ethanol extract and essential oil prepared from *R. damascena* flowers on ulcerative colitis induced by acetic acid in rats was investigated. Male Wistar rats were treated orally/intraperitoneally for 4 days with different doses of extract (250, 500, and 1000 mg/kg) and essential oil (100, 200, and 400 μL/kg) 2 hours before colitis. Prednisolone (4 mg/kg) was given orally and dexamethasone (1 mg/kg) intraperitoneally to reference groups. Macroscopic, histopathological, and biochemical examinations were performed by measuring the weight and length of the wet tissue and the myeloperoxidase (MPO) activity of the tissues. The results are reported as follows: at all doses (250, 500, and 1000 mg/kg) after oral extract treatment; ulcer score 0–4 (2.83 ± 0.48, 2.50 ± 0.43, and 1.50 ± 0.34), ulcer area (4.68 ± 0.63, 4.15 ± 0.55, and 2.70 ± 0.44 cm^2), ulcer index 0–12 (7.52 ± 1.06, 6.65 ± 0.64, and 4.20 ± 0.66), weight/length ratio (1.28 ± 0.07, 1.12 ± 0.07, and 1.08 ± 0.08 g/cm). After oral essential oil treatment at the lowest dose (100 μL/kg), ulcer score 0–4 (2.33 ± 0.42), ulcer area (3.42 ± 0.54 cm^2), ulcer index 0–12 (5.75 ± 0.95), weight/length rate (1.34 ± 0.06 g/cm). At the lowest dose (125 mg/kg) after the 70% ethanol extract treatment intraperitoneally; ulcer score 0–4 (2.67 ± 0.33), ulcer area (5.05 ± 0.66 cm^2), ulcer index 0–12 (7.77 ± 0.84), weight/length ratio (1.17 ± 0.12 g/cm). In summary, *R. damascena* reduced all colitis indices as well as MPO activity [70].

Anticonvulsant and Neuroprotective Effects

The effects of aqueous, ethanolic, and chloroform extracts prepared from *R. damasecena* dried flowers on pentylenetetrazole (PTZ)-induced seizures in mice were investigated. The mice were divided into two groups: the normal saline control group and the diazepam group (3 mg/kg). The mice were injected intraperitoneally with all three extracts at doses of 100, 500, and 1000 mg/kg 30 minutes before PTZ injection. Latency values of the minimal clonic seizure (MCS) and generalized tonic clonic seizure (GTCS) and mortality rates in mice were calculated. The aqueous extract showed significant increases ($p<0.05$ or $p<0.01$) in the MCS and GTCS latency values at all three doses compared to the normal saline group. The MCS latency of ethanolic extract at 1000 mg/kg dose and GTCS latency at 500 and 1000 mg/kg doses were observed higher than normal saline control groups ($p<0.05$, $p<0.01$). According to these results, *R. damascena* aqueous and ethanol extracts have been reported to have anticonvulsant effects and are effective in reducing epileptic seizures [71]. The anti-seizure and neuroprotective effects of 70% ethanol extract prepared from *R. damascena* dried flowers on PTZ-induced seizures in rats were investigated. The rats to be used in the study were divided into five groups: Group 1: Control (given saline), Group 2: PTZ 100 mg/kg, intraperitoneal, Group 3: PTZ-Extract 50 mg/kg, Group 4: PTZ-Extract 100 mg/kg, and Group 5: PTZ-Extract 200 mg/kg. Rats were treated with *R. damascena* extracts one week before PTZ injection. After examining the electrocorticography (ECoG) record of the animals, their brains were removed for histological examination. The extract significantly prolonged the delay of seizure attacks and decreased the frequency of epileptiform discharge caused by PTZ injection ($p<0.01$ - $p<0.001$) [72].

ANTIPARASITIC EFFECT

The effect of methanol extract prepared from *R. damascena* flowers on Nobeh fever and fever caused by *Plasmodium berghei* in albino mice was investigated by Peters' suppression test. It was observed that the extract injected into mice intraperitoneally at 10 mg/kg daily for 4 days reduced parasitemia by 57.7% in mice compared to the control group [73].

EFFECT ON MEMORY PERFORMANCE

The effect of *R. damascena* standardized extract on neurogenesis and synaptogenesis in rats was investigated. In the Morris water maze behavior test used to determine the effect of the extract on spatial learning and memory, 60 rats were randomly divided into six equal groups: Group 1: control, Group 2: imitation, Group 3: Alzheimer's disease + normal saline, Group 4: Alzheimer's disease + 300 mg *R. damascena* extract, Group 5: Alzheimer's disease + 600 mg *R. damascena* extract, Group 6: Alzheimer's disease + 1200 mg *R. damascena* extract. Before the *R. damascena* extract was given, learning performance was studied in rats using spatial memory. Before the rats were treated with amyloid-β, no significant difference was observed between the mean latency time ($F_{5,54}=0.03$, $p=1>0.05$), mean swimming distance ($F_{5,54}=0.12$, $p= 0.9>0.05$), mean swimming speed ($F_{5,54}= 0.53$, $p= 0.7>0.05$), and spatial acquisition blocks (1–5) to find hidden platforms in all groups. However, significant differences in escape latency, swimming distance, and swimming speed between blocks in all groups (Block effect $F_{4.54}=629.26$, $p<0.001$; $F_{4.54}=718.47$, $p<0.001$; $F_{4.54}=54.55$, $p<0.001$) have been observed. This showed that the spatial acquisition performance in all groups improved significantly during the training days. Multivariate ANOVA analysis showed that there were significant differences ($p<0.001$) between groups in the first probe test. Post hoc Tukey's HSD analysis showed that 21 days after the operation, the average time percentage spent in the target quadrant in the amyloid-β1-42 injected groups (Group 3 and Group 6) decreased significantly ($p<0.001$) compared to the control and imitation groups. However, there was no significant difference between the control and the simulated groups in terms of the average percentage of time spent in the target quadrant ($p=0.9$). These results showed that amyloid-β1-42 caused learning and memory impairment in rats. To compare the proliferation of adult neural stem cells and neural progenitor cells among all groups, the proliferation marker BrdU was injected twice daily for 7 days (50 mg/kg body weight) and examinations were performed 24 hours after the last injection. According to post hoc Tukey's HSD analysis, a significant (1.987 ± 21.46, $p<0.001$) increase in the number of BrdU-positive cells was detected in the groups treated with the extract (Groups 4–6 respectively; 2.107 ± 19.68, 2.432 ± 2.92, 2.703 ± 10.41) compared to Group 3. Proliferation of neural stem cells and neural progenitor cells in the hippocampus was observed in all groups treated with *R. damascena* extract. In conclusion, *R. damascena* has been reported to increase neurogenesis, hippocampal volume, and synaptic plasticity in adults, and reverse memory abnormalities in amyloid-β induced Alzheimer's disease in rats [74]. The effects of methanol extract prepared from dried fresh flowers of *R. damascena* on behavioral function in rats with amyloid-β induced Alzheimer's disease were investigated. Morris water maze and passive avoidance tests were used to evaluate basic education behavior performance. Amyloid-β was injected bilaterally into the hippocampus CA1 area. Twenty-one days after amyloid injection, the first probe test of behavioral tests was used to detect learning and memory impairment. In order to examine the potential effects of the extract on behavioral tasks, the second probe test was performed after one month application of *R. damascena* extract. It was determined that the groups treated with *R. damascena* extract significantly ($p<0.001$) developed spatial and long-term memories at medium (600 mg/kg) and high (1200 mg/kg) doses, depending on the dose [75]. The effect of 50% ethanol extract prepared from *R. damascena* petals on memory performance in scopolamine-induced memory impairment in rats was investigated. Thirty-two adult male rats were used for the study and the rats were divided into four equal groups. Control group (Group 1) was treated with saline without *R. damascena* extract and scopolamine. Scopolamine group (Group 2) was given saline for 2 weeks and treated with scopolamine (1.5 mg/kg) 30 minutes before each experiment in the Morris water maze test. Scopolamine + 50 mg/kg *R. damascena* extract and scopolamine + 250 mg/kg *R. damascena* extract were given to the treatment groups (Groups 3 and 4) for 2 weeks, respectively. Scopolamine was injected intraperitoneally before each experiment in the Morris water maze test. The brains of the rats were then removed for biochemical measurements. Delay time and path length in the scopolamine group were higher than in the control group ($p<0.01$-0.001). Both treatment groups showed shorter distance and time delay compared to the scopolamine group ($p<0.05$–0.001). Although the time spent by the scopolamine group in the target quadrant was less than the control group, it was observed that the time spent in the target quadrant of Group 4 was longer than the scopolamine group ($p<0.05$). While the malondialdehyde concentrations in the hippocampal and cortex tissues of the scopolamine group were higher, thiol concentrations were found to be lower than the control group ($p<0.001$). It was observed that extracts treated at both doses decreased the malondialdehyde concentration while increasing the thiol concentration ($p<0.05$–0.001) [76]. The effect of chloroform extract prepared from *R. damascena* buds and long-chain polyunsaturated fatty acid with the molecular formula $C_{37}H_{64}O_2$ isolated from the extract was investigated on atrophy induced by amyloid-β (25-35). It has been observed that the chloroform extract significantly triggers neurite growth activity and inhibits amyloid-β (25-35) induced atrophy and cell death. The protective effect of the isolated fatty acid on amyloid-β (25-35) induced dendritic atrophy was compared with nerve growth factor and docosahexaenoic acid (DHA). It has been found that it has a stronger effect than DHA at a dose of 1 μM [77].

Hepatoprotective Effect

Hepatoprotective efficacy of the aqueous extract prepared from *R. damascena* flowers at different oral doses (250, 500, and 1000 mg/kg body weight) in rats exposed to toxicity with acetaminophen (2 g/kg oral N-acetyl-p-aminophenol) was investigated. Acetaminophen treatment caused changes in several biochemical parameters, including serum transaminases, serum alkaline phosphatase, lactate dehydrogenase, albumin, bilirubin, urea and creatine, hepatic lipid peroxidation, and low glutathione levels. Adenosine triphosphatase and glucose-6-phosphatase activity in the liver was significantly decreased in acetaminophen-treated animals. These changes were found to be significantly reversed ($p\leq0.05$) depending on the dose when treated with three separate doses of the extract. In addition, histopathological changes reveal the protective property of the extract against acetaminophen-induced necrotic damage of hepatic tissues [78]. To investigate the hepatoprotective effects of aqueous extract prepared from *R. damascena* flowers on CCl_4-induced hepatocellular damage in rats, 0.15 mL/kg CCl_4 was administered intraperitoneally to rats for 21 days. As a result of these applications, important biochemical changes occurred in the blood and the levels of enzymes such as ALT and AST increased significantly. While hepatic and renal lipid peroxidation increased significantly, enzymatic antioxidants (glutathione reductase and glutathione peroxidase) and glutathione levels were decreased. In addition, a significant decrease was observed in aniline hydroxylase, adenosine triphosphatase, and glucose-6-phosphatase activities, while a slight increase in protein content was observed. Pathological changes caused by CCl_4 were evaluated by histopathological studies. These values were measured again after oral administration of *R. damascena* extract at doses of 513 and 1026 mg/kg for 5 days in rats. It has been found that the extract given at a dose of 513 mg/kg significantly restored blood values to almost normal [79]. The protective effect of the acetone fraction purified from lyophilized rose water obtained from fresh petals of *R. damascena* on carbon tetrachloride (CCl_4)-induced hepatotoxic effect in rats was investigated. Fresh petals were crushed in a mortar, the rose water was filtered and diluted with distilled water, lyophilized, and the acetone fraction was obtained with a silica gel column. Six rats were selected for use in *in vivo* studies and they were divided into three equal groups. The rats in group 1 were determined as the control group and no treatment was applied. The rats in group 2 were treated orally with CCl_4 (0.25 mL/100 g) (1:1) in liquid paraffin. In addition to CCl_4, the animals in group 3 were administered the acetone fraction (50 mg/kg body weight/day) orally. Two hours after the 4th dose of acetone fraction, all animals except the normal group were treated with CCl_4 and two more doses of acetone fraction intraperitoneally. Animals were killed 48 hours after the acute dose of CCl_4. Serum alkaline phosphatase (ALP), glutamine pyruvate transaminase (GPT), and glutamine oxaloacetate transaminase (GOT) activity and lipid peroxide levels were significantly decreased ($p<0.001$) in animals in group 3 [59].

Hypnotic Effect

The hypnotic effect of ethanol, aqueous, and chloroform extracts prepared from dried flowers of *R. damascena* on mice was investigated by determining the potentiation of pentobarbital-induced sleep time by the extracts. Three doses of extract (100, 500, and 1000 mg/kg) were injected intraperitoneally into the mice and the results were evaluated in comparison with the diazepam (3 mg/kg) group as positive control and the saline group as negative control. Pentobarbital (30 mg/kg) was injected 30 minutes after the extract was injected, and the extracts were noted to increase sleep time. The results revealed that ethanolic and aqueous extracts significantly ($p <0.001$) increased pentobarbital-induced sleep time at doses of 500 and 1000 mg/kg compared with diazepam [80]. The hypnotic effects of 50% ethanolic extract from *R. damascena* flowers and some fractions (aqueous, ethyl acetate, and n-butanol) prepared by partitioning from this extract on pentobarbital-induced extended sleep time in mice were investigated. The extract and fractions were injected into mice intraperitoneally in two doses (250 and 500 mg/kg). The results were evaluated by comparing the positive control diazepam (3 mg/kg) group with the negative control saline (10 mL/kg) group. Thirty minutes after the injection of the extract and fractions, pentobarbital (30 mg/kg) was injected. Increases in sleep time were noted due to extracts and fractions. It was observed that *R. damascena* ethanolic extract and its fractions prolonged pentobarbital-induced sleep time in mice at both doses ($p<0.05$–0.001). Among all fractions, the aqueous fraction showed the lowest hypnotic effect, while the ethyl acetate fraction showed the best hypnotic effect at a dose of 500 mg/kg [81].

Cardiovascular Effect

The acute cardiovascular effect of 70% ethanol extract prepared from *R. damascena* flowers on rats was investigated. Thirty-two Wistar rats used in the study were equally divided into four groups randomly. After anesthesia, a catheter was placed in the femoral artery, and blood pressure and heart rate were recorded regularly. Animals were treated intraperitoneally with three doses of extracts

(250, 500, and 1000 mg/kg) and after 30 minutes systolic blood pressure, mean arterial pressure, and heart rate values were recorded and compared with the control group. When the extract and the control group were compared, no significant change was observed in heart rate at all doses. It was observed that the extract decreased systolic blood pressure depending on the dose at all doses and the maximum response was significant when compared with the saline group ($p<0.01–0.001$). It was found that the extract applied in different doses decreased the maximum changes in mean arterial pressure depending on the dose compared to the control group. The effect of the extract on systolic blood pressure and mean arterial pressure at higher doses was more significant when compared with lower doses ($p<0.05–0.001$). As a result, it was determined that the hypotensive effect of *R. damascena* extract did not have a significant effect on heart rate. Therefore, *R. damascena* has been reported to have a beneficial effect on controlling blood pressure [82]. Effects of aqueous ethanolic extract prepared from dried *R. damascena* flowers on heart rate and contractility in guinea pig hearts were investigated. Isolated guinea pig hearts were perfused through the aorta with the Langendorff model. Heart rate and contractility were determined by comparing the baseline values in the presence and absence of propranolol in 4 different concentrations of extract (0.1%, 0.2, 0.4, and 1.0 mg) and isoprenaline (1, 10, 100 nM, and 1 μM). The extract caused an increase in isoprenaline, heart rate, and contractility ($p<0.05–0.001$). In the absence of propranolol, the percentage increase in heart rate due to the final concentration of isoprenaline was significantly ($p<0.01$) higher than that of the extract. Propranolol caused a significant decrease in both heart rate and contractility ($p<0.05$ in both cases). However, this effect was significantly reversed with isoprenaline and extract ($p<0.05–0.001$). The percent increase in cardiac contractility due to the final concentration of the extract was significantly higher than isoprenaline in the absence and presence of propranolol ($p<0.05$ in both cases). A significant correlation was observed between both heart rate and heart contractility and isoprenaline and extract concentration ($p<0.05–0.001$). As a result, it has been stated that the aqueous ethanolic extract prepared from *R. damascena* flowers has a strong inotropic and chronotropic effect on isolated guinea pig hearts [83].

Laxative Effect

The laxative and prokinetic effects of the extract prepared by boiling *R. damascena* dried flowers in hot water were investigated. The 14 rats to be used in the study were divided into two equal groups, and while the experimental group was extracted by gavage, seven rats in the other group were determined as the control group to investigate the placebo effect. The number of feces, weight, and water percentage were measured for up to 24 hours. To assess the possible osmotic laxative effects of the extract, the jejunum was randomly divided into 4 cm segments in anesthetized rats ($n = 7$) and 0.5 mL of extract, lactulose, or saline was injected into each section. The volume of the contents of each segment was measured after 1 hour. To evaluate the intestinal transit time, the extract was treated by gavage, in fasting hungry rats, with the extract or placebo. After 30 minutes after the last treatment, all rats were given phenol red and methyl cellulose (1.5 mL) by gavage. The rats in the experimental group and the control group were killed in groups of 4 after 30 minutes, 1, 2, and 4 hours, and the amount of phenol red in various parts of the gastrointestinal system was measured. *R. damascena* extract significantly increased the number of feces and the percentage of water. However, it had no effect on the transit time to the intestine. It was observed that the content volume in the jejunum segments significantly increased with the extract compared to the placebo. It has been reported that *R. damascena* extract has a mild laxative effect in the intestines with osmotic infiltration of fluids (24). The laxative effects of the extract prepared from *R. damascena* dried flowers on dogs were investigated. Five groups of animals ($n = 5$) were given boiled extract prepared from *R. damascena* flowers at doses 0.5–8 times the recommended dose for conventional human use (180 mg/kg). The animals in the 6th group ($n = 4$) were selected as the negative control group and given 12 mL/kg of placebo (distilled water). The animals in the 7th group ($n = 4$) were selected as the positive control group and given daily 300 mg/kg lactulose. The fecal water content and the incidence of diarrhea were measured in dogs over 10 days. In addition, weights, ECG values, heart rate, fever, and respiration of the dogs were recorded. The extract has been found to cause diarrhea depending on the dose. No significant change was observed in the fecal water content. It has been observed that the extract caused sedation and slight weight loss in animals at the highest dose. It was stated that no other side effects were observed in animals other than these symptoms [84].

Relaxing and Antidepressant Effects

The relaxant effects of four cumulative concentrations (0.25%, 0.5, 0.75, and 1.0 g) of *R. damascena* ethanolic extract and its essential oil (0.25%, 0.5, 0.75, and 1.0 v/v) on guinea pig tracheal chains were evaluated. Results are given by comparing the positive control anhydrous theophylline (0.25, 0.5, 0.75, and 1.0 mM) and the negative control saline (1 mL). The guinea pigs to be used in the study were divided into three groups: Tracheal chains contracted with 60 mM KCl in Group 1, unincubated tracheal chains contracted with 10 μM methacholine hydrochloride in Group 2, and tissues incubated with 1 μM propranolol and 1 μM chlorpheniramine in Group 3. In group 1, a relaxant effect was observed when the last two concentrations of essential oil and theophylline and the final concentration of ethanolic extract were compared with saline ($p<0.01–0.001$). In group 2, the three high concentrations of ethanolic extract and theophylline and all concentrations of essential oil showed a dose-dependent relaxant effect when compared to saline ($p<0.05–0.001$). Additionally, the effect of 0.25% and 0.5 v/h essential oil was found to be significantly higher when compared with theophylline and ethanol extracts ($p<0.01$). However, extract and essential oil did not show a significant relaxant effect in Group 3. Significant correlations were observed between ethanol extract, essential oil, theophylline concentrations, and relaxant effect in Group 1 and Group 2. As a result, it was stated that *R. damascena* showed a strong relaxant effect on guinea pig tracheal chains compared to theophylline at the specified concentrations [85]. Aqueous, ethyl acetate, and n-butanol fractions obtained by partitioning from 50% ethanolic extract prepared from *R. damascena* dried flowers were examined in terms of their relaxant effects on guinea pig tracheal smooth muscles. Saline and three cumulative theophylline concentrations (0.1, 0.2, and 0.4 mM) and fractions (0.1%, 0.2, and 0.4 g) were investigated for relaxant effects on guinea pig tracheal chains precontracted with 60 mM KCl (Group 1, $n = 5$) and 10 μM methacholine (Group 2, $n = 8$). In group 1, all concentrations of the theophylline and ethyl acetate fraction and the last two concentrations of the *n*-butanol fraction showed a significant relaxant effect when compared with saline ($p<0.05–0.001$). In group 2, all concentrations of theophylline, ethyl acetate, and aqueous fractions showed a concentration-dependent relaxant effect compared to saline ($p<0.01–0.001$). In addition, the effect of the ethyl acetate fraction in Group 1 was significantly higher than that of theophylline ($p<0.05–0.001$). However, the effects of other fractions were significantly lower than theophylline in both groups ($p<0.01–0.001$). Significant correlations were observed between the relaxant effect and concentrations for theophylline and all fractions (except aqueous fractions in Group 1) in both groups. As a result, the ethyl acetate fraction of *R. damascena* has been reported to have a strong relaxant effect on tracheal smooth muscle compared to theophylline [86].

Other Effects

The inhibitory effect of ethanolic extract prepared from *R. damascena* petals on intraabdominal adhesion in rats was investigated. Thirty healthy rats were divided into three groups. The rats in Group A were determined as the animals to be treated with 1% extract and the rats in Group B 5% *R. damascena* extract. Group C was determined as the control group. After anesthesia, the abdominal walls of the rats were opened and a 2 cm shallow incision was made on the right wall. Then, a 2x2 part of the peritoneal surface was removed from the left side of the abdominal wall and 3 mL of 1% and 5% *R. damascena* extract was applied to the abdominal cavity. 3 mL of distilled water was given to the control group. The abdominal cavity was sutured and a second laparotomy was performed according to the Canbaz scale for the adhesions formed after 14 days. After histopathological examinations, all data were evaluated with SPSS software, and the results were considered statistically significant ($p<0.05$). The amount of adhesion in group A (1.4 ± 1.265) was significantly lower than the amount of adhesion in group C (3 ± 0.816) ($p = 0.007$). In histological examinations, a significant difference was observed between groups A and C between the severity of fibrosis ($p = 0.029$) and inflammation ($p = 0.009$). All rats in group B were found dead. As a result, it was stated that the use of 1% *R. damascena* extract caused a significant decrease in intraabdominal adhesion after laparotomy in rats [87]. The nephroprotective effect of the aqueous extract (250 and 500 mg/kg) prepared from *R. damascena* flowers was investigated in comparison with silymarin (200 mg/kg) on albino rabbits exposed to toxicity with gentamicin (80 mg/kg). Biochemical analysis of blood samples collected on days 0, 7, 14, and 21 was performed using standard protocols and kit methods. Results were evaluated by two-way analysis of variance and Duncan multi-range test. While gentamicin caused a significant increase in serum urea, creatine, and blood urea nitrogen levels, it caused a significant decrease in serum albumin levels ($p \leq 0.01$). The extract resulted in a statistically significant decrease in serum urea (80 ± 8.37; 56±4.59 mg/dL), creatine, and blood urea nitrogen (37.38 ± 3.91; 26.17 ± 2.14 mg/dL) levels depending on the dose at doses of 250 and 500 mg/kg, respectively, compared to the gentamicin-free, gentamicin-induced toxicity and silymarin-treated (control groups. On the other hand, a statistically significant increase in albumin (0.30 ± 0.30; 0.19 ± 0.19 g/dL) levels was observed [88].

EX VIVO BIOLOGICAL EFFECTS OF *ROSA DAMASCENA*

Effect on the Digestive System

The effect and possible mechanisms of 70% ethanol extract prepared from *R. damascena* petals on ileum contraction in Wistar rats were investigated. Forty-eight male rats were divided into six groups. On the day of the experiment, rats were anesthetized with chloroform, and segments isolated from the distal of the ileum (excluding the last 2 cm) were placed vertically in tissue baths (50 mL) containing tyrode solution between two stainless steel hooks. The contraction of the ileum was achieved with the application of KCl and when the intensity of the contractions increased, cumulative concentrations of *R. damascena* extract (100, 500, and 1000 mg/kg) were added to the organ bath. In order to understand the mechanism of action of the extract on the ileum, propranolol at a concentration of 1 μM, naloxone at a concentration of 1 μM and L-NAME (N (ω)-nitro-L-arginine methyl ester) at a concentration of 100 μM were used. It was observed that *R. damascena* extract decreased ileum contractions depending on the dose ($p>0.0001$). While propranolol and naloxone significantly reduced the inhibitory effect of *R. damascena* extract on KCl-induced contractions in the ileum ($p<0.001$), L-NAME did not cause a significant change. At the same time, calcium caused contraction of depolarized tissue, and this effect was significantly reduced with the cumulative concentrations of the extract ($p<0.001$) [89]. The effect of 80% ethanol extract prepared from *R. damascena* petals on ileum motility in rats was investigated. A part of rat ileum was suspended in organ bath containing tyrode solution. Tissue was stimulated with electrical field stimulation, KCl, and acetylcholine. It was kept under 1 g tension at 37 °C and continuously exposed to O_2 gas. The effect of *R. damascena* extract on ileum contraction was compared with atropine. *R. damascena* extract (10–100 μg/mL) induced a contraction in the ileum isolated from the rat, while at a bath concentration of 1 mg, it had a relaxant effect on the rat ileum. *R. damascena* extract (1–8 mg/mL), KCl (IC50 = 3.3 ± 0.9 mg/mL), acetylcholine (IC50 = 1.4 ± 0.1 mg/mL), and electrical field stimulation (IC50 = 1.5 ± 0.3 mg/mL) inhibited ileum contraction in a concentration-dependent manner. The carrier had no significant effect on ileum contractions. It was concluded that *R. damascena* extract had a stimulating effect on the smooth muscle of the ileum in microgram concentrations. However, it shows an inhibitory effect at milligram concentrations [90]. The effect of the aqueous fraction prepared from *R. damascena* flowers on the contraction of isolated guinea pig ileum was investigated. After removing the ethyl acetate and n-butanol fractions, the aqueous fraction of the plant was obtained from the ethanol extract. In order to determine the effect of the aqueous fraction on the contraction of the ileum, the guinea pig ileums were removed, placed in the organ bath, and its contraction was recorded. The effect of the aqueous fraction on contraction of the ileum at various concentrations (0.66, 0.83, and 1.3 mg/mL) was evaluated in comparison with acetylcholine in the presence/absence of atropine. The response of ileum to 1 μg/mL acetylcholine was accepted as 100%. As a result, *R. damascena* aqueous fraction was observed to increase basal guinea pig ileum contraction depending on the dose ($p<0.05–0.001$). Acetylcholine gave a maximum response of 23.4%, versus the maximal contraction of the fraction (1.3 mg/mL). Contraction of the ileum with the aqueous fraction was significantly reduced in the presence of 0.001 μg/mL atropine [91].

CLINICAL STUDIES OF *ROSA DAMASCENA*

EFFECT ON APHTHOUS STOMATITIS

A randomized, double-blind, placebo-controlled clinical study was conducted to evaluate the clinical efficacy of mouthwash prepared from *R. damascena* aqueous extract in the treatment of recurrent aphthous stomatitis. The mouthwash to be used for the experiment was prepared with 0.025% polysorbate, 2% glycerin, 0.5% sodium carboxymethylcellulose, 0.05% saccharin sodium and up to 20 mL of distilled water, and 20% *R. damascena* extract as active ingredient. Except for the active substance, the other substances are the same in the mouthwash prepared for the placebo group. In addition, 10% caramel was added to make the color the same. The experiment and placebo groups were asked to rinse their mouths four times a day with 5 mL of mouthwash given after performing oral hygiene procedures. Within the scope of the study, the clinical effectiveness of mouthwash on ulcers in terms of pain, size, and number was evaluated on 50 patients for 2 weeks in comparison with the placebo group on days 4, 7, 11, and 14. While there was no difference between the two groups for initial ulcer size, aphthous number, and pain median values, a significant difference was observed between the experimental group and the placebo group on the 4th and 7th days ($p<0.05$). As a result, it has been reported that mouthwashing with a mouthwash containing *R. damascena* extract is more effective than placebo in the treatment of recurrent aphthous stomatitis [92].

EFFECT ON DYSMENORRHEA

The effect of capsules (200 mg) obtained from 70% ethanol extract prepared from *R. damascena* fruits on primary dysmenorrhea was investigated in a double-blind, cross-randomized clinical study. Ninety-two single students, aged between 18 and 24, with a body mass index between 19 and 25, and pain intensity between 5 and 8 on a visual analog scale, were selected and randomly divided into two equal groups. Participants were given the same physical properties Mefenamic Acid (250 mg) capsule and *R. damascena* capsule (200 mg) in a cross-over every 6 hours for 3 days. In the first cycle, the drugs were given to the participants at random, without knowing which capsule they were taking. In the second cycle, drugs were replaced with the other drugs and continued to be given under the same conditions. The pain intensity of the participants was measured when there was menstrual pain before drug administration and 1-2-3-6-12-24-48-72 hours after drug administration. Data were collected through demographic characteristics questionnaire and visual analog scale checklist. Descriptive statistics, repeated measurement test, and independent sample t test were used via SPSS software to determine and compare the effects of the pain intensity of dysmenorrhea in both drugs in the groups. In each group, after the end of the first and second cycles, a significant difference was observed between the mean pain intensity in the measurements made at different hours ($p<0.001$). There was no significant difference between the mean pain intensity of the two groups in the first (p=0.35) and second cycles (p=0.22). As a result, it was observed that *R. damascena* and Mefenamic acid had similar effects on the severity of pain in primary dysmenorrhea [93].

Effect on Postoperative Pain

The effects of capsules containing 70% extract of *R. damascena* fruits on pain after elective cesarean operations were investigated in a double-blind placebo-controlled clinical study. Ninety-two patients who had cesarean section were selected for the study. The patients were divided into two equal groups: the experimental group (Group A) receiving capsules containing *R. damascena* extract and the placebo group (Group B). While the patients in Group A were given two capsules of 400 mg *R. damascena* 15 minutes before spinal anesthesia, the patients in Group B were given the same-looking capsules containing starch. Pain intensity was measured using a visual analog scale at various hours after the operation and the results were evaluated statistically. The total dosage and pain intensity of analgesics used to relieve post-operative pain were lower in group A than in group B. No significant side effects were observed in either group. The initiation of breastfeeding and its effects on newborns were the same in both groups [94].

Aromatherapeutic Effect on Post-dressing Pain in Burn Patients

The aromatherapeutic effects of essential oil obtained from *R. damascena* plant on pain after dressing in patients with burns were investigated in a randomized controlled clinical study. Fifty patients between the ages of 18–65 who were able to speak, have the ability to communicate effectively, and had second- and third-degree burns were selected for the study. Initial pains of the patients were measured 30 minutes before entering the dressing room on the first and second day. While five drops of 40% rose oil were dropped into distilled water for inhalation to the patients in the experimental group, five drops of distilled water were dropped to the control group as placebo. The pain intensity of the patients 15 and 30 minutes after leaving the dressing room was measured using a visual analog scale. The data were evaluated using descriptive and inferential statistics via SPSS software. A significant difference was observed between the mean pain intensity before and after the intervention at 15 and 30 minutes after dressing ($p<0.001$). In addition, a significant difference was observed in the reduction of pain intensity before and after aromatherapy in the experimental group ($p<0.05$). At the same time, it was reported that there was a significant decrease in pain severity after dressing in the experimental group compared to the control group ($p<0.05$) [95].

CONCLUSION AND FUTURE PERSPECTIVES

Damask rose is a very valuable plant that has traditional uses for various purposes in different parts of the world and has commercial importance in cosmetics. It has been determined that they have different effects in *in vitro* and *in vivo* biological activity studies on *Rosa damascena*. However, in addition to all these activity investigations, clinical studies and meta-analysis researches are quite inadequate. Therefore, the biological effects of *R. damascena* have not come to the fore yet. In order for damask rose to be marketed as a tea or herbal medicine in the future, it is important to evaluate its effects, especially by conducting clinical studies.

REFERENCES

1. M.J.M. Christenhusz, J.W. Byng, The number of known plants species in the world and its annual increase, *Phytotaxa*, 261 (2016) 201–217.
2. APG III, An update of the Angiosperm Phylogeny Group classification for the orders and families of flowering plants: APG III, *Botanical Journal of the Linnean Society*, 161 (2009) 105–121.

3. Z.M. Zhu, X.F. Gao, M. Fougere-Danezan, Phylogeny of Rosa sections Chinenses and Synstylae (Rosaceae) based on chloroplast and nuclear markers, *Molecular Phylogenetics and Evolution*, 87 (2015) 50–64.
4. NGBB Elektronik Herbaryumu, *Nezahat Gökyiğit Botanik Bahçesi*, Ataşehir, İstanbul (2021). [http://bizimbitkiler.org.tr; Date of access: 25/05/2021].
5. R. Dilmen, N. Baydar, Yağ Gülü (Rosa damascena Mill.)'nde Doku Kültürü Uygulamaları, *SDÜ Ziraat Fakültesi Dergisi*, 1 (2016) 134–141.
6. M. Mahboubi, Rosa damascena as holy ancient herb with novel applications, *Journal of Traditional and Complementary Medicine*, 6 (2016) 10–16.
7. P.K. Pal, Evaluation, genetic diversity, recent development of distillation method, challenges and opportunities of Rosa damascena: A review, *Journal of Essential Oil Bearing Plants*, 16 (2013) 1–10.
8. S. Baziar, A. Zakerin, V. Rowshan, Antioxidant properties and polyphenolic compounds of rosa damascene collected from different altitudes, *International Journal of Agronomy and Plant Production*, 4 (2013) 2937–2942.
9. S. Andalib, A. Vaseghi, G. Vaseghi, A.M. Naeini, Sedative and hypnotic effects of Iranian traditional medicinal herbs used for treatment of insomnia, *EXCLI Journal*, 10 (2011) 192–197.
10. M.H. Boskabady, M.N. Shafei, Z. Saberi, S. Amini, Pharmacological effects of Rosa damascena, *Iranian Journal of Basic Medical Sciences*, 14 (2011) 295–307.
11. M. Aslam, A. Siddiqui, Rosa damascena: An overview, *European Journal of Biomedical and Pharmaceutical Sciences*, 3 (2016) 73–79.
12. N. Mahmood, S. Piacente, C. Pizza, A. Burke, A.I. Khan, A.J. Hay, The anti-HIV activity and mechanisms of action of pure compounds isolated from Rosa damascena, *Biochemical and Biophysical Research Communications*, 229 (1996) 73–79.
13. N. Kumar, P. Bhandari, B. Singh, A.P. Gupta, V.K. Kaul, Reversed phase-HPLC for rapid determination of polyphenols in flowers of rose species, *Journal of Separation Science*, 31 (2008) 262–267.
14. N. Yassa, F. Masoomi, S.E.R. Rankouhi, A. Hadjiakhoondi, Chemical composition and antioxidant activity of the extract and essential oil of Rosa damascena from Iran, population of Guilan, *Daru*, 17 (2009) 175–180.
15. N. Kumar, B. Singh, V.K. Kaul, Flavonoids from Rosa damascena Mill., *Natural Product Communications*, 1 (2006) 623–626.
16. G. Papanov, P. Malakov, Tomova K., Aromatic compounds, flavones, and glycosides from extracted flowers of Rosa damascena, *Nauchni Trudove*, 22 (1984) 221–226.
17. E.K. Kwon, D.Y. Lee, H. Lee, D.O. Kim, N.I. Baek, Y.E. Kim, H.Y. Kim, Flavonoids from the buds of Rosa damascena inhibit the activity of 3-hydroxy-3-methylglutaryl-coenzyme a reductase and angiotensin I-converting enzyme, *Journal of Agricultural and Food Chemistry*, 58 (2010) 882–886.
18. Y.S. Velioglu, G. Mazza, Characterization of flavonoids in petals of Rosa-Damascena by Hplc and spectral-analysis, *Journal of Agricultural and Food Chemistry*, 39 (1991) 463–467.
19. A. Schieber, K. Mihalev, N. Berardini, P. Mollov, R. Carle, Flavonol glycosides from distilled petals of Rosa damascena Mill., *Zeitschrift für Naturforschung C*, 60 (2005) 379–384.
20. P. Bhandari, N. Kumar, A.P. Gupta, B. Singh, V.K. Kaul, A rapid RP-HPTLC densitometry method for simultaneous determination of major flavonoids in important medicinal plants, *Journal of Separation Science*, 30 (2007) 2092–2096.
21. S.N. Li, L.P. Guo, C.M. Liu, Y.C. Zhang, Application of supercritical fluid extraction coupled with counter-current chromatography for extraction and online isolation of unstable chemical components from Rosa damascena, *Journal of Separation Science*, 36 (2013) 2104–2113.
22. X.M. Gao, L.Y. Yang, X.Z. Huang, L.D. Shu, Y.Q. Shen, Q.F. Hu, Z.Y. Chen, Aurones and Isoaurones from the flowers of Rosa Damascena and their biological activities, *Heterocycles*, 87 (2013) 583–589.
23. Y.K. Li, J.Q. Sun, X.M. Gao, C. Lei, New Isoprenylated Aurones from the flowers of Rosa damascena, *Helvetica Chimica Acta*, 97 (2014) 414–419.
24. H. Wang, J.-X. Yang, L.L. Jie Lou, G.-Y. Liu, Q.-F. Hu, Y.-Q. Ye, X.-M. Gao, A new Isoprenylated Aurone from the flowers of Rosa damascene and its cytotoxicities, *Asian Journal of Chemistry*, 26 (2014).
25. M. Ali, S. Sultana, M. Jameel, Phytochemical investigation of flowers of Rosa damascena Mill., *International Journal of Herbal Medicine*, 179 (2016) 179–183.
26. M. Kurkcuoglu, A. Abdel-Megeed, K.H.C. Baser, The composition of Taif rose oil, *Journal of Essential Oil Research*, 25 (2013) 364–367.
27. B. Gong, D. Chen, X. Zhao, X. Yang, GC-MS analysis and antioxidant activity of chemical constituents of volatile oil from Rosa damascena Mill., *Huaxue Yu Shengwu Gongcheng*, 33 (2016) 62–65.

28. J. Vilaplana, C. Romaguera, F. Grimalt, Contact-dermatitis from geraniol in Bulgarian rose oil, *Contact Dermatitis*, 24 (1991) 301–319.
29. S. Erbas, H. Baydar, Variation in scent compounds of oil-bearing Rose (Rosa damascena Mill.) produced by headspace solid phase microextraction, hydrodistillation and solvent extraction, *Records of Natural Products*, 10 (2016) 555–565.
30. M. Mohamadi, T. Shamspur, A. Mostafavi, Comparison of microwave-assisted distillation and conventional hydrodistillation in the essential oil extraction of flowers Rosa damascena Mill., *Journal of Essential Oil Research*, 25 (2013) 55–61.
31. J. Krupcik, R. Gorovenko, I. Spanik, P. Sandra, D.W. Armstrong, Enantioselective comprehensive two-dimensional gas chromatography. A route to elucidate the authenticity and origin of Rosa damascena Miller essential oils, *Journal of Separation Science*, 38 (2015) 3397–3403.
32. T. Shamspur, M. Mohamadi, A. Mostafavi, The effects of onion and salt treatments on essential oil content and composition of Rosa damascena Mill, *Industrial Crops and Products*, 37 (2012) 451–456.
33. A. Karami, M. Khosh-Khui, H. Salehi, M.J. Saharkhiz, P. Zandi, Essential oil chemical diversity of forty-four Rosa damascena accessions from Iran, *Journal of Essential Oil Bearing Plants*, 17 (2014) 1378–1388.
34. A. Gorji-Chakespari, A.M. Nikbakht, F. Sefidkon, M. Ghasemi-Varnamkhasti, J. Brezmes, E. Llobet, Performance comparison of fuzzy ARTMAP and LDA in qualitative classification of Iranian Rosa damascena essential oils by an electronic nose, *Sensors (Basel)*, 16 (2016).
35. H. Batooli, J. Safaei-Ghomi, Comparison of essential oil composition of flowers of three Rosa damascena Mill. genotypes from Kashan, *Faslnamah-i Giyahan-i Daruyi*, 11 (2021) 241.
36. S.O.S. Bahaffi, Volatile oil composition of Taif rose, *Journal of Saudi Chemical Society*, 9 (2005).
37. Y. Li, X. Xue, L. Yu, L. Kong, L. L, Analysis of Changbai mountain Rosa damascena oil by GC/MS, *Shipin Yanjiu Yu Kaifa*, 34 (2013) 65–68.
38. M. Straubinger, H. Knapp, N. Watanabe, N. Oka, H. Washio, P. Winterhalter, Three novel eugenol glycosides from rose flowers, Rosa damascena Mill., *Natural Product Letters*, 13 (1999) 5–10.
39. P. Khadzhieva, B. Stoyanova-Ivanova, Callitrisic acid and sterols in the flowers of Bulgarian oil-bearing rose (Rosa damascena Mill), *Advances in Mass Spectrometry in Biochemistry and Medicine*, 1 (1976) 303–307.
40. Hadjieva P., Stoyanova-Ivanova B., Butchvarova, Z, Dascalow, R , Composition and structure of chemical constituents of rose flower stamen (Rosa damascena Miller), *J Global*, 61 (1979) 283–286.
41. M. Paridhavi, S.S. Agrawal, Isolation and characterization of flowers of Rosa damascena, *Asian Journal of Chemistry*, 19 (2007) 2751–2756.
42. A. Ait, H. Said, S. Derfoufi, I. Sbai, A. Benmoussa, Research article Ethnopharmacological survey of traditional medicinal plants used for the treatment of infantile colic in Morocco, *Journal of Chemical and Pharmaceutical Research*, 7 (2015) 664–671.
43. H. Krajian, A. Odeh, Polycyclic aromatic hydrocarbons in medicinal plants from Syria, *Toxicological & Environmental Chemistry*, 95 (2013) 942–953.
44. W.H. Talib, A.M. Mahasneh, Antimicrobial, cytotoxicity and phytochemical screening of Jordanian plants used in traditional medicine, *Molecules*, 15 (2010) 1811–1824.
45. M.D. Carmona, R. Llorach, C. Obon, D. Rivera, "Zahraa", a Unani multicomponent herbal tea widely consumed in Syria: Components of drug mixtures and alleged medicinal properties, *Journal of Ethnopharmacology*, 102 (2005) 344–350.
46. Victoria, J., & Arunmozhi, V., Antibacterial activity of hibiscus rosa-sinensis and rosa- damascena petals against dental pathogens, *International Journal of Integrated Science Innovative Technology*, 3 (2014) 1–6.
47. M. Shohayeb, E.S.S. Abdel-Hameed, S.A. Bazaid, I. Maghrabi, Antibacterial and antifungal activity of Rosa damascena MILL. essential oil, different extracts of rose petals, *Global Journal of Pharmacology*, 8 (2014) 1–7.
48. M.H. Eman, Antimicrobial activity of Rosa damascena petals extracts and chemical composition by gas chromatography-mass spectrometry (GC/MS) analysis, *African Journal of Microbiology Research*, 8 (2014) 2359–2367.
49. N. Shokouhinejad, M. Emaneini, M. Aligholi, F. Jabalameli, Antimicrobial effect of Rosa damascena extract on selected endodontic pathogens, *Journal of the California Dental Association*, 38 (2010) 123–126.
50. G. Özkan, O. Sağdiç, N.G. Baydar, H. Baydar, Note: Antioxidant and antibacterial activities of Rosa damascena flower extracts, *Food Science and Technology International*, 10 (2004) 277–281.

51. H. Subeki, M. Matsuura M. Yamasaki, O. Yamato, Y. Maede, K. Katakura, M. Suzuki, Trimurningish, Chairul, T. Yoshihara, Effects of central Kalimantan plant extracts on intraerythrocytic Babesia gibsoni in culture, *The Journal of Veterinary Medical Science*, 66 (2004) 871–874.
52. A. Elkhateeb, H. Matsuura, M. Yamasaki, Y. Maede, K. Katakura, K. Nabeta, Anti-babesial compounds from Rosa damascena mill., *Natural Product Communications*, 2 (2007) 765–769.
53. S.F. Zaidi, J.S. Muhammad, S. Shahryar, K. Usmanghani, A.H. Gilani, W. Jafri, T. Sugiyama, Anti-inflammatory and cytoprotective effects of selected Pakistani medicinal plants in Helicobacter pylori-infected gastric epithelial cells, *Journal of Ethnopharmacology*, 141 (2012) 403–410.
54. F.T. Artun, A. Karagoz, G. Ozcan, G. Melikoglu, S. Anil, S. Kultur, N. Sutlupinar, In vitro anticancer and cytotoxic activities of some plant extracts on HeLa and Vero cell lines, *J Buon*, 21 (2016) 720–725.
55. X.M. Gao, R.R. Wang, D.Y. Niu, C.Y. Meng, L.M. Yang, Y.T. Zheng, G.Y. Yang, Q.F. Hu, H.D. Sun, W.L. Xiao, Bioactive Dibenzocyclooctadiene Lignans from the stems of Schisandra neglecta, *Journal of Natural Products*, 76 (2013) 1052–1057.
56. A. Karagoz, F.T. Artun, G. Ozcan, G. Melikoglu, S. Anil, S. Kultur, N. Sutlupinar, In vitro evaluation of antioxidant activity of some plant methanol extracts, *Biotechnology & Biotechnological Equipment*, 29 (2015) 1184–1189.
57. M.D. Kalim, D. Dutta, P. Das, S. Chattopadhyay, Evaluation of phytochemicals interrelated to antioxidant potential of Unani plants, *International Journal of Pharma and Bio Sciences*, 6 (2015) 330–342.
58. N. Kumar, P. Bhandari, B. Singh, S.S. Bari, Antioxidant activity and ultra-performance LC-electrospray ionization-quadrupole time-of-flight mass spectrometry for phenolics-based fingerprinting of Rose species: Rosa damascena, Rosa bourboniana and Rosa brunonii, *Food and Chemical Toxicology*, 47 (2009) 361–367.
59. C.R. Achuthan, B.H. Babu, J. Padikkala, Antioxidant and hepatoprotective effects of Rosa damascena, *Pharmaceutical Biology*, 41 (2003) 357–361.
60. P. Wetwitayaklung, T. Phaechamud, C. Limmatvapirat, S. Keokitichai, The study of antioxidant activities of edible flower extracts, *Acta Hortic*, 786 (2008) 185–+.
61. N.G. Baydar, H. Baydar, Phenolic compounds, antiradical activity and antioxidant capacity of oil-bearing rose (Rosa damascena Mill.) extracts, *Industrial Crops and Products*, 41 (2013) 375–380.
62. B.L. Diffey, J. Robson, A new substrate to measure sunscreen protection factors throughout the ultraviolet-spectrum, *Journal of the Society of Cosmetic Chemists*, 40 (1989) 127–133.
63. H. Tabrizi, S.A. Mortazavi, M. Kamalinejad, An in vitro evaluation of various Rosa damascena flower extracts as a natural antisolar agent, *International Journal of Cosmetic Science*, 25 (2003) 259–265.
64. A. Gholamhoseinian, B. Shahouzehi, F. Sharifi-Far, Inhibitory activity of some plant methanol extracts on 3-Hydroxy-3-Methylglutaryl coenzyme a reductase, *International Journal of Pharmacology*, 6 (2010) 705–711.
65. D.A. Kleinsek, S. Ranganathan, J.W. Porter, Purification of 3-hydroxy-3-methylglutaryl-coenzyme A reductase from rat liver, *PNAS*, 74 (1977) 1431–1435.
66. D.W. Cushman, H.S. Cheung, Spectrophotometric assay and properties of the angiotensin-converting enzyme of rabbit lung, *Biochemical Pharmacology*, 20 (1971) 1637–1648.
67. S.B. Jazayeri, A. Amanlou, N. Ghanadian, P. Pasalar, M. Amanlou, A preliminary investigation of anticholinesterase activity of some Iranian medicinal plants commonly used in traditional medicine, *Daru*, 22 (2014) 17.
68. A. Gholamhoseinian, B. Shahouzehi, F. Sharifi-far, Inhibitory effect of some plant extracts on pancreatic lipase, *International Journal of Pharmacology*, 6 (2010) 18–24.
69. A. Gholamhoseinian, H. Fallah, F.S. Far, Inhibitory effect of methanol extract of Rosa damascena Mill. flowers on alpha-glucosidase activity and postprandial hyperglycemia in normal and diabetic rats, *Phytomedicine*, 16 (2009) 935–941.
70. G. Latifi, A. Ghannadi, M. Minaiyan, Anti-inflammatory effect of volatile oil and hydroalcoholic extract of Rosa damascena Mill. on acetic acid-induced colitis in rats, *Research in Pharmaceutical Sciences*, 10 (2015) 514–522.
71. M. Hosseini, M.G. Rahbardar, H.R. Sadeghnia, H. Rakhshandeh, Effects of different extracts of Rosa damascena on pentylenetetrazol-induced seizures in mice, *Zhongxiyi Jiehe Xuebao*, 9 (2011) 1118–1124.
72. M. Homayoun, M. Seghatoleslam, M. Pourzaki, R. Shafieian, M. Hosseini, A.E. Bideskan, Anticonvulsant and neuroprotective effects of Rosa damascena hydro-alcoholic extract on rat hippocampus, *Avicenna Journal of Phytomedicine*, 5 (2015) 260–270.

73. S. Esmaeili, A. Ghiaee, F. Naghibi, M. Mosaddegh, Antiplasmodial activity and cytotoxicity of plants used in traditional medicine of Iran for the treatment of fever, *Iranian Journal of Pharmaceutical Research*, 14 (2015) 103–107.
74. E. Esfandiary, M. Karimipour, M. Mardani, H. Alaei, M. Ghannadian, M. Kazemi, D. Mohammadnejad, N. Hosseini, A. Esmaeili, Novel effects of Rosa damascena extract on memory and neurogenesis in a rat model of Alzheimer's disease, *Journal of Neuroscience Research*, 92 (2014) 517–530.
75. E. Esfandiary, M. Karimipour, M. Mardani, M. Ghanadian, H.A. Alaei, D. Mohammadnejad, A. Esmaeili, Neuroprotective effects of Rosa damascena extract on learning and memory in a rat model of amyloid-beta-induced Alzheimer's disease, *Advanced Biomedical Research*, 4 (2015) 131.
76. T. Mohammadpour, M. Hosseini, A. Naderi, R. Karami, H.R. Sadeghnia, M. Soukhtanloo, F. Vafaee, Protection against brain tissues oxidative damage as a possible mechanism for the beneficial effects of Rosa damascena hydroalcoholic extract on scopolamine induced memory impairment in rats, *Nutritional Neuroscience*, 18 (2015) 329–336.
77. S. Awale, C. Tohda, Y. Tezuka, M. Miyazaki, S. Kadota, Protective effects of Rosa damascena and its active constituent on Abeta(25–35)-induced neuritic atrophy, *Evidence-Based Complementary and Alternative Medicine*, 2011 (2011) 131042.
78. M. Saxena, A.K. Shakya, N. Sharma, S. Shrivastava, S. Shukla, Therapeutic efficacy of Rosa damascena Mill. on acetaminophen-induced oxidative stress in Albino rats, *Journal of Environmental Pathology, Toxicology*, 31 (2012) 193–201.
79. S. Monika, S. Yamini, S. Sangeeta, Antioxidant and hepatoprotective activity of aqueous extract of rosa damascena against CCL4-induced oxidative damage, *Indian Journal of Pharmacology*, 40 (2008) 162–162.
80. H. Rakhshandah, M. Hosseini, Potentiation of pentobarbital hypnosis by Rosa damascena in mice, *Indian Journal of Experimental Biology*, 44 (2006) 910–912.
81. H. Rakhshandah, M.T. Shakeri, M.R. Ghasemzadeh, Comparative hypnotic effect of Rosa damascena fractions and diazepam in mice, *Iranian Journal of Pharmaceutical Research*, 6 (2007) 193–197.
82. A. Baniasad, A. Khajavirad, M. Hosseini, M.N. Shafei, S. Aminzadah, M. Ghavi, Effect of hydro-alcoholic extract of Rosa damascena on cardiovascular responses in normotensive rat, *Avicenna Journal of Phytomedicine*, 5 (2015) 319–324.
83. M.H. Boskabady, A. Vatanprast, H. Parsee, M. Ghasemzadeh, Effect of aqueous-ethanolic extract from Rosa damascena on guinea pig isolated heart, *Iranian Journal of Basic Medical Sciences*, 14 (2011) 116–121.
84. M. Abbaszadeh, H.R. Kazerani, A. Kamrani, Laxative effects of Rosa damascena Mill in dogs, *Journal of Applied Animal Research*, 38 (2010) 89–92.
85. M.H. Boskabady, S. Mani, H. Rakhshandah, Relaxant effects of Rosa damascena on guinea pig tracheal chains and its possible mechanism(s), *Journal of Ethnopharmacology*, 106 (2006) 377–382.
86. H. Rakhshandah, M.H. Boskabady, Z. Mossavi, M. Gholami, Z. Saberi, The differences in the relaxant effects of different fractions of Rosa damascena on guinea pig tracheal smooth muscle, *Iranian Journal of Basic Medical Sciences*, 13 (2010) 126–132.
87. M. Karimi, S.Y. Asadi, P. Parsaei, M. Rafieian-kopaei, H. Ghaheri, S. Ezzati, The effect of ethanol extract of Rose (Rosa damascena) on intra-abdominal adhesions after laparotomy in rats, *Wounds*, 28 (2016) 167–174.
88. T. Khaliq, F. Mumtaz, Zia-ur-Rahman, I. Javed, A. Iftikhar, Nephroprotective potential of Rosa damascena Mill flowers, Cichorium intybus Linn roots and their mixtures on gentamicin-induced toxicity in albino rabbits, *Pakistan Veterinary Journal*, 35 (2015) 43–47.
89. M. Sedighi, M. Rafieian-Kopaei, M. Noori-Ahmadabadi, I. Godarzi, A. Baradaran, In vitro impact of hydro-alcoholic extract of Rosa damascena Mill. on rat Ileum contractions and the mechanisms involved, *International Journal of Preventive Medicine*, 5 (2014) 767–775.
90. H. Sadraei, G. Asghari, S. Emami, Effect of Rosa damascena Mill. flower extract on rat ileum, *Research in Pharmaceutical Sciences*, 8 (2013) 277–284.
91. K. Dolati, H. Rakhshandeh, M.N. Shafei, Effect of aqueous fraction of Rosa damascena on ileum contractile response of guinea pigs, *Avicenna Journal of Phytomedicine*, 3 (2013) 248–253.
92. H. Hoseinpour, S.A.F. Peel, H. Rakhshandeh, A. Forouzanfar, M. Taheri, O. Rajabi, M. Saljoghinejad, K. Sohrabi, Evaluation of Rosa damascena mouthwash in the treatment of recurrent aphthous stomatitis: A randomized, double-blinded, placebo-controlled clinical trial, *Quintessence International*, 42 (2011) 483–491.

93. S. Bani, S. Hasanpour, Z. Mousavi, P.M. Garehbaghi, M. Gojazadeh, The effect of Rosa Damascena extract on primary dysmenorrhea: A double-blind cross-over clinical trial, *Iranian Red Crescent Medical*, 16 (2014).
94. P.M. Gharabaghi, F. Tabatabei, S.A. Fard, M. Sayyah-Melli, E. Ouladesahebmadarek, A. Del Azar, S.A. Khoei, M.M. Gharabaghi, M. Ghojazadeh, O. Mashrabi, Evaluation of the effect of preemptive administration of Rosa damascena extract on post-operative pain in elective cesarean sections, *African Journal of Pharmacy and Pharmacology*, 5 (2011) 1950–1955.
95. A. Bikmoradi, M. Harorani, G. Roshanaei, S. Moradkhani, G.H. Falahinia, The effect of inhalation aromatherapy with damask rose (Rosa damascena) essence on the pain intensity after dressing in patients with burns: A clinical randomized trial, *Iranian Journal of Nursing and Midwifery Research*, 21 (2016) 247–254.

21 *Salvia* sp.

Alper Gökbulut

INTRODUCTION

Salvia genus is a member of the Lamiaceae family which comprises over a thousand species distributed throughout the world. Species of this genus are widely localized in temperate, subtropical, and tropical regions (Zengin *et al.* 2018). The origin of word *Salvia* comes from the Latin word "salvare" meaning "to heal". In Turkey, there are over 100 *Salvia* species, 53 of them being endemic (Aydın Kıran *et al.* 2021; Güzel *et al.* 2020). Anatolia is an important localization for the diversity of the genus in Asia. *Salvia* plants have traditionally been used for the treatment of colds, aches, infections, wounds, bronchitis, and insomnia. Furthermore, *Salvia* species also have economic importance as flavor, spices, teas, and food preservatives in the food industry. Phytochemical analysis revealed that terpenoids, flavonoids, phenolic acids, and steroids are the main secondary metabolites of the *Salvia* species contributing to the anti-inflammatory, anticholinesterase, antiviral, antihepatotoxic, cytotoxic, antitumor, antioxidant, and antimicrobial effects. The phytochemical content and biological activities of the species can vary due to the growing regions, season of collection, genetical origins, and physiological properties (Zengin *et al.* 2019; Karatoprak *et al.* 2020; Gürbüz *et al.* 2021).

Standardization is one of the most important processes to benefit entirely from the herbals and their preparations. For this purpose, various analytical techniques such as liquid chromatography (LC) and gas chromatography (GC) equipped with different detectors are commonly used. In order to present the secondary metabolites qualitatively and quantitatively, and to add value to the chemical composition of the *Salvia* species, the aforementioned techniques have also been commonly used.

In this chapter, studies on the chemical constituents and biological activities of the species growing in Turkey are summarized.

CHEMISTRY

SALVIA HYPARGEIA

Bakır *et al.* (2020) performed the GC-MS and LC-MS/MS analysis of the *Salvia hypargeia* Fisch. & C.A. Mey. GC-MS results revealed that the roots of *S. hypargeia* were found to be rich in ferruginol (30,787.97 μg/g extract) and lupenone (23,276.21 μg/g extract). According to the LC-MS/MS analysis results, rosmarinic acid (**1**) (38,035.7 μg/g extract) and isoquercitrin (4136.91 μg/g extract) were the important constituents of *S. hypargeia* leaf extract. It was underlined that these developed and validated methods could be easily applied to determine major/active/toxic secondary metabolites of *Salvia* species.

Eliuz (2021) investigated the essential oil obtained from the aerial parts of *Salvia hypargeia* using GC-MS. β-pinene, 1,8-cineole, camphor, *α* pinene, 4-terpineol, and 4-thujanol were found as the major constituents.

SALVIA MARASHICA

Aydın Kıran *et al.* (2021) investigated the dichloromethane, acetone, and methanol extracts of the aerial parts of *Salvia marashica* which is an endemic species to Anatolia. Various di- and triterpenoids

 DOI: 10.1201/9781003146971-21

were isolated and their structures were elucidated. Two abietane diterpenes, abieta-8,11,13-triene and 18-acetoxymethylene-abieta-8,11,13-triene, were obtained from the acetone extract, which were isolated for the first time from *Salvia* genus. Also, rosmarinic acid (Figure 21.1) was found as a significant constituent of the methanol extract of *S. marashica*.

SALVIA CERINO-PRUINOSA

The ethanol extracts of the aerial parts and roots of an endemic species, *Salvia cerino-pruinosa* Rech. f. var. *cerino-pruinosa*, were chromatographed on silica gel columns, and 20 known secondary metabolites were isolated from the antioxidant fractions. The main compounds were determined as rosmarinic, chlorogenic and caffeic acids, luteolin 7-*O*-glucoside, salvianolic acids A and B, and inuroyleanol (Ertaş *et al.* 2021).

SALVIA ABSCONDITIFLORA

Koysu *et al.* (2019) performed phytochemical analysis and evaluated the antioxidative properties of the secondary metabolites isolated from *Salvia absconditiflora*. By using spectrophotometric techniques including 1D, 2D NMR, and LC-TOF/MS, ursolic acid, crismaritin, luteolin, rosmarinic acid methyl ester, 3,4-dihydroxyl benzaldehyde (protocatechuic aldehyde), caffeic acid, apigenin-7-*O*-*β*-glucoside, rosmarinic acid, and luteolin-7-*O*-β-glucoside were isolated and their structures were elucidated.

Drying method is a key step in the preparation of plant materials for experimental evaluation. In this regard, the effects of three drying methods (oven, freeze, and shade drying) on phytochemical components of *Salvia absconditiflora* Greuter & Burdet collected from Konya, Turkey, were examined. Shade drying provided the highest total phenolic content (TPC) (99.33 mg GAE (gallic acid equivalent)/g) and total flavonoid content (TFC) (46.88 mg RE (rutin equivalent)/g) followed by oven drying (58.15 mg GAE/g for TPC, 40.65 mg RE/g for TFC) and freeze-drying (43.73 mg GAE/g for TPC, 36.68 mg RE/g for TFC). The main compound determined by HPLC was rosmarinic acid. Shade drying presented the highest total bioactive compounds and possessed the strongest antioxidant features. Also, it was suggested that shade drying was the most suitable drying method for *S. absconditiflora* due to the presence of the highest rosmarinic acid content and biological activities (Uysal 2019).

SALVIA ALBIMACULATA, SALVIA POTENTILLIFOLIA, AND SALVIA NYDEGGERI

Aerial parts of *Salvia albimaculata* Hedge & Hub.-Mor., *Salvia potentillifolia* Boiss & Heldr. ex Bentham., and *Salvia nydeggeri* Hub.-Mor. from Southwest Anatolia, Turkey, were examined for their phenolic compounds. UPLC-ESI–MS/MS analysis results revealed that caffeic acid (3582.8 ± 2.5 μg/g, 2956.5 ± 4.6 μg/g, and 2457.7 ± 3.1 μg/g) and 3,4-dihydroxy benzoic acid (1846.2 ± 3.1 μg/g, 2019.1 ± 2.2 μg/g, and 1901.3 ± 1.5 μg/g) were determined as major phenolic compounds in

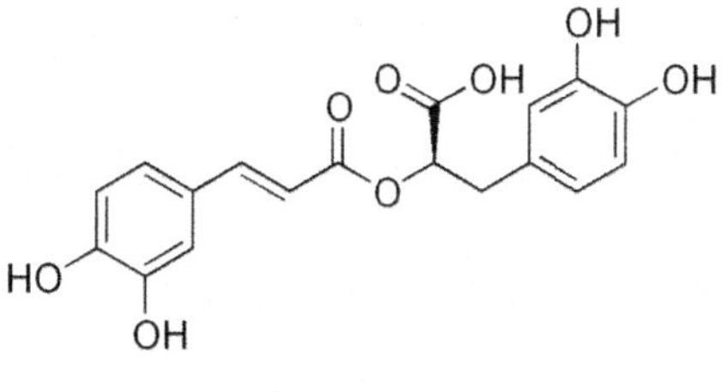

FIGURE 21.1 Rosmarinic acid, Drowen by Alper Gokbulut

S. potentillifolia, *S. albimaculata*, and *S. nydeggeri*, respectively, which were believed to be responsible for the antioxidant potentials of the species (Kıvrak *et al.* 2019).

Salvia virgata, Salvia verticillata, Salvia trichoclada, and Salvia kronenburgii

The Turkish name of *S. virgata* is "yılancık" or "fatmanaotu", which is used for the cure of wounds and various skin problems. A decoction of the aerial parts of *S. virgata* has been used for preventing blood cancer. Also, it was reported to be consumed as a strong tea in Iran. The essential oil composition of the species was analyzed by GC-MS. The main essential oil constituents of the aerial parts of *S. virgata* were detected and identified as pentacosane (20.09%), caryophyllene oxide (6.90%), phytol (6.83%), spathulenol (6.09%), and nonacosane (5.15%) (Şenkal 2019).

Salvia virgata and *Salvia verticillata* ssp. *amasiaca* collected from Ankara, Turkey, together with five commercial *Salvia* tea bag samples were investigated for their phenolic acid contents. HPLC-PDA analysis allowed the identification and quantification of rosmarinic, chlorogenic, and caffeic acids in the methanol extracts. It was indicated that the studied *Salvia* species were found to be rich in rosmarinic acid, and tea bag samples belonging to different companies have a similar phenolic acid profile (Gökbulut 2013).

Reversed-phase high-performance liquid chromatography (RP-HPLC) coupled with UV spectrometry has been used for the quantification of phenolic acids in *Salvia vertisillata* L. subsp. *vertisillata*, *Salvia trichoclada* Benth., and *Salvia kronenburgii* Rech.f. which were collected from the Van-Gurpinar-Catak region, Turkey in their flowering times. It was concluded that higher amounts of rosmarinic, caffeic, and chlorogenic acids (0.4617%, 0.0192%, and 0.1165%, respectively) were detected in *Salvia vertisillata* L. subsp. *vertisillata*, while gallic acid (0.5463%) was found in particular amount in *Salvia trichoclada*. It was also reported that the endemic species *Salvia kronenburgii* Rech.f. had less amount of phenolics compared with the other two *Salvia* species (Gökbulut *et al.* 2010).

Salvia ekimiana

Kaya *et al.* (2021) studied the essential oils of endemic *Salvia ekimiana* collected from two different regions of Anatolia. The obtained oils were tested using gas chromatography (GC-FID) and gas chromatography-mass spectrometry (GC-MS). Oxygenated sesquiterpenes were determined in almost equal amounts in the oils, while sesquiterpene hydrocarbons were found in high amounts in the oil of Yozgat locality. Also, oxygenated monoterpenes in Kayseri oil and diterpenes in Yozgat oil of *S. ekimiana* were detected in high amounts.

Salvia candidissima, Salvia heldreichiana, and Salvia tomentosa

The essential oil composition of *Salvia candidissima* ssp. *occidentalis* and *Salvia heldreichiana* from Konya and *Salvia tomentosa* from Eskisehir, Turkey, were analyzed using GC-MS. Obtained results revealed alpha-pinene/borneol/caryophyllene oxide/terpinen-4-ol chemotype present in *S. heldreichiana* and alpha-pinene/borneol chemotype present in *S. tomentosa* (Bardakçı *et al.* 2019).

Salvia divaricata, S. eriophora, S. longipedicellata, and S. pilifera

Characterization of the essential oils of four endemic species of Turkey, *Salvia divaricata* Montbret & Aucher ex Bentham, *S. eriophora* Boiss. & Kotschy, *S. longipedicellata* Hedge, and *S. pilifera* Montbret & Aucher ex Bentham, was analyzed by GC and GC-MS. The major constituents of the essential oils were detected as 1,8-cineole (34.4%), caryophyllene oxide (13.7%), β-caryophyllene (47.9%), and α -pinene (9.4%) in *Salvia divaricata*, *Salvia eriophora*, *Salvia longipedicellata*, and *Salvia pilifera*, respectively (Kürkçüoğlu *et al.* 2019).

Chemical compound characterization of the leaves of *Salvia eriophora* collected from Sivas, Turkey, was examined by LC–MS/MS (Liquid Chromatography Tandem Mass Spectrometry). Salvigenin (158.64 ± 10.8 mg/kg), fumaric acid (123.09 ± 8.54 mg/kg), and quercetagetin-3.6-dimethylether (37.85 ± 7.09 mg/kg) were found as main compounds in the ethanol extract, while fumaric acid (555.96 ± 38.56 mg/kg), caffeic acid (103.62 ± 20.51 mg/kg), and epicatechin (83.19 ± 8.43 mg/kg) were the major constituents of the water extract (Bursal *et al.* 2019).

Mericarps of the *Salvia longipedicellata* Hedge which is a perennial endemic species found in East Anatolia, were investigated for fatty acid, sterol, tocol compositions, mineral, total phenolic, and flavonoid contents. *α* -Linolenic acid, β-sitosterol, γ-tocopherol, and potassium were determined in significant amounts. Also, it was indicated that total phenolic and flavonoid contents (1.04 µg GAE/mg extract and 0.32 µg QE (quercetin equivalent)/mg extract, respectively) contributed to the antioxidant potential (Güzel *et al.* 2020).

Salvia fruticosa, Salvia officinalis, Salvia sclarea, and Salvia dichroantha

The essential oil compositions of *Salvia fruticosa* Mill., *Salvia officinalis* L., *Salvia sclarea* L., hybrid (*Salvia fruticosa* Mill. x *Salvia officinalis* L.), and *Salvia dichroantha* L. cultivated and collected from İzmir, Turkey, were investigated by using GC and GC-MS, and the main constituents of the essential oils were detected as 1,8-cineole (57.18%), β-thujone (34.59%), linalyl acetate (46.77%), 1,8-cineole (21.42%), and β-caryophyllene (23.11%), respectively (Karik *et al.* 2018).

Süzgeç-Selçuk *et al.* (2021) studied the volatile constituents from the leaves and the galls of *Salvia fruticosa* Mill. by GC and GC-MS using both the conventional hydrodistillation and micro-steam distillation-solid-phase microextraction methods. Qualitative distinctions were observed among the volatiles obtained by two different extraction methods. The oxygenated monoterpenes (62.4–69.3%) were detected as the dominant secondary metabolite group with eucalyptol and camphor as the main constituents in all the tested samples with the exception of the gall oil in which oxygenated sesquiterpenes (25.6%) and diterpenes (17.3%) were found in high amount.

Salvia montbretii

The essential oil obtained from the aerial parts of *Salvia montbretii* Benth collected from Şanlıurfa, Turkey, was analyzed by using GC and GC-MS. The main constituents of the essential oil were detected as β-caryophyllene (32.8%), β-pinene (9.8%), *α* -humulene (8.2%), 12-hydroxy-β-caryophyllene acetate (6.6%), germacrene D (4.9%), and *α* -pinene (4.5%) (Abak *et al.* 2018).

Salvia poculata

Chemical constituents of *Salvia poculata* were investigated by Yener (2020). GC-MS results depicted that germacrene D (20.07%), caryophyllene (17.57%), and tricosanoic acid (17.01%) were determined as the main components of the essential oil, aroma, and fatty acid compositions, respectively. LC-MS/MS results also showed a significant amount of ferruginol (9292.9 µg analyte/g extract) and rosmarinic acid (15,612.08 µg analyte/g extract) which should be responsible for the antioxidant potential.

Miscellaneous

Rosmarinic acid and total phenolic contents of 14 *Salvia* species collected from seven regions of Anatolia were studied. It was indicated that all the samples contained a significant amount of rosmarinic acid which contributed to the antioxidant and *α* -glucosidase enzyme inhibitory activities (Adımcılar 2019).

A new validated HPLC-PDA method was developed for the determination of rosmarinic, caffeic, chlorogenic, and gallic acids in 12 *Salvia* species growing naturally in Anatolia. Considering the botanical parts (flowers, leaves, stems, and roots), distinctions in the content of gallic and rosmarinic acids were detected. It was indicated that the rosmarinic acid and gallic acid contents in the leaves and flowers were found much higher than those of the branch or root parts of ten species of *Salvia*. Nevertheless, it was reported that the rosmarinic and gallic acid contents in the roots of *S. halophila* Hedge and *S. syriaca* L. were found to be in higher amounts among the investigated species (Kan *et al.* 2007).

Seeds of six *Salvia* species cultivated and grown in Cumra/Konya (Turkey) were investigated using headspace gas chromatography mass spectroscopy (GC-MS) and Fourier transform infrared spectroscopy-attenuated total reflectance (FTIR-ATR) combined with chemometrics of hierarchical cluster analysis (HCA) and principal component analysis (PCA). The dominant (%) volatile compounds were determined as α -pinene, camphene, β-pinene, and eucalyptol (Tulukçu *et al.* 2019).

BIOLOGICAL ACTIVITIES

Since *Salvia* species have diverse secondary metabolites, it is not surprising that they have various biological activities such as antioxidant, antimicrobial, wound healing, and enzyme inhibitory.

Antioxidant and Enzyme Inhibitory Activities

Salvia sclarea L. (clary sage) is traditionally used for various disorders including stress, asthma, amenorrhea, dysmenorrhea, sore throat, colics, depression, fatigue, nervousness, migraine, varicose veins, and hemorrhoids, and is also used as a food product. Phytochemical profile, antioxidant potential, α -amylase, α -glucosidase, acetylcholinesterase, butyrylcholinesterase, and tyrosinase enzyme inhibitory activities of *S. sclarea* collected from Kastamonu, Turkey, were studied. Reversed-phase high-performance liquid chromatography with diode array detector analysis results showed that *S. sclarea* was rich in rosmarinic acid. The water extract showed the lowest inhibitory activity against α -amylase, although it exhibited the best activity against α -glucosidase and tyrosinase. It was also indicated that only the aqueous extract (8.86 mg KAE (kojic acid equivalent)/g extract) predominantly inhibited tyrosinase enzyme. Docking studies revealed that quercetin was supposed to be responsible for the antityrosinase effect (Zengin *et al.* 2018).

Salvia viridis L. (synonym: *Salvia horminum* L.), commonly known as "Red topped sage", naturally occurs in the Mediterranean region. In Turkey, aerial parts of *S. viridis* have been used as an infusion against sore throat, cough, and gastrointestinal complications. Phytochemical characterization and enzyme inhibitory and antioxidant activities of the ethanolic root extracts of *S. viridis* obtained by several extraction methods such as microwave-assisted extraction, maceration, supercritical fluid extraction, Soxhlet extraction, and ultrasonic-assisted extraction were investigated. Secondary metabolite characterization of these extracts was examined by HPLC–MS/MS, and the profiles (23 components) of the supercritical fluid extract were found to be different from the other extraction methods. It was reported for the first time that the inhibitory activity of ethanolic root extract of *S. viridis* on acetylcholinesterase, butyrylcholinesterase, α -amylase, α -glucosidase, and tyrosinase was promising, so further investigations are needed which might lead to the development of new pharmaceutical preparations for the treatment of diabetes, Alzheimer's disease, and skin hyperpigmentation problems (Zengin *et al.* 2019).

Salvia aramiensis Rech. f. is a species that only grows in Hatay, Turkey, and is traditionally used as a tea for gastrointestinal problems. Collagenase and elastase enzyme inhibitory activities of 70% methanol extract of S. *aramiensis* were analyzed. The collagenase enzyme inhibition of the different concentrations of the *S. aramiensis* extract was found to be between 66.64% and 72.66 %. Furthermore, nanoliposomes which were formed with 70% methanol extract were evaluated for the aforementioned activities. It was concluded that the researchers successfully developed a long-term antioxidant and enzyme inhibitory formulation containing *S. aramiensis* (Karatoprak *et al.* 2020).

Anticholinergic, antidiabetic, and antioxidant activities of *Salvia pilifera* Montbret & Aucher ex Bentham were evaluated. Methanol and lyophilized water extracts possessed significant inhibitory activities against AChE (IC_{50}: 94.93 and 138.61 µg/mL), BChE (IC_{50}: 60.05 and 99.13 µg/mL), α -glucosidase (IC_{50}: 23.28 and 36.47 µg/mL), and α -amylase (IC_{50}: 46.21 and 97.67 µg/mL) enzymes. Both extracts also showed radical scavenging activity (Gülçin *et al.* 2019).

α -Glucosidase enzyme inhibitory activities of 14 Anatolian *Salvia* species were investigated. Among the investigated extracts, *S. aucheri* var. *aucheri* and *S. adenocaulon* possessed the best α -glucosidase inhibitory activities with the lowest IC_{50} values, 17.6 and 25.9 µg/mL, respectively. A positive correlation was found between both ursolic and oleanolic acid contents and α -glucosidase inhibitory activities of the extracts (Kalaycıoğlu *et al.* 2018).

Anticholinesterase, antiurease, antityrosinase and antielastase inhibitory, antioxidant, and cytotoxic activities of *S. hypargeia* were studied. Butyrylcholinesterase and elastase inhibitory potential of ferruginol, the main component of the roots of *S. hypargeia*, was found to be promising (Bakır *et al.* 2020).

Acetylcholinesterase, α -amylase, butyrylcholinesterase, and α -glycosidase enzyme inhibitory activities of *S. eriophora* were studied. It was indicated that the aqueous and methanol extracts of *S. eriophora* significantly inhibited all the investigated enzymes (Bursal *et al.* 2019).

Antioxidant and antimicrobial activities of the main extract and subfractions of *Salvia heldreichiana*, an endemic plant from Turkey, were examined. The obtained results have indicated a positive correlation between the phenolic composition and the antioxidant profile as well as the antimicrobial activities of the extracts. Ethyl acetate subfraction showed the highest antioxidant and antimicrobial activities which also had the highest phenolic and rosmarinic acid content (Bardakci *et al.* 2019).

Uysal *et al.* (2021) examined the chemical characterization, antioxidant, cytotoxic, antimicrobial, and enzyme inhibitory effects of different extracts of *Salvia ceratophylla* L. from Turkey. The hydro-methanolic extract of the aerial parts showed promising DPPH radical scavenging activity (193.40 +/-0.27 mg TE (trolox equivalent)/g), and the highest reducing potential against CUPRAC (CUPric Reducing Antioxidant Capacity) (377.93 +/- 2.38 mg TE/g). Significant tyrosinase inhibitory activity was observed with dichloromethane extract obtained from the underground parts (125.45 +/- 1.41 mg KAE/g). Phytochemical analysis showed the presence of phenolics such as luteolin, gallic acid, and rosmarinic acid which should be responsible for the observed activities.

Kocakaya *et al.* (2020) investigated both *in vitro* and *in silico* acetylcholinesterase, butyrylcholinesterase, urease, and tyrosinase enzymes inhibitory potentials of secondary metabolites of *Salvia* species such as rosmarinic acid, salvigenin, salvianolic acids A and B, tanshinones I and IIA, cyrtotanshinone, dihydrotanshinone I, carnosic acid, carnosol, and danshensu sodium salt. Inhibitory activities on acetyl- and butyryl-cholinesterase of dihydrotanshinone I, carnasol, and carnosic acid were found to be promising. Also, all other secondary metabolites inhibited butyrylcholinesterase in different concentrations. Tanshinone I possessed significant anti-urease (42.41 +/- 0.85%) and anti-tyrosinase (39.82 +/- 1.16%) activities. According to Dock score analysis and Lipinski parameters, these ligands were thought to be potential inhibitors of the aforementioned enzymes.

Among the plants that have been investigated for dementia and cancer treatment, *Salvia* (sage) is one of the most promising taxa in the family of Lamiaceae in the world. Karakaya *et al.* (2020) studied the enzyme inhibitory activity against acetylcholinesterase/butyrylcholinesterase, total phenolics content, and antioxidant potential of polar and apolar extracts together with the essential oils of flowers and aerial parts of *S. verticillata* subsp. *amasiaca*. Caryophyllene oxide indicated significant cholinesterase inhibitory and antioxidant activities.

Antimicrobial Activity

Yazgan (2020) investigated the antimicrobial activity of sage essential oil and its emulsified form on the growth of important food-related microorganisms (*Enterococcus faecalis* ATCC29212, *Klebsiella pneumoniae* ATCC700603, *Salmonella paratyphi A* NCTC13, and *Staphylococcus aureus* ATCC29213) and spoilage microorganisms (*Proteus mirabilis*, *Photobacterium damselae*, *Vibrio vulnificus*, *Enterococcus faecalis*, *Pseudomonas luteola*, and *Serratia liquefaciens*). GC-MS analysis results indicated that the volatile compounds of sage essential oil were eucalyptol, (+)-2-bornanone, α -pinene, borneol, camphene, β-pinene with percentages of 14.46, 14.33, 14.00, 8.15, 7.53, and 6.27, respectively. The essential oil was found to be tremendously effective on *K. pneumoniae* and *P. damselae* with inhibition zones of 17.25 and 19.00 mm, respectively. The nanoemulsion was found to be active on *S. aureus*, *S. paratyphi*, and *P. mirabilis* with inhibition zones of 12.63, 12.00, and 12.25 mm, respectively.

Yılar *et al.* (2020) examined the antifungal and bioherbicidal properties of essential oils of *Salvia absconditiflora* which were collected from ten locations in Kirsehir province, Turkey. GC-MS analysis results indicated that the main constituents of the oil were camphor (10.52–58.64%) and viridiflorol (3.42–25.2%). The essential oil of *S. absconditiflora* (a dose of 10 μL/petri dishes) inhibited the mycelium growth of *Sclerotinia sclerotiorum* and *Alternaria solani* pathogens by 9.3 and 54.40%, respectively. It was concluded that the essential oil of the plant should be used as an alternative to synthetic fungicide and herbicide chemicals.

Eliuz (2021) studied the antimicrobial activity of the essential oil of *Salvia hypargeia* against *A. baumannii*, *S. aureus*, and *C. tropicalis* using agar well diffusion assay and microdilution methods. It was concluded that an increase in DNA, protein, and K+ leakage was observed due to the interaction of microorganisms with the essential oil. On the other hand, the triphenyltetrazolium chloride-dehydrogenase of the treated microorganism cells was also significantly decreased due to slowing the respiration.

Cytotoxic and Antiproliferative Activities

Among noncommunicable diseases, cancer with various types is the second leading cause of death. The mortality rate due to cancer is estimated to be 11.5 million in 2030. To evaluate the anticancer potential of herbs, the cytotoxic and photo-cytotoxic activities of different medicinal plants including *Salvia cedronella*, *Salvia chionantha*, and *Salvia adenophylla* collected from South West Anatolia were studied against HT 144 (human malignant melanoma) cancer cell lines using the MTT assay and flow cytometry with annexin V/PI dual staining technique. Results showed that *Salvia* species induced apoptosis with the intracellular reactive oxygen species (ROS) generation secreted by TNF-α , and have the potential in the treatment of melanoma (Aydoğmuş-Öztürk *et al.* 2020).

The cytotoxic and genotoxic effects of *Salvia kronenburgii* were investigated on breast cancer cell lines, MCF-7 and MDA-MB-231. *S. kronenburgii* possessed antiproliferative effect in a dose-dependent manner on MCF-7 and MDA-MB-231 cell lines by inducing apoptosis-like cell death. Furthermore, it was concluded that meaningful augmentation was determined in the genetic damage index and frequencies in the damaged cells. Consequently, *S. kronenburgii* was offered as a promising natural source for cancer therapy, although *in vivo* studies are needed for understanding the possible mechanism of activity (Çebi *et al.* 2019).

Phenolic profile, antioxidant, antiproliferative, and glutathione S-transferase (GST) inhibitory activities of aqueous extract of *Salvia fruticosa* Mill. collected from İzmir, Turkey, were investigated. According to HPLC analysis results, the main bioactive constituents of the plant were rosmarinic and caffeic acids. The extract showed a dose-dependent antiproliferative activity on colorectal adenocarcinoma cell lines (Caco-2 and HT-29). Also, the extract inhibited GST with an IC_{50} value of 6.8 μg/mL. It was suggested that *S. fruticosa* may be used in the prevention and treatment of colon cancer (Altay *et al.* 2019).

Anti-inflammatory Activity

Aqueous, methanol, *n*-butanol, acetone, and chloroform extracts of *S. fruticosa*, *S. verticillata*, and *S. trichoclada* collected from different locations of Anatolia were investigated for their anti-inflammatory activity using rats as *in vivo* experimental model. A carrageenan-induced inflammatory paw edema model was employed. Results indicated that all the extracts possessed anti-inflammatory activities, while *n*-butanol extract of *Salvia fruticosa* was found to be the most active (Çadırcı *et al.* 2012).

Wound Healing Activity

Salvia species are traditionally used for their wound healing features. For this reason, the effects of ointment prepared with ethanol extract obtained from the aerial parts of *Salvia hypargeia*, an endemic plant from Turkey, were investigated on diabetic rat incisional and excisional skin wounds. In both excisional and incisional wounds, significant healing ratios (up to 99%) were determined. According to histopathological results, re-epithelialization and formation of granulation tissue in all *S. hypargeia* groups were observed. It was concluded that *S. hypargeia* should potentially be used for the treatment and management of diabetic wounds with topical application (Yusuf *et al.* 2021).

In vivo wound healing activity, *in vitro* antimicrobial and antioxidant potential, and total phenolic and flavonoid contents of the aerial parts of *Salvia kronenburgii* Rech. f. and *Salvia euphratica* Montbret, Aucher & Rech. f. var. *euphratica* collected from Eastern Anatolia were studied. 0.5% and 1% (w/w) concentrations of ethanol extracts were examined in incision and excision wound models on streptozotocin-induced diabetic rats using biomechanical, biochemical, histopathological, macroscopic, and genotoxic methods for 1 and 2 weeks. It was observed that *S. kronenburgii* and *S. euphratica* possessed significant wound healing activities, while *S. kronenburgii* was found to be more active at both 7 and 14 days. Also, both species possessed tremendous antibacterial activities against *A. baumannii*, and *S. euphratica* showed strong antimycobacterial activity against *M. tuberculosis* (0.24 μg/mL MIC value) (Güzel *et al.* 2019).

Antiamnesic Activity

Antiamnesic activity of *Salvia multicaulis* essential oil on scopolamine-induced amnesia in rats was examined *using in vivo* and *in silico* approaches. *S. multicaulis* essential oil (1% and 3%) applied for 21 days attenuated cognitive deficits and exhibited anxiolytic and antidepressant-profile in the scopolamine-induced amnesia in rats. Behavioral test results demonstrated that inhalation of *S. multicaulis* essential oil exhibited significant antiamnesic activity together with anxiolytic-antidepressant-like effects in the scopolamine-treated rats (Bağcı *et al.* 2019).

Salvia species have been proven as potential natural therapeutics for Alzheimer's disease. *Salvia fruticosa* which is known as Turkish sage or Greek sage has been shown to possess anticholinergic effects. Gürbüz *et al.* (2021) studied the protective effects of the *S. fruticosa* infusion and its main constituent rosmarinic acid on amyloid beta 1-42-induced cytotoxicity on SH-SY5Y cells together with p-GSK-3β activation. *S. fruticosa* infusion exhibited activity to prevent amyloid beta 1-42-induced neurotoxicity by the regulation of *p*-GSK-3β protein.

Analgesic Activity

The aerial parts of *Salvia wiedemannii* Boiss. have been used to treat peptic ulcers and to relieve pain in Turkey. The analgesic effect of *S. wiedemannii* was evaluated using tail flick and acetic acid-induced writhing tests in mice. It was determined that the chloroform extract (500 mg/kg, i.p.) showed significant analgesic activity on tail flick assay, while water, ethanol, and butanol extracts had no activity. Chloroform extract (500 mg/kg, i.p.) also inhibited writhings induced by acetic acid (Ustun and Sezik 2011).

CONCLUSIONS

During the last decades, the role of natural products as protecting and healing agents has become a key subject all over the world. The aforementioned studies on various *Salvia* species indicated that Turkish *Salvia* species have been a promising natural source to combat health disorders and manage common human ailments due to their wide range of biological activities. Therefore, usage of the effective and unique secondary metabolites and preparations of *Salvia* plants in the pharmaceutical, cosmetical, and food industries would contribute significantly to mankind's benefit.

REFERENCES

Abak F. et al. (2018). Composition of the essential oil of *Salvia montbretii* Benth. from Turkey. *Records of Natural Products*, 12(5), 426–431.

Adımcılar V. (2019). Rosmarinic and carnosic acid contents and correlated antioxidant and antidiabetic activities of 14 *Salvia* species from Anatolia. *Journal of Pharmaceutical and Biomedical Analysis*, 175, 112763.

Altay A. et al. (2019). Anatolian sage *Salvia fruticosa* inhibits cytosolic glutathione-s-transferase activity and colon cancer cell proliferatıon. *Food Measure*, 13, 1390–1399.

Aydın Kıran S. et al. (2021). Di-, and triterpenoids isolation and LC-MS analysis of *Salvia marashica* extracts with bioactivity studies. *Records of Natural Products*, 15(6), 463–475.

Aydoğmuş-Öztürk F. et al. (2020). Preclinical study of the medicinal plants for the treatment of malignant melanoma. *Molecular Biology Reports*, 47, 5975–5983.

Bağcı E. et al. (2019). Evaluation of antiamnesic activity of *Salvia multicaulis* essential oil on scopolamine-induced amnesia in rats: *In vivo* and *in silico* approaches. *Heliyon*, 5(8), e02223.

Bakır D. et al. (2020). GC-MS method validation for quantitative investigation of some chemical markers in *Salvia hypargeia* Fisch. & CA Mey. of Turkey: Enzyme inhibitory potential of ferruginol. *Journal of Food Biochemistry*, 44(9), e13350.

Bardakci H. et al. (2019a). A comparative investigation on phenolic composition, antioxidant and antimicrobial potentials of *Salvia heldreichiana* Boiss. ex Bentham extracts. *South African Journal of Botany*, 125, 72–80.

Bardakçı H. et al. (2019b). Essential oil composition of *Salvia candidissima* Vahl. *occidentalis* Hedge, *S. tomentosa* Miller and *S. heldreichiana* Boiss. ex Bentham from Turkey. *Journal of Essential Oil Bearing Plants*, 22(6), 1467–1480.

Bursal E. et al. (2019). Phytochemical content, antioxidant activity, and enzyme inhibition effect of *Salvia eriophora* Boiss. & Kotschy against acetylcholinesterase, α-amylase, butyrylcholinesterase, and α-glycosidase enzymes. *Journal of Food Biochemistry*, 43(3), e12776.

Çadırcı E. et al. (2012). Anti-inflammatory effects of different extracts from three *Salvia* species. *Turkish Journal of Biology*, 36(1), 59–64.

Çebi A. et al. (2019). Cytotoxic and genotoxic effects of an endemic plant of Turkey *Salvia kronenburgii* on breast cancer cell lines. *Journal of Cancer Research and Therapeutics*, 15(5), 1080–1086.

Eliuz, EAE. (2021). Antimicrobial activity and mechanism of essential oil of endemic *Salvia hypargeia* Finc. & Mey. in Turkey. *Indian Journal of Microbiology*. https://doi.org/10.1007/s12088-021-00939-1.

Ertaş A. et al. (2021). Bioguided isolation of secondary metabolites from *Salvia cerino-pruinosa* Rech. f. var. *cerino-pruinosa*. *Records of Natural Products*, 15(6), 585–592.

Gökbulut A. (2013). Investigations on rosmarinic, chlorogenic and caffeic acid contents of *Salvia virgata*, *Salvia verticillata* ssp. *amasiaca* and five commercial *Salvia* tea bag samples using HPLC-DAD method. *Fabad Journal of Pharmaceutical Sciences*, 38, 49–53.

Gökbulut A. et al. (2010). Simultaneous determination of selected phenolic acids in Turkish *Salvia* species by HPLC-DAD. *Chemistry of Natural Compounds*, 46(5), 805–806.

Gülçin İ. et al. (2019). Sage (*Salvia pilifera*): Determination of its polyphenol contents, anticholinergic, antidiabetic and antioxidant activities. *Food Measure*, 13, 2062–2074.

Gürbüz P. et al. (2021). In vitro biological activity of *Salvia fruticosa* Mill. infusion against amyloid β-peptide-induced toxicity and inhibition of GSK-3β, CK-1δ, and BACE-1 enzymes relevant to Alzheimer's disease. *Saudi Pharmaceutical Journal*, 29(3), 236–243.

Güzel S. et al. (2019). Wound healing properties, antimicrobial and antioxidant activities of *Salvia kronenburgii* Rech. f. and *Salvia euphratica* Montbret, Aucher & Rech. f. var. *euphratica* on excision and incision wound models in diabetic rats. *Biomedicine & Pharmacotherapy*, 111, 1260–1276.

Güzel S. et al. (2020). Chemical composition and some biological activities of *Salvia longipedicellatahedge* mericarps. *Chemistry of Natural Compounds*, 56(5), 788–792.

Kalaycıoğlu Z. et al. (2018). α-Glucosidase enzyme inhibitory effects and ursolic and oleanolic acid contents of fourteen Anatolian *Salvia* species. *Journal of Pharmaceutical and Biomedical Analysis*, 155, 284–287.

Kan Y. et al. (2007). Development and validation of a LC method for the analysis of phenolic acids in Turkish *Salvia* species. *Chromatographia*, 66, 147–152.

Karakaya S et al. (2020). A caryophyllene oxide and other potential anticholinesterase and anticancer agent in *Salvia verticillata* subsp. *amasiaca* (Freyn & Bornm.) Bornm. (Lamiaceae). *Journal of Essential Oil Research*, 32(6), 512–525.

Karatoprak GS. et al. (2020). Potential antioxidant and enzyme ınhibitory effects of nanoliposomal formulation prepared from *Salvia aramiensis* Rech. f. extract. *Antioxidants*, 9(4), 293.

Karik Ü. et al. (2018). Essential oil composition of some sage (*Salvia* spp.) species cultivated in İzmir (Turkey) ecological conditions. *Indian Journal of Pharmaceutical Education and Research*, 52(4), 102–107.

Kaya A. et al. (2021). Composition of the essential oils of endemic *Salvia ekimiana* growing in two different areas of Turkey. *Chemistry of Natural Compounds*, 57, 563–565.

Kıvrak Ş. et al. (2019). Investigation of phenolic profiles and antioxidant activities of some *Salvia* species commonly grown in Southwest Anatolia using UPLC-ESI-MS/MS. *Food Science and Technology*, 39(2), 423–431.

Kocakaya SO. et al. (2020). Selective *in-vitro* enzymes' inhibitory activities of fingerprints compounds of Salvia species and molecular docking simulations. *Iranian Journal of Pharmaceutical Research*, 19(2), 187–198.

Koysu P. et al. (2019). Isolation, identification of secondary metabolites from *Salvia absconditiflora* and evaluation of their antioxidative properties. *Natural Product Research*, 33(24), 3592–3595.

Kürkçüoğlu M. et al. (2019). The essentıal oils of four endemic *Salvia* species in Turkey. *Chemistry of Natural Compounds*, 55(2), 354–358.

Şenkal BC. (2019). Determination of essential oil components, mineral matter, and heavy metal content of *Salvia virgata* Jacq. grown in culture conditions. *Turkish Journal of Agriculture and Forestry*, 43, 395–404.

Süzgeç-Selçuk S. et al. (2021). The leaf and the gall volatiles of *Salvia fruticosa* Miller from Turkey: Chemical composition and biological activities. *Records of Natural Products*, 15(1), 10–24.

Tulukçu E. et al. (2019). Chemical fingerprinting of seeds of some *Salvia* species in Turkey by using GC-MS and FTIR. *Foods*, 8(4), 118.

Ustun O., Sezik E. (2011). Analgesic activity of *Salvia wiedemannii* Boiss. used in Turkish folk medicine. *Records of Natural Products*, 5(4), 328–331.

Uysal S. (2019). A comparative study of three drying methods on the phenolic profile and biological activities of *Salvia absconditiflora*. *Food Measure*, 13, 162–168.

Uysal S. et al. (2021). Chemical characterization, cytotoxic, antioxidant, antimicrobial, and enzyme inhibitory effects of different extracts from one sage (*Salvia ceratophylla* L.) from Turkey: Open a new window on industrial purposes. *RSC Advances*, 11(10), 5295–5310.

Yazgan H. (2020). Investigation of antimicrobial properties of sage essential oil and its nanoemulsion as antimicrobial agent. *LWT-Food Science and Technology*, 130, 109669.

Yener İ. (2020). Determination of antioxidant, cytotoxic, anticholinesterase, antiurease, antityrosinase, and antielastase activities and aroma, essential oil, fatty acid, phenolic, and terpenoid-phytosterol contents of *Salvia poculata*. *Industrial Crops and Products*, 155, 112712.

Yılar M. et al.(2020). Allelopathic and antifungal potentials of endemic *Salvia absconditiflora* Greuter & Burdet collected from different locations in Turkey. *Allelopathy Journal*, 49(2), 243–255.

Yusuf O. et al. (2021) Biochemical, histopathologic, and genotoxic effects of ethanol extract of *Salvia hypargeia* (fisch. & mey.) on incisional and excisional wounded diabetic rats. *Journal of Investigative Surgery*, 34(1), 7–19.

Zengin G. et al. (2018). New insights into the *in vitro* biological effects, *in silico* docking and chemical profile of clary sage-*Salvia sclarea* L. *Computational Biology and Chemistry*, 75, 111–119.

Zengin G. et al. (2019). Metabolomic profile of *Salvia viridis* L. root extracts using HPLC–MS/MS technique and their pharmacological properties: A comparative study. *Industrial Crops and Products*, 131, 266–280.

22 *Satureja* sp.

Dudu Altıntaş, Zehra Bektur, and Ufuk Koca Çalışkan

INTRODUCTION

The species belonging to the *Satureja* genus, which is a plant of the Lamiaceae family, are known by different local names in various regions of our country Turkey. They are also known by their local names such as "kekik", "dağ kekiği", and "sivri kekik" [1]. A total of 50 species of *Satureja* genus were determined all over the world, 15 of them naturally grown in Turkey [2]. A study in 2019 that focused on *Satureja metastasiantha* led to the first-ever identification of the species in the flora of Turkey, which led to a total of 16 different species of *Satureja* in the country [3]. *Satureja cilicica*, *S. amani*, *S. wiedemanniana*, *S. parnassica* subsp. *sipylea*, and *S. aintabensis* (Figure 22.1) are examples of endemic species of these 16 taxa [1, 4].

It spreads in rocky or eroded slopes, mountain foothills, pebbly and stony, coastal areas, sand dunes, roadsides, fallow areas, and up to 1920 m above sea level [5]. The species belonging to the *Satureja* genus are distributed mostly in Mediterranean countries of Europe, North Africa, Morocco and Libya, Saudi Arabia, South America, the Canary Islands, Caucasus, Iran, Iraq, and Turkiye [4]. About 200 species of *Satureja* are grown throughout the Mediterranean basin. *Satureja* species have a strong odor [6]. The Aegean and Mediterranean are where (Figure 22.2) most *Satureja*

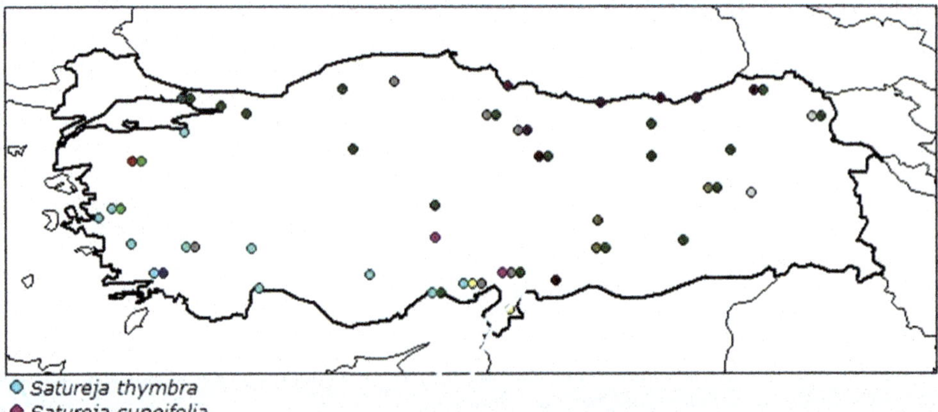

FIGURE 22.1 *Satureja* species in Turkiye (http://194.27.225.161/yasin/tubives/index.php?sayfa=karsilastir).

DOI: 10.1201/9781003146971-22

FIGURE 22.2 *S. hortensis* (https://identify.plantnet.org/tr/the-plant-list/species/Satureja%20hortensis%20L./data).

FIGURE 22.3 *S. thymbra* (https://identify.plantnet.org/tr/the-plant-list/species/Satureja%20thymbra%20L./data).

species are traded. (Figure 22.3) *S. wiedemanniana* (Avé-Lall.) Velen, *S. spicigera*, *S. hortensis* L., *S. cuneifolia* Ten., *S. thymbra* L., and *S. cilicica* P H Davis are examples of species [7]. This study is a summary of research on the *Satureja* species. Information about the chemical profile of *Satureja* species is provided in addition to the botanical properties. In addition to the botanical characteristics of *Satureja* species, information is also given about the chemical profile. Additionally, biological activity studies depending on chemical components are also included in this chapter.

BOTANICAL FEATURES/ETHNOBOTANICAL USAGE

Satureja plants are usually 10–35 cm tall, multi-branched, single or perennial aromatic herbaceous plants. The leaves are dark or gray-green, 1–3 cm long and half cm wide, sessile, and hairy. The flower color is lilac or white colored, the petals are tubular, with two lips at the top and three parts

FIGURE 22.4 *S. spicigera* (https://powo.science.kew.org/taxon/urn:lsid:ipni.org:names:457813-1).

at the bottom, and the calyx with five-pointed teeth, the teeth are approximately equal to the tube part [8]. For instance, *S. spicigera* is a perennial woody herb at the base with 25–60 cm in height (Figure 22.4). Leaf: numerous branches at the base, herbaceous, green light on both surface, densely hairs and sessile, secretory in the lower surface, attenuate at the base into petiole, acute, median nerve slender, lower leaves 15–20 mm in length and 2–4 mm in weight, upper leaves and leaves of branches small, 1 mm; Flower: three flowered, spike, slender peduncle, calyx 3–4.5 mm, campanulate-funnel, sparse canescent or sessile, subbilabiate, lower slightly longer, lanceolate, upper teeth 1/3 length of calyx; Corolla: 8–10 mm, almost white- pink [9]. *S. boissieri* is perennial herb with 40–60 cm height. Leaf: 10–26 mm length, 2–5 mm weight, linear, obtuse, upper linear, canaliculated; Corolla: 9 mm, corolla tube exerted, stamens exerted [9].

The *Satureja* genus is widely used all over the world as an herbal tea, spice, food additive, and in cosmetics industry. They have historically been used as carminatives, tonics, and muscle relaxants to treat cardiovascular problems as well as stomach and gastrointestinal ailments such as cramps, nausea, indigestion, and diarrhea [4, 10, 11]. Traditionally, *S. hortensis* seeds have been used for inflammatory illnesses [9]. Traditional remedies for colds and bronchitis is used the essential oil of *S. spicigera*, *S. hortensis*, and *S. cilicia* [12]. *S. montana* is used as a carminative to treat stomach and intestinal diseases such as nausea, indigestion, and diarrhea, tonic, and as a muscle pain reliever to treat cramps [15]. Aerial parts of *S. hortensis* have historically been utilized for their pharmacological characteristics, which include antispasmodic, antioxidant, antibacterial, antidiarrheal, and sedative activities [18]. *S. thymbra* aerial parts are used to prepare tea, which has traditionally been used to treat stomachaches and related problems [20]. According to a report, a decoction made out of the herbaceous part of *Satureja thymbra* is used to treat gingivitis [21].

PHYTOCHEMICALS

Essential Oils

Thymol and carvacrol are found to be the primary constituents in most of the species followed by P-cymene and terpinene. Linalool, borneol, β-caryophyllene, α-pinene, limonene, α-terpinene, terpineol-4-ol, spathulenol, and myrcene are also reported as major components in some of the *Satureja* sp. [10]. The phytochemical and volatile oil content varies depending on the season it is grown, geographical conditions and soil conditions, harvest period, different drying methods,

TABLE 22.1
Components of *Satureja* Species Volatile Oil

Components	Species	References
Carvacrol	***S. hortensis***	[11]
	S. cuneifolia	
Thymol	***S. hortensis***	[12]
γ-Terpinene	***S. thymbra***	[13]
p-Cymene	*S. boissieri*	[13]
ß-Caryophyllene	*S. coerulea*	[10]
Caryophyllene oxide	*S. coerulea*	[10]
α-Terpinene	***S. hortensis***	[14]
	S. spicigera	
Germakren-D	*S. coerula*	[2]
Myrcene	***S. hortensis***	[14]
	S. spicigera	
Linalool	*S. cuneifolia*	[15]
Borneol	*S. cuneifolia*	[16]
α-Pinene	***S. thymbra***	[10]
Spathulenol	*S. metastasiantha*	[12]
Terpinen-4-ol	***S. thymbra***	[16]
Limonene	***S. hortensis***	[8]
α-Thujene	***S. hortensis***	[17]

TABLE 22.2
Phenolic Components of *Satureja* Species

Components	Species	References
Rosmarinic acid	***S. thymbra***	[16]
Ferulic acid	***S. thymbra***	[16]
Caffeic acid	***S. hortesensis***	[10]
p-Coumaric	*S. cilicica*	[10]
Vanillic acid	***S. thymbra***	[16]
Protocatechuic acid	***S. thymbra***	[16]
Gallic acid	***S. thymbra***	[16]
Chlorogenic acid	*S. montana*	[10]

and distillation techniques [10, 24]. The best harvest time for volatile oil yield is the beginning of flowering. The highest rate of carvacrol was obtained from the harvest made during the flowering period [25].

Phenolic Acids

Rosmarinic acid is the phenolic acid that is most frequently discovered in *Satureja* species. Caffeic acid is a different phytochemical that has been found in *Satureja* species. These species also contain the phenolic acids p-coumaric, ferulic, gallic, chlorogenic, protocatechuic, vanillic, and carnosic [10].

TABLE 22.3
Flavonoids in *Satureja* Species

Species	Components	References
S. hortensis	Luteolin	[10]
S. aintabensis	Apigenin	[10]
S. thymbra	Quercetin	[16]
	Kaempferol	
	Myricetin	
	Naringenin	
	Catechin	
	Epicatechin	
S. parvifolia	Epigallocatechine	[10]

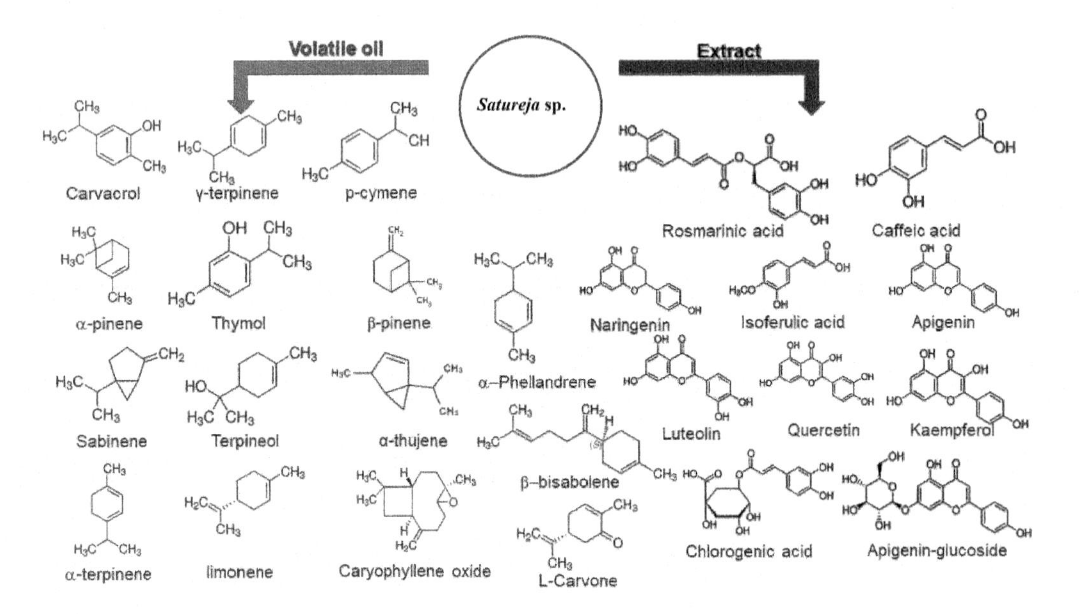

FIGURE 22.5 Main phytochemical components in *Satureja* sp. [21].

Flavanoids and Related Compounds

The subgroups of flavonoids include flavones, flavonols, flavanones, flavan-3-ols, and flavanonols. Additionally, glucosides such as flavanone glucosides, flavone glucosides, and flavonol glucosides are presented. The two flavones that are most frequently found in *Satureja* species are luteolin and apigenin [10].

Five distinct flavonols, including quercetin, epicatechin-3-O-gallate, epigallocatechin-3-O-gallate, kaempferol, and myrcetin, were found in *Satureja* species, according to the previous studies. The most prevalent flavonoid aglycone detected in *Satureja* species is quercetin, which was identified in *S. aintabensis* P. H. Davis, *S. cilicica*, *S. corulea* L., *S. hortensis*, and *S. icerica* [17, 18], as well as flavanonols (aromadendrin, taxifolin) [19]. Flavonoid glucoside derivatives have also been extensively researched, in addition to flavonoid aglycones. The literature research indicates that flavonoid glucosides, flavanone glucosides, flavone glucosides, and flavonol glucosides are the major chemical groups (Figure 22.5) found in *Satureja* sp. [10].

Moreover, vitexin, apigenin-7-glucoside, lutein glucoside, apigenin glucoside, naringin, naringenin glucoside, hesperidin, and rutin (a flavonol glusocide) were also reported in *S. cilicica* by Cetojevic-Simin *et al.* (2012) [10].

Various Phytochemicals and Enzymes

The fatty acids of *S. cuneifolia* and *S. thymbra* that have been described by Goren *et al.* (2003) including 9-octadecenoic acid methyl ester, hexadecanoic acid methyl ester (methyl palmitate), 9,12,15-octadecatrienoic acid methyl ester, and octadecanoic acid methyl ester [10].

So far, only *S. hortensis* had its enzyme content assessed, based on a literature review [21]. The enzymes found in this species include lignin peroxidase, ascorbate peroxidase, o-dianiside peroxidases, flavocytochrome b(2), cytochrome b(2) (NAD-independent lactate dehydrogenase), peroxidase isoenzymes, and superoxide dismutase [22, 23].

The phytochemical and volatile oil content varies depending on the season it is grown, geographical conditions and soil conditions, harvest period, different drying methods, and distillation techniques [10, 24]. The best harvest time for volatile oil yield is the beginning of flowering. The highest rate of carvacrol was obtained from the harvest made during the flowering period [25].

BIOLOGICAL ACTIVITIES AND EFFECTS

Antimicrobial Activity

In various studies, it has been discovered that the primary components of *Satureja* species, volatile oils, carvacrol and thymol, coupled with flavonoids, have potent antibacterial effects on Gram-positive and Gram-negative bacteria as well as potent antifungal effects [16, 26].

In a study conducted with *Satureja aintabensis*, its antimicrobial effect was found to be strong [27]. Sharifi *et al.* showed that *S. hortensis* volatile oil has an antibiofilm and antiquorum sensing effect [28].

In the study by Ahmadi *et al.*, green-synthesized Fe nanoparticles possessed higher antimicrobial properties than *Satureja hortensis* volatile oil against selected pathogenic microorganisms, especially Gram-negative bacteria [29]. Seyedtaghiya *et al.* (2021) showed that *S. hortensis* volatile oil prevented biofilm formation at concentrations lower than MIC, and this effect on *Salmonella* was seen to be stronger than *E. coli* [30].

Antioxidant Activity

Data from the literature suggested that two methods, such as b-carotene bleaching test and determination of free radical scavenging potential, are used to determine the antioxidant activity, and as a result, especially carvacrol and thymol, which are oxygen monoterpenes in *Satureja* species, showed great antioxidant effect. However, it is emphasized that polar phytochemicals are also biologically active compounds responsible for this activity. *Satureja* species are considered reliable sources for consumption due to their great antioxidant potentials [10]. Among the *Satureja* species with the highest carvacrol concentration, *Satureja hortensis* has the highest antioxidant effect [31]. In addition, Bimbiraite-Surviliene *et al.* demonstrated that rutin and rosmarinic acid found in *Satureja* species also possessed antioxidant activity [32].

Cytotoxic Activity

In studies using plant extract or volatile oil, *Satureja* species have been shown to have cytotoxic activity and both methanolic extract and rosmarinic acid have the best cytotoxic activity. Saab *et al.* found that *Satureja thymbra* is the species with the highest selective index (SI) [33]. Bimbiraite-Surviliene *et al.* observed that rosmarinic acid increases lipid peroxidation in melanoma cells [32].

Anti-leukemic Activity

According to Asadipour *et al.*'s research, the dichloromethane and hexane extracts of *S. bachtiarica* have anti-leukemic properties because they induce apoptosis and stop the advancement of the cell cycle [34].

Insecticidal Activity/Insect Repellant Activity/Fumigant Toxicity

Studies have shown that *S. hortensis, S. thymbra,* and *S. spinosa* L. species grown in Turkey have insecticidal activity. In numerous research, *S. hortensis* and *S. thymbra,* in particular, draw attention. Volatile oil was used as a target agent in the studies because volatile oil has a stronger insecticidal effect compared to the plant extract [10]. In the study conducted by Gokturk, the insecticidal effect of the volatile oil of *S. spicigera* was proven [14]. In addition to insecticidal activity, *S. hortensis* and *S. thymbra* also have insect-repellant activities [10].

Acetylcholinesterase (AChE) and Butyrylcholinesterase (BChE) Inhibitory Activities

The two main enzymes associated with Alzheimer's disease are known as AChE and BChE. It was tested for *S. cuneifolia* (Ten.) and *S. thymbra*'s ability to inhibit AChE and BChE [35].

Orhan *et al.* tested the AChE and BChE inhibitory activities of volatile oil from *S. cuneifolia.* The volatile oil showed a very high inhibitory activity (over 80%) against AChE [35].

The cholinesterase inhibitory effect against thymol AChE in the volatile oil of *Satureja thymbra* was evaluated by Öztürk *et al.* This investigation found that the presence of volatile oil had a promising enzyme inhibitory activity [36].

Antiviral Activity

As a result of the studies on the antiviral activity potential of the volatile oil of *Satureja thymbra*, it has been proven that it has a moderate inhibitory effect against HSV-1 and SARS-CoV replication and shows a highly selective index (SI) for HSV-1 [36]. In addition, in another study, it has been proven that the volatile oil of *S. cuneifolia* has a strong antiviral effect [38].

In the study by Wang *et al.*, carvacrol, the major component of *Satureja* volatile oil, inhibits the HSV-2 proliferation process and increases HSV-2-induced TNF-α levels, RIP3 and MLKL via intracellular RIP3 (Receptor interacting proteins)-mediated programmed cell necrosis. Carvacrol inhibits the proliferation of HSV-2 and has high antiviral activity. The IC_{50} for Carvacrol was calculated as 0.49 mmol/L [39].

Antinociceptive/Analgesic Activity

Aydın *et al.* showed analgesic activity associated with carvacrol found in *S. cuneifolia* volatile oil [39]. Hydroalcoholic and polyphenolic extracts and volatile oil of *S. hortensis* [41] and volatile oil of *S. thymbra* were found to have analgesic activity [42].

Antileishmanial Activity

Carvacrol and thymol, which make up the majority of *Satureja* sp., exhibit potent antileishmanial activity in the study conducted by Mirzaei *et al.* [43]. High antileishmanial activity was observed in the ethanolic extract of *S. hortensis* (IC_{50} value 15.6251 μM). Active chemicals found in *S. hortensis* have the potential to replace alternative treatments for cutaneous leishmaniasis [44].

Anti-inflammatory Activity

Studies have shown that the seeds of *Satureja hortensis*, hydroalcoholic and polyphenolic extracts, and its volatile oil have anti-inflammatory activity [40]. *S. hortensis* volatile oil had an immunomodulatory effect as seen by an uptick in the gene production of both pro-inflammatory (IL-1, IL-6, and TNF) and anti-inflammatory (IL-10) cytokines, according to Popa *et al.* [45]. *Satureja* species contain rosmarinic acid, which also has anti-inflammatory properties [10]. Karabay-Yavaşoğlu *et al.* showed that *Satureja thymbra* also has an anti-inflammatory effect [42].

Herbicidal Activity

Satureja thymbra encouraged and increased parasite infestation so some plants are important as a type of trap or capture for parasites. It has been shown that it has herbicidal activity [10]. Taban *et al.* reported that the microencapsulation of *S. hortensis* volatile oil showed irreversible herbicidal activity [46].

Antifungal Activity

S. hortensis essential oil has been investigated for *Aspergillus flavus*. MIC value has been found to be 6.25 μg/mL. Therefore essential oil can also be used as a potential source of antifungal [9].

Antihyperlipidemia Activity

Twelve men and nine women with type 2 diabetes were randomly divided into two groups in the study. Of these, 11 patients received tablets containing 250 mg of flowering aerial parts of *S. khuzestanica*, while the other ten patients received a placebo. Both total cholesterol and low-density lipoprotein cholesterol (LDL-C) levels have decreased significantly, from 202.8 mg/dL to 153.2 mg/dL and 120 mg/dL to 103 mg/dL, respectively [9].

Other Activities

S. hortensis and *S. thymbra* are the two *Satureja* species that have attracted the most investigation. Below activities were recorded for *S. hortensis* that were summarized.

For acaricidal activity, the LD_{50} value was calculated by using leaf-dipping method in the study on *Tetranychus urticae*. It was found that the essential oil of *S. hortensis* has an LD_{50} value of 0.876 $\mu L/cm^2$ [47].

For antispasmodic activity, the aqueous ethanolic extract of *S. hortensis* in an organ bath was investigated capability to relax muscles. Depending on the dose, an increase in the antispasmodic effect was shown. At a concentration of 0.15 mg/mL, there is 13.54% inhibition, whereas a dose of 0.9 mg/mL has an inhibitory effect of 89.90% [48].

For antihepatoma activity, *S. hortensis* aerial part water extract showed 71% inhibition in HA22T/VGH cell line at 2000 μg/mL concentration [49].

For anthelmintic activity, the ethanol extract prepared from *S. hortensis* aerial part was investigated for anthelmintic activity. In the study with *Ascaris suum* eggs, a decrease in the effect was observed depending on the concentration. At 62 μg/mL dosage, ovicidal activity was 90.69%; at 1000 μg/mL, it was 69% [50].

For inhibition of blood platelet adhesion, *S. hortensis* methanol extract was investigated for platelet adhesion, and a 48% adhesion inhibition was found at a dosage of 200 μg/mL [51].

CONCLUSION

Numerous medicinal plants belonging to the Lamiaceae family have been used medicinally for a long time. There are too many biologically active chemicals in *Satureja* species, to which the activities are devoted. *Satureja* is rich in volatile oils. The information gathered here offers a way to comprehend the most current changes to the pharmacology and phytochemistry of the genus. Recent researches have focused on separating and purifying a variety of secondary metabolites from this genus, particularly identifying phenolic acids like rosmarinic acid, caffeic acid, and chlorogenic acid, and subsequently flavonoids and triterpenoids. *Satureja* species' essential oil composition includes monoterpenes compounds such as carvacrol, thymol, and p-cymene.

REFERENCES

1. Paşa, C., Kılıç, T., Selvi, S., Özer sağır, Z., Determination of essential oil ratio and essential oil components by applying different drying methods of Satureja cuneifolia (Lamiaceae) species, *J. Inst. Sci. Technol.*, 9, 2330–2335, 2019.
2. Baser, K. H. C., Özek, T., Kirimer, N., Tümen, G., A comparative study of the essential oils of wild and cultivated Satureja hortensis L, *J. Essent. Oil Res.*, 16, 422–424, 2004.
3. Dirmenci, T., Yıldız, B., Öztekin, M., Türkiye Florası İçin Yeni Bir Tür Kaydı: Satureja metastasiantha Rech.f. (Ballıbabagiller / Lamiaceae), *Bağbahçe Bilim Derg.*, 6, 54–58, 2017.
4. Momtaz, S., Abdollahi, M., An update on pharmacology of satureja species, *Int. J. Pharmacolojy.*, 6, 454–461, 2010.
5. Cantino, P. D., Harley, R. M., Wagstaff, S. J., *Genera of Labiatae Status and Classification*, Advances in Labiatae Science, 1992.
6. Bukvički, D. vd., Satureja horvatii essential oil: In vitro antimicrobial and antiradical properties and in situ control of Listeria monocytogenes in pork meat, *Meat Sci.*, 96, 1355–1360, 2014.
7. Satil, F., Dirmenci, T., Tumen, G., Turan, Y., Commercial and ethnic uses of Satureja (Sivri kekik) species in Turkey, *Ekoloji.*, 17, 1–7, 2008.
8. Jafari, F., Ghavidel, F., Zarshenas, M. M., A critical overview on the pharmacological and clinical aspects of popular satureja species, *JAMS J. Acupunct. Meridian Stud.*, 9, 118–127, 2016.
9. Saeidnia, S., Gohari, A. R., Manayi, A., Kourepaz-Mahmoodabadi, M., *Satureja: Ethnomedicine, Phytochemical Diversity and Pharmacological Activities*, Springer International Publishing, 2016.
10. Tepe, B., Cilkiz, M., A pharmacological and phytochemical overview on Satureja, *Pharm. Biol.*, 54, 375–412, 2016.
11. Arabaci, T., Uzay, G., Keleştemur, U., Karaaslan, M. G., Balcioğlu, S., Ateş, B., Türkiye'de yetişen satureja cilicica P.H. Davis türünün uçucu yağinin sitotoksisitesi, antioksidan aktivitesi ve kimyasal kompozisyonu, *Marmara Pharm. J.*, 21, 500–505, 2017.
12. Carikci, S., Goren, A. C., Dirmenci, T., Yalcinkaya, B., Erkucuk, A., Topcu, G., Composition of the essential oil of Satureja metastasiantha: A new species for the flora of Turkey, *Zeitschrift fur Naturforsch. - Sect. C J. Biosci.*, 75, 271–277, 2020.
13. Başer, K. H. C., Kırımer, N., Essential oils of anatolian lamiaceae-An update, *Nat. Volatiles Essent. Oils.*, 5, 1–28, 2018.
14. Gokturk, T., Chemical composition of Satureja spicigera essential oil and its insecticidal effectiveness on Halyomorpha halys nymphs and adults, *Zeitschrift fur Naturforsch. - Sect. C J. Biosci.*, 76, 451–457, 2021.
15. Kan, Y., Uçan, U. S., Kartal, M., Altun, M. L., Aslan, S., Sayar, E., Ceyhan, T., GC-MS analysis and antibacterial activity of cultivated Satureja cuneifolia Ten. essential oil, *Turkish J. Chem.*, 30, 253–259, 2006.
16. Skendi, A., Katsantonis, D. N., Chatzopoulou, P., Irakli, M., Papageorgiou, M., Antifungal activity of aromatic plants of the lamiaceae family in bread, *Foods.*, 9, 8–12, 2020.
17. Popovici, R. vd., A comparative study on the biological activity of essential oil and total hydro-alcoholic extract of Satureja hortensis L., *Exp. Ther. Med.*, 18, 932–942, 2019.

18. Alizadeh, A., Moghaddam, M., Asgharzade, A., Sourestani, M. M., Phytochemical and physiological response of Satureja hortensis L. to different irrigation regimes and chitosan application, *Ind. Crops Prod.*, 158, 1–10, 2020.
19. Palavra, A. M. F. vd., Supercritical carbon dioxide extraction of bioactive compounds from microalgae and essential oils from aromatic plants, *J. Supercrit. Fluids.*, 60, 21–27, 2011.
20. Malmir, M., Gohari, A. R., Saeidnia, S., Silva, O., A new bioactive monoterpene-flavonoid from Satureja khuzistanica, *Fitoterapia.*, 105, 107–112, 2015.
21. Fierascu, I., Dinu-Pirvu, C. E., Fierascu, R. C., Velescu, B. S., Anuta, V., Ortan, A., Jinga, V., Phytochemical profile and biological activities of satureja hortensis l.: A review of the last decade, *Molecules.*, 23, 1–19, 2018.
22. Keyhani, J., Keyhani, E., Peroxidase activity in Satureja hortensis L. roots, *Acta Hortic.*, 723, 209–214, 2006.
23. Arzi, L., Keyhani, J., Keyhani, E., Flavocytochrome b2 Activity in Satureja hortensis L. leaves, *Acta Hortic.*, 853, 369–378, 2010.
24. Baydar, H., Sağdiç, O., Özkan, G., Karadoğan, T., Antibacterial activity and composition of essential oils from Origanum, Thymbra and Satureja species with commercial importance in Turkey, *Food Control.*, 15, 169–172, 2004.
25. Katar, D., Arslan, Y., Subaşı, İ., Bülbül, A., Ankara Ekolojík KoçuUarinda Sater (Satureja hortensis L) Bitkisinde Uçucu Yag ve Bileşenlerinin Ontogenetik Varyabilitesinin Belirlenmesi, *Journal of Tekirdag Agriculture Faculty*, 8, 29–36, 2011.
26. Dikbas, N., Dadasoglu, F., Kotan, R., Cakir, A., Influence of summer savory essential oil (satureja hortensis) on decay of strawberry and grape, *J. Essent. Oil-Bearing Plants.*, 14, 151–160, 2011.
27. Askun, T., Tekwu, E. M., Satil, F., Modanlioglu, S., Aydeniz, H., Preliminary antimycobacterial study on selected Turkish plants (Lamiaceae) against Mycobacterium tuberculosis and search for some phenolic constituents, *BMC Complement. Altern. Med.*, 13, 1–11, 2013.
28. Sharifi, A., Mohammadzadeh, A., Zahraei Salehi, T., Mahmoodi, P., Antibacterial, antibiofilm and anti-quorum sensing effects of Thymus daenensis and Satureja hortensis essential oils against Staphylococcus aureus isolates, *J. Appl. Microbiol.*, 124, 379–388, 2018.
29. Ahmadi, S., Fazilati, M., Nazem, H., Mousavi, S. M., Green synthesis of magnetic nanoparticles using Satureja hortensis Essential oil toward superior antibacterial/fungal and anticancer performance, *Biomed Res. Int.*, 2021, 1–14, 2021.
30. Seyedtaghiya, M. H., Fasaei, B. N., Peighambari, S. M., Antimicrobial and antibiofilm effects of satureja hortensis essential oil against escherichia coli and salmonella isolated from poultry, *Iran. J. Microbiol.*, 13, 74–80, 2021.
31. Chambre, D. R., Moisa, C., Lupitu, A., Copolovici, L., Pop, G., Copolovici, D. M., Chemical composition, antioxidant capacity, and thermal behavior of Satureja hortensis essential oil, *Sci. Rep.*, 10, 1–12, 2020.
32. Bimbiraitė-Survilienė, K. vd., Evaluation of chemical composition, radical scavenging and antitumor activities of satureja hortensis L. Herb extracts, *Antioxidants.*, 10, 1–15, 2021.
33. Saab, A. M. vd., In vitro evaluation of the biological activity of Lebanese medicinal plants extracts against herpes simplex virus type 1, *Minerva Biotec.*, 147, 135–140, 2012.
34. Asadipour, M., Malek-Hosseini, S., Amirghofran, Z., Anti-leukemic activity of Satureja bachtiarica occurs by apoptosis in human cells, *Biotech. Histochem.*, 95, 506–513, 2020.
35. Orhan, I., Şener, B., Kartal, M., Kan, Y., Activity of essential oils and individual components against Acetyl-and Butyrylcholinesterase, *Zeitschrift fur Naturforsch. – Sect. C J. Biosci.*, 63, 547–553, 2008.
36. Öztürk, M., Anticholinesterase and antioxidant activities of Savoury (Satureja thymbra L.) with identified major terpenes of the essential oil, *Food Chem.*, 134, 48–54, 2012.
37. Loizzo, M. R., Saab, A. M., Tundis, R., Statti, G. A., Menichimi, F., Lampronti, D., Gambari, R., Cinatl, J., Doerr, H. W., Phytochemical analysis and in vitro antiviral activities of the essential oils of seven Lebanon species, *Chem. Biodivers.*, 5, 461–470, 2008.
38. Erdoğan Orhan, I., Özçelik, B., Kartal, M., Kan, Y., Antimicrobial and antiviral effects of essential oils from selected Umbelliferae and Labiatae plants and individual essential oil components, *Turkish J. Biol.*, 36, 239–246, 2012.
39. Wang, L., Wang, D., Wu, X., Xu, R., Li, Y., Antiviral mechanism of carvacrol on HSV-2 infectivity through inhibition of RIP3-mediated programmed cell necrosis pathway and ubiquitin-proteasome system in BSC-1 cells, *BMC Infect. Dis.*, 20, 1–16, 2020.
40. Aydın, S., Öztürk, Y., Beis, R., Başer, K. H. C., Investigation of Origanum onites, Sideritis congesta and Satureja cuneifolia essential oils for analgesic activity, *Phyther. Res.*, 10, 342–344, 1996.

41. Hajhashemi, V., Ghannadi, A., Pezeshkian, S. K., Antinociceptive and anti-inflammatory effects of Satureja hortensis L. extracts and essential oil, *J. Ethnopharmacol.*, 82, 83–87, 2002.
42. Karabay-Yavasoglu, N. U., Baykan, S., Ozturk, B., Apaydin, S., Tuglular, I., Evaluation of the antinociceptive and anti-inflammatory activities of Satureja thymbra L. essential oil, *Pharm. Biol.*, 44, 585–591, 2006.
43. Mirzaei, F., Bafghi, A. F., Mohaghegh, M. A., Jaliani, H. Z., Faridnia, R., Kalani, H., In vitro antileishmanial activity of Satureja hortensis and Artemisia dracunculus extracts on Leishmania major promastigotes, *J. Parasit. Dis.*, 40, 1571–1574, 2016.
44. Kheiri Manjili, H., Jafari, H., Ramazani, A., Davoudi, N., Anti-leishmanial and toxicity activities of some selected Iranian medicinal plants, *Parasitol. Res.*, 111, 2115–2121, 2012.
45. Popa, M., Măruţescu, L., Oprea, E., Bleotu, C., Kamerzan, C., Chifiriuc, M. C., Pircalabioru, G. G., In vitro evaluation of the antimicrobial and immunomodulatory activity of culinary herb essential oils as potential perioceutics, *Antibiotics.*, 9, 1–14, 2020.
46. Taban, A., Saharkhiz, M. J., Khorram, M., Formulation and assessment of nano-encapsulated bioherbicides based on biopolymers and essential oil, *Ind. Crops Prod.*, 149, 1–11, 2020.
47. Ebadollahi, A., Sendi, J. J., Aliakbar, A., Razmjou, J., Acaricidal activities of essential oils from Satureja hortensis (L.) and Teucrium polium (L.) against the two spotted spider mite, Tetranychus urticae koch (Acari: Tetranychidae), *Egypt. J. Biol. Pest Control.*, 25, 171–176, 2015.
48. Boskabady, M. H., Aslani, M. R., Mansuri, F., Amery, S., Relaxant effect of Satureja hortensis on guinea pig tracheal chains and its possible mechanism(s), *Daru.*, 15, 199–204, 2007.
49. Lin, L. T., Liu, L. T., Chiang, L. C., Lin, C. C., In vitro anti-hepatoma activity of fifteen natural medicines from Canada, *Phyther. Res.*, 16, 440–444, 2002.
50. Urban, J., Kokoska, L., Langrova, I., Matejkova, J., In vitro anthelmintic effects of medicinal plants used in Czech Republic, *Pharm. Biol.*, 46, 808–813, 2008.
51. Yazdanparast, R., Shahriyary, L., Comparative effects of Artemisia dracunculus, Satureja hortensis and Origanum majorana on inhibition of blood platelet adhesion, aggregation and secretion, *Vascul. Pharmacol.*, 48, 32–37, 2008.

23 *Scorzonera* sp.

Özlem Bahadır Acıkara

INTRODUCTION

BOTANICAL FEATURES

The genus *Scorzonera* L., which belongs to the Asteraceae family, originates mainly from the Mediterranean Region, is extensively distributed from Central Europe to Central Asia and northern Africa, and is represented by more than 160 species in the World (Çoşkunçelebi et al., 2015). In Europe, 28 species are widely distributed in the region extending from the north of Russia to Spain and Greece (Zidorn et al., 2003), 23 species are grown in China (Zhu et al., 2010), and 11 species have been recorded in Mongolia two of which are endemic (Wang et al., 2009). Anatolia is considered as one of the diversity centers for *Scorzonera* species due to the many rare species found naturally and the number of growing species as well as the endemism rate. Forty-two species of *Scorzonera* have been recorded in The Flora of Turkey, according to the Chamberlain (1975). Together with the new species this genus is represented by 52 species (59 taxa) in Turkey currently, and among these 31 are endemics (Çoşkunçelebi et al., 2015).

Scorzonera species are annual, biennial, or perennial herbaceous, rarely suffructicose, subscapigerous, or caulescent. The stem is usually leafless. The underground part is thick, cylindrical, rhizome, or tuberous. Leaves at the base or on the stem, simple and entire, linear to broadly ovate lanceolate or deeply divided, petiolate or sessile. The capitulum is homogamous and all flowers are ligulate, solitary, or several per stem. Involucrum is ovate or cylindrical, phyllaries 2-seriate, inner ones are longer and herbaceous than outer ones. The inflorescence is glabrous, and the flowers are white to yellow, violet to purple, or purple-yellow. Achenes are cylindrical, smooth, or ribbed, sometimes transversely lamellate-rugulose, glabrous or hairy, beakless, stipitate, or sessile. Papus 3-seriate, sessile, hairs plumose, sometimes with barbellate on the upper parts and sometimes completely (Chamberlain, 1975).

TRADITIONAL USE

The importance of *Scorzonera* species as vegetable as well as medicinal plants in Turkey, China, Mongolia, and in some European countries has been recorded. *S. hispanica* known as black salsify, Spanish salsify, or serpent's root is the most recognized species that grows naturally and widely in Europe. The species, which was consumed as food by the ancient Romans and Greeks, has been cultivated since the 1500s and has been used as an ornamental plant and for medicinal purposes. The current part of the plant is the carrot-shaped roots with black skin, which gives the plant its name. In the Middle Ages, it was used as a strong tonic and curative for snake bites, and it was even called gentian due to its usage. *S. hispanica* has appetizing, mucolytic, stomachic, diuretic, and antipyretic effects. Young leaves of the plant are also used in salads, and the roots are consumed as a cooked vegetable in European cuisine after the removal of black color skin (Zidorn et al., 2000; Wang et al., 2009; Granica et al., 2015).

S. hispanica, which is recorded in The Flora of Turkey, does not grow naturally in Turkey. However, it has been stated that it was cultivated in Istanbul and its surroundings during the Ottoman Empire and was used against snake bites in folk medicine, and it is used as a diaphoretic and pectoral in folk medicine (Baytop, 1999).

DOI: 10.1201/9781003146971-23

In Turkey, the roots and fresh shoots of some *Scorzonera* species are consumed as a vegetable, fresh or cooked. For this purpose the plant should be collected while it is fresh; old and flowering plants are not used. Species consumed as vegetables in Turkey are *S. cana* (C.A. Mey.) Hoffm. (*karakök* or *teke sakalı*), *S. latifolia* (Fisch. and Mey.) DC. (*geniş yapraklı karakök* or *mesdek*), *S. mollis* Bieb. (*goftigoda*) and *S. suberosa* C. Koch. (wild carrot) (Baytop, 1999; Turan et al., 2003), *S. cinerea* Boiss. (*at yemliği, boz kanak, kıllı yemlik, sakız otu, yemlik*), *S. lacera* Boiss. et Bal. (*dedesakalı, tekesakalı*), *S. rigida* Aucher (*yemlik*), *S. semicana* DC. (*kıvrım, yemlik*), and *S. tomentosa* L. (*arvent, çitlembik, karasakız otu, neraband, yersakızı*) (Şenkardeş et al., 2019). In addition to their use as vegetable, the usage of the *Scorzonera* species is recorded in a variety of illnesses including arteriosclerosis, kidney diseases, hypertension, diabetes mellitus, and rheumatism, as well as for pain relief and wound healing (Sezik et al., 1997; Baytop, 1999). *S. latifolia* has been known for its special usage in Turkish folk medicine. A gum is prepared by drying the latex obtained from the roots of *S. latifolia* and other similar species. This latex, which is initially white in color, turns brown and then black when dried in the sun, is poured into molds when it is soft, and shaped and strung on. Commercially prepared gum is circular in shape and has a hole in the middle. The gum prepared and used in this way is known by different names such as "*beniş* , *çingene sakızı, kandil sakızı, karasakız, merkosakızı, selsepet sakızı, yerleme sakızı*" in local markets. This preparation is used for its antihelmintic activity internally, as a pain reliever, and against infertility externally. To prepare a patch for pain relieving, 100 g of latex is beaten with 10 g of *Viscum album* and applied to a cloth, for applying to the aching part and to the abdomen of women for infertility and left on the skin for a week (Baytop, 1999). *S. phaeopappa, S. sosnowskyii, S. mirabilis*, and *S. rigida* are used in the treatment of headaches, and *S. mollis* is used as a diuretic and against kidney stones (Yıldırım et al., 2008; Altundağ and Öztürk, 2011), *S. tomentosa* roots have hemostatic activity (Karakaya et al., 2020) and are used in stomach ailments, *S. cinerea* latex is useful in the treatment of cough and backache, both aerial parts and roots of *S. suberosa* subsp. *suberosa* have galactogogue and appetizing effects (Altundağ and Öztürk, 2011), and aerial parts of *S. latifolia are* used against nausea (Karakaya et al., 2020) in Turkish folk medicine.

Chemistry

Earlier studies have examined *Scorzonera* species and have revealed that *Scorzonera* species display promising biological activities and interesting chemistry from the chemical point of view. A number of compounds such as dihydroisocoumarines, bibenzyl derivatives, flavonoids, lignans, stilbene derivatives, quinic and caffeic acid derivatives, sesquiterpene, sesquiterpene lactones, and triterpenes have been isolated from the *Scorzonera* species.

Stigmasterol 3β-glucoside, β-sitosterol, lupeol, and lupeol acetate (Öksüz et al., 1990); two new dihydroisocoumarins, (±)-scorzotomentosin, (–)-scorzotomentosin 4′ - *O*- β-glucoside; a new phthalide (±)-scorzophthalide; and a new stilbene derivative, scorzoerzincanin together with four known compounds, (±)-hydrangenol, (–)-hydrangenol 4′ - *O*- β-glucoside, (±)-hydramacrophyllol A, and (±)-hydramacrophyllol B were isolated from *S. tomentosa* L. (Sarı et al., 2007).

Taraxasteryl myristate, taraxasteryl acetate, fern-7-en-3-β-one (Bahadir et al., 2010), motiol, β-sitosterol (Bahadır Acıkara et al., 2014), a novel triterpene (3-β-hydroxy-fern-7-en-6-one-acetate) together with urs-12-en-11-one-3-acetyl, 3-β-hydroxy-fern-8-en-7-one-acetate, olean-12-en-11-one-3-acetyl and leucodin (Bahadir Acikara et al., 2012), (±) hydrangenol and (-)-scorzotomentosin 4'-*O*-β-glucopyranoside (Saltan et al., 2010), a new 3-benzylphthalide, scorzoveratrin 4'-*O*-β-glucoside, together with the known 3-benzylphthalides, scorzoveratrin and scorzoveratrozit, the caffeoyl derivatives, chlorogenic acid methyl ester, 4,5-di-*O*-caffeoylquinic acid, 4,5-di-*O*-caffeoylquinic acid methyl ester, 3,5-di-*O*-caffeoylquinic acid and caffeic acid (Sarı, 2012) and quercetin-3-*O*-β-glucoside, hyperoside, hydrangenol-8-*O*-glucoside, swertisin, 7-*O*-methylisoorientin, 4,5-di-*O*-caffeoyl-quinic acid, 3,5-di-*O*-caffeoyl-quinic acid, and chlorogenic acid (Bahadir Acikara et al., 2016), isoorientin, quercetin-3-*O*-β-apiofuranosyl-(1′ ′ ′ →2′ ′)-β-D-glucopyranoside, and quercetin-3-*O*-α-rhamnopyranosyl-(1→6)-β-D-galactopyranoside were isolated from *S. latifolia* (Küpeli Akkol et al., 2019).

A phytochemical investigation of *S. veratrifolia* Fenzl revealed that the plant contains two new 3-benzylphthalides, scorzoveratrin and scorzoveratrozit, together with five known compounds, chlorogenic acid, chlorogenic acid methyl ester, cryptochlorogenic acid, 4,5-di-*O*-caffeoylquinic acid, and 3,5-di-*O*-caffeoylquinic acid (Sarı, 2010).

New natural compounds (6-*trans*-*p*-coumaroyl)-3-*O*-β-D-glucopyranosyl-2-deoxy-D-riburonic acid, (6-cis-*p*-coumaroyl)-3-*O*-β-D-glucopyranosyl-2-deoxy-D-riburonic acid, (6-*trans*-*p*-coumaroyl)-3-*O*-β-D-glucopyranosyl-2-deoxy-D-riburonic acid methyl ester, and (6-*trans*-*p*-coumaroyl)-3-*O*-β-D-glucopyranosyl-(5-acetyl)-2-deoxy-D-riburonic acid, having the rare deoxy-D-riburonic acid moiety as well as known compounds thunberginol G, isoorientin, orientin, isoschaftoside, and swertiajaponin were isolated from *S. papposa* DC (Milella et al., 2014).

S. hieraciifolia Hayek subaerial part ethanol extract gives quinic acid derivatives 5-*O*-feruloyl quinic acid methyl ester, 1,5-di-*O*-feruloylquinic acid, chlorogenic acid methyl ester, 3-*O*-caffeoylquinic acid methyl ester, 1,3-di-*O*-caffeoylquinic acid methyl ester, 3,5-di-*O*-caffeoylquinic acid methyl ester, 4,5-di-*O*-caffeoylquinic acid methyl ester, caffeic acid, and 3-(4′-hydroxyphenyl)-2-propenoic acid (4″-carboxyl)-phenyl ester (Sarı et al., 2019).

Scorzopygmaecoside, scorzonerol, cudrabibenzyl A, thunberginol C, scorzocreticoside I and II, chlorogenic acid, chlorogenic acid methyl ester, and 3,5-di-*O*-caffeoylquinic acid were isolated from the roots of *S. pygmaea* Sibth. and Sm. (Şahin et al., 2020a); 3,5-di-*O*-caffeoylquinic acid, chlorogenic acid, chlorogenic acid methyl ester, cudrabibenzyl A, scorzocreticoside I and scorzocreticoside II, scorzonerol, scorzopygmaecoside, thunberginol C, and protocatechuic acid were isolated from the ethanol extract of *S. pygmaea* aerial parts (Şahin et al., 2020b).

Methanol extract of *S. aucheriana* aerial parts led to the isolation of chlorogenic acid derivatives as methyl 1-(2-methylcyclopropyl-1-carbonyloxy) chlorogenate and 3,4-bis [(3′,4-dioxo-1′,3′,5′,6′-tetrahydrospiro [cyclohexa-2,5-diene-1,4′-cyclopenta-[c]-furan]-1′-yl)] chlorogenic acid, triterpenoids (taraxasterol, taraxasterol acetate, taraxasterol oleate, lupeol, lupeol acetate, and ptiloepoxide), and β-sitosterol (Erik et al., 2021a), dihydroisocoumarin derivatives including scorzopygmaecoside, scorzocreticoside II, iso-scorzopygmaecoside, scorzoaucherioside I and II, quinic acid derivatives; 3,5-*O*-dicaffeoyl-*epi*-quinic acid, 3,5-*O*-dicaffeoylquinic acid, and 3,4-dihydroxyphenyl caffeate (Erik et al., 2021b).

Biological Activity

Analgesic Activity

S. tomentosa, *S. latifolia*, *S. mollis* ssp. *szowitsii*, and *S. suberosa* ssp. *suberosa* roots were investigated for their analgesic activities by writhing test and tail-flick test at 100 mg/kg, 50 and 25 mg/kg dosage in order to prove their ethnopharmacological usage in Turkish folk medicine (Bahadır et al., 2012). *n*-hexane fraction of *S. latifolia* roots methanol extract displayed significant analgesic activity and inhibited abdominal writhings by 88.70%, 77.41%, and 76.45% at 100 mg/kg, 50 mg/kg, and 25 mg/kg dosage, respectively, on mice. In tail-flick test, administration of 50 mg/kg body weight *n*-hexane fraction of *S. latifolia* resulted in significant analgesic activity. Biological activity-guided fractionation revealed that taraxasteryl acetate (Bahadır et al., 2010), motiol, and β-sitosterol (Bahadır Acıkara et al., 2014a) were active components. The antinociceptive activity of *S. latifolia* was reported as significant and due to the synergistic activity of the compounds found in *n*-hexane fraction (Bahadır et al., 2010).

Antidiabetic Activity

S. hieraciifolia aerial parts and roots' different extracts prepared by using n-hexane, dichloromethane, ethyl acetate, methanol, and water were determined to have moderate inhibitory activity on α-amylase and α-glucosidase (Dall'Acqua et al., 2020)

S. tomentosa, S. mollis ssp. *szowitsii, S. suberosa* ssp. *suberosa, S. eriophora, S. acuminata, S. sublanata,* and *S. cana* var. *jacquiniana* aerial part extracts and rutin as one of the constituents of *Scorzonera* species were tested for their antidiabetic activities on alloxan-induced diabetic mice test model at 100 mg/kg dose. Significant reduction in blood glucose levels was observed after 1, 2, and 4 hours of treatments of *S. sublanata* extract, and the activity was determined as the highest when compared to other species. *S. cana* var. *jacquiniana* extract also displayed a notable decrease after 4 hours of treatment. Rutin, which is one of the constituents of *Scorzonera* species, was also evaluated for its antidiabetic activity, and a significant lowering effect on blood glucose level was observed by treatment with rutin at all tested times at 100 mg/kg i.p. injection. However, the rutin content of the tested *Scorzonera* species and the antidiabetic activity were not correlated (Sakul et al., 2021).

Ethanolic extract of *S. cinerea* leaves was evaluated for its therapeutic activities on streptozotocin-induced diabetic rats at 100 mg/kg dosage. Dried and frozen plant materials were used for preparing two different extracts. Both extracts prepared from the leaves induced a decrease in blood glucose and HbA1c levels and induced an increase in insulin levels. The inhibitory effects on α-glucosidase and α-amylase activities were also evaluated. While dried *Scorzonera* extract was more effective in α-amylase inhibitory activity, frozen *Scorzonera* extract displayed higher activity in α-glucosidase inhibitory activity (Temiz, 2021).

Anti-inflammatory Activity

S. acuminata, S. cana var. *alpina, S. cana* var. *jacquiniana, S. cana* var. *radicosa, S. cinerea, S. eriophora, S. incisa, S. laciniata* ssp. *laciniata, S. parviflora, S. sublanata, S. latifolia, S. mollis* ssp. *szowitsii, S. suberosa, and S. tomentosa* aerial part and root extracts were investigated for their anti-inflammatory potential by carrageenan, PGE2, and serotonin- induced hind paw edema and 12-*O*-tetradecanoyl-13-acetate (TPA)-induced mouse ear edema models on mice. Extracts prepared from the aerial parts of *S. cana* var. *jacquiniana, S. cinerea, S. eriophora, S. incisa* and *S. parviflora* showed significant inhibitory effect on carrageenan and PGE(2)-induced hind paw edema test model. The extracts did not show any remarkable activity on serotonin-induced hind paw edema and TPA-induced mouse ear edema models. *S. latifolia, S. mollis* ssp. *szowitsii*, and *S. tomentosa* aerial parts exhibited notable inhibition in carrageenan-induced hind paw edema model, while *S. latifolia* and *S. tomentosa* showed potent activity against PGE2-induced hind paw edema model as well as in (TPA)-induced mouse ear edema model. The highest activity was observed with *S. latifolia* aerial parts by 29.4% inhibition at 270th min and 33.3% inhibition at 45th min on carrageenan and PGE(2)-induced hind paw edema model test, respectively (Küpeli Akkol et al., 2011; Bahadır Acıkara et al., 2014b). *S. latifolia* and its major constituents (quercetin-3-*O*-β-glucoside, hyperoside, hydrangenol-8-*O*-glucoside, swertisin, 7-*O*-methylisoorientin, 4,5-*O*-dicaffeoyl-quinic acid, 3,5-di-*O*-caffeoyl-quinic acid, and chlorogenic acid) as well as *S. cana* var. *jacquiniana, S. tomentosa, S. mollis* ssp. *szowitsii, S. eriophora, S. incisa, S. cinerea*, and *S. parviflora* were also evaluated for their inhibitory activities of TNF-α and IL-1β production, and NF-κB nuclear translocation in THP-1 macrophages in order to reveal their anti-inflammatory mechanisms. *S. tomentosa* and *S. latifolia* aerial part extracts were established as the most active in the inhibition of TNF-α production. Additionally, the ability of *S. tomentosa* in decreasing IL-1β production after inflammatory stimulation displayed as notable. The greatest NF-κB nuclear translocation inhibitory activity by *S. latifolia* and *S. tomentosa* was observed. On the other hand none of the isolated compounds was found to be active (Bahadır Acıkara et al., 2016).

Anti-inflammatory activity of *S. hieraciifolia* roots was evaluated by measuring its inhibitory activity on COX enzymes. The fractions prepared from the ethanol extract with petroleum ether, chloroform, ethyl acetate, and n-butanol were investigated for their COX inhibitory properties. Inhibitory percentage of all fractions obtained from *S. hieraciifolia* roots indicated that the plant and its fractions have an insignificant effect on both COX-1 and COX-2 enzymes when compared with indomethacin (Sarı et al., 2019). Similar results were obtained for *S. pygmaea* of which subaerial parts and roots' anti-inflammatory activities were evaluated also by COX (cyclooxygenase) inhibition. The inhibitory activity of ethanol extract and its fractions against COX-1 and COX-2 (cyclooxygenase 2) was not found to be notable (Şahin et al., 2020a; Şahin et al., 2020b).

Antimicrobial Activity

The antibacterial and antifungal activity of the aerial part and root extracts of *S. papposa* which were collected from two different places, Turkey and Iraq, were investigated for their antimicrobial activities. The extracts were found to be active against bacteria at the concentrations of 50–800 µg/mL and display antifungal activity at 50–100 µg/mL, while the reference compounds (ampicillin, amikacin, ciprofloxacin, fluconazole, and amphotericin B) were active at lower concentrations (1.56–3.12 µg/mL). All tested extracts displayed mild antimicrobial activity, and Turkey samples were slightly more effective against both bacteria and fungi (50 µg/mL against *Pseudomonas aeruginosa*) than Iraq samples (Mohammed et al., 2020).

S. hieraciifolia roots fractions: petroleum ether, chloroform, ethyl acetate, and n-butanol were tested against *Staphylococcus aureus, Staphylococcus epidermidis, Escherichia coli, Klebsiella pneumoniae, Proteus mirabilis, Pseudomonas aeruginosa, Enterococcus feacalis,* and *Candida albicans* using micro broth dilution technique in order to evaluate their antimicrobial activities. While remarkable antimicrobial activity was not observed against tested bacteria or fungi by none of the tested fractions of *S. hieraciifolia* roots, only weak antimicrobial activity against *Enterococcus feacalis* with 1250 mg/l MIC value was found (Sarı et al., 2019). No antimicrobial activity against the same microorganism which were used in the study carried out by Sarı et al. (2019) was also determined by *S. pygmaea* roots ethanolic extract (Şahin et al., 2020a). On the other hand *S. pygmaea* aerial parts exhibited a weak activity against *Staphylococcus aureus* and *S. epidermidis* as Gram-positive bacteria, but no antimicrobial activity against Gram-negative bacteria and yeast strain was determined (Şahin et al., 2020b).

Isolated compounds: iso-scorzopygmaecoside, scorzoaucherioside I, scorzoaucherioside II, scorzopygmaecoside, scorzocreticoside II, 3,5-*O*-dicaffeoyl-*epi*-quinic acid, 3,5-*O*-dicaf-feoylquinic acid, and 3,4-dihydroxyphenyl caffeate from the aerial parts of *S. aucheriana* were evaluated for their antimicrobial activities against *Escherichia coli, Yersinia pseudotuberculosis, Pseudomonas aeruginosa, Enterococcus faecalis, Staphylococcus aureus, Listeria monositogenes, Bacillus cereus, Mycobacterium smegmatis, Candida albicans,* and *Saccharomyces cerevisiae*. Anti-tuberculosis activity was observed by iso-scorzopygmaecoside and scorzoaucherioside I with 25.6 µg/mL, and 21.2 µg/mL MIC values respectively. 3,4-Dihydroxyphenyl caffeate and scorzoaucherioside I exhibited antimicrobial activity against *Pseudomonas aeruginosa* as Gram-negative bacteria strain (MIC = 290 and 377.5 µg/mL, respectively), while scorzopygmaecoside and scorzocreticoside II were found to be active against *Enterococcus faecalis*, a Gram-positive strain by 135 and 200 µg/mL MIC values, respectively (Erik et al., 2021b).

Antioxidant Capacity

S. suberosa, S. laciniata, and *S. latifolia* methanol extracts were investigated for their antioxidant activities by measuring DPPH radical scavenging abilities. The concentration-dependent antioxidant activity was observed with IC_{50} values of 29.36 mg/mL for *S. latifolia*, 42.33 mg/mL for *S. suberosa,* and 77.07 mg/mL for *S. laciniata* (Erden et al., 2013).

Another study revealed that the dried and powdered aerial parts and roots of the *S. cana* var. *alpina, S. cana* var. *jacquiniana, S. cana* var. *radicosa, S. cinerea, S. eriophora, S. incisa,* and *S. laciniata ssp. laciniata* were found to possess significant antioxidant activities against superoxide anion radicals. The extracts of *Scorzonera* aerial parts and roots displayed notable anti-superoxide anion formation with IC_{50} values ranging from 3.0 to 6.0 mg/mL and 2.25 to 9.0 mg/mL respectively. *S. parviflora* root extract was determined as the most active species with an IC_{50} value of 2.25 mg/mL among the tested *Scorzonera* species. The extracts of the aerial parts of *S. cinerea* and *S. incisa* also exhibited scavenging activity significantly with 3.0 mg/mL of IC_{50} value, and these are followed by the extracts of *S. cana* var. *jacquiniana* and *S. eriophora* aerial parts with 3.3 and 3.5 mg/mL of IC_{50} values respectively (Bahadır Acıkara et al., 2013).

Aerial parts and roots of *Scorzonera papposa* and isolated compounds from methanol extracts were evaluated for their antioxidant properties. In the study, the antioxidant activity was measured by four different assays: the DPPH assay, the FRAP (ferric reducing antioxidant power) assay, the BCB (β-Carotene bleaching) assay, and the TPC (total phenolic content) assay. *n*-Hexane, chloroform, chloroform–methanol (9:1), and methanol extract that was partitioned between *n*-butanol and water were prepared from the aerial parts and the roots. *n*-Butanol extracts of the roots and the aerial parts as well as chloroform and chloroform-methanol extracts of aerial parts displayed significant antioxidant activities while no remarkable activities were observed with the remaining extracts. Bioassay-guided fractionation led to the isolation of a rare deoxy-d-riburonic moiety containing phenolic structures as new secondary metabolites and the known compounds as thunberginol G, isoorientin, orientin, isoschaftoside, and swertiajaponin. Among the isolated compounds, thunberginol G showed the highest β-carotene bleaching inhibitory activity and FRAP value (46.1% and 82.6 mg TE/g, respectively) followed by isoorientin (47.7% and 60.9 mg TE/g, respectively). According to the research, the antioxidant activity of the extract was suggested as a result of the synergistic effect of the compounds found in the extract (Milella et al., 2014). Further study on *Scorzonera papposa* revealed the antioxidant activity of ethanol extracts from aerial parts and roots of *S. papposa* from Iraq and Turkey. The results indicated that *Scorzonera papposa* aerial parts exhibited notable antioxidant activity at 50–800 μg/mL concentrations. Iraq samples (6.328 ± 0.141 mmol/L and 4.817 ± 0.073 mmol/L for aerial parts and roots, respectively) displayed a higher level of Total Antioxidant Status than samples collected in Turkey (5.314 ± 0.100 mmol/L and 4.504 ± 0.042 mmol/L for aerial parts and roots respectively) and a lower level of TOS (Total Oxidant Status). Therefore, the OSI (Oxidative Stress Index – The TAS/TOS ratio) parameter of the Iraq samples was lower than the Turkey samples (Mohammed et al., 2020).

S. hieraciifolia root ethanolic extract fractions which were prepared by petroleum ether, chloroform, ethyl acetate, and n-butanol, respectively, were evaluated for their antioxidant activities by measuring the inhibitory activity on lipid peroxidation, scavenging ability of DPPH, superoxide, and ABTS radicals, as well as ferric reducing antioxidant power (FRAP). The ethyl acetate fraction of the roots displayed the highest antioxidant activity in preventing lipid peroxidation, scavenging superoxide, DPPH, and ABTS radicals by 1.26 mg/mL, 1.47 mg/mL, 0.45 mg/mL EC_{50} values, respectively, as well as in reducing power assay with 2.69 mM ferrous ions equivalents expression (Sarı et al., 2019). Further studies on aerial parts of the *S. hieraciifolia* revealed that *S. hieraciifolia* dichloromethane, ethyl acetate, and n-hexane extracts exhibited low anti-radical activity, against DPPH and ABTS radicals, while they displayed considerable reducing potencies for CUPRAC and FRAP assay. Infusion of the plant and methanol extract antioxidant activities were determined as significant (DPPH: 44.51 ± 0.37 mg Trolox equivalent (TE)/g and 85.52 ± 0.62 mg TE/g; ABTS: 48.07 ± 0.05 mg TE/g and 85.11 ± 2.20 mg TE/g respectively). The same study also revealed that the underground parts of *S. hieraciifolia* have similar antioxidant activity for its infusion (DPPH: 18.87 ± 0.38 mg TE/g and ABTS: 43.38 ± 0.88 mg TE/g) and methanol extracts (DPPH: 43.38 ± 0.88 mg TE/g and ABTS: 66.55 ± 1.52 mg TE/g). Remarkable reducing power effects were determined on CUPRAC and FRAP, by mean values ranging from 96.72 ± 3.49 to 52.27 ± 2.48 mg TE/g, and 47.76 ± 1.70 to 21.25 ± 0.86 mg TE/g for the root extracts, respectively. Regarding metal chelating activity, infusion of the aerial parts displayed the highest metal chelating activity (15.93 ± 0.10 mg EDTAE/g) (Dall'Acqua et al., 2020).

S. pygmaea aerial parts and roots' fractions prepared by using petroleum ether, chloroform, ethyl acetate, and n-butanol respectively were also evaluated for their antioxidant potentials with five different methods as following: scavenging activity against ABTS radical, DPPH radical, and superoxide radical as well as by the determination of inhibitory activities on lipid peroxidation and the ferric reducing antioxidant power assay (FRAP) (Şahin et al., 2020a and 2020b). The EtOAc fraction of both the aerial parts and roots was found to be the highest active fraction and the most scavenged radical by all fractions was determined as superoxide. EC_{50} values were determined as 1.58 mg/mL, 0.49 mg/mL, 1.21 mg/mL, and 2.12 mg/mL, for the roots, and 0.92 mg/mL, 0.46 mg/mL, 1.24 mg/mL, and 1.09 mg/mL, for the aerial parts in lipid peroxidation assay and scavenging activity against superoxide, DPPH, ABTS radicals, respectively. Furthermore, in reducing power assay *S. pygmaea* aerial parts and roots exhibited antioxidant activity with 2.77 mM and 1.76 mM ferrous ions equivalent expression, respectively. EtOAc fraction of the aerial parts displayed comparable antioxidant activity with the rutin as a standard compound on lipid peroxidation and the ferric reducing antioxidant power assay as well as superoxide radical scavenging activity and the activity was determined as stronger than the same fractions of the roots in all the methods except the DPPH radical scavenging activity (Şahin et al., 2020b).

Ethanolic extract of *Scorzonera cinerea* leaves was evaluated for its DPPH and ABTS radical scavenging capacities in order to clarify antioxidant activity. Moreover, superoxide dismutase (SOD), glutathione peroxidase (GPx), and catalase (CAT) activities, glutathione (GSH) concentration, malondialdehyde (MDA) levels, total antioxidant status (TAS), and total oxidant status (TOS) were analyzed in the liver tissues of streptozotocin-induced diabetic rats which were treated with *S. cinerea* leaf extract. An increased level of liver MDA and TOS were determined in diabetic rats while *S. cinerea* treated group values were lowered significantly. On the other hand, GSH, TAS, and antioxidant enzyme activities which were found to be in low amounts in the diabetic group were determined remarkably enhanced in *S. cinerea* leaf extract treated groups (Temiz, 2021).

Hepatoprotective Activity

Several *Scorzonera* species root extracts (*S. cana* var. *jacquiniana, S. latifolia, S. mollis* ssp. *szowitsii, S. parviflora, S. tomentosa*) at 100 mg/kg dosage were investigated for their hepatoprotective activities, along with compounds isolated from the *S. latifolia* root extract (chlorogenic acid, hydrangenol-8-*O*-*β*-glucoside, and scorzotomentosin-4′-*O*-*β*-glucoside) at 5 mg/kg dosage by measuring their protective activities against CCl_4-induced liver toxicity on animals. *Scorzonera* extract treatments did not induce significant changes in ALT and AST levels after seven days of treatment, while the histopathological results displayed that all tested groups have less damage when compared to the carbon tetrachloride group except scorzotomentosin-4'-*O*-*β*-glucoside and hydrangenol-8-*O*-*β*-glucoside groups. Although *Scorzonera* species displayed moderate hepatoprotective activities against carbon tetrachloride-induced acute toxicity, chlorogenic acid as one of the main constituents of the investigated species exhibited significant recovery on cellular damage and less sinusoidal dilatation, vascular congestion, and ballooning which were proved by histopathological examinations (Özbek et al., 2017).

Wound Healing Activity

S. cinerea, S. latifolia, S. incisa, S. parviflora, S. mollis ssp. *szowitsii, S. tomentosa,* and yakı sakızı that were prepared traditionally from *S. latifolia* root latex, aerial parts and root aqueous methanolic extracts were tested for their wound healing properties by *in vivo* linear incision and circular excision wound models. Hydroxyproline content of the treated tissues was also evaluated. The highest wound healing activity was observed with the treatment of *S. latifolia, S. mollis* ssp. *szowitsii* aerial parts by the values of 30.6% and 25.4% wound tensile strength on day 10 in the linear incision wound model. In the same study *S. tomentosa* was also determined to be effective (23.2%). In circular excision wound model, the wound contractions were determined as 45.82% and 61.44% for *S. latifolia*, 33.48% and 59.88% for *S. mollis* ssp. *szowitsii*, 42.13% and 37.62% for *S. tomentosa,* and 42.13% and 37.62% for yakı sakızı treated group on days 10 and 12, which were comparable to reference drug Madecassol® (90.78–100%). High hydroxyproline levels that correlate with enhanced collagen synthesis have an important role in hemostasis and are required for strength and integrity of the tissue and were also determined in *S. latifolia, S. mollis* ssp. *szowitsii* and *S. tomentosa* treated groups, and these groups demonstrated faster re-modeling compared to the others (Küpeli Akkol et al., 2011). Another study on *S. acuminata, S. cana* var. *alpina, S. cana* var. *jacquiniana,*

S. cana var. *radicosa, S. eriophora, S. laciniata* ssp. *laciniata, S. suberosa* ssp. *Suberosa,* and *S. sublanata* revealed aqueous methanolic extracts of *S. cana* var. *jacquiniana* and *S. eriophora* aerial parts displayed significant tensile strength by 40.5% and 34.3% values respectively on day 10. In the circular excision wound model the same extracts caused the contraction of the wound area by up to 46.27% and 39.44% on day 12 respectively, and significant enhancement of hydroxyproline levels in the regenerated tissue was also determined (Süntar et al., 2012).

Further studies on *S. latifolia* and *S. cana* var. *jacquiniana* aerial parts which were found to have significant wound healing activities in previous researches report that ethyl acetate fractions of aerial parts methanol extract of both plants have wound healing properties *in vitro.* Bioactivity guided fractionation resulted in the isolation of isoorientin, 7-*O*-methylisoorientin, quercetin-3-*O*-*β*-apiofuranosyl-(1′ ′ ′ →2′ ′)-*β*-D-glucopyranoside, and quercetin-3-*O*-*α*-rhamnopyranosyl-(1→6)-*β*-D-galactopyranoside, from the active fraction of *S. latifolia.* 7-*O*-methylisoorientin displayed inhibitory activity on collagenase and elastase enzymes, while quercetin-3-*O*-*β*-apiofuranosyl-(1′ ′ ′ →2′ ′)-*β*-D-glucopyranoside inhibited only collagenase enzymes. No activity was observed on hyaluronidase enzyme by none of the fractions or isolated compounds. The activity of the *S. latifolia* methanolic extract was considered by the synergistic interaction of the compounds found in the methanolic extract (Küpeli Akkol et al., 2019). A similar study on *S. cana* var. *jacquiniana* (Syn *P. canum*) aerial parts gave 11 known compounds: arbutin, 6′-*O*-caffeoylarbutin, cichoriin, 3,5-di-*O*-caffeoylquinic acid methyl ester, apigenin-7-*O*-*β*-glucoside, luteolin-7-*O*-*β*-glucoside, apigenin-7-*O*-*β*-rutinoside, isoorientin, orientin, vitexin, procatechuic acid, and new compound 4-hydroxy-benzoic acid-4-(6-*O*-*α*-rhamnopyranosyl-*β*-glucopyranosyl)-benzyl ester as a result of bioactivity-guided fractionation. Apigenin-7-*O*-*β*-glucoside, apigenin-7-*O*-*β*-rutinoside, and isoorientin inhibited elastase enzyme by 43.04%, 40.36%, and 30.66%, and apigenin-7-*O*-*β*-glucoside, luteolin7-*O*-*β*-glucoside, and apigenin 7-*O*-*β*-rutinoside displayed inhibitory activity on collagenase enzymes with values of 26.48%, 29.96%, and 24.28%. However, this activity was not found to be higher than the total methanolic extract. Therefore the inhibitory activity of the plant is considered by the synergistic interaction of the compounds in the extract (Bahadır Acıkara et al., 2019).

REFERENCES

Acikara O. B., Çitoğlu, G. S., & Çoban, T., Phytochemical Screening and Antioxidant Activities of Selected Scorzonera Species., *Turk J Pharm Sci*, 10(3), 453–462 (2013).

Altundag E, Ozturk M, Ethnomedicinal studies on the plant resources of east Anatolia, Turkey, *Procedia. Soc. Behav. Sci.*, 19, 756–777 (2011).

Bahadır Ö, Saltan HG, Özbek H, Antinociceptive activity of some *Scorzonera* L. species, *Turk. J. Med. Sci.*, 42, 861–866 (2012).

Bahadır Ö, Saltan Çitoğlu G, Smejkal K, Dall'Acqua S, Özbek H, Cvacka J, Zemlicka M, Analgesic compounds from *Scorzonera latifolia* (Fisch. and Mey.) DC., *J Ethnopharmacol.*, 131, 83–87 (2010).

Bahadır Acıkara Ö, Saltan Çitoğlu G, Dall'Acqua S, Smejkal K, Cvacka J, Zemlicka M, A new triterpene from *Scorzonera latifolia* (Fisch. and Mey.) DC., *Nat. Prod. Res.*, 26, 1892–1897 (2012).

Bahadır Acıkara Ö, Hosek J, Babula P, Cvacka J, Budesinsky M, Dracinsky M, Saltan G, Smejkal K, Turkish *Scorzonera* species extracts attenuate cytokine secretion via NF-κB inhibition showing promising anti-inflammatory effect, *Molecules*, 21, 43 (2016).

Bahadır Acıkara Ö, Saltan Çitoğlu G, Dall'Acqua S, Özbek H, Cvacka J, Zemlicka M, Smejkal K, Bioassay-guided isolation of the antinociceptive compounds motiol and beta-sitosterol from *Scorzonera latifolia* root extract, *Pharmazie*, 69, 711–714 (2014a).

Bahadır Acıkara Ö, Süntar İ, Saltan Çitoğlu G, Keleş H, Ergene B, Akkol Küpeli E, Determination of phenolic acids and flavonoids and anti-inflammatory activity of *Scorzonera* species, *IJPPR*, 6, 59–65 (2014b).

Bahadır Acıkara Ö, Ilhan M, Kurtul E, Šmejkal K, Küpeli Akkol E, Inhibitory activity of *Podospermum canum* and its active components on collagenase, elastase and hyaluronidase enzymes, *Bioorg. Chem.*, 93, 103330 (2019).

Baytop T, *Therapy with Medicinal Plants in Turkey Past and Present*, 2nd Edition, Nobel Tıp Kitabevleri, İstanbul (1999).

Chamberlain DF, Scorzonera L. In: Davis, P.H. (Ed.) *Flora of Turkey and The East Aegean Islands*. Edinburgh: University Press, Vol. 5, pp: 632–657 (1975).
Coşkunçelebi K, Makbul S, Gültepe M, Okur S, Güzel ME, A conceptus of *Scorzonera* s.l. in Turkey, *Turk. J. Bot.*, 38, 1401–1410 (2015).
Dall'Acqua S, Ak G, Sut S, Zengin G, Yıldıztugay E, Mahomoodally MF, Sinan KI, Lobine D, Comprehensive bioactivity and chemical characterization of the endemic plant *Scorzonera hieraciifolia* Hayek extracts: A promising source of bioactive compounds, *Food Res. Int.*, 137, 109371 (2020).
Erden Y, Kırbağ S, Yılmaz Ö, Phytochemical composition and antioxidant activity os some *Scorzonera* species, *Proc. Natl. Acad. Sci. India, Sect. B Biol. Sci.*, 83, 271–276 (2013).
Erik I, Coskuncelebi K, Makbul S, Yayli N, New chlorogenic acid derivatives and triterpenoids from *Scorzonera aucheriana*, *Turk. J. Chem.*, 45, 199–209 (2021a).
Erik İ, Yaylı N, Coşkunçelebi K, Makbul S, Alpay Karaoğlu Ş, Three new dihydroisocoumarin glycosides with antimicrobial activities from *Scorzonera aucheriana*, *Phytochem. Lett.*, 43, 45–52 (2021b).
Granica S, Lohwasser U, Jöhrer K, Zidorn C, Qualitative and quantitative analyses of secondary metabolites in aerial an subaerial of *Scorzonera hispanica* L. (black salsify), *Food Chem.*, 173, 321–331 (2015).
Karakaya S, Polat A, Aksakal Ö, Sümbüllü YZ, 'Incekara Ü, Ethnobotanical study of medicinal plants in aziziye district (Erzurum, Turkey), *Turk. J. Pharm. Sci.*, 17, 211–220 (2020).
Küpeli Akkol E, Bahadır Acıkara Ö, Süntar İ, Saltan Çitoğlu G, Keleş M, Ergene B, Enhancement of wound healing by topical application of *Scorzonera* species: Determination of the constituents by HPLC with new validated reverse phase method, *J. Ethnopharmacol.*, 137, 1018–1027 (2011).
Küpeli Akkol E, Šmejkal K, Kurtul E, Ilhan M, Güragac FT, Saltan İşcan G, Bahadır Acıkara Ö, Cvačka J, Buděšínský M, Inhibitory activity of *Scorzonera latifolia* and its components on enzymes connected with healing process, *J. Ethnopharmacol.*, 245 (2019).
Milella L, Bader A, De Tommasi N, Russo D, Braca A, Antioxidant and free radical-scavenging activity of constituents from two *Scorzonera* species, *Food Chem.*, 160, 298–304 (2014).
Mohammed FS, Günal S, Sabik AE, Akgül H, Sevindik M, Antioxidant and antimicrobial activity of *Scorzonera papposa* collected from Iraq and Turkey, *KSU J. Agric Nat.*, 23(5), 1114–1118 (2020).
Özbek H, Bahadir O, Keskin I, Kırmızı NI, Yigitbasi T, Sayin Sakul A, Iscan G, Preclinical evaluation of *Scorzonera* sp. root extracts and major compounds against acute hepatotoxicity induced by carbon tetrachloride, *Indian J. Pharm. Sci.*, 79 (2017).
Öksüz S, Gören N, Ulubelen A, Terpenoids from *Scorzonera tomentosa*, *Fitoterapia*, 61, 92–93 (1990).
Saltan Çitoğlu G, Bahadır G, Dall'Acqua S, Dihydroisocoumarin derivatives isolated from the roots of *Scorzonera latifolia*, *Turk. J. Pharm. Sci.*, 7, 205–211 (2010).
Sarı A, Zidorn C, Spitaler R, Ellmerer EP, Ozgokce F, Orgama KH, Stuppner H, Phenolic compounds from *Scorzonera tomentosa* L., *Helv. Chim. Acta*, 90, 311–317 (2007).
Sarı A, Two new 3-benzylphyhalides from Scorzonera veratrifolia Fenzl, *Nat. Prod. Res.*, 24, 56–62 (2010).
Sarı A, Phenolic compounds from *Scorzonera latifolia* (Fisch. & Mey.) DC, *Nat. Prod. Res.*, 26, 50–55 (2012).
Sarı A, Şahin H, Özsoy N, Özbek Çelik B, Phenolic compounds and in vitro antioxidant, anti-inflammatory, antimicrobial activities of *Scorzonera hieraciifolia* Hayek roots, *S. Afr. J. Bot.*, 125, 116–119 (2019).
Süntar, I, Acikara, OB, Citoglu, GS, Keles, H, Ergene, B, Akkol, EK. (2012) *In vivo* and *in vitro* evaluation of the therapeutic potential of some Turkish Scorzonera species as wound healing agent. *Curr Pharm Des*,18(10):1421–33.
Şahin H, Sarı A, Özsoy N, Özbek Çelik B, Koyuncu O, Two new phenolic compounds and some biological activities of *Scorzonera pygmaea* Sibth. & Sm. subaerial parts, *Nat. Prod. Res.*, 34(5), 621–628 (2020a).
Sahin H, Sari A, Ozsoy N, Ozbek Celik B, Phenolic compounds and bioactivity of *Scorzonera pygmaea* Sibth. & Sm. aerial parts: In vitro antioxidant, anti-inflammatory and antimicrobial activities, *J. Pharm. Istanbul Univ.*, 50(3), 294–299 (2020b).
Şakul AA, Kurtul E, Özbek H, Kırmızı Nİ, Bahtiyar BC, Saltan H, Bahadır Acıkara Ö, Evaluation of antidiabetic activities of *Scorzonera* species on alloxan-induced diabetic mice, *Clin. Exp. Health. Sci.*, 11(1), 74–80 (2021).
Şenkardeş İ, Bulut G, Doğan A, Tuzlacı E, An ethnobotanical analysis on wild edible plants of the Turkish Asteraceae tax, *Agric. Conspec. Sci.*, 84, 17–28 (2019).
Sezik E, Yeşilada E, Tabata M, Honda G, Takaishi Y, Fujita T, Tanaka T, Takeda Y, Traditional medicine in Turkey VIII. Folk medicine in East Anatolia; Erzurum, Erzincan, Ağrı, Kars, Iğdır provinces, *Econ. Bot.*, 51, 195–211 (1997).
Temiz MA, Antioxidant and antihyperglycemic activities of *Scorzonera cinerea* radical leaves in streptozocin-induced diabetic rats, *Acta Pharm.*, 71, 603–617 (2021).

Turan M, Kordali S, Zengin H, Dursun A, Sezen Y, Macro and micro mineral content of some wild edible leaves consumed in Eastern Anatolia, *Acta. Agr. Scand. B-SP*, 53, 129–137 (2003).

Wang Y, Edrada-Ebel RA, Tsevegsuren N, Sendker J, Braun M, Wray V, Lin W, Proksch P, Dihydrostilbene derivatives from the Mongolian medicinal plant *Scorzonera radiata*, *J. Nat. Prod.*, 72, 671–675 (2009).

Yildirim B, Terzioglu Ö, Özgökçe F, Türközü D, Ethnobotanical and pharmacological uses of some plants in the districts of Karpuzalan and Adigüzel (Van-Turkey), *J. Anim. Vet. Adv.*, 7, 873–878 (2008).

Zhu Y, Hu PZ, He ZW, Wu QX, Li J, Wu WS, Sesquiterpene lactones, from *Scorzonera austriaca*, *J. Nat. Prod.*, 73, 237–241 (2010).

Zidorn C, Ellmerer-Müller EP, Stuppner H, Sesquiterpenoids from *Scorzonera hispanica* L., *Pharmazie*, 55, 550–551 (2000).

Zidorn C, Ellmerer EP, Sturm S, Stuppner H, Tyrolobibenzyls E and F from Scorzonera humilis and distribution of caffeic acid derivatives, lignans and tyrolobibenzyls in European taxa of the subtribe Scorzonerinae (Lactuceae, Asteraceae), *Phytochemistry*, 63, 61–67 (2003).

24 *Sideritis* sp.

Tuğba Günbatan and İlhan Gürbüz

INTRODUCTION

Sideritis L. is a large genus that grows in wide region and is represented by more than 150 species. Particularly, Mediterranean countries such as Spain, Macedonia, Greece, and Bulgaria are rich in *Sideritis* species (1) and Turkey is also among the richest countries with 45 wildly grown *Sideritis* species (54 taxa, 40 of them are endemic) (2). *Sideritis* species grown in Turkey are given in Table 24.1. In recent years, the genus has attracted attention with its high endemicity (74 %) and rich phytochemical diversity and has been researched in detail in terms of botany, pharmacology, and phytochemistry. In this chapter, ethnobotanical, phytochemical, and biological activity studies on *Sideritis* species growing in Turkey were summarized in order to provide an overall perspective.

BOTANICAL DESCRIPTION

Sideritis species are annual or perennial plants that can be in small shrub or herbaceous form. They have pilose or tomentose indumentum which can bear glandular hair, but can rarely be glabrous. Flowers form usually 6-flowered verticillaster, covered with leaf-like bracts (Figure 24.1). Tubular-campanulate calyx has 5 teeth and all calyx teeth are equal or the upper tooth could be larger. The corolla is usually yellow colored, but may be white, purple, or red in some species. The corolla tube is shorter than the calyx. The lower corolla lip is three-lobed; while the upper corolla lip could be flat, patent, entire, or bifit. Flowers bear 4 stamens, and the lengths of the stamens and style do not exceed the corolla tube. The fruit type is nutlet (1, 3).

ETHNOBOTANICAL INFORMATION

The origin of the genus name is "*sideros*" which means iron in Greek. This name was given to the genus because it was used in the treatment of injuries caused by iron materials such as knives and swords. In European countries *Sideritis* species are named "Olympus tea", "Mountain tea", "Mursala Tea", "Rabo de gato", "Zajarena" etc. and are traditionally used for disorders in a wide spectrum like respiratory system disorders, common cold, fever, urinary system infections, peptic ulcer, wounds, and rheumatism (4).

In Anatolia, although the members of the *Sideritis* genus are mostly known as "dağ çayı"; names such as "adaçayı, çay otu, yayla çayı, çay çalbası, altınbaş, şıltık, yaraotu, elduran, tepeli çay, cavva, fenerli çay, fincan çayı, kandil çayı, kandilli çay" were are also given (4-9). It is generally used for stomachache, as an appetite stimulant, and as carminative (9), but very common ethnobotanical usages have been identified in relation to its richness in the genus *Sideritis*. Ethnobotanical records about *Sideritis* species in Turkey are compiled in detail and presented in alphabetical order according to their scientific names in Table 24.1. As can be seen in this table, different *Sideritis* species can be known with the same name and used for the same or similar purposes, and it is understood that Turkish *Sideritis* species are generally used in upper respiratory tract diseases such as common cold, flu, and gastrointestinal complaints, e.g., stomach pain and abdominal pain (Table 24.1).

DOI: 10.1201/9781003146971-24

PHYTOCHEMISTRY

Sideritis species have been studied in detail in terms of phytochemistry. Compounds isolated from Turkish *Sideritis* species, their essential oil and fixed oil analyses are summarized in Table 24.1.

With more than 160 identified compounds, *Sideritis* species are very rich in diterpenes (4). As can be seen in Table 24.1, terpenes found in Turkish *Sideritis* species are generally kaurane diterpenes. Sidol, sideridiol, linearol, and isolinearol are the most common kaurane diterpene derivatives; however, diterpenes with labdane and beyerane skeleton have also been isolated.

Genus is also rich in flavonoid, especially hypolaetin, isoscutellarein, apigenin and chrysoeriol derivatives, ozturcosides, xanthomicrol, and salvigenin have been isolated from different Turkish *Sideritis* species (Table 24.1).

The members of the Lamiaceae family are rich in essential oil, but *Sideritis* genus may be excluded from this generalization. Essential oil analyses have revealed that *Sideritis* species yields essential oil at low rates (4, 10). Turkish *Sideritis* species were grouped under six classes, according to the main components of their essential oil: rich in monoterpene hydrocarbons, rich in oxygenated monoterpenes, rich in sesquiterpene hydrocarbons, rich oxygenated sesquiterpenes, rich in diterpenes, and others (4, 11). More than half of the Turkish *Sideritis* species are in the monoterpene hydrocarbon-rich group with their high α-pinene, β-pinene, and phellandrene content (e.g., *S. erythrantha, S. stricta,* and *S. vuralii*). The sesquiterpene-rich group, which includes majorly β-caryophyllene, bicyclogermacrene, germacrene D, calamen, etc., corresponds to 27% (e.g., *S. caesarea, S. ozturkii, S. tmolea, S. vulcanica*) (4, 11, 12).

Although iridoids are not common in the *Sideritis* genus, some iridoids such as ajugoside, ajugol, melittoside, and stachysosides E and G have been isolated from Turkish *Sideritis* species. On the other hand, phenylethanoid derivatives such as verbascoside, lavandulifolioside, leucosceptoside A, and martynoside are also found in members of the genus (Table 24.1).

The composition of fixed oils obtained from seeds of some *Sideritis* species has also been elucidated. As could be seen in Table 24.1, the seeds of the Turkish *Sideritis* species mainly contain linoleic, oleic, palmitic, and stearic acid.

PHARMACOLOGICAL ACTIVITY

Sideritis species have been studied extensively in terms of pharmacological activity. Pharmacological activity studies on Turkish *Sideritis* species are summarized in Table 24.1 and the details are provided under separate headings below.

Analgesic and Anti-inflammatory Activity

Akçoş et al. (1999) evaluated the anti-inflammatory activity of *S. lycia* on mice by carrageenan-induced paw edema test. According to results, petroleum ether, chloroform, ethyl acetate, and water fractions obtained from methanol extract of *S. lycia* exhibited significant anti-inflammatory activity at 1500 mg/kg doses with inhibition rates of 69%, 80%, 58%, and 69%, respectively. But on the contrary of extracts and fractions, isolated compounds lavandulifolioside, verbascoside, martinoside + leucoceptoside A and 4'-*O*-methyl-hypolaetin-7-*O*-[6'''-*O*-acetyl-β-D-allopyranosil-(1→2)-β-D-glucopyranoside + 4'-*O*-methyl-hypolaetin-7-*O*-[6'''-*O*-acetyl-β-D-allopyranosyl–(1→2)-6''-*O*-acetyl-β-D-glucopyranoside mixture determined to cause weak inhibition (in the range of 5–13%) at a dose of 100 mg/kg (13).

In another research, methanol extract of *S. brevibracteata*, chloroform, *n*-butanol, and water fractions obtained from methanol extract; hypolaetin-7-*O*-[6'''-*O*-acetyl-β-D-allopyranosyl-(1→2)]-β-D-glucopyranoside, isoscutellarein-7-*O*-[6'''-*O*-acetyl-β-D-allopyranosyl-(1→2)]-β-D-glucopyranoside, 3'-hydroxy-4'-*O*-methylisoscutellarein-7-*O*-[6'''-*O*-acetyl-β-D-allopyranosyl-(1→2)]-β-D-glucopyranoside,

TABLE 24.1
Ethnobotanical Usages, Pharmacological Investigations, and Phytochemical Findings on Turkish *Sideritis* Species

	Ethnobotanical Records				
***Sideritis* Species**	**Locality**	**Local Name**	**Usage**	**Pharmacological Activity**	**Phytochemical Findings**
**S. akmanii* Aytaç, Ekici & Dönmez	-	-	-	Antioxidant (18) Anticholinesterase (18) Antiglucosidase (18) Antiamylase (18) Antityrosinase (18)	*Diterpenes:* Foliol, isofoliol, linearol, isolinearol, sideridiol, sideroxol (43) *Essential oil:* Major compounds are hexadecanoic acid, tetradecanoic acid, spathulenol (18)
**S. albiflora* Hub.-Mor.	-	-	-	Antimicrobial (44, 45) Antioxidant (21, 46-48) Anticholinesterase (21, 48) Antityrosinase (21, 48) Antiurease (21)	*Essential oil*: Major compounds are *trans*-caryophyllene, α-pinene, β-pinene, γ-cadinene, pulegone (49) Major compounds are β-caryophyllene, γ-elemene, caryophyllene oxide, aromadendrene, γ-cadinene (47) Major compounds are germacrene D, β-caryophyllene, caryophylene oxide (50) Major compounds are β-caryophyllene, germacrene D, τ-gurjunene (48) *Fixed oil:* Major compounds are palmitic acid, linoleic acid, oleic acid (21) *Flavonoids*: 4'-*O*-methylisoscutellarein-7-*O*-allosyl-(1→2) glucoside, apigenin, apigenin glucoside, isoscutellarein 7-*O*-[6'''-*O*-acetyl]- allosyl-(1→2)-glucoside, luteolin, naringin (21, 44, 50) *Other phenolics*: Chlorogenic acid, 5-caffeoylquinic acid, gallic acid, caffeic acid, 2,4-dihydroxy benzoic acid, *p*-coumaric acid, ferulic acid, *trans*-2-hydroxycinnamic acid, rosmarinic acid, (21, 50) *Others*: *trans*-Cinnamic acid (21) *Phenylethanoids*: Verbascoside, forsythoside B (50)
**S. amasiaca* Bornm.	-	-	-	-	*Essential oil*: Major compounds are β-pinene, bicyclogermacrene, β-caryophyllene (51)

(*Continued*)

TABLE 24.1 (CONTINUED)
Ethnobotanical Usages, Pharmacological Investigations, and Phytochemical Findings on Turkish *Sideritis* Species

	Ethnobotanical Records				
***Sideritis* Species**	**Locality**	**Local Name**	**Usage**	**Pharmacological Activity**	**Phytochemical Findings**
**S. arguta* Boiss. & Heldr.	Karaman	Dağ çayı	Aerial parts; for stomachache (52)	Anticholinesterase (19, 53) Cytotoxicity (27) Antioxidant (19, 27, 46, 53, 54)	*Diterpenes*: Diacetyldistanol, *ent*-7α-15β-16β-18-trihydroxykaur-16-ene, *ent*-7α-18-diacetoxy-15β,16β-epoxykauran, *ent*-7α-acetoxy-15β,18-dihydroxykaur-16-ene, epoxysiderol, eubol, eubotriol, sideroxol, siderol, 7-epicandicandiol, 15-*epi*-eubol (27, 53) *Flavonoids*: Apigenin, kaempferol, quercetin, myricetin (54) *Other phenolics*: Ferulic acid, rosmarinic acid, caffeic acid, chlorogenic acid, *p*-coumaric acid (54)
	Antalya	Dağ çayı, yayla çayı	Flower, leaf; as appetizer and stomachic (9)		
**S. argyrea* P.H. Davis	Antalya	Eşek çayı	Flower, leaf; as appetizer and stomachic (9)	Anticholinesterase (55) Antityrosinase (55)	*Diterpenes*: *Ent*-6β,8α-dihydroxylabda-13(16),14-dien, *ent*-7α-acetoxy-18-hydroxykaur-16-ene, 7-epicandicandiol, 7-epicandicandiol-18-monoacetate, foliol, candol B, linearol, sideridiol, siderol, sidol (56, 57) *Essential oil*: Major compounds are β-pinene, limonene, α-pinene (58)
**S. armeniaca* Bornm.	-	-	-	-	*Essential oil*: Major compounds are α-pinene, β-pinene (59)
S. athoa Papan. & Kokkini	Balıkesir	Kandil çayı, tilki kuyruğu çayı, yüzüklü çay, Kazdağı çayı	Aerial parts; for common cold and flu (60)	Central nervous system depressant/ stimulant activity (32)	*Diterpenes*: Athonolone,7-epicandicandiol, *ent*-3α-18-dihydroxykaur-16-ene, *ent*-3β,7α-dihydroxykaur-l6-ene, *ent*-3β-hydroxykaur-16-ene, *ent*-7α,18-dihydroxybeyer-15-ene, canadiol, foliol, linearol, sidol (57, 61-63) *Fixed oil*: Major compounds are linoleic acid, oleic acid, palmitic acid, stearic acid (64)

(*Continued*)

TABLE 24.1 (CONTINUED)
Ethnobotanical Usages, Pharmacological Investigations, and Phytochemical Findings on Turkish *Sideritis* Species

Sideritis Species	Ethnobotanical Records			Pharmacological Activity	Phytochemical Findings
	Locality	Local Name	Usage		
S. bilgerana P. H. Davis	Karaman	Dağ çayı	Aerial parts; for stomachache (52)	Antimicrobial (65, 66) Antinociceptive (15) Anti-inflammatory (15) Antioxidant (15) Anticholinesterase (15) Antityrosinase (15) Antiamylase (15) Antiglucosidase (15)	*Essential oil*: Major compounds are β-pinene, α-pinene, β-phellandren (65, 67) *Flavonoids*: Apigenin, quercetin, luteolin (68) *Other phenolics:* Ferulic acid, caffeic acid (68)
	Niğde	Boz şalba, kekik çayı, yayla çayı	Aerial parts; for heartburn, common cold, hemorrhoids (69)		
	Konya	-	Flower, leaf; as appetizer and stomachic (9)		
*S. brevibracteata P. H. Davis	-	-	-	Anti-inflammatory, Antinociceptive (14) Antialdose reductase inhibitory (14) Antimicrobial (66) Glutathione reductase inhibitory (38) Antioxidant (14, 46, 70) Anticholinesterase (70)	*Essential oil*: α-Cadinene β-caryophyllene, germacrene D (70) *Diterpenoids:*7-Acetyl sideroxol, athanolone, eubotriol, linearol, sideridiol, siderol (70) *Flavonoids*: Hypolaetin-7-*O*-[6‴-*O*-acetyl-β-D-allopyranosyl-(1→2)]-6″-*O*-acetyl-β-D-glucopyranoside, hypolaetin-7-*O*-[6‴-*O*-acetyl-β-D-allopyranosyl-(1→2)]-6"-β-D-glucopyranoside, isorhamnetin, isoquercetin, isoscutellarein-7-*O*-[6‴-*O*-acetyl-β-D- allopyranosyl-(1→2)]-β-D-glucopyranoside, isoscutellarein-7-*O*-[6‴-*O*-acetyl-β-D-allopyranosyl-(1→2)]-6″-*O*-acetyl-β-D-glucopyranoside, 3′-hydroxy-4′-*O*-methylisoscutellarein-7-*O*-[6‴-*O*-acetyl-β-D-allopyranosyl-(1⟶2)]-6″-*O*-acetyl-β-D-glycopyranoside, 3′-hydroxy-4′-*O*-metylisoscutellarein-7-*O*-[6‴-*O*-acetyl-β-D-allopyranosyl-(1→2)]-β-D-glycopyranoside, luteolin-5-glucoside, luteolin-7-glucoside, kaempferol, penduletin, quercetagetin-3,6-dimethylether, quercitrin, salvigenin (14, 70)

(*Continued*)

TABLE 24.1 (CONTINUED)
Ethnobotanical Usages, Pharmacological Investigations, and Phytochemical Findings on Turkish *Sideritis* Species

Sideritis Species	Ethnobotanical Records			Pharmacological Activity	Phytochemical Findings
	Locality	Local Name	Usage		
					Other phenolics: *p*-Coumaric acid, *p*-hyroxy benzoic acid, caffeic acid, chlorogenic acid, *t*-ferulic acid, gallic acid, rosmarinic acid, pelargonin chloride, pyrogallol, salicylic acid, syringic acid, vanillin (70) *Phenylethanoids:* Verbascoside (14) *Steroids*: Stigmasterol (70)
**S. brevidens* P. H. Davis	-	--		Antimicrobial (71) Antioxidant (72) Anticholinesterase (55) Antityrosinase (55)	*Diterpenes*: Epicandicandiol, linearol, siderol, sidol (43, 72) *Essential oil*: Major compounds are *β*-pinene, *epi*-cubebol, *α*-pinene, cubebol (73) *Fixed oil*: Major compounds are linoleic acid, oleic acid, palmitic acid, stearic acid (64) *Flavonoids*: Apigenin, kaempferol-3-*O*-glucoside, quercetin, luteolin-7-*O*-glucoside (72) *Other phenolics*: Caffeic acid, ferulic acid, gallic acid, *p*-hydroxybenzoic acid, *p*-coumaric acid, syringic acid, vanillin (72) *Steroids*: *β*-Sitosterol, stigmasterol (72) *Tannins*: Ellagic acid (72)
**S. caesarea* H. Duman, Aytaç & Başer	Kayseri	Dağ çayı	Aerial parts; for stomachache, intestinal spasm, as sedative (74)	Antimicrobial (75, 76) Antioxidant (77) Anticholine esterase inhibitory (78) Anti-ulcerogenic (30, 31)	*Diterpenes*: Epoxysiderol, siderol, eubol, eubotriol ve *ent-7α*,18-dihydroxy-15-oxokaur-16-ene (62, 79) *Essential oil*: Major compounds are hexadecanoic acid, caryophyllene oxide, *β*-caryophyllene, spathulenol (22) Major compounds are *β*-caryophyllene, caryophyllene oxide, spathulenol (80)

(*Continued*)

TABLE 24.1 (CONTINUED)
Ethnobotanical Usages, Pharmacological Investigations, and Phytochemical Findings on Turkish *Sideritis* Species

Sideritis Species	Ethnobotanical Records			Pharmacological Activity	Phytochemical Findings
	Locality	Local Name	Usage		
					Fixed oil: Major compounds are linoleic acid, oleic acid, palmitic acid (64) *Flavonoids*: Apigenin, eriodictyol, hypolaetin-7-*O*-[6"'-*O*-acetyl-β-D-allopyranosyl-(1→2)]-β-D-glucopyranoside, (+)-luteolin, naringin, naringenin, penduletin, rutin, 4'-*O*-methylhypolaetin-7-*O*-[6"'-*O*-acetyl-β-D-allopyranosyl-(1→2)]-β-D-glucopyranoside, 4'-*O*-methylhypolaetin-7-*O*-[6"'-*O*-acetyl-β-D-allopyranosyl-(1→2)]-6"-*O*-acetyl-β-D-glucopyranoside, isoscutellarein-7-*O*-[6"'-*O*-acetyl-β-D-allopyranosyl-(1→2)]-β-D-glucopyranoside, isosccutellarein-7-*O*-[6"'-*O*-acetyl-β-D-allopyranosyl-(1→2)]-6"-*O*-acetyl-β-D-glucopyranoside, 4'-*O*-methyl-isoscutellarein-7-*O*-[6"'-*O*-acetyl-β-D-allopyranosyl-(1→2)]-6"-*O*-acetyl-β-D-glucopyranoside (78, 81) *Other phenolics*: Caffeic acid, gallic acid, rosmarinic acid (81) *Tannins*: (-)-Epicatechin, catechin (81)
**S. cilicica* Boiss. & Balansa	-	-	-	Antimicrobial (65, 71)	*Essential oil*: Major compounds are β-pinene, α-pinene, β-phellandrene (65)
**S. condensata* Boiss. & Heldr.	Karaman	Dağ çayı	Aerial parts; for stomachache (52)	Antimicrobial (66, 82, 83) Antioxidant (46, 84, 85)	*Diterpenes:* Isolinearol, candol B, linearol, sideridiol, sideroxol, siderol, 7-acetylsideroxol (86) *Essential oil*: Major compounds are β-caryophyllene, β-pinene, α-pinene (58) Major compounds are β-caryophyllene, germacrene D (87) Major compounds are caryophllene, germacrene, muurola-3,5-diene (88)

(Continued)

TABLE 24.1 (CONTINUED)
Ethnobotanical Usages, Pharmacological Investigations, and Phytochemical Findings on Turkish *Sideritis* Species

Sideritis Species	Ethnobotanical Records			Pharmacological Activity	Phytochemical Findings
	Locality	Local Name	Usage		
					Major compounds are β-pinene, 3-octanol, limonene (89) *Fixed oil*: Major compounds are linoleic acid, oleic acid, palmitic acid, stearic acid (64) *Flavonoids*: Apigenin-7-*O*-glucoside, luteolin, naringenin, naringin, isoscutellarein 7-*O*-(6‴-*O*-acetyl)-β-allopyranosyl-(1 → 2)-β-glucopyranoside, rutin, stachyspinoside, vitexin (81, 85) *Iridoids*: Ajugoside (85) *Monoterpenes*: Betulalbuside A, 1-hydroxylinaloyl 6-*O*-β-D-glucopyranoside (85) *Other phenolics:* Chlorogenic acid, gallic acid, caffeic acid, rosmarinic acid (81, 85) *Phenylethanoids*: Verbascoside (85)
**S. congesta* P. H. Davis & Hub.- Mor.	Karaman Mersin Antalya	Dağ çayı - Dağ çayı, yayla çayı	Aerial parts; for stomachache (52) Aerial parts; as tonic (93) Flower, leaf; as appetizer and stomachic (9)	Analgesic (90) Anticholinesterase (19, 20) Cytotoxicity (27) Antioxidant (19, 20, 27, 46, 54) Anti-inflammatory (16)	*Flavonoids*: Chrysoeriol-7-diglucoside, chrysoeriol-7-glucoside, salvigenin, stachyspinoside (16, 91) *Diterpenes*: Epoxyisolinearol, foliol, linearol, sideridiol, sideroxol, siderol, sidol, 7-acetyldistanol, 7-epicandicandiol, siderol-18-palmitat, *ent*-3β,7α-dihydroxy-18-asetoxy-15β,16β-epoxykaurane (16, 20, 27, 91, 92) *Essential oil*: Major compounds are β-pinene, α-pinene, δ-cadinen, linalool (58) Major compounds are β-pinene, α-pinene, muurola-3,5-diene (88) Major compounds are muurol-5-en-4β-ol, murol-5-en-4α-ol, α-cadinol (67) *Fixed oil*: Major compounds are linoleic acid, oleic acid, palmitic acid, stearic acid (64)

(*Continued*)

TABLE 24.1 (CONTINUED)

Ethnobotanical Usages, Pharmacological Investigations, and Phytochemical Findings on Turkish *Sideritis* Species

Sideritis Species	Ethnobotanical Records			Pharmacological Activity	Phytochemical Findings
	Locality	Local Name	Usage		
**S. dichotoma* Huter	-	-	-	Antimicrobial (66) Antioxidant (46, 94)	*Diterpenes*: Siderol, sideridiol, *ent-7*α- acetoxy-15,18-trihydroxykaur-16-ene, *ent-7a*,15,18-trihydroxykaur-16-ene, *ent-7a*,18-dihidroxy -15β,16β-epoxykaurane, *ent-7a*,18-dihydroxybeyer-15-ene, *ent-7*β,15*a*,18-trihydroxykaur-16-ene, *ent-7*β-acetoxy-15α,18-trihydroxy-kaur-16-ene, *ent-7a*-acetoxy-18-hydroxy-15β,16β-epoxykaurane (57, 95) *Fixed oil*: Major compounds are linoleic acid, oleic acid, palmitic acid, stearic acid (64)
S. erythrantha Boiss. & Heldr.	Mersin	Adaçayı	Aerial parts; for common colds, flu, pharyngitis, and bronchitis (96)	-	-
**S. erythrantha* Boiss. & Heldr. subsp. *cedretorum* (P. H. Davis) H. Duman	Mersin	Adaçayı	Aerial parts; for common cold, flu, blurred vision (97)	Antimicrobial (25) Antioxidant (25, 46, 94)	*Essential oil*: Major compounds are α-pinene, α-bisabolol, β-pinene (25) Major compounds are myrcene, α-pinene, β-caryophyllene (98, 99) *Fixed oil*: Major compounds are linoleic acid, oleic acid, palmitic acid (64)
**S. erythrantha* Boiss. & Heldr .subsp. *erythrantha*	Mersin	Adaçayı	Aerial parts; for common cold, flu, pharyngitis, blurred vision (97)	Antimicrobial (6, 25) Antioxidant (25, 46, 84)	*Diterpenes*: Sideridiol (100) *Essential oil*: Major compounds are α-pinene, β-caryophyllene, sabinene (6) Major compounds are α-pinene, eucalyptol, linalool (101) Major compounds are α-pinene, sabinene, β-phellandrene (98) Major compounds are α-pinene, β-caryophyllene, β-pinene (25, 99)

(Continued)

TABLE 24.1 (CONTINUED)
Ethnobotanical Usages, Pharmacological Investigations, and Phytochemical Findings on Turkish *Sideritis* Species

Sideritis Species	Ethnobotanical Records: Locality	Local Name	Usage	Pharmacological Activity	Phytochemical Findings
**S. galatica* Bornm.	-	-	-	α-Amylase, α-glucosidase inhibitory (28, 102) Anticholinesterase (28, 102) Antimicrobial (66) Antioxidant (28, 102)	*Diterpenes*: Galaticat (103) *Essential oil*: Major compounds are β-pinene, α-pinene, β-caryophyllene (102)
S. germanicopolitana Bornm.	Ankara	Adaçayı	Aerial parts; for common cold (104)	Anti-inflammatory (105) Anticholinesterase (105) Antifungal (106)	*Essential oil*: Major compounds are α-pinene, α-limonen, β-pinene (106) *Flavonoids*: 3'-Hydroxy-4'-*O*-methylisoscutellarein 7-O-[6'''-*O*-acetyl-β-allopyranosyl-(1→2)]-β-glucopyranoside, 4'-*O*-methylisoscutellarein 7-*O*-[6'''-*O*-acetyl-β-allopyranosyl-(1→2)]-β-glucopyranoside), isoscutellarein 7-*O*-[6'''-*O*-acetyl-β-allopyranosyl-(1→2)]-β-glucopyranoside, xanthomicrol (105) *Iridoids*: 5-Allosyloxy-aucubine, ajugol, melittoside (105) *Lignans*: Dehydrodiconiferylalcohol 4-*O*-β-D-glucopyranose, pinoresinol 4'-*O*-β-glucopyranoside (105) *Phenylethanoids*: Decaffeoylverbascoside, lamalboside, leucoseptoside A, martynoside, verbascoside (105)
**S. germanicopolitana* Bornm. subsp. *germanicopolitana*	-	-	-	-	*Fixed oil*: Major compounds are linoleic acid, oleic acid, palmitic acid (64)

(*Continued*)

TABLE 24.1 (CONTINUED)
Ethnobotanical Usages, Pharmacological Investigations, and Phytochemical Findings on Turkish *Sideritis* Species

	Ethnobotanical Records				
***Sideritis* Species**	**Locality**	**Local Name**	**Usage**	**Pharmacological Activity**	**Phytochemical Findings**
*S. *germanicopolitana* Bornm. subsp. *viridis* Hausskn. ex Bornm.	-	-	--		*Essential oil*: Major compound is myrcene (107)
*S. *gulendamii* H. Duman & Karavel.	-	-	-	-	*Diterpenes*: Athonolone, epicandicandiol, linearol, siderol (43, 108)
*S. *hispida* P. H. Davis	Karaman	Dağ çayı	Aerial parts; for stomachache (52)	-	*Essential oil*: Major compounds are (*E*)-2-hexenal, β-myrcene, caryophyllene (89)
	Konya	-	Flower, leaf; as appetizer and stomachic (9)		
*S. *hololeuca* Boiss. & Heldr.	-	-	-	Antimicrobial (109) Antioxidant (110)	*Diterpenes*: Eubol, siderol, 7-acetylsideroxol, *ent*-7α-acetyl-18-hydroxy-kaur-16-ene (110) *Fixed oil*: Major compounds are linoleic acid, oleic acid (64)
*S. *huber-morathii* Greuter & Burdet	-	-	-	Antioxidant (46)	*Diterpenes*: Foliol acetonide, candicandiol, linearol, siderol, sideridiol, sidol, 3,7,18-triacetylfoliol (79)

(*Continued*)

TABLE 24.1 (CONTINUED)
Ethnobotanical Usages, Pharmacological Investigations, and Phytochemical Findings on Turkish *Sideritis* Species

Sideritis Species	Ethnobotanical Records: Locality	Local Name	Usage	Pharmacological Activity	Phytochemical Findings
S. lanata L.	-	-		Antimicrobial (24, 111) Central nervous system depressant/ stimulant activity (32)	*Essential oil*: Major compounds are hexadecanoic acid, spathulenol, heptacosane (112) Major compounds are spathulenol, *ent*-2α-hydroxy-8(14),5-pimaradien, heptacosane (111) *Fixed oil*: Major compounds are linoleic acid, oleic acid, palmitic acid (64) *Flavonoids*: 3'-*O*-Methylhypolaetin-7-*O*-[6‴-*O*-acetyl]-allosyl-(1→2)-glucoside, 7-*O*-(6″-*O*-acetyl-)-β-glucopyranosylchrysoeriol, 7-*O*-β-D-glucopyranosylchrysoeriol, 4'-*O*-methylhypolaetin-7-*O*-[6‴-*O*-acetyl]-allosyl-(1→2)-[6″-*O*-acetyl]-glucoside, 4'-*O*-methylisoscutellarein-7-*O*-allosyl-(1→2)-glucoside,7-*O*-[(6″'-*O*-acetyl)-β-D-allopyranosyl-(1→2)-β-D-glucopyranosyl]-hypolaetin, 7-*O*-[(6″'-*O*-acetyl)-β-D-allopyranosyl-(1→2)-β-D-glucopyranosyl]-hypolaetin, 7-*O*-[(6″'-*O*-acetyl)-β-D-allopyranosyl (1→2)-β-D-glucopyranosyl]-hypolaetin-3'-me thylether, 7-*O*-[(6″'-*O*-acetyl)-β-D-allopyranosyl (1→2)-β-D-glucopyranosyl]-isoscutellarein, 7-*O*-(6″-*O*-acetyl-)-β-glucopyranosyl-chrysoeriol, 7-*O*-β-D-glucopyranosyl-chrysoeriol, hypolaetin-7-*O*-allosyl-(1→2)-glucoside, chrysoeriol-7-*O*-[6‴-*O*-acetyl]- allosyl-(1→2)-glucoside, hypolaetin 7-*O*-allosyl-(1→2)-[6″-*O*-acetyl]-glucoside, isoscutellarein-7-*O*-[6‴-*O*-acetyl]-allosyl-(1→2)-glucoside, isoscutellarein-7-*O*-allosyl-(1→2)-glucoside (113, 114) *Iridoid*: 10-*O*-(*E*)-*p*-Coumaroylmelittoside (113) *Other phenolics*: Chlorogenic acid (114) *Phenylethanoids*: Allysonoside, verbascoside (114)

(*Continued*)

TABLE 24.1 (CONTINUED)
Ethnobotanical Usages, Pharmacological Investigations, and Phytochemical Findings on Turkish *Sideritis* Species

Sideritis Species	Ethnobotanical Records			Pharmacological Activity	Phytochemical Findings
	Locality	Local Name	Usage		
*S. leptoclada O. Schwarz & P.H. Davis	Afyon	Kırk boğum, ulama otu, çay otu	Flower, leaf; for fatigue, common cold, as expectorant (115)	Antimicrobial (44) Antioxidant (21, 46-48, 116) Cytotoxic (37) Anticholinesterase (21, 48) Antityrosinase (21, 48) Antiurease (21)	*Diterpenes*: Linearol, sideroxol, sidol, 7-epicandicandiol, *ent*-7α,18-dihydroxy-15β,16β-epoxykaurane, *ent*-7α,15β,18-trihydroxykaur-16-ene, *ent*-7α-acetoxy-15β,18-dihydroxykaur-16-ene, *ent*-7α-acetoxy-18-hydroxy-15β,16β-epoxykaurane, *ent*-7α-hydroxy, 18-acetoxy-15β,16β-epoxykaur-16-ene (117, 118) *Essential oil*: Major compounds are germacrene D, β-caryophyllene, caryophyllene oxide (47) Major compounds are β-caryophyllene, germacrene D, τ-gurjunene (48) *Flavonoids*: Apigenin, apigetrin, hesperidin, isotrifoliin, luteolin, naringenin, rutin (37, 44) *Other phenolics*: 2,4-Dihydroxy benzoic acid, 4-hydroxy-benzoic acid, caffeic acid, chlorogenic acid, *p*-coumaric acid, protocatechuic acid, rosmarinic acid, salicylic acid, sinapinic acid, syringic acid, *trans*-2-hydroxycinnamic acid, *trans*-ferulic acid, vanillic acid, vanillin (21, 37) *Others*: Malic acid, quinic acid, *trans*-cinnamic acid (37)
	Muğla	Kızlan çayı, anababakokusu	Aerial parts; for hoarseness, common cold (119)		
S. libanotica Labill.	Malatya	Dağ çayı	Flower, leaf; for common cold (120)	Anticholinesterase (121) Genotoxicity (122)	-
	Mersin, Antalya	Dağ çayı, altınbaş, adaçayı	Aerial parts; as tonic (93)		

(*Continued*)

TABLE 24.1 (CONTINUED)
Ethnobotanical Usages, Pharmacological Investigations, and Phytochemical Findings on Turkish *Sideritis* Species

Sideritis Species	Ethnobotanical Records: Locality	Local Name	Usage	Pharmacological Activity	Phytochemical Findings
	Niğde	Dağ çayı, Toros çayı, yayla çayı	Aerial parts; for diarrhea, digestive, and stomach problems (69)		
	Güney Anadolu Bölgesi	Dağ çayı	Flower, leaf; as appetizer and stomachic (9)		
S. libanotica Labill. subsp. *libanotica*	Antakya	Dağ çayı, dağ kekiği	Aerial parts; as sedative, carminative, and appetizer (123)	Antioxidant (46)	*Diterpenes*: Sideridiol, siderol (100)
S. libanotica Labill. subsp. *linearis* (Benth.) Bornm.	Muğla	Boz çay, diken çayı, anababakokusu	Aerial parts; for common colds, hoarseness, and as laxative (119)	Anticholineterase (19) Antimicrobial (124, 125) Cytotoxicity (126, 127) Antioxidant (19, 125, 128-132) Anticarbonic anhydrase (133)	*Diterpenes*: Sideridiol, sideroksol (129, 130) *Essential oil*: Major compounds are β-phellandrene, β-pinene, germacrene D (126) Major compounds are (*E*)-2-hexenal, 3-octanole, limonene (89) *Fixed oil*: Major compounds are palmitic acid, oleic acid, lineloic acid (129) *Flavonoids*: 3'-*O*-Methylhypolaetin-7-*O*-[6'''-*O*-acetyl-β-D-allopyranosyl-(1⟶2)]-6''-*O*-acetyl-β-D-glucopyranoside, 3'-*O*-methylhypolaetin-7-*O*-β-D-allopyranosyl-β-D-glucopyranoside, 3′-*O*-methylhypolaetin, 4'-*O*-methylisoscutellarein-7-*O*-[6'''-*O*-acetyl]-allosyl-(1→2)-glucoside, 4'-*O*-methylisoscutellarein 7-*O*-[6'''-O-acetyl]-allosyl-(1→2)-[6''-*O*-acetyl]-glucoside, apigenin, hesperetin, kaempferol, luteolin, morin, myricetin, quercetin, rutin (126, 130, 132, 133) *Iridoid*: Ajugoside (124)
	Mersin, Konya	Acemarpası, çay otu	Aerial parts; as tonic (93)		
	Malatya	Dağ çayı	Aerial parts; for common colds (134)		
	Karaman	Dağ çayı	Aerial parts; for stomachache (52)		

(*Continued*)

TABLE 24.1 (CONTINUED)
Ethnobotanical Usages, Pharmacological Investigations, and Phytochemical Findings on Turkish *Sideritis* Species

***Sideritis* Species**	Ethnobotanical Records: Locality	Ethnobotanical Records: Local Name	Ethnobotanical Records: Usage	Pharmacological Activity	Phytochemical Findings
					Other phenolics: Caffeic acid, chlorogenic acid, ferulic acid, gallic acid, p-coumaric acid, rosmarinic acid, vanillic acid (132)
S. libanotica Labill. subsp. *kurdica* (Bornm.) Hub.-Mor.	-	-	-	Central nervous system depressant/ stimulant activity (32)	-
S. libanotica Labill. subsp. *microchlamys*(Hand.-Mazz.) Hub.-Mor.	Şanlıurfa	Kara gekme	Aerial parts; for common cold and flu (135)	-	-
**S. libanotica* Labill. subsp. *violascens* (P.H. Davis) P. H. Davis	Karaman	Dağ çayı	Aerial parts; for stomachache (52)	Antimicrobial (109)	*Fixed oil*: Major compounds are linoleic acid, oleic acid, palmitic acid (64)
**S. lycia* Boiss. & Heldr.	-	-	-	Anti-inflammatory, antinociceptive (13) Antimicrobial (26, 136) Anticholinesterase (55) Antityrosinase (55) Antioxidant (132) Cytotoxic (26) Insecticidal activity (26)	*Diterpenes*: Foliol, isolinearol, isosidol, linearol, sideridiol, siderol, sidol, 7-epicandicandiol (26, 136) *Flavonoids*: 4'-*O*-Methyl-hypolaetin-7-*O*-[6'''-*O*-acetyl-*β*-D-allopyranosil-(1→2)-*β*-D-glucopyranoside], 4'-*O*-methyl-hypolaetin-7-*O*-[6'''-*O*-acetyl-*β*-D-allopyranosyl–(1→2)-6''-*O*-acetyl-*β*-D-glucopyranoside], apigenin, hesperetin, kaempferol, luteolin, morin, myricetin, quercetin, rutin (13, 132) *Fixed oil*: Major compounds are linoleic acid, oleic acid, palmitic acid (64) *Iridoids*: Ajugol, ajugoside (136) *Other phenolics*: Caffeic acid, chlorogenic acid, ferulic acid, gallic acid, *p*-coumaric acid, rosmarinic acid, vanillic acid (132) *Phenylethanoids:* Lavandulifolioside, leucosceptoside A, martynoside, verbascoside (13) *Tannins*: (+)-Catechin, (–)-epicatechin (132)

(*Continued*)

TABLE 24.1 (CONTINUED)
Ethnobotanical Usages, Pharmacological Investigations, and Phytochemical Findings on Turkish *Sideritis* Species

	Ethnobotanical Records				
***Sideritis* Species**	**Locality**	**Local Name**	**Usage**	**Pharmacological Activity**	**Phytochemical Findings**
S. montana L.	Niğde	Dağ çayı, yayla çayı	Flower, leaf; for stomach pain, as carminative and stimulant (69)	Antimicrobial (137, 138) Antioxidant (138-140) Antiproliferative (34)	*Diterpenes*: 9α,13α-Epi-dioxyabiet-8(14)-en-18-ol, sideritins A and B (34) *Essential oil*: Major compounds are bicyclogermacrene, germacrene D, isocaryophyllene, β-pinene, *trans*-β-farnesene, viridiflorol (141) *Flavonoids*: 6-Methoxysakuranetin, apigenin, chrysoeriol, hypolaetin-3'-methyl ester, hypolaetin-4'-methyl ester, hypolaetin-4'-methylether-7-*O*-[6'''-*O*-acetyl-β-D-allopyranosyl-(1→2)-β-D-glucopyranoside], hypolaetin-4'-methylether-7-*O*-[6'''-*O*-acetyl-β-D-allopyranosyl-(1→2)-6''-*O*-acetyl-β-D-glucopyranoside], kaempferol-3-*O*-glucoside, kaempferol-3-*O*-rutinoside, luteolin-7-*O*-glucoside, pomiferin E (34, 140, 142, 143) *Lignans*: Paulownin (34) *Methoxystigmanes*: 3-Oxo-α-ionol (34) *Other phenolics*: 4-Allyl-2,6-dimethoxyphenol glucoside, caffeic acid, ferulic acid, rosmarinic acid (34, 140)
S. montana L. subsp. *montana*	Kırklareli	Tilki kuyruğu	Aerial parts; for common cold, flu, and cough (144)	Antioxidant (145)	*Essential oil*: Major compounds are germacrene D, bicyclogermacrene, (*E*)-β-farnesene (112) Major compounds are β-caryophyllene, α-pinene, β-pinene (146) *Fixed oils:* Major compounds are α-linolenic acid, oleic acid, palmitic acid (145) *Flavonoids*: Kaempferol, quercetin, myricetin, morin, naringenin (145) *Other phenolics*: Resveratrol (145) *Steroids*: Major compounds are ergosterol, stigmasterol, β-sitosterol (145) *Tannins*: Catechin (145)
	Tekirdağ	Karaçay	Aerial parts; for cough (147)		

(*Continued*)

TABLE 24.1 (CONTINUED)
Ethnobotanical Usages, Pharmacological Investigations, and Phytochemical Findings on Turkish *Sideritis* Species

Sideritis Species	Ethnobotanical Records			Pharmacological Activity	Phytochemical Findings
	Locality	Local Name	Usage		
S. montana L. subsp. *remota* (d'Urv.) P.W. Ball	-	-	-	-	*Essential oil*: Major compounds are germacrene D, bicyclogermacrene, β-pinene (112)
**S. niveotomentosa* Hub.-Mor.	-	-	-	Antioxidant (72, 148) Cytotoxic (148)	*Diterpenes*: Athonolone, linearol, eubol, eubotriol, sidol, sideridiol, siderol, 7-epicandicandiol, foliol, linearol (43, 72) *Fixed oils*: Major compounds are linoleic acid, oleic acid, palmitic acid (64) *Steroids*: β-Sitosterol, stigmasterol (72)
**S. ozturkii* Aytaç & Aksoy	-	-	-	Anti-inflammatory, Antinociceptive (149) Antimicrobial (76, 150) Antioxidant (35, 76) Cytotoxic (35, 36) Antiamylase (35) Antiglucosidase (35) Anticholinesterase (35) Antityrosinase (35)	*Diterpenes*: 7-Epicandicandiol, epoxyisolinearol, linearol, sideroxol, sidol, (151) *Essential oil*: Major compounds are α-pinene, β-pinene, bicyclogermacrene isosteviol (10, 35) *Flavonoids*: 5,7-Dihydroxy-8,2'-dimethoxyflavone, 5-*O*-demethylnobiletin, apigenin, apigenin-7-*O*-glucoside, cirsilineol, cirsimaritin, kaempferol, martynoside, myricetin, naringenin, naringenin chalcone 4'-*O*-glucoside, naringenin-7-*O*-glucoside, osturcoside A, B, C, quercetin, quercetin-3-*O*-glucoside, rutin trihydrate, xanthomicrol, vicenin-2 (35, 36, 151) *Glycosides*: Citric acid, galactonic acid, gluconic acid, glucosamine (or galactosamine), saccharic acid (35) *Iridoids*: Loganic acid, mussaenosidic acid (35) *Other phenolics*: 1-*O*-Caffeoylquinic acid, 3-*p*-coumaroylquinic acid, 4-hydroxybenzoic acid, caffeic acid, chlorogenic acid, ferulic acid, gallic acid, protocatechuic acid, syringic acid, *trans*-*p*-coumaric acid, *trans*-ferulic acid, *trans*-resveratrol, (35, 36) *Phenylethanoids*: Forsythoside E, leucoseptoside A, martynoside, verbascoside (35, 151) *Others*: γ-Aminobutyric acid, glutamine, indole-4-carbaldehyde, phenylalanine, proline, quinic acid, *trans*-cinnamic acid, tryptophan (35) *Tannins*: Catechin (36)

(*Continued*)

TABLE 24.1 (CONTINUED)
Ethnobotanical Usages, Pharmacological Investigations, and Phytochemical Findings on Turkish *Sideritis* Species

Sideritis Species	Ethnobotanical Records: Locality	Local Name	Usage	Pharmacological Activity	Phytochemical Findings
S. perfoliata L.	Çanakkale	Fenerli çay, fincan çayı, kandil çayı	Aerial parts; for cough, bronchitis, stomach diseases (8)	α-Glucosidase inhibitory (39) Anticholinesterase (19, 152) Antimicrobial (153) Cytotoxicity (154, 155) ACE inhibition (39) Central nervous system depressant/stimulant activity (32) Antioxidant (19, 152, 155) Antityrosinase (152) Antiamylase (152) Scolicidal activity (156)	*Diterpenes*: 2-Oxo-13-epi-manoyl oxide, sideridiol, sideritriol, siderol (100, 157) *Essential oil*: Major compounds are α-pinene, β-phellandrene, β-pinene (158) Major compounds are limonene, *cis*-ocimene, α-pinene (58) Major compounds are limonene, sabinene, manoyloxide (153) Major compounds are β-phellandrene, sabinene, β-pinene (154) Major compounds are α-pinene, β-pinene, limonene (89) *Fixed oil*: Major compounds are linoleic acid, oleic acid, palmitic acid (64) *Flavonoids*: 3'-Hydroxy-4'-*O*-methylisoscutellarein-7-*O*-(2"-*O*-6"'-*O*-acetyl-β-D-allopyranoside)-β-D-glucopyranoside, 3'-hydroxy-4'-*O*-methylisoscutellarein-7-*O*-[6'''-*O*-acetyl-β-D-allopyranosyl-(1$\rightarrow$2)-β-D-glucopyranoside, 4'-*O*-methylisoscutellarein-7-*O*-(2"-*O*-6"'-*O*-acetyl-β-D-allopyranosyl)-β-D-glucopyranoside, 4'-*O*-methylisoscutellarein-7-*O*-[6'''-*O*-acetyl-β-D-allopyranosyl-(1$\longrightarrow$2)]-β-D-glucopyranoside, apigenin, apigenin-7-glucoside, apigenin-7-*O*-(4"-*O*-*p*-coumaroyl)-β-D-glucopyranoside, hesperidin, hyperoside, isoscutellarein-7-*O*-[6'''-*O*-acetyl-β-D-allopyranosyl-(1$\longrightarrow$2)]-β-D-glucopyranoside, isoscutellarein-7-*O*-[6'''-*O*-acetyl-β-D-allopyranosyl-(1$\rightarrow$2)]-6"-*O*-acetyl-β-D-glucopyranoside, kaempferol, luteolin, luteolin-7-glucoside, myricetin, naringenin, quercetin (152, 155, 156, 159, 160) *Iridoids*: Ajugoside, oleuropein (156, 160) *Lignans*: Pinoresinol (152) *Other phenolics*: Caffeic acid, chlorogenic acid, curmin, hydroxybenzoic acid, gallic acid, hydroxycinnamic acid, 2,5-dihydroxybenzoic acid, 3,4-dihydroxyphenylacetic acid, 4- hydroxybenzoic acid, hydroxybenzoic acid ferulic acid, prothocatechuic acid, *p*-coumaric acid, resveratrol, rosmarinic acid, salicylic acid, sinapic acid, syringic acid, vanillic acid (152, 155, 156, 160) *Phenylethanoids*: Lavandulifolioside, leucoseptoside A, martynoside, verbascoside (152, 159, 160) *Others*: Acetohydroxamic acid, alizarin, fumaric acid (155, 156) *Tannins*: (-)-Epicatechin, catechin hydrate, ellagic acid (152, 156)
	Alanya	Dağ çayı	Flower, leaf; as appetizer and stomachic (9)		

(Continued)

TABLE 24.1 (CONTINUED)
Ethnobotanical Usages, Pharmacological Investigations, and Phytochemical Findings on Turkish *Sideritis* Species

Sideritis Species	Ethnobotanical Records			Pharmacological Activity	Phytochemical Findings
	Locality	Local Name	Usage		
**S. phlomoides* Boiss. & Balansa	Kayseri	Dağ çayı	For abdominal pain (161)	-	*Essential oil*: Major compounds are β-caryophllene, α-bisabolol, caryophyllene oxide (80)
**S. phrygia* Bornm.	-	-	-	Antioxidant (46)	*Fixed oil*: Major compounds are linoleic acid, oleic acid, palmitic acid (64)
**S. pisidica* Boiss. & Heldr.	Muğla	Dağ çayı	Flower, leaf; for stomach disorders (162)	Antimicrobial (45) Antioxidant (19, 163) Anticholinesterase (163)	*Essential oil:* Major compounds are β-caryophyllene, α-pinene, sabinene (164) Major compounds are δ-cadinene, *T*-cadinol, β-cubebene (163)
	Antalya	Eldiven çayı	Aerial parts; as tonic (93)	Urease inhibitory (163)	*Fixed oil*: Major compounds are linoleic acid, oleic acid, palmitic acid (64)
	Konya	Hava otu, dallı adaçayı	Aerial parts; for abdominal pain (165)		Major compounds are linolenic, palmitic, linoleic acids (163)
	Muğla, Antalya	Çay şalbası	Flower, leaf; as appetizer and stomachic (9)		
S. romana L	-	-	-	-	*Essential oils*: Major compounds are germacrene D, *trans*-caryophyllene, limonene, bicyclogermacrene, δ-cadinene (141)
S. romana L. subsp. *curvidens* (Stapf) Holmboe	-	-	-	Antimicrobial (24)	
S. romana L. subsp. *romana*	-	-	-	-	*Essential oils*: Major compounds are 1-octen-3-ol, carvacrol, γ-terpinene (112)
**S. rubriflora* Hub.-Mor.	Mersin	Kazıklı çayı, adaçayı, kazıklı çayı	Aerial parts; for common cold, flu, bronchitis pharyngitis (96, 97)	Antimicrobial (66) Antioxidant (46)	*Diterpenes*: Epicandicandiol, linearol, sideroxol, sidol (43) *Essential oil*: Major compounds are β-pinene, α-pinene, cubebol (166) Major compounds are β-pinene, α-pinene, epicubebol (73)

(Continued)

TABLE 24.1 (CONTINUED)
Ethnobotanical Usages, Pharmacological Investigations, and Phytochemical Findings on Turkish *Sideritis* Species

	Ethnobotanical Records				
***Sideritis* Species**	**Locality**	**Local Name**	**Usage**	**Pharmacological Activity**	**Phytochemical Findings**
S. scardica Griseb.	-	-	-	Antihyaluronidase (167) Antielastase (167) Antioxidant (167-171) Antityrosinase (167) Peptidase activity (167) Antimatrix metalloproteinases (167) Cytotoxicity (42, 167, 172) Antidiabetic (173) Effect on cognitive performance and blood pressure (174) Acute and subchronic toxicity (41) Effect on cognition and β-amyloidosis (175) Antimicrobial (171, 176) Monoamine reuptake inhibitory (177) Anti-inflammatory (172) Gastroprotective (172)	*Essential oil:* Major compounds are *trans*-caryophyllene, *β*-pinene, myrtenal, *α*-copaene, bornyl acetate, *α*-pinene, 1-octen-3-ol (178) Major compounds are hexadecanoic acid, myristicin, menthol, dodecanoic acid, (E)-coniferyl alcohol, cyclopentadecanolide (176) *Flavonoids*: 3'-*O*-Methylhypolaetin-7-*O*-allosyl-(1→2)-glucoside, 4'-*O*-methylhypolaetin 7-*O*-[6'''-*O*-acetyl] –allosyl-(1→2)-glucoside, 4'-*O*-methylisoscutellarein 7-*O*-[6'''-*O*-acetyl]-allosyl(1→2)-glucoside, apigenin, apigenin-7-*O*-[6'''-*O*-acetyl]-allosyl-(1→2)-glucoside, apigenin 7-*O*-allosyl(1→2)-glucoside, apigenin 7-*O*-glucoside, hypolaetin 7-*O*-[6'''-*O*-acetyl]-allosyl-(1→2)-glucoside, hypolaetin 7-*O*-allosyl-(1→2)-glucoside, isoscutellarein 7-*O*-[6'''-*O*-acetyl]-allosyl(1→2)-glucoside, isoscutellarein 7-*O*-allosyl-(1→2)-glucoside, isoscutellarein 7-*O*-allosyl-(1→2)-[6''-*O*-acetyl]-glucoside, isoscutellarein 7-*O*-[6'''-*O*-acetyl]-allosyl-(1→2)-[6''-*O*-acetyl]-glucoside (167) *Other phenolics*: 5-Caffeoylquinic acid, echinacoside (167) *Phenylethanoids*: Allysonoside, lavandulifolioside, leucosceptoside A, amioside, verbascoside (167)
S. scardica Griseb. subsp. *scardica*	Kırklareli	Kırçayı, taşlık çayı, başak çayı, pazlak, çayı	Aerial parts; for common cold, flu, cough, bronchitis (144)	-	-
S. serratifolia Hub.-Mor.	-	-	-	Antioxidant (46)	-

(*Continued*)

TABLE 24.1 (CONTINUED)
Ethnobotanical Usages, Pharmacological Investigations, and Phytochemical Findings on Turkish *Sideritis* Species

	Ethnobotanical Records				
***Sideritis* Species**	**Locality**	**Local Name**	**Usage**	**Pharmacological Activity**	**Phytochemical Findings**
**S. sipylea* Boiss.	Manisa	Dağçayı, adaçayı, calba, şabla, ca otu, ca şalbası	Aerial parts; for dyspepsia, diarrhea, respiratory tract diseases, gallstone, common cold, flu, athlete's foot, foot odor hair care (179, 180)	Antimicrobial (23, 66, 181, 182) Antioxidant (46, 183)	*Diterpenes*: Epoxyisolinearol, isolinearol, isosidol, linearol, sideridiol, siderol, 7-epicandicandiol (57) *Essential oil*: Major compounds are α-pinene, β-pinene, 1,8-cineole (181) *Flavonoids*: 4'-*O*-methylisoscutellarein-7-*O*-[6'''-*O*-acetyl-β-D-allopyranosyl-(1→2)]-β-D-glucopyranoside, 4'-*O*-methylisoscutellarein-7-*O*-[6'''-*O*-acetyl-β-D-allopyranosyl-(1→2)]-6''-*O*-acetyl-β-D-glucopyranoside, apigenin, apigenin-7-*O*-β-D-glucopyranoside, apigenin-7-*O*-[4''-*O*-*trans*-*p*-coumaroyl]-β-D-glucopyranoside, apigenin-7-*O*-[6''-*O*-*trans*-*p*-coumaroyl]-β-D-glucopyranoside, isoscutellarein-7-*O*-[6'''-*O*-acetyl-β-D-allopyranosyl-(1→2)]-β-D-glucopyranoside, isoscutellarein-7-*O*-[6'''-*O*-acetyl-β-D-allopyranosyl-(1→2)]-6''-*O*-acetyl-β-D-glucopyranoside (184) *Iridoids*: Ajugol, ajugoside, melittoside, 5-allosyloxy-aucubin (184) *Other phenolics*: Caffeic acid, chlorogenic acid (184) *Phenylethanoids*: Acteoside, martynoside, lamalboside, leucosceptoside A, lavandulifolioside (184)
**S. stricta* Boiss. & Heldr.	Niğde	Dağ çayı	Aerial parts; for common cold, flu (69)	Anti-inflammatory, antinociceptive (185) Antimicrobial (83, 186, 187) Cytotoxic (188) Antioxidant (189) Anticholinesterase (189) Antityrosinase (189)	*Essential oil*: Major compounds are β-pinene, α-pinene, α-terpinene (82) *Diterpenes*: 7-Acetylsideroxol, 7-epicandicandiol, athonolone, *ent*-1β-hydroxy-7α-acetyl-15β,16β-epoxykaurane, *ent*-7α,15β,18-trihydroxy-kaur-16-ene,*ent*-7α-acetyl,15,18-dihydroxy-kaur-16-ene foliol, linaerol, isolinearol, isosidol, sideridiol, sideroxol, siderol (187, 190-192) *Flavonoids*: Isoscutellarein-7-*O*-[6'''-*O*-acetyl-β-D-allopyranosyl-(1→2)]-β-D-glucopyranoside, isoscutellarein-7-*O*-[6'''-*O*-acetyl-β-D-allopyranosyl-(1→2)]-6''-*O*-acetyl-β-D-glucopyranoside, xanthomicrol (189, 191)
	Antalya	Adaçayı	Leaf; as carminative, as mouthwash in tonsillitis and gum diseases (193)		

(Continued)

TABLE 24.1 (CONTINUED)
Ethnobotanical Usages, Pharmacological Investigations, and Phytochemical Findings on Turkish *Sideritis* Species

Sideritis Species	Ethnobotanical Records			Pharmacological Activity	Phytochemical Findings
	Locality	Local Name	Usage		
					Other phenolics: Caffeic acid, gallic acid, *p*-coumaric acid, *p*-hydroxybenzoic acid, rosmarinic acid, *trans*-2-hydroxycinnamic acid, vanillin (189) *Others*: *trans*-Cinnamic acid (189) *Phenylethanoids*: Verbascoside (191) *Tannins*: Catechin hydrate, ellagic acid (189)
S. syriaca L.	Güneydoğu Anadolu, Antalya	-	Flower, leaf; as appetizer, stomachic (9)	Antioxidant (139)	-
S. syriaca L. subsp. *nusariensis* (Post) Hub.-Mor.	Antakya	Adaçayı, ana-baba bohuru	Aerial parts; for bronchitis, cough, common cold, flu, diabetes, as aphrodisiac (123)	Antioxidant (46)	-
S. taurica Stephan ex Willd.	-	-	-	Anti-inflammatory, analgesic (17) Antidiabetic (17) Anti-ulcerogenic, acute toxicity (17) Central nervous system inhibitory activity (17) Antimicrobial (194) Antitumor (194)	*Essential oil*: Major compounds are α-cadinol, β-pinene, curcumenol (195)
S. tmolea P. H. Davis	Manisa	Balbaşı, balşalbası, yakı otu, yakı şalbası,	Aerial parts; for respiratory tract diseases, flu, common colds, diarrhea, indigestion, gastrointestinal diseases, bowel spasm, as carminative (179, 180)	Antimicrobial (196) Antioxidant (46)	*Diterpenes*: Athonolon, 7-acetoxyisideroxol, *ent*-7α,15β-18-trihydroxykaur-16-ene, siderol, *ent*-7α-acetoxy-15β,18-dihydroxykaur-16-ene (196, 197) *Essential oil*: Major compounds are β-caryophyllene, calamenene, muurol-5-en-4-β-ol (67)

(*Continued*)

TABLE 24.1 (CONTINUED)

Ethnobotanical Usages, Pharmacological Investigations, and Phytochemical Findings on Turkish *Sideritis* Species

	Ethnobotanical Records				
***Sideritis* Species**	**Locality**	**Local Name**	**Usage**	**Pharmacological Activity**	**Phytochemical Findings**
	İzmir	Balbaşı, sivri çay	Flower, leaf; as appetizer, stomachic (9)		
**S. trojana* Bornm.	Çanakkale	Kazdağı çayı, tüylü çay	Aerial parts; for stomach diseases, abdominal pain, kidney diseases, sore throat, as laxative (8)	Antimicrobial (66, 153, 198) Cytotoxicity (33) Antioxidant (33, 199, 200) Antiamylase (200) Antiglucosidase (200)	*Diterpenes*: Isocandol B, candol A acetate, sideridiol, siderol, 7-epicandicandiol, *ent*-2α-hydroxy-8(14),15-pimaradiene, *ent*-7α-15β,16β-epoxykaurane, *ent*-7α-acetoxy-15β,16β-epoxykaurane, *ent*-7α-acetoxykaur-15-ene (57, 95) *Essential oil*: Major compounds are β-pinene, α-pinene, germacrene D (153)
	Balıkesir	Sarız çayı, dağ çayı, cılbak çayı	Aerial parts; for common colds, stomachache, and flu (60)		Major compounds are α-bisabolol, valeranone, 4-terpineol, caryophyllene, 3-methyl nonane, copaene (201) Major compounds are valeranone, α-bisabolol, β-caryophyllene (198) *Flavonoids*: 3′-Hydroxy-4′-*O*-methylisoscutellarein-7-*O*-[6‴-*O*-acetyl-β-allopyranosyl-(1→2)]-β-glucopyranoside], 4′-*O*-methylisoscutellarein-7-*O*-[6‴-*O*-acetyl-β-allopyranosyl-(1→2)]-β-glucopyranoside, isoscutellarein-7-*O*-[6‴-*O*-acetyl-β-allopyranosyl-(1→2)]-β-D-glucopyranoside (33) *Iridoids*: 10-*O*-(*E*)-Feruloylmelittoside, melittoside, 10-*O*-(*E*)-*p*-coumaroylmelittoside, stachysosides E and G (33) *Other phenolics*: Di-*O*-methylcrenatin (33) *Phenylethanoids*: Isoacteoside, isolavandulifolioside, lamalboside, leonoside A, verbascoside (33)
**S. vulcanica* Hub.-Mor.	-	-	-	-	*Essential oil*: Major compounds are α-pinene, β-caryophyllene, 1,8-cineole (146) Major compounds are β-caryophyllene, hexadecanoic acid, germacrene D (80)
**S. vuralii* H. Duman & Başer	-	-	-	Antimicrobial (71, 75) Antioxidant (94)	*Essential oil*: Major compounds are β-pinene, α-pinene, 1,8-cineole (80)

*Endemic species

FIGURE 24.1 An endemic species for Turkey: *Sideritis caesarea.*

hypolaetin-7-*O*-[6‴-*O*-acetyl-*β*-D-allopyranosyl-(1→2)]-6″-*O*-acetyl-*β*-D-glucopyranoside, isoscutellarein-7-*O*-[6‴-*O*-acetyl-*β*-D-allopyranosyl-(1→2)]-6″-*O*-acetyl-*β*-D-glucopyranoside, 3′-hydroxy-4′-*O*-methylisoscutellarein-7-*O*-[6‴-*O*-acetyl-*β*-D-allopyranosyl-(1→2)]-6″-*O*-acetyl-*β*-D-glucopyranoside, and verbascoside isolated from *n*-butanol fraction were administered to mice at doses of 100 mg/kg. The butanol extracts showed significant anti-inflammatory activity in carrageenan-induced hind paw edema. In addition, some of the isolated compounds have different activities in the different *in vivo* models (12-*O*-tetradecanoyl phorbol-13-acetate-induced ear edema, prostaglandin E_2-induced paw edema, and p-benzoquinone-induced abdominal constriction tests) (14).

Cavalcanti et al. (2020) investigated the anti-inflammatory and anti-neuropathic activity of aerial parts of *S. bilgeriana* methanolic extract on mice. Results revealed that *S. bilgeriana* methanolic extract causes significant reduction in carrageenan-induced myeloperoxidase activity and inhibition of hyperalgesia. Likewise, a significant reduction was observed in paw licking/lifting time in capsaicin test. The extract treatment reduced tumor necrosis factor-α, interleukin (IL)-1β, and leukocyte levels in the pleural cavity after carrageenan administration. In partial sciatic nerve ligation model, one hour after extract application and 7-day assessment, a significant reduction in hyperalgesia was observed. Also, latency time in hot plate test significantly increased 1 h after the extract application. The extract reversed the loss of strength in legs caused by sciatic nerve ligation. As the level IL-6 and factor nuclear factor (NF)-κB in medulla were analyzed, a decrease in these cytokine levels was observed in *S. bilgeriana* methanolic extract group (15).

Anti-inflammatory properties of aqueous extract from *S. congesta* and its fractions obtained by centrifugal partition chromatography were investigated. The fraction I (92.3%) and the crude extract (90.2%) caused strong inhibition on cyclooxygenase-2, but the isolated flavonoid (stachyspinoside 1) showed lower inhibitory activity (45.4%). Moreover, fraction III (99.9%) and crude extract (44.8%) exhibited significant inhibitory activity on NF-κB transnucleation in human embryonic kidney 293 cells (16).

Antidiabetic Activity

Antihyperglycemic activity of petroleum ether and ethanol (70%) extract of *S. taurica* aerial parts; butanol and dichloromethane fraction of its ethanol extract were investigated in alloxan diabetic rats. Obtained results revealed that petroleum ether (300 and 400 mg/kg) and ethanol extract (350 and 450 mg/kg) caused a significant decrease in serum glucose level (changes between 56% and 61%). Additionally, a similar decrease in serum glucose levels (between 48% and 62%) was observed with dichloromethane and *n*-butanol fractions at doses of 200 and 300 mg/kg (17).

Besides animal experiments, there are studies investigating the antidiabetic activity of *Sideritis* species by *in vitro* methods based on different mechanisms. In one of these studies, the inhibitory effect of flavonoids and a phenylpropanoid, which were isolated from the methanol extract of *S. brevibracteata* aerial parts, on aldose reductase, one of the enzymes associated with diabetes, was researched. As a result, hypolaetin-7-*O*-[6‴-*O*-acetyl-*β*-D-allopyranosyl-(1→2)]-*β*-D-glucopyranoside, isoscutellarein-7-*O*-[6‴-*O*-acetyl-*β*-D-allopyranosyl-(1→2)]-*β*-D-glucopyanoside, 3′-hydroxy-4′-*O*-methylisoscutellarein-7-*O*-[6‴-*O*-acetyl-*β*-D-allopyranosyl-(1→2)]-*β*-D-glucopyranoside, hypolaetin-7-*O*-[6‴-*O*-acetyl-*β*-D-allopyranosyl-(1→2)]-6″-*O*-acetyl-*β*-D-glucopyranoside, isoscutellarein-7-*O*-[6‴-*O*-acetyl-*β*-D-allopyranosyl-(1→2)]-6″-*O*-acetyl-*β*-D-glucopyranoside, 3′-hydroxy-4′-*O*-methylisoscutellarein-7-*O*-[6‴-*O*-acetyl-*β*-D-allopyranosyl-(1→2)]-6″-*O*-acetyl-*β*-D-glucopyranoside, and verbascoside were found to inhibit aldose reductase with an half maximal inhibitory concentration (IC_{50}) value of 0.61, 1.16, 1.25, 0.47, 1.02, >1000, and 2.10 µg/mL, respectively (14). In another research, acetone and methanol extracts of *S. akmanii* were determined to inhibit α-glucosidase (76% inhibition) and α-amylase (~45% inhibition) close to acarbose (18).

Anticholinesterase Activity

A lot of *Sideritis* species have been investigated for anticholinesterase activity *in vitro*. In one of these studies, infusions from *S. arguta, S. congesta, S. libanotica* subsp. *linearis, S. perfoliata,* and *S. pisidica* var. *termessi* samples obtained from the market were determined to have no inhibitory effect on acetylcholinesterase (19). But in another investigation, ent-kaurane diterpenoids (*ent*-7*α*-acetoxy-16*β* ,18-dihydroxy-kaurane, epoxyisolinearol, sideroxol, sideridiol, siderol, 7-epicandicandiol, linearol, sidol) obtained from petroleum ether and acetone extracts whole plant of *S. congesta* were determined to inhibit acetyl choline esterase with IC_{50} values between 0.23 and 8.04 mM; while they inhibit butyryl choline esterase with IC_{50} values between 0.022 and 3.67 mM. Sideroxol and 7-epicandicandiol have attracted attention by inhibiting the butyryl choline esterase with lower IC_{50} values than galantamine (20).

Acetylcholinesterase and butyrylcholinesterase inhibitory activity of *n*-hexane, acetone, and methanol extracts of *S. albiflora* and *S. leptoclada* were investigated. Only hexane extracts of *S. albiflora* and *S. leptoclada* demonstrated moderate acetylcholinesterase inhibitory activity (162.78 and 104.44 µg/mL IC_{50} values, respectively), while hexane and acetone extracts exhibited moderate butyrylcholinesterase inhibitory activity (IC_{50} values between 119.02 and 191.18) (21).

Antimicrobial Activity

Nowadays, development of resistance to commercially available antibiotics has become an important problem, and in this context, plants have come to the fore with their chemical richness and have been extensively studied in terms of antimicrobial activity. The effects of *Sideritis* species on many different microorganisms have also been studied in detail and one of the most studied activities on the *Sideritis* genus is antimicrobial activity. Most of these researches focused on the activity of the methanol extract and essential oil obtained from the aerial parts. It is not possible to describe all of the studies done in this section. For this reason, only some of them have been discussed here and the rest are given in Table 24.1.

Antimicrobial activities of essential oils acquired from four *S. caesarea* samples obtained from different localities in Kayseri were studied against six pathogenic microorganisms (*Escherichia coli, Staphylococcus aureus, Pseudomonas aeruginosa, Bacillus subtilis, Staphylococcus epidermidis,* and *Candida albicans*). None of the studied samples were shown to be active against *Candida albicans*, but they were found to inhibit the growth of other tested microorganisms with varying degrees (inhibition zones between 0.7 and 1.3 mm) (22). Loğoğlu et al. (2006) isolated five diterpenes (linearol, siderol, epicandicandiol, linearol diacetate, and epicandicandiol diacetate) from petroleum ether extract of *S. sipylea* and evaluated their antimicrobial activity on *Staphylococcus aureus,*

Bacillus subtilis, Escherichia coli, Pseudomonas aeruginosa, and *Candida albicans*. Among the studied diterpenes, only epicandicandiol showed antimicrobial activity with an inhibition zone of 10 mm against *S. aureus, B. subtilis,* and *C. albicans* (23).

In another research, *S. curvidens* and *S. lanata* essential oils were found to exhibit moderate to high antibacterial effect against *Escherichia coli, Enterobacter aerogenes, Shigella sonnei, Salmonella typhimurium, Pseudomonas aeruginosa, Streptococcus mutans, Staphylococcus aureus, Staphylococcus epidermidis, Micrococcus luteus, Bacillus subtilis,* and *B. cereus* at 5, 10, and 25 μL/disk concentrations with inhibition zones in the range of 8 and 38 mm (24).

Antimicrobial activity of *S. erythrantha* var. *erythrantha* and *S. erythrantha* var. *cedretorum* essential oils has been investigated against *Staphylococcus aureus, S. epidermidis, Enterococcus faecalis, Streptococcus pyogenes, Escherichia coli, Klebsiella pneumoniae, Enterobacter cloacae, Serratia marcescens, Proteus vulgaris, Salmonella typhimurium, Pseudomonas aeruginosa*, and *Haemophilus influenza* with diffusion method. The essential oils were found to have no activity against studied Gram (-) bacteria. But they demonstrated moderate activity on Gram (+) bacteria and *H. influenza* with inhibition zones in the range of 8.00 and 18.00 mm. Additionally, researchers noted that both essential oils are as effective as oxacylline and vancomycin against vancomycin-resistant *E. faecalis* and ampicillin-resistant *H. influenzae.* In addition essential oil of *S. erythrantha* var. *cedretorum* showed effective against methicillin-resistant *S. aureus* (25).

Antiviral activity of aerial parts of *S. lycia* acetone extract and isolated compounds (linearol, sidol, and isosidol) was determined against human parainfluenza virus type 2 in Vero cells. Although the most potent activity was observed with the acetone extract [2.91 μg/mL, 50% cytotoxic dose (CD_{50})], the isolated compounds exhibited antiviral activity with CD_{50} values in the range of 14.64–29.32 μg/mL (26).

Antioxidant Activity

Another pharmacological activity that has been studied in detail on *Sideritis* species is the antioxidant. The antioxidant activity of different extracts, essential oil, and isolated compounds obtained from *Sideritis* species was mostly investigated by DPPH (2,2-diphenyl-1-picrylhydrazyl), β-carotene bleaching test and ABTS (2,2'-azino-bis(3-ethylbenzothiazoline-6-sulfonic acid) methods. However, antioxidant activity determination models such as CUPRAC (cupric reducing antioxidant capacity) and FRAP (ferric ion reducing antioxidant power) were also used. Since there has been a lot of research on the antioxidant activities of Turkish *Sideritis* species, a limited number of studies are included in this section and the rest is presented in Table 24.1.

The petroleum ether, acetone, and methanol extracts of *S. arguta* and *S. congesta* were determined to cause inhibition between 15.5% and 65.7% in the *β*-carotene bleaching assay; between 0% and 89.8% in the DPPH test; between 1.09% and 13.04% in the ferrous ion chelating assay. In the superoxide radical scavenging activity test, studied extracts caused inhibition between 3.56% and 42.7%, respectively (27). Petroleum ether and acetone extracts of *S. congesta* were studied once again by Topçu et al. (2011) with the same methods (DPPH, β-carotene bleaching assay, and copper oxide radical scavenging activity tests); unlike the other study, isolated compounds were also included. In these research, petroleum ether and acetone extracts were determined to have 25.47 and 79.44 μM IC_{50} values, respectively, in DPPH method. In the β-carotene bleaching assay petroleum ether, acetone extract, and isolated compounds (*ent*-7*α*-acetoxy-16*β*,18-dihydroxy-kaurane, epoxyisolinearol, 7-epicandicandiol, linearol, and sidol) determined to have IC_{50} values ranging from 4.5 to 355.0 μM; while in the superoxide radical scavenging activity test, petroleum ether, acetone extract, and isolated compounds (*ent*-7*α*-acetoxy-16*β*,18-dihydroxy-kaurane, epoxyisolinearol, 7-epicandicandiol, linearol, sideroxol, sideridiol, siderol, and sidol) were determined to have IC_{50} values ranging from 107.27 to 571.05 μM (20).

Petroleum ether, ethyl acetate, methanol, and water extracts of *S. galatica* displayed important antioxidant activity in DPPH (0.0028, 0.1290, 0.3900, and 0.3520 mM trolox equivalent/g extract, respectively), ABTS (0.0045, 0.1630, 0.6750, and 0.7560 mM trolox equivalent/g extract), nitric oxide radical scavenging activity (6.668, 1.799, 5.569, and 0.773 mM trolox equivalent/g extract, respectively), CUPRAC (74.6, 83.7, 195.6, and 199.0 mg trolox equivalent/g extract respectively), FRAP (41.2, 49.4, 132.8, and 142.7 mg trolox equivalent/g extract, respectively), ferrous ion chelating assay (52.4, 16.4, 7.9, and 6.6 mg disodium edetate equivalent/g extract, respectively), phosphomolybdenum bleaching assay (1.3, 1.5, 2.3, and 2.3 mM trolox equivalent/g extract, respectively), and β-carotene bleaching assay (74.5%, 59.6%, 0% and 30.2%, respectively) (28).

The essential oil of *S. perfoliata* showed high antioxidant activity (higher than positive kontrol) in ABTS, β-carotene bleaching, and reducing power tests (29).

Antioxidant activity of different extracts (*n*-hexane, acetone, and methanol) of *S. albiflora* and *S. leptoclada* were investigated with DPPH, ABTS, CUPRAC, and metal chelating assay. In DPPH method, acetone extract of *S. albiflora* and acetone and methanol extracts of *S. leptoclada* exhibited strong antioxidant activity, higher than α-tocopherol (IC_{50} values were in the range of 28.08 and 42.92 µg/mL). Similarly, strong antioxidant activities, higher than α-tocopherol, were observed with acetone extract of *S. albiflora* and acetone and methanol extracts of *S. leptoclada* in ABTS method (IC_{50} values were in the range of 15.18 and 28.42 µg/mL). In CUPRAC test, acetone extract of *S. albiflora* showed higher antioxidant activity than an α-tocopherol (A0.05: 66.72 µg/mL) (21).

Anti-ulcerogenic Activity

In a research evaluating the anti-ulcerogenic effects of plants, which was used as folk medicine against peptic ulcer symptoms in Pınarbaşı district (Kayseri, Turkey), 80% ethanol extracts of aerial parts of *S. caesarea* were determined to prevent completely visible lesions in half of the animals and inhibit peptic ulcer formation significantly (95.8%) on ethanol-induced gastric ulcer method in rats (30). In the continuation of this work, the anti-ulcerogenic activity of *n*-hexane, dichloromethane, ethyl acetate, *n*-butanol, and remaining water fractions prepared from ethanol (80%) extract of *S. caesarea* was evaluated with the same *in vivo* experimental ulcer method. The ethyl acetate fraction attracted attention with its significant anti-ulcerogenic activity (91.4% inhibition), and two flavonoids {4′-*O*-methylhypolaetin-7-*O*-[6‴-*O*-acetyl-*β*-D-allopyranosyl-(1→2)]-6″-*O*-acetyl-*β*-D-glucopyranoside and isoscutellarein-7-*O*-[6‴-*O*-acetyl-*β*-D-allopyranosyl-(1→2)]-6″-*O*-acetyl-*β*-D-glucopyranoside} were isolated from the ethyl acetate fraction as active compounds (70.0% and 69.0% inhibition at 250 mg/kg dose, respectively) (31).

Aboutabl et al. investigated the anti-ulcerogenic activity of petroleum ether and ethanol (70%) extracts of *S. taurica*, and *n*-butanol and dichloromethane fractions of its ethanol extract with indomethacin-induced gastric ulcer model. The researchers stated that they detected significant anti-ulcerogenic activity at different doses between 400 and 200 mg/kg in all studied extracts and fractions (17).

Effect on the Central Nervous System

Öztürk et al. administered infusions of aerial parts from four different *Sideritis* species (*S. athoa*, *S. lanata*, *S. libanotica* subsp. *kurdica,* and *S. perfoliata*) to mice (250 and 500 mg/kg) for examining their swimming performance. The researchers have stated that studied *Sideritis* extracts displayed biphasic (not dose-dependent) effect on the central nervous system. The swimming time is shortened at lower dose, which indicates the depressant effect on the central nervous system; but at higher dose the extracts displayed a central nervous system stimulating effect by causing an increase in swimming time (32).

Petroleum ether and ethanol (70%) extracts of aerial parts of *S. taurica*, *n*-butanol, and dichloromethane fractions of its ethanol extract were determined not to prevent convulsions induced by electricity in experimental animals at different doses between 200 and 450 mg/kg (17).

Cytotoxic and Genotoxic Activity

The known diterpene 7-epicandedicandiol isolated from two different *Sideritis* species (*S. arguta* and *S. congesta*) and determined to be a cytotoxic agent against COL-2 (human colon cancer) and A2780 (human ovarian cancer) cells [median effective dose (ED_{50}) values: 11.8 and 9.0 μg/mL, respectively]. These two compounds also showed a weak cytotoxic effect (ED_{50} values between 13.3 and 17.9 μg/mL) on different cancer cells [KB (human epidermoid cancer), LU1 (human liver cancer), and LNCaP (human prostate cancer)]. In addition, it did not show any effect on the P-388 (mouse leukemia) and hTERTRPE (human retinal pigment epithelial cell) ($ED_{50} \geq 20$ μg/mL). Linearol and sideridiol did not display effect against the studied cell lines, while sidol had only weak cytotoxicity (ED_{50} 20 μg/mL) against A2780 cell (27).

Cytotoxicity of verbascoside, izoacteoside, lamalboside, leonoside A, isolavandulifolioside, isoscutellarein-7-*O*-[6‴-*O*-acetyl-*β*-allopyranosyl-(1→2)]-*β*-glucopyranoside, 4′-*O*-methyisoscutellarein-7-*O*-[6‴-*O*-acetyl-*β*-allopyranosyl-(1→2)]-*β*-glucopyranoside, 3′-hydroxy-4′-*O*-methylisoscutellarein-7-*O*-[6‴-*O*-acetyl-*β*-allopyranosyl-(1→2)]-*β*-glucopyranoside], and di-*O*-methylcrenatin that isolated from *S. trojana* root was assessed by the MTT [3- (4,5-dimethylthiazol-2-yl)-2,5-diphenyltetrazolium bromide] test against human prostate cancer cells (PC3). Results of this research revealed that evaluated compounds, except verbascoside, did not have cytotoxic effect against PC3 cells (in the range of 1 and 50 μM concentrations). Verbascoside exhibited cytotoxic activity with 42.5 μM IC_{50} value (33).

Four abietane diterpenoids (sideritin A and B, pomiferin E, 9α,13α-epi-dioxyabiet-8(14)-en-18-ol), a lignan (paulownin), flavanone (6-methoxysakuranetin), methoxystigmane (3-oxo-α-ionol), and phenolic glucoside (4-allyl-2,6-dimethoxyphenol glucoside) were isolated from *S. montana* methanol extract and their antiproliferative activity were evaluated on human cervical cancer cell lines (HeLa, SiHa, and C33A) with MTT assay. According to obtained results, pomiferin E was determined to show important activity on HeLa cell lines (46.93% inhibition), but a moderate inhibitory effect on SiHa cells (24.49 % inhibition). 6-Methoxysakuranetin was determined to show important activity on C33A cell lines (51.52% inhibition, respectively), while it showed moderate inhibitory activity on HeLA and SiHA (39.70% and 35.49%, respectively). Sideritin A represented moderate activity on HeLa and SiHa cells (28.34% and 26.87%, respectively) (34).

Cytotoxic effect of *S. ozturkii* water, ethyl acetate, and methanol extracts was evaluated on human breast cancer cell lines (MDA-MB-231 cells). MTT test results revealed that water extract (1 mg/mL, 48 h incubation) faintly inhibited (22%) MDA-MB-231 cells growth, but ethyl acetate and methanol extracts significantly inhibited the growth of same cells with 65.4 and 32.2 μg/mL IC_{50} values, respectively. In addition, ethyl acetate and methanol extracts caused a significant increase in Bax (apoptosis promoter) but a reduction in Bcl-2 (apoptosis inhibitor) gene expression (35). Cytotoxic activity of the same *Sideritis* species on different cell lines was also investigated. Methanol extracts prepared with flower and leaf were determined to exhibit cytotoxic activity on DLD-1 human colorectal cancer cells, and their IC_{50} values were calculated as 0.53 and 0.33 mg/mL for 48 h, respectively (36).

Cytotoxic effect of 7-epicandicandiol, linearol, sidol, siderol, and isosidol isolated from *S. lycia* against KB, P-388, COL-2, hTERTRPE, LU1, LNCaP, and A2780 cell lines was investigated. Highest cytotoxic activity was observed with 7-epi-candicandiol against KB, COL-2, LU1, LNCaP, and A2780 with ED_{50} values between 9.0 and 17.9 μg/mL. Sidol exhibited moderate cytotoxic activity against only A2780 cells (ED_{50}:15.6 μg/m) while the other studied compounds exhibited important activity on the aforementioned cell lines (26).

Cytotoxic activity of two different extracts (ethyl acetate and 96% ethanol) of *S. leptoclada* against the HT-144 (malignant melanoma) cancer cell line was evaluated with MTT test. The ethanol extracts were determined to show important inhibitory activity (83.49% inhibition at 100 μg/mL concentration), but ethyl acetate extract was determined to have weak cytotoxic activity. In illuminated MTT assay (~ ≥400 nm), the ethanol extract almost preserved its activity (77.46% inhibition at 100 μg/mL concentration), therefore ethanol extract was used in further experiments. Afterwards, the ethanol extract did not show cytotoxicity on normal cells (3T3), while it caused an increase in reactive oxygen species and tumor necrosis factor-α levels (37).

Other Pharmacological Activities

Hypolaetin-7-*O*-[6‴-*O*-acetyl-*β*-D-allopyranosyl-(1→2)]-*β*-D-glucopyranoside, hypolaetin-7-*O*-[6‴-*O*-acetyl-*β*-D-allopyranosyl-(1→2)]-6"-*O*-acetyl-*β*-D-glucopyranoside, isoscutellarein-7-*O*-[6‴-*O*-acetyl-*β*-D-allopyranosyl-(1→2)]-*β*-D-glucopyranoside, isoscutellarein 7-*O*-[6‴-*O*-acetyl-β-D-allopyranosyl-(1→2)]-6"-*O*-acetyl-*β*-D-glucopyranoside, and 3′-hydroxy-4′-*O*-methylisoscutellarein-7-*O*-[6‴-*O*-acetyl-*β*-D-allopyranosyl-(1→2)]-*β*-D-glucopyranoside isolated from methanol extract of *S. brevibracteata* were found to inhibit glutathione reductase enzyme, and IC_{50} values were 0.177, 0.297, 0.372, 0.368, and 0.295 mM, respectively (38).

In another investigation, methanol extract of *S. perfoliata* has been determined to inhibit angiotensin-converting enzyme (ACE) with 150 μg/mL IC_{50} values (39).

Ethanol extract of *S. perfoliata* was determined to have significant elastase inhibitory activity with 57.18 μg/mL IC_{50} value (29).

Urease and tyrosinase inhibitory activity of three different extracts (*n*-hexane, acetone, and methanol) of *S. albiflora* and *S. leptoclada* was investigated. All studied extracts of *S. albiflora* and hexane extracts of *S. leptoclada* caused higher urease inhibitory activity compared with thiourea (IC_{50} values between 1.02 and 5.10 μg/mL). All studied extracts, except the *n*-hexane of *S. albiflora*, were determined to have moderate tyrosinase inhibitory activity (% inhibition ratios were between 15.25 and 42.52) (21).

TOXICOLOGICAL DATA

In a repeated dose toxicity study, after 28 days of exposure, no observable adverse effect level (NOAEL) for *S. scardica* dry extract was determined as 1000 mg/kg body weight on Sprague Dawley rats. Besides, there was no evidence that the dry extracts from *S. scardica* were mutagenic in *Salmonella typhimurium* reverse mutation assay (AMES) on five *Salmonella typhimurium* strains (40).

Acute and repeated dose oral toxicity of 20% ethanol extract of *S. scardica* was evaluated on Sprague Dawley rats, while its water, *n*-heptane, and 20% and 50% ethanol extracts mutagenicity were determined using the Ames test. In acute and repeated dose oral toxicity tests, no evidence of toxicity or mortality was observed. At the end of repeated dose toxicity investigation, NOAEL was calculated as 1000 mg/kg body weight in rats. In Ames test, mutations in the *Salmonella* strains were not detected (41).

Cytotoxic activity of *S. scardica* 70% ethanol extract and its diethyl ether, ethyl acetate, and *n*-butanol fractions were investigated against rat glioma C6 line and rat astrocytes. Diethyl ether fraction was determined to exhibit cytotoxic effect by increasing reactive oxygen species production, and leading to apoptotic and autophagic cell death, while the ethyl acetate fraction exhibited cytotoxic effect by inducing G_2M cell cycle arrest and autophagy. But studied extracts did not present cytotoxic activity in rat astrocytes (42).

HERBAL PRODUCTS

In Turkey, an infusion is prepared from the flowers of *Sideritis* species and used as tea. Different preparation techniques are not very common. In addition, some herbal products in pharmaceutical form containing *Sideritis* species are also encountered. In one of these products, which is released to market as a food supplement, the extract of *S. congesta* has been combined with extracts of different plants (such as ginger and linden) into a syrup formulation and is recommended for cough.

S. angustifolia and *S. scardica* are also included in the content of food supplements in the syrup formulation that is recommended for kidney stones and as a sexual performance enhancer, although they are not the components responsible for the effect.

Aqueous extract and tincture of *S. scardica* are commercially available in the market in some European countries and these products are recommended for conditions such as lung disease, cough, anemia, liver disease, nervous system disorders, and hypertension.

SUGGESTED USAGE AND POSOLOGY

According to a report by European Medicines Agency, two Turkish species *S. scardica* and *S. syriaca* could be used as herbal tea for cough caused by cold and mild gastrointestinal discomfort, by preparing infusion with 2–4 g grinded herb in a glass of boiling water (2–3 times in a day) and daily dose up to 12 g. However, due to insufficient data, it is not recommended for use in children younger than 18 years of age (40).

CONCLUSION

Due to having high endemism rate (74%) and hosting 54 taxa, Turkey can be regarded as one of the gene centers for *Sideritis* (26). The Turkish *Sideritis* genus provides a wide research opportunity in terms of folk medicinal usage, chemical constituents, and pharmacological activity due to its diversity of species and the high rate of endemism. Its traditional usages recorded in ethnobotanical studies also guide pharmacological studies.

The essential oil content of many *Sideritis* species has been evaluated and a great number of terpenoid, flavonoid, iridoid, and phenylethanoid derivatives have been isolated and their structures have been clarified. Different extracts or compounds obtained from *Sideritis* species have been shown to exhibit different activities such as antimicrobial, anti-inflammatory, anti-ulcer, antioxidant, and cytotoxicity. However, it is possible to isolate new compounds with phytochemical studies, and identifying new drug candidates may be possible through pharmacological studies associated with or without their traditional usages. Considering all these issues and its widespread use as tea among the people, it is understood that *Sideritis* species grown in Turkey also have serious economic potential.

REFERENCES

1. Duman H, Kırımer N, Ünal F, Güvenç A, Şahin P. *Türkiye Sideritis L. türlerinin revizyonu.* Turkey: Türkiye Bilimsel ve Teknolojik Araştırma Kurumu; 2005. Report No.: TBAG-1853 (199T090).
2. Güner A, Aslan S, Ekim T, Vural M, Babaç MT. *Türkiye Bitkileri Listesi (Damarlı Bitkiler).* İstanbul: Nezahat Gökyiğit Botanik Bahçesi ve Flora Araştırmaları Derneği Yayını; 2012.
3. Huber-Morath A. Sideritis L. In: Davis PH, editor. *Flora of Turkey and the East Eagean Islands.* 7. 1st ed. Edinburgh: Edinburgh University Press; 1982. pp. 178–99.
4. González-Burgos E, Carretero ME, Gómez-Serranillos MP. *Sideritis* spp.: Uses, chemical composition and pharmacological activities-a review. *Journal of Ethnopharmacology.* 2011;135:209–25.
5. Akgul A, Akgul A, Senol SG, Yildirim H, Secmen O, Dogan Y. An ethnobotanical study in Midyat (Turkey), a city on the silk road where cultures meet. *Journal of Ethnobiology and Ethnomedicine.* 2018;14(1):12.
6. Altundağ E, Öztürk M. Ethnomedicinal studies on the plant resources of east Anatolia, Turkey. *Procedia - Social and Behavioral Sciences.* 2011;19:756–77.
7. Bulut G, Haznedaroğlu MZ, Doğan A, Koyu H, Tuzlacı E. An ethnobotanical study of medicinal plants in Acipayam (Denizli-Turkey). *Journal of Herbal Medicine.* 2017;10:64–81.
8. Bulut G, Tuzlacı E. An ethnobotanical study of medicinal plants in Bayramiç. *Marmara Pharmaceutcal Journal.* 2015;19(3):268–82.
9. Baytop T. *Türkiye'de Bitkiler ile Tedavi.* 2. ed. İstanbul: Nobel Tıp Kitabevleri; 1999.
10. Kırımer N, Tabanca N, Demirci B, Başer KHC, Duman H, Aytaç Z. The essential oil of a new *Sideritis* species: *Sideritis Ozturkii* Aytaç and Aksoy. *Chemistry of Natural Compounds.* 2001;37(3).

11. Başer KHC. Aromatic biodiversity among the flowering plant taxa of Turkey. *Pure and Applied Chemistry*. 2002;74:527–45.
12. Kırımer N, Başer KHC, Demirci B, Duman H. Essential oils of *Sideritis* species of Turkey belonging to the section Empedoclia. *Chemistry of Natural Compounds*. 2004;40:19–23.
13. Akcoş Y, Ezer N, Çalış İ, Demirdamar R, Tel BC. Polyphenolic compounds of *Sideritis lycia* and their anti-inflammatory activity. *Pharmaceutical Biology*. 1999;37(2):118–22.
14. Güvenç A, Okada Y, Küpeli-Akkol E, Duman H, Okuyama T, Çalış İ. Investigations of anti-inflammatory, antinociceptive, antioxidant and aldose reductase inhibitory activities of phenolic compounds from *Sideritis brevibracteata*. *Food Chemistry*. 2010;118:686–92.
15. Cavalcanti MRM, Passos FRS, Monteiro BS, Gandhi SR, Heimfarth L, Lima BS, et al. HPLC-DAD-UV analysis, anti-inflammatory and anti-neuropathic effects of methanolic extract of Sideritis bilgeriana (lamiaceae) by NF-kappaB, TNF-alpha, IL-1beta and IL-6 involvement. *Journal of Ethnopharmacology*. 2021;265:113338.
16. Osthoff J, Carola C, Salazar A, von Hagen J, Llorent Mart nez EJ. Bioactivity-guided single-step isolation of Stachyspinoside from Sideritis congesta by centrifugal partition chromatography. *Journal of Chemistry*. 2021;2021:1–12.
17. Aboutabl EA, Nassar MI, Elsakhawy FM, Maklad YA, Osman AF, El-Khrisy EAM. Phytochemical and pharmacological studies on *Sideritis taurica* Stephan ex Wild. *Journal of Ethnopharmacology*. 2002;82:177–84.
18. Aksoy L, Guzey I, Duz M. Essential oil content, antioxidative characteristics and enzyme inhibitory activity of Sideritis akmanii Aytac, Ekici & Donmez. *Turkish Journal of Pharmaceutical Sciences*. 2022;19(1):76–83.
19. Orhan İE, Baki E, Şenol S, Yılmaz G. Sage-called plant species sold in Turkey and their antioxidant activities. *Journal of the Serbian Chemical Society*. 2010;75(11):1491–501.
20. Topçu G, Ertaş A, Öztürk M, Dinçel D, Kılıç T, Halfon B. Ent-kaurane diterpenoids isolated from *Sideritis congesta*. *Phytochemistry Letters*. 2011;4:436–9.
21. Deveci E, Tel- Çayan G, Duru ME, Öztürk M. Phytochemical contents, antioxidant effects, and inhibitory activities of key enzymes associated with Alzheimer's disease, ulcer, and skin disorders of *Sideritis albiflora* and *Sideritis leptoclada*. *Journal of Food Biochemistry*. 2019;43(12).
22. Günbatan T, Demirci B, Gürbüz İ, Demirci F, Özkan AMG. Comparison of volatiles of *Sideritis caesarea* specimens collected from different localities in Turkey. *Natural Product Communications*. 2017;12(10):1639–42.
23. Loğoğlu E, Arslan S, Öktemer A, Şakiyan I. Biological activities of some natural compounds from *Sideritis sipylea* Boiss. *Phytotherapy Research*. 2006;20:294–7.
24. Uğur A, Varol Ö, Ceylan Ö. Antibacterial activity of *Sideritis curvidens* and *Sideritis lanata* from Turkey. *Pharmaceutical Biology*. 2005;43(1):47–52.
25. Köse EO, Deniz İG, Sarıkürkçü C, Aktaş Ö, Yavuz M. Chemical composition, antimicrobial and antioxidant activities of the essential oils of *Sideritis erythrantha* Boiss. and Heldr. (var. erythrantha and var. cedretorum P.H. Davis) endemic in Turkey. *Food and Chemical Toxicology*. 2010;48:2960–5.
26. Kılıç T, Topcu G, Goren AC, Aydogmus Z, Karagoz A, Yildiz YK, et al. Ent-kaurene diterpenoids from Sideritis lycia with antiviral and cytotoxic activities. *Records of Natural Products*. 2020;14(4):256–68.
27. Ertaş A. *Endemik İki Sideritis türü S. arguta ve S. congesta'nın Diterpenik Bileşiklerinin İzolasyonu ve Biyolojik Aktivitelerinin incelenmesi [Yüksek Lisans]*. İstanbul: İstanbul Üniversitesi; 2005.
28. Zengin G, Sarikurkcu C, Aktumsek A, Ceylan R. Sideritis galatica Bornm.: A source of multifunctional agents for the management of oxidative damage, Alzheimer's and diabetes mellitus. *Journal of Functional Foods*. 2014;11:538–47.
29. Lall N, Chrysargyris A, Lambrechts I, Fibrich B, Blom Van Staden A, Twilley D, et al. Sideritis Perfoliata (Subsp. Perfoliata) nutritive value and its potential medicinal properties. *Antioxidants (Basel)*. 2019;8(11).
30. Gürbüz İ, Özkan AMG, Yeşilada E, Kutsal O. Anti-ulcerogenic activity of some plants used in folk medicine of Pınarbaşı (Kayseri, Turkey). *Journal of Ethnopharmacology*. 2005;101:313–8.
31. Günbatan T, Gürbüz İ, Bedir E, Gençler Özkan AM, Özçınar Ö. Investigations on the anti-ulcerogenic activity of *Sideritis caesarea* H. Duman, Aytac & Baser. *Journal of Ethnopharmacology*. 2020;258:112920.
32. Öztürk Y, Aydın S, Öztürk N, Başer KHC. Effects of extracts from certain *Sideritis* species on swimming performance in mice. *Phytotherapy Research*. 1996;10:70–3.
33. Kırmızıbekmez H, Arıburnu E, Masullo M, Festa M, Capasso A, Yesilada E, et al. Iridoid, phenylethanoid and flavonoid glycosides from *Sideritis trojana*. *Fitoterapia*. 2012;83:130–6.

34. Toth B, Kusz N, Forgo P, Bozsity N, Zupko I, Pinke G, et al. Abietane diterpenoids from Sideritis montana L. and their antiproliferative activity. *Fitoterapia.* 2017;122:90–4.
35. Zengin G, Ugurlu A, Baloglu MC, Diuzheva A, Jeko J, Cziaky Z, et al. Chemical fingerprints, antioxidant, enzyme inhibitory, and cell assays of three extracts obtained from Sideritis ozturkii Aytac & Aksoy: An endemic plant from Turkey. *Journal of Pharmaceutical and Biomedical Analysis.* 2019;171:118–25.
36. Demirelma H, Gelinci E. Determination of the cytotoxic effect on human colon cancer and phenolic substance content of the endemic species Sideritis Ozturkii AytaÇ & Aksoy. *Applied Ecology and Environmental Research.* 2019;17(4).
37. Aydogmus-Ozturk F, Gunaydin K, Ozturk M, Jahan H, Duru ME, Choudhary MI. Effect of Sideritis leptoclada against HT-144 human malignant melanoma. *Melanoma Research.* 2018;28(6):502–9.
38. Tandoğan B, Güvenç A, Çalış İ, Ulusu NN. In vitro effects of compounds isolated from *Sideritis brevibracteata* on bovine kidney cortex glutathione reductase. *Acta Biochimica Polonica.* 2011;58(4):471–5.
39. Loizzo MR, Saab AM, Tundis R, Menichini F, Bonesi M, Piccolo V, et al. In vitro inhibitory activities of plants used in Lebanon traditional medicine against angiotensin converting enzyme (Ace) and digestive enzymes related to diabetes. *Journal of Ethnopharmacology.* 2008;119:109–16.
40. EMA. *Assessment Report on Sideritis scardica Griseb.; Sideritis clandestina (Bory & Chaub.) Hayek; Sideritis raeseri Boiss. & Heldr.; Sideritis syriaca L., herba.* The Netherlands; 2016.
41. Feistel B, Wegener T, Rzymski P, Pischel I. Assessment of the acute and subchronic toxicity and mutagenicity of Sideritis scardica Griseb. extracts. *Toxins (Basel).* 2018;10(7).
42. Jeremic I, Tadic V, Isakovic A, Trajkovic V, Markovic I, Redzic Z, et al. The mechanisms of in vitro cytotoxicity of mountain tea, Sideritis scardica, against the C6 glioma cell line. *Planta Medica.* 2013;79(16):1516–24.
43. Bondi ML, Bruno M, Piozzi F, Başer KHC, Simmonds MSJ. Diversity and antifeedant activity of diterpenes from Turkish species of *Sideritis. Biochemical Systematics and Ecology.* 2000;28:299–303.
44. Askun T, Tümen G, Satıl F, Ateş M. Characterization of the phenolic composition and antimicrobial activities of Turkish medicinal plants. *Pharmaceutical Biology.* 2009;47(7):563–71.
45. Dülger B, Gönüz A, Bican T. Antimicrobial studies on three endemic species of *Sideritis* from Turkey. *Acta Biologica Cracoviensia Series Botanica.* 2005;47(2):153–6.
46. Güvenç A, Houghton PJ, Duman H, Coşkun M, Şahin P. Antioxidant activity studies on selected *Sideritis* species native to Turkey *Pharmaceutical Biology.* 2005;43(2):173–7.
47. Usluer Ö. *Sideritis albiflora Hub.-Mor. ve Sideritis leptoclada O. Schwarz & P. H. Davis türlerinin uçucu bileşenlerinin izolasyonları, karakterizasyonları ve antioksidant aktivitelerinin incelenmesi [Yüksek Lisans].* Muğla: Muğla Universitesi; 2005.
48. Deveci E, Tel-Cayan G, Usluer O, Emin Duru M. Chemical composition, antioxidant, anticholinesterase and anti-tyrosinase activities of essential oils of two sideritis species from Turkey. *Iranian Journal of Pharmaceutical Research.* 2019;18(2):903–13.
49. Topçu G, Barla A, Gören AC, Bilsel G, Bilsel M, Tümen G. Analysis of the essential oil composition of *Sideritis albiflora* using direct thermal desorption and headspace GC-MS techniques. *Turkish Journal of Chemistry.* 2005;29:525–9.
50. Kırcı D, Saltan N, Göger F, Köse Y, Demirci B. Chemical compositions of Sideritis albiflora Hub. – Mor. *Istanbul Journal of Pharmacy.* 2021;51(3):378–85.
51. Tümen G, Baser KHC, Kirimer N, Ermin N. Essential oil of Sideritis amasiaca Bornm. *Journal of Essential Oil Research.* 1995;7(6):699–700.
52. Özhatay N, Koçak S. Plants used for medicinal purposes in Karaman province (Southern Turkey). *İstanbul Üniversitesi Eczacılık Fakültesi Dergisi.* 2011;41:75–89.
53. Ertaş A, Öztürk M, Boğa M, Topçu G. Antioxidant and anticholinesterase activity evaluation of ent-kaurane diterpenoids from *Sideritis arguta. Journal of Natural Products.* 2009;72:500–2.
54. Erkan N, Çetin H, Ayrancı E. Antioxidant activities of *Sideritis congesta* Davis et Huber-Morath and *Sideritis arguta* Boiss et Heldr: Identification of free flavonoids and cinnamic acid derivatives. *Food Research International.* 2011;44:297–303.
55. Ceylan R, Demirbas A, Ocsoy I, Aktumsek A. Green synthesis of silver nanoparticles using aqueous extracts of three Sideritis species from Turkey and evaluations bioactivity potentials. *Sustainable Chemistry and Pharmacy.* 2021;21.
56. Topçu G, Gören AC, Kılıç T, Yıldız YK, Tümen G. Diterpenes from *Sideritis argyrea. Fitoterapia.* 2001;72:1–4.
57. Kılıç T, Yıldız YK, Gören AC, Tümen G, Topçu G. Phytochemical analysis of some *Sideritis* species of Turkey. *Chemistry of Natural Compounds.* 2003;39(5):453–6.

58. Ezer N, Vila R, Canigueral S, Adzet T. Essential oil composition of four Turkish species of *Sideritis*. *Phytochem*. 1996;41(1):203–5.
59. Kirimer N, Tabanca N, Ozek T, Baser K, Tumen G, Duman H. Composition of essential oils from five endemic Sideritis species. *Journal of Essential Oil Research*. 2003;15(4):221–5.
60. Polat R, Satıl F. An ethnobotanical survey of medicinal plants in Edremit Gulf (Balıkesir-Turkey). *Journal of Ethnopharmacology*. 2012;139(2):626–41.
61. Topçu G, Gören AC, Yıldız YK, Tümen G. Ent-kaurene diterpenes from *Sideritis athoa*. *Natural Product Letters*. 1999;14(2):123–9.
62. Halfon B, Gören AC, Ertaş A, Topçu G. Complete ^{13}C NMR assignments for ent-kaurane diterpenoids from *Sideritis* species. *Magnetic Resonance in Chemistry*. 2011;49:291–4.
63. Fraga BM, Hernandez MG, Diaz CE. On the Ent-Kaurene diterpenes from *Sideritis Athoa*. *Natural Product Research*. 2003;17(2):141–4.
64. Ertan A, Azcan N, Demirci B, Başer KHC. Fatty acid composition of *Sideritis* species. *Chemistry of Natural Compounds*. 2001;37(4):301–3.
65. İşcan G, Kırımer N, Kürkçüoğlu M, Başer KHC. Composition and antimicrobial activity of the essential oils of two endemic species from Turkey: *Sideritis cilicica* and *Sideritis bilgerana*. *Chemistry of Natural Compounds*. 2005;41(6):679–82.
66. Dülger B, Gönüz A, Aysel V. Inhibition of clotrimazole-resistant *Candida albicans* by some endemic *Sideritis* species from Turkey. *Fitoterapia*. 2006;77:404–5.
67. Özcan M, Chalchat JC, Akgül A. Essential oil composition of Turkish mountain tea (*Sideritis* spp.). *Food Chemistry*. 2001;75:459–63.
68. Tekeli Y. Antioxidant activities and phenolic compounds of two endemic taxa of Labiatae *Sideritis*. *Revista de chimie*. 2012;63(5):465–9.
69. Özdemir E, Alpınar K. An ethnobotanical survey of medicinal plants in western part of central Taurus Mountains: Aladağlar (Niğde - Turkey). *Journal of Ethnopharmacology*. 2015;166:53–65.
70. Sagir ZO, Carikci S, Kilic T, Goren AC. Metabolic profile and biological activity of Sideritis brevibracteata P. H. Davis endemic to Turkey. *International Journal of Food Properties*. 2017;20(12):2994–3005.
71. Dülger B, Uğurlu E, Akı C, Süerdem TB, Çamdeviren A, Tazeler G. Evaluation of antimicrobial activity of some endemic *Verbascum*, *Sideritis*, and *Stachys* species from Turkey. *Pharmaceutical Biology*. 2005;43(3):270–4.
72. Çarıkçı S, Kılıç T, Azizoğlu A, Topçu G. Chemical constituents of two endemic *Sideritis* species from Turkey with antioxidant activity. *Records of Natural Products*. 2012;6(2):101–9.
73. Kırımer N, Tabanca N, Özek T, Başer KHC, Tümen G. Composition of essential oils from two endemic *Sideritis* species of Turkey. *Chemistry of Natural Compounds*. 1999;35(1):61–4.
74. Özkan AMG, Koyuncu M. Traditional medicinal plants used in Pınarbaşı area (Kayseri-Turkey). *Turkish Journal of Pharmaceutical Sciences*. 2005;2(2):63–83.
75. Askun T, Tümen G, Satıl F, Kılıç T. Effects of some Lamiaceae species methanol extracts on potential mycotoxin producer fungi. *Pharmaceutical Biology*. 2008;46:688–94.
76. Sağdıç O, Aksoy A, Özkan G, Ekici L, Albayrak S. Biological activities of the extracts of two endemic *Sideritis* species in Turkey. *Innovative Food Science & Emerging Technologies*. 2008;9:80–4.
77. Çelik İ, Kaya MS. The antioxidant role of *Sideritis caesarea* infusion against TCA toxicity in rats. *British Journal of Nutrition*. 2011;105:663–8.
78. Halfon B, Çiftçi E, Topçu G. Flavonoid constituents of *Sideritis caesarea*. *Turkish Journal of Chemistry*. 2013;37:464–72.
79. Başer KHC, Bondi ML, Bruno M, Kırımer N, Piozzi F, Tümen G, et al. An ent-kaurane from *Sideritis huber-morathii*. *Phytochemistry*. 1996;43(6):1293–5.
80. Kırımer N, Tabanca N, Tümen G, Duman H, Başer KHC. Composition of the essential oils of four endemic *Sideritis* species from Turkey. *Flavour and Fragrance Journal*. 1999;14:421–5.
81. Özkan G. Comparison of antioxidant phenolics of ethanolic extracts and aqueous infusions from *Sideritis* species. *Asian Journal of Chemistry*. 2009;21(2):1024–8.
82. Dülgeroğlu C. *Sideritis stricta Boiss. & Heldr. (Lamiaceae) Türünün Doğal ve Kültür Formlarının Morfolojik, Anatomik, Ekolojik ve Uçucu Yağ İ çerikleri Yönünden Karş ılaştırılması [Yüksek Lisans]*. Antalya: Akdeniz Üniversitesi; 2013.
83. Koyuncu İ. *Bazi Sideritis Türlerinin Antimikrobiyal Aktivitelerinin Araştırılması [Yüksek Lisans]*. Antalya: Akdeniz Üniversitesi; 2009.
84. Özkan G, Sağdıç O, Özcan M, Özçelik H, Ünver A. Antioxidant and antibacterial activities of Turkish endemic *Sideritis* extracts. *Grasas y Aceites*. 2005;56(1):16–20.

85. Bardakci H, Cevik D, Barak TH, Gozet T, Kan Y, Kirmizibekmez H. Secondary metabolites, phytochemical characterization and antioxidant activities of different extracts of Sideritis congesta P.H. Davis et Hub.-Mor. *Biochemical Systematics and Ecology.* 2020;92.
86. Kılıç T, Çarıkçı S, Topçu G, Aslan I, Gören AC. Diterpenoids from *Sideritis condensata.* Evaluation of chemotaxonomy of *Sideritis* species and insecticidal activity. *Chemistry of Natural Compounds.* 2009;45(6):918–20.
87. Kırımer N, Kürkçüoğlu M, Özek T, Başer KHC. Composition of the essential oil of *Sideritis condensata* Boiss. et Heldr. *Flavour and Fragrance Journal.* 1996;11:315–20.
88. Gümüşçü A, Tugay O, Kan Y. Comparison of essential oil compositions of some natural and cultivated endemic *Sideritis* species. *Advances in Environmental Biology.* 2011;5(2):222–6.
89. Gul Sarıkaya A, Canıs S. Volatile components of leaf and flowers of natural mountain sage (Sideritis spp.) taxa from Davraz Mountain, Isparta-Turkey. *International Journal of Biology and Chemistry.* 2019;12(2).
90. Aydın S, Oztürk Y, Beis R, Başer KHC. Investigation of *Origanum onites, Sideritis congesta* and *Satureja cuneifolia* essential oils for analgesic activity. *Phytotherapy Research.* 1996;10:342–4.
91. Ezer N. *Sideritis congesta Davis et Huber-Morath Üzerinde Farmakognozik Araştırmalar [Doktora Tezi].* Ankara: Hacettepe Üniversitesi; 1980.
92. Öktemer A, Loğoğlu E. Isolation of a new diterpene from *Sideritis congesta. Communications, Faculty Of Science, University of Ankara Series B: Chemistry and Chemical Engineering.* 2003;49:1–3.
93. Yeşilada E, Honda G, Sezik E, Tabata M, Goto K, Ikeshiro Y. Traditional medicine in Turkey IV. folk Medicine in the Mediterranean subdivision. *Journal of Ethnopharmacology.* 1993;39:31–8.
94. Dorman HJD, Koşar M, Başer KHC, Hiltunen R. Iron (III) reducing and antiradical activities of three *Sideritis* from Turkey. *Pharmaceutical Biology.* 2011;49(8):800–4.
95. Topçu G, Gören AC. Diterpenes from *Sideritis sipylea* and *S. dichotoma. Turkish Journal of Chemistry.* 2002;26:189–94.
96. Sargın SA, Selvi S, Büyükcengiz M. Ethnomedicinal plants of Aydıncık District of Mersin, Turkey. *Journal of Ethnopharmacology.* 2015;174:200–16.
97. Ahmet Sargın S. Ethnobotanical survey of medicinal plants in Bozyazı district of Mersin, Turkey. *Journal of Ethnopharmacology.* 2015;173:105–26.
98. Tabanca N, Kırımer N, Başer KHC. The composition of essential oils from two varieties of *Sideritis erythrantha* var. *erythrantha* and var. *cedretorum. Turkish Journal of Chemistry.* 2001;25:201–8.
99. Nermin N. *Sideritis erthrantha'nın iki varyetesi: Var. erythrantha ve var. cedretorum uçucu yağlarının bileşimi [Yüksek Lisans].* Eskişehir: Anadolu Üniversitesi; 1995.
100. Bruno M, Piozzi F, Arnold NA, Başer KHC, Tabanca N, Kırımer N. Kaurane diterpenoids from three *Sideritis* species. *Turkish Journal of Chemistry.* 2005;29:61–4.
101. Chalchat J-C, Özcan M. Constituents of the essential oil of *Sideritis erythrantha* Boiss. & Heldr. var. *erythrantha. General and Applied Plant Physiology.* 2005;31(1–2):65–70.
102. Zengin G, Sarıkürkçü C, Aktümsek A, Ceylan R. Antioxidant potential and inhibition of key enzymes linked to Alzheimer's diseases and diabetes mellitus by monoterpene-rich essential oil from *Sideritis galatica* Bornm. endemic to Turkey. *Records of Natural Products.* 2016;10(2):195–206.
103. Dişli A, Yıldırır Y, Yaşar A. Galaticat, a new diterpene from *Sideritis galatica. Ankara Üniversitesi Eczacılık Fakültesi Dergisi.* 2002;31(2):83–9.
104. Günbatan T, Gürbüz İ, Özkan AMG. The current status of ethnopharmacobotanical knowledge in Çamlıdere (Ankara, Turkey). *Turkish Journal of Botany.* 2016;40:241–9.
105. Kirmizibekmez H, Erdogan M, Kusz N, Karaca N, Erdem U, Demirci F, et al. Secondary metabolites from the aerial parts of Sideritis germanicopolitana and their in vitro enzyme inhibitory activities. *Natural Product Research.* 2021:1–4.
106. Bayan Y, Akşit H. Antifungal activity of essential oils and plant extracts from Sideritis germanicopolitana Bornm. growin in Turkey. *Egyptian Journal of Biological Pest Control.* 2016;26(2):333–7.
107. Kirimer N, Koca F, Baser KHC, Özek T, Tanriverdi H, Kaya A. Composition of the Essential Oils of Two Subspecies of Sideritis germanicopolitana Bornm. *Journal of Essential Oil Research.* 1992;4(5):533–4.
108. Karahan A. *Sideritis gülendami H. Duman & F. A. Karavelioğlu Bitkisinin Diterpen Bileşiklerinin Izolasyonu ve Yapılarının Tayini [Yüksek Lisans].* Balıkesir: Balıkesir Üniversitesi; 2007.
109. Ayaz A. *Sideritis hololeuca Boiss. & Heldr. apud Bentham ve Sideritis libanotica Labill. subsp. violascens ekstraktlarının antibakteriyel aktivitelerinin belirlenmesi [Yüksek Lisans Tezi].* Konya: Selçuk Üniversitesi; 2008.
110. Çarıkçı S. *Bazi Sideritis (Sideritis niveotomentosa, Sideritis hololeuca, Sideritis brevidens) Türlerinin Diterpenik Bileşenlerinin İzolasyonu ve Yapılarının Tayini [Doktora Tezi].* Balıkesir: Balıkesir Üniversitesi; 2010.

111. Koutsaviti A, Bazos I, Milenkovic M, Pavlovic-Drobac M, Tzakou O. Antimicrobial activity and essential oil composition of five *Sideritis* taxa of Empedoclia and Hesiodia Sect. from Greece. *Records of Natural Products.* 2013;7(1):6–14.
112. Kırımer N, Tabanca N, Özek T, Tümen G, Başer KHC. Essential oils of annual *Sideritis* Species Growing in Turkey. *Pharmaceutical Biology.* 2000;38(2):106–11.
113. Alipieva KI, Kostadinova EP, Evstatieva LN, Stefova M, Bankova VS. An iridoid and a flavonoid from *Sideritis lanata* L. *Fitoterapia.* 2009;80:51–3.
114. Stanoeva JP, Stefova M, Stefkov G, Kulevanova S, Alipieva K, Bankova V, et al. Chemotaxonomic contribution to the *Sideritis* species dilemma on the Balkans. *Biochemical Systematics and Ecology.* 2015;61:477–87.
115. Kargıoğlu M, Cenkçi S, Serteser A, Evliyaoğlu N, Konuk M, Kök MŞ, et al. An ethnobotanical survey of Inner-west Anatolia, Turkey. *Human Ecology.* 2008;36:763–77.
116. Kayalı HA, Urek RÖ, Nakiboğlu M, Tarhan L. Antioxidant activities of endemic *Sideritis leptoclada* and *Mentha dumetorum* aqueous extracts used in Turkey folk medicine. *Journal of Food Processing and Preservation.* 2009;33:285–95.
117. Kılıç T. *Sidertis lycia ve Sideritis leptoclada Türlerinin Diterpen Bileşiklerinin Izolasyonu ve Karakterizasyonu [Doktora Tezi].* Balıkesir: Balıkesir Üniversitesi; 2002.
118. Kılıç T, Yıldız YK, Topçu G, Gören AC, Ay M, Bodige SG, et al. X-ray analysis of sideroxol from *Sideritis leptoclada. Journal of Chemical Crystallography.* 2005;35(8).
119. Gürdal B, Kültür Ş. An ethnobotanical study of medicinal plants in Marmaris. *Journal of Ethnopharmacology.* 2013;146:113–26.
120. Tetik F, Civelek S, Çakılcıoğlu U. Traditional uses of some plants in Malatya (Turkey). *Journal of Ethnopharmacology.* 2013;146:331–46.
121. *Korkmaz S. İnsan Plazma ve Eritrosit Hemolizatı Asetilkolinesteraz (AChE, E.C.3.1.1.7) Enzimi Üzerine Balbaş ı Otunun (Sideritis libanotica) Polar ve Nonpolar Ekstraktının İn Vitro Etkilerinin Araştırılması [Yüksek lisans].* Van: Yüzüncü Yıl Üniversitesi; 2015.
122. Köker E. *Sideritis libanotica Bitkisinin Genotoksik Etkisinin Değerlendirilmesi [Yüksek Lisans Tezi].* KahramanMaraş: KahramanMaraş Sütçü İmam Üniversitesi; 2014.
123. Güzel Y, Güzelsemme M, Miski M. Ethnobotany of medicinal plants used in Antakya: A multicultural district in Hatay province of Turkey. *Journal of Ethnopharmacology.* 2015;174:118–52.
124. Ezer N, Akçoş Y, Rodriguez B, Abbasoğlu U. *Sideritis libanotica* Labill. subsp. *linearis* (Bentham) Bornm.'den elde edilen iridoit heteroziti ve antimikrobiyal aktivitesi. *Hacettepe Üniv Ecz Fak* Der. 1995;15(1):15–21.
125. Güven UM. Design of microemulsion formulations loaded Scutellaria salviifolia Benth, Sideritis libanotica Labill. subsp. linearis (Bentham) Bornm, and Ziziphora clinopodioides Lam. extracts from Turkey and in vitro evaluation of their biological activities. *Turkish Journal of Botany.* 2021;45(SI-2):789–99.
126. Ayhan B. *Sideritis libanotica linearis bitkisinden sekonder metabolitlerinin saflaştırılması, karakterizasyonu ve bazı biyolojik aktivitelerinin incelenmesi [Yüksek Lisans Tezi].* Tokat: Gaziosmanpaşa Üniversitesi; 2008.
127. Demirtaş İ, Şahin A, Ayhan B, Tekin S, Telci İ. Antiproliferative effects of the methanolic extracts of *Sideritis libanotica* Labill. subsp. *linearis. Records of Natural Products.* 2009;3(2):104–9.
128. Şahin M. *Sideritis libanotica Labill. ssp. linearis (Bentham) Bornm'in methanol Ekstraktının Antioksidan Etkilerinin Araştırılması [Yüksek Lisans Tezi].* Elazığ: Fırat Universitesi; 2010.
129. Demirtaş İ, Ayhan B, Şahin A, Aksit H, Elmastaş M, Telci İ. Antioxidant activity and chemical composition of *Sideritis libanotica* Labill. ssp. *linearis* (Bentham) Borm. (Lamiaceae). *Natural Product Research.* 2011;25(16):1512–23.
130. Çabuk N. *Sideritis libanotica linearis subsp. linearis Bitkisinden Sekonder Metabolitlerin İzolasyonu, Yapi Tayini, Antioksidan Aktivitelerinin İncelenmesi [Yüksek Lisans Tezi].* Tokat: Gaziosmanpaşa Üniversitesi; 2012.
131. Tepe B, Sökmen M, Akpulat HA, Yumrutaş Ö, Sökmen A. Screening of antioxidative properties of the methanolic extracts of *Pelargonium endlicherianum* Fenzl., *Verbascum wiedemannianum* Fisch. & Mey., *Sideritis libanotica* Labill. subsp. *linearis* (Bentham) Borm., Centaurea *mucronifera* DC. and *Hieracium cappadocicum* Freyn from Turkish flora. *Food Chemistry.* 2006;98:9–13.
132. Dincer C, Torun M, Tontul I, Topuz A, Sahin-Nadeem H, Gokturk RS, et al. Phenolic composition and antioxidant activity of Sideritis lycia and Sideritis libanotica subsp. linearis : Effects of cultivation, year and storage. *Journal of Applied Research on Medicinal and Aromatic Plants.* 2017;5:26–32.
133. Adem S, Akkemik E, Aksit H, Guller P, Tüfekci AR, Demirtas İ, et al. Activation and inhibition effects of some natural products on human cytosolic CAI and CAII. *Medicinal Chemistry Research.* 2019;28(5):711–22.

134. Yeşil Y, Akalın E. Folk medicinal plants in Kürecik area (Akçadağ/Malatya-Turkey). *Turkish Journal of Pharmaceutical Sciences.* 2009;6(3):207–20.
135. Balos MM, Akan H. Zeytinbahçe - Akarçay (Birecik, Şanlı Urfa) arasmda kalan bolgenin etnobotanik ozellikleri. *S 0 Fen Ed Fak Fen Derg.* 2007;29:155–71.
136. Akcoş Y, Ezer N, Özçelik B, Abbasoğlu U. Iridoid glucosides from *Sideritis Lycia* Boiss. & Heldr. and its antimicrobial activities. *Journal of Pharmaceutical Sciences.* 1998;23:99–103.
137. Radojevic ID, Stanković MS, Stefanović OD, Topuzović MD, Čomić LR, Ostojić AM. Anti-Aspergillus properties of different extracts from selected plants. *African Journal of Microbiology Research.* 2011;5(23):3986–90.
138. Radojević ID, Stanković MS, Stefanović OD, Topuzović MD, Čomić LR. Antioxidative and antimicrobial properties of different extracts from *Sideritis montana* L. *Romanian Biotechnological Letters.* 2012;17(2):7160–8.
139. Koleva II, Linssen JP, van-Beek TA, Evstatieva LN, Kortenska V, Handjieva N. Antioxidant activity screening of extracts from *Sideritis* species (Labiatae) grown in Bulgaria. *Journal of Science of Food and Agriculture.* 2003;83:809–19.
140. Bilgin M, Elhussein EAA, Ozyurek M, Guclu K, Sahin S. Optimizing the extraction of polyphenols from Sideritis montana L. using response surface methodology. *Journal of Pharmaceutical and Biomedical Analysis.* 2018;158:137–43.
141. Maric T, Friscic M, Marijanovic Z, Males Z, Jerkovic I. Comparison of volatile organic compounds of Sideritis romana L. and Sideritis montana L. from Croatia. *Molecules.* 2021;26(19).
142. Sattar AA, Bankova V, Spassov S. Flavonoid glycosides from *Sideritis* species. *Fitoterapia.* 1993;64(3):278–9.
143. Tomas-Barberan FA, Rejdali M, Harborne JB, Heywood VH. External and vacuolar flavonoids from İbero-North African *Sideritis* species. A chemosystematic approach. *Phytochem.* 1988;27(1):165–70.
144. Kültür Ş. Medicinal plants used in Kırklareli province (Turkey). *Journal of Ethnopharmacology.* 2007;111:341–64.
145. Emre İ, Kurşat M, Yılmaz Ö, Erecevit P. Some biological compounds, radical scavenging capacities and antimicrobial activities in the seeds of *Nepeta italica* L. and *Sideritis montana* L. subsp. *montana* from Turkey. *Grasas y aceites.* 2011;62(1):68–75.
146. Kılıç Ö. Essential oil composition of two *Sideritis* L. taxa from Turkey: A chemotaxonomic approach. *Asian Journal of Chemistry.* 2014;26(8).
147. Akalın E, Alpınar K. Tekirdağ'ın tıbbi ve yenen yabani bitkileri hakkında bir araştırma. *Ege Üniv Ecz Fak Derg.* 1994;2(1):1–11.
148. Simsek Sezer EN, Uysal T. Phytochemical analysis, antioxidant and anticancer potential of Sideritis niveotomentosa: Endemic wild species of Turkey. *Molecules.* 2021;26(9).
149. Küpeli E, Şahin FP, Çalış İ, Yeşilada E, Ezer N. Phenolic compounds of *Sideritis ozturkii* and their in vivo anti-inflammatory and antinociceptive activities. *Journal of Ethnopharmacology.* 2007;112:356–60.
150. Gelinci E, Maçin S, Demirelma H, Türk Dağı H. Antimicrobial effect of Sideritis öztürkii Aytaç & Aksoy Species. *Flora the Journal of Infectious Diseases and Clinical Microbiology.* 2020;25(1):84–90.
151. Şahin FP, Ezer N, Çalış İ. Three acylated flavone glycosides from *Sideritis ozturkii* Aytaç & Aksoy. *Phytochemistry Letters.* 2004;65:2095–9.
152. Sarikurkcu C, Locatelli M, Mocan A, Zengin G, Kirkan B. Phenolic profile and bioactivities of Sideritis perfoliata L.: The plant, its most active extract, and its broad biological properties. *Frontiers in Pharmacology.* 2020;10:1642.
153. Kırımer N, Demirci B, İşcan G, Başer KHC, Duman H. Composition of the essential oils of two *Sideritis* species from Turkey and antimicrobial activity. *Chemistry of Natural Compounds.* 2008;44(1).
154. Loizzo MR, Tundis R, Menichini F, Saab AM, Statti GA, Menichini F. Cytotoxic activity of essential oils from Labiatae and Lauraceae families against in vitro human tumor models. *Anticancer Research.* 2007;27:3293–300.
155. Cocelli G, Pehlivan M, Yumrutas O. Sideritis perfoliata inhibits cell proliferation, induces apoptosis and exhibits cellular antioxidant activity in cervical cancer cells. *Boletin Latinoamericano y del Caribe de Plantas Medicinales y Aromaticas.* 2021;20(4):394–405.
156. Celik T, Onderci M, Pehlivan M, Yumrutas O, Uckardes F. In vitro scolicidal effects of Sideritis perfoliata extract against Echinococcus granulosus. *International Journal of Clinical Practice.* 2021;75(10):e14498.
157. Celik I, Atioglu Z, Aksit H, Demirtas I, Erenler R, Akkurt M. Crystal structure and Hirshfeld surface analysis of 2-oxo-13-epi-manoyl oxide isolated from Sideritis perfoliata. *Acta Crystallographica Section E: Crystallographic Communications.* 2018;74(Pt 5):713–7.

158. Karabörklü S. Chemical characterization of *Sideritis perfoliata* L. essential oil and its fumigant toxicity against two pest insects. *Journal of Food, Agriculture & Environment.* 2014;12(2):434–7.
159. Ezer N, Sakar MK, Rodriguez B, Torre MCDL. Flavonoid glycosides and a phenylpropanoid clycoside from *Sideritis perfoliata. International Journal of Pharmacognosy.* 1992;30(1):61–6.
160. Charami M-T, Lazari D, Karioti A, Skaltsa H, Hadjipavlou-Litina D, Souleles C. Antioxidant and anti-inflammatory activities of *Sideritis perfoliata* subsp. *perfoliata* (Lamiaceae). *Phytotherapy Research.* 2008;22:450–4.
161. Bağcı Y. Aladağlar (Yahyalı, Kayseri) ve çevresinin etnobotanik özellikleri. *Ot Sistematik Botanik Dergisi.* 2000;7(1):89–94.
162. Sayar A, Güvensen A, Özdemir F, Öztürk M. Muğla (Türkiye) ilindeki bazı türlerin etnobotanik özellikleri. *Ot Sistematik Botanik Dergisi.* 1995;2(1):151–60.
163. Deveci E, Tel-Çayan G, Yıldırım H, Duru ME. Chemical composition, antioxidant, anticholinesterase and anti-urease activities of *Sideritis pisidica* Boiss. & Heldr. endemic to Turkey. *Marmara Pharmaceutical Journal.* 2017;21(4):898–905.
164. Münevver A, Özek G, Özek T. Essential oil compositions and site characteristics of Sideritis pisidica in natural habitat. *Contemporary Problems of Ecology.* 2021;14(6):675–89.
165. Yeşilada E, Honda G, Sezik E, Tabata M, Fujita T, Tanaka T, et al. Traditional medicine in Turkey. V. folk medicine in the Inner Taurus Mountains. *Journal of Ethnopharmacology.* 1995;46:133–52.
166. Chalchat JC, Özcan MM, Figueredo G. The composition of essential oils of different parts of laurel, mountain tea, sage and ajowan. *Journal of Food Biochemistry.* 2011;35:484–99.
167. Sklirou AD, Angelopoulou MT, Argyropoulou A, Chaita E, Boka VI, Cheimonidi C, et al. Phytochemical study and in vitro screening focusing on the anti-aging features of various plants of the Greek Flora. *Antioxidants (Basel).* 2021;10(8).
168. Korompokis K, Igoumenidis PE, Mourtzinos I, Karathanos VT. Green extraction and simultaneous inclusion complex formation of Sideritis scardica polyphenols *International Food Research Journal.* 2017;24(3):1233–8.
169. Vassilevska-Ivanova R, Shtereva L, Stancheva I, Geneva M, Hristozkova M. Determination of the antioxidant capacity of Sideritis scardica specimens collected at different regions in Bulgaria. *Comptes rendus de l'Acad´emie bulgare des Sciences.* 2016;69(10):1307–14.
170. Georgieva L, Mihaylova D. Screening of total phenolic content and radical scavenging capacity of Bulgarian plant species. *International Food Research Journal.* 2015; 22(1):240–5.
171. Mihaylova DS, Lante A, Krastanov IA. A study on the antioxidant and antimicrobial activities of pressurized-liquid extracts of Clinopodium vulgare and Sideritis scardica. *Agro Food Industry Hi Tech.* 2014;25(6).
172. Tadic VM, Jeremic I, Dobric S, Isakovic A, Markovic I, Trajkovic V, et al. Anti-inflammatory, gastroprotective, and cytotoxic effects of *Sideritis scardica* extracts. *Planta Medica.* 2012;78(5):415–27.
173. Jeremic I, Petricevic S, Tadic V, Petrovic D, Tosic J, Stanojevic Z, et al. Effects of Sideritis scardica extract on glucose tolerance, triglyceride levels and markers of oxidative stress in Ovariectomized rats. *Planta Medica.* 2019;85(6):465–72.
174. Wightman EL, Jackson PA, Khan J, Forster J, Heiner F, Feistel B, et al. The acute and chronic cognitive and cerebral blood flow effects of a Sideritis scardica (Greek Mountain Tea) extract: A double blind, randomized, placebo controlled, parallel groups study in healthy humans. *Nutrients.* 2018;10(8).
175. Hofrichter J, Krohn M, Schumacher T, Lange C, Feistel B, Walbroel B, et al. Sideritis spp. extracts enhance memory and learning in Alzheimer's beta-amyloidosis mouse models and aged C57Bl/6 mice. *Journal of Alzheimer's Disease.* 2016;53(3):967–80.
176. Tadic V, Bojovic D, Arsic I, Dordevic S, Aksentijevic K, Stamenic M, et al. Chemical and antimicrobial evaluation of supercritical and conventional Sideritis scardica Griseb., Lamiaceae extracts. *Molecules.* 2012;17(3):2683–703.
177. Knorle R. Extracts of Sideritis scardica as triple monoamine reuptake inhibitors. *Journal of Neural Transmission (Vienna).* 2012;119(12):1477–82.
178. Qazimi B, Stefkov G, Karapandzova M, Cvetkovikj I, Kulevanova S. Aroma compounds of mountain tea (Sideritis scardica and S. raeseri) from Western Balkan. *Natural Product Communications.* 2014 9(9):1369–72.
179. Sargın SA, Selvi S, Lopez V. Ethnomedicinal plants of Sarıgöl district (Manisa), Turkey. *Journal of Ethnopharmacology.* 2015;171:64–84.
180. Sargın SA, Akçiçek E, Selvi S. An ethnobotanical study of medicinal plants used by the local people of Alaşehir (Manisa) in Turkey. *Journal of Ethnopharmacology.* 2013;150:860–74.

181. Aligiannis N, Kalpoutzakis E, Chinou IB, Mitakou S, Gikas E, Tsarbopoulos A. Composition and antimicrobial activity of the essential oils of five taxa of *Sideritis* from Greece. *Journal of Agricultural and Food Chemistry*. 2001;49:811–5.
182. Gergis V, Spiliotis V, Poulos C. Antimicrobial activity of essential oils from Greek *Sidertis* species. *Pharmazie*. 1990;45:70.
183. Nakiboğlu M, Ürek RÖ, Kayalı HA, Tarhan L. Antioxidant capacities of endemic *Sideritis sipylea* and *Origanum sipyleum* from Turkey. *Food Chemistry*. 2007;104:630–5.
184. Tomou EM, Lytra K, Chrysargyris A, Christofi MD, Miltiadous P, Corongiu GL, et al. Polar constituents, biological effects and nutritional value of Sideritis sipylea Boiss. *Natural Product Research*. 2021:1–5.
185. Küpeli E, Şahin FP, Yeşilada E, Çalış İ, Ezer N. In vivo anti-inflammatory and antinociceptive activity evaluation of phenolic compounds from *Sideritis stricta*. *Zeitschrift für Naturforschung*. 2007;62c:519–25.
186. Aslan İ, Kılıç T, Gören AC, Topçu G. Toxicity of acetone extract of *Sideritis trojana* and 7-epicandicandiol, 7-epicandicandiol diacetate and 18-acetylsideroxol against stored pests *Acanthoscelides obtectus* (Say), *Sitophilus granarius* (L.) and *Ephestia kuehniella* (Zell.). *Industrial Crops and Products*. 2006;23:171–6.
187. Kösedağ A. *Sideritis stricta Boiss. & Heldr. bitkisinin fitokimyasal analizi [Yüksek Lisans Tezi]*. Balıkesir: Balıkesir Üniversitesi; 2005.
188. Erdoğan A, Özkan A, Ünal O, Dülgeroğlu C. Parental ve epirubicin-HCl dirençli H1299 hücrelerinde dağ çayı (Sideritis stricta Boiss & Heldr.) uçucu yağının sitotoksik ve membran hasar verici etkilerinin değerlendirilmesi. *Cukurova Medical Journal*. 2018;43(3):669–77.
189. Deveci E, Tel-Çayan G, Duru ME. Phenolic profile, antioxidant, anticholinesterase, and anti-tyrosinase activities of the various extracts of Ferula elaeochytris and Sideritis stricta. *International Journal of Food Properties*. 2018;21(1):771–83.
190. Kılıç T. Isolation and biological activity of new and known diterpenoids from *Sideritis stricta* Boiss. & Heldr. *Molecules*. 2006;11:257–62.
191. Şahin FP, Ezer N, Çalış İ. Terpenic and phenolic compounds from *Sideritis stricta*. *Turkish Joıurnal of Chemistry*. 2006;30:495–504.
192. Kilic T, Sagir ZO, Carikci S, Azizoğlu A. Experimental and theoretical vibrational spectra of Sideridiol isolated from Sideritis species. *Russian Journal of Physical Chemistry A*. 2018;91(13):2608–12.
193. Tuzlacı E. *Şifa Niyetine, Türkiye'nin Bitkisel Halk İlaçları*. İstanbul: Alfa Basım Yayım Dağıtım Ltd. Şti; 2006.
194. Turker AU, Yildirim AB, Karakas FP, Turker H. In vitro antibacterial and antitumor efficiency of some traditional plants from Turkey. *Indian Journal of Traditional Knowledge*. 2018;17(1):50–8.
195. Aboutabl EA, Meselhy KM, El-Azzouny AM. Composition and antiwormal activity of essential oil from *Sideritis taurica* Stephan ex Willd grown in Egypt. *Journal of Essential Oil Research*. 2009;21:94–6.
196. Çarıkçı S, Çöl Ç, Kılıç T, Azizoğlu A. Diterpenoids from *Sideritis tmolea* P. H. Davis. *Records of Natural Products*. 2007;1(4):44–50.
197. Çöl Ç. *Sideritis tmolea P. H. Davis Bitkisinin Diterpen Bileşenlerinin İzolasyonu ve Yapılarının Tayini [Yüksek Lisans Tezi]*. Balıkesir: Balıkesir Üniversitesi; 2007.
198. Kirmizibekmez H, Karaca N, DemİRcİ B, DemİRcİ F. Characterization of Sideritis trojana Bornm. essential oil and its antimicrobial activity. *Marmara Pharmaceutical Journal*. 2017;21(4):860–5.
199. Kırca A, Arslan E. Antioxidant capacity and total phenolic content of selected plants from Turkey. *International Journal of Food Science and Technology*. 2008;43:2038–46.
200. Celep E, Seven M, Akyüz S, İnan Y, Yesilada E. Influence of extraction method on enzyme inhibition, phenolic profile and antioxidant capacity of Sideritis trojana Bornm. *South African Journal of Botany*. 2019;121:360–5.
201. Paşa C, Selvi S, Özer Z, Kılıç T. Kazdağlarında (Edremit-Balıkesir) yayılış gösteren Sideritis trojana türünün Uçucu Yağ Oranı ve Bileşenlerinin Diurnal ve Ontogenetik Varyasyonunun Belirlenmesi Üzerine Bir Araştırma. Kahramanmaraş Sütçü İmam Üniversitesi Tarım ve Doğa Dergisi. 2019.

25 *Teucrium sp.*

Fatma Ayaz

INTRODUCTION

Teucrium L. (Lamiaceae) is a wide and diverse genus that is generally characterized by perennial plants that grow in sunny environments. *Teucrium* genus, belonging to the subfamily Ajugoideae, consists of ten sections and grows in mild climate zones, such as Europe, North Africa, and Asia, mainly in the Mediterranean basin and Central Asia (Dinç & Doğu, 2016; Harley et al., 2004; McClintock & Epling, 1946; Salmaki et al., 2016). The Plant List includes almost 1000 scientific plant names of species order for the genus *Teucrium*, more than 300 of which were accepted to verify the scientific names of the species (*Kew and Missouri Botanical Garden*, 2021). The genus *Teucrium* is recorded by 37 species and 49 taxa in Turkey with 17 endemic taxa (the endemism ratio, approximately 35%) (Aksoy et al., 2020; Dinç & Doğu, 2016).

According to the literature, there has been no recent comprehensive research on *Teucrium* species in Turkey for their medicinal importance up to date. In this work, significant knowledge about some of *Teucrium* species in Turkey has been summarized in the hopes of being useful to guide scientists in the future. Therefore, it is revealed that this work evaluates the botanical descriptions, phytochemical constituents, traditional usages, medicinal administration doses, and biological activities of the selected species, *T. chamaedrys* L. (Wall Germander), *T. polium* L. (Poley), and *T. scordium* L. (Water Germander), growing in Turkey. These plants have also been the subject of many printed and online publications in pharmacognosy discipline, as well as regulatory controls in the world (Duke, 2002; Gruenwald et al., 2007; Khare & Plants, 2007).

TEUCRIUM SPECIES

The genus *Teucrium*, derived from in Greek word "Teucer", a Trojan king, who utilized the plant as a remedy or a botanist and physician, may have been the inspiration for the name. The first mention of these plant species for their medicinal properties can also be recorded in the tenth century BC in Greek mythology as Teucer, a son of Telamon, king of Salamis (Charters, 2021).

Teucrium species are generally perennial, rarely annual or biannual herbs or tiny shrubs, and their flowers are in the axils of upper leaves or in the bom in racemes, panicles, or heads, with entire dentate or deeply dissected leaves (Ulubelen et al., 2000).

The selected species, *T. chamaedrys*, *T. polium*, and *T. scordium*, and their botanical descriptive parameters are presented in Table 25.1 (Davis, 1982; Güner & Aslan, 2012; Güner et al., 2000).

PHYTOCONSTITUENTS

The medicinal plants belonging to *Teucrium* genus mainly contain clerodane or neo-clerodane diterpenoids, known as chemotaxonomic markers for the genus, terpenoids such as monoterpenes, abietane diterpenes, sesquiterpenes, and triterpenes, as well as steroids and flavonoids. They also include polyphenols, iridoids, fatty acid esters, and volatile constituents (Boulila et al., 2008; Candela et al., 2020; Hachicha et al., 2009; Piozzi et al., 2005; Ulubelen et al., 2000). The records on chemical compounds of *T. chamaedrys*, *T. polium*, and *T. scordium* are given in Table 25.2, according to the PDR for Herbal Medicines (Gruenwald et al., 2007; Khare & Plants, 2007).

DOI: 10.1201/9781003146971-25

TABLE 25.1
Morphological Comparison of the Species *T. chamaedrys*, *T. polium*, and *T. scordium*

Properties	*T. chamaedrys*	*T. polium*	*T. scordium*
English name	Wall Germander	Poley	Water Germander
Perennial	Suffrutescent herb, often rhizomatous	Suffruticose herb	Stoloniferous herb, shortly villous
Stems	5–50 cm, with very variable indumentum	stems 10–40 cm, prostrate to erect, with white, gray lanuginose-tomentose or crisped indumentum	erect or sometimes ascending, branched, 5–75 cm
Leaves	Oblong or obovate-oblong, cuneate at base, usually crenate-dentate or shortly lobulate, lobes entire or toothed	Oblong to narrowly obovate or linear, obtuse, crenate to the base or middle, flat or revolute-margined, usually tomentose	Oblong to ovate-oblong, crenate to dentate, cordate to cuneate at base, usually amplexicaul on main stem
Flowers	Verticillasters 4–8-flowered, in dense or ± lax terminal racemes, floral leaves usually distinct from cauline leaves, entire or toothed; pedicels equaling or usually shorter than calyx	Flowers very shortly pedicellate, borne in heads. Bracts linear-spathulate, crenate, or entire, shorter to slightly longer than calyces	Lateral branches bearing smaller dentate leaves, upper ones bearing 1–4 flowers in each axil and overtopping the flowers. Pedicels 3–5 mm
Calyx	Calyx tubular-campanulate, subgibbous at base, teeth lanceolate-triangular, ± subulate-tipped, margins usually densely short-haired and sparsely long-hairy, or only long-hairy, or glabrous, usually purplish	Calyx 3–5 mm, tubular-obconical, usually densely lanuginose or adpressed-canescent, divided to 1/4-1/3 into subequal, obtuse teeth	Calyx obconical-campanulate, 3–5 mm, divided to 1/3 into equal triangular acute teeth
Corolla	Corolla reddish-purple, 2 x calyx	Corolla whitish, proximal lobes occasionally glabrous. Nutlets 2 mm	Corolla purplish-pink or mauve, 2 x calyx
Flowering period	Fl. 6–9	Fl. 6–9	Fl. 5–9
Habitat	Open forests, cliffs, slopes, steppe, s.I.-1600 m	Dry places, Quercus scrub, rocky slopes; steppe, dunes, field margins, etc., s.l. 2050 m	Damp places, in fresh or slightly saline marshes, by streams, in forest, moist dunes, and fallow fields, 50–2350 m.

In addition to this Table 25.2, these species have extensively been investigated on their essential oil chemical compositions until now. It was reported that the essential oil of *T. polium* species and subspecies mainly contain different kinds of substances such as farnesol, α-curcumene, cis-verbenol, carvone, (E)-β-farnesene, germacrene D, nerolidol, alloaromadendrene, aromadendrene, sabinol, 11-acetoxyeudesman-4-α-ol, α-bisabolol, β-bisabolol, caryophylleneoxide, α-bisabololoxide B, bicyclogermacrene, trans-β-guaiene, camphor, pinocarvone, pinene oxide, α-terpineol, spathulenol, β-bourbonene, γ-muurolene, α-gurjunene, rosifoliol, 3-carene, carvacrol, and dehydro-sesquicineole, depending on their different localities and seasonal variations (Basudan & Abu-Gabal, 2018; Belmekki et al., 2013; Benali et al., 2021; BOUKHEBTI et al., 2019; Cakir et al., 1998; Chabane et al., 2021; Yassine El Atki et al., 2020b; Farahbakhsh et al., 2020, 2021; Hayta et al., 2017; Kovacevic

TABLE 25.2
Phytochemical Ingredients of *T. chamaedrys*, *T. polium*, and *T. scordium*

	T. chamaedrys	*T. polium*	*T. scordium*
Volatile oil	0.07%, major components beta-caryophyllene (20%), humulene (15%)	0.1 to 1%, major components, varying according to chemical race, alpha-pinene and beta-pinene, alpha-cadinol, alpha-humulene, beta-caryophyllene, caryophyllene oxide, cedrol, gamma-cadinene, delta-cadinene, limonene, linalool, menthofurane, myrcene, ocimene, T-cadinol, terpine-4-ol	-
Iridoids	Harpagide, acetyl harpagide	8-O-acetyl harpagide, harpagide, teucardoside	Harpagide and acetyl harpagide
Diterpenes	Teugin, teuflin, teuflidin, dihydroteugin, teucrin A, B, E, F, G, marrubiin	Picropolin, picropolinol, picropolinon, teucrin A, teucrin PI, teucrin H3, montanines B and C, teupolins I to V, gnaphalidin	6,20-bisdeacetylteupyreinidin, 6-deacetylteupyreinidin, 2beta, 6beta-dihydroxyteuscordin, 2beta,6beta-dihydroxyteuscordin, dihydroteugin, teuflidin, teucrin E, teugin, 2-keto-19-hydroxyteuscordin
Caffeic acid derivatives	Teucroside	-	-
Flavonoids	Cirsiliol, cirsimaritin, luteolin	Apigenin-7-O-glucoside, luteolin-7-Oglucoside, acacetine, apigenin, cirsiliol, cirsimaritin. eupatorin, luteolin, salvigenin	Rutin, quercetin, iso-quercetin
Others	-	-	Choline, stigmasterol, beta-sitosterol, beta-amyrin, chlorogenic and ursolic acids

-: not reported in PDR for Herbal Medicines

et al., 2001; Maizi et al., 2019; Reaisi et al., 2019; Roukia et al., 2013; Sadeghi et al., 2014; Saltan et al., 2019; Sarer et al., 1987; Sevindik et al., 2016). The chemical composition of essential oil of *T. chamaedrys* was less examined than *T. polium*. Major constituents of the essential oil of *T. chamaedrys* were also determined as germacrene D, δ-cadinene, bicyclogermacrene, β-farnesene, caryophyllene oxide, linalool, nonacosane, α-pinene, caryophyllenol II, δ-cadinene, and carvacrol (Bagci et al., 2010; Kaya et al., 2009; Kovacevic et al., 2001; Küçük et al., 2006; Morteza-Semnani et al., 2005; Muselli et al., 2009; Sajjadi & Shookohinia, 2010). Otherwise, *T. scordium* has recently been explored for essential oil components. Essential oil composition was found to be rich with trans-a-bergamotene, (Z)-a-trans-bergamotol, linalool, piperitenone oxide, caryophyllene oxide, α-pinene, β-pinene, pulegone, β-caryophyllene, β-farnesene, menthofuran, 1,8-cineole, α-humulene, and β-eudesmol (Ebrahimi Anaraki et al., 2020; Gagliano Candela et al., 2020; Kovacevic et al., 2001; Morteza-Semnani et al., 2007; Radulović et al., 2012; Sharififar et al., 2010).

As for isolated secondary metabolites of *T. polium*, the neo-clerodane diterpenoids, polivincins A-C, teulamifin B, 20-O-acetyl-teucrasiatin, syrapolin II, teupolins VI-XII, teulolin A and B, and abeo-abietanes were isolated (Bedir et al., 1999; Bozov & Penchev, 2019; Fiorentino et al., 2011; Fiorentino et al., 2010; Moustapha, 2011; Venditti et al., 2017). Other phytochemicals, such as saponin glycosides, triterpene glycoside (poliusaposide C), iridoid derivatives (6'-O-Caffeoyl-8-O-acetylharpagide), phenylethanol glycosides (poliumoside B), phenylpropanoid glycoside, monoterpenoid, sesquiterpenes (β-eudesmol and α-cadinol), flavonoids, and clerosterol-3-O-glucosid, were also identified in *T. polium* (D'Abrosca et al., 2013; De Marino et al., 2012; Elmasri et al., 2014; Elmasri, Hegazy, et al., 2015; Elmasri, Yang, et al., 2015; Kamel, 1995; Moustapha, 2011). On the other hand, iridoid glycosides and phenolic compounds, such as cinaroside, nuchensin, and (E)-p-coumaroyl-O-β-D-glucoside, verbascoside, forsythoside b, samioside, lamiuside A alyssonoside, and β-arbutin, as well as neo-clerodane diterpenoids (teuchamaedryn A, D, chamaedryoside A, B, and C, syspirensins A and B) were isolated from *T. chamaedrys* (Çalis et al., 1996; Elmastas et al., 2016; Fiorentino et al., 2009; Frezza et al., 2018; Gecibesler et al., 2019; Gori et al., 2011; Prescott et al., 2011). Phytoconstituents of *T. scordium* were defined as diterpenoids, such as scordidesin A, teucrin A, 6-ketoteuscordin, 6β-hydroxyteuscordin, 6-ketoteuscordin, 6α-hydroxy-teuscordin, and teucrin H4 (Bozov et al., 2020; Papanov & Malakov, 1981, 1982, 1985).

TRADITIONAL USE

Teucrium species have been used as herbal medicines for over 2000 years. The many species from *Teucrium* are traditionally utilized against stomach aches, insect as antifeedant agents, as well as for wound healing, antimicrobial, and antitumor properties. These plants are also used as diuretic, hypoglycemic, diaphoretic, carminative, appetizer, astringent, stimulant, expectorant, tonic, antiseptic, antipyretic, and anti-inflammatory in folk medicine (Ulubelen et al., 2000). They are also extensively utilized in folk medicine now in Turkey (Candela et al., 2020; Charters, 2021). Among one of the most well-known medicinal *Teucrium* species, *T. polium*, commonly used in ethnomedicine in several cultures, has frequently been investigated until now (Bahramikia & Yazdanparast, 2012; Jaradat, 2015). It is shown as an overview of the traditional usage of *T. polium*, and *T. chamaedrys* in Turkey in Table 25.3.

POSOLOGY AND ADMINISTRATION

Teucrium species, known with principal bioactive constituents as diterpenoids, have been the subject of a lot of research, although they are not detected including in Pharmacopoeial and Other Monographs, such as British Pharmacopoeia 2009, European Pharmacopoeia 9th Edition, European Scientific Cooperative on Phytotherapy (ESCOP), Martindale 35th edition, Committee on Herbal Medicinal Products (HMPC) in European Medicines Agency (EMA). In the literature, there have been several investigations on the chemical components and biological activities of *Teucrium* species (Bahramikia & Yazdanparast, 2012; Jaradat, 2015; Ke et al., 2009; Piozzi et al., 2005; Ulubelen et al., 2000). Handbook of Medicinal Herbs, and PDR for Herbal Medicine present three *Teucrium* species, *T. chamaedrys*, *T. polium*, and *T. scordium*, growing in Turkey. Therefore, the medicinal administrations and the description of the medicinal parts of these plants are summarized in Table 25.4 with doses, and other knowledge such as effects, indications, contraindications precautions, and side effects (Duke, 2002; Gruenwald et al., 2007).

BIOLOGICAL ACTIVITY

Teucrium genus is known to have diverse volatile components, which are responsible for various biological effects, including antioxidant, antimicrobial, antitumor, anti-inflammatory, phytotoxic, antiprotozoal, and insecticidal, and which have been recorded, supporting the extensive utilization in folk medicine in many regions (Candela et al., 2020). Moreover, the extracts and isolates of *Teucrium* species have mostly been investigated on their biological activities in *in vitro* and *in vivo* assays up to date. Among *Teucrium* species, *T. polium* is one of the most investigated species in the world, similarly in Turkey. Other important species can be *T. chamaedrys* and *T. scordium* in terms

TABLE 25.3
Traditional Usage of *T. chamaedrys* and *T. polium* Growing in Turkey*

Species	Vernacular Names, Locality	Usages	References
T. chamaedrys	Kisa mahmut, Sancti otu	Hemorrhoids	(Erbay & Sarı, 2018)
	Uzun mahmut, Balikesir	Abdominal pain, kidney stones	(Polat & Satıl, 2012)
	Kısacıkmahmut, Central	Stomachache	(Sezik et al., 2001)
T. chamaedrys ssp. *lydium*	Mayasil otu, Ayvacik	Hemorrhoids	(Uysal et al., 2012)
	Mayasil otu, Bayramiç	Hemorrhoids, eczema [49]	(Bulut & Tuzlacı, 2015)
T. chamaedrys ssp. *chamaedris*	Bodurca mahmut, Kinin otu, Acipayam	Hemorrhoids, hearth diseases, malaria	(Bulut et al., 2017)
	Kisamahmut otu, Bilecik	Ulcer in mouth, kidney infection	(Ünsal et al., 2010)
	Kisamahmut otu, Kirklareli	Diuretic, kidney stones, diarrhea, abdominal pain	(Kültür, 2007)
	Kisacik mahmut, Sancti otu, Bodurmahmut, Ulukışla	Hemorrhoids	(Erbay & Sarı, 2018; Paksoy et al., 2016)
T. polium	Kısamahmut	Diabetes, kidney stone	(Polat & Satıl, 2012)
	Several vernacular names	Hemorrhoid	(Erbay & Sarı, 2018)
	Kisa Mahmut Otu, Ayvacik	Antipyretic, cough, tonic	(Uysal et al., 2012)
	Kokar yavşan, Peryavşanotu, Sırçanotu, Yavşan, Central	Colds, antipyretic, rheumatic pain, stomachache	(Sezik et al., 2001)
	Cay kekigi, Kekikmisi, Acipayam	Diabetes, abdominal pain	(Bulut et al., 2017)
	Merwend (Kurdish), Batman	Abdominal pain, digestive, colds, diabetes, stomachache, antitussive	(Bulut et al., 2019)
	Mayasil otu, Bayramiç	Hemorrhoids, eczema	(Bulut & Tuzlacı, 2015)
	Aci yavşan, Bilecik	Stomach diseases, wounds, carminative	(Ünsal et al., 2010)
	Periyavşan, Kahramanmaraş	Lung inflammations, stomach ulcers, diabetes disease, fever lowering	(Demirci & Özhatay, 2012)
	Kefen otu, Karaisalı	Menstruation, common cold	(Güneş et al., 2017)
	Kekik, Kirkalareli	Flu, colds, abdominal pain	(Kültür, 2007)
	Urper, Maden	Diabetes, stomachache, antipyretic, colds, liver disorders, inflammation, stomachic, wounds	(Cakilcioglu et al., 2011; Cakilcioglu et al., 2010)
	Cığde, Midyat	Stomachache	(Akgul et al., 2018)
	Kısamahmut, Silopi	Stomach diseases, wounds	(Akin et al., 2010)
	Meyremxort, Solhan	Antihypertensive, colds and flu, diabetes, diarrhea, headache, stomachache	(Polat et al., 2013)

* There is no information on the traditional usage of *T. scordium* in the literature.

of the amounts of accessed detailed data in the literature. Therefore, the biological effects of the extracts and the phytoconstituents as well as the nanoformulations of these products of *T. chamaedrys*, *T. polium*, and *T. scordium* are compiled in Table 25.5.

According to a review of the literature, *Teucrium* species, *T. chamaedrys*, *T. polium*, and *T. scordium*, are common plants with a variety of traditional uses. Otherwise, there is no experimental proof enough that they are safe herbal medicines. As a result, it is important to emphasize that biological testing and clinical trials have been conducted to confirm their potential medical benefits.

TABLE 25.4
Description of the Medicinal Parts and Their Dosages of *T. chamaedrys*, *T. polium*, and *T. scordium*

Data	*T. chamaedrys*	*T. polium*	*T. scordium*
Medicinal parts	the herb collected during the flowering season.	the whole herb.	the herb harvested during or shortly before the flowering season and the fresh flowering herb.
Dosage (Duke, 2002)	liquid herb extract (2–4 ml)	herb/cup tea (1.5 g)	herb/cup (4 teaspoon, ~7.2 g); liquid herb extract (2–4 ml)
Dosage: Mode of Administration (Gruenwald et al., 2007)	Used in tea mixtures on occasionally. Toxic symptoms can occur at doses of more than 600 mg per day.	Internal usage of cut drug and liquid extract; 1.5 gm drug per cup in a single dosage for infusion	Internally and externally, the same preparation can be applied. As an infusion, four teaspoonful of the herb (7.2 grams) are orally administered daily.
Effects (Gruenwald et al., 2007)	The drug, which includes a lot of strong amaroids, is said to have a cholagogic effect, but this hasn't been verified scientifically. As a result, the poisonous principle is unknown. Hepatitis-like symptoms, including liver cell necrosis, are caused by higher doses or poisoning.	he drug's antidiabetic and anti-ulcer efficacy has yet to be proven in comprehensive clinical trials. In animal studies, a decline in the ulcer index was observed, as well as a significant decrease in blood sugar levels following the intravenous injection of a 4 percent decoction of the dried plant. In addition, the drug has antimicrobial, antipyretic, and maybe anti-edema and antiexudative properties.	Reported as other *Teucrium* species
Indications And Usage Contraindications (Gruenwald et al., 2007)	Germander is used as a digestive support, a gout rinse, a weight-loss aid, and a fever reducer. The drug is extremely dangerous and should not be used.	Poley is utilized for diabetes in Israel, gastrointestinal symptoms in North Africa, fever in Italy, and as a vulnerary in Spain.	The plant is used to heal festering and inflammatory wounds, respiratory problems, diarrhea, fever, hemorrhoids, and intestinal parasites.
Precautions And Adverse Reactions (Gruenwald et al., 2007)	Following ingestion of the medication, liver cell necrosis has been seen. Jaundice and a high level of aminotransferase in the blood are two symptoms. There has been one death reported. As a result, the medicine will not be given.	There are no known health risks associated with the correct administration of prescribed therapeutic amounts.	There are no known health risks or side effects associated with the correct administration of prescribed therapeutic amounts.

TABLE 25.5
Biological Properties of *T. chamaedrys*, *T. polium*, and *T. scordium* in the Literature

Species	Bioassays	Biological effects	Reference
T. polium	*In vivo*	Hypoglycemic effect	(Afifi et al., 2005; Alkofahi et al., 2017; Ardestani et al., 2008; Asghari et al., 2020; Esmaeili & Yazdanparast, 2004; Gharaibeh et al., 1988; Hasanein & Shahidi, 2012; Ireng et al., 2016; Monfared & Pournourmohammadi, 2010; Qujeq et al., 2013; Shahraki et al., 2007; Tatar et al., 2012; Yazdanparast et al., 2005; Zal et al., 2001)
		Antioxidant activity	(Ardestani et al., 2008; Baali et al., 2016; Esmaeili et al., 2009; Hasani et al., 2007; Krache et al., 2018; Mahmoudabady et al., 2018; Matsuda et al., 2005; Nargesi, 2018; Salimnejad et al., 2017; Sharififar et al., 2009)
		Cardiac function	(Mahmoudabady et al., 2018)
		Anti-inflammatory activity	(Aghazadeh & Yazdanparast, 2010; Krache et al., 2018; Rahmouni et al., 2017; Tariq et al., 1989)
		Antinociceptive and antispasmodic activity	(Abdollahi et al., 2003; Baluchnejadmojarad et al., 2005; Parsaee & Shafiee-Nick, 2006; Zendehdel et al., 2011)
		Hepatoprotective effect	(Aghazadeh & Yazdanparast, 2010; Amini et al., 2009; Amini & Yazdanparast, 2009; Amini et al., 2011; Bahramikia et al., 2009; Forouzandeh et al., 2013; Khleifat et al., 2002; Mehdineya et al., 2013; Panovska et al., 2007)
		Hypolipidemic effect	(Nargesi, 2018; Niazmand et al., 2007; Rasekh et al., 2001; Shahraki et al., 2007)
		Anticancer activity	(Movahedi et al., 2014)
		Anticonvulsant activity	(Khoshnood-Mansoorkhani et al., 2010)
		Antiamnesic activity	(Hasanein & Shahidi, 2012; Mousavi et al., 2015; Orhan & Aslan, 2009; Simonyan et al., 2019)
		Hypotensive effect	(Dehghani et al., 2005; Suleiman et al., 1988)
		Anti-ulcerogenic effect	(Alkofahi & Atta, 1999; Fallah Huseini et al., 2020; Mehrabani et al., 2009; Rasik & Shukla, 2000; Twaij et al., 1987)
		Antimicrobial activity	(Capasso et al., 1984)
		Wound healing	(Meguellati et al., 2019)
		Analgesic effect	(Verdi et al., 2010)
		Vascular effect	(Feridoni et al., 2012; Niazmand et al., 2007)
		Effects on renal disorders	(Rafieian-Kopaei & Baradaran, 2013)
		Anti-hemolytic and antihyperuricemic effects	(Krache et al., 2018)
		Toxic effects on serum levels of liver enzymes, lipoproteins and blood sugar	(Shahraki et al., 2006)
	In vitro	Antioxidant activity	(Al-Mustafa & Al-Thunibat, 2008; Alali & Khazem, 2019; Ali et al., 2018; Atki et al., 2019; Chioibas et al., 2019; Couladis et al., 2003; Djeridane et al., 2006; Y El Atki et al., 2020a; El Atki et al., 2019; Elmasri et al., 2016; Ghara & Ghadi, 2016; Gülçin et al., 2003; Kerbouche et al., 2015; Krache et al., 2018; Ljubuncic et al., 2005; Ljubuncic et al., 2006; Malki et al., 2017; Noumi et al., 2020; Orhan & Aslan, 2009; Panovska et al., 2005; Qabaha et al., 2021; Qabaha, 2013; Yazdanparast & Ardestani, 2009)

(*Continued*)

TABLE 25.5 (CONTINUED)
Biological Properties of *T. chamaedrys*, *T. polium*, and *T. scordium* in the Literature

Species	Bioassays	Biological effects	Reference
		Anti-inflammatory activity	(Qabaha et al., 2021)
		Cytotoxic activity	(Alachkar & Al Moustafa, 2011; Alreshidi et al., 2020; Kandouz et al., 2010; Khodaei et al., 2018; Line, 2007; Ljubuncic et al., 2005; Nematollahi-Mahani et al., 2007; Nematollahi-Mahani et al., 2012; Rajabalian, 2008; Stankovic et al., 2011)
		Antidiabetic activity	(Kadan et al., 2018)
		Anticancer activity	(Al-Hamwi et al., 2021; Al-Shalabi et al., 2020; Elmasri et al., 2016; Hashemi et al., 2020; Noumi et al., 2020; Sheikhbahaei et al., 2018)
		Antimicrobial activity	(Aggelis et al., 1998; Ali et al., 2018; Alreshidi et al., 2020; Bakari et al., 2015; Basudan & Abu-Gabal, 2018; Chioibas et al., 2019; Essawi & Srour, 2000; Ghojavand et al., 2020; Gülçin et al., 2003; Jaradat, 2015; Kerbouche et al., 2015; Kunduhoglu et al., 2011; Malki et al., 2017; Mosadegh et al., 2002; Motamedi et al., 2015; Othman et al., 2017; Rojas et al., 1992; Sarac & Uğur, 2007; Tabatabaei Yazdi et al., 2014)
		Anti-quorum-sensing	(Alreshidi et al., 2020)
		Antiviral activity	(Alwan et al., 1988)
		Anticholinesterase activity	(Orhan & Aslan, 2009)
		Antimutagenic activity	(Khader et al., 2010; Khader et al., 2007)
		Scolicidal activity	(Al-Nakeeb et al., 2015)
	Clinical trial	Antimicrobial activity	(Akbarzdeh et al., 2019; Tusi et al., 2020)
		Effects on hepatic disorders	(Laliberte & Villeneuve, 1996; Mattéi et al., 1995; Mazokopakis et al., 2004; Polymeros et al., 2002; Savvidou et al., 2007)
T. chamaedrys	*In vitro*	Antiproliferative and cytotoxic activity	(Gecibesler et al., 2019; Milutinović et al., 2019; STANKOVIĆ et al., 2015)
		Antimicrobial activity	(Ceyhan et al., 2012; Gursoy & Tepe, 2009; Haziri et al., 2017; Kunduhoglu et al., 2011; Stanković et al., 2012; Vlase et al., 2014)
		Antioxidant activity	(Antognoni et al., 2012; Gursoy & Tepe, 2009; Nencini et al., 2014; Oalđe et al., 2020; Piccolella et al., 2021; Stankovic et al., 2010; Vlase et al., 2014; Zlatić et al., 2017)
		Antiherpes and cytotoxic activity	(Todorov et al., 2015)
		Immunomodulatory activity	(Prescott et al., 2011)
		Amoebicidal activity	(Tepe et al., 2012)
	In vivo	Hepatotoxicity	(Kouzi et al., 1994)
T. scordium	*In vitro*	Antimicrobial activity	(Bozov et al., 2020; Ebrahimi Anaraki et al., 2020; Stanković et al., 2012)
		Cytotoxic activity	(STANKOVIĆ et al., 2015)
		Antioxidant activity	(Rahimifard et al., 2014)

REFERENCES

Abdollahi, M., Karimpour, H., & Monsef-Esfehani, H. R. (2003). Antinociceptive effects of Teucrium polium L. total extract and essential oil in mouse writhing test. *Pharmacological Research*, *48*(1), 31–35.

Afifi, F., Al-Khalidi, B., & Khalil, E. (2005). Studies on the in vivo hypoglycemic activities of two medicinal plants used in the treatment of diabetes in Jordanian traditional medicine following intranasal administration. *Journal of Ethnopharmacology*, *100*(3), 314–318.

Aggelis, G., Athanassopoulos, N., Paliogianni, A., & Komaitis, M. (1998). Effect of a Teucrium polium L. extract on the growth and fatty acid composition of Saccharomyces cerevisiae and Yarrowia lipolytica. *Antonie van Leeuwenhoek*, *73*(2), 195–198.

Aghazadeh, S., & Yazdanparast, R. (2010). Inhibition of JNK along with activation of ERK1/2 MAPK pathways improve steatohepatitis among the rats. *Clinical Nutrition*, *29*(3), 381–385.

Akbarzdeh, M., Bonyadpour, B., Pakshir, K., & Mohagheghzadeh, A. A. (2019). Comparative investigation of the sensitivity of Candida fungi isolated from vulvovaginal candidiasis to nystatin and Teucrium polium smoke product. *International Journal of Women's Health and Reproduction Science*, *7*, 508–514.

Akgul, A., Akgul, A., Senol, S. G., Yildirim, H., Secmen, O., & Dogan, Y. (2018). An ethnobotanical study in Midyat (Turkey), a city on the silk road where cultures meet. *Journal of Ethnobiology and Ethnomedicine*, *14*(1), 1–18.

Akin, M., Oguz, D., & Saracoglu, H. (2010). Antibacterial activity of essential oil from Thymbra spicata var. spicata L. and Teucrium polium (Stapf Brig.). *Interventions*, *8*(9), 53–58.

Aksoy, A., Özcan, T., Girişken, H., Celik, J., & Dirmenci, T. (2020). A phylogenetic analysis and biogeographical distribution of Teucrium Sect. Teucrium (Lamiaceae) and taxonomic notes for a new species from southwest Turkey. *Turkish Journal of Botany*, *44*(3), 322–337.

Al-Hamwi, M., Aboul-Ela, M., El-Lakany, A., & Nasreddine, S. (2021). Anticancer activity of Micromeria fruticosa and Teucrium polium growing in Lebanon. *Pharmacognosy Journal*, *13*(1).

Al-Mustafa, A. H., & Al-Thunibat, O. Y. (2008). Antioxidant activity of some Jordanian medicinal plants used traditionally for treatment of diabetes. *Pakistan Journal of Biological Sciences*, *11*(3), 351–358.

Al-Nakeeb, S. A., Al-Taae, A.-R. A., & Kadir, M. A. (2015). Evaluation of scolicidal effect of Teucrium polium, Zingiber officinale and Nigella sativa in-vitro on Echinococcus granulosus. *Pharmacie Globale*, *6*(2), 1.

Al-Shalabi, E., Alkhaldi, M., & Sunoqrot, S. (2020). Development and evaluation of polymeric nanocapsules for cirsiliol isolated from Jordanian Teucrium polium L. as a potential anticancer nanomedicine. *Journal of Drug Delivery Science and Technology*, *56*, 101544.

Alachkar, A., & Al Moustafa, A.-E. (2011). Teucrium polium plant extract provokes significant cell death in human lung cancer cells. *Health*, *3*(6), 366.

Alali, R. G., & Khazem, M. R. (2019). Total phenolic content and antioxidant activity of two Teucrium species from Syria. *Journal of Pharmacy and Nutrition Sciences*, *9*(6).

Ali, F., Jan, A. K., Khan, N. M., Ali, R., Mukhtiar, M., Khan, S., Khan, S. A., & Aziz, R. (2018). Selective biological activities and phytochemical profiling of two wild plant species, Teucrium polium and Capsicum annum from Sheringal, Pakistan. *Chiang Mai Journal of Science*, *45*, 881–887.

Alkofahi, A., & Atta, A. (1999). Pharmacological screening of the anti-ulcerogenic effects of some Jordanian medicinal plants in rats. *Journal of Ethnopharmacology*, *67*(3), 341–345.

Alkofahi, A. S., Abdul-Razzak, K. K., Alzoubi, K. H., & Khabour, O. F. (2017). Screening of the anti-hyperglycemic activity of some medicinal plants of Jordan. *Pakistan Journal of Pharmaceutical Sciences*, *30*(3).

Alreshidi, M., Noumi, E., Bouslama, L., Ceylan, O., Veettil, V. N., Adnan, M., Danciu, C., Elkahoui, S., Badraoui, R., & Al-Motair, K. A. (2020). Phytochemical screening, antibacterial, antifungal, antiviral, cytotoxic, and anti-quorum-sensing properties of Teucrium polium l. aerial parts methanolic extract. *Plants*, *9*(11), 1418.

Alwan, A., Jawad, A.-L. M., Albana, A., & Ali, K. (1988). Antiviral activity of some Iraqi indigenous plants. *International Journal of Crude Drug Research*, *26*(2), 107–111.

Amini, R., Nosrati, N., Yazdanparast, R., & Molaei, M. (2009). Teucrium polium in prevention of steatohepatitis in rats. *Liver International*, *29*(8), 1216–1221.

Amini, R., & Yazdanparast, R. (2009). Suppression of hepatic TNF-α and TGF-β gene expressions in rats with induced nonalcoholic steatohepatitis. *Pharmacology Online*, *3*, 340–350.

Amini, R., Yazdanparast, R., Aghazadeh, S., & Ghaffari, S. H. (2011). Teucrium polium reversed the MCD diet-induced liver injury in rats. *Human & Experimental Toxicology*, *30*(9), 1303–1312.

Antognoni, F., Iannello, C., Mandrone, M., Scognamiglio, M., Fiorentino, A., Giovannini, P. P., & Poli, F. (2012). Elicited Teucrium chamaedrys cell cultures produce high amounts of teucrioside, but not the hepatotoxic neo-clerodane diterpenoids. *Phytochemistry*, *81*, 50–59.

Ardestani, A., Yazdanparast, R., & Jamshidi, S. (2008). Therapeutic effects of Teucrium polium extract on oxidative stress in pancreas of streptozotocin-induced diabetic rats. *Journal of Medicinal Food*, *11*(3), 525–532.

Asghari, A. A., Mokhtari-Zaer, A., Niazmand, S., Mc Entee, K., & Mahmoudabady, M. (2020). Anti-diabetic properties and bioactive compounds of Teucrium polium L. *Asian Pacific Journal of Tropical Biomedicine*, *10*(10), 433.

Atki, Y., Aouam, I., Kamari, F., Taroq, A., Lyoussi, B., & Abdellaoui, A. (2019). Antioxidant activity of two wild teucrium species from Morocco. *International Journal of Pharmaceutical Sciences and Research*, *10*(6), 2723–2739.

Baali, N., Belloum, Z., Baali, S., Chabi, B., Pessemesse, L., Fouret, G., Ameddah, S., Benayache, F., Benayache, S., & Feillet-Coudray, C. (2016). Protective activity of total polyphenols from Genista quadriflora Munby and Teucrium polium geyrii Maire in acetaminophen-induced hepatotoxicity in rats. *Nutrients*, *8*(4), 193.

Bagci, E., Yazgın, A., Hayta, S., & Cakılcıoglu, U. (2010). Composition of the essential oil of Teucrium chamaedrys L.(Lamiaceae) from Turkey. *Journal of Medicinal Plants Research*, *4*(23), 2588–2590.

Bahramikia, S., Ardestani, A., & Yazdanparast, R. (2009). Protective effects of four Iranian medicinal plants against free radical-mediated protein oxidation. *Food Chemistry*, *115*(1), 37–42.

Bahramikia, S., & Yazdanparast, R. (2012). Phytochemistry and medicinal properties of Teucrium polium L.(Lamiaceae). *Phytotherapy Research*, *26*(11), 1581–1593.

Bakari, S., Ncir, M., Felhi, S., Hajlaoui, H., Saoudi, M., Gharsallah, N., & Kadri, A. (2015). Chemical composition and in vitro evaluation of total phenolic, flavonoid, and antioxidant properties of essential oil and solvent extract from the aerial parts of Teucrium polium grown in Tunisia. *Food Science and Biotechnology*, *24*(6), 1943–1949.

Baluchnejadmojarad, T., Roghani, M., & Roghani-Dehkordi, F. (2005). Antinociceptive effect of Teucrium polium leaf extract in the diabetic rat formalin test. *Journal of Ethnopharmacology*, *97*(2), 207–210. https://doi.org/10.1016/j.jep.2004.10.030

Basudan, N., & Abu-Gabal, N. (2018). Phytochemistry and biological properties investigation of Teucrium polium L. *International Journal of Pharmacy and Biological Sciences*, *8*, 660–670.

Bedir, E., Tasdemir, D., Çalis, I., Zerbe, O., & Sticher, O. (1999). Neo-clerodane diterpenoids from Teucrium polium. *Phytochemistry*, *51*(7), 921–925.

Belmekki, N., Bendimerad, N., & Bekhechi, C. (2013). Chemical analysis and antimicrobial activity of Teucrium polium L. essential oil from Western Algeria. *Journal of Medicinal Plants Research*, *7*(14), 897–902.

Benali, T., Habbadi, K., Bouyahya, A., Khabbach, A., Marmouzi, I., Aanniz, T., Chtibi, H., Mrabti, H. N., Achbani, E. H., & Hammani, K. (2021). Phytochemical analysis and study of antioxidant, anticandidal, and antibacterial activities of Teucrium polium subsp. polium and Micromeria graeca (Lamiaceae) essential oils from Northern Morocco. *Evidence-Based Complementary and Alternative Medicine*, *2021*.

Boukhebti, H., Ramdani, M., Lasmi, I., Katfi, F., Chaker, A. N., & Lograda, T. (2019). Chemical composition, antibacterial activity, and anatomical study of teucrium Polium L. *Asian Journal of Pharmaceutical and Clinical Research*, *12*(6), 337–341.

Boulila, A., Béjaoui, A., Messaoud, C., & Boussaid, M. (2008). Variation of volatiles in Tunisian populations of Teucrium polium L.(Lamiaceae). *Chemistry & Biodiversity*, *5*(7), 1389–1400.

Bozov, P. I., & Penchev, P. N. (2019). Neo-clerodane diterpenoids from Teucrium polium subsp. vincentinum (rouy) D. Wood. *Phytochemistry Letters*, *31*, 237–241.

Bozov, P. I., Penchev, P. N., Girova, T. D., & Gochev, V. K. (2020). Diterpenoid constituents of Teucrium scordium L. Subsp. s cordioides (Shreb.) Maire Et Petitmengin. *Natural Product Communications*, *15*(9), 1934578X20959525.

Bulut, G., Doğan, A., Şenkardeş, İ., Avci, R., & Tuzlaci, E. (2019). The medicinal and wild food plants of Batman City and Kozluk district (Batman-Turkey). *Agriculturae Conspectus Scientificus*, *84*(1), 29–36.

Bulut, G., Haznedaroğlu, M. Z., Doğan, A., Koyu, H., & Tuzlacı, E. (2017). An ethnobotanical study of medicinal plants in Acipayam (Denizli-Turkey). *Journal of Herbal Medicine*, *10*, 64–81.

Bulut, G., & Tuzlacı, E. (2015). An ethnobotanical study of medicinal plants in Bayramiç (Çanakkale-Turkey). *Marmara Pharmaceutical Journal*, *19*(1), 269–282.

Cakilcioglu, U., Khatun, S., Turkoglu, I., & Hayta, S. (2011). Ethnopharmacological survey of medicinal plants in Maden (Elazig-Turkey). *Journal of Ethnopharmacology*, *137*(1), 469–486.

Cakilcioglu, U., Sengun, M., & Turkoglu, I. (2010). An ethnobotanical survey of medicinal plants of Yazikonak and Yurtbasi districts of Elazig province, Turkey. *Journal of Medicinal Plants Research*, *4*(7), 567–572.

Cakir, A., Duru, M. E., Harmandar, M., Ciriminna, R., & Passannanti, S. (1998). Volatile constituents of Teucrium polium L. from Turkey. *Journal of Essential Oil Research, 10*(1), 113–115.

Çalis, I., Bedir, E., Wright, A. D., & Sticher, O. (1996). Neoclerodane diterpenoids from Teucrium chamaedrys subsp. syspirense. *Journal of Natural Products, 59*(4), 457–460.

Candela, R. G., Rosselli, S., Bruno, M., & Fontana, G. (2020). A review of the phytochemistry, traditional uses and biological activities of the essential oils of Genus Teucrium. *Planta Medica.*

Capasso, F., De Fusco, R., Fasulo, M., Lembo, M., Mascolo, N., & Menghini, A. (1984). Antipyretic and antibacterial actions of Teucrium polium (L.). *Pharmacological Research Communications, 16*(1), 21–29.

Ceyhan, N., Keskin, D., & Ugur, A. (2012). Antimicrobial activities of different extracts of eight plant species from four different family against some pathogenic microoorganisms. *Journal of Food, Agriculture and Environment, 10*, 193–197.

Chabane, S., Boudjelal, A., Napoli, E., Benkhaled, A., & Ruberto, G. (2021). Phytochemical composition, antioxidant and wound healing activities of Teucrium polium subsp. capitatum (L.) Briq. essential oil. *Journal of Essential Oil Research, 33*(2), 143–151.

Charters, M. L. (2021). *California plant names: Latin and Greek meanings and derivations, a dictionary of botanical and biographical etymology.* Retrieved July 11 from http://www.calflora.net/botanicalnames/pageT.html

Chioibas, R., Susan, R., Susan, M., Mederle, O., Vaduva, D. B., Radulescu, M., Berceanu, M., Danciu, C., Khaled, Z., & Draghici, G. (2019). Antimicrobial activity exerted by total extracts of germander. *Revista de Chimie, 70*(9), 3242–3244.

Couladis, M., Tzakou, O., Verykokidou, E., & Harvala, C. (2003). Screening of some Greek aromatic plants for antioxidant activity. *Phytotherapy Research: An International Journal Devoted to Pharmacological and Toxicological Evaluation of Natural Product Derivatives, 17*(2), 194–195.

D'Abrosca, B., Pacifico, S., Scognamiglio, M., D'Angelo, G., Galasso, S., Monaco, P., & Fiorentino, A. (2013). A new acylated flavone glycoside with antioxidant and radical scavenging activities from Teucrium polium leaves. *Natural Product Research, 27*(4–5), 356–363.

Davis, P. H. (1982). *Flora of Turkey and the east Aegean Islands. Vol. 7.* Edinburgh University Press.

De Marino, S., Festa, C., Zollo, F., Incollingo, F., Raimo, G., Evangelista, G., & Iorizzi, M. (2012). Antioxidant activity of phenolic and phenylethanoid glycosides from Teucrium polium L. *Food Chemistry, 133*(1), 21–28.

Dehghani, F., Khozani, T. T., Panjehshahin, M., & Karbalaedoost, S. (2005). Effect of Teucrium polium on histology and histochemistry in rat stomach. *Indian Journal of Gastroenterology, 24*(3), 126.

Demirci, S., & Özhatay, N. (2012). An ethnobotanical study in Kahramanmaraş (Turkey); wild plants used for medicinal purpose in Andirin, Kahramanmaraş. *Turkish Journal of Pharmaceutical Sciences, 9*(1), 75–92.

Dinç, M., & Doğu, S. (2016). Teucrium pruinosum var. aksarayense var. nov.(Lamiaceae) from Central Anatolia, Turkey. *Modern Phytomorphology, 9*, 13–17.

Djeridane, A., Yousfi, M., Nadjemi, B., Boutassouna, D., Stocker, P., & Vidal, N. (2006). Antioxidant activity of some Algerian medicinal plants extracts containing phenolic compounds. *Food Chemistry, 97*(4), 654–660.

Duke, J. A. (2002). *Handbook of medicinal herbs.* CRC Press.

Ebrahimi Anaraki, P., Abocc-Mehrizi, F., Dehghani Ashkezary, M., & Sedighi, S. (2020). Chemical composition and biological activities of essential oil and methanol extract of Teucrium scordium. *Iranian Journal of Chemistry and Chemical Engineering, 39*(2).

El Atki, Y., Aouam, I., El Kamari, F., Taroq, A., Lyoussi, B., Oumokhtar, B., & Abdellaoui, A. (2020a). Phytochemistry, antioxidant and antibacterial activities of two Moroccan Teucrium polium L. subspecies: Preventive approach against nosocomial infections. *Arabian Journal of Chemistry, 13*(2), 3866–3874.

El Atki, Y., Aouam, I., El Kamari, F., Taroq, A., Zejli, H., Taleb, M., Lyoussi, B., & Abdellaoui, A. (2020b). Antioxidant activities, total phenol an flavonoid contents of two Teucrium polium subspecies extracts. *Moroccan Journal of Chemistry, 8*(2), 2446–2455.

El Atki, Y., Aouam, I., Taroq, A., Lyoussi, B., Taleb, M., & Abdellaoui, A. (2019). Total phenolic and flavonoid contents and antioxidant activities of extracts from Teucrium polium growing wild in Morocco. *Materials Today: Proceedings, 13*, 777–783.

Elmasri, W. A., Hegazy, M.-E. F., Aziz, M., Koksal, E., Amor, W., Mechref, Y., Hamood, A. N., Cordes, D. B., & Paré, P. W. (2014). Biofilm blocking sesquiterpenes from Teucrium polium. *Phytochemistry, 103*, 107–113.

Elmasri, W. A., Hegazy, M.-E. F., Mechref, Y., & Paré, P. W. (2015). Cytotoxic saponin poliusaposide from Teucrium polium. *RSC Advances*, *5*(34), 27126–27133.

Elmasri, W. A., Hegazy, M.-E. F., Mechref, Y., & Paré, P. W. (2016). Structure-antioxidant and anti-tumor activity of Teucrium polium phytochemicals. *Phytochemistry Letters*, *15*, 81–87.

Elmasri, W. A., Yang, T., Tran, P., Hegazy, M.-E. F., Hamood, A. N., Mechref, Y., & Paré, P. W. (2015). Teucrium polium phenylethanol and iridoid glycoside characterization and flavonoid inhibition of biofilm-forming Staphylococcus aureus. *Journal of Natural Products*, *78*(1), 2–9.

Elmastas, M., Erenler, R., Isnac, B., Aksit, H., Sen, O., Genc, N., & Demirtas, I. (2016). Isolation and identification of a new neo-clerodane diterpenoid from Teucrium chamaedrys L. *Natural Product Research*, *30*(3), 299–304.

Erbay, M. Ş., & Sarı, A. (2018). Plants used in traditional treatment against hemorrhoids in Turkey. *Marmara Pharmaceutical Journal*, *22*(2), 110–132.

Esmaeili, M. A., & Yazdanparast, R. (2004). Hypoglycaemic effect of Teucrium polium: Studies with rat pancreatic islets. *Journal of Ethnopharmacology*, *95*(1), 27–30.

Esmaeili, M. A., Zohari, F., & Sadeghi, H. (2009). Antioxidant and protective effects of major flavonoids from Teucrium polium on β-cell destruction in a model of streptozotocin-induced diabetes. *Planta Medica*, *75*(13), 1418–1420.

Essawi, T., & Srour, M. (2000). Screening of some Palestinian medicinal plants for antibacterial activity. *Journal of Ethnopharmacology*, *70*(3), 343–349.

Fallah Huseini, H., Abdolghaffari, A. H., Ahwazi, M., Jasemi, E., Yaghoobi, M., & Ziaee, M. (2020). Topical application of Teucrium polium can improve wound healing in diabetic rats. *The International Journal of Lower Extremity Wounds*, *19*(2), 132–138.

Farahbakhsh, J., Najafian, S., Hosseinifarahi, M., & Gholipour, S. (2020). The effect of time and temperature on shelf life of essential oil composition of Teucrium polium L. *Natural Product Research*, 1–5.

Farahbakhsh, J., Najafian, S., Hosseinifarahi, M., & Gholipour, S. (2021). Essential Oil composition and phytochemical properties from leaves of Felty Germander (Teucrium polium L.) and Spearmint (Mentha spicata L.). *Journal of Essential Oil Bearing Plants*, *24*(1), 147–159.

Feridoni, E., Niazmand, S., Harandizadeh, F., Hosseini, S., & Mahmodabadi, M. (2012). Vasorelaxant effect of hydroalcoholic extract of Teucrium Polium L. on isolated rat aorta.

Fiorentino, A., D'Abrosca, B., Ricci, A., Pacifico, S., Piccolella, S., & Monaco, P. (2009). Structure determination of chamaedryosides A—C, three novel nor-neo-clerodane glucosides from Teucrium chamaedrys, by NMR spectroscopy. *Magnetic Resonance in Chemistry*, *47*(11), 1007–1012.

Fiorentino, A., D'Abrosca, B., Pacifico, S., Scognamiglio, M., D'Angelo, G., Gallicchio, M., Chambery, A., & Monaco, P. (2011). Structure elucidation and hepatotoxicity evaluation against HepG2 human cells of neo-clerodane diterpenes from Teucrium polium L. *Phytochemistry*, *72*(16), 2037–2044.

Fiorentino, A., D'Abrosca, B., Pacifico, S., Scognamiglio, M., D'Angelo, G., & Monaco, P. (2010). Abeo-Abietanes from Teucrium polium roots as protective factors against oxidative stress. *Bioorganic & Medicinal Chemistry*, *18*(24), 8530–8536.

Forouzandeh, H., Azemi, M. E., Rashidi, I., Goudarzi, M., & Kalantari, H. (2013). Study of the protective effect of Teucrium polium L. extract on acetaminophen-induced hepatotoxicity in mice. *Iranian Journal of Pharmaceutical Research: IJPR*, *12*(1), 123.

Frezza, C., Venditti, A., Matrone, G., Serafini, I., Foddai, S., Bianco, A., & Serafini, M. (2018). Iridoid glycosides and polyphenolic compounds from Teucrium chamaedrys L. *Natural Product Research*, *32*(13), 1583–1589.

Gagliano Candela, R., Ilardi, V., Badalamenti, N., Bruno, M., Rosselli, S., & Maggi, F. (2020). Essential oil compositions of Teucrium fruticans, T. scordium subsp. scordioides and T. siculum growing in Sicily and Malta. *Natural Product Research*, 1–10.

Gecibesler, I. H., Demirtas, I., Koldas, S., Behcet, L., Gul, F., & Altun, M. (2019). Bioactivity-guided isolation of compounds with antiproliferative activity from Teucrium chamaedrys L. subsp. sinuatum (Celak.) Rech. f.

Ghara, A. R., & Ghadi, F. E. (2016). Effect of Extraction Solvent/technique on the Antioxidant activity of Selected 10 herb native species south of Iran. *Advances in Bioresearch*, *7*(4).

Gharaibeh, M. N., Elayan, H. H., & Salhab, A. S. (1988). Hypoglycemic effects of Teucrium polium. *Journal of Ethnopharmacology*, *24*(1), 93–99.

Ghojavand, S., Madani, M., & Karimi, J. (2020). Green synthesis, characterization and antifungal activity of silver nanoparticles using stems and flowers of felty germander. *Journal of Inorganic and Organometallic Polymers and Materials*, *30*(8), 2987–2997.

Gori, L., Galluzzi, P., Mascherini, V., Gallo, E., Lapi, F., Menniti-Ippolito, F., Raschetti, R., Mugelli, A., Vannacci, A., & Firenzuoli, F. (2011). Two contemporary cases of hepatitis associated with Teucrium chamaedrys L. decoction use. Case reports and review of literature. *Basic & Clinical Pharmacology & Toxicology*, *109*(6), 521–526.

Gruenwald, J., Brendler, T., & Jaenicke, C. (2007). *PDR for herbal medicines*. Thomson, Reuters.

Gursoy, N., & Tepe, B. (2009). Determination of the antimicrobial and antioxidative properties and total phenolics of two "endemic" Lamiaceae species from Turkey: Ballota rotundifolia L. and Teucrium chamaedrys C. Koch. *Plant Foods for Human Nutrition*, *64*(2), 135–140.

Gülçin, I., Uguz, M., Oktay, M., Beydemir, S., & Küfrevioglu, Ö. (2003). Antioxidant and antimicrobial activities of Teucrium polium. L. *Journal of Food Technology*, *1*, 9–17.

Güner, A., & Aslan, S. (2012). *Türkiye bitkileri listesi:(damarlı bitkiler)*. Nezahat Gökyiğit Botanik Bahçesi Yayınları.

Güner, A., Özhatay, N., Ekim, T., & Başer, K. H. C. (2000). *Flora of Turkey and the east Aegean Islands* (Vol. 11).

Güneş, S., Savran, A., Paksoy, M. Y., Koşar, M., & Çakılcıoğlu, U. (2017). Ethnopharmacological survey of medicinal plants in Karaisalı and its surrounding (Adana-Turkey). *Journal of Herbal Medicine*, *8*, 68–75.

Hachicha, S., Barrek, S., Skanji, T., Zarrouk, H., & Ghrabi, Z. (2009). Fatty acid, tocopherol, and sterol content of three Teucrium species from Tunisia. *Chemistry of Natural Compounds*, *45*(3), 304–308.

Harley, R., Atkins, S., Budantsev, A., Cantino, P., Conn, B., Grayer, R., Harley, M., de Kok, R., Krestovskaja, T., & Morales, R. (2004). Labiatae. In: J. W. Kaderit (Ed.), Kubitzki, K. (Ed. in chief). *The families and genera of vascular plants*, vol 7. *Flowering plants. Dycotiledons: Lamiales (except Acanthaceae including Avicenniaceae)*. Berlin: Springer.

Hasanein, P., & Shahidi, S. (2012). Preventive effect of Teucrium polium on learning and memory deficits in diabetic rats. *Medical Science Monitor: International Medical Journal of Experimental and Clinical Research*, *18*(1), BR41.

Hasani, P., Yasa, N., Vosough-Ghanbari, S., Mohammadirad, A., Dehghan, G., & Abdollahi, M. (2007). In vivo antioxidant potential of Teucrium polium, as compared to α-tocopherol. *Acta Pharmaceutica*, *57*(1), 123–129.

Hashemi, S. F., Tasharrofi, N., & Saber, M. M. (2020). Green synthesis of silver nanoparticles using Teucrium polium leaf extract and assessment of their antitumor effects against MNK45 human gastric cancer cell line. *Journal of Molecular Structure*, *1208*, 127889.

Hayta, Ş., Yazgın, A., & Bağcı, E. (2017). Constituents of the volatile oils of two teucrium species from Turkey. *Bitlis Eren University Journal of Science & Technology*, *7*(2).

Haziri, A., Faiku, F., Berisha, R., Mehmeti, I., Govori, S., & Haziri, I. (2017). Evaluation of antibacterial activity of different solvent extracts of Teucrium chamaedys (L.) growing wild in Kosovo. *Chemistry*, *26*, 431–441.

Ireng, A., Helmerhorst, E., Parsons, R., & Caccetta, R. (2016). Teucrium polium significantly lowers blood glucose levels acutely in normoglycemic male Wistar rats: A comparative to insulin and metformin. *Advancement in Medicinal Plant Research*, *4*(1), 1–10.

Jaradat, N. A. (2015). Review of the taxonomy, ethnobotany, phytochemistry, phytotherapy and phytotoxicity of germander plant (Teucrium polium L.). *Asian Journal of Pharmaceutical and Clinical Research*, 13–19.

Kadan, S., Sasson, Y., Abu-Rcziq, R., Saad, B., Benvalid, S., Linn, T., Cohen, G., & Zaid, H. (2018). Teucrium polium extracts stimulate GLUT4 translocation to the plasma membrane in L6 muscle cells. *Advancement in Medicinal Plant Research*, *6*(1), 1–8.

Kamel, A. (1995). 7-epi-Eudesmanes from Teucrium polium. *Journal of Natural Products*, *58*(3), 428–431.

Kandouz, M., Alachkar, A., Zhang, L., Dekhil, H., Chehna, F., Yasmeen, A., & Al Moustafa, A.-E. (2010). Teucrium polium plant extract inhibits cell invasion and motility of human prostate cancer cells via the restoration of the E-cadherin/catenin complex. *Journal of Ethnopharmacology*, *129*(3), 410–415.

Kaya, A., Demirci, B., & Başer, K. H. C. (2009). Compositions of essential oils and Trichomes of Teucrium chamaedrys L. subsp. trapezunticum Rech. fil. and subsp. syspirense (C. Koch) Rech. fil. *Chemistry & Biodiversity*, *6*(1), 96–104.

Ke, P., Qian-jun, Z., Ming-yue, M., Qing, C., & Rong-jun, Y. (2009). Study on chemical constituents and bioactivities of plants from Teucrium. *Natural Product Research & Development*, *21*(5).

Kerbouche, L., Hazzit, M., Ferhat, M.-A., Baaliouamer, A., & Miguel, M. G. (2015). Biological activities of essential oils and ethanol extracts of Teucrium polium subsp. capitatum (L.) Briq. and Origanum floribundum Munby. *Journal of Essential Oil Bearing Plants*, *18*(5), 1197–1208.

Kew and Missouri botanical garden. (2021). Royal Botanic Gardens. Retrieved July 7 from http://www.theplantlist.org/tpl1.1/search?q=Teucrium

Khader, M., Bresgen, N., & Eckl, P. (2010). Antimutagenic effects of ethanolic extracts from selected Palestinian medicinal plants. *Journal of Ethnopharmacology*, *127*(2), 319–324.

Khader, M., Eckl, P., & Bresgen, N. (2007). Effects of aqueous extracts of medicinal plants on MNNG-treated rat hepatocytes in primary cultures. *Journal of Ethnopharmacology*, *112*(1), 199–202.

Khare, C., & Plants, I. M. (2007). *An illustrated dictionary*. Vol. 812. Springer New Delhi.

Khleifat, K., Shakhanbeh, J., & Tarawneh, K. A. (2002). The chronic effects of Teucrium polium on some blood parameters and histopathology of liver and kidney in the rat. *Turkish Journal of Biology*, *26*(2), 65–71.

Khodaei, F., Ahmadi, K., Kiyani, H., Hashemitabar, M., & Rezaei, M. (2018). Mitochondrial effects of Teucrium polium and Prosopis farcta extracts in colorectal cancer cells. *Asian Pacific Journal of Cancer Prevention: APJCP*, *19*(1), 103.

Khoshnood-Mansoorkhani, M. J., Moein, M. R., & Oveisi, N. (2010). Anticonvulsant activity of Teucrium polium against seizure induced by PTZ and MES in mice. *Iranian Journal of Pharmaceutical Research: IJPR*, *9*(4), 395.

Kouzi, S. A., McMurtry, R. J., & Nelson, S. D. (1994). Hepatotoxicity of germander (Teucrium chamaedrys L.) and one of its constituent neoclerodane diterpenes teucrin A in the mouse. *Chemical Research in Toxicology*, *7*(6), 850–856.

Kovacevic, N. N., Lakusic, B. S., & Ristic, M. S. (2001). Composition of the essential oils of seven Teucrium species from Serbia and Montenegro. *Journal of Essential Oil Research*, *13*(3), 163–165.

Krache, I., Boussoualim, N., Trabsa, H., Ouhida, S., Abderrahmane, B., & Arrar, L. (2018). Antioxidant, antihemolytic, antihyperuricemic, anti-inflammatory activity of algerian germander methanolic extract. *Annual Research & Review in Biology*, 1–14.

Kunduhoglu, B., Pilatin, S., & Caliskan, F. (2011). Antimicrobial screening of some medicinal plants collected from Eskisehir, Turkey. *Fresenius Environmental Bulletin*, *20*(4), 945–952.

Küçük, M., Güleç, C., Yaşar, A., Üçüncü, O., Yaylı, N., Coşkunçelebi, K., Terzioğlu, S., & Yaylı, N. (2006). Chemical composition and antimicrobial activities of the essential oils of Teucrium chamaedrys. subsp. chamaedrys., T. orientale. var. puberulens., and T. chamaedrys. subsp. lydium. *Pharmaceutical Biology*, *44*(8), 592–599.

Kültür, Ş. (2007). Medicinal plants used in Kırklareli province (Turkey). *Journal of Ethnopharmacology*, *111*(2), 341–364.

Laliberte, L., & Villeneuve, J.-P. (1996). Hepatitis after the use of germander, a herbal remedy. *CMAJ: Canadian Medical Association Journal*, *154*(11), 1689.

Line, M. C. (2007). Evaluation of cytotoxic effect of teuerium polium on a new glioblastoma multiforme cell line (reyf-1) using mtt and soft agar clonogenic assays. *International Journal of Pharmacology*, *3*(5), 435–437.

Ljubuncic, P., Azaizeh, H., Portnaya, I., Cogan, U., Said, O., Saleh, K. A., & Bomzon, A. (2005). Antioxidant activity and cytotoxicity of eight plants used in traditional Arab medicine in Israel. *Journal of Ethnopharmacology*, *99*(1), 43–47.

Ljubuncic, P., Dakwar, S., Portnaya, I., Cogan, U., Azaizeh, H., & Bomzon, A. (2006). Aqueous extracts of Teucrium polium possess remarkable antioxidant activity in vitro. *Evidence-Based Complementary and Alternative Medicine*, *3*(3), 329–338.

Mahmoudabady, M., Haghshenas, M., & Niazmand, S. (2018). Extract from Teucrium polium L. protects rat heart against oxidative stress induced by ischemic–reperfusion injury. *Advanced Biomedical Research*, *7*.

Maizi, Y., Meddah, B., Tir Touil Meddah, A., & Gabaldon Hernandez, J. A. (2019). Seasonal variation in essential oil content, chemical composition and antioxidant activity of Teucrium polium L. growing in Mascara (North West of Algeria). *Journal of Applied Biotechnology Reports*, *6*(4), 151–157.

Malki, S., Abidi, L., Hioun, S., & Yahia, A. (2017). Variability of phenolic contents in ethanolic extracts of Teucrium polium L. populations and effect on antioxidant and antimicrobial activities. *Journal of Microbiology and Biotechnology Research*, *5*(4), 21–27.

Matsuda, T., Ferreri, K., Todorov, I., Kuroda, Y., Smith, C. V., Kandeel, F., & Mullen, Y. (2005). Silymarin protects pancreatic β-cells against cytokine-mediated toxicity: Implication of c-Jun NH2-terminal kinase and janus kinase/signal transducer and activator of transcription pathways. *Endocrinology*, *146*(1), 175–185.

Mattéi, A., Rucay, P., Samuel, D., Feray, C., Reynes, M., & Bismuth, H. (1995). Liver transplantation for severe acute liver failure after herbal medicine (Teucrium polium) administration. *Journal of Hepatology*, *22*(5), 597–597.

Mazokopakis, E., Lazaridou, S., Tzardi, M., Mixaki, J., Diamantis, I., & Ganotakis, E. (2004). Acute cholestatic hepatitis caused by Teucrium polium L. *Phytomedicine*, *11*(1), 83–84.

McClintock, E., & Epling, C. (1946). A revision of Teucrium in the new world, with observations on its variation, geographical distribution and history. *Brittonia*, *5*(5), 491–510.

Meguellati, H., Ouafi, S., Saad, S., & Djemouai, N. (2019). Evaluation of acute, subacute oral toxicity and wound healing activity of mother plant and callus of Teucrium polium L. subsp. geyrii Maire from Algeria. *South African Journal of Botany*, *127*, 25–34.

Mehdineya, Z., Eftekhar-Vaghefi, R., Mehrabani, M., Nabipoor, F., & Mehdiniya, N. (2013). The effect of petroleum ether fraction of Teucrium polium extract on laboratory mouse liver. *Journal of Kerman University of Medical Sciences*, *20*(3), 279–291.

Mehrabani, D., Rezaee, A., Azarpira, N., Fattahi, M. R., Amini, M., Tanideh, N., Panjehshahin, M. R., & Saberi-Firouzi, M. (2009). The healing effects of Teucrium polium in the repair of indomethacin-induced gastric ulcer in rats. *Saudi Medical Journal*, *30*(4), 494–499.

Milutinović, M. G., Maksimović, V. M., Cvetković, D. M., Nikodijević, D. D., Stanković, M. S., Pešić, M., & Marković, S. D. (2019). Potential of Teucrium chamaedrys L. to modulate apoptosis and biotransformation in colorectal carcinoma cells. *Journal of Ethnopharmacology*, *240*, 111951.

Monfared, S. S. M. S., & Pournourmohammadi, S. (2010). Teucrium polium complex with molybdate enhance cultured islets secretory function. *Biological Trace Element Research*, *133*(2), 236–241.

Morteza-Semnani, K., Saeedi, M., & Akbarzadeh, M. (2007). Essential oil composition of Teucrium scordium L. *Acta Pharmaceutica*, *57*(4), 499.

Morteza-Semnani, K., Akbarzadeh, M., & Rostami, B. (2005). The essential oil composition of Teucrium chamaedrys L. from Iran. *Flavour and Fragrance Journal*, *20*(5), 544–546.

Mosadegh, M., Dehmoubed, S. A., Nasiri, P., Esmaeili, S., & Naghibi, F. (2002). The study of phytochemical, antifungal and antibacterial effects of Teucrium polium and Cichourium intybus. *Scientific Journal of Kurdistan University of Medical Sciences* 2002;7(1):1–6.

Motamedi, H., Aalivand, S., Ebrahimian, M., & Moosavian, S. M. (2015). The antibacterial properties of methanolic extract of Teucrium polium against MRSA. *Journal of Kermanshah University of Medical Sciences*, *18*(10).

Mousavi, S. M., Niazmand, S., Hosseini, M., Hassanzadeh, Z., Sadeghnia, H. R., Vafaee, F., & Keshavarzi, Z. (2015). Beneficial effects of Teucrium polium and metformin on diabetes-induced memory impairments and brain tissue oxidative damage in rats. *International Journal of Alzheimer's Disease*, *2015*.

Moustapha, C. (2011). Chemical constituents of Teucrium polium L. var. mollissimum Hand-Mazz. *Jordan Journal of Chemistry (JJC)*, *6*(3), 339–345.

Movahedi, A., Basir, R., Rahmat, A., Charaffedine, M., & Othman, F. (2014). Remarkable anticancer activity of Teucrium polium on hepatocellular carcinogenic rats. *Evidence-Based Complementary and Alternative Medicine*, *2014*.

Muselli, A., Desjobert, J.-M., Paolini, J., Bernardini, A.-F., Costa, J., Rosa, A., & Dessi, M. A. (2009). Chemical composition of the essential oils of Teucrium chamaedrys L. from Corsica and Sardinia. *Journal of Essential Oil Research*, *21*(2), 138–143.

Nargesi, S. (2018). Studying the effect of hydroalcoholic extract of Teucrium polium L. leaves on antioxidant activity and lipid profile alterations. *Asian Journal of Pharmaceutics (AJP)*, *12*(2).

Nematollahi-Mahani, S., Rezazadeh-Kermani, M., Mehrabani, M., & Nakhaee, N. (2007). Cytotoxic effects of teucrium polium. On some established cell lines. *Pharmaceutical Biology*, *45*(4), 295–298.

Nematollahi-Mahani, S. N., Mahdinia, Z., Eftekharvaghefi, R., Mehrabani, M., Hemayatkhah-Jahromi, V., & Nabipour, F. (2012). In vitro inhibition of the growth of glioblastoma by teucrium polium crude extract and fractions. *International Journal of Phytomedicine*, *4*(4), 582.

Nencini, C., Galluzzi, P., Pippi, F., Menchiari, A., & Micheli, L. (2014). Hepatotoxicity of Teucrium chamaedrys L. decoction: Role of difference in the harvesting area and preparation method. *Indian Journal of Pharmacology*, *46*(2), 181.

Niazmand, S., Hajizade, M., & Keshavarzipour, T. (2007). The effects of aqueous extract from Teucrium polium L. on rat gastric motility in basal and vagal-stimulated conditions. *Iran J Basic Med Sci.*. 2007;10(1):60–5.

Noumi, E., Snoussi, M., Anouar, E. H., Alreshidi, M., Veettil, V. N., Elkahoui, S., Adnan, M., Patel, M., Kadri, A., & Aouadi, K. (2020). HR-LCMS-based metabolite profiling, antioxidant, and anticancer properties of Teucrium polium L. methanolic extract: Computational and in vitro study. *Antioxidants*, *9*(11), 1089.

Oalđe, M. M., Kolarević, S. M., Živković, J. C., Vuković-Gačić, B. S., Marić, J. M. J., Kolarević, M. J. K., Đorđević, J. Z., Aradski, A. Z. A., Marin, P. D., & Šavikin, K. P. (2020). The impact of different extracts of six Lamiaceae species on deleterious effects of oxidative stress assessed in acellular, prokaryotic and eukaryotic models in vitro. *Saudi Pharmaceutical Journal*, *28*(12), 1592–1604.

Orhan, I., & Aslan, M. (2009). Appraisal of scopolamine-induced antiamnesic effect in mice and in vitro antiacetylcholinesterase and antioxidant activities of some traditionally used Lamiaceae plants. *Journal of Ethnopharmacology*, *122*(2), 327–332.

Othman, M. B., Salah-Fatnassi, K. B. H., Ncibi, S., Elaissi, A., & Zourgui, L. (2017). Antimicrobial activity of essential oil and aqueous and ethanol extracts of Teucrium polium L. subsp. gabesianum (LH) from Tunisia. *Physiology and Molecular Biology of Plants*, *23*(3), 723–729.

Paksoy, M. Y., Selvi, S., & Savran, A. (2016). Ethnopharmacological survey of medicinal plants in Ulukışla (Niğde-Turkey). *Journal of Herbal Medicine*, *6*(1), 42–48.

Panovska, T., Gjorgoski, I., Bogdanova, M., & Petrushevska, G. (2007). Hepatoprotective effect of the ethyl acetate extract of Teucrium polium L. against carbontetrachloride-induced hepatic injury in rats. *Acta Pharmaceutica*, *57*(2), 241.

Panovska, T. K., Kulevanova, S., & Stefova, M. (2005). In vitro antioxidant activity of some Teucrium species (Lamiaceae). *Acta Pharmaceutica*, *55*, 207–214.

Papanov, G. Y., & Malakov, P. Y. (1981). New Furanoid Diterpenes from Teucrium scordium L. *Zeitschrift für Naturforschung B*, *36*(1), 112–113.

Papanov, G. Y., & Malakov, P. Y. (1982). 6β-Hydroxyteuscordin and 2β-, 6β-Dihydroxyteuscordin two new diterpenoids from Teucrium scordium L. *Zeitschrift für Naturforschung B*, *37*(4), 519–520.

Papanov, G. Y., & Malakov, P. Y. (1985). 2-keto-19-hydroxyteuscordin, a neo-clerodane diterpene from Teucrium scordium. *Phytochemistry*, *24*(2), 297–299.

Parsaee, H., & Shafiee-Nick, R. (2006). Anti-spasmodic and anti-nociceptive effects of Teucrium polium aqueous extract. *Iranian Biomedical Journal*, *10*(3), 145–149.

Piccolella, S., Scognamiglio, M., D'Abrosca, B., Esposito, A., Fiorentino, A., & Pacifico, S. (2021). Chemical fractionation joint to in-mixture NMR analysis for avoiding the hepatotoxicity of Teucrium chamaedrys L. subsp. chamaedrys. *Biomolecules*, *11*(5), 690.

Piozzi, F., Bruno, M., Rosselli, S., & Maggio, A. (2005). Advances on the chemistry of furanoditerpenoids from Teucrium genus. *Heterocycles-Sendai Institute of Heterocyclic Chemistry*, *65*(5), 1221–1234.

Polat, R., Cakilcioglu, U., & Satıl, F. (2013). Traditional uses of medicinal plants in Solhan (Bingöl—Turkey). *Journal of Ethnopharmacology*, *148*(3), 951–963.

Polat, R., & Satıl, F. (2012). An ethnobotanical survey of medicinal plants in Edremit Gulf (Balıkesir–Turkey). *Journal of Ethnopharmacology*, *139*(2), 626–641.

Polymeros, D., Kamberoglou, D., & Tzias, V. (2002). Acute cholestatic hepatitis caused by Teucrium polium (golden germander) with transient appearance of antimitochondrial antibody. *Journal of Clinical Gastroenterology*, *34*(1), 100–101.

Prescott, T. A., Veitch, N. C., & Simmonds, M. S. (2011). Direct inhibition of calcineurin by caffeoyl phenylethanoid glycosides from Teucrium chamaedrys and Nepeta cataria. *Journal of Ethnopharmacology*, *137*(3), 1306–1310.

Qabaha, K., Hijawi, T., Mahamid, A., Mansour, H., Naeem, A., Abbadi, J., & Al-Rimawi, F. (2021). Anti-inflammatory and antioxidant activities of Teucrium polium leaf extract and its phenolic and flavonoids content. *Asian Journal of Chemistry*; 33(4), 881–884

Qabaha, K. I. (2013). Antimicrobial and free radical scavenging activities of five Palestinian medicinal plants. *African Journal of Traditional, Complementary and Alternative Medicines*, *10*(4), 101–108.

Qujeq, D., Tatar, M., Feizi, F., Parsian, H., & Halalkhor, S. (2013). Effect of Teucrium polium leaf extracts on AMPK level in isolated rat pancreases. *Research in Molecular Medicine*, 1(3), 29–33.

Radulović, N., Dekić, M., Joksović, M., & Vukićević, R. (2012). Chemotaxonomy of Serbian Teucrium species inferred from essential oil chemical composition: The case of Teucrium scordium L. ssp. scordioides. *Chemistry & Biodiversity*, *9*(1), 106–122.

Rafieian-Kopaei, M., & Baradaran, A. (2013). Teucrium polium and kidney. *Journal of Renal Injury Prevention*, *2*(1), 3–4.

Rahimifard, M., Navaei-Nigjeh, M., Mahroui, N., Mirzaei, S., & Siahpoosh, Z. (2014). Improvement in the function of isolated rat pancreatic islets through reduction of oxidative stress using traditional Iranian medicine. *Cell Journal (Yakhteh)*, *16*(2), 147.

Rahmouni, F., Hamdaoui, L., & Rebai, T. (2017). In vivo anti-inflammatory activity of aqueous extract of Teucrium polium against carrageenan-induced inflammation in experimental models. *Archives of Physiology and Biochemistry*, *123*(5), 313–321.

Rajabalian, S. (2008). Methanolic extract of Teucrium polium L potentiates the cytotoxic and apoptotic effects of anticancer drugs of vincristine, vinblastine and doxorubicin against a panel of cancerous cell lines. *Experimental Oncology*, 30(2),133–8.

Rasekh, H., Khoshnood-Mansourkhani, M., & Kamalinejad, M. (2001). Hypolipidemic effects of Teucrium polium in rats. *Fitoterapia*, *72*(8), 937–939.
Rasik, A. M., & Shukla, A. (2000). Antioxidant status in delayed healing type of wounds. *International Journal of Experimental Pathology*, *81*(4), 257–263.
Reaisi, Z., Yadegari, M., & Shirmardia, H. A. (2019). Effects of phenological stage and elevation on phytochemical characteristics of essential oil of Teucrium polium L. and Teucrium orientale L. *International Journal of Horticultural Science and Technology*, *6*(1), 89–99.
Rojas, A., Hernandez, L., Pereda-Miranda, R., & Mata, R. (1992). Screening for antimicrobial activity of crude drug extracts and pure natural products from Mexican medicinal plants. *Journal of Ethnopharmacology*, *35*(3), 275–283.
Roukia, H., Mahfoud, H. M., & Ould El Hadj, M. D. (2013). Chemical composition and antioxidant and antimicrobial activities of the essential oil from Teucrium polium geyrii (Labiatae). *Journal of Medicinal Plants Research*, *7*(20), 1506–1510.
Sadeghi, H., Jamalpoor, S., & Shirzadi, M. H. (2014). Variability in essential oil of Teucrium polium L. of different latitudinal populations. *Industrial Crops and Products*, *54*, 130–134.
Sajjadi, S. E., & Shookohinia, Y. (2010). Composition of the essential oil of Teucrium chamaedrys L. subsp. syspirense (C. Koch) rech. fil. growing wild in Iran. *Journal of Essential Oil Bearing Plants*, *13*(2), 175–180.
Salimnejad, R., Sazegar, G., Mousavi, S. M., Borujeni, M. J. S., Shokoohi, M., Allami, M., Zarezade, M. M., Boldaji, H. N., & Dashty, M. M. (2017). Effect of Teucriumpolium on oxidative damages and sperm parameters in diabetic rat induced with streptozotocin. *International Journal of Advanced Biotechnology and Research*, *8*(3), 1909–1917.
Salmaki, Y., Kattari, S., Heubl, G., & Bräuchler, C. (2016). Phylogeny of non-monophyletic Teucrium (Lamiaceae: Ajugoideae): Implications for character evolution and taxonomy. *Taxon*, *65*(4), 805–822.
Saltan, N., Kose, Y. B., Iscan, G., & Demirci, B. (2019). Essential oil composition and anticandidal activity of Teucrium polium L. *Fresenius Environmental Bulletin*, *28*(2A), 1174–1178.
Sarac, N., & Uğur, A. (2007). Antimicrobial activities and usage in folkloric medicine of some Lamiaceae species growing in Mugla, Turkey.
Sarer, E., Konuklugil, B., & Eczacilik, F. (1987). Investigation of the volatile oil of Teucrium polium L. *Turkish Tipve Eczacillk Dergisi*, *11*, 317–325.
Savvidou, S., Goulis, J., Giavazis, I., Patsiaoura, K., Hytiroglou, P., & Arvanitakis, C. (2007). Herb-induced hepatitis by Teucrium polium L.: Report of two cases and review of the literature. *European Journal of Gastroenterology & Hepatology*, *19*(6), 507–511.
Sevindik, E., Abacı, Z. T., Yamaner, C., & Ayvaz, M. (2016). Determination of the chemical composition and antimicrobial activity of the essential oils of Teucrium polium and Achillea millefolium grown under North Anatolian ecological conditions. *Biotechnology & Biotechnological Equipment*, *30*(2), 375–380.
Sezik, E., Yeşilada, E., Honda, G., Takaishi, Y., Takeda, Y., & Tanaka, T. (2001). Traditional medicine in Turkey X. Folk medicine in central Anatolia. *Journal of Ethnopharmacology*, *75*(2–3), 95–115.
Shahraki, M., MIRI, M. E., Palan, M., Mirshekari, H., & Shahraki, E. (2006). The survey of Teucrium Polium toxicity effect on liver and serum lipoproteins in normoglycemic male rats. *Zahedan Journal of Research in Medical Sciences*, 8(3).
Shahraki, M. R., Arab, M. R., Mirimokadam, E., & Palan, M. J. (2007). The effect of Teucrium polium (Calpoureh) on liver function, serum lipids and glucose in diabetic male rats. *Iranian Biomedical Journal*, 11(1), 65–68.
Sharififar, F., Dehghn-Nudeh, G., & Mirtajaldini, M. (2009). Major flavonoids with antioxidant activity from Teucrium polium L. *Food Chemistry*, *112*(4), 885–888.
Sharififar, F., Mahdavi, Z., Mirtajaldini, M., & Purhematy, A. (2010). Volatile constituents of aerial parts of Teucrium scordium L. from Iran. *Journal of Essential Oil Research*, *22*(3), 202–204.
Sheikhbahaei, F., Khazaei, M., & Nematollahi-Mahani, S. N. (2018). Teucrium polium extract enhances the anti-angiogenesis effect of tranilast on human umbilical vein endothelial cells. *Advanced Pharmaceutical Bulletin*, *8*(1), 131.
Simonyan, K., Galstyan, H., & Chavushyan, V. (2019). Post-tetanic potentiation and depression in hippocampal neurons in a rat model of Alzheimer's disease: Effects of Teucrium Polium extract. *Neurophysiology*, *51*(5), 332–343.
Stankovic, M. S., Curcic, M. G., Zizic, J. B., Topuzovic, M. D., Solujic, S. R., & Markovic, S. D. (2011). Teucrium plant species as natural sources of novel anticancer compounds: Antiproliferative, proapoptotic and antioxidant properties. *International Journal of Molecular Sciences*, *12*(7), 4190–4205.
Stanković, M. S., Mitrović, T. L., Matić, I. Z., Topuzović, M. D., & Stamenković, S. M. (2015). New values of Teucrium species: In vitro study of cytotoxic activities of secondary metabolites. *Notulae Botanicae Horti Agrobotanici Cluj-Napoca*, *43*(1), 41–46.

Stanković, M. S., Stefanović, O., Čomić, L., Topuzović, M., Radojević, I., & Solujić, S. (2012). Antimicrobial activity, total phenolic content and flavonoid concentrations of Teucrium species. *Central European Journal of Biology*, *7*(4), 664–671.

Stankovic, M. S., Topuzovic, M., Solujic, S., & Mihailovic, V. (2010). Antioxidant activity and concentration of phenols and flavonoids in the whole plant and plant parts of Teucrium chamaerdys L. var. glanduliferum Haussk. *Journal of Medicinal Plants Research*, *4*(20), 2092–2098.

Suleiman, M.-S., Abdul-Ghani, A.-S., Al-Khalil, S., & Amin, R. (1988). Effect of Teucrium polium boiled leaf extract on intestinal motility and blood pressure. *Journal of Ethnopharmacology*, *22*(1), 111–116.

Tabatabaei Yazdi, F., Alizadeh Behbahani, B., Heidari Sureshjani, M., & Mortazavi, S. (2014). The in vitro study of antimicrobial effect of teucrium polium extract on infectious microorganisms. *Avicenna Journal of Clinical Medicine (Scientific Journal of Hamadan University of Medical Sciences and Health Services)*, *21*(1 (SN 71)), 16–24.

Tariq, M., Ageel, A., Al-Yahya, M., Mossa, J., & Al-Said, M. (1989). Anti-inflammatory activity of Teucrium polium. *International Journal of Tissue Reactions*, *11*(4), 185–188.

Tatar, M., Qujeq, D., Feizi, F., Parsian, H., Faraji, A. S., Halalkhor, S., Abassi, R., Abedian, Z., Pourbagher, R., & Mir, S. M. A. (2012). Effects of Teucrium Polium aerial parts extract on oral glucose tolerance tests and pancreas histopathology in Streptozocin-induced diabetic rats. *International Journal of Molecular and Cellular Medicine*, *1*(1), 44.

Tepe, B., Malatyali, E., Degerli, S., & Berk, S. (2012). In vitro amoebicidal activities of Teucrium polium and T. chamaedrys on Acanthamoeba castellanii trophozoites and cysts. *Parasitology Research*, *110*(5), 1773–1778.

Todorov, D., Pavlova, D., Hinkov, A., Shishkova, K., Dragolova, D., Kapchina-Toteva, V., & Shishkov, S. (2015). Effect of Teucrium chamaedrys l. extracts on herpes simplex virus type 2. *Comptes rendus de l'Académie bulgare des Sciences*, *68*(12).

Tusi, S. K., Jafari, A., Marashi, S. M. A., Niknam, S. F., Farid, M., & Ansari, M. (2020). The effect of antimicrobial activity of Teucrium Polium on Oral Streptococcus Mutans: A randomized cross-over clinical trial study. *BMC Oral Health*, *20*(1), 1–8.

Twaij, H. A., Albadr, A. A., & Abul-Khail, A. (1987). Anti-ulcer activity of Teucrium polium. *International Journal of Crude Drug Research*, *25*(2), 125–128.

Ulubelen, A., Topu, G., & Sönmez, U. (2000). Chemical and biological evaluation of genus Teucrium. *Studies in Natural Products Chemistry*, *23*, 591–648.

Uysal, I., Guecel, S., Tütenocakli, T., & Öztürk, M. (2012). Studies on the medicinal plants of Ayvacik-Çanakkale in Turkey. *Pakistan Journal of Botany*, *44*(Supp. 1), 239–244.

Ünsal, Ç., Vural, H., Sariyar, G., Özbek, B., & Ötük, G. (2010). Traditional medicine in Bilecik Province (Turkey) and antimicrobial activities of selected species. *Turkish Journal of Pharmaceutical Sciences*, *7*(2).

Venditti, A., Frezza, C., Trancanella, E., Zadeh, S. M. M., Foddai, S., Sciubba, F., Delfini, M., Serafini, M., & Bianco, A. (2017). A new natural neo-clerodane from Teucrium polium L. collected in Northern Iran. *Industrial Crops and Products*, *97*, 632–638.

Verdi, J., Komeilizadeh, H., Kamalinejad, M., & Sh, S. (2010). Analgesic effects of aqueous extract of Teucriun polium L. in experimental models of pain in male rats. *Iranian Journal of Pharmaceutical Research*, *3*(2), 62–62.

Vlase, L., Benedec, D., Hanganu, D., Damian, G., Csillag, I., Sevastre, B., Mot, A. C., Silaghi-Dumitrescu, R., & Tilea, I. (2014). Evaluation of antioxidant and antimicrobial activities and phenolic profile for Hyssopus officinalis, Ocimum basilicum and Teucrium chamaedrys. *Molecules*, *19*(5), 5490–5507.

Yazdanparast, R., & Ardestani, A. (2009). Suppressive effect of ethyl acetate extract of Teucrium polium on cellular oxidative damages and apoptosis induced by 2-deoxy-D-ribose: Role of de novo synthesis of glutathione. *Food Chemistry*, *114*(4), 1222–1230.

Yazdanparast, R., Esmaeili, M. A., & Helan, J. A. (2005). Teucrium polium extract effects pancreatic function of streptozotocin diabetic rats: A histopathological examination. *Iranian Biomedical Journal*, *9*(2): 81–85.

Zal, F., Rasti, M., Vesal, M., & Vaseei, M. (2001). Hepatotoxicity associated with hypoglycemic effects of Teucrium polium in diabetic rats. *Archives of Iranian Medicine*, 4(4), 188–192.

Zendehdel, M., Taati, M., Jadidoleslami, M., & Bashiri, A. (2011). Evaluation of pharmacological mechanisms of antinociceptive effect of Teucrium polium on visceral pain in mice. *Z Iranian Journal of Veterinary Research*, 12(4), 292–297.

Zlatić, N. M., Stanković, M. S., & Simić, Z. S. (2017). Secondary metabolites and metal content dynamics in Teucrium montanum L. and Teucrium chamaedrys L. from habitats with serpentine and calcareous substrate. *Environmental Monitoring and Assessment*, *189*(3), 110.

26 *Thymus* sp.

Methiye Mancak and Ufuk Koca Çalışkan

INTRODUCTION

The genus *Thymus*, commonly used English name is "thyme", belongs to the Lamiaceae family. It has 250–350 taxa scattered throughout Asia, North Africa, and Southern Europe (Casiglia et al., 2019). *Thymus* in Turkiye has 60 taxa belonging to 39 species with an endemism of 45% (Baser, 2002). Endemic *Thymus* species growing in Turkiye and their places are given in Figure 26.1 and Table 26.1.

Thymus sp. are included in various international monographs with their rich content, pharmacological properties, and important economic value (Basch et al., 2004; WHO Monographs, 1999; EMA 2013; ESCOP, 2003; PDR for Herbal Medicines, 2000). Above-ground parts (herba) of the plant and its essential oil are used. Thymus vulgaris L., Thymus zygis L., or a hybrid of both species' complete leaves and flowers separated from their previously dried stems are known as thymi herba. (EMA, 2013).

Thyme has been used traditionally for many years. In addition to its use as a spice, it is also preferred because of its traditionally known medicinal effects. Due to its antioxidant and antimicrobial effects, thyme is used for storing foods and extending their shelf life (Jayari et al., 2018; Gonçalves et al., 2017). Thymol and carvacrol, the active ingredients of thyme, have antitussive, antioxidative, antimicrobial, expectorant, antispasmodic, and antibacterial effects. They are used as antiseptic, flavorings, food preservatives, and wound healing. Scientific studies demonstrated that thymol also has anticarcinogenic, anti-inflammatory, immunomodulatory properties with the application in the treatment of disorders affecting the cardiovascular, nervous, and respiratory systems (Salehi et al., 2018). Thymol is included in the content of antiseptic mouthwashes, in order to control supragingival plaque and gingivitis along with oral hygiene (Alshehri, 2018). Although carvacrol has antibacterial, antifungal, insecticidal, antioxidant, and antihistaminic effects, its use in the pharmaceutical industry is limited because it cannot be crystallized (Bozdemir, 2019). *Thymus* sp. have antibacterial, antifungal, antiviral, anti-inflammatory, antihypertensive, antihyperlipidemic, antidiabetic, antiproliferative, and anticancer effects (Salehi et al., 2018).

BOTANICAL FEATURES/ETHNOBOTANICAL USAGE

It is an aromatic perennial sub-shrub, 20–30 cm in height, with ascending, quadrangular, grayish brown to purplish brown lignified and twisted stems bearing oblong-lanceolate to ovate-lanceolate greyish green leaves that are pubescent on the lower surface. The flowers are borne in verticillasters. Corolla is bilobate, pinkish or whitish, and calyx is pubescent. The fruit has four brown ovoid nutlets (WHO Monographs, 1999).

Thyme is a symbol of nobility, courage, and wealth in ancient times, and it is known that Roman soldiers bathed with thyme water. Thyme oil was used as a battlefield antiseptic until World War I (Bozdemir, 2019). Thyme has been associated with death in many cultures. For instance, the ancient Greeks used thyme as incense in funerals and temples, whereas the Ancient Etruscans and Egyptians used thyme oil to embalm their dead (Johnson, 2012). *Thymus* sp. are traditionally used

DOI: 10.1201/9781003146971-26

FIGURE 26.1 Distribution of endemic *Thymus* species in Turkiye. (Modified from Turkish Plants Data Service-TÜBİVES.)

for its medicinal effects and culinary properties. Thyme leaves are used to give taste and flavor, especially in meat dishes. Moreover, they also have been used as an insect repellent in the form of incense (Salehi et al., 2018). The use of thyme oil in the field of health began in the first century AD. They have been exploited for oral hygiene and mixed with wine for cough; thyme-filled pillow has been used to treat melancholy. In Europe, incense was used as a purifier of the home air; the bodies were covered with thyme to protect against infectious diseases such as plague and leprosy (Bozdemir, 2019). *Thymus* sp. have traditionally been used to treat respiratory ailments, gastrointestinal problems, and hypertension. Additionally, antihelmintic, antimicrobial, antiseptic, antirheumatic, and calming effects were also recorded for the plant species (Salehi et al., 2018).

In Turkiye, herbs of *T. longicaulis* C. are consumed on an empty stomach for stomachache as well as a decoction of the aerial parts to treat stomach chills (Fujita et al., 1995; Kargıoglu et al., 2010). Similarly, *T. zygioides* Griseb. var. *lycaonicus* is also used for stomachache (Fujita et al., 1995). The herbs of *T. transcaucasicus* Ronniger, leaves of *T. haussknechtii* Velen. and aerial parts of *T. kotschyanus* Boiss. & Hohen. are applied for common colds (Polat et al., 2013; Tetik et al., 2013; Fujita et al., 1995). *T. argaeus*, endemic to Turkiye, is used as a spice and as a wild tea in Central Anatolia (Sagdıc et al., 2009). Decoction of aerial parts of *T. zygioides* Griseb. var. *lycaonicus* and aerial parts of *T. longicaulis* subsp. *longicaulis* var. *subisophyllus* is used in the treatment of diabetes (Honda et al., 1996; Tuzlacı et al., 2001). Decoction and infusion of aerial parts of *T. kotschyanus* Boiss. & Hohen. subsp. *glabrescens* Boiss. are applied in the treatment of colds, diabetes, and hypertension. Infusion of aerial parts of *T. sipyleus* Boiss. subsp. *rosulans* is consumed in the treatment of abdominal pain, cold, and diabetes (Altındag and Ozturk, 2011).

PHYTOCHEMICALS

Thymus sp. contain essential oil, caffeic acid derivatives, flavonoids, and triterpenes (PDR for Herbal Medicines, 2000). The principal components are thymol, carvacrol, apigenin, α-pinene, cymene, linalool, luteolin, p-cymol, thymene, and 6-hydroxyluteolin glycosides, as well as di-, tri-, and tetramethoxylated flavones (WHO Monographs, 1999). The amount and content of essential oil varies depending on the species, chemotype, and environmental factors. The main component of the essential oils of *Thymus* sp. grown in Turkiye are usually thymol, carvacrol, borneol, or linalool (Azaz et al., 2004; Ozcan et al., 2004). The major components in *T. haussknechtii* oil were linalool (19.91%) and borneol (10.35%), while in *T. cariensis* oil were borneol (13.42%), 1,8-cineole (12.65%), and α-pinene (12.15%), two endemic species growing in Turkiye (Baser et al., 1992). The major composition of essential oil of *Thymus* species is given in Table 26.2.

TABLE 26.1
Endemic *Thymus* Species Growing in Turkiye and Their Places

	Species name	Regions
	Thymus argaeus BOISS ET BAL	Middle Anatolia (Kayseri, Aksaray)
	*Thymus aznavourii VELEN.*	Thrace (Istanbul)
-	*Thymus bornmuelleri* VELEN.	Northwest Anatolia (Bursa)
	*Thymus brachychilus* JALAS	Southern and Continental Anatolia (Adana, Erzincan, Icel, Nigde, Sivas, Tunceli, Van)
-	*Thymus canoviridis* JALAS	East Anatolia (Erzurum, Sivas)
	*Thymus cappadocicus* BOISS. (var. *pruinosus* (BOISS.) BOISS. var. *cappadocicus* BOISS. var. *globifer* JALAS)	East Anatolia (Sivas) East Anatolia (Sivas, Erzincan) East Anatolia (Sivas, Erzincan)
	Thymus cariensis HUB.-MOR. ET JALAS	South West Anatolia (Mugla)
	*Thymus cherlerioides* VIS (var. *cherlerioides* VIS var. *oxydon* JALAS var. *isauricus* JALAS)	West and South Anatolia (Antalya, Balıkesir, Isparta, Icel) South Anatolia (Antalya) South Anatolia (Antalya, Sivas)
	*Thymus convolutus* KLOKOV	East Anatolia (Erzincan)
	*Thymus fedtschenkoi* RONNIGER var. *handelii* (RONNIGER) JALAS	East Anatolia (Batman, Bitlis, Agri, Van)
	*Thymus haussknechtii* VELEN.	East Anatolia (Elazig, Erzincan, Erzurum, Sivas, Tunceli)

(*Continued*)

TABLE 26.1 (CONTINUED)
Endemic *Thymus* Species Growing in Turkiye and Their Places

	Species name	Regions
	Thymus leucostomus HAUSSKN. ET VELEN.	Middle and North Anatolia (Cankiri, Kastamonu, Ankara, Çorum, Konya)
	var. *leucostomus* HAUSSKN. ET VELEN.	
	var. *argillaceus* JALAS	Middle Anatolia (Ankara, Eskisehir)
	var. *gypsaceus* JALAS	Middle Anatolia (Cankiri)
-	*Thymus longicaulis* C. PRESL subsp. *chaubardii* (BOISS. ET HELDR. EX REICHB. FIL.) JALAS var. *antalyanus* (KLOKOV) JALAS	Middle and South West Anatolia (Antalya, Denizli, Konya, Kutahya)
-	*Thymus pectinatus* FISCH. ET MEY.	Middle Anatolia (Adana, Kayseri, Malatya, Sivas)
	(var. *pectinatus* FISCH. ET MEY.	
	var. *pallasicus* (HAYEK ET VELEN.) JALAS	Middle Anatolia (Kayseri)
	*Thymus pubescens* BOISS. ET KOTSCHY EX CELAK var. *crateicola* JALAS	East Anatolia (Bitlis, Tunceli, Van)
	*Thymus pulvinatus* CELAK	West Anatolia (Balıkesir)
	Thymus revolutus CELAK	, Southern Anatolia (Antalya, İcel)
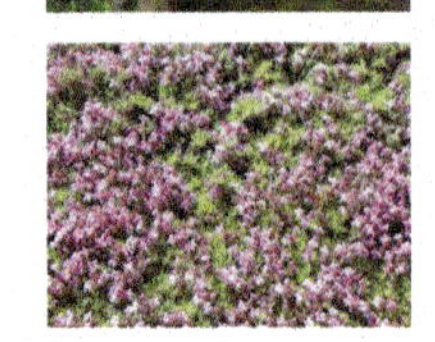	*Thymus sipyleus* BOISS subsp. *sipyleus* BOISS. var. *davisianus* RONNIGER	South West Anatolia (Antalya, Burdur, Mugla)
-	*Thymus spathulifolius* HAUSSKN. ET VELEN.	East Anatolia (Sivas, Erzincan)

(Turkish Plants Data Service-TÜBİVES)

BIOLOGICAL ACTIVITIES/EFFECTS

Anticancer and Antiproliferative Activity

The effects of ten essential oils, including *T. vulgaris*, on three human cancer cells [A-549 (human lung carcinoma cells), PC- 3 (human prostate carcinoma cells), and MCF-7 (human breast cancer cells)] were examined *in vitro*. *T. vulgaris* oil had the strongest cytotoxicity among all cancer cells tested (Zu et al., 2010). In another study, *T. vulgaris* essential oil had an antiproliferative effect on A549, HCT116 (colon), MCF-7, and PC3 but not on HepG2 (liver) cells. Moreover, thyme oil was more effective than doxorubicin in breast and lung cells (Ramadan et al., 2015). Time-dependent cytotoxic effect on A549 and MDA-MD-231 (breast cancer cells) has been shown in different studies (Miladi et al., 2013; Al-Shahrani et al., 2017). In a different study, hexane extract

TABLE 26.2
Major Composition of Essential Oil of *Thymus* Species

Species name		Composition	Reference
Thymus cappadocicus		Thymol (70.82%); p-cymene (9.52%); γ-terpinene (9.27%); carvacrol (4.65%)	Albayrak and Aksoy, 2013
Thymus cariensis		Borneol (13.42%); 1,8-cineole (12.65%), α-pinene (12.15%)	Baser et al., 1992
Thymus canoviridis	I	Thymol (60.44%); β-caryophyllene (8.49%)	Azaz and Celen, 2012
	II	Thymol (64.79%); β-caryophyllene (6.58%)	
Thymus cilicicus		Thymol (34.03%); carvacrol (12.11 %); terpinolen (8.29%)	Azaz and Celen, 2012
Thymus convolutus		Camphor (16.6%); (E)-β-ocimene (8.90%); 1,8-cineole (7.72%)	Celen Yuceturk et al., 2021
Thymus fallax		Carvacrol (66.1%); *p*-cymene (7.1%); (E)-β-ocimene (5.5%); γ-terpinene (4.6%)	Kucukbay et al., 2014
Thymus haussknechtii		Linalool (19.91%); borneol (10.35%)	Baser et al., 1992
Thymus leucostomus		o-Cymene (30.6%); carvacrol (9.6%); thymol methyl ether (7.2%); limonene (6.8%)	Elkiran and Avsar, 2020
Thymus longicaulis subsp. *chaubardii* var. *chaubardii*	Chemotype I	Thymol (56.6%); *p*-cymene (9.8%); γ-terpinene (7.1%)	Azaz et al., 2004
	Chemotype II	Thymol (42.8%); *p*-cymene (14.3%); γ-terpinene (11.3%)	
Thymus longicaulis subsp. *longicaulis* var. *subisophyllus*		Carvacrol (60.0%); thymol (7.0%); β-bisabolene (4.8%)	Azaz et al., 2004
Thymus migricus		α-Terpineol (30.6%); thymol (20.7%); α-terpinyl acetate (14.9%); borneol (5.5%)	Kucukbay et al., 2014
Thymus pectinatus		Thymol (48.77%); *m*-cymene (9.15%); isoborneol (5.19%); trans-caryophyllene (4.43%); carvacrol (3.91%); ɣ-terpinene (3.54%)	Saygınl et al., 2018
Thymus pubescens		*cis*-Carveol (29.6%); α-terpinol (10.8%); (E)-β-ocimene (9.5%)	Kucukbay et al., 2014
		Thymol (53.33%); 1,8 cineol (7.40%); *o*-cymene (7.09%); endo-borneol (5.17%)	Yigitkan et al., 2022
Thymus pulvinatus		Borneol (27.9%); camphene (9.3%); camphor (8.7%)	Azaz et al., 2004
Thymus revolutus		Thymol (66.96%); carvacrol (10.12%); γ-terpinene (8.13%)	Azaz and Celen, 2012
		Carvacrol (24.53%); thymol (15.39%); p-cimene (15.50%); borneol (14.66%)	Gokturk et al., 2013
		Cymene (32.57%); γ-terpinene (17.18%); carvacrol (11.89%)	Erdogan and Ozkan, 2013
Thymus sipyleus BOISS. subsp. *sipyleus* BOISS. var. *davisianus* RONNIGER		Thymol (38.31%); carvacrol (37.95%)	Ceylan and Ugur, 2015
Thymus spathulifolius		Thymol (50.5%); borneol (16.7%); carvacrol (7.7%)	Ceylan et al., 2016
Thymus vulgaris		Thymol (46.2%); γ-terpinene (14.1%); *p*-cymene (9.9%); linalool (4.0%)	Ozcan et al., 2004
Thymus zygioides var. *lycaonicus*		Thymol (36.9%); geraniol (22.8%); borneol (6.4%)	Azaz et al., 2004

of *T. vulgaris* inhibited the mitogen-stimulated proliferation of lymphocytes (Amirghofran et al., 2011). Antiproliferative effect of ethanol extract of *T. spathulifolius* was detected with MTT (3-(4,5-dimethylthiazol-2-yl)-2, 5-diphenyltetrazolium bromide) (ED_{50}: 12.0 µg/mL) (Eruygur et al., 2017).

Antimicrobial Activity

Thymus sp. and its active ingredients, particularly thymol, have a broad antimicrobial spectrum. They are effective against respiratory and foodborne pathogens.

A thyme extract preparation used as oral drops showed antiviral activity against human rhinoviruses (HRV) and influenza A (Lenz et al., 2018). *T. vulgaris* essential oil (3.1 µL/mL) had a strong anti-influenza effect in the liquid phase (Vimalanathan and Hudson, 2014). In a different study, thyme essential oil had antibacterial activity against respiratory pathogens in both liquid and vapor phases, but this effect was lower than reference antibiotics (Ács et al., 2018). *T. vulgaris* extract at 5%, 10%, and 20% w/v concentrations showed a time and concentration-dependent reduction in *Streptococcus mutans*. Maximum inhibition of bacterial growth (96%) occurred as a result of exposure to 20% thyme extract for 48 hours. At the same time, thyme extracts reduced the adhesion of *S. mutans* to buccal epithelial cells. After rinsing with 20% aqueous thyme extract, there was an 89% reduction in the adhesion of bacterial cells to buccal epithelial cells, while the reduction in rinsing with chlorhexidine digluconate was 45% (Hammad et al., 2007). *T. vulgaris* essential oil [0.25% (v/v)] presented strong bactericidal activity against *Propionibacterium acnes* and killed it completely after 5 minutes (Zu et al., 2010). In another study, its essential oil demonstrated an inhibitory effect against herpes simplex virus type 2 (HSV-2) (Koch et al., 2008).

Thymol showed antifungal effects against *Aspergillus*, *C. albicans*, and *Cryptococcus* (Braga et al., 2007; Faria et al., 2011; Shen et al., 2016). Thymol inhibited the hyphae formation and viability of *Candida* by affecting the fungal enzymes that have roles in the cell wall and then disrupt the cell membranes and metabolism of the fungus (Braga et al., 2007). Moreover, the thymol and eugenol combinations showed a synergistic effect against *C. albicans in vitro* (Braga et al., 2007). Thymol-induced reactive oxygen species with nitric oxide showed antifungal effect against *Aspergillus flavus* spores (Shen et al., 2016). Thymol was added to collagen-containing wound dressings and then their antimicrobial properties were compared. The results have shown that thymol inhibited the formation of biofilms on the collagen surface and then the growth of bacteria, mostly *Staphylococcus aureus* (Michalska-Sionkowska et al., 2017). Thymol was found to be effective on oral bacteria *Porphyromonas gingivalis*, *Selenomonas artemidis*, and *Streptococcus sobrinus* (Shapiro and Guggenheim, 1995). The preparation containing 1% chlorhexidine/thymol suppressed *S. mutans*, which is associated with caries development (Twetman and Petersson, 1999). Multicomponent mouthwash containing thymol was effective against influenza A, herpes simplex type-1 and type-2, *S. mutans* (Dennison et al., 1995; Fine et al., 2000).

Some *Thymus* sp. showed an antifungal effect on *Candida albicans in vitro* (Giordani et al., 2008). *Thymus* sp. (*T. canoviridis*, *T. cilicicus*, *T. comptus*, *T. revolutus*) grown in Turkiye on *Enterobacter aerogenes*, *Escherichia coli*, *Listeria monocytogenes*, *Pseudomonas aeruginosa*, *Proteus vulgaris*, *Serratia marcescens*, *Staphylococcus aureus*, *Candida albicans*, and *Aspergillus niger* antimicrobial effect detected (Azaz and Celen, 2012). *T. kotschyanus* essential oil (MIC: 0.5 to 1 µg/mL) showed antifungal effect on *Alternaria alternata*, *Aspergillus flavus*, and *Fusarium oxysporum*. Antifungal activity was approximately the same as that of amphotericin B (Mohammadi et al., 2014). *T. spathulifolius* hexane extract was highly effective on *C. albicans* and *S. aureus* (MIC values 0.62 and 0.31 mg/mL, respectively) (Eruygur et al., 2017). Antibacterial activities of thyme essential oils and their main components, vapor and liquid phases, have been investigated in studies. Thyme oil and thymol in gaseous state showed antibacterial activity against *Staphylococcus aureus*, *Streptococcus pneumoniae*, *S. pyogenes*, and *Haemophilus influenzae* being the most. The antibacterial effect is maximum when applied in a short time at high vapor concentration (Inouye et al., 2001). *T. longicaulis* essential oil showed antimicrobial activity against respiratory pathogens. It was most effective on *Haemophilus influenzae* and *Streptococcus pneumoniae* (Vladimir-Knežević et al., 2012). Both essential oil and methanolic extracts of *T. argaeus* had antimicrobial activity. The extract was effective mostly against *Pseudomonas aeruginosa*, whereas the essential oil was effective mostly against *Aeromonas hydrophila* and *P. aeruginosa* (Sagdıc et al., 2009). *T. haussknechtii* Velen infusion at a concentration of 4% showed antibacterial and antifungal effects (Kılıcgun et al., 2016).

Anti-Inflammatory Activity

Anti-inflammatory activity was associated mostly with thymol content; *thymus* extracts suppressed the gene expression and production of proinflammatory mediators TNF-α, IL-1β, and IL-6 (Ocaña and Reglero, 2012). Thymol (2.5–20 μg/mL) inhibited neutrophil elastase release and exhibited anti-inflammatory activity (Braga et al., 2006). *T. vulgaris* essential oil and carvacrol showed anti-inflammatory effects through the inhibition of inflammatory edema and leukocyte migration (Fachini-Queiroz et al., 2012). Thymol ameliorated inflammation in asthmatic mice by inhibiting NF-κB activation (Zhou et al., 2014). Gel containing thymol exhibited anti-ınflammatory effects in rats and reduced gingivitis (Botelho et al., 2016).

Antioxidant Activity

The extract and essential oil of *Thymus* sp. and their compounds (e.g., carvacrol eriodictyol p-cymene, rosmarinic acid, thymol) have antioxidant effects (Haraguchi et al., 1996; Dapkevicius et al., 2002; Chizzola et al., 2008). The antioxidant activity of the essential oil mainly depends on the content of phenolic compounds (e.g., carvacrol, thymol) and the antioxidant activity of the extracts depends on the rosmarinic acid content (EMA).

In one study, the methanol extract of *T. argaeus* had strong antioxidant activity, but this activity of the essential oil was quite low (Sagdıc et al., 2009). Aqueous extracts of *T. longicaulis* subsp. *chaubardii* (Rchb.f.) Jalas and *T. longicaulis* subsp. *longicaulis* C. presl. had a strong antioxidant effect. The extracts prepared with decoction had higher antioxidant capacity than the extracts prepared with infusion (Zuleyha, 2020). In the study of *T. zygioides* Griseb. with ethanol, methanol, acetone, and water extracts, all the extracts indicated strong antioxidant activity. They served as metal chelators and free radical scavengers. The methanol extract revealed the highest amount of free radical scavenging (ABTS, 11.44 μg/mL and DPPH, 20.82 μg/mL) and antioxidant activities (phosphomolybdenum, 45.63 μg/mg). In the metal chelating activity, acetone extract was statistically different from the other extracts (54.26%, $P<0.05$). The water extract exhibited the highest antioxidant capacity (β-carotene/linoleic acid test system, 82.20%) and the ethanol extract presented the highest amount of ferric-reducing power activity (0.141 mg/mL) (Kaska et al., 2018).

Metabolic Activity

Thymol supplementation (3 or 6 mg/kg/day) for 8 weeks prevented the progression of high-fat diet-induced hyperlipidemia and atherosclerosis in rabbits. It reduced inflammation, serum lipids, aortic intimal lipid lesion, and oxidative stress (Yu et al., 2016). Administration of thymol (14 mg/kg, twice per day) to obese rats for 4 weeks improved insulin and leptin sensitivity. It decreased body weight gain, visceral fat accumulation, lipid, ALT, AST, BUN, LDH, and glucose levels (Haque et al., 2014).

20 g of *T. kotschyanus* aqueous extract was administered orally daily for 3 months to type II diabetes patients. While FBG, HbA1C, and LDL levels were decreased in the group that took the

extract in addition to normal antidiabetic drugs, beta cell function index (HOMA-Beta) increased (Taleb et al., 2017). *T. zygioides* var. *lycaonicus* showed high α-glucosidase inhibitory activity (85.28 ± 0.89%), while α-amylase inhibitory activity (8.93 ± 1.73%) was quite low, whereas Its butyrylcholinesterase inhibitory activity was moderate (30.92 ± 1.44%) (Ekin et al., 2019).

Other Activities

Topical essential oil of 2% lavender-thymol in jojoba oil was applied in a placebo-controlled, single-blind, randomized clinical study, conducted on 60 women, with mediolateral episiotomy. Topically applied lavender-thymol aromatherapy has been found to be effective, suitable, and safe for episiotomy wound care (Marzouk et al., 2015). Large doses of oral thymol found to be effective in dermatomyositis (Buccellato, 1965). The application of a topical cream containing thyme extract improved vulval lichen sclerosis successfully (Hagedorn, 1989).

Other studies reported that *T. vulgaris* extracts and flavonoids have spasmolytic activity on ileal smooth muscles and trachea (Van Den Broucke and Lemli, 1981; Van Den Broucke and Lemli, 1983; Meister et al., 1999). This was achieved by inhibition of acetylcholine, histamine, L- noradrenaline receptors, and Ca^{2+}. Thyme oil showed spasmolytic effect by inhibiting phasic contractions (Reiter and Brandt, 1985). Nasal administration of thymol (0.025 mL, 10^{-3} M) modulated cough in healthy individuals (Gavliakova et al., 2013). In a study of patients with productive cough, thyme syrup showed similar results with bromhexine in relieving cough complaints (Knols et al., 1994). Thyme honey nasal spray prepared with thymol and honey was applied as an additional drug to patients with chronic rhinosinusitis after functional endoscopic sinus surgery. Thyme honey promoted healing of patients' mucosal and reduced inflammation and polyp formation (Hashemian et al., 2015). In another study, thyme honey was used for the management of oral mucositis in head and neck cancer patients. It positively affected radiation-induced oral mucositis, overall health, and quality of life (Charalambous et al., 2018).

Preparations containing the combination of thyme with different herbs are used in the treatment of bronchitis. A preparation containing a mixture of thyme and evening primrose was found to be as effective as synthetic drugs on acute bronchitis (Ernst et al., 1997). Similarly, the combination of thyme and ivy extracts suppressed bronchoalveolar inflammation and goblet cell hyperplasia in experimental bronchoalveolitis (Seibel et al., 2015; Mastalerz-Migas et al., 2017).

T. vulgaris hydroalcoholic extract showed modulatory effects on acute and chronic pain (Taherian et al., 2009). Application of *T. vulgaris* aqueous extract (100 mg/kg/day, orally) for 8 weeks presented antihypertensive effect and a marked improvement in hypertension-related biochemical changes and aortic vascular damage in rats (Kensara et al., 2013). *T. vulgaris* extract administered to hypertensive rats at 100 mg/kg daily for 8 weeks revealed significant antihypertensive effect. This practice showed improvement in hypertension-related biochemical changes and aortic vascular damage (Kensara et al., 2013).

CONCLUSION

Turkiye is very rich in terms of thyme diversity and the endemism rate of thyme is significantly high. *Thymus* sp. have varying proportions of thymol and carvacrol; this ratio is quite high in Turkish thyme. *Thymus* sp. are generally used as spice in Turkiye, by being a natural preservative to extend the shelf life of foods with its antimicrobial and antioxidant properties. Studies have been conducted based on the traditional use of *Thymus* sp. with most of the studies on *Thymus vulgaris*; the research should be extended to the rest of *Thymus* sp. grown in Turkiye. The fact that thyme has natural herbicide and insecticide properties increases the importance and cultivability of the plant. Furthermore, only a few clinical studies were recorded on *Thymus* sp., their promising efficacy on respiratory system disorders, oral health, and diabetes should be considered to carry on the clinical research.

REFERENCES

Ács, K., Balázs, V. L., Kocsis, B., Bencsik, T., Böszörményi, A., & Horváth, G. (2018). Antibacterial activity evaluation of selected essential oils in liquid and vapor phase on respiratory tract pathogens. *BMC Complementary and Alternative Medicine*, 18(1), 1–9.

Albayrak, S., & Aksoy, A. (2013). Essential oil composition and *in vitro* antioxidant and antimicrobial activities of *Thymus cappadocicus* boiss. *Journal of Food Processing and Preservation*, 37(5), 605–614.

Al-Shahrani, M. H., Mahfoud, M., Anvarbatcha, R., Athar, M. T., & Al Asmari, A. (2017). Evaluation of antifungal activity and cytotoxicity of Thymus vulgaris essential oil. *Pharmacognosy Communications*, 7(1), 34.

Alshehri, F. A. (2018). The use of mouthwash containing essential oils (LISTERINE®) to improve oral health: A systematic review. *The Saudi Dental Journal*, 30(1), 2–6.

Altundağ, E., & Öztürk, M. (2011). Ethnomedicinal studies on the plant resources of East Anatolia, Turkiye. *Procedia Social and Behavioral Sciences*, 19, 756–777.

Amirghofran, Z., Hashemzadeh, R., Javidnia, K., Golmoghaddam, H., & Esmaeilbeig, A. (2011). In vitro immunomodulatory effects of extracts from three plants of the Labiatae family and isolation of the active compound (s). *Journal of Immunotoxicology*, 8(4), 265–273.

Azaz, A. D., Irtem, H. A., Kurkcuoğlu, M., & Baser, K. H. (2004). Composition and the in vitro antimicrobial activities of the essential oils of some *Thymus* species. *Zeitschrift für Naturforschung. C, A Journal of Biosciences*, 59(1–2), 75–80. doi: 10.1515/znc-2004-1-216. PMID: 15018057.

Azaz, A. D., & Celen, S. (2012). Composition and in vitro antimicrobial and antioxidant activities of the essential oils of four *Thymus* species in Turkiye. *Asian Journal of Chemistry*, 24(5), 2082–2086.

Basch, E., Ulbricht, C., Hammerness, P., Bevins, A., & Sollars, D. (2004). Thyme (*Thymus vulgaris* L.), thymol. *Journal of Herbal Pharmacotherapy*, 4(1), 49–67.

Baser, K. H. C. (2002). Aromatic biodiversity among the flowering plant taxa of Turkiye. *Pure and Applied Chemistry*, 74(4), 527–546.

Baser, K. H. C., Özek, T., & Tümen, G. (1992). Essential oils of *Thymus cariensis* and *Thymus haussknechtii*, two endemic species in Turkiye. *Journal of Essential Oil Research*, 4(6), 659–661.

Botelho, M. A., Barros, G., Queiroz, D. B., Carvalho, C. F., Gouvea, J., Patrus, L., ... Araújo-Filho, I. (2016). Nanotechnology in phytotherapy: Antiinflammatory effect of a nanostructured thymol gel from *Lippia sidoides* in acute periodontitis in rats. *Phytotherapy Research*, 30(1), 152–159.

Bozdemir, Ç. (2019). Türkiye'de yetişen kekik türleri, ekonomik önemi ve kullanım alanları. *Yüzüncü Yıl Üniversitesi Tarım Bilimleri Dergisi*, 29(3), 583–594.

Braga, P. C., Alfieri, M., Culici, M., & Dal Sasso, M. (2007). Inhibitory activity of thymol against the formation and viability of *Candida albicans* hyphae. *Mycoses*, 50(6), 502–506. doi: 10.1111/j.1439-0507.2007.01412.x. PMID: 17944714.

Braga, P. C., Sasso, M. D., Culici, M., & Alfieri, M. (2007). Eugenol and thymol, alone or in combination, induce morphological alterations in the envelope of *Candida albicans*. *Fitoterapia*, 78(6), 396–400. doi: 10.1016/j.fitote.2007.02.022. Epub 2007 May 23. PMID: 17590533.

Braga, P. C., Dal Sasso, M., Culici, M., Bianchi, T., Bordoni, L., & Marabini, L. (2006). Anti-inflammatory activity of thymol: Inhibitory effect on the release of human neutrophil elastase. *Pharmacology*, 77(3), 130–136.

Buccellato, G. (1965). Dermatomyositis cured by administration of para-methyl-isopropyl-phenol (thymol). *Giornale italiano di dermatolotia. Minerva dermatologica*, 106(1), 89–94.

Casiglia, S., Bruno, M., Scandolera, E., Senatore, F., & Senatore, F. (2019). Influence of harvesting time on composition of the essential oil of *Thymus capitatus* (L.) Hoffmanns. & Link. growing wild in northern Sicily and its activity on microorganisms affecting historical art crafts. *Arabian Journal of Chemistry*, 12(8), 2704–2712.

Celen Yuceturk, S., Aydogan Turkoglu, S., Kockar, F., Kucukbay, F. Z., & Azaz, A. D. (2021). Essential oil chemical composition, antimicrobial, anticancer, and antioxidant effects of *Thymus convolutus* Klokov in Turkiye. *Zeitschrift für Naturforschung C*, 76(5–6), 193–203.

Ceylan, R., Zengin, G., Uysal, S., Ilhan, V., Aktumsek, A., Kandemir, A., & Anwar, F. (2016). GC-MS analysis and *in vitro* antioxidant and enzyme inhibitory activities of essential oil from aerial parts of endemic *Thymus spathulifolius* Hausskn. et Velen. *Journal of Enzyme Inhibition and Medicinal Chemistry*, 31(6), 983–990.

Ceylan, O., & Ugur, A. (2015). Chemical composition and anti-biofilm activity of *Thymus sipyleus* BOISS. subsp. *sipyleus* BOISS. var. *davisianus* RONNIGER essential oil. *Archives of Pharmacal Research*, 38, 957–965.

Charalambous, M., Raftopoulos, V., Paikousis, L., Katodritis, N., Lambrinou, E., Vomvas, D., Georgiou, M., & Charalambous, A. (2018). The effect of the use of thyme honey in minimizing radiation- induced oral mucositis in head and neck cancer patients: A randomized controlled trial. *European Journal of Oncology Nursing*, 34, 89–97. doi: 10.1016/j.ejon.2018.04.003. Epub 2018 Apr 30. PMID: 29784145.

Chizzola, R., Michitsch, H., & Franz, C. (2008). Antioxidative properties of *Thymus vulgaris* leaves: Comparison of different extracts and essential oil chemotypes. *Journal of Agricultural and Food Chemistry*, 56(16), 6897–6904.

Dapkevicius, A., van Beek, T. A., Lelyveld, G. P., van Veldhuizen, A., de Groot, A., Linssen, J. P., & Venskutonis, R. (2002). Isolation and structure elucidation of radical scavengers from *Thymus vulgaris* leaves. *Journal of Natural Products*, 65(6), 892–896.

Dennison, D. K., Meredith, G. M., Shillitoe, E. J., & Caffesse, R. G. (1995). The antiviral spectrum of Listerine antiseptic. *Oral Surgery, Oral Medicine, Oral Pathology, Oral Radiology, and Endodontology*, 79(4), 442–448.

Ekin, H. N., Orhan, D. D., Orhan, İ. E., Orhan, N., & Aslan, M. (2019). Evaluation of enzyme inhibitory and antioxidant activity of some Lamiaceae plants. *Journal of Research in Pharmacy*, 23(4), 749–758.

Elkiran, O., & Avsar, C. (2020). Chemical composition and biological activities of essential oils of endemic *Thymus leucostomus* Hausskn. et Velen. *Bangladesh Journal of Botany*, 49(4), 957–965.

Elmasulu, S., Karaca, M. (2020). Genetic diversity of some Thymus L. taxa naturally occurring in flora of Antalya based on Td-DAMD-PCR markers. 10th International Ecology Symposium, 26-28 November, Bursa, Turkey.

Erdogan, A., & Ozkan, A. (2013). Effects of *Thymus revolutus* Célak essential oil and its two major components on Hep G2 cells membrane. *Biologia*, 68, 105–111.

Ernst, E., Marz, R., & Sieder, C. (1997). A controlled multi-centre study of herbal versus synthetic secretolytic drugs for acute bronchitis. *Phytomedicine*, 4, 287–293.

Eruygur, N., Ataş, M., Çevir, Ö., & Tekin, M. (2017). Investigating of phytochemicals, antioxidant, antimicrobial and proliferative properties of different extracts of *Thymus spathulifolius* Hausskn. and Velen. endemic medicinal plant from Sivas, Turkiye. *International Journal of Secondary Metabolite*, 4(3, Special Issue 1), 155–166.

ESCOP Monographs, Second Edition, 2003.

Fachini-Queiroz, F. C., Kummer, R., Estevao-Silva, C. F., Carvalho, M. D. D. B., Cunha, J. M., Grespan, R., ... Cuman, R. K. N. (2012). Effects of thymol and carvacrol, constituents of *Thymus vulgaris* L. essential oil, on the inflammatory response. *Evidence-Based Complementary and Alternative Medicine*, 1–10.

Faria, N. C., Kim, J. H., Gonçalves, L. A., Martins Mde, L., Chan, K. L., & Campbell, B. C. (2011). Enhanced activity of antifungal drugs using natural phenolics against yeast strains of *Candida* and *Cryptococcus*. *Letters in Applied Microbiology*, 52(5), 506–513. doi: 10.1111/j.1472-765X.2011.03032.x. Epub 2011 Mar 14. PMID: 21332761.

Fine, D. H., Furgang, D., Barnett, M. L., Drew, C., Steinberg, L., Charles, C. H., & Vincent, J. W. (2000). Effect of an essential oilcontaining antiseptic mouthrinse on plaque and salivary *Streptococcus mutans* levels. *Journal of Clinical Periodontology*, 27(3), 157–161.

Fujita, T., Sezik, E., Tabata, M., Yeşilada, E., Honda, G., Takeda, Y., et al. (1995). Traditional medicine in Turkey VII. Folk medicine in middle and west Black Sea regions. *Economic botany*. 406–422.

Gavliakova, S., Biringerova, Z., Buday, T., Brozmanova, M., Calkovsky, V., Poliacek, I., & Plevkova, J. (2013). Antitussive effects of nasal thymol challenges in healthy volunteers. *Respiratory Physiology & Neurobiology*, 187(1), 104–107. doi: 10.1016/j.resp.2013.02.011. Epub 2013 Feb 21. PMID: 23438788.

Giordani, R., Hadef, Y., & Kaloustian, J. (2008). Compositions and antifungal activities of essential oils of some Algerian aromatic plants. *Fitoterapia*, 79(3), 199–203. doi: 10.1016/j.fitote.2007.11.004. Epub 2007 Nov 29. PMID: 18164558.

Gokturk, R. S., Sagdic, O., Ozkan, G., Unal, O., Aksoy, A., Albayrak, S., ... Durak, M. Z. (2013). Essential oil compositions and bioactivities of *Thymus revolutus* and *Cyclotrichium origanifolium*. *Journal of Essential Oil Bearing Plants*, 16(6), 795–805.

Gonçalves, N. D., de Lima Pena, F., Sartoratto, A., Derlamelina, C., Duarte, M. C. T., Antunes, A. E. C., & Prata, A. S. (2017). Encapsulated thyme (*Thymus vulgaris*) essential oil used as a natural preservative in bakery product. *Food Research International*, 96, 154–160.

Hagedorn, M. (1989). Genitaler vulvärer Lichen sclerosus bei Geschwistern [Genital vulvar lichen sclerosis in 2 siblings]. *Zeitschrift für Hautkrankheiten*, 64(9), 810, 813–814. German. PMID: 2815905.

Hammad, M., Sallal, A. K., & Darmani, H. (2007). Inhibition of *Streptococcus mutans* adhesion to buccal epithelial cells by an aqueous extract of Thymus vulgaris. *International Journal of Dental Hygiene*, 5(4), 232–235.

Haque, M. R., Ansari, S. H., Najmi, A. K., & Ahmad, M. A. (2014). Monoterpene phenolic compound thymol prevents high fat diet-induced obesity in murine model. *Toxicology Mechanisms and Methods*, 24(2), 116–123. doi: 10.3109/15376516.2013.861888. Epub 2013 Dec 5. PMID: 24175857.

Haraguchi, H., Saito, T., Ishikawa, H., Date, H., Kataoka, S., Tamura, Y., & Mizutani, K. (1996). Antiperoxidative components in *Thymus vulgaris*. *Planta Medica*, 62(3), 217–221.

Hashemian, F., Baghbanian, N., Majd, Z., Rouini, M. R., Jahanshahi, J., & Hashemian, F. (2015). The effect of thyme honey nasal spray on chronic rhinosinusitis: A double-blind randomized controlled clinical trial. *European Archives of Oto-Rhino-Laryngology*, 272(6), 1429–1435. doi: 10.1007/s00405-014-3233-x. Epub 2014 Aug 9. PMID: 25106547.

https://davesgarden.com/guides/pf/go/250578/#b.

https://davesgarden.com/guides/pf/go/198356/#b.

http://www.cappadociaexplorer.com/en/detay.php?id=311&cid=179.

http://www.kazdaglari.com/bitkiler/kazdag/FOT-1/thymus/thymus-3.html.

https://turkiyebitkileri.com/tr/foto%C4%9Fraf-galerisi/lamiaceae-ball%C4%B1babagiller/thymus-kekik/thymus-argaeus.html.

https://turkiyebitkileri.com/tr/foto%C4%9Fraf-galerisi/lamiaceae-ball%C4%B1babagiller/thymus-kekik/thymus-cariensis.html.

https://turkiyebitkileri.com/tr/foto%C4%9Fraf-galerisi/lamiaceae-ball%C4%B1babagiller/thymus-kekik/thymus-convolutus.html.

https://turkiyebitkileri.com/tr/foto%C4%9Fraf-galerisi/lamiaceae-ball%C4%B1babagiller/thymus-kekik/thymus-haussknechtii.html.

https://www.ema.europa.eu/en/documents/herbal-report/final-assessment-report-thymus-vulgaris-l-vulgaris-zygis-l-herba_en.pdf.

http://www.vanherbaryum.yyu.edu.tr/flora/famgenustur/la/th/fe/pages/Thymus%20fedtschenkoi%20RONNIGER%20var_%20handelii%20%28RONNIGER%29%20JALAS%20%20%20%20%20_jpg.htm.

Inouye, S., Takizawa, T., & Yamaguchi, H. (2001). Antibacterial activity of essential oils and their major constituents against respiratory tract pathogens by gaseous contact. *Journal of Antimicrobial Chemotherapy*, 47(5), 565–573. doi: 10.1093/jac/47.5.565. PMID: 11328766.

Jayari, A., El Abed, N., Jouini, A., Mohammed Saed Abdul-Wahab, O., Maaroufi, A., & Ben Hadj Ahmed, S. (2018). Antibacterial activity of *Thymus capitatus* and *Thymus algeriensis* essential oils against four food-borne pathogens inoculated in minced beef meat. *Journal of Food Safety*, 38(1), e12409.

Johnson, R. L. (2012). *36 Healing Herbs: The World's Best Medicinal Plants*. Random House Publisher Services, 89 pages.

Kargıoğlu, M., Cenkci, S., Serteser, A., Konuk, M., & Vural, G. (2010). Traditional uses of wild plants in the middle Aegean region of Turkiye. *Human Ecology*, 38(3), 429–450.

Kaska, A., Çiçek, M., Deniz, N., & Mammadov, R. (2018). Investigation of phenolic content, antioxidant capacities, anthelmintic and cytotoxic activities of *Thymus zygioides* Griseb. *Journal of Pharmaceutical Research International*, 1–13.

Kensara, O. A., ElSawy, N. A., El-Shemi, A. G., & Header, E. A. (2013). *Thymus vulgaris* supplementation attenuates blood pressure and aorta damage in hypertensive rats. *Journal of Medicinal Plants Research*, 7(11), 669–676.

Kensara, O. A., ElSawy, N. A., El-Shemi, A. G., & Header, E. A. (2013). *Thymus vulgaris* supplementation attenuates blood pressure and aorta damage in hypertensive rats. *Journal of Medicinal Plants Research*, 7(11), 669–676.

Kılıçgün, H., & Korkmaz, M. (2016). Dose-dependent medicinal effects of *Thymus haussknechtii* Velen grown wild in Turkiye. *Pakistan Journal of Pharmaceutical Sciences*, 29(1), 179–183. PMID: 26826832.

Knols, G., Stal, P. C., & Van Ree, J. W. (1994). Productive coughing complaints: Sirupus Thymi or Bromhexine? A double-blind randomized study. *Huisarts en Wetenschap*, 37, 392–394.

Koch, C., Reichling, J., Schneele, J., & Schnitzler, P. (2008). Inhibitory effect of essential oils against herpes simplex virus type 2. *Phytomedicine*, 15(1–2), 71–78.

Kucukbay, F. Z., Kuyumcu, E., Azaz, A. D., Arabacı, T., & Yuceturk, S. Ç. (2014). Chemical composition of the essential oils of three *Thymus* taxa from Turkiye with antimicrobial and antioxidant activities. *Records of Natural Products*, 8(2), 110–120.

Lenz, E., Müller, C., Mostafa, A., Dzieciolowski, J., Kanrai, P., Dam, S., ... Pleschka, S. (2018). Authorised medicinal product Aspecton® Oral Drops containing thyme extract KMTv24497 shows antiviral activity against viruses which cause respiratory infections. *Journal of Herbal Medicine*, 13, 26–33.

Marzouk, T., Barakat, R., Ragab, A., Badria, F., & Badawy, A. (2015). Lavender-thymol as a new topical aromatherapy preparation for episiotomy: A randomized clinical trial. *Journal of Obstetrics and Gynaecology*, 35(5), 472–475. doi: 10.3109/01443615.2014.970522. PMID: 25384116.

Mastalerz-Migas, A., Doniec, Z., & Płusa, T. (2017). Bronchipret TE® in therapy of acute infections of the respiratory tract. *Polski Merkuriusz Lekarski: Organ Polskiego Towarzystwa Lekarskiego*, 43(258), 255–257.

Meister, A., Bernhardt, G., Christoffel, V., & Buschauer, A. (1999). Antispasmodic activity of *Thymus vulgaris* extract on the isolated guinea-pig trachea: Discrimination between drug and ethanol effects. *Planta Medica*, 65(6), 512–516.

Michalska-Sionkowska, M., Walczak, M., & Sionkowska, A. (2017). Antimicrobial activity of collagen material with thymol addition for potential application as wound dressing. *Polymer Testing*, 63, 360–366.

Miladi, H., Slama, R. B., Mili, D., Zouari, S., Bakhrouf, A., & Ammar, E. (2013). Essential oil of *Thymus vulgaris* L. and *Rosmarinus officinalis* L.: Gas chromatography-mass spectrometry analysis, cytotoxicity and antioxidant properties and antibacterial activities against foodborne pathogens. *Natural Science*, 5, 729–739.

Mohammadi, A., Nazari, H., Imani, S., & Amrollahi, H. (2014). Antifungal activities and chemical composition of some medicinal plants. *Journal de Mycologie Medicale*, 24(2), e1–e8.

Ocaña, A., & Reglero, G. (2012). Effects of thyme extract oils (from *Thymus vulgaris*, *Thymus zygis*, and *Thymus hyemalis*) on cytokine production and gene expression of oxLDL-stimulated THP-1-macrophages. *Journal of Obesity*, 1–11.

Ozcan, M., & Chalchat, J. C. (2004). Aroma profile of *Thymus vulgaris* L. growing wild in Turkiye. *Bulgarian Journal of Plant Physiology*, 30(3–4), 68–73.

PDR for Herbal Medicines. (2000). Fleming T, ed. Physicians' Desk Reference (PDR) for Herbal Medicines. Published by Medical Economics Company, Inc. at Montvale, NJ 07645-1742.

Polat, R., Cakilcioglu, U., & Satıl, F. (2013). Traditional uses of medicinal plants in Solhan (Bingöl-Turkiye). *Journal of Ethnopharmacology*, 148(3), 951–963. doi: 10.1016/j.jep.2013.05.050. Epub 2013 Jun 14. PMID: 23770029.

Ramadan, M. M., Ali, M. M., Ghanem, K. Z., & El-Ghorabe, A. H. (2015). Essential oils from Egyptian aromatic plants as antioxidant and novel anticancer agents in human cancer cell lines. *Grasas y Aceites*, 66(2), e080–e080.

Reiter, M., & Brandt, W. (1985). Relaxant effects on tracheal and ileal smooth muscles of the guinea pig. *Arzneimittel-Forschung*, 35(1A), 408–414.

Sagdic, O., Ozkan, G., Aksoy, A., & Yetim, H. (2009). Bioactivities of essential oil and extract of *Thymus argaeus*, Turkish endemic wild thyme. *Journal of the Science of Food and Agriculture*, 89(5), 791–795.

Salehi, B., Mishra, A. P., Shukla, I., Sharifi-Rad, M., Contreras, M. D. M., Segura-Carretero, A., Fathi, H., Nasrabadi, N. N., Kobarfard, F., Sharifi-Rad, J. (2018). Thymol, thyme, and other plant sources: Health and potential uses. *Phytotherapy Research*, 32(9), 1688–1706. doi: 10.1002/ptr.6109. Epub 2018 May 22. PMID: 29785774.

Saygınl, A. G., Göze, İ., Alim, A., Ercan, N., Durmuş, N., Vural, N., & Alim, B. A. (2018). Essential oil of *Thymus pectinatus* Fisch&Mey. var. *pectinatus*: Chemical formation, antimicrobial, antioxidant, antispasmodic and angiogenic activities. *African Journal of Traditional, Complementary and Alternative Medicines*, 15(1), 34–41.

Seibel, J., Pergola, C., Werz, O., Kryshen, K., Wosikowski, K., Lehner, M. D., & Haunschild, J. (2015). Bronchipret® syrup containing thyme and ivy extracts suppresses bronchoalveolar inflammation and goblet cell hyperplasia in experimental bronchoalveolitis. *Phytomedicine*, 22(13), 1172–1177.

Selma, K. Ö. S. A. (2021) *Thymus revolutus* C. Türünün Sert Odun Çeliklerinin Köklenme Özellikleri Üzerine Yetiştirme Ortamlarının ve IBA Konsantrasyonlarının Etkilerinin Belirlenmesi. *Bartın Orman Fakültesi Dergisi*, 23(2), 595–605.

Shapiro, S., & Guggenheim, B. (1995). The action of thymol on oral bacteria. *Oral Microbiology and Immunology*, 10(4), 241–246.

Shen, Q., Zhou, W., Li, H., Hu, L., & Mo, H. (2016). ROS involves the fungicidal actions of Thymol against spores of *Aspergillus flavus* via the induction of nitric oxide. *PLoS One*, 11(5), e0155647. doi: 10.1371/journal.pone.0155647. PMID: 27196096; PMCID: PMC4872997.

Taherian, A. A., Babae, M., Vafaei, A. A., Jarrahi, M., Jadidi, M., & Sadeghi, H. (2009). Antinociceptive effects of hydroalcoholic extract of *Thymus vulgaris*. *Pakistan Journal of Pharmaceutical Sciences*, 22(1): 83–89.

Taleb, A. M., Qannadi, F., Changizi-Ashtiyani, S., Zarei, A., Rezvanfar, M. R., Akbari, A., & Hekmatpou, D. (2017). The effect of aqueous extract *Thymus kotschyanus* boiss. Et hohen on glycemic control and dyslipidemia associated with type II diabetes: A randomized controlled trial. *Iranian Journal of Endocrinology and Metabolism*, 19(4), 234–243.

Tetik, F., Civelek, S., & Cakilcioglu, U. (2013). Traditional uses of some medicinal plants in Malatya (Turkiye). *Journal of Ethnopharmacology*, 146(1), 331–346. doi: 10.1016/j.jep.2012.12.054. Epub 2013 Jan 17. PMID: 23333750.

Tuzlacı, E., & Eryaşar-Aymaz, P. (2001). Turkish folk medicinal plants, part IV: Gönen (Balıkesir). *Fitoterapia*, 72, 323–343.

Twetman, S., & Petersson, L. G. (1999). Interdental caries incidence and progression in relation to mutans streptococci suppression after chlorhexidine-thymol varnish treatments in schoolchildren. *Acta Odontologica Scandinavica*, 57(3), 144–148.

Van den Broucke, C. O., & Lemli, J. A. (1981). Pharmacological and chemical investigation of thyme liquid extracts. *Planta Medica*, 41(2), 129–135.

Van Den Broucke, C. O., & Lemli, J. A. (1983). Spasmolytic activity of the flavonoids from *Thymus vulgaris*. *Pharmaceutisch Weekblad*, 5(1), 9–14.

Vimalanathan, S., & Hudson, J. (2014). Anti-influenza virus activity of essential oils and vapors. *American Journal of Essential Oils and Natural Products*, 2(1), 47–53.

Vladimir-Knežević, S., Kosalec, I., Babac, M., Petrović, M., Ralić, J., Matica, B., & Blažeković, B. (2012). Antimicrobial activity of *Thymus longicaulis* C. Presl essential oil against respiratory pathogens. *Central European Journal of Biology*, 7(6), 1109–1115.

World Health Organization. (1999). *WHO Monographs on Selected Medicinal Plants*, Vol. 1. Geneva. Typeset in Hong Kong, Printed in Malta 97/11795-Best-set/Interprint-6500, ISBN 92-4-154517-8

Yigitkan, S., Akdeniz, M., Yener, I., Seker, Z., Yilmaz, M. A., Firat, M., ... Orhan, I. E. (2022). Comprehensive study of chemical composition and biological activity of *Thymus pubescens* Boiss. et Kotschy ex Celak. *South African Journal of Botany*, 149, 425–434.

Yu, Y. M., Chao, T. Y., Chang, W. C., Chang, M. J., & Lee, M. F. (2016). Thymol reduces oxidative stress, aortic intimal thickening, and inflammation-related gene expression in hyperlipidemic rabbits. *Journal of Food and Drug Analysis*, 24(3), 556–563. doi: 10.1016/j.jfda.2016.02.004. Epub 2016 Apr 12. PMID: 28911561.

Zhou, E., Fu, Y., Wei, Z., Yu, Y., Zhang, X., & Yang, Z. (2014). Thymol attenuates allergic airway inflammation in ovalbumin (OVA)-induced mouse asthma. *Fitoterapia*, 96, 131–137.

Zu, Y., Yu, H., Liang, L., Fu, Y., Efferth, T., Liu, X., & Wu, N. (2010). Activities of ten essential oils towards *Propionibacterium acnes* and PC-3, A-549 and MCF-7 cancer cells. *Molecules*, 15(5), 3200–3210. doi: 10.3390/molecules15053200. PMID: 20657472; PMCID: PMC6263286.

Zuleyha, O. (2020). Antioxidant activities of aqueous extracts from two *Thymus* species growing in Turkiye. *GSC Biological and Pharmaceutical Sciences*, 10(3), 16–22.

27 *Trigonella foenum-graecum* L.

Erkan Yılmaz

INTRODUCTION

Trigonella foenum-graecum, commonly known as fenugreek in English, is an aromatic and medicinal plant belonging to the family Fabaceae (Hilles and Mahmood 2021). The name of the plant, *Trigonella*, originates from the Greek word *trigonou* meaning triangle, because of the triangular-shaped leaflets. The term foenum-graecum owes its name to its extensive use, which is referred to with expressions such as "Greek hay" or "Greek grass" in ancient Greece (Bahmani et al. 2016). The history of fenugreek can be traced back to Egyptian papyrus dated to 1500 B.C., where the plant was used as embalming material for mummies. From ancient times, it has been utilized in the treatment of many diseases such as diabetes, inflammation, cancer, reproductive problems, microbial infection, respiratory diseases, etc. (Singh et al. 2022). In addition to the medicinal properties of fenugreek, it is also known for its culinary values. The herb is widely used as a spice, which not only improves the taste of food but also has the potential to significantly improve overall health (Żuk-Gołaszewska and Wierzbowska 2017). Phytochemical assessments have revealed that fenugreek includes many secondary metabolites. Among them, saponins, alkaloids, and flavonoids are the major classes of phytochemicals (Chaudhary et al. 2018). Modern pharmacological studies have shown that fenugreek possesses antidiabetic, anti-inflammatory, antilipidemic, antimicrobial, antioxidant, anticarcinogenic, anti-obesity, immunomodulator, gastroprotective effects, and acting on reproductive system disorders (Nagulapalli Venkata et al. 2017; Syed et al. 2020; Almatroodi et al. 2021; Singh et al. 2022).

PLANT CHARACTERISTICS

Morphologically, it is herbaceous and annual plant reaching up to 50 cm in height with sparingly pubescent to glabrescent, branched stem (Huber-Morath 1972; Bahmani et al. 2016). Leaves have dark green, serrated, oval, consisting of three small (10–30 x 5–15 mm) obovate to oblanceolate, dentate, or sometimes incised leaflets distributed from one point (Huber-Morath 1972; Bahmani et al. 2016; Hilles and Mahmood 2021). Flowers are clustered in leafily. Tubular calyx is 7–8 mm long. Corolla is yellowish-white, sometimes tinged with lilac, 12–18 mm long (Huber-Morath 1972). Fruit morphology is legumes. Seed-bearing pods, which are greenish or reddish, turning brown on ripening, contain 10–20 oblong (3–4 x 2–3 mm) seeds (Huber-Morath 1972; Hilles and Mahmood 2021). Maturation period takes four to seven months for this fast-growing plant (Hilles and Mahmood 2021). The leaves and seeds of *T. foenum-graecum* are depicted in Figure 27.1.

ORIGIN AND DISTRIBUTION

T. foenum-graecum is a member of genus *Trigonella* consisting of 260 species in the world (Hilles and Mahmood 2021). Origin of fenugreek is still doubtful and thought that probably mostly the plant escaped from cultivation (Huber-Morath 1972). It is widely cultivated throughout many parts of the world, including Egypt, India, Pakistan, Russia, Canada, Afghanistan, Iran, Morocco, Ethiopia, China, Turkey, Chile, Australia, Argentina, and the USA because of its great economic importance

DOI: 10.1201/9781003146971-27

FIGURE 27.1 The leaves and seeds photos of *Trigonella foenum-graecum*. (Photos: Erkan Yılmaz.)

(Hilles and Mahmood 2021; Singh et al. 2022). Some countries such as India, Lebanon, France, Egypt, and Argentina take the lead in the trade of the plant (Hilles and Mahmood 2021).

TRADITIONAL USES OF FENUGREEK

T. foenum-graecum is one of the important and old medicinal plants, which has a long medicinal history in the world. Fenugreek has always been part of the traditional healing system (Basch et al. 2003). Its traditional uses can be encountered in Ayurveda, Iranian traditional medicinal system, and many pharmacopeias including Arabic, Chinese, Greek, Latin, and Korean. According to writings, many uses such as curing different ailments such as inflammation, migraine, joint pain, mucosal, digestion, and other gastric issues are mentioned. Moreover, lactation stimulant, easing childbirth, curing metabolic disorders, and vaginal ailments are documented in these writings (Singh et al. 2022). More recent ethnopharmacological studies show that fenugreek has been used in many parts of the world for curing a number of diseases due to its wide distribution over the world (Tabata et al. 1994; Miara et al. 2019; Idm'Hand et al. 2020; Prabhu et al. 2021). Ethnobotanical uses regarding *T. foenum-graecum* with the geographic region where the plant is used, its therapeutic effect, and plant parts used are summarized in Table 27.1. These data gathered from ethnobotanical survey display that the most frequent traditional use is antidiabetic utilization (Ocvirk et al. 2013; Bhatia et al. 2014; Ullah et al. 2020). Meanwhile, it has been used to treat the symptoms related to inflammation (Ouelbani et al. 2016; Idm'Hand et al. 2020), wounds (Ugulu et al. 2009; Malik et al. 2019), and problems in the reproductive system of women (Ghorbani 2005; Kılıç et al. 2020). Many countries including Algeria (Miara et al. 2019), Saudi Arabia (Abdelhalim and Saleem 2021), and Syria (Alachkar et al. 2011) accepted this plant as a galactagogue. Moreover, seeds have been employed to cure cancer in some countries like Algeria (Taïbi et al. 2020) and Morocco (Merrouni and Elachouri 2021). Traditionally, Arabs and Persian women utilized seeds of fenugreek to increase lean muscle mass. It is reported that in societies where fenugreek is consumed as food, menopausal symptoms are milder (Srinivasan 2006). Flowers, fruits, leaves, and especially seeds are used for curing many symptoms and illnesses (AbouZid and Mohamed 2011; Alalwan et al. 2019; Malik et al. 2019). The seeds of fenugreek also function as a natural food preservative with a broad range and are added to chutneys, pickles, and other similar products (Moradi Kor and Zadeh 2013).

TABLE 27.1
Ethnobotanical Uses Regarding *Trigonella foenum-graecum* with Geographic Region, Its Medicinal Uses and Plant Parts Used.

Geographic regions	Therapeutic effects, ailments treated	Plant parts	References
Algeria	Anti-inflammatory, diuretic, wounds, appetite stimulant, immune system diseases, galactagogue, respiratory system diseases, digestive disorders, metabolic system diseases, anemia	–	(Ouelbani et al. 2016)
Algeria	Cancer therapy	Seeds	(Taïbi et al. 2020)
Algeria	Stomach pains, bronchitis, diuretic, galactagogue, antidiabetic, slimming, appetite stimulant	Seeds	(Miara et al. 2019)
Bahrain	Antidiabetic, litholitic, oxytocic, emmenagogue, expectorant	Seeds, fruits, leaves	(Alalwan et al. 2019)
Bangladesh	Diarrhea, dyspepsia, joint pain	Seeds	(Islam et al. 2014)
Bangladesh	Antidiabetic	Seeds, whole plants	(Ocvirk et al. 2013)
Egypt	Diuretic, colic, nutritive, antidiabetic	Seeds	(AbouZid and Mohamed 2011)
Saudi Arabia	Antidiabetic, antipyretic, diuretic	Seeds	(Ullah et al. 2020)
Saudi Arabia	Galactagogue, antidiabetic, menstrual colic, helping delivery of newborn	Seeds	(Abdelhalim and Saleem 2021)
India	Antidiabetic, gonorrhea, strangury, mucous diarrhea	Seeds	(Prabhu et al. 2021)
India	Antidiabetic, arthralgia	Seeds	(Bhatia et al. 2014)
Iran	Menstrual cramps	Fruits	(Khajoei Nasab and Khosravi 2014)
Iran	Respiratory complaints, infection, anti-hyperlipidemic, women infertility	Aerial parts	(Ghorbani 2005)
Iraq	Antidiabetic, antihypertensive, anti-hyperlipidemic, hemostatic for diabetics, galactagogue, appetite stimulant, sedative, catarrh, vulnerary	Seeds	(Mati and De Boer 2011)
Italy	Sprain pains	Aerial parts	(Menale et al. 2016)
Morocco	Antidiabetic	Seeds	(Skalli et al. 2019)
Morocco	Cancer therapy	Seeds	(Merrouni and Elachouri 2021)
Morocco	Emmenagogue, rheumatism, stomach pain, antidiabetic, blood purification, appetite stimulant	Seeds	(Idm'Hand et al. 2020)
Pakistan	Wound healing	Leaves, flowers	(Malik et al. 2019)
Palestine	Urinary system and stones	Seeds	(Ali-Shtayeh et al. 2000)
Sudan	Stomach ache	Seeds	(Adam et al. 2020)
Syria	Antidiabetic, antitussive, anti-hyperlipidemic, galactagogue, bruises	Seeds	(Alachkar et al. 2011)
Turkey	Bronchitis, expectorant, stomachic, wounds, furuncle	Seeds	(Ugulu et al. 2009)
Turkey	Antidiabetic	Seeds	(Tabata et al. 1994)
Turkey	Prostate, gynecological diseases	Leaves, stem, seeds	(Kılıç et al. 2020)
Yemen	Diuretic, laxative	Seeds	(Hussein and Dhabe 2018)

PHYTOCHEMISTRY AND NUTRITIONAL PROPERTIES

The seeds and leaves of fenugreek are very important as they possess valuable ingredients for food and pharmaceutical industries (Niknam et al. 2021). Fenugreek seeds have a rich source of secondary metabolites and a significant role in the cure of many diseases. Management through regulation of various biological activities may be attributed to these constituents (Almatroodi et al. 2021). Fenugreek is dominated by three main classes of secondary metabolites including saponins, alkaloids, and flavonoids (Chaudhary et al. 2018). In the study in which metabolomic profiling of seeds derived from three *Trigonella* species including *T. foenum-graecum*, *T. corniculata*, and *T. caerulea* were assessed, 93 metabolites including 5 peptides, 22 flavonoid conjugates, 2 phenolic acids, 26 saponins, and 9 fatty acids were identified by using UPLC-MS. Among the three species examined, *T. foenum-graecum* was found to be a better source in terms of secondary bioactive metabolites (Farag et al. 2016). The fenugreek is particularly rich in saponins which is the most abundant class of phytoconstituents (Nagulapalli Venkata et al. 2017). The most diverse and characteristic steroidal saponins of fenugreek are diosgenin and yamogenin. Other sapogenins including gitogenin, smilagenin, yuccagenin, and sarsapogenin are also present in the seeds (Bahmani et al. 2016). Apart from the seeds, leaves also contain some saponins, known as graecunins, glycosides of diosgenin (Varshney et al. 1984). The growing biological and economic interest in diosgenin is because of its role as advanced product intermediate for the synthesis of important drugs like oral contraceptives and steroid hormone and also due to its wide range of pharmacological effects as anticancer and anti-inflammatory agents (Nagulapalli Venkata et al. 2017). Other important group of phytochemicals in fenugreek is alkaloids. Phytochemical investigations of the plant demonstrates its richness in various pyridine-type alkaloids. Trigonelline is known as the most substantial among them (Zandi et al. 2015). Several other alkaloids such as choline, gentianine, and carpaine have been identified from this plant (Bahmani et al. 2016). What makes trigonellin special from other substances is that it has hypoglycemic, neuroprotective, and ability of estrogen receptor activity (Nagulapalli Venkata et al. 2017). Another important compound which may be useful for the management of diabetes by increasing secretion of insulin is a free amino acid namely 4-hydroxyisoleucine (Zandi et al. 2015). Moreover, fenugreek is rich in dietary fibers. There are two types of dietary fibers in fenugreek seeds, insoluble (30%) and soluble fraction (20%), which is mostly made of galactomannan (Srinivasan 2006). Furthermore, several molecules belonging to flavonoids, such as quercetin, luteolin, isovitexin, quercitrin, afroside, vitexin, vicenin-1, and vicenin-2 were detected in fenugreek (Nagulapalli Venkata et al. 2017). Thirty-two phenolic compounds among which various phenolic acid and flavonoid glycosides were characterized from crude seeds by HPLC–DAD–ESI/MS analysis (Benayad et al. 2014). Additionally, fenugreek seeds contain small amount of volatile oils and around 7% fixed oil (Zandi et al. 2015).

Several parts of the plant, fresh or dried leaves, tender stems, and seeds are edible. The characteristic bitterness of fenugreek is mainly due to the oil along with alkaloids and steroidal saponins (Srinivasan 2006). It is considered as an important source of dietary fiber, proteins, amino acids, and fatty acids (Chaudhary et al. 2018). Being rich in a number of vitamins (A, B1, B2, C, nicotinic acid, niacin), a good source of major dietary elements (Ca, Fe, S, P, Mg), and having valuable trace dietary elements (Zn, Cu, Mn, Co, Br) appear to be important in terms of nutritional composition (Nagulapalli Venkata et al. 2017).

ANTIDIABETIC ACTIVITY

The possible hypoglycemic properties of fenugreek have been investigated in humans and several animal models such as rats (streptozotocin-induced or alloxanized rats), diabetic obese KK-Ay mice, and db/db mice (Mooventhan and Nivethitha 2017). In a streptozotocin-induced diabetes rat model, fenugreek seed extract at a dose of 100 mg/kg was given either orally or intraperitoneally (daily or every other day) for four weeks. Generally, fenugreek administration in both ways significantly reduced blood glucose, triglycerides, and AST and ALT levels, while fenugreek treatment caused a significant rise in protein levels. Increased catalase and glutathione S-transferase levels were detected with treatment, while peroxidase antioxidant enzyme levels were observed to be reduced. Significant changes in cholesterol levels were not seen after administration. When evaluated in terms of differences between the two application ways, increased high-density lipoprotein and reduced urea levels were achieved by daily injections. It was observed that rats administered orally demonstrated a reduction in creatinine levels. Increased glutathione peroxidase levels were also seen only after daily injection. This study revealed that although improvements in the antioxidant enzymes and other diabetic parameters were observed in all groups, daily injection of fenugreek has better antidiabetic activity with better serum values than other groups (Baset et al. 2020). In a clinical study, fenugreek seed powder was administered orally in sachet form (20 g/day) to type 2 diabetes (T2DM) patients divided into two groups (control, treated). Demographic profiles of patients and their laboratory values including fasting blood glucose level, random blood glucose level, and HbA1c level were screened at baseline, after one month and then at the end of the clinical study. Sachet form receiving group of T2DM patients along with dietary guidelines displayed a gradual decrease in fasting blood glucose, random blood glucose, and HbA1c levels, significantly compared to the control group (Rehman et al. 2021). In another clinical trial, multicenter, randomized, placebo-controlled, double-blind, add-on clinical study was planned to investigate the efficacy of a patented fenugreek seed extract enriched in approximately 40% furostanolic saponins at doses of 500 mg bid. This study was conducted with 154 subjects with T2DM over a period of 90 days consecutively. It was reported that significant fasting plasma and post-prandial blood sugar lowering activity were observed with the administration of the seed extract. Reduced HbA1c levels in both placebo and treatment groups were also seen. A significant increase in values of fasting and post-prandial C-peptide compared to the respective baseline values was reported, while no significant changes in fasting and post-prandial C-peptide values were seen between the two groups. No significant adverse events were seen by analysis of blood parameters. Moreover, reduced dosages of antidiabetic therapy in the extract-treated group and the placebo group were reported in 48.8% and 18.05% of subjects, respectively (Verma et al. 2016). A meta-analysis of ten clinical trials revealed that use of fenugreek seeds showed a significant reduction in fasting blood glucose, 2 hours post load glucose, and HbA1c levels. Moreover, no major harmful side effects caused by fenugreek treatment were reported in all studies (Neelakantan et al. 2014).

ANTI-INFLAMMATORY ACTIVITY

Anti-inflammatory effects of the *T. foenum-graecum* were examined *in vivo* and *in vitro* methods (Almatroodi et al. 2021). *In vitro* anti-inflammatory assessment and COX-2 inhibitory actions of fenugreek seed extracts were evaluated using the human red blood cell membrane stabilization and COX-2 enzyme immunoassay kit, respectively. The aqueous and ethanol extracts of seeds at a concentration of 200 μg/mL were found to display a significant protection against inflammation at the percentage of 61.4% and 56.9%, respectively. COX-2 inhibition potency was detected as 59.2% for aqueous extract and as 57.9% for ethanol extract in comparison to indomethacin (78.3%) (Ramadhan 2020). In other research, ethanolic extract of seeds was used for evaluating anti-inflammatory effects on carrageenan-induced inflammation in rats. The extract at the dose of 200 mg/kg did not exhibit a significant anti-inflammatory effect. However, a moderate anti-inflammatory effect at the dose of 400 mg/kg was seen at the end of two hours when compared to Aspirin used as a standard drug (Narapogu et al. 2021). In other study, anti-arthritic activity of ethanol extract derived from fenugreek seeds was tested against Freund's complete adjuvant-induced arthritic rats. Volume of paw, hematological properties, tumour necrosis factor (TNF)-α and interleukins (IL-1α, IL-1β, IL-2, IL-6) in blood, and antioxidant parameters in cartilage were monitored. Administration of extract to the rats with both doses of 200 and 400 mg/kg significantly reduced the paw edema, restored loss in body weight, and improved hematological parameters. Fenugreek was found to decrease the levels of inflammatory cytokines significantly. Decreasing levels of lipid peroxidation and increasing levels of superoxide dismutase and glutathione in cartilage tissue were also seen. The higher dose of extract was found to display more prominent results when compared to the lower dose of extract (Suresh et al. 2012). In another study investigating anti-inflammatory activity of methanolic extract of seeds, carrageenan-induced edema method was used and the efficiency of the extracts was compared with dexamethasone and ibuprofen. Different doses of the extract at doses of 100, 200, and 400 mg/kg were injected intraperitoneally to left hind paw. The edema inhibition activity of the plant at doses of 100 and 200 was found to show no significant difference when compared to ibuprofen- and dexamethasone-mediated inhibitions. Moreover, several cream formulation of the plant extract (2–5%) were tested for anti-inflammatory effects compared with reference drug (1% hydrocortisone ointment). Among them, 3% and 5% creams prepared from plant extrats were found to exhibit the most inhibition of edema. No significant differences between the 1% hydrocortisone ointment and the formulations of cream at concentrations of 3% and 5% were observed (Sharififara et al. 2009).

ANTI-HYPERLIPIDEMIC EFFECT

Hypocholesterolemic activity of fenugreek has been extensively studied in experimental animals and clinical studies by researchers (Nagulapalli Venkata et al. 2017). The fenugreek ethanol extract derived from seeds has been investigated in diet-induced hypercholesterolemic rats. Thirty or fifty gram ethanol extract/kg administered to the rats for four week caused reductions in cholesterol levels in plasma ranging from 18% to 26% along with a tendency for lower liver cholesterol levels. It was considered that hypocholesterolemic effects of extracts were caused by the saponin in the extracts that interacted with bile salts in the digestive tract (Stark and Madar 1993). In another study investigating the effect of ethyl acetate extract of seeds on the lipoprotein profile in rats taking a high cholesterol diet, administration of ethyl acetate extract was found to display anti-hyperlipidemic action along with a decreased rate of lipid peroxidation and regulated antioxidant system (Belguith-Hadriche et al. 2010). In another research, the fenugreek leaves powder (2 g and 8 g/100 g chow) has been investigated for its effects on the serum lipid profile, aspartate transaminases, alanine transaminases, and liver LDL receptor gene expression in male hamsters fed with a high cholesterol diet. Total cholesterol, serum triglyceride, LDL and VLDL, and aspartate and alanine transaminases' levels notably decreased in animals taking high cholesterol diet and 8 g powder group. Moreover, increased high-density lipoprotein cholesterol level in serum was observed. Although no significant change in the liver LDL receptor gene expression was observed in high cholesterol diet and 2 g powder receiving group, gene expression in high cholesterol diet and 8 g powder receiving group increased 7.8-fold (Kassaee et al. 2021). In a clinical trial conducted on 20 hypocholesterolemic adults of both sexes between 50 and 65 years old, they consumed 12.5 g and 18.0 g germinated seed powder for 30 days. Although both doses of seed powder consumption showed a hypocholesterolemic effect, a significant decrease in total cholesterol and LDL levels were seen with a higher dose. No significant changes were detected in VLDL, HDL, and triglyceride levels in all subjects (Sowmya and Rajyalakshmi 1999). A meta-analysis was carried out to evaluate lipid-lowering effects of fenugreek supplementation on the human serum lipid profile. Data sources on this analysis collected from 2000 up to 2019 revealed that fenugreek supplementation improved lipid parameters including LDL cholesterol, triglyceride, total cholesterol, and HDL cholesterol levels. At the end of the analysis, it was said that fenugreek can be considered an effective lipid-lowering agent (Heshmat-Ghahdarijani et al. 2020).

ANTIOXIDANT ACTIVITY

Aqueous ethanol (50%) and methanol extracts prepared from the seeds of some fenugreek cultivars were assessed for antioxidant potentials using different types of tests. Aqueous ethanol extracts of seeds exhibited high levels of total phenolic content, total flavonoids content, and antioxidant activity according to three model systems including 2,2′-Azino-bis (3-ethylbenzothiazoline-6-sulfonate) ($ABTS^+$), total antioxidant capacity (TAC), and reducing power activity (RPA) as compared to methanol extract. However, methanolic extracts were found to show higher condensed tannin content and 1,1-diphenyl-2-picrylhydrazyl (DPPH) radical scavenging activity. Additionally, aqueous ethanol extracts were evaluated for DNA damage protection activity against *in vitro* plasmid deoxyribonucleic acid (DNA) (pBR322). The extracts showed great potential for DNA protection (Dhull et al. 2020). In other research, the aqueous extract prepared from seeds was found to exhibit potent $ABTS^+$ and DPPH radical scavenging activity with IC_{50} values of 71.44 μg/mL and 62.67 μg/mL, respectively (Kumar et al. 2021). In one of the studies conducted on different plant parts other than seeds, antioxidant properties of the leaves were studied using DPPH test. The finding proved that ethanolic leaf extract has excellent free radical scavenging activities (Akhlaghi and Najafpour-Darzi 2021). In another study aiming to determine *in vivo* antioxidant effect of leaf and seed extracts, a mice model of oxidative stress generated by treatment with mercuric chloride was used to evaluate methanol extract of both parts. Biomarkers of oxidative stress including catalase, super oxide dismutase, xanthine oxidase, and glutathione reductase enzyme activities were measured after feeding of these extracts orally. The antioxidant potential of the two extracts was also determined using DPPH assay. Seed extract was found to be a more potent antioxidant agent than the extract of leaves. Supplementation with methanolic seed extract was found to decrease lipid peroxidation and increase the levels of antioxidant enzymes significantly and brought them close to the normal range. Leaf extract was determined having the capacity to augment the antioxidant defense mechanism but to a lesser extent than the seed extract (Jha and Srivastava 2012).

ANTICANCER ACTIVITY

Several reported studies have exhibited the anticancer potential of *T. foenum-graecum* in experimental models using cancer cell lines or animals (El Bairi et al. 2017). The management of different types of cancers including breast, hematological, lung, prostate, skin gastrointestinal tract, and associated cancers has been published (Nagulapalli Venkata et al. 2017). A study conducted by Khoja (2022) evaluated the cytotoxic potential and antiproliferative activities of fenugreek seeds and sprouts methanol extracts in MCF-7 breast cancer cells. The cell viability assay on MCF-7 cells was measured using 3-(4,5-dimethylthiazol-2-yl)-2,5-diphenyltetrazolium bromide to observe the effects of fenugreek extracts. Time and dose-dependent effects on cell viability were reported. The MCF-7 cancer cells treated with both extracts of fenugreek also exhibited increased relative mitochondrial DNA damage and suppressed metastasis and proliferation. Another study was performed to investigate *in vivo* anticancer activity. The seed powder given in the diet to 1,2-dimethylhydrazine treated rats was found to reduce the colon tumor incidence to 16.6%, decrease hepatic lipid peroxidation, and increase the activities of glutathione S-transferase, catalase, superoxide dismutase, and glutathione peroxidase in the liver (Devasena and Menon 2007). In another *in vivo* study conducted on active constituents of fenugreek, intra-tumoral administration of diosgenin exhibited significant regression in human breast cancer cells MCF-7 and MDA-MB-231 tumor xenografts in nude female mice (Srinivasan et al. 2009).

ANTIMICROBIAL ACTIVITY

The antibacterial activity of ethanol and water extracts of fenugreek seeds was evaluated against six clinically isolated pathogenic bacteria using the diffusion agar method. The ethanol extract of fenugreek seeds exhibited considerable activity against *Staphylococcus aureus* and *Pseudomonas aeruginosa*. It did not exhibit antibacterial activity against *Vibrio parahaemolyticus* and *Escherichia coli*. Besides, the aqueous extract was found to possess low to moderate activities against all bacterial strains except *Staphylococcus aureus* and *Escherichia coli*. MIC value ranging from 50 to

500 μg/mL was detected for ethanol extract except *Vibrio parahaemolyticus* which was found to be the most resistant strain (Al-timimi 2019). In another study different extracts prepared from leaves, seeds, and stem of fenugreek using various solvents (methanol, acetone, water) were tested against *Escherichia coli* and *Staphylococcus*. The methanol extract was found to be susceptible to *Escherichia coli* and *Staphylococcus* with inhibition zone 20 mm and 19 mm, respectively. Acetone extract was found to display an equal 16 mm inhibition zone for both bacteria, while the aqueous extract did not show any inhibition zone. When evaluated in terms of plant parts, leaf extract was found to be much more effective than seeds and stems extracts (Sharma et al. 2017). In a different study conducted on popular nanoscience technic, silver nanoparticles using fenugreek aqueous extract was found to have good inhibition potential against *Escherichia coli* and *Bacillus cereus*. However, low antibacterial activity was seen against *Staphylococcus aureus*. Another important point was that the aqueous extract did not act on all bacteria (Awad et al. 2021). In another study using leaves of fenugreek for synthesizing silver nanoparticles, the synthesized silver nanoparticles were found to have higher antibacterial effect against *Pseudomonas aeruginosa* and *Escherichia coli* as compared to other bacteria *Staphylococcus aureus* and *Vibrio cholera*. It was also seen that effects of silver nanoparticles on *Pseudomonas aeruginosa* and *Escherichia coli* were higher than antibiotics, Rifampicin (Ghoshal and Singh 2022).

ANTI-OBESITY EFFECT

Administration of aqueous extract of seeds to high-fat diet-induced obese rats was seen to prevent fat accumulation and dyslipidemia. Indicators of obesity like blood glucose, serum insulin, weight gain, body mass index, white adipose tissue weights, lipids, leptin, lipase reduction, and elevation in apolipoprotein-B levels were observed. Besides, some hepatic and cardiac parameters were restored to normal levels. In addition to the increased antioxidant system, some hepatic and cardiac risk factors were improved (Kumar et al. 2014). In a clinical trial conducted by Chevassus et al. (2009) aiming to evaluate the effects of a repeated administration of hydro-alcoholic seed extracts (588 and 1176 mg) on eating behavior and energy intake in healthy human volunteers, double-blind randomized placebo-controlled clinical study lasting six weeks showed that daily fat consumption was significantly reduced by 1176 mg seed extract with a tendency towards a reduction in total energy intake. Moreover, fenugreek seed extract did not increase appetite or food consumption.

GASTROPROTECTIVE AND ANTI-LITHOGENIC EFFECTS

Gastroprotective efficacy of seed extract was tested *in vitro* and *in vivo* models. Moreover, *in silico* assessment on H^+/K^+ ATPase receptor was used for elucidation of the possible mechanism of fenugreek seeds. Cell line of human gastric carcinoma epithelial was utilized as an *in vitro* model to understand the gastro-protective effect of aqueous seed extract. The gastroprotective effect of the extract *in vivo* was investigated using Sprague Dawley rats. In cell lines, fenugreek extract was found to protect against the damage induced by ethanol at 5 μg/mL, while in the animal studies a protection of 67% was seen at the dose of 1000 mg/kg. It was also observed in *in silico* analysis that H^+/K^+ ATPase interacted well with the flavonoid derivatives including vicenin-2, vitexin-7-O-glucoside, luteolin, and orientin, whereas the saponins lacked good interaction (Figer et al. 2017).

A recently published study has examined the potential of fenugreek seeds as an anti-lithogenic effect in gallstone-induced Swiss mice. It has been demonstrated that the inclusion of fenugreek in the high cholesterolemic diet at levels 5%, 10%, and 15% significantly reduced the incidence of gallstones in mice by 63%, 40%, and 10%, respectively. The biliary cholesterol saturation index fell from 2.57 to 0.77–0.99 during fenugreek treatment. Moreover, 10 weeks of feeding of fenugreek seeds with 6% and 12% caused regression of pre-established gallstone by 61% and 64%. At the end of the study, it was shown that fenugreek seeds not only prevent the formation of stones but also eliminate the preexisted gallstone in mice (Reddy et al. 2019).

ROLE IN REPRODUCTIVE SYSTEM

The efficacy and safety of a patented seed extract were evaluated in 100 healthy volunteers aged 35 to 60 years over a period of 12 consecutive weeks in terms of potential benefits on reproductive systems. Significant improvements in free and bound testosterone levels were seen at 12 weeks of treatment with seed extract. An increase in sperm motility was seen significantly at 8 and 12 weeks of treatment, while abnormal sperm morphology was found to be decreased significantly at 12 weeks of treatment with seed extract. Furthermore, remarkable alleviation of pain mental alertness, mood, and reflex erection score was reported. It has been observed that these effects increase with age. Moreover, cardiovascular health and libido were also significantly improved. Hematological values were found to display broad-spectrum safety. No adverse events in clinical trials were observed (Sankhwar et al. 2021).

TOXICOLOGY AND SAFETY PROFILE

Fenugreek has been used as a traditional herbal remedy since ancient times. It has been reported as a safe plant and a widely used food additive (Singh et al. 2022). The available data demonstrate some adverse and toxic effects apart from the beneficial effects of fenugreek. The reported adverse events are transient diarrhea, flatulence, dizziness, and allergic reactions (Syed et al. 2020). Patients taking oral antidiabetic agents or using insulin should be monitored closely, because of the hypoglycemic activity. Altered levels of thyroid hormones might be seen after fenugreek consumption (Yadav and Baquer 2014). Coumarin derivatives found in fenugreek may increase the risk of bleeding complications (Syed et al. 2020). Neurodevelopmental, neurobehavioral, and neuropathological side effects can be observed (Ouzir et al. 2016). Seeds of fenugreek are recommended for increasing milk production traditionally in nursing women and are widely preferred during pregnancy (Taloub et al. 2013). Consumption of fenugreek seeds may lead to some serious toxicological side effects. Literature reports have found a relationship between seed consumption during pregnancy and a range of congenital malformations, including anencephaly, hydrocephalus, and spina bifida (Bernstein et al. 2021). Moreover, a current publication obtained from studies on rodent models indicated testicular toxicity and anti-fertility effects in males, as well as anti-fertility, anti-implantation, and abortifacient activity in females related to saponins of the plant. The use of fenugreek seeds seems not safe during pregnancy (Ouzir et al. 2016).

CONCLUSION AND FUTURE PERSPECTIVES

Fenugreek has been one of the most valuable plants for disease prevention and remediation since ancient times. It has been used by a growing number of people in the treatment of many diseases including diabetes, hyperlipidemia, obesity, gastrointestinal system ailments, inflammation, reproductive problems, and cancer because of its traditional reputation and many studies conducted on it. Traditional uses of fenugreek have been confirmed by several studies pertaining to pharmacological functions. However, a comprehensive analysis should be conducted to understand some unexplained mechanisms of action of the fenugreek. Scientific reports available on phytochemical studies of the fenugreek have focused on some important molecules, especially 4-hydroxyisoleucine, diosgenin, trigonellin, and galactomannan; thus, more scientific efforts are required to isolate active moieties from other class of secondary metabolites for different therapeutic potential. In addition to the presence of many preclinical models on the efficacy of fenugreek, clinical studies are predominantly concentrated on diabetes, hyperlipidemia, obesity, and reproductive health. However, clinical research on its pharmacologically active constituents have been scarce. At the same time, although clinical studies generally give an idea about short-term results in terms of toxicology and safety profile, there are still uncertainties about long-term use. Studies on fenugreek do not give substantial clarity about toxicological evaluation. Moreover, the applications to special groups including children, elderly population, pregnant women, and people with chronic illnesses also need to be concerned. For example, many questions are waiting to be clarified about the results of the use of many fenugreek-containing products by pregnant women. In summary, fenugreek is a worthy medicinal resource with valuable pharmacological effects. Providing better scientific evidence for fenugreek might provide new opportunities for drug discovery.

REFERENCES

Abdelhalim A, Saleem N (2021) Medicinal plants used for women's healthcare in Al-Madinah Al-Munawarah, Saudi Arabia. *Indian J Tradit Knowl* 20:132–140.

AbouZid SF, Mohamed AA (2011) Survey on medicinal plants and spices used in Beni-Sueif, Upper Egypt. *J Ethnobiol Ethnomed* 7:18. https://doi.org/10.1186/1746-4269-7-18

Adam M, Ahmed AA, Yagi A, Yagi S (2020) Ethnobotanical investigation on medicinal plants used against human ailments in Erkowit and Sinkat areas, eastern Sudan. *Biodiversitas* 21:3255–3262. https://doi.org/10.13057/biodiv/d210748

Akhlaghi N, Najafpour-Darzi G (2021) Phytochemical analysis, antioxidant activity, and pancreatic lipase inhibitory effect of ethanolic extract of Trigonella foenum graceum L. leaves. *Biocatal Agric Biotechnol* 32:101961. https://doi.org/10.1016/j.bcab.2021.101961

Al-timimi LAN (2019) Antibacterial and anticancer activities of fenugreek seed extract. *Asian Pacific J Cancer Prevetiontion* 20:3771–3776. https://doi.org/10.31557/APJCP.2019.20.12.3771

Alachkar A, Jaddouh A, Elsheikh MS, et al (2011) Traditional medicine in Syria: Folk medicine in Aleppo Governorate. *Nat Prod Commun* 6:79–84. https://doi.org/10.1177/1934578x1100600119

Alalwan TA, Alkhuzai JA, Jameel Z, Mandeel QA (2019) Quantitative ethnobotanical study of some medicinal plants used by herbalists in Bahrain. *J Herb Med* 17–18:100278. https://doi.org/10.1016/j.hermed.2019.100278

Ali-Shtayeh MS, Yaniv Z, Mahajna J (2000) Ethnobotanical survey in the Palestinian area: A classification of the healing potential of medicinal plants. *J Ethnopharmacol* 73:221–232. https://doi.org/10.1016/S0378-8741(00)00316-0

Almatroodi SA, Almatroudi A, Alsahli MA, Rahmani AH (2021) Fenugreek (Trigonella foenum-graecum) and its active compounds: A review of its effects on human health through modulating biological activities. *Pharmacogn J* 13:813–821. https://doi.org/10.5530/pj.2021.13.103

Awad MA, Hendi AA, Ortashi KM, et al (2021) Biogenic synthesis of silver nanoparticles using Trigonella foenum-graecum seed extract: Characterization, photocatalytic and antibacterial activities. *Sensors Actuators A Phys* 323:112670. https://doi.org/10.1016/j.sna.2021.112670

Bahmani M, Shirzad H, Mirhosseini M, et al (2016) A review on ethnobotanical and therapeutic uses of fenugreek (Trigonella foenum-graceum L). *J Evidence-Based Complement Altern Med* 21:53–62. https://doi.org/10.1177/2156587215583405

Basch E, Ulbricht C, Kuo G, et al (2003) Therapeutic applications of fenugreek. *Altern Med Rev* 8:20–27.

Baset ME, Ali TI, Elshamy H, et al (2020) Anti-diabetic effects of fenugreek (Trigonella foenum-graecum): A comparison between oral and intraperitoneal administration – an animal study. *Int J Funct Nutr* 1:1. https://doi.org/10.3892/ijfn.2020.2

Belguith-Hadriche O, Bouaziz M, Jamoussi K, et al (2010) Lipid-lowering and antioxidant effects of an ethyl acetate extract of fenugreek seeds in high-cholesterol-fed rats. *J Agric Food Chem* 58:2116–2122. https://doi.org/10.1021/jf903186w

Benayad Z, Gómez-Cordovés C, Es-Safi NE (2014) Characterization of flavonoid glycosides from fenugreek (Trigonella foenum-graecum) crude seeds by HPLC-DAD-ESI/MS analysis. *Int J Mol Sci* 15:20668–20685. https://doi.org/10.3390/ijms151120668

Bernstein N, Akram M, Yaniv-Bachrach Z, Daniyal M (2021) Is it safe to consume traditional medicinal plants during pregnancy? *Phyther Res* 35:1908–1924. https://doi.org/10.1002/ptr.6935

Bhatia H, Sharma YP, Manhas RK, Kumar K (2014) Ethnomedicinal plants used by the villagers of district Udhampur, J&K, India. *J Ethnopharmacol* 151:1005–1018. https://doi.org/10.1016/j.jep.2013.12.017

Chaudhary S, Chaudhary PS, Chikara SK, et al (2018) Review on fenugreek (Trigonella foenum-graecum L.) and its important secondary metabolite diosgenin. *Not Bot Horti Agrobot Cluj-Napoca* 46:22–31. https://doi.org/10.15835/nbha46110996

Chevassus H, Molinier N, Costa F, et al (2009) A fenugreek seed extract selectively reduces spontaneous fat consumption in healthy volunteers. *Eur J Clin Pharmacol* 65:1175–1178. https://doi.org/10.1007/s00228-009-0733-5

Devasena T, Menon PV (2007) Fenugreek seeds modulate 1,2-dimethylhydrazine-induced hepatic oxidative stress during colon carcinogenesis. *Ital J Biochem* 56:28–34.

Dhull SB, Kaur M, Sandhu KS (2020) Antioxidant characterization and in vitro DNA damage protection potential of some Indian fenugreek (Trigonella foenum-graecum) cultivars: Effect of solvents. *J Food Sci Technol* 57:3457–3466. https://doi.org/10.1007/s13197-020-04380-y.

El Bairi K, Ouzir M, Agnieszka N, Khalki L (2017) Anticancer potential of Trigonella foenum graecum: Cellular and molecular targets. *Biomed Pharmacother* 90:479–491. https://doi.org/10.1016/j.biopha.2017.03.071

Farag MA, Rasheed DM, Kropf M, Heiss AG (2016) Metabolite profiling in Trigonella seeds via UPLC-MS and GC-MS analyzed using multivariate data analyses. *Anal Bioanal Chem* 408:8065–8078. https://doi.org/10.1007/s00216-016-9910-4

Figer B, Pissurlenkar R, Ambre P, et al (2017) Treatment of gastric ulcers with fenugreek seed extract; in vitro, in vivo and in silico approaches. *Indian J Pharm Sci* 79:724–730. https://doi.org/10.4172/PHARMACEUTICAL-SCIENCES.1000285

Ghorbani A (2005) Studies on pharmaceutical ethnobotany in the region of Turkmen Sahra, north of Iran (Part 1): General results. *J Ethnopharmacol* 102:58–68. https://doi.org/10.1016/j.jep.2005.05.035

Ghoshal G, Singh M (2022) Characterization of silver nano-particles synthesized using fenugreek leave extract and its antibacterial activity. *Mater Sci Energy Technol* 5:22–29. https://doi.org/10.1016/j.mset.2021.10.001

Heshmat-Ghahdarijani K, Mashayekhiasl N, Amerizadeh A, et al (2020) Effect of fenugreek consumption on serum lipid profile: A systematic review and meta-analysis. *Phyther Res* 34:2230–2245. https://doi.org/10.1002/ptr.6690

Hilles AR, Mahmood S (2021) Historical background, origin, distribution, and economic importance of fenugreek. In: Naeem M, Aftab T, Khan MMA (eds) *Fenugreek*. Springer, Singapore, pp 3–11.

Huber-Morath A (1972) Trigonella L. In: Davis PH (ed) *Flora of Turkey and the East Aegean Islands Volume 3*. Edinburgh University Press, Edinburgh, pp 452–482.

Hussein S, Dhabe A (2018) Ethnobotanical study of folk medicinal plants used by villagers in Hajjah district-Republic of Yemen. *J Med Plants Stud* 6:24–30.

Idm'Hand E, Msanda F, Cherifi K (2020) Ethnobotanical study and biodiversity of medicinal plants used in the Tarfaya province, Morocco. *Acta Ecol Sin* 40:134–144. https://doi.org/10.1016/J.CHNAES.2020.01.002

Islam MK, Saha S, Mahmud I, et al (2014) An ethnobotanical study of medicinal plants used by tribal and native people of Madhupur forest area, Bangladesh. *J Ethnopharmacol* 151:921–930. https://doi.org/10.1016/j.jep.2013.11.056

Jha SK, Srivastava SK (2012) Determination of antioxidant activity of Trigonella foenum-graecum Linn. in mice. *Ann Phytomedicine* 1:69–74.

Kassaee SM, Goodarzi MT, Kassaee SN (2021) Ameliorative effect of Trigonella foenum graecum L. on lipid profile, liver histology and LDL-receptor gene expression in high cholesterol-fed hamsters. *Acta Endocrinol (Copenh)* 17:7–13. https://doi.org/10.4183/aeb.2021.7

Khajoei Nasab F, Khosravi AR (2014) Ethnobotanical study of medicinal plants of Sirjan in Kerman Province, Iran. *J Ethnopharmacol* 154:190–197. https://doi.org/10.1016/j.jep.2014.04.003

Khoja KK, Howes MJR, Hider R, et al (2022) Cytotoxicity of fenugreek sprout and seed extracts and their bioactive constituents on MCF-7 breast cancer cells. *Nutrients* 14:784. https://doi.org/10.3390/nu14040784

Kılıç M, Yıldız K, Mungan Kılıç F (2020) Traditional uses of medicinal plants in Artuklu, Turkey. *Hum Ecol* 48:619–632. https://doi.org/10.1007/s10745-020-00180-2

Kumar N, Ahmad A, Singh S, et al (2021) Phytochemical analysis and antioxidant activity of Trigonella foenum-graecum seeds. *J Pharmacogn Phytochem* 10:23–26.

Kumar P, Bhandari U, Jamadagni S (2014) Fenugreek seed extract inhibit fat accumulation and ameliorates dyslipidemia in high fat diet-induced obese rats. *Biomed Res Int.* 2014:606021. https://doi.org/10.1155/2014/606021

Malik K, Ahmad M, Zafar M, et al (2019) An ethnobotanical study of medicinal plants used to treat skin diseases in northern Pakistan. *BMC Complement Altern Med* 19:210. https://doi.org/10.1186/s12906-019-2605-6

Mati E, De Boer H (2011) Ethnobotany and trade of medicinal plants in the Qaysari Market, Kurdish Autonomous Region, Iraq. *J Ethnopharmacol* 133:490–510. https://doi.org/10.1016/j.jep.2010.10.023

Menale B, De Castro O, Cascone C, Muoio R (2016) Ethnobotanical investigation on medicinal plants in the Vesuvio National Park (Campania, Southern Italy). *J Ethnopharmacol* 192:320–349. https://doi.org/10.1016/j.jep.2016.07.049

Merrouni IA, Elachouri M (2021) Anticancer medicinal plants used by Moroccan people: Ethnobotanical, preclinical, phytochemical and clinical evidence. *J Ethnopharmacol* 266:113435. https://doi.org/10.1016/j.jep.2020.113435

Miara MD, Bendif H, Rebbas K, et al (2019) Medicinal plants and their traditional uses in the highland region of Bordj Bou Arreridj (Northeast Algeria). *J Herb Med* 16:100262. https://doi.org/10.1016/j.hermed.2019.100262

Mooventhan A, Nivethitha L (2017) A narrative review on evidence-based antidiabetic effect of fenugreek (Trigonella foenum-graecum). *Int J Nutr Pharmacol Neurol Dis* 7:84–87. https://doi.org/10.4103/ijnpnd.ijnpnd_36_17

Moradi Kor Z, Zadeh JB (2013) Fenugreek (Trigonella foenum-graecum L.) as a valuable medicinal plant. *Int J Adv Biol Biomed Res* 1:922–931.

Nagulapalli Venkata KC, Swaroop A, Bagchi D, Bishayee A (2017) A small plant with big benefits: Fenugreek (Trigonella foenum-graecum Linn.) for disease prevention and health promotion. *Mol Nutr Food Res* 61:1600950. https://doi.org/10.1002/mnfr.201600950

Narapogu V, Swathi C, Polsani MR, Premendran J (2021) Evaluation of anti-diabetic and anti-inflammatory activities of fenugreek (Trigonella foenum-graecum) seed extracts on Albino Wistar rats. *Natl J Physiol Pharm Pharmacol* 11:1277–1282. https://doi.org/10.5455/njppp.2021.11.07221202114092021

Neelakantan N, Narayanan M, De Souza RJ, Van Dam RM (2014) Effect of fenugreek (Trigonella foenum-graecum L.) intake on glycemia: A meta-analysis of clinical trials. *Nutr J* 13:7. https://doi.org/10.1186/1475-2891-13-7

Niknam R, Kiani H, Mousavi ZE, Mousavi M (2021) Extraction, detection, and characterization of various chemical components of Trigonella foenum-graecum L. (fenugreek) known as a valuable seed in agriculture. In: Naeem M, Aftab T, Khan MMA (eds) *Fenugreek*. Springer, Singapore.

Ocvirk S, Kistler M, Khan S, et al (2013) Traditional medicinal plants used for the treatment of diabetes in rural and urban areas of Dhaka, Bangladesh – An ethnobotanical survey. *J Ethnobiol Ethnomed* 9:1–8. https://doi.org/10.1186/1746-4269-9-43

Ouelbani R, Bensari S, Mouas TN, Khelifi D (2016) Ethnobotanical investigations on plants used in folk medicine in the regions of Constantine and Mila (North-East of Algeria). *J Ethnopharmacol* 194:196–218. https://doi.org/10.1016/j.jep.2016.08.016

Ouzir M, El Bairi K, Amzazi S (2016) Toxicological properties of fenugreek (Trigonella foenum graecum). *Food Chem Toxicol* 96:145–154. https://doi.org/10.1016/j.fct.2016.08.003

Prabhu S, Vijayakumar S, Morvin Yabesh JE, et al (2021) An ethnobotanical study of medicinal plants used in pachamalai hills of Tamil Nadu, India. *J Herb Med* 25:100400. https://doi.org/10.1016/j.hermed.2020.100400

Ramadhan U (2020) In vitro assessment of anti-inflammatory and COX-2 inhibitory action of some medicinal plants. *J Biol Res* 93:8723. https://doi.org/10.4081/jbr.2020.8723

Reddy RLR, Gowda AN, Srinivasan K (2019) Antilithogenic and hypocholesterolemic effect of dietary fenugreek seeds (Trigonella foenum-graecum) in experimental mice. *Med Plants-International J Phytomedicines Relat Ind* 11:117–126. https://doi.org/10.5958/0975-6892.2019.00018.2

Rehman MHU, Ahmad A, Amir RM, et al (2021) Ameliorative effects of fenugreek (Trigonella foenum-graecum) seed on type 2 diabetes. *Food Sci Technol* 41:349–354. https://doi.org/10.1590/fst.03520

Sankhwar SN, Kumar P, Bagchi M, et al (2021) Safety and efficacy of Furosap®, a patented Trigonella foenum-graecum seed extract, in boosting testosterone level, reproductive health and mood alleviation in male volunteers. *J Am Coll Nutr*: 1–9. https://doi.org/10.1080/07315724.2021.1978348

Sharififara F, Khazaelia P, Allib N (2009) In vivo evaluation of anti-inflammatory activity of topical preparations from fenugreek (Trigonella foenum-graecum L.) seeds in a cream base. *Iran J Pharm Sci* 5:157–162.

Sharma V, Singh P, Rani A (2017) Antimicrobial activity of Trigonella foenum-graecum L. (Fenugreek). *Eur J Exp Biol* 7:1–4. https://doi.org/10.21767/2248-9215.100004

Singh N, Yadav SS, Kumar S, Narashiman B (2022) Ethnopharmacological, phytochemical and clinical studies on fenugreek (Trigonella foenum-graecum L.). *Food Biosci* 46:101546. https://doi.org/10.1016/j.fbio.2022.101546

Skalli S, Hassikou R, Arahou M (2019) An ethnobotanical survey of medicinal plants used for diabetes treatment in Rabat, Morocco. *Heliyon* 5:e01421. https://doi.org/10.1016/j.heliyon.2019.e01421

Sowmya P, Rajyalakshmi P (1999) Hypocholesterolemic effect of germinated fenugreek seeds in human subjects. *Plant Foods Hum Nutr* 53:359–365. https://doi.org/10.1023/A:1008021618733

Srinivasan K (2006) Fenugreek (Trigonella foenum-graecum): A review of health beneficial physiological effects. *Food Rev Int* 22:203–224. https://doi.org/10.1080/87559120600586315

Srinivasan S, Koduru S, Kumar R, et al (2009) Diosgenin targets Akt-mediated prosurvival signaling in human breast cancer cells. *Int J Cancer* 125:961–967. https://doi.org/10.1002/ijc.24419

Stark A, Madar Z (1993) The effect of an ethanol extract derived from fenugreek (Trigonella foenum-graecum) on bile acid absorption and cholesterol levels in rats. *Br J Nutr* 69:277–287. https://doi.org/10.1079/bjn19930029

Suresh P, Kavitha CN, Babu SM, et al (2012) Effect of ethanol extract of Trigonella foenum graecum (fenugreek) seeds on Freund's adjuvant-induced arthritis in albino rats. *Inflammation* 35:1314–1321. https://doi.org/10.1007/s10753-012-9444-7

Syed QA, Rashid Z, Ahmad MH, et al (2020) Nutritional and therapeutic properties of fenugreek (Trigonella foenum-graecum): A review. *Int J Food Prop* 23:1777–1791. https://doi.org/10.1080/10942912.2020.1825482

Tabata M, Sezik E, Honda G, et al (1994) Traditional medicine in Turkey III. Folk medicine in East Anatolia, Van and Mitlis provinces. *Int J Pharmacogn* 32:3–12. https://doi.org/10.3109/13880209409082966

Taïbi K, Abderrahim LA, Ferhat K, et al (2020) Ethnopharmacological study of natural products used for traditional cancer therapy in Algeria. *Saudi Pharm J* 28:1451–1465. https://doi.org/10.1016/j.jsps.2020.09.011

Taloub LM, Rhouda H, Belahcen A, et al (2013) An overview of plants causing teratogenicity: Fenugreek (Trgonella foenum graecum). *Int J Pharm Sci Res* 4:516–519.

Ugulu I, Baslar S, Yorek N, Dogan Y (2009) The investigation and quantitative ethnobotanical evaluation of medicinal plants used around Izmir province, Turkey. *J Med Plants Res* 3:345–367.

Ullah R, Alqahtani AS, Noman OMA, et al (2020) A review on ethno-medicinal plants used in traditional medicine in the Kingdom of Saudi Arabia. *Saudi J Biol Sci* 27:2706–2718. https://doi.org/10.1016/j.sjbs.2020.06.020

Varshney IP, Jain DC, Srivastava HC (1984) Saponins from Trigonella foenum-graecum leaves. *J Nat Prod* 47:44–46. https://doi.org/10.1021/np50031a002

Verma N, Usman K, Patel N, et al (2016) A multicenter clinical study to determine the efficacy of a novel fenugreek seed (Trigonella foenum-graecum) extract (Fenfuro™) in patients with type 2 diabetes. *Food Nutr Res* 60:32382. https://doi.org/10.3402/fnr.v60.32382

Yadav UCS, Baquer NZ (2014) Pharmacological effects of Trigonella foenum-graecum L. in health and disease. *Pharm Biol* 52:243–254. https://doi.org/10.3109/13880209.2013.826247

Zandi P, Basu SK, Khatibani LB, et al (2015) Fenugreek (Trigonella foenum-graecum L.) seed: A review of physiological and biochemical properties and their genetic improvement. *Acta Physiol Plant* 37:1714. https://doi.org/10.1007/s11738-014-1714-6

Żuk-Gołaszewska K, Wierzbowska J (2017) Fenugreek: Productivity, nutritional value and uses. *J Elem* 22:1067–1080. https://doi.org/10.5601/jelem.2017.22.1.1396

28 *Viburnum* sp.

Burçin Ergene and Gülçin Saltan İşcan

INTRODUCTION

In Turkey, the genus *Viburnum* L. belongs to the Caprifoliaceae family and is represented by four species, namely, *Viburnum opulus* L., *Viburnum lantana* L. (Syn: *Viburnum tomentosa* Lam. in Lam. & DC), *Viburnum orientale* Pallas, and *Viburnum tinus* L. (Davis, 1972; Altun and Yilmaz, 2007).

Viburnum opulus, which is known as European cranberry bush, guelder-rose, snowball tree, crampbark, water elder, and locally known as "*gilaburu*" in Turkey, is a deciduous shrub of 2–4 m in height. The plant is widespread in Eastern, Northeastern, Western, and Central regions of Europe, as well as in North and Central Asia, North America, and North Africa. It also grows naturally in Turkey, Central Anatolia, and Black Sea region (Baytop, 1999; Velioglu *et al.*, 2006; Altun and Yilmaz, 2007; Yilmaztekin and Sislioglu, 2015; Kraujalis *et al.*, 2017; Kajszczak *et al.*, 2020; Ozrenk *et al.*, 2020).

The monograph of *V. opulus* is recorded in Turkish Pharmacopoeia (2017), American Herbal Pharmacopoeia (2011), and British Herbal Pharmacopoeia (1996) (Altun, 2017; TF, 2017). The monographs for the bark and fruits of *V. opulus* were also found in Russian Pharmacopoeia (The State Pharmacopoeia of USSR, 11th Edition) (Shikov *et al.*, 2014).

Viburnum lantana, which is also called "*wayfaring tree*" throughout the world and "*germişek*" in Turkey, is widespread in Western, Central, and Southern Europe as well as South of England and the Northern regions of Asia. It is rarely found in North Africa and northeastern parts of the USA (Kollmann and Grubb, 2002; Turker *et al.*, 2012). In Turkey, it is distributed in East, Central, and North Anatolia (Davis, 1972).

Viburnum orientale is mainly distributed in the Northern regions of Anatolia (Davis, 1972) and is known as "*kartopu*" in Turkey (Orhan *et al.*, 2018).

Viburnum tinus, which naturally grows in the Mediterranean basin, is known as "laurustinus" or "spring bouquet" in English and "*kartopu*" or "*defne yapraklı kartopu*" in Turkish (Sever Yilmaz *et al.*, 2013). The presence of *V. tinus* started with cultivation in New Zealand, but afterward the plant has become neutralized in the same region (Kollmann and Grubb, 2002).

BOTANICAL FEATURES

Viburnum opulus has three-lobed dark green palmate leaves with opposite arrangement and the leaves turn red in autumn. The twigs are grayish and naked, and the buds are scaly. It has white flowers and sheds its leaves in winter. The inflorescence is at 5–10 cm diameter with large sterile flowers at a diameter of 1.5–2.5 cm, around tiny fertile flowers at a diameter of 4–5 mm form the bloom at the top of the stems. The cultivar, which is especially known as "snowball", has only sterile flowers. The plant has dark red, globose, edible fruits (Figure 28.1) at approximately 8 mm diameter, with astringent taste. The twigs are grayish and glabrous. The barks of the plant are green-brown, and the color of the inner surface of the bark is green-yellow to red-brown (Davis, 1972; Yilmaztekin and Sislioglu, 2015; Kraujalis *et al.*, 2017; Kajszczak *et al.*, 2020).

Viburnum lantana is a deciduous shrub, branching near the base, and with a height of 2–6 m. It has gray-brown hairy twigs, and naked buds are in the same color. Older branches of the plant

DOI: 10.1201/9781003146971-28

FIGURE 28.1 *Viburnum opulus* leaves (USDA) and fruits (Zaklos-Szyda *et al.*, 2019).

FIGURE 28.2 *Viburnum lantana* fruits (Konarska and Domaciuk, 2018).

have reddish-brown, smooth barks. Ovate to obovate leaves are arranged opposite and do not have stipules. It has terminal umbel-like inflorescences with flat-shaped tops at 6–10 cm in diameter. Creamy white flowers are fertile with two small bracts and no pedicel. The smell of the flowers is unpleasant. The fruits (Figure 28.2) are flattened oval drupes, which have leathery, shiny skin, and stellate hairs. The fruits are 7–11 mm long, 5–8 mm wide, and 4–7 mm in thickness, first red and then become black (Davis, 1972; Kollmann and Grubb, 2002).

Viburnum orientale is a shrub 1–3 m tall. It has palmate, three-lobbed greenish leaves, which become red in autumn. The leaves are 8–15 cm, dentate-mucronate. The inflorescence, at a diameter of 5–10 cm, is fertile and the flowers are white. The fruits are red, oval drupe with a diameter of 10 mm (Davis, 1972).

Viburnum tinus is a small tree or shrub, with odorous, tiny, white flowers. The fruits are drupe and black-dark blue colored. It has leaves resembling *Laurus nobilis* (bay tree), which the name "*laurustinus*" refers to (Sever Yilmaz *et al.*, 2013).

TRADITIONAL USE

The fruits of *Viburnum opulus* are edible and have been used for the preparation of jelly, sauce, marmalade, and beverage (Kraujalis *et al.*, 2017). Apart from the cultivation of *V. opulus* as an ornamental plant, there are several medicinal uses of different parts of the plant. Its flowers and barks have been used as diuretic, purgative, and sedative in Europe; the juice of squeezed fruits has been used against gallbladder and liver disorders in the region of Middle Anatolia (Baytop, 1999; Dienaite *et al.*, 2021). It is also known that the decoction prepared from barks was used as diuretic by North American Indians. In America, barks of *V. opulus* are used as a remedy to ease uterine irritation and cramps and against neuralgia and dysmenorrhea (Youngken, 1932). The bark of the plant has been used against bleeding in stomach and uterus as well as hemorrhoids (Kajszczak *et al.*, 2020). In Russia, the fruits of the plant were reported to be used as tonic, diuretic, hypotensive, choleretic, anti-inflammatory, and sedative agents, while the infusion prepared using fruits was used for the treatment of hemorrhoids. In Russian folk medicine, fruit juice of *V. opulus* was also consumed as a laxative and against cold. The recommended dose for the infusion is reported as 1/3 glass of infusion prepared using 10 g of fruits in 200 mL water, 3–4 times daily (Shikov *et al.*, 2014). The fruits of *V. opulus* have been used for the treatment of kidney and bladder infections, digestive disorders, duodenal ulcers, heart and lung disease, cough, cold and shortness of breath, tuberculosis, rheumatic aches, menstrual cramps, dysmenorrhea, diabetes as well as for the regulation of blood pressure (Altun *et al.*, 2010; Kraujalyte *et al.*, 2013; Dietz *et al.*, 2016; Saltan *et al.*, 2016; Moldovan *et al.*, 2017; Kajszczak *et al.*, 2020). It is also reported that the infusion prepared using barks was used against scrofula, asphyxia, cold as well as uterine, gastric, and hemorrhoidal bleeding in Russian folk medicine. The recommended dose of decoction of the bark is 18–36 mL, 3–4 times daily. Besides, the decoction, prepared with 10 g of bark in 200 mL water and 1–2 tablespoons of decoction is recommended 3–4 times a day for diuretic and antiseptic effect (Shikov *et al.*, 2014). It is reported that 2–4 g of dried bark are used for the preparation of liquid extract in 25% ethanol (1:1) and tincture in 45% ethanol (1:5). The recommended doses of extract and tincture are 2–4 mL and 5–10 mL, respectively (Altun, 2017).

A beverage, also called "*gilaburu*", has been prepared by squeezing the brined fruits in Central Anatolia and this drink has also been used as an antihyperglycemic agent (Altun *et al.*, 2008) and against renal disorders (Yilmaztekin and Sislioglu, 2015). The beverage is prepared by leaving the fruits in water at room temperature for 2–3 months for fermentation (Baschali *et al.*, 2017). This fermented drink is also considered a potential probiotic due to its lactic acid bacteria microbiota (Sagdic *et al.*, 2014).

The bark of *V. lantana* has been used due to its analgesic and rubefacient properties in folk medicine (Baytop, 1999).

PHYTOCHEMICALS OF *VIBURNUM* SPECIES

The genus is reported to comprise terpenoids, sterols as well as polyphenolic substances such as flavonoids, anthocyanins, tannins, iridoids, and their glycosides (Sever Yılmaz *et al.*, 2007; Shafaghat and Shafaghatlonbar, 2019).

It is determined that the fruit juice of *V. opulus* is a rich source of vitamin C and dietary fibers. Besides, the phytochemical studies have revealed that the fruits contained various compounds such as phenolic acids, hydroxybenzoic acids, flavonoids, anthocyanins, coumarins, iridoids, and tannins (Altun *et al.*, 2008; Yilmaztekin and Sislioglu, 2015; Saltan *et al.*, 2016; Polka *et al.*, 2019; Kajszczak *et al.*, 2020).

Zaklos-Szyda *et al.* (2015) evaluated the amount of phenolic compounds in the *V. opulus* fruit extract prepared using 70% acetone. According to the results of HPLC analyses, content of hydroxybenzoic acids and flavonols, hydroxy-cinnamic acids, flavonols, anthocyanins, and total polyphenolics were found to be 8.813 ± 0.219 mg gallic acid equivalents (GAE)/g dry weight (DW), 54.205 ± 0.817 mg chlorogenic acid equivalents/g DW, 0.557 ± 0.020 rutin equivalents/g DW, 1.066 ± 0.072 cyanidin equivalents (CYE)/g DW, and 64.641 ± 1.155 mg polyphenolics/g dry weight, respectively.

Among these phenolics, chlorogenic acid was found to be the major constituent of *V. opulus* fruits (Kraujalyte *et al.*, 2013; Zaklos-Szyda *et al.*, 2019). In the study of Kraujalyte *et al.* (2013), the content of the fruit juice was evaluated using UPLC/ESI-QTOP-MS and the juice was shown to comprise chlorogenic acid, neochlorogenic acid, quinic acid, catechin, epicatechin, and procyanidin.

In other studies evaluating the chemical composition of fruit juice, cyaniding-3-glucoside, cyaniding-3-rutinoside, rutin (quercetin-3-rutinoside), quercitrin (quercetin-3-rhamnoside), isoquercitrin, (quercetin-3-β-D-glucoside), quercetin-3-arabinoside, and quercetin-3-xyloside (Velioglu *et al.*, 2006) were determined. Along with these phenolic compounds; isorhamnetin, isorhamnetin-3-*O*-rutinoside, gallocathecin gallate, cyaniding-3-sambubioside, B-typeprocyanidin dimer derivative, cyaniding dimers and trimers, feruloylquinic acid I-II, quercetin-3-vicianoside, and quercetin-3-galactoside were also detected in fruit extracts (Zaklos-Szyda *et al.*, 2019; Zaklos-Szyda *et al.*, 2020c). The study of Perova *et al.* (2014), in the content of which bioactive components of *V. opulus* fruit extracts were evaluated, showed that water extract contained cyaniding glycosides, such as cyanidin-3-vivianoside, cyanisin-3-xylosyl-rutinoside, cyaniding-3-sambubioside, cyaniding-3-glucoside, and cyaniding-3-rutinoside, and ethanol extract (50%) comprised flavonoids, namely, quercetin-3-sambubioside, rutin, isorhamnetin-3-sambubioside, isoquercetin, isorhamnetin-3-rutinoside, and quercitrin. In the same study, the iridoid content (opuloside I-IV, X) of 50% ethanol extract and the carbohydrate content (fructose, glucose, and sucrose) of water extract were determined using HPLC-MS/MS.

Chlorogenic acid was also determined as the major component of ethyl acetate and methanol extracts obtained from *V. opulus* fruits (Saltan *et al.*, 2016). In the methanol extracts of *V. opulus* leaves, branches, and fruits, salicin was also determined as well as chlorogenic acid (Altun and Yilmaz, 2007). Amentoflavone was only detected in the methanolic leaf extract of *V. opulus* (Altun *et al.*, 2007). Yilmaztekin and Sislioglu (2015) compared the chemical compositions of *V. opulus* fruits, traditionally prepared juices that were fermented for one, two, three, and four months in their study. The results revealed that the total phenol content of fermented juice (for four months) was the highest with a value of 483 ± 4.2 mg/100 mg gallic acid. Total phenol contents of other samples were found as 457 ± 4.2; 453 ± 3.5; and 464 ± 1.4 mg/100 mg gallic acid for one, two, and three months of fermented juices, respectively, while it was found 413 ± 3.5 mg/100 mg gallic acid for raw fruit. Dienaite *et al.* (2020) determined the chemical composition of ethanol and water fractions of defatted *V. opulus* fruit pomace and determined 45 compounds using the UPLC-QTOF-MS system combined with an online antioxidant capacity screening method. Among detected components, phenolic acids (chlorogenic acid, cryptochlorogenic acid, neochlorogenic acid, ethyl chlorogenate, dihydroferulic acid, hydroxybenzoic acid, dihydroxybenzoic acid derivatives, 1-caffeoylquinic acid), iridoids (secologanate, virbutinoside derivatives, opulus iridoids), flavalignans (cinchonain derivatives), catechins (procyanidin derivatives, procyanidin dimers, epicatechin hexoside, epicatechin-dihexoside, catechin), quercetin derivatives (quercetin pentoside hexoside derivatives, rutin, quercetin-3-O-glucoside), and other phenolic compounds (scopoletin-7-O-sophoroside, gambiriin, dihydromyrcetin) determined were found to be responsible for the radical scavenging activity of the extracts.

The acidity of *V. opulus* fruits especially rises from the organic acid content, which is particularly higher than that of bark and flowers. Succinic acid and malic acid were found to be the major organic acids of bark and flowers. Malic acid and oxalic acid were also determined in *V. opulus* seeds. Various studies revealed that the organic acid content of the fruits was comprised of malic acid, citric acid, quinic acid, shikimic acid, oxalic acid, tartaric acid, fumaric acid, succinic acid, acetic acid, gallic acid, syringic acid, caffeic acid, ferulic acid, *p*-coumaric acid, and maleic acid as well as ascorbic acid (Cam *et al.*, 2007; Kajszczak *et al.*, 2020; Karakurt *et al.*, 2020). In the study of Cam *et al.* (2007), the organic acid content of fruit flesh and seeds was determined separately and fruit flesh was found to contain malic acid as a major component, along with oxalic acid, ascorbic acid, and citric acid; while oxalic acid was the predominant organic acid of the seeds, in which malic acid was determined as well.

It was also reported that bark, flowers, leaves, stalks, and fruits of *V. opulus* contained proteins, as the protein content of flowers was 2–3 times higher than that of bark and fruits. Glutamic acid, aspartic acid, arginine, and leucine were determined as the major amino acids in *V. opulus* seeds (Kajszczak *et al.*, 2020). A coumarin derivative scopoletin and a sesquiterpene derivative viopudial were isolated from the barks of *V. opulus* (Jarboe *et al.*, 1967; Nicholson *et al.*, 1972). Triterpene alcohols, namely α-amyrin, β-amyrin, and lupeol were determined in seeds (Karimova *et al.*, 2000).

The lipophilic extract of *V. opulus* fruits was found to comprise triacylglycerols, sterols, triterpenes, mono- and diacylglycerides, tocopherols, carotenoids, and phytosterols. It was shown that the pomace of *V. opulus* fruits was extracted using a supercritical CO_2 extraction technique; the lipophilic extract contained α-, β-, γ-, and δ-tocopherols, along with β-sitostrerol, campesterol, stigmasterol, ergosterol, and cholestan-3-ol. β-sitostrerol was found to be the major phytosterol, as it constitutes about 85% of all sterols determined in the oil of fruit pomace (Kraujalis *et al.*, 2017; Kajszczak *et al.*, 2020; Dienaite *et al.*, 2021). Tocopherol content of the extracts obtained from dried berries was found to be higher than that of the pomace (Kraujalis *et al.*, 2017). In the study of Dienaite *et al.* (2021), the fatty acid composition of the oil obtained from *V. opulus* fruit pomace was also determined. Linoleic acid (49.91% ± 0.30) and oleic acid (46.50% ± 1.30) were found to be the predominant constituents along with other minor fatty acids, namely, linolenic acid, myristoleic acid, palmitic acid, stearic acid, and myristic acid (Dienaite *et al.*, 2021). While Kraujalis *et al.* (2017) found that the fatty acid contents of dried berries and pomace were mainly comprised of oleic, linoleic acids, as well as palmitic, stearic, eicosenoic, and linolenic acids. Briefly, according to the literature, the fatty acid content of *V. opulus* fruit is comprised of oleic acid, linoleic acid, arachidic acid, gondoic acid, palmitic acid, lauric acid, myristic acid, behenic acid, and stearic acid (Zarifikhosroshahi *et al.*, 2020). Yilmaz *et al.* (2008) investigated the content of the essential oil obtained from *V. opulus*, reporting the presence of 40 components, among which phytol was found to be the predominant component, along with trans-β-damascenone, α-cadinol, γ-cadinene, Δ-cadinene, and methyl pentanoate. In another study by Yilmaz N. (2011), 21 fatty acids were determined in the seed oil of *V. opulus*, and the major fatty acids were found to be oleic and linoleic acids.

The fruits of *V. opulus* have an undesirable odor, which is shown to be rising from the volatile constituents. Kraujalyte *et al.* (2012) evaluated the aromatic components of *V. opulus* fruits by solid phase microextraction-gas chromatography-olfactometry, yielding 41 compounds among which 3-methyl butanoic acid and 2-octanone were found to be predominant for different cultivars. The main odor active components were determined as 3-methyl butanoic acid, 2-methyl butanoic acid, linalool, and ethyl decanoate. In another study, more than 40 volatile compounds were determined in the fruits, and the predominant compounds responsible for the characteristic odor of the fruit were reported as 3-methylbutanoic acid, 2-methylbutanoic acid, linalool, and ethyl decanoate. Gilaburu, which is a traditional fermented beverage prepared using *V. opulus* fruits, was shown to contain 3-methylbutanoic acid major volatile component, among 58 volatile components that were determined in the study of Yilmaztekin and Sislioglu (2015). In this study, it was also determined that the amounts of acids, ketones, and alcohols increased proportionally to extending the fermentation period. Sariozkan *et al.* (2017) also determined the chemical constituents of the volatile oil of *V. opulus* fruits obtained by hydrodistillation. α and β-pinene were found to be the predominant components of the essential oil, along with butanoic acid, dl-limonene, α-terpineol, and germacrene D.

It is reported that the fruits of *V. lantana* contain essential oil, steroids, saponins, carotenoids, flavonoids, tannins, and anthocyanins, as well as lipids, protein, pectin, and mucilage (Konarska and Domaciuk, 2018). Iridoid glucosides, namely, lantanoside, dihydropenstemide, betulalbuside A, acetylpatrinoside, 2'-*O*-acetyldihydropenstemide, decapetaloside (Calis *et al.*, 1995), and 3'-acetylpatrinoside (Tomassini and Brkic, 1997), were isolated from the methanolic extract of *V. lantana* leaves as well as chalcone glycosides, *trans*-3-ethoxy-4-*O*-(glucopyranoside)-2',3',4',5',6'-pentah ydroxy chalcone and *trans*-3-methoxy-4-*O*-(glucopyranoside)-2',3',4',5',6'-pentahydroxy chalcone (Shafaghat and Shafaghatlonbar, 2019). The iridoid structures, 2'-*O*-acetyldihydropenstemide, 2'-*O*-acetylpatritunoside, and decapetaloside, were also isolated from the bark of *V. lantana* (Handjieva *et al.*, 1988; Handjieva *et al.*, 1991).

Salicin and chlorogenic acid were also determined in methanol extracts of *V. lantana* leaves as well as branches and fruits of the plant (Altun and Yilmaz, 2007). Amentoflavone was detected only in leaf and branch extracts (Altun *et al.*, 2007). Chlorogenic acid and caffeic acid were reported to be found in the buds of the plant (Orodan *et al.*, 2016).

In the content of the essential oil obtained from *V. lantana,* 53 components, among which occidecanol was found to be the major component, along with α-cadinol, γ-cadinene, 2E,4E-decadienal, n-heptanal, and Δ-cadinene, were detected by GC-MS (Yilmaz *et al.*, 2008).

An ester iridoid glycoside viborientoside, acyclic monoterpene glycosides anatolioside, and anatoliasides A-D were isolated from the water-soluble part of methanol extract prepared for *V. orientale* leaves (Calis *et al.*, 1993). Monoterpenes, betulalbusides A and B and anatolioside E were also determined as phytochemical components of *V. orientale* (Wang *et al.*, 2010).

The studies revealed that *V. tinus* contained iridoids, coumarins, saponins, and flavonoids (Gao *et al.*, 2014). Iridoid glycosides, namely, suspensolide A, viburtinosides I-V, were isolated from the methanolic extract of *V. tinus* leaves and branches (Tomassini *et al.*, 1995; Mohamed *et al.*, 2005). The ethyl acetate fraction of 80% methanolic extract of *V. tinus* leaves was shown to contain acylated iridoid glucosides, which are viburtinoside A and B, a coumarin diglucoside, and scopoletin 7-*O*-D-sophoroside along with suspensolides A and F. Furthermore, 2,6-di-*C*-methyl nicotinic acid, 3,5-diethyl ester, and oleanolic acid were obtained from chloroform fraction of total extract (Mohamed *et al.*, 2005; Wang *et al.*, 2010). Diterpenoid structures, vibsatins A and B, were isolated along with vibsanins B, C, and K from the twigs and leaves of a cultivated species of *V. tinus* (Gao *et al.*, 2014). *V. tinus* was reported to contain some compounds, which have flavonoid structure, namely, asoquercitroside, kaempferol-3-*O*-β-D-galactopyranoside, quercetine, nobiletin, rutin, and afzelin (Wang *et al.*, 2010).

Nazir *et al.* (2011) reported that depside galactoside and depsitinuside along with ergosterol and (22E,24S)-24-methyl-5-α-cholesta-7,22-diene-3β,5,6β-triol were obtained from the endophytic fungi culture, isolated from the leaves of *V. tinus*.

BIOLOGICAL ACTIVITIES

In Vitro Studies

Antioxidant Capacity

Cam and Hisil (2007) investigated the antioxidant capacity of *V. opulus* fruit juice and compared the activities of fresh fruit juice (FFJ) and pasteurized fruit juice (PFJ). FFJ and PFJ possessed antioxidant capacity with EC_{50} values of 25.06 ± 1.36 μL/mg and 30.87 ± 0.93 μL/mg, respectively, while total phenolic contents were determined as 351.26 ± 27.73 mg gallic acid equivalent (GAE)/100 mL for FFJ and 330.40 ± 29.46 mg GAE/100 mL for PFJ. In another study, EC_{50} values for seeds and fruit flesh were found as 2.35 ± 0.56 mg/mg DPPH and 24.26 ± 2.38 mg/mg DPPH in $DPPH^{\bullet}$ radical scavenging assay, respectively. The results revealed higher antioxidant capacity and phenolic content of the seeds, with total phenol contents of 1231.03 ± 8.93 mg GAE/100 g for seeds and 355.59 ± 3.72 mg GAE/100 g for fruit flesh.

In a study, evaluating the antioxidant capacity of water extracts of *V. opulus* fruits, leaves, and branches, the extract prepared using branches was found to be promising with IC_{50} values of 0.014 mg/mL and 3.7 mg/mL in $DPPH^{\bullet}$ radical and superoxide anion scavenging activity tests, respectively (Altun *et al.*, 2008).

Rop *et al.* (2010) evaluated the antioxidant capacity of the fruits of different *V. opulus* cultivars. The total phenolic content of the fruit extracts, which were prepared using hydrochloric acid:methanol:water mixture (2:80:18), was found in the range of 6.80–8.29 g GAE/kg fresh weight, while total flavonoid content was between 3.14 and 4.89 g/kg fresh weight. Antioxidant capacities of the extracts were determined by $ABTS^{\bullet+}$, $DPPH^{\bullet}$, hydroxyl radical, nitric oxide, and superoxide anion scavenging activity assays as well as lipid peroxidation inhibition test. The results were found in the ranges of 8.58–9.79 g ascorbic acid equivalent (AAE)/kg fresh weight for $DPPH^{\bullet}$ radical scavenging test and 9.10–11.12 g AAE/kg fresh weight for $ABTS^{\bullet+}$ radical scavenging test, while the determined ranges were 21.89–25.44%; 25.13–28.50%; 19.49–23.98%, and 11.20–13.90% for nitric oxide, superoxide anion, hydroxyl radical scavenging, and lipid peroxidation inhibition assays, respectively.

In the study of Erdogan-Orhan *et al.* (2011), the antioxidant capacity of the extracts prepared by successive extraction of *V. opulus* leaves, fruits, and branches with ethyl acetate, methanol, and water was evaluated as well as the total phenolic and flavonoid contents of the extracts. Methanol extract of the branches was found to have the highest phenolic content (217.95 ± 1.29 mg gallic acid equivalent (GAE)/g extract), while the highest total flavonoid content was determined in the ethyl acetate extract of leaves (1.38 ± 0.02 mg rutin equivalents/g extract).

Moldovan *et al.* (2012) assessed the total phenolic content and antioxidant capacity of the ethanol extract prepared from *V. opulus* fruits. The phenolic content of the extract was found to be 4.22 g GAE/kg frozen mass, while the $ABTS^{\bullet+}$ activity was determined as 7.05 g AAE/kg frozen fruits.

Kraujalyte *et al.* (2013) evaluated the antioxidant capacity of the fruit juice of six *V. opulus* genotypes using $ABTS^{\bullet+}$ radical scavenging activity, ferric reducing antioxidant power (FRAP), and oxygen radical absorbance capacity (ORAC) tests and Folin–Ciocalteu test for the determination of total phenolic content. In this study, the antioxidant capacity of *V. opulus* genotypes was found to be in the range of 31.9–109.8 µmol TE/g for $ABTS^{\bullet+}$ radical scavenging activity assay, 32.3–61.8 µmol TE/g for FRAP assay, and 141.6–260.4 µmol TE/g for ORAC assay, where the range for total phenol content was 5.4–10.6 mg GAE/g.

In the study of Karacelik *et al.* (2015), the juice, seeds, and skin of the fruits were obtained and extracted using methanol, acetonitrile, and water in order to evaluate the antioxidant capacity of *V. opulus* fruits. Total phenolic contents were calculated as 755 µg/mL for fruit juice (FJ); 673 µg/mL for methanol extract of the seeds (SM); 389 µg/mL for methanol extract of the skin (KM); 406 µg/mL for acetonitrile extract of the seeds (SA); 150 µg/mL for acetonitrile extract of the skin (KA); 577 µg/mL for water extract of the seeds (SW); and 189 µg/mL for water extract of the skin (KW). The highest antioxidant capacity was determined for the fruit juice with a TEAC value of 1297 µM in FRAP assay and SC_{50} value (dilution ratio) of 0.00033 in ABTS assay.

Zaklos-Szyda *et al.* (2015) compared the bioactivities of 20 edible plants containing polyphenolic compounds and determined $ABTS^{\bullet+}$ radical scavenging capacity of the extract of *V. opulus* fruits prepared using 70% acetone as 0.112 ± 0.009 µmol TE/g DW and total phenolic content of the extract was found to be 70.992 ± 1.724 mg GAE/g DW.

In the study of Kraujalis *et al.* (2017), the antioxidant capacity of dried berries and pomace was evaluated for the extracts obtained by supercritical CO_2 extraction and Soxhlet extraction as well as for the homogeneous solid ground particles and more polar fractions prepared using acetone, ethanol, and water successively. All of the extracts were found to possess antioxidant activity in $ABTS^{\bullet+}$ and $DPPH^{\bullet}$ radical scavenging activity and oxygen radical absorbance capacity (ORAC) tests. The highest activity was observed for the supercritical CO_2 extract of berry pomace at $ABTS^{\bullet+}$ radical scavenging activity test (1414 ± 178 µmol trolox equivalent (TE)/g dry weight) and ORAC test (2576 ± 324 µmol TE/g dry weight). This extract was also found to have the highest phenolic content with a value of 54.9 ± 6.7 mg GAE/g dry weight. Among more polar fractions which were obtained subsequent to the supercritical CO_2 extraction of the pomace, water fraction showed the highest antioxidant activity with the values of 267.4 ± 1.17 mg TE/dry weight for $DPPH^{\bullet}$ free radical scavenging activity test; 602.3 ± 6.62 mg TE/dry weight for $ABTS^{\bullet+}$ radical scavenging activity test; and 8.72 ± 2.28 mmol TE/dry weight for ORAC test. The total phenol content of this extract was determined as 174.9 ± 0.40 GAE/g dry weight.

Ersoy *et al.* (2017) investigated different genotypes of *V. opulus* growing in Turkey, in terms of some properties such as total phenolics, flavonoids, anthocyanins, and antioxidant capacity. The study revealed that the fruit extracts prepared using acetone:water:acetic acid (70:29.5:0.5) comprised 621–987 mg GAE/100 g fruit weight (FW) of total phenolics; 15–51 mg cyaniding-3-rutinoside equivalents/100 g FW of total anthocyanins; and 202–318 mg rutin equivalents/100 g FW of total flavonoids, while the antioxidant capacity determined by FRAP assay was found in the range of 21.02–34.90 μM TE/g. In another study, different genotypes collected from another province were found to contain 703–911 mg GAE/100 g FW of total phenolics; 22–48 mg cyaniding-3-rutinoside equivalents/100 g FW of total anthocyanins; and 187–302 mg rutin equivalents/100 g FW of total flavonoids. Ferric reducing antioxidant power (FRAP) of the extracts was determined between 23.41 and 32.70 μM TE/g (Ersoy *et al.*, 2018). Total carotenoid, polyphenol, and flavonoid contents of the fruits of *V. opulus* were also determined by Konarska and Domaciuk (2018), revealing the values of 1.50 ± 0.41 mg/100 g fresh weight for carotenoids; 12.30 ± 0.35 g caffeic acid equivalents/100 g dry extract for polyphenol content; and 0.18 ± 0.04 g QE/100 g dry extract for the total flavonoid content of acetone (80%):petroleum ether (1:1) extracts of the fruits.

Dienaite *et al.* (2020) evaluated the antioxidant capacity of *V. opulus* fruit pomace defatted using a supercritical CO_2 extraction technique. The residue was further extracted with ethanol and water successively. Antioxidant capacity of ethanol and water extracts were found as 3085 ± 55.53 and 2544 ± 33.30 μM TE/g in ORAC test; 1667 ± 32.12 and 1282 ± 19.33 μM TE/g in $ABTS^{\bullet+}$ radical scavenging activity test; 667.5 ± 20.74 and 496.18.58 μM TE/g in $DPPH^{\bullet}$ free radical scavenging activity test, respectively. The total phenolic content of ethanol extract was 154.2 ± 3.33 mg GAE/g and that of water extract was 131.6 ± 2.18 mg GAE/g. Among the compounds determined in the extracts, chlorogenic acid was found to be the most potent structure, in terms of $DPPH^{\bullet}$ radical scavenging activity. In the same study, the cellular antioxidant activity of the extracts was also investigated using Caco-2 cells, and the results revealed that ethanol and water extracts had high antioxidant capacity with the values of 30.36 ± 4.82 and 27.23 ± 9.01 μM quercetin equivalent(QE)/mg dry weight of extract.

The antioxidant capacities of fruit juice, phenolic rich fraction obtained from fruit juice, the extract prepared using methanol:acetone:water, and 70% acetone extract were evaluated using $ABTS^{\bullet+}$ radical scavenging and oxygen radical absorbance capacity tests. The results were found in the range of 38.36–2619.56 μmol TE/g for ABTS assay and 103.50–7810.29 μmol TE/g for ORAC. The phenolic content of the extracts was determined between 5.98 and 361.52 mg GAE/g (Zaklos-Szyda *et al.*, 2019).

Polka *et al.* (2019) evaluated the amounts of the components with antioxidant properties in the content of *V. opulus* bark, flowers, and fruits. Carotenoid, total phenolic, flavonoid, and proantocyanidin contents of the extracts were found as 1.12 ± 0.06 mg β-carotene/100 g DW; 3.51 ± 0.13 g GAE/100 g DW; 1.67 ± 0.07 g (+)-catechin equivalents (CE)/100 g DW; 0.22 ± 0.00 g CYE/100 g DW for flowers; 1.13 ± 0.03 mg β-carotene/100 g DW; 3.98 ± 0.04 g GAE/100 g DW; 2.25 ± 0.12 g CE/100g DW; 1.03 ± 0.03 g CYE/100 g DW for barks and 2.70 ± 0.07 mg β-carotene/100 g DW; 3.73 ± 0.16 GAE/100 g DW; 2.01 ± 0.11 g CE/100g DW; 0.52 ± 0.02 g CYE/100 g DW for fruits, respectively. The antioxidant capacity of the extracts prepared using 70% methanol was evaluated by $ABTS^{\bullet+}$, hydroxyl radical scavenging, superoxide anion radical scavenging, ORAC, and FRAP assays. The results showed that all parts of *V. opulus* possessed significant antioxidant capacity except for the superoxide anion radical scavenging activity of fruits and flowers. The results were found in the range of 16.18–40.21 mM TE/100 g DW for ABTS assay; 5.91–10.05 mM TE/100 g DW for hydroxyl radical scavenging assay; 10.93–108.17 mM TE/100 g DW for ORAC assay; 13.65–23.47 mM TE/100 g DW for FRAP assay; 89.77–115.44 mM CE/100 g DW for superoxide anion radical scavenging assay, revealing that the bark possesses the highest antioxidant capacity.

In the study of Karakurt *et al.* (2020), the antioxidant capacity of methanol extract prepared from *V. opulus* fruits was determined using $DPPH^{\bullet}$ radical scavenging method (IC_{50} = 156.51 ± 1.12 μg/mL) and cupric ion reducing antioxidant capacity (CUPRAC) assay (44.06 ± 1.55 mg GAE/g extract) as well as total phenolic content (310.07 ± 2.20 mg GAE/g extract) and total flavonoid content (5.86 ± 0.40 mg QE/g extract).

Bujor *et al.* (2020) determined the $DPPH^{\bullet}$ free radical scavenging capacity of *V. opulus* fruits, which were extracted using acetone:water:acetic acid (80:19.5:0.5) prior to the partition with diethyl ether, with EC_{50} value of 47.18 ± 0.12 μg/mL. In the content of the same study, total phenol content of the extract, which was prepared using acetone:water:acetic acid (80:19.5:0.5) prior to the liquid-liquid partition with diethyl ether, was determined as 213.15 ± 1.54 mg chlorogenic acid equivalents/g extract, while total flavonoid and total proanthocyanidin contents were calculated as 95.33 ± 0.95 mg catechin equivalents/g extract and 18.40 ± 0.87 mg cyaniding/g extract, respectively.

In the study of Ozrenk *et al.* (2020), the antioxidant capacities of different genotypes growing in Turkey were evaluated using FRAP assay, and the effects of the extracts, which were prepared using acetone:water:acetic acid (70:29.5:0.5) mixture, were found in the range of 27.67–35.65 μmol trolox/g. In the same study, the total phenolic content and total anthocyanin content of the genotypes were determined between the ranges of 690–830 mg GAE/100 g and 27.6–54.3 mg cyaniding-3-glucoside/100 g, respectively.

The results of the studies evaluating the antioxidant capacity of *V. opulus* are summarized in Table 28.1.

In order to determine the influence of the digestion process on the antioxidant activity of *V. opulus* fruits, Barak *et al.* (2019) conducted a study evaluating the changes in total phenol, phenolic acid, and flavonoid contents as well as the content of chlorogenic acid, the predominant bioactive compound of the fruits using a digestion stimulation *in vitro*. In the content of the study, antioxidant capacities of methanol and aqueous extracts were also evaluated by DPPH• and dimethyl-4-phenyl-enediamine (DMPD) free radical scavenging activity tests as well as CUPRAC and FRAP assays. The total phenolic content of methanol and water extracts was found as 40.17 ± 1.11 and 25.64 ± 1.06 mg/g GAE before digestion process and those values were found to reduce to 14.77 ± 0.18 and 15.23 ± 0.31 mg/g GAE, respectively. Total phenolic acid contents were decreased from 50.98 ± 1.46 mg/g caffeic acid equivalents (CAE) to 36.87 ± 2.21 mg/g CAE for methanol extract and from 42.54 ± 1.13 to 38.56 ± 0.26 mg/g CAE for water extract. Minor decreases were observed in terms of total flavonoid contents of methanol (from 25.09 ± 0.77 to 23.35 ± 0.37 mg/g QE) and water (from 24.57 ± 0.47 to 22.02 ± 0.44 mg/g QE) extracts. The chlorogenic acid content of the methanol extract was altered from 34.42 ± 1.22 to 25.50 ± 0.45 mg/g dry extract and that of water extract from 26.76 ± 0.91 to 18.61 ± 0.07 mg/g dry extract. The effect of digestion process on the antioxidant capacity of the extracts was not significant.

In the study of Altun *et al.* (2008), the antioxidant capacity of water extracts of *V. lantana* fruits, leaves, and branches was evaluated. In DPPH• radical scavenging activity tests, IC_{50} values were found as 0.085, 0.052, and 0.035 mg/mL for fruit, leaf, and branch extracts, respectively. The IC_{50} value for branches in superoxide anion scavenging capacity assay was 3.1 mg/mL, while other extracts were found ineffective.

In the study of Erdogan-Orhan *et al.* (2011), the antioxidant capacity of the extracts prepared by successive extraction of *V. lantana* leaves, fruits, and branches with ethyl acetate, methanol, and water were evaluated as well as the total phenolic and flavonoid contents of the extracts. Methanol extract of the leaves was found to have the highest phenolic content (368.78 ± 1.71 mg GAE/g extract) and total flavonoid content (1.56 ± 0.41 mg rutin equivalents/g extract). Antioxidant capacity assays were conducted at the concentrations of 500 μg/mL, 1000 μg/mL, and 2000 μg/mL. Ferrous ion chelating capacities of branch, leaf, and fruit extracts were found in the ranges of 3.0–22.6%; 2.21–17.10%, and 1.25–79.50%, respectively. Methanol extract of the fruits was found to be the most potent extract. The effect of the extracts in the FRAP assay was determined between 0.296% and 0.583% for branches; 0.278% and 3.401% for leaves; and 0.205% and 0.743% for fruits. In this assay, the highest capacity was observed for the methanol extract of leaves.

In the study of Orodan *et al.* (2016), total phenolic and flavonoid contents of *V. lantana* buds were determined as 12.07 ± 1.05 mg GAE/mL and 6.18 ± 0.52 mg rutin equivalent/mL, respectively. Besides, the IC_{50} value of the extract was found as 30.98 ± 2.14 μg/mL in DPPH• radical scavenging assay, revealing a low antioxidant capacity.

The antioxidant capacity of *V. lantana* leaf extract prepared using methanol was determined using DPPH• radical scavenging method in the study of Shafaghat and Shafaghatlonbar (2019). The IC_{50} value of the extract was found as 52 μg/mL.

TABLE 28.1
Antioxidant Capacity of *V. opulus*

Assays	Used Parts	Extraction Method/ Solvent	Antioxidant Capacity	References
DPPH• free radical scavenging capacity	Fruit juice (fresh)	-	EC_{50}= 25.06 μL/mg	Cam and Hisil (2007)
	Fruit juice (pasteurized)	-	EC_{50}= 30.87 μL/mg	Cam and Hisil (2007)
	Seed	-	EC_{50}= 2.35 ± 0.56 mg/mg DPPH	Cam *et al.* (2007)
	Fruit flesh	-	EC_{50}= 24.26 ± 2.38 mg/mg DPPH	Cam *et al.* (2007)
	Branch	Water	IC_{50}= 0.014 mg/mL	Altun *et al.* (2008)
	Leaf	Water	IC_{50}= 0.25 mg/mL	Altun *et al.* (2008)
	Fruit	Water	IC_{50}= 0.057 mg/mL	Altun *et al.* (2008)
	Fruit	Hydrochloric acid:methanol:water	8.58–9.79 g AEE/kg FW	Rop *et al.* (2010)
	Unwashed fruit pomace	Acetone	121.8 ± 0.85 μmol TE/g DW	Kraujalis *et al.* (2017)
	Unwashed fruit pomace	Ethanol	106.9 ± 0.48 μmol TE/g DW	Kraujalis *et al.* (2017)
	Unwashed fruit pomace	Water	267.4 ± 1.17 μmol TE/g DW	Kraujalis *et al.* (2017)
	Fruit	Methanol	103.59 ± 5.26 mg BHTE/g	Barak *et al.* (2019)
	Fruit	Water	96.74 ± 4.151 mg BHTE/g	Barak *et al.* (2019)
	Fruit	Diethyl ether fraction of acetone:water:acetic acid extract	EC_{50} = 47.18 ± 0.12 μg/mL	Bujor *et al.* (2020)
	Fruit pomace (defatted)	Ethanol	667.5 ± 20.74 μM TE/g DW	Dienaite *et al.* (2020)
	Fruit pomace (defatted)	Water	496.0±18.58 μM TE/g DW	Dienaite *et al.* (2020)
	Fruit	Methanol	IC_{50} = 156.0 ± 1.12 μg/mL	Karakurt *et al.* (2020)
ABTS•+ cation radical scavenging capacity	Fruit	hydrochloric acid:methanol:water	9.10–11.12 g AEE/kg FW	Rop *et al.* (2010)
	Fruit	Ethanol	7.05 g AAE/kg frozen fruit	Moldovan *et al.* (2012)
	Fruit juice	-	31.9–109.8 μmol TE/g	Kraujalyte *et al.* (2013)
	Fruit juice	-	SC_{50} = 0.00033	Karacelik *et al.* (2015)
	Seed	Methanol	SC_{50} = 0.00034	Karacelik *et al.* (2015)
	Seed	Acetonitrile	SC_{50} = 0.00413	Karacelik *et al.* (2015)
	Seed	Water	SC_{50} = 0.00180	Karacelik *et al.* (2015)
	Skin of fruit	Methanol	SC_{50} = 0.00387	Karacelik *et al.* (2015)
	Skin of fruit	Acetonitrile	SC_{50} = 0.01342	Karacelik *et al.* (2015)
	Skin of fruit	Water	SC_{50} = 0.01020	Karacelik *et al.* (2015)
	Fruit	Aqueous acetone (70%)	0.112 ± 0.009 μmol TE/g DW	Zaklos-Szyda *et al.* (2015)
	Fruit	-	643 ± 87 μmol TE/g DW	Kraujalis *et al.* (2017)
	Unwashed fruit pomace	-	1007 ± 85 μmol TE/g DW	Kraujalis *et al.* (2017)

(Continued)

TABLE 28.1 (CONTINUED)
Antioxidant Capacity of *V. opulus*

Assays	Used Parts	Extraction Method/ Solvent	Antioxidant Capacity	References
	Washed fruit pomace	-	863 ± 80 µmol TE/g DW	Kraujalis *et al.* (2017)
	Fruit	Supercritical CO_2 extraction	707 ± 84 µmol TE/g DW	Kraujalis *et al.* (2017)
	Unwashed fruit pomace	Supercritical CO_2 extraction	1414 ± 178 µmol TE/g DW	Kraujalis *et al.* (2017)
	Washed fruit pomace	Supercritical CO_2 extraction	791 ± 81 µmol TE/g DW	Kraujalis *et al.* (2017)
	Fruit	Soxhlet extraction (hexane)	687 ± 110 µmol TE/g DW	Kraujalis *et al.* (2017)
	Unwashed fruit pomace	Acetone	376.8 ± 2.85 µmol TE/g DW	Kraujalis *et al.* (2017)
	Unwashed fruit pomace	Ethanol	331.0 ± 1.22 µmol TE/g DW	Kraujalis *et al.* (2017)
	Unwashed fruit pomace	Water	602.3 ± 6.62 µmol TE/g DW	Kraujalis *et al.* (2017)
	Fruit	Methanol:acetone:water	1230.81 ± 26.1 µmol TE/g DW	Zaklos-Szyda *et al.* (2019)
	Fruit	Acetone	1227.73 ± 32.0 µmol TE/g DW	Zaklos-Szyda *et al.* (2019)
	Fruit juice	-	38.36 ± 2.7 µmol TE/g DW	Zaklos-Szyda *et al.* (2019)
	Fruit juice	Methanol	2619.59 ± 123.1 µmol TE/g DW	Zaklos-Szyda *et al.* (2019)
	Flower	Ethanol	16.18 ± 1.35 mM TE/100 g DW	Polka *et al.* (2019)
	Bark	Ethanol	40.21 ± 0.67 mM TE/100 g DW	Polka *et al.* (2019)
	Fruit	Ethanol	26.57 ± 1.71 mM TE/100 g DW	Polka *et al.* (2019)
	Fruit pomace (defatted)	Ethanol	1667 ± 32.12 µM TE/g DW	Dienaite *et al.* (2020)
	Fruit pomace (defatted)	Water	1282 ± 19.33 µM TE/g DW	Dienaite *et al.* (2020)
DMPD free radical scavenging capacity	Fruit	Methanol	52.55 ± 4.87 mg TE/g	Barak *et al.* (2019)
	Fruit	Water	55.00 ± 2.81 mg TE/g	Barak *et al.* (2019)
Hydroxyl radical scavenging capacity	Fruit	Hydrochloric acid:methanol:water	19.49–23.98%	Rop *et al.* (2010)
	Flower	Ethanol	8.23 ± 0.39 mM TE/100 g DW	Polka *et al.* (2019)
	Bark	Ethanol	5.91 ± 0.33 mM TE/100 g DW	Polka *et al.* (2019)
	Fruit	Ethanol	10.05 ± 0.31 mM TE/100 g DW	Polka *et al.* (2019)
Superoxide anion scavenging capacity	Branch	Water	IC_{50}= 3.7 mg/mL	Altun *et al.* (2008)

(*Continued*)

TABLE 28.1 (CONTINUED)
Antioxidant Capacity of *V. opulus*

Assays	Used Parts	Extraction Method/ Solvent	Antioxidant Capacity	References
	Leaf	Water	IC_{50}= 8.3 mg/mL	Altun *et al.* (2008)
	Fruit	Hydrochloric acid:methanol:water	25.13–28.50%	Rop *et al.* (2010)
	Flower	Ethanol	91.13 ± 3.40 mM CE/100 g DW	Polka *et al.* (2019)
	Bark	Ethanol	115.44 ± 5.28 mM CE/100 g DW	Polka *et al.* (2019)
	Fruit	Ethanol	89.77 ± 2.56 mM CE/100 g DW	Polka *et al.* (2019)
Nitric oxide scavenging capacity	Fruit	Hydrochloric acid:methanol:water	21.89–25.44%	Rop *et al.* (2010)
Lipid peroxidation inhibition assay	Fruit	Hydrochloric acid:methanol:water	11.20–13.90%	Rop *et al.* (2010)
Antioxidant activity coefficient for β-carotene bleaching assay	Branch	Ethyl acetate (0.5–2 mg/mL dose)	2.0–6.50	Erdogan-Orhan *et al.* (2011)
	Branch	Methanol (0.5–2 mg/mL dose)	2.0–23.0	Erdogan-Orhan *et al.* (2011)
	Branch	Water (0.5–2 mg/mL dose)	12.5–22.5	Erdogan-Orhan *et al.* (2011)
	Leaf	Ethyl acetate (0.5–2 mg/mL dose)	6.0–21.0	Erdogan-Orhan *et al.* (2011)
	Leaf	Methanol (0.5–2 mg/mL dose)	7.0–13.0	Erdogan-Orhan *et al.* (2011)
	Fruit	Ethyl acetate (0.5–2 mg/mL dose)	26.5–60.5	Erdogan-Orhan *et al.* (2011)
	Fruit	Water (0.5–2 mg/mL dose)	4.0–11.0	Erdogan-Orhan *et al.* (2011)
Cupric ion reducing antioxidant capacity (CUPRAC)	Fruit	Methanol	208.87 ± 9.32 mg AAE/g	Barak *et al.* (2019)
	Fruit	Water	156.49 ± 4.32 mg AAE/g	Barak *et al.* (2019)
	Fruit	Methanol	44.06 ± 1.55 mg GAE/g extract	Karakurt *et al.* (2020)
Ferric reducing antioxidant power (FRAP)	Branch	Ethyl acetate (0.5–2 mg/mL dose)	0.365–0.414	Erdogan-Orhan *et al.* (2011)
	Branch	Methanol (0.5–2 mg/mL dose)	0.366–3.321	Erdogan-Orhan *et al.* (2011)
	Branch	Water (0.5–2 mg/mL dose)	0.363–3.396	Erdogan-Orhan *et al.* (2011)
	Leaf	Ethyl acetate (0.5–2 mg/mL dose)	0.319–0.432	Erdogan-Orhan *et al.* (2011)
	Leaf	Methanol (0.5–2 mg/mL dose)	0.349–0.508	Erdogan-Orhan *et al.* (2011)
	Leaf	Water (0.5–2 mg/mL dose)	0.355–0.458	Erdogan-Orhan *et al.* (2011)

(Continued)

TABLE 28.1 (CONTINUED)
Antioxidant Capacity of *V. opulus*

Assays	Used Parts	Extraction Method/ Solvent	Antioxidant Capacity	References
	Fruit	Ethyl acetate (0.5–2 mg/mL dose)	0.345–0.739	Erdogan-Orhan *et al.* (2011)
	Fruit	Methanol (0.5–2 mg/mL dose)	0.324–1.988	Erdogan-Orhan *et al.* (2011)
	Fruit	Water (0.5–2 mg/mL dose)	0.345–0.390	Erdogan-Orhan *et al.* (2011)
	Fruit juice	-	32.3–61.8 μmol TE/g	Kraujalyte *et al.* (2013)
	Fruit juice	-	1297 μM TE/mL	Karacelik *et al.* (2015)
	Seed	Methanol	1227 μM TE/mL	Karacelik *et al.* (2015)
	Seed	Acetonitrile	276 μM TE/mL	Karacelik *et al.* (2015)
	Seed	Water	520 μM TE/mL	Karacelik *et al.* (2015)
	Skin of fruit	Methanol	260 μM TE/mL	Karacelik *et al.* (2015)
	Skin of fruit	Acetonitrile	99 μM TE/mL	Karacelik *et al.* (2015)
	Skin of fruit	Water	95 μM TE/mL	Karacelik *et al.* (2015)
	Fruit	Acetone:water:acetic acid	21.02–34.90 μmol TE/g	Ersoy *et al.* (2017)
	Fruit	Acetone:water:acetic acid	23.41–32.70 μmol TE/g	Ersoy *et al.* (2018)
	Fruit	Methanol	0.46 ± 0.05 mM $FESO_4$ equivalents/g	Barak *et al.* (2019)
	Fruit	Water	0.41 ± 0.09 mM $FESO_4$ equivalents/g	Barak *et al.* (2019)
	Flower	Ethanol	13.65 ± 0.64 mM TE/100 g DW	Polka *et al.* (2019)
	Bark	Ethanol	23.47 ± 1.50 mM TE/100 g DW	Polka *et al.* (2019)
	Fruit	Ethanol	19.29 ± 0.83 mM TE/100 g DW	Polka *et al.* (2019)
	Fruit	Acetone:water:acetic acid	27.67–35.65 μmol TE/g	Ozrenk *et al.* (2020)
Ferrous ion chelating capacity	Leaf	Ethyl acetate (0.5–2 mg/mL dose)	33.01–44.62%	Erdogan-Orhan *et al.* (2011)
	Fruit	Ethyl acetate (0.5–2 mg/mL dose)	17.75–36.36%	Erdogan-Orhan *et al.* (2011)
Oxygen radical absorbance capacity (ORAC)	Fruit juice	-	141.6–260.4 μmol TE/g	Kraujalyte *et al.* (2013)
	Fruit	-	1277 ± 204 μmol TE/g DW	Kraujalis *et al.* (2017)
	Unwashed fruit pomace	-	2576 ± 324 μmol TE/g DW	Kraujalis *et al.* (2017)
	Washed fruit pomace	-	659 ± 80 μmol TE/g DW	Kraujalis *et al.* (2017)
	Fruit	Supercritical CO_2 extraction	1515 ± 167 μmol TE/g DW	Kraujalis *et al.* (2017)
	Unwashed fruit pomace	Supercritical CO_2 extraction	2174 ± 220 μmol TE/g DW	Kraujalis *et al.* (2017)

(*Continued*)

TABLE 28.1 (CONTINUED)
Antioxidant Capacity of *V. opulus*

Assays	Used Parts	Extraction Method/ Solvent	Antioxidant Capacity	References
	Washed fruit pomace	Supercritical CO_2 extraction	612 ± 75 μmol TE/g DW	Kraujalis *et al.* (2017)
	Fruit	Soxhlet extraction (hexane)	459 ± 50 μmol TE/g DW	Kraujalis *et al.* (2017)
	Unwashed fruit pomace	Acetone	5.75 ± 0.99 μmol TE/g DW	Kraujalis *et al.* (2017)
	Unwashed fruit pomace	Ethanol	5.32 ± 0.37 μmol TE/g DW	Kraujalis *et al.* (2017)
	Unwashed fruit pomace	Water	8.72 ± 2.28 μmol TE/g DW	Kraujalis *et al.* (2017)
	Fruit	Methanol:acetone:water	2388.58 ± 150.5 μmol TE/g DW	Zaklos-Szyda *et al.* (2019)
	Fruit	Acetone	2092.65 ± 48.3 μmol TE/g DW	Zaklos-Szyda *et al.* (2019)
	Fruit juice	-	103.50 ± 6.4 μmol TE/g DW	Zaklos-Szyda *et al.* (2019)
	Fruit juice	Methanol	7810.29 ± 342.3 μmol TE/g DW	Zaklos-Szyda *et al.* (2019)
	Flower	Ethanol	61.82 ± 2.04 mM TE/100 g DW	Polka *et al.* (2019)
	Bark	Ethanol	108.17 ± 3.38 mM TE/100 g DW	Polka *et al.* (2019)
	Fruit	Ethanol	10.93 ± 0.39 mM TE/100 g DW	Polka *et al.* (2019)
	Fruit pomace (defatted)	Ethanol	3085 ± 55.53 μM TE/g DW	Dienaite *et al.* (2020)
	Fruit pomace (defatted)	Water	2544 ± 33.30 μM TE/g DW	Dienaite *et al.* (2020)
Cellular antioxidant activity (CAA)	Fruit pomace (defatted)	Ethanol	30.36 ± 4.82 μM QE/mg DW	Dienaite *et al.* (2020)
	Fruit pomace (defatted)	Water	27.23 ± 9.01 μM QEm/g DW	Dienaite *et al.* (2020)
Total antioxidant capacity	Fruit	Methanol	56.89 ± 5.14 mg AAE/g	Barak *et al.* (2019)
	Fruit	Water	49.07 ± 6.20 mg AAE/g	Barak *et al.* (2019)

AAE: ascorbic acid equivalents; BHTE: butylated hydroxytoluene equivalents; CE: (+)-catechin equivalents; DW: dry weight; EC_{50}: the half maximal effective concentration; FW: fresh weight; GAE: gallic acid equivalents; IC_{50}: the half maximal inhibitory concentration; QE: quercetin equivalents; SC_{50}: dilution ratio of 50% scavenging activity of $ABTS^{•+}$ radical present in the test medium; TE: Trolox equivalents

Sever Yilmaz *et al.* (2013) evaluated the antioxidant capacity of *V. tinus* by $DPPH^•$, N,N-dimethyl-p-phenylendiamine ($DMPD^+$), superoxide (SO), nitric oxide radical scavenging methods as well as metal chelating capacity, FRAP, and phosphomolybdenum reducing antioxidant power assays. Ethyl acetate, methanol, and water extracts were prepared sequentially from the leaves, branches, and fruits of the plant and the assays were conducted at concentrations of 500, 1000, and 2000 μg/

mL. Methanol extracts of fruits, branches, and leaves as well as aqueous extract of leaves exhibited remarkable DPPH• scavenging activity (over 89%) compared to the reference gallic acid (93.1% ± 0.27). Methanol extract of fruits was the most potent extract against DMPD radical (67.1% ± 0.33), while the inhibition value of the reference was 68.3% ± 0.67. It also exhibited the highest activity in FRAP and PRPA assays. Only the ethyl acetate extract of fruit possessed SO radical scavenging activity (38.4% ± 1.01). Water extracts prepared from fruits, branches, and leaves were found to be active in metal chelation capacity test with the values of 69.4 ± 5.54; 75.4 ± 3.22; and 62.3 ± 5.38%, respectively. All of the extracts showed NO radical scavenging activity by over 50% inhibition. In further studies, the antioxidant potential of ethyl acetate, methanol, and water extracts, which were prepared sequentially from the leaves, branches, and fruits of *V. orientale*, was evaluated. Methanol extracts exerted activity against DPPH• by over 90%, while the activities of aqueous extracts of leaves (86.25% ± 0.62) and branches (73.64% ± 3.29) were also remarkable. The scavenging capacities of tested extracts on DMPD radical were insignificant or low with the inhibition values below 35%. NO radical scavenging capacity values were found in the range of 37–75%, aqueous extract of the leaves possessed the highest activity (75.00% ± 1.22) against this radical. Methanol extracts showed the highest activity in FRAP and PRAP assays. The aqueous extract of the leaves was found to have the best metal chelating capacity with a value of 54.66% ± 3.56 (Orhan *et al.*, 2018).

Antimicrobial Activity

Antimicrobial effect of *V. opulus* fruit extract (chloroform:methanol, 1:1) against two Gram-positive bacteria (*Bacillus cereus* (ATCC 31430), *Staphylococcus aureus* (ATCC 6538P)); Gram-negative bacteria (*Helicobacter pylori* (ATCC 43504), *Pseudomonas aeruginosa* (ATCC 27858), and *Escherichia coli* (ATCC 128)); and two yeasts (*Saccharomyces cerevisiae* (ATCC 287) and *Candida albicans* (ATCC 0028)) was evaluated in the study of Laux *et al.* (2007). The extract was found to possess antimicrobial activity against all tested strains except from *C. albicans* with the average inhibition zone of 3 mm.

Yilmaz *et al.* (2008) evaluated the antimicrobial activity of essential oils obtained from *V. opulus*, *V. lantana*, and *V. orientale* using broth dilution assay. Weak antibacterial activity of *V. orientale* against *Enterococcus faecalis* ATCC29212, *S. aureus* ATCC25923, and *B. cereus* 709 Roma was observed with MIC value of 1000 μg/mL, while *V. opulus* and *V. lantana* did not possess activity.

In the study of Cesoniene *et al.* (2012), the fruit juice of six genotypes of *V. opulus* was evaluated for antimicrobial activity using the agar well diffusion test. The fruit juice of *V. opulus* exhibited strong antimicrobial activity against Gram-negative (*Salmonella typhimurium, S. agona*) and Gram-positive (*Staphylococcus aureus, Listeria monocytogenes, Enterococcus faecalis*) bacteria, while no effect was observed against the yeast strains tested in the content of study. In the following studies conducted using the same method, the antimicrobial effects of fruit juice and ethanol extract prepared using fruits were compared. Significant antibacterial activity was observed against the strains of *S. typhimurium*, *S. agona*, and *L. monocytogenes*. The inhibition zones of the fruit juice, ethanol extract, and the reference ceftazimidime/clavulanic acid (30/10 μg) were determined as 27.9 mm, 20.7 mm, and 34.3 mm for *S. typhimurium*; 26.3, 20.7, and 35.7 mm for *S. agona*; and 26.5 mm, 19.1 mm, and 32.0 mm for *L. monocytogenes*, respectively. The results revealed that the fruit juice was more effective than the ethanol extract of the fruits.

Bubulica *et al.* (2012) evaluated the bactericidal and antibiofilm effect of *V. opulus* fruit extract prepared using water on *Staphylococcus epidermidis* and *S. aureus* strains isolated from wound and ear infections using broth dilution and microtiter assays. The extract possessed antimicrobial activity with the minimum inhibition concentration (MIC) value of 250 μL/mL. In another study, the antibacterial activity of *V. opulus* fruits was assessed using disc diffusion test. The inhibition zones observed for the extract prepared using ethanol were found in the range of 8.8–10.3 mm, revealing a moderate activity against *S. aureus, Streptococcus pyogenes*, and *S. epidermidis*. Ethanolic extract prepared from *V. lantana* fruits exerted higher activity against these strains with an inhibition zone range of 9.4–13.3 mm (Turker *et al.*, 2012).

Eryilmaz *et al.* (2013) investigated the antimicrobial effect of ethanol (75%) and water extracts prepared from stems, leaves, and fruits of *V. opulus, V. lantana,* and *V. tinus* using disc diffusion test. The minimum inhibition concentrations (MIC) of the active extracts were also calculated as 125 μg/mL by broth dilution assay. The inhibition zones for ethanol extracts of *V. opulus* leaves were determined as 10 mm against *Staphylococcus aureus* ATCC 25923; 10 mm against *Staphylococcus aureus* ATCC 43300 (MRSA); 8 mm against *Escherichia coli* ATCC 25922; 8 mm against *Pseudomonas aeruginosa* ATCC 27853; 8 mm against *Klebsiella pneumoniae* RSKK 574. A 7 mm inhibition zone was observed for water extract of *V. opulus* leaves against *K. pneumoniae.* Ethanol extract of *V. opulus* fruits exhibited activity against *E. coli, P. aeruginosa,* and *K. pneumoniae* with inhibition zones of 9 mm, 10 mm, and 8 mm, respectively. *S. aureus* and MRSA were also inhibited by the ethanol extract of *V. opulus* stems, with the values for inhibition zones of 8 mm for *S. aureus* and 7 mm for MRSA. The inhibition zones of *V. orientale* were found as 10 mm against *S. aureus*; 8 mm against MRSA, *B. subtilis,* and *E. coli*; 9 mm against *P. aeruginosa* and 10 mm against *K. pneumoniae* for ethanol extract of fruits; 8 mm against *P. aeruginosa* and *K. pneumoniae*; 9 mm against *E. coli* and *C. albicans* for ethanol extracts of stems; 7 mm against *K. pneumoniae* for water extract of leaves; 8 mm against *S. aureus* and MRSA for ethanol extract; and 8 mm against *E. coli* and *K. pneumoniae* for ethanol extract of leaves. The effect of *V. tinus* ethanol extracts was observed with the inhibition zone values of 8 mm against *S. aureus* and MRSA for stems; 8 mm against *S. aureus, P. aeruginosa*, and *K. pneumoniae* and 9 mm against *E. coli* for fruits; and 8 mm against MRSA and *K. pneumoniae* and 9 mm *S. aureus* and *E. coli* for leaves. The inhibition zone was calculated as 8 mm against *K. pneumoniae* for water extract of leaves. The ethanol extract of *V. lantana* leaves inhibited *E. coli, P. aeruginosa,* and *K. pneumoniae* by 8 mm of inhibition zone and the inhibition zone against *K. pneumoniae* for water extract of the leaves was 7 mm.

In the study of Zaklos-Szyda *et al.* (2020b), antimicrobial effect of the fruit juice (FJ) and the juice purified via solid phase extraction (PJ) were studied using the agar well diffusion method. The highest activity was observed against *S. aureus* ATTC 6538 with inhibition zones of 12.5 mm for FJ and 9.0 mm for PJ. The inhibition zones for FJ and PJ against *S. aureus* ATTC 25923 were found as 2.0 and 7.00 mm, respectively. PJ possessed the highest inhibition capacity against *Enterococcus faecalis* ATCC 29212 (inhibition zone = 10.0 mm), and the inhibition zone of FJ against this strain was found to be 6.0 mm. Only PJ exhibited antimicrobial activity on *L. monocytogenes* ATCC19115 with an inhibition zone of 8.7 mm. The results showed that tested Gram-positive bacteria were more sensitive to the extracts than Gram-negative bacteria and yeast (*Candida albicans* ATC 10231). The extract did not possess antimicrobial activity against the tested strains of Gram-negative bacteria and the yeast, as well as Gram-positive lactic acid bacteria.

In the study of Yildirim *et al.* (2013), it was shown that methanol extract prepared from the leaves and fruits of *V. lantana* possessed weak antibacterial activity against *S. aureus* ATTC 25923, *S. epidermidis* ATTC 12228, and *S. pyogenes* ATTC 19615 in disc diffusion test, with the inhibition zones of 7.0 mm, 7.5 mm, and 7.7 mm, respectively.

In the study of Orodan *et al.* (2016), the antimicrobial activity of *V. lantana* buds was evaluated by the microdilution method. The buds were found to possess low antimicrobial activity with MIC values of 7.5 mg/mL against *S. aureus* and *L. monocytogenes*; 1.87 mg/mL against *P. aeruginosa*; and 3.75 mg/mL against *E. coli.*

Anticarcinogenic Effect

The influence of *V. opulus* fruit extract on gastric carcinoma cell line (CRL-5971) was assessed in the study of Laux *et al.* (2007). It was found that the aldehyde bioactive fraction of the fruit extract prepared using chloroform:methanol (1:1) induced the level of apoptosis of gastric cancer cells at the concentration of 27 μM. IC_{50} values were found in the range of 0.9–1.7 μM.

Turker *et al.* (2012) evaluated the antitumor activity of ethanolic and aqueous extracts prepared from *V. lantana* fruits using the potato disc method. The hot water and hot ethanol extracts were found to possess high activity with inhibition values of 90.5% and 95.2%, respectively. In further studies on the antitumor activities of water, ethanol, and methanol extracts of *V. lantana* leaves and fruits investigated with the same method, a strong antitumor activity was revealed with inhibition values of 86.4% for aqueous extract, 90.9% for ethanol extract, and 100% for methanol extract (Yildirim *et al.*, 2013).

In the study of Ceylan *et al.* (2018), trypan blue exclusion test was conducted to evaluate the cytotoxic effect of *V. opulus* fruit juice on Ehrlich ascites carcinoma (EAC) cells. The IC_{50} value was found to be 199.58 μg/mL.

Dienaite *et al.* (2020) evaluated the antiproliferative activity and cytotoxicity of ethanol and water extracts which were successively extracted from *V. opulus* fruit pomace defatted using supercritical CO_2 extraction. Ethanol extract showed high antiproliferative activity against HT29 cells with EC_{50} concentration of 0.39 ± 0.03 mg/mL, while it did not possess cytotoxic effect on Caco-2 cell line. Water extract did not exhibit antiproliferative activity at the concentration of 2 mg/mL, which was the highest dose applied in the study.

In the study of Karakurt *et al.* (2020), anticarcinogenic properties of the methanol extract of *V. opulus* fruits were assessed on human colon cancer cells (DLD-1, HT-29, SW620, Caco-2) and CCD-18Co colon epithelial cells. The results revealed that the fruit extract selectively inhibited the proliferation of human colon cancer cells. IC_{50} values were determined as 254.3 μg/mL for DLD-1 cells, 553.3 μg/mL for HT-29 cells, 327.4 μg/mL for SW-620 cells, and 714.6 μg/mL for Caco-2 cells. The inhibition observed on DLD-1 and HT-29 cells was found to be dose-dependent. The extract did not possess a significant effect on healthy colon epithelial cells. Since the *p*-coumaric acid content and the activity was found to be in positive correlation with antiproliferative activity, the effect of the extract was mainly attributed to the active component *p*-coumaric acid.

Antidiabetic and Antiobesity Activity

The inhibitory effect of *V. opulus* fruit on α-amylase, α-glucosidase, and protein tyrosine phosphatase (PTP) was investigated. IC_{50} values of the extract prepared using 70% acetone were found as 0.970 ± 0.058; 2.194 ± 0.077; 4.053 ± 0.483; and 0.151 ± 0.015 mg/mL for α-amylase (in DNS reagent assay and iodine assay), α-glucosidase, and PTP-1B, respectively (Zaklos-Szyda *et al.*, 2015). Further study on *V. opulus* fruits was conducted to evaluate the effect to decrease lipid accumulation and to lower glucose and free fatty acid uptake by the human colon adenocarcinoma cell line (Caco-2), as well as their protective effect against oxidative stress induced by *tert*-butylhydroxy peroxide by determination of cellular viability and the effect on intracellular generation of reactive oxygen species. For this purpose, the pomace of the fruits was extracted with methanol:acetone:water (MAE) at the ratio of 2:2:1, prior to the extraction with 70% acetone (AE). Besides, phenolic-rich fraction (PRF) was obtained by eluting dried fruit juice (FJ) with methanol. The activities of PRF, MAE, and AE were tested. IC_{50} values of the extracts were calculated as 450, 250, 275, and 400 μg/mL for FJ, PRF, MAE, and AE, respectively. Cell viability values were determined at maximal non-toxic concentrations (IC_0) of the samples and found to be 95.64% ± 8.78 for FJ; 96.12% ± 10.12 for PRF; 91.70% ± 8.31 for MAE; and 90.50% ± 3.02 for AE. The highest antidiabetic activity was observed for PRF in terms of decrease in free fatty acids and glucose intake as well as accumulation of lipids in Caco-2 cells. The lipid accumulation in the cells was found to be 17% less in PRF-treated cells than that of untreated group (Zaklos-Szyda *et al.*, 2019). In addition, the effects of FJ and PRF were studied using mice insulinoma MIN6 cells. In order to determine the cell viability IC_0 (the concentration, which did not possess any effect on cell viability) and IC_{50} values of FJ (IC_0=75 μg/mL; IC_{50}=180 μg/mL) and PRF (IC_0=50 μg/mL; IC_{50}=135 μg/mL) were calculated for MIN6 cells. With respect to the occurrence of oxidative stress due to the increase in fatty acid and glucose levels in the case of obesity and type 2 diabetes, the effects of FJ and PRF on intracellular oxidative stress were investigated. The extracts were found to decrease the intracellular level of reactive oxygen species by 10–20%. Besides, PRF used at IC_0 concentration decreased glutathione peroxidase enzyme activity and hydrogen peroxide production by about 15% and 7%, respectively. On the contrary, the extracts induced intracellular oxidative stress above IC_0 concentration. The effect of the extracts on fatty acid intake was evaluated using MIN6 cells in which lipid accumulation was induced by oleic acid, but the extract did not decrease the cellular lipid levels, instead, a slight increase was observed. FJ and PRF were found to decrease the secretion of glucagon-like peptide-1 by 30% and 50%, respectively, and to inhibit glucose-induced insulin secretion by 15–20% at IC_{50} concentration. Apart from this, in order to test the bioavailability, the affinity of PRF to bind human serum albumin was detected as a parameter showing drug bioavailability. These findings demonstrated the influence of *V. opulus* fruits on type 2 diabetes and obesity as well as the toxicity potential and safety limits of *V. opulus* fruit extracts (Zaklos-Szyda *et al.*, 2020a). In another study by Zaklos-Szyda *et al.* (2020c), the effect of fruit juice (FJ) and the juice exposed to solid phase extraction (PJ) on adipogenesis with murine 3T3-L1 preadipocyte cell and their pancreatic lipase activity in a triolein emulsion by spectrophotometry were evaluated. Both the tested extracts were found to possess pancreatic lipase activity dose-dependently. IC_{50} value was 55.26 ± 2.54 mg/mL for PJ and 261.94 ± 2.00 mg/mL for FJ, while the value for the reference compound orlistat was 0.380 ± 0.004 μg/mL. The tests to evaluate the effect of the extracts on the metabolic activity of 3T3-L1 cells were conducted between the doses of 10 μg/mL and 200 μg/mL. FJ did not exhibit cytotoxic effects on cells under the concentration of 100 μg/mL, while the IC_{50} value of PJ was about 85 μg/mL. The tests for the investigation of adipogenesis were conducted at maximum nontoxic concentrations (IC_0) of the extracts, which were found as 100 μg/mL for FJ and 25 μg/mL for PJ. At these concentrations, FJ significantly decreased the lipid accumulation, the decrease provided by PJ was 25%. On the other hand, PJ was found to induce lipolysis by 20% at the dose of 25 μg/mL, as the value was 7% for FJ at 100 μg/mL concentration. The effect of the extracts on adipocyte differentiation was also investigated at the gene level. The extracts were shown to significantly suppress some mRNA expression of master adipogenic regulators at their IC_0 concentrations. FJ and PJ samples also inhibited fatty acid synthase, which is an enzyme involved in the synthesis of long-chain fatty acids. Based on the formation of reactive oxygen species during adipogenesis, the effect of the extracts on intracellular reactive oxygen species formation was investigated in the content of the same study, revealing that both extracts inhibited the formation of reactive oxygen species at the level of 10–15% at IC_0 concentrations. PJ also showed a reducing effect (40%) on the activity of PPARγ, which is a regulator in the differentiation process of adipocytes.

In the study of Podsedek *et al.* (2020), *V. opulus* fruits were evaluated with respect to the pancreatic lipase inhibitory effect and the effect on adipogenesis in 3T3-L1 cells. 70% acetone was used for extraction consecutively, and crude extract (CE) was purified with methanol using solid phase extraction method in order to obtain a partially purified phenolic extract (SPPE). Orlistat was used as a reference for the determination of pancreatic lipase activity. IC_{50} value of SPPE, which exhibited three times stronger activity than CE against pancreatic lipase was found to be 3.29 ± 0.06 mg/mL, where this value for orlistat was 0.38 ± 0.00 mg/mL. The cell viability of 3T3-L1 cells was also investigated at 25–125 μg/mL concentration of the extracts. SPPE exhibited toxicity by 11% at the dose of 125 μg/mL, while CE did not possess any toxic effect. The extracts decreased the lipid accumulation by 22% at the dose of 75 μg/mL. The extracts, especially SPPE, also affected the size of the lipid droplets as well as the number. Leptin level of 3T3-L1 cells was reduced by the extracts at 75 μg/mL dose by 21–30%. However, CE and SPPE did not influence adiponectin level of the cells significantly. Furthermore, the synergistic effect of the combination of the extracts and orlistat was revealed.

Anti-inflammatory Activity

Zaklos-Szyda *et al.* (2020c) evaluated the effect of fruit juice (FJ) and the juice exposed to solid phase extraction (PJ) on the secretion of tumor necrosis factor (TNFα) and interleukin-1β (IL-1β) and IL-6, which are among the mediators involved in the inflammation process, in mouse 3T3-L1 preadipocyte cells. FJ (100 μg/mL) and PJ (25 μg/mL) were found to decrease the expression of TNF-α mRNA by 20% and 70%, respectively. The extracts also reduced IL-6 secretion at the level of 40–60%, without any effect on the mRNA level. In another study by Zaklos-Szyda *et al.* (2020b), the effect of the fruit juice (FJ) and the juice purified via solid phase extraction (PJ) on IL-6, TNF-α, and vascular endothelial growth factor (VEGF) levels in human osteosarcoma Saso-2 cells was investigated. The extracts were applied at IC_0 concentrations, which were found to be 100 μg/mL for FJ and 25 μg/mL for PJ. It was shown that PF reduced TNF-α secretion by 40%, FJ reduced the level of cytokine to 90%, and both of the extracts decreased the IL-6 gene expression (15–30%) and secretion of IL-6 protein (55–80%). With respect to IL-6 inhibition, PJ was found to be more effective than FJ.

Cytoprotective Activity

Zaklos-Szyda *et al.* (2015) evaluated the cytoprotective activity of *V. opulus* fruit extract prepared using 70% acetone against oxidative stress induced by *tert*-butylhydroxy peroxide in βTC3 cells. The cytoprotective activity (108.435% of cells viability) of *V. opulus* fruit extracts was observed at the concentration of 0.365 mg/mL, which was determined as IC_0 value of the extract.

In another study by Zaklos-Szyda *et al.* (2019), the protective effect of the extracts against DNA damage induced by methylnitronitrosoguanidine (MNNG) or hydrogen peroxide on human colon adenocarcinoma cell line (Caco-2) was also established. The pomace of the fruits was extracted with methanol:acetone:water (MAE) at the ratio of 2:2:1, prior to the extraction with 70% acetone (AE). Besides, phenolic-rich fraction (PRF) was obtained by eluting dried fruit juice (FJ) with methanol. IC_0 values were determined as 95.64% ± 8.78 for FJ; 96.12% ± 10.12 for PRF; 91.70% ± 8.31 for MAE; and 90.50% ± 3.02 for AE. The highest cytoprotective activity against *tert*-butylhydroxy peroxide (500 μM) was observed for PRF at the dose of 50 μg/mL. The DNA repair efficacy of PRF after 120 minute was found as 93%, while that of MAE was 88%.

In a further study by Zaklos-Szyda *et al.* (2020b) on human osteosarcoma Saos-2 cell lines, the cytoprotective effect of the fruit juice (FJ) and the juice purified via solid phase extraction (PJ) were investigated. The cytoprotective activity of extracts at IC_0 concentrations (100 μg/mL for FJ and 25 μg/mL for PJ) against cellular DNA damage induced by methylnitronitrosoguanidine mutagen was evaluated on Saos-2 cells. FJ and PJ induced the repair of DNA in 60 minutes by 40% and 55%, respectively. The effect of PF was found to be about 50% over 120 minutes.

Anticarcinogenic Activity

In the study of Zaklos-Szyda *et al.* (2020b), the fruit juice (FJ) and the juice purified via solid phase extraction (PJ) were investigated in terms of their effect on osteosarcoma in human osteosarcoma Saos-2 cell lines. In the content of the study, the effect of extracts on cell metabolism was assessed at the concentration range of 10–200 μg/mL. In this range, no cytotoxicity on Saos-2 cells was observed for FJ, while PJ possessed a cytotoxic potential by reducing the metabolic activity by 90% at 200 μg/mL. IC_0 values were determined as 100 μg/mL for FJ and 25 μg/mL for PJ in order to evaluate the activity of the extracts. The activity of alkaline phosphatase, which is involved in the cell mineralization process of osteoblasts, was determined and a significant increase (35%) was observed with the treatment of PJ due to the gene expression. No influence of FJ was seen on ALP gene. Alizarin red S staining was also conducted in order to assess cell matrix mineralization and it was shown that FJ enhanced calcium deposition in the extracellular matrix by 15%, and the increase provided by PJ was found to be 65% in comparison to the untreated cells. The expression of some bone marker genes was evaluated. It was found that FJ induced the expression of RUNX2 mRNA levels by 40%, osteonectin expression was observed only for PJ (75%) and both extracts induced the expression of COL1A1 by 80–95%. A decrease by 15% with PJ treatment was observed in RANKL mRNA level.

Other Bioactivity Studies

In the study of Jarboe *et al.* (1967), aqueous and alcoholic extracts of *V. opulus* barks were found to possess antispasmodic effect, which was attributed to the active component scopoletin. *In vitro* barium-stimulated rat uterus model was used for the investigation of the activity, and the results revealed that the extracts provided an inhibition of uterus construction by 50% at the dose of 0.09 mg/mL. In further studies, the effect of viopudial isolated from the aqueous *V. opulus* fruit extract, on barium-stimulated rat uterus, was investigated. ED_{50} value of the compound was determined as 24 µg/kg (Nicholson *et al.*, 1972).

Ovodova *et al.* (2000) determined the immunostimulant activity of polysaccharide fractions obtained from *V. opulus* fruit pomace by ion-exchange chromatography. The fractions were found to induce the lysosomal enzyme secretion and stimulate phagocytic peritoneal macrophages obtained from rats.

It was reported that the extract prepared from *V. opulus* barks using ethanol:water (1:1) mixture exhibited hypotensive activity due to its inhibitory effect of angiotensin-converting enzyme (Barbosa-Filho *et al.*, 2006).

Erdogan-Orhan *et al.* (2011) assessed the anti-acetylcholinesterase activity of *V. opulus* and *V. lantana* using enzyme-linked immunosorbent assay. The extracts were prepared from the leaves, branches, and fruits using ethyl acetate, methanol, and water consecutively. This study revealed that methanolic extract of *V. opulus* leaves possessed high activity with the inhibition capacities of 57.63 ± 1.23%, 87.41 ± 0.99%, and 93.19 ± 0.87% at the concentrations of 50 µg/mL, 100 µg/mL, and 200 µg/mL, respectively, while those of the reference, galanthamine was found as 90.45 ± 0.83 and 98.97 ± 0.24 at the doses of 50 and 100 µg/mL. A moderate activity was observed for water extracts of *V. lantana* leaves and fruits. The inhibition capacities of water extract prepared from leaves were found as 1.44 ± 1.09, 42.40 ± 1.23, and 58.39 ± 1.00 at 50, 100, and 200 µg/mL doses, respectively. Water extract of the fruits was effective only at the doses of 100 and 200 µg/mL, with inhibition values of 13.79 ± 0.96 and 61.80 ± 0.76, respectively. In the same study, acetylcholinesterase inhibitory activity of the compounds amentoflavone, salicin, and chlorogenic acid, which were formerly detected in the extracts of *V. opulus* and *V. lantana*, was also investigated, revealing a mild effect of salicin with inhibition value of 34.54% at 200 µg/mL concentration. In further studies, the inhibitory effect of *V. tinus* extracts on acetylcholinesterase (AChE), tyrosinase, and butyrylcholinesterase (BChE) was investigated in order to determine the neuroprotective activity. For this purpose, leaves, branches, and fruits were extracted using ethyl acetate, methanol, and water, consecutively. The assays were conducted with the concentrations of the extracts at 50, 100, and 200 µg/mL, revealing a dose-dependent inhibition of the enzymes AChE and BChE. Ethyl acetate extract of branches exhibited the highest inhibition of AChE (81.7% ± 1.60), while methanol extract of fruits possessed the best BChE inhibition by 97.7% ± 0.47 at the concentration of 200 µg/mL. The highest tyrosinase inhibition (47.0% ± 0.68) was observed for the methanol extract of the fruits at the dose of 200 µg/mL (Sever Yilmaz *et al.*, 2013). In another study, the enzyme inhibitory effect of *V. orientale* leaves, branches, and fruits, which were sequentially extracted using ethyl acetate, methanol, and water, was investigated. Methanolic extract of the fruits was found to exert the highest inhibition against AChE (73.29% ± 2.73) and BChE (88.62 ± 0.61). Methanolic extracts of the branches and fruits exhibited the highest activity against tyrosinase enzyme with inhibition values of 49.75% ± 1.41 and 44.82 ± 2.49, respectively (Orhan *et al.*, 2018).

Ho *et al.* (2011) investigated the herb-drug interactions by evaluating the inhibitory effects of some herbal drugs on CYP1A1, which is an enzyme responsible for activating and detoxifying some carcinogens and on major human drug metabolizing enzymes (CYP1A2, CYP2C9, CYP2C19, CYP2D6, CYP3A4) to determine the CYP450 interaction kinetics. For this purpose, fluorogenic substrates and cDNA-expressed CYP450 isoforms were used. It was shown that the commercial preparation containing *V. opulus* extract was effective against CYP1A2 and CYP2C19 with IC_{50} values of 0.94 and 1.1 µg/mL, respectively. The concentrations to inhibit the enzymes by 50% were found to be higher than the predicted single dose concentration of the preparation which is assumed to possess 100% bioavailability.

In the content of the study of Cometa *et al.* (1998), spasmolytic effect of *V. tinus* methanol extract and ethyl acetate fraction on rabbit jejunum was also evaluated. Viburtinoside I, which was isolated from the ethyl acetate fraction, exhibited a weak activity, while a dose-dependent, reversible inhibition of spontaneous activity was observed for viburtinoside II/III, revealing a slight reduction in spontaneous contraction.

In Vivo Studies

Analgesic and Anti-inflammatory Activity

Altun *et al.* (2009) evaluated the analgesic activity of the water extract of *V. opulus* leaves using acetic acid-induced writhing test and tail flick tests in mice, and anti-inflammatory activity using acetic acid-induced rat paw edema test. In this study, the extract exhibited analgesic activity at the doses of 100 mg/kg and 200 mg/kg, with the inhibition value of 56.59% and 63.32% in acetic acid-induced writhing test, where the reference compound, acetylsalicylic acid inhibited the pain at the dose of 300 mg/kg with the value of 44.37%. In tail flick test, the extract possessed a comparable effect (45.29% ± 17.52 and 57.71% ± 15.65 for the doses of 100 and 200 mg/kg, respectively) with acetylsalicylic acid (23.55% ± 08.87 at the dose of 300 mg/kg) and morphine (69.38% ± 12.28 at the dose of 10 mg/kg) at 90 minutes. LD_{50} value of the extract was found to be 5.447 g/kg in mice.

It was reported that amentoflavone, which is isolated from *V. opulus* leaves, reduced IL-6, TNF-α, IL-1β, and prostaglandin E2 production, yielding a protective effect on hippocampal neurons of Kunming mice with epilepsy at the dose of 25 mg/mL (Kajszczak *et al.*, 2020).

Sever Yılmaz *et al.* (2007) evaluated the analgesic and anti-inflammatory activity of aqueous extract prepared from *V. lantana* leaves. Acetic acid-induced writhing test and tail flick test were conducted for the evaluation of analgesic activity, while carrageenan-induced hind paw edema test was used for anti-inflammatory activity. The extract exhibited a significant analgesic effect at the dose of 100 mg/kg in both models. The inhibitory effect of 200 mg/kg dose of the extract on abdominal stretching was higher than that of acetylsalicylic acid. The anti-inflammatory activity of the extract was found to be weak in comparison to the reference indomethacin. In the content of the study, LD_{50} value of the extract was calculated as 2.169 g/kg.

Anticarcinogenic Effect

In the study of the influence of *V. opulus* fruit juice on colon cancer induced by 1,2-dimethylhydrazine (DMH) injection for 12 weeks, the mice in test groups received fruit juice for 30 weeks (Ulger et al., 2013) (starting with the first DMH injection) and for 18 weeks (starting at the end of DMH injections). At the end of the experiment, the average number of tumor lesions were decreased in mice that received fruit juice for 30 weeks (8.5) and for 18 weeks (8.3) in comparison to DMH received group (11.3). The reduction in the incidence of invasive carcinoma in the test group of 30 weeks of treatment was significant, showing the preventive effect of *V. opulus* fruit juice against the progress of colon cancer.

Ceylan *et al.* (2018) investigated the effect of *V. opulus* extract on mice with Ehrlich ascites carcinoma. The extract was obtained from the juice of the fruits, and the treatment was conducted at doses of 1, 2, and 4 g/kg. The changes in weight and number of viable/non-viable cells were evaluated for the determination of anticancer activity. It was shown that *V. opulus* fruit juice reduced the weight of the tumor. The values exhibiting the viability of tumor cells were found to be 88.72%, 69.02%, and 51.87% for 1, 2, and 4 g/kg doses, respectively.

Antioxidant Activity

Zayachkivska *et al.* (2006) evaluated the effect of *V. opulus* fruits on antioxidant enzymes of rats with stress-induced gastrointestinal mucosal damage. The activity of superoxide dismutase (SOD), catalase (CAT), and glutathione peroxidase (GPx) was assessed, as well as malondialdehyde (MDA) to determine the effect on the lipid peroxidation. The samples were applied at doses of 25, 50, and 75 mg/kg, and it was observed that the fruits exhibited a dose-dependent activity, causing a decrease in MDA levels, an increase in CAT and GPx, and activation of SOD.

In the study of Ilhan *et al.* (2014), the antioxidant activity of squeezed and lyophilized fruit juice, lyophilized commercial juice, and fruit extracts which were extracted using *n*-hexane, ethyl acetate, and methanol consecutively were evaluated using thiobarbituric acid reactive substances (TBARs) assay and determination of the levels of total thiols (TSH) and glutathione (GSH) in kidney tissues of the rats, in which urolithiasis was induced by sodium oxalate. The results revealed the significant antioxidant activity of squeezed fruit juice and commercial fruit juice. The results in the TBARs assay were found as 134.5 ± 6.5 nmol/g and 121.1 ± 5.1 nmol/g for fruit juice and commercial fruit juice, while the value for the control group was 266.2 ± 11.8 nmol/g. GSH levels of fruit juice, commercial juice, and control group were determined as 21.7 ± 1.2 μmol/g, 24.8 ± 1.1 μmol/g, and 25.2 ± 1.3 μmol/g, respectively. TSH levels were measured as 15.5 ± 1.7 μmol/g for fruit juice, 13.65 ± 1.9 μmol/g for commercial juice, and 7.7 ± 1.1 μmol/g for the control group.

Eken *et al.* (2017) evaluated the influence of the methanol extract prepared from the fruits of *V. opulus* on oxidative stress induced by ischemia/reperfusion (I/R) after lung transplantation in rats. Superoxide dismutase (SOD), glutathione peroxidase (GPx), catalase (CAT) enzyme activities, the levels of glutathione, and total antioxidant status (TAS), which were the parameters decreased by ischemia/reperfusion, and the levels of malondialdehyde (MDA), total oxidant status (TOS), and protein carbonyl, which were increased in I/R group, were assessed in the content of the study. The total phenolic content of the extract was found as 67.73 mg GAE/g extract. The SOD enzyme activity was determined as 3.6 ± 0.37, 1.65 ± 0.33, and 3.31 ± 3.31 ± 0.43 U/mg protein for the control, I/R, and test group, respectively. The activity of GPx in the I/R group (1.42 ± 0.29 U/mg protein) was significantly lower than the control group (3.18 ± 0.25 U/mg protein); however, the fruit extract (2.84 ± 0.28 U/mg protein) was seen to enhance the GPx activity. CAT activity was decreased in the I/R group (17.97 ± 1.62 U/mg protein) in comparison to the control group (27.02 ± 1.41 U/mg protein) and test group (24.5 ± 1.72 U/mg protein). Reduced levels of glutathione by I/R (1.32 ± 0.43 μmol/g tissue) compared to the control group (2.20 ± 0.09 μmol/g tissue) were ameliorated by fruit extract (2.13 ± 0.47 μmol/g tissue). MDA levels of the I/R group (0.74 ± 0.09 nmol/mg protein) were elevated, while those of the control group (0.48 ± 0.05 nmol/mg protein) and test group (0.53 ± 0.08 nmol/mg protein) were significantly lower. Carbonyl levels were found as 0.71 ± 0.06 nmol/mg protein and 0.76 ± 0.03 nmol/mg protein for the control group and test group, respectively, while higher levels were determined in the I/R group (0.85 ± 0.05 nmol/mg protein). TAS levels of the I/R group (43.02 ± 4.75 μmol TE/mg protein) were found to be reduced compared to those of the control (66.98 ± 3.8 μmol TE/mg protein) and test (69.59 ± 8.9 μmol TE/mg protein) groups. TOS levels of the I/R group (0.32 ± 0.05 μmol H_2O_2 eq/mg protein) were elevated, while those values for control (0.17 ± 0.05 μmol H_2O_2 eq/mg protein) and test (0.18 ± 0.04 μmol H_2O_2 eq/mg protein) groups were lower. The oxygenation index was also determined for each group by blood gas analysis. The results revealed that I/R decreased PaO_2/FiO_2 level by 22% as compared to the control group. However, a significant increase in PaO_2/FiO_2 level was observed by 9% compared to the I/R group.

Erdem *et al.* (2016) evaluated the effect of the ethanol (90%) extract prepared from *V. opulus* fruits on antioxidant enzymes of rats with ethylene glycol-induced urolithiasis model. In the experiment, the low-dose group received 0.5 mL of 10 mg/mL extract solution for five days a week, while the high-dose group received 1 mL of the extract solution during a period of one month. Following the sacrification, the kidneys of the rats were removed, and homogenate analyses of the kidney samples were performed in order to determine MDA, GPx, and SOD levels. It was observed that the extract decreased the urinary excretion of phosphate and calcium. Besides, Cystine and oxalate levels in urine were lowered and urine citrate levels were increased along with the urine volume in these groups. The investigation of kidney tissue revealed that the crystal deposits were also lowered by the extract. GPx levels were found to be increased, and MDA levels were decreased by the high-dose application of the extract.

Sariozkan *et al.* (2017) determined the influence of water extract prepared from *V. opulus* fruits on the lipid peroxidation level and SOD, GPx, and CAT activity in testis and epididymis tissues of rats. Fruit extract was shown to lower the MDA level increased by chemotherapeutics (docetaxel and paclitaxel) and induced SOD and GPx activities, which were reduced by the administration of chemotherapeutics. MDA levels were found to be 0.788 ± 0.029 nmol/mL for the docetaxel and fruit juice administered group; 0.726 ± 0.053 nmol/mL paclitaxel and fruit juice administered group, while 1.251 ± 0.078 nmol/mL and 1.139 ± 0.057 nmol/mL for docetaxel and paclitaxel administered groups, respectively. SOD and GPx levels of the docetaxel group were determined as 1.647 ± 0.106 and 1.425 ± 0.065 U/mL, and those of the paclitaxel group were 1.835 ± 0.189 and 1.575 ± 0.084 U/mL, respectively. The influence of the treatment with fruit extract on these parameters was observed with SOD and GPx levels of 2.439 ± 0.102 and 2.202 ± 0.103 U/mL for docetaxel and 2.588 ± 0.067 and 2.160 ± 0.080 U/mL for paclitaxel. Besides, the CAT activity of the docetaxel group (0.012 ± 0.001 U/mL) was lower than that of fruit juice (0.016 ± 0.001 U/mL) treated group.

Antiurolithiatic Activity

Antiurolithiatic activity of squeezed and lyophilized fruit juice and lyophilized commercial juice was evaluated in the sodium oxalate-induced urolithiasis rat model. The urinary output after seven days of experiment was determined as 11.75 ± 1.21 mL/rat in 24 hours for only sodium oxalate treated group (70 mg/kg), while the values for squeezed juice (100 mg/kg), commercial juice (100 mg/kg), and the reference cystone (500 mg/kg) were found to be 25.21 ± 0.19 mL, 24 ± 1.28 mL, and 19.56 ± 1.53 mL, respectively. The results revealed that urinary output significantly increased in the test and cystone groups, as no significant alteration was observed in the sodium oxalate group. Squeezed juice decreased oxalate, creatinine, and uric acid levels in urine to 0.69 ± 0.10 mg/dL, 2.95 ± 0.24 mg/dL, and 1.96 ± 0.95 mg/dL from the increased levels by sodium oxalate; 2.15 ± 0.74 mg/dL, 4.25 ± 0.17 mg/dL, and 2.97 ± 0.56 mg/dL, respectively. The increased urine parameters were also reduced in the group of commercial juice with the values of 0.61 ± 0.09 mg/dL for oxalate, 2.25 ± 0.1 9mg/dL for creatinine, and 1.79 ± 0.81 mg/dL for uric acid, while those of cystone group were 0.59 ± 0.03 mg/dL, 1.39 ± 0.10mg/dL, and 1.63 ± 0.75mg/dL, respectively. Decreased excretion of magnesium was also normalized in test and cystone groups. The test samples did not change the pH of urine, which was increased by sodium oxalate. The increased oxalate level in urine, which was induced by sodium oxalate administration, caused crystal deposition in urine samples; however, no crystal deposition was observed in cystone and test groups. The levels of uric acid, sodium, potassium, and creatinine in serum were also decreased by fruit juice samples as well as cystone. The results showing the effect of the fruit juice samples were verified by histopathological investigations (Ilhan *et al.*, 2014).

Erdem *et al.* (2016) evaluated the effect of the ethanol (90%) extract prepared from *V. opulus* fruits on ethylene glycol-induced urolithiasis model in rats. The experiment was conducted for 28 days, and after urine and blood samples of the rats were taken, various parameters such as pH, volume, cyrstalluria, levels of urea, uric acid, creatinine, calcium, magnesium, sodium, potassium, and chloride were determined. Furthermore, histopathological investigations and homogenate analyses of the kidney samples were performed. Fruit extract was found to decrease the cysteine and oxalate levels in urine as well as the crystal deposition in tubules, revealing the preventive effect against stone formation.

Hepatoprotective Activity

Altun *et al.* (2010) investigated the hepatoprotective activity of water extract prepared from *V. opulus* leaves using CCl_4-induced hepatotoxicity model in rats. The extract exhibited low activity at the dose of 100 mg/kg in comparison to silibinin (50 mg/kg), which was used as the control. The serum ALT, AST, ALP, and total bilirubin levels of the rats treated with the extract were 500.8 ± 82.7, 1158.2 ± 217.3, 174.8 ± 11.3 U/L, and 0.05 ± 0.01 mg/dL, respectively, while those values for silibinin were found as 205.8 ± 66.4, 708.8 ± 183.4, 316.3 ± 35.7 U/L, and 0.26 ± 0.51 mg/dL, respectively. The results were corroborated by histopathological investigations. LD_{50} value of the extract was found as 5.447 g/kg.

In order to determine the hepatoprotective effect of aqueous extract of *V. lantana* leaves, Sever Yilmaz *et al.* (2006) conducted a study on rats with carbon tetrachloride (CCl_4)-induced hepatotoxicity. Bilirubin, aspartate aminotransferase (AST), and alanine aminotransferase (ALT) levels in the serum were determined to see the influence of the extract on hepatic damage. The extract was administered at the dose of 100 mg/kg, while the dose of the reference compound silibinin was 50 mg/kg. Serum bilirubin, ALT, and AST levels, which were increased by CCl_4 (0.33 ± 0.17 mg/dL, 959.4 ± 152.1 U/L, 1931.9 ± 303.6 U/L, respectively), were reduced in extract (0.28 ± 0.12 mg/dL, 244.2 ± 28.7 U/L, 913.3 ± 271.9 U/L, respectively) and silibinin (0.26 ± 0.51 mg/dL, 205.8 ± 66.4 U/L, 708.8 ± 183.4 U/L, respectively) treated groups. The results were confirmed by histopathological examinations. In the content of the same study, LD_{50} dose of the extract was calculated as 2.169 g/kg in mice.

Mohamed *et al.* (2005) assessed the hepatoprotective effect of *V. tinus* leaves in rats with CCl_4-induced hepatotoxicity. The administered doses were 25 and 50 mg/kg for the 80% methanolic extract and 25 mg/kg for silymarin, the reference compound. LD_{50} value of the extract was determined as 500 mg/kg. Decreased serum ALT, AST, lipid peroxide, and nitric oxide levels by CCl_4 were not significantly influenced by 25 mg/kg dose of extract, while a significant reduction was observed at the dose of 50 mg/kg with the values of 118.6 ± 3.67 U/mL for ALT, 192.4 ± 4.73 U/mL for AST, 4.60 ± 0.18 nmole/mL for lipid peroxide and 2.38 ± 0.25 μg/L for nitric oxide. Those values for silymarin were found as 78.2 ± 3.10 U/mL, 130.7 ± 3.20 U/mL, 3.70 ± 0.16 nmole/mL, and 1.06 ± 0.11 μg/L, respectively.

Other Biological Activities

Zayachkivska *et al.* (2006) evaluated the gastroduodenoprotective activity of *V. opulus* fruit proanthocyanidins on rats, which were exposed to non-topical ulcerogens. It was observed that the mean areas of gastroduodenal lesions were reduced and the nitric oxide (NO) system was activated in the test group, in which a 50 mg/kg dose was applied. However, gastroprotective effect was not seen at this dose, on the rats with sensory denervation by capsaicin. Furthermore, it was shown that *V. opulus* fruit enhanced the restoration of the content of gastroduodenal glycoproteins, providing an increase in mucus gel layer and gastric mucosal resistance.

In the study of Saltan *et al.* (2016), dried fruits of *V. opulus* were extracted using *n*-hexane, ethyl acetate, and methanol successively and the effect of these extracts on surgically induced endometriosis in rats was investigated. The results of this study showed that ethyl acetate and methanol extracts reduced the endometriotic volume significantly at the dose of 100 mg/kg. The results were supported by histopathological investigations.

In the study of Sariozkan *et al.* (2017), a rat model with sperm damage induced by docetaxel and paclitaxel was used for the evaluation of the potential of *V. opulus* fruit extract as a protective agent. The reproductive system was damaged by weekly use of docetaxel (5 mg/kg) or paclitaxel (4 mg/kg) for 10 weeks. Aqueous extract of *V. opulus* fruits was administered *per os* at the dose of 100 mg/kg/day. A significant decrease in the serum levels of testosterone was observed in the docetaxel and paclitaxel groups, while treatment with fruit extract without chemotherapeutic administration provided an increase in testosterone levels. However, no significant change was determined in these groups, to which chemotherapeutic was administered along with fruit extract. *V. opulus* fruit extract was found to enhance the motility and the concentration of the sperms, which were reduced by the influence of docetaxel and paclitaxel. Sperm motility and concentration values were found as 14.33% ± 1.22 and 47.40 ± 1.16 million for the docetaxel group; 38.00% ± 3.27 and 51.20 ± 1.64 million for the paclitaxel group; 37.00% ± 2.60 and 70.60 ± 1.34 million for the group treated with fruit extract along with docetaxel administration; 65.41% ± 1.99 and 106.25 ± 5.40 million for the group treated with fruit extract along with paclitaxel administration. Fruit juice treatment was also shown to ameliorate the histopathological findings in the chemotherapeutic administered groups.

The study of Nicholson *et al.* (1972) showed that viopudial isolated from the aqueous *V. opulus* fruit extract exerted a hypotensive effect on rats at i.v. dose of 250 μg/mL. A significant decrease in blood pressure and heartbeat was observed at the dose of 1 μg/mL. These effects were also observed in dogs and cats at 2 mg/kg dose.

Bujor *et al.* (2019) evaluated the vasorelaxant and the arginase inhibitory activity of *V. opulus* fruit extracts in phenylephrine precontracted aortic rings of rats. The extract was prepared using acetone:water:acetic acid (80:19.5:0.5) prior to the liquid-liquid partition with diethyl ether. The extract exhibited a remarkable vasorelaxant activity with an endothelium-dependent mechanism and the EC_{50} value was determined as 6.31 ± 1.61 μg/mL. The IC_{50} value for the inhibition of arginase was found to be 71.02 ± 3.06 μg/mL. The toxicity of the extract was also assessed by brine shrimp acute toxicity assay, revealing no toxic effect up to the concentration of 5 mg/mL.

Cometa *et al.* (1998) investigated the sedative effect of *V. tinus* methanol extract, ethyl acetate fraction and the isolated compounds viburtinosides I and II/III using the behavioral Irwing test on mice. The results revealed that the extract at the doses of 100–300 mg/kg and fraction at the doses of 25–150 mg/kg provided a sedative effect by reducing curiosity and spontaneous activity as well as by muscle relaxant activity. In the dose range of 25–100 mg/kg, the isolated compounds exhibited a dose-dependent activity, where viburtinoside I was found to be more potent.

Clinical Studies

Antiurolithiatic Activity

Kizilay *et al.* (2019) investigated the effect of *V. opulus* extract on distal ureteral stones at diameters of 5–10 mm. In the content of the study, 53 patients were given 1000 mg (at the dose of two tablets, three times a day) *V. opulus* extract along with 50 mg diclofenac sodium and 50 patients were given only 50 mg diclofenac sodium. *V. opulus* extract provided a significant increase in the rate of expulsion rate of distal ureteral stones by 82% in comparison to diclofenac sodium group in which that value was determined as 66%. The elapsed time to stone expulsion was found to be 9 ± 1.8 days, while that of the diclofenac sodium group was 14 ± 2.3 days. No significant difference was observed between the two groups with respect to the admission to emergency service and the rate of complication; however, the analgesic need of the patients was 24.5% less for the patients treated with *V. opulus* extract, while a decrease in the analgesic need of the patients in diclofenac sodium group was 44%.

CONCLUSIONS

According to the phytochemical studies, the genus *Viburnum* contains terpenoids, sterols, and polyphenolic substances such as flavonoids, anthocyanins, tannins, iridoids, and their glycosides.

Among the four *Viburnum* species grown in Turkey, *V. opulus* is the most known and widely used as a folk remedy. In folk medicine, *V. opulus* is used for various purposes such as diuretic, sedative, antispasmodic, hemorrhagic, tonic, hypotensive, choleretic, and for the treatment of cold, flu, and lung diseases.

V. opulus is known to contain carbohydrates, lipids, and proteins, as well as secondary metabolites such as phenolic acids, anthocyanins, iridoids, flavonoids, coumarins, and tannins. Fruits are reported as a good source of vitamin C and dietary fibers, and organic acids. Chlorogenic acid was determined as the major bioactive component of the fruits. Most of the studies were focused on the fruits of *V. opulus*, yielding various activities such as antioxidant, antimicrobial, cytoprotective, anticarcinogenic, anti-inflammatory, antiobesity, antispasmodic, and antidiabetic.

The barks of *V. lantana* are used as analgesic and rubefacient as a folk remedy. It is reported that the plant contains essential oil, steroids, saponins, carotenoids, flavonoids, tannins, and anthocyanins as well as lipids. According to the studies, *V. lantana* exhibited antioxidant, antimicrobial, antitumor, anti-acetylcholinesterase, anti-inflammatory, analgesic, and hepatoprotective activities.

There are fewer studies that reveal the phytochemical contents and bioactivities of *V. tinus* and *V. orientale*. The phytochemical studies reported that *V. tinus* contained iridoids, coumarins, saponins, and flavonoids, while terpenoid and iridoid structures were isolated from *V. orientale*. Biological activity studies revealed *in vitro* antioxidant and spasmolytic and *in vivo* hepatoprotective and sedative effects of *V. tinus*. *V. orientale* was found to possess antioxidant and antimicrobial activity *in vitro*. Besides, inhibitory effects of *V. tinus* and *V. orientale* on acetylcholinesterase, butyrylcholinesterase, and tyrosinase enzymes were also reported.

REFERENCES

Altun, M., Sever Yılmaz, B., & Saltan Citoğlu, G. Viburnum opulus L. ve Viburnum lantana L.'da amentoflavon YBSK analizi. *Ankara Ecz. Fak. Derg.(Journal of Faculty of Pharmacy of Ankara University)*, **36**: 161–169 (2007).

Altun, M. L. Viburnum opulus, Gilaburu. In L. Ö. E. Demirezer, T.; Saraçoğlu, İ.; Şener, B.; Köroğlu, A.; Yalçın, F. N. (Ed.), *FFD Monografları Bitkiler ve Etkileri* (pp. 1103–1106). Ankara: Akademisyen Kitabevi (2017).

Altun, M. L., Citoglu, G. S., Yilmaz, B. S., & Coban, T. Antioxidant properties of Viburnum opulus and Viburnum lantana growing in Turkey. *International Journal of Food Sciences and Nutrition*, **59**(3): 175–180 (2008).

Altun, M. L., Citoglu, G. S., Yilmaz, B. S., & Ozbek, H. Antinociceptive and anti-inflammatory activities of Viburnum opulus. *Pharmaceutical Biology*, **47**(7): 653–658 (2009).

Altun, M. L., Özbek, H., Çitoğlu, G. S., Yilmaz, B. S., Bayram, I., & Cengiz, N. Hepatoprotective and hypoglycemic activities of Viburnum opulus L. *Turk J Pharm Sci*, **7**: 35–48 (2010).

Altun, M. L., & Yilmaz, B. S. HPLC method for the analysis of salicin and chlorogenic acid from Viburnum opulus and V-lantana. *Chemistry of Natural Compounds*, **43**(2): 205–207 (2007).

Barak, T. H., Celep, E., Ivan, Y., & Yesilada, E. Influence of in vitro human digestion on the bioavailability of phenolic content and antioxidant activity of Viburnum opulus L. (European cranberry) fruit extracts. *Industrial Crops and Products*, **131**: 62–69 (2019).

Barbosa-Filho, J. M., Martins, V. K., Rabelo, L. A., Moura, M. D., Silva, M. S., Cunha, E. V., Souza, M. F., Almeida, R. N., & Medeiros, I. A. Natural products inhibitors of the angiotensin converting enzyme (ACE): A review between 1980–2000. *Revista Brasileira de Farmacognosia*, **16**(3): 421–446 (2006).

Baschali, A., Tsakalidou, E., Kyriacou, A., Karavasiloglou, N., & Matalas, A. L. Traditional low-alcoholic and non-alcoholic fermented beverages consumed in European countries: A neglected food group. *Nutrition Research Reviews*, **30**(1): 1–24 (2017).

Baytop, T. *Türkiye'de Bitkilerle tedavi Geçmişte ve Bugün*. İstanbul: Nobel Tıp Kitabevi (1999).

Bubulica, M. V., Anghel, I., Grumezescu, A. M., Saviuc, C., Anghel, G. A., Chifiriuc, M. C., Gheorghe, I., Lazar, V., & Popescu, A. In vitro evaluation of bactericidal and antibiofilm activity of Lonicera tatarica and Viburnum opulus plant extracts on Staphylococcus strains. *Farmacia*, **60**(1): 80–91 (2012).

Bujor, A., Miron, A., Luca, S. V., Skalicka-Wozniak, K., Silion, M., Ancuceanu, R., Dinu, M., Girard, C., Demougeot, C., & Totoson, P. Metabolite profiling, arginase inhibition and vasorelaxant activity of Cornus mas, Sorbus aucuparia and Viburnum opulus fruit extracts. *Food and Chemical Toxicology*, **133**: 10 (2019).

Bujor, A., Ochiuz, L., Sha'at, M., Stoleriu, I., Iliuta, S. M., Luca, S. V., & Miron, A. Chemical, antioxidant and in vitro permeation and penetration studies of extracts obtained from Viburnum opulus and Crataegus pentagyna. *Farmacia*, **68**(4): 672–678 (2020).

Calis, I., Yuruker, A., Ruegger, H., Wright, A. D., & Sticher, O. Anatoliosides – 5 Novel acyclic monoterpene glycosides from Viburnum orientale. *Helvetica Chimica Acta*, **76**(1): 416–424 (1993).

Calis, I., Yuruker, A., Ruegger, H., Wright, A. D., & Sticher, O. Lantanoside, a monocyclic C-10 iridoid glucoside from Viburnum lantana. *Phytochemistry*, **38**(1): 163–165 (1995).

Cam, M., & Hisil, Y. Comparison of chemical characteristics of fresh and pasteurised juice of gilaburu (Viburnum opulus L.). *Acta Alimentaria*, **36**(3): 381–385 (2007).

Cam, M., Hisil, Y., & Kuscu, A. Organic acid, phenolic content, and antioxidant capacity of fruit flesh and seed of Viburnum opulus. *Chemistry of Natural Compounds*, **43**(4): 460–461 (2007).

Cesoniene, L., Daubaras, R., Viskelis, P., & Sarkinas, A. Determination of the total phenolic and anthocyanin contents and antimicrobial activity of Viburnum Opulus fruit juice. *Plant Foods for Human Nutrition*, **67**(3): 256–261 (2012).

Ceylan, D., Aksoy, A., Ertekin, T., Yay, A. H., Nisari, M., Karatoprak, G. S., & Ulger, H. The effects of gilaburu (Viburnum opulus) juice on experimentally induced Ehrlich ascites tumor in mice. *Journal of Cancer Research and Therapeutics*, **14**(2): 314–320 (2018).

Cometa, M. F., Mazzanti, G., & Tomassini, L. Sedative and spasmolytic effects of Viburnum tinus L. and its major pure compounds. *Phytotherapy Research*, **12**: S89–S91 (1998).

Davis, P. H. *Flora of Turkey and The East Aegean Islands* (Vol. 4). Edinburgh: Edinburgh University Press (1972).

Dienaite, L., Baranauskiene, R., & Venskutonis, P. R. Lipophilic extracts isolated from European cranberry bush (Viburnum opulus) and sea buckthorn (Hippophae rhamnoides) berry pomace by supercritical CO2 – Promising bioactive ingredients for foods and nutraceuticals. *Food Chemistry*, **348**: 10 (2021).

Dienaite, L., Pukalskiene, M., Pereira, C. V., Matias, A. A., & Venskutonis, P. R. Valorization of European cranberry bush (Viburnum opulus L.) Berry Pomace extracts isolated with pressurized ethanol and water by assessing their phytochemical composition, antioxidant, and antiproliferative activities. *Foods*, **9**(10): 23 (2020).

Dietz, B. M., Hajirahimkhan, A., Dunlap, T. L., & Bolton, J. L. Botanicals and their bioactive phytochemicals for women's health. *Pharmacological Reviews*, **68**(4): 1026–1073 (2016).

Eken, A., Yucel, O., Bosgelmez, II, Baldemir, A., Cubuk, S., Cermik, A. H., Unlu Endirlik, B. U., Bakir, E., Yildizhan, A., Guler, A., & Kosar, M. An investigation on protective effect of Viburnum opulus L. fruit extract against Ischemia/Reperfusion-Induced oxidative stress after lung transplantation in rats. *Kafkas Universitesi Veteriner Fakultesi Dergisi*, **23**(3): 437–444 (2017).

Erdem, G., Kesik, V., Honca, T., Ozcan, A., Uguz, S., Akgul, E. O., Aykutlug, O., Alp, B. F., Korkmazer, N., Saldir, M., & Bayrak, Z. Antinephrolithiatic activity of Persea americana (avocado) and Viburnum opulus (guelder rose) against ethylene glycol-induced nephrolithiasis in rats. *African Journal of Traditional Complementary and Alternative Medicines*, **13**(2): 110–119 (2016).

Erdogan-Orhan, I., Altun, M. L., Sever-Yilmaz, B., & Saltan, G. Anti-Acetylcholinesterase and antioxidant assets of the major components (Salicin, Amentoflavone, and Chlorogenic Acid) and the extracts of Viburnum opulus and *Viburnum lantana* and their total phenol and flavonoid contents. *Journal of Medicinal Food*, **14**(4): 434–440 (2011).

Ersoy, N., Ercisli, S., Akin, M., Gundogdu, M., Colak, A. M., & Ben Ayed, R. Agro-morphological and biochemical characteristics of European cranberry bush (Viburnum opulus L.). *Comptes Rendus De L Academie Bulgare Des Sciences*, **71**(4): 491–499 (2018).

Ersoy, N., Ercisli, S., & Gundogdu, M. Evaluation of European Cranberrybush (*Viburnum opulus* L.) genotypes for agro-morphological, biochemical and bioactive characteristics in Turkey. *Folia Horticulturae*, **29**(2): 181–188 (2017).

Eryilmaz, M., Ozbilgin, S., Ergene, B., Yilmaz, B. S., Altun, M. L., & Saltan, G. Antimicrobial activity of Turkish *Viburnum* species. *Bangladesh Journal of Botany*, **42**(2): 355–360 (2013).

Gao, X., Shao, L.-D., Dong, L.-B., Cheng, X., Wu, X.-D., Liu, F., Jiang, W.-W., Peng, L.-Y., He, J., & Zhao, Q.-S. Vibsatins A and B, two new tetranorvibsane-type diterpenoids from Viburnum tinus cv. variegatus. *Organic Letters*, **16**(3): 980–983 (2014).

Handjieva, N., Baranovska, I., Mikhova, B., & Popov, S. Iridoids from *Viburnum-lantana*. *Phytochemistry*, **27**(10): 3175–3179 (1988).

Handjieva, N., Saadi, H., Popov, S., & Baranovska, I. Separation of Iridoids by vacuum liquid chromatography. *Phytochemical Analysis*, **2**(3): 130–133 (1991).

Ho, S. H. Y., Singh, M., Holloway, A. C., & Crankshaw, D. J. The effects of commercial preparations of herbal supplements commonly used by women on the biotransformation of fluorogenic substrates by human cytochromes P450. *Phytotherapy Research*, **25**(7): 983–989 (2011).

Ilhan, M., Ergene, B., Suntar, I., Ozbilgin, S., Citoglu, G. S., Demirel, M. A., Keles, H., Altun, L., & Akkol, E. K. Preclinical evaluation of antiurolithiatic activity of Viburnum opulus L. on sodium oxalate-induced urolithiasis rat model. *Evidence-Based Complementary and Alternative Medicine*, **2014**: 1–10 (2014).

Jarboe, C. H., Zirvi, K. A., Nicholso, J. A., & Schmidt, C. M. Scopoletin an antispasmodic component of Viburnum opulus and prunifolium. *Journal of Medicinal Chemistry*, **10**(3): 488–& (1967).

Kajszczak, D., Zaklos-Szyda, M., & Podsedek, A. Viburnum opulus L.-A review of phytochemistry and biological effects. *Nutrients*, **12**(11): 30 (2020).

Karacelik, A. A., Kucuk, M., Iskefiyeli, Z., Aydemir, S., De Smet, S., Miserez, B., & Sandra, P. Antioxidant components of Viburnum opulus L. determined by on-line HPLC-UV-ABTS radical scavenging and LC-UV-ESI-MS methods. *Food Chemistry*, **175**: 106–114 (2015).

Karakurt, S., Abusoglu, G., & Arituluk, Z. C. Comparison of anticarcinogenic properties of Viburnum opulus and its active compound p-coumaric acid on human colorectal carcinoma. *Turkish Journal of Biology*, **44**(5): 252–263 (2020).

Karimova, A. R., Yunusova, S. G., Maslennikov, S. I., Galkin, E. G., Yunusov, T. S., Shereshovets, V. V., & Yunusov, M. S. Lipids, lipophilic components, and biologically active fractions of Viburnum opulus L. seeds. *Chemistry of Natural Compounds*, **36**(6): 560–564 (2000).

Kizilay, F., Ulker, V., Celik, O., Ozdemir, T., Cakmak, O., Can, E., & Nazli, O. The evaluation of the effectiveness of Gilaburu (Viburnum opulus L.) extract in the medical expulsive treatment of distal ureteral stones. *Turkish Journal of Urology*, **45**: 63–69 (2019).

Kollmann, J., & Grubb, P. J. Viburnum lantana L. and Viburnum opulus L. (V-lobatum Lam., Opulus vulgaris Borkh.). *Journal of Ecology*, **90**(6): 1044–1070 (2002).

Konarska, A., & Domaciuk, M. Differences in the fruit structure and the location and content of bioactive substances in Viburnum opulus and Viburnum lantana fruits. *Protoplasma*, **255**(1): 25–41 (2018).

Kraujalis, P., Kraujaliene, V., Kazernaviciute, R., & Venskutonis, P. R. Supercritical carbon dioxide and pressurized liquid extraction of valuable ingredients from Viburnum opulus pomace and berries and evaluation of product characteristics. *Journal of Supercritical Fluids*, **122**: 99–108 (2017).

Kraujalyte, V., Leitner, E., & Venskutonis, P. R. Chemical and sensory characterisation of aroma of Viburnum opulus fruits by solid phase microextraction-gas chromatography-olfactometry. *Food Chemistry*, **132**(2): 717–723 (2012).

Kraujalyte, V., Venskutonis, P. R., Pukalskas, A., Cesoniene, L., & Daubaras, R. Antioxidant properties and polyphenolic compositions of fruits from different European cranberrybush (Viburnum opulus L.) genotypes. *Food Chemistry*, **141**(4): 3695–3702 (2013).

Laux, M. T., Aregullin, M., & Rodriguez, E. Inhibition of Helicobacter pylori and gastric cancer cells by lipid aldehydes from Viburnum opulus (Adoxaceae). *Natural Product Communications*, **2**(10): 1015–1018 (2007).

Mohamed, M. A., Marzouk, M. S. A., Moharram, F. A., El-Sayed, M. M., & Baiuomy, A. R. Phytochemical constituents and hepatoprotective activity of Viburnum tinus. *Phytochemistry*, **66**(23): 2780–2786 (2005).

Moldovan, B., David, L., Vulcu, A., Olenic, L., Perde-Schrepler, M., Fischer-Fodor, E., Baldea, I., Clichici, S., & Filip, G. A. In vitro and in vivo anti-inflammatory properties of green synthesized silver nanoparticles using Viburnum opulus L. fruits extract. *Materials Science & Engineering C-Materials for Biological Applications*, **79**: 720–727 (2017).

Moldovan, B., Ghic, O., David, L., & Chisbora, C. The influence of storage on the total phenols content and antioxidant activity of the Cranberrybush (Viburnum opulus L.) fruits extract. *Revista De Chimie*, **63**(5): 463–464 (2012).

Nazir, M., Sultan, M., Riaz, N., Hafeez, M., Hussain, H., Ahmed, I., Schulz, B., Draeger, S., Jabbar, A., Krohn, K., Ashraf, M., & Saleem, M. Depsitinuside: A new depside galactoside from an endophytic fungus isolated from Viburnum tinus. *Journal of Asian Natural Products Research*, **13**(11): 1056–1060 (2011).

Nicholson, J. A., Darby, T. D., & Jarboe, C. H. Viopudial, a hypotensive and smooth muscle antispasmodic from Viburnum opulus. *Proceedings of the Society for Experimental Biology and Medicine*, **140**(2): 457–+ (1972).

Orhan, I. E., Senol, F. S., Yilmaz, B. S., Altun, M. L., Ozbilgin, S., Yazgan, A. N., Yuksel, E., & Iscan, G. S. Neuroprotective potential of Viburnum orientale Pallas through enzyme inhibition and antioxidant activity assays. *South African Journal of Botany*, **114**: 126–131 (2018).

Orodan, M., Vodnar, D. C., Toiu, A. M., Pop, C. E., Vlase, L., Viorica, I., & Arsene, A. L. Phytochemical analysis, antimicrobial and antioxidant effect of some gemmotherapic remedies used in respiratory diseases. *Farmacia*, **64**(2): 224–230 (2016).

Ovodova, R. G., Golovchenko, V. V., Popov, S. V., Shashkov, A. S., & Ovodov, Y. S. The isolation, preliminary structural studies, and physiological activity of water-soluble polysaccharides from the squeezed berries of snowball tree Viburnum opulus. *Bioorganicheskaya Khimiya*, **26**(1): 61–67 (2000).

Ozrenk, K., Ilhan, G., Sagbas, H. I., Karatas, N., Ercisli, S., & Colak, A. M. Characterization of European cranberrybush (Viburnum opulus L.) genetic resources in Turkey. *Scientia Horticulturae*, **273**: 109611 (2020).

Perova, I. B., Zhogova, A. A., Cherkashin, A. V., Eller, K. I., Ramenskaya, G. V., & Samylina, I. A. Biologically active substances from European Guelder berry fruits. *Pharmaceutical Chemistry Journal*, **48**(5): 332–339 (2014).

Podsedek, A., Zaklos-Szyda, M., Polka, D., & Sosnowska, D. Effects of Viburnum opulus fruit extracts on adipogenesis of 3T3-L1 cells and lipase activity. *Journal of Functional Foods*, **73**: 10 (2020).

Polka, D., Podsedek, A., & Koziolkiewicz, M. Comparison of chemical composition and antioxidant capacity of fruit, flower and bark of Viburnum opulus. *Plant Foods for Human Nutrition*, **74**(3): 436–442 (2019).

Rop, O., Reznicek, V., Valsikova, M., Jurikova, T., Mlcek, J., & Kramarova, D. Antioxidant properties of European Cranberrybush fruit (Viburnum opulus var. edule). *Molecules*, **15**(6): 4467–4477 (2010).

Sagdic, O., Ozturk, I., Yapar, N., & Yetim, H. Diversity and probiotic potentials of lactic acid bacteria isolated from gilaburu, a traditional Turkish fermented European cranberrybush (Viburnum opulus L.) fruit drink. *Food Research International*, **64**: 537–545 (2014).

Saltan, G., Suntar, I., Ozbilgin, S., Ilhan, M., Demirel, M. A., Oz, B. E., Keles, H., & Akkol, E. K. Viburnum opulus L.: A remedy for the. treatment of endometriosis demonstrated by rat model of surgically-induced endometriosis. *Journal of Ethnopharmacology*, **193**: 450–455 (2016).

Sariozkan, S., Turk, G., Eken, A., Bayram, L. C., Baldemir, A., & Dogan, G. Gilaburu (Viburnum opulus L.) fruit extract alleviates testis and sperm damages induced by taxane-based chemotherapeutics. *Biomedicine & Pharmacotherapy*, **95**: 1284–1294 (2017).

Sever Yilmaz, B., Altun, M. L., Orhan, I. E., Ergene, B., & Saltan Citoglu, G. S. Enzyme inhibitory and antioxidant activities of Viburnum tinus L. relevant to its neuroprotective potential. *Food Chemistry*, **141**(1): 582–588 (2013).

Sever Yilmaz, B., Saltan Citoglu, G., Altun, M. L., Özbek, H., Bayram, İ., Cengiz, N., & Altinyay, Ç. Hepatoprotective and hypoglycemic activity of Viburnum lantana L. *Turkish J. Pharm. Sci*, **3**(3): 151–165 (2006).

Sever Yılmaz, B., Saltan Çitoğlu, G. S., Altun, M., & Özbek, H. Antinociceptive and anti-inflammatory activities of Viburnum lantana. *Pharmaceutical Biology*, **45**(3): 241–245 (2007).

Shafaghat, A., & Shafaghatlonbar, M. Two new chalcone glycoside compounds from Viburnum lantana (Family Caprifoliaceae) and antioxidant activity of its hydroalcoholic extract. *Letters in Organic Chemistry*, **16**(2): 93–98 (2019).

Shikov, A. N., Pozharitskaya, O. N., Makarov, V. G., Wagner, H., Verpoorte, R., & Heinrich, M. Medicinal plants of the Russian pharmacopoeia; their history and applications. *Journal of Ethnopharmacology*, **154**(3): 481–536 (2014).

TF. Viburni Fructus. In *Türk Farmakopesi 2017* (Vol. Cilt 3, pp. 2412–2414). Ankara: T.C. Sağlık Bakanlığı Türkiye İlaç ve Tıbbi Cihaz Kurumu Yayın Komisyonu (2017).

Tomassini, L., & Brkic, D. Iridoid glucosides from Viburnum lantana var. discolor. *Planta Medica*, **63**(5): 485–486 (1997).

Tomassini, L., Cometa, M. F., Foddai, S., & Nicoletti, M. Iridoid glucosides from Viburnum tinus. *Phytochemistry*, **38**(2): 423–425 (1995).

Turker, A. U., Yildirim, A. B., & Karakas, F. P. Antibacterial and antitumor activities of some wild fruits grown in Turkey. *Biotechnology & Biotechnological Equipment*, **26**(1): 2765–2772 (2012).

Ulger, H., Ertekin, T., Karaca, O., Canoz, O., Nisari, M., Unur, E., & Elmali, F. Influence of gilaburu (Viburnum opulus) juice on 1,2-dimethylhydrazine (DMH)-induced colon cancer. *Toxicology and Industrial Health*, **29**(9): 824–829 (2013).

USDA. United States Department of Agriculture.

Natural Resources Conservation Service. Retrieved from https://plants.sc.egov.usda.gov/home/plantProfile?symbol=VIOP.

Velioglu, Y. S., Ekici, L., & Poyrazoglu, E. S. Phenolic composition of European cranberrybush (Viburnum opulus L.) berries and astringency removal of its commercial juice. *International Journal of Food Science and Technology*, **41**(9): 1011–1015 (2006).

Wang, X. Y., Shi, H. M., & Li, X. B. Chemical constituents of plants from the genus Viburnum. *Chemistry & Biodiversity*, **7**(3): 567–593 (2010).

Yildirim, A. B., Karakas, F. P., & Turker, A. U. In vitro antibacterial and antitumor activities of some medicinal plant extracts, growing in Turkey. *Asian Pacific Journal of Tropical Medicine*, **6**(8): 616–624 (2013).

Yilmaz, N., Yayli, N., Misir, G., Coskuncelebi, K., Karaoglu, S., & Yayli, N. Chemical composition and antimicrobial activities of the essential oils of Viburnum opulus, Viburnum lantana and Viburnum orientala. *Asian Journal of Chemistry*, **20**(5): 3324–3330 (2008).

Yilmaz, N., Beyhan, Ö., Gerçekçioğlu, R., & Kalayci, Z. Determination of fatty acid composition in seed oils of some important berry species and genotypes grown in Tokat Provinceof Turkey. *African Journal of Biotechnology*, **10**(41): 8070–8073 (2011).

Yilmaztekin, M., & Sislioglu, K. Changes in volatile compounds and some physicochemical properties of European Cranberrybush (Viburnum opulus L.) during ripening through traditional fermentation. *Journal of Food Science*, **80**(4): C687–C694 (2015).

Youngken, H. W. The pharmacognosy, chemistry and pharmacology of viburnum. III. History, botany and pharmacognosy of viburnum opulus L. var. Americanum (miller) ait. *Journal of the American Pharmaceutical Association*, **21**(5): 444–462 (1932).

Zaklos-Szyda, M., Kowalska-Baron, A., Pietrzyk, N., Drzazga, A., & Podsedek, A. Evaluation of Viburnum opulus L. fruit phenolics cytoprotective potential on insulinoma MIN6 cells relevant for Diabetes mellitus and obesity. *Antioxidants*, **9**(5): 26 (2020a).

Zaklos-Szyda, M., Majewska, I., Redzynia, M., & Koziolkiewicz, M. Antidiabetic effect of polyphenolic extracts from selected edible plants as alpha-amylase, alpha-glucosidase and PTP1B inhibitors, and beta pancreatic cells cytoprotective agents – A comparative study. *Current Topics in Medicinal Chemistry*, **15**(23): 2431–2444 (2015).

Zaklos-Szyda, M., Nowak, A., Pietrzyk, N., & Podsedek, A. Viburnum opulus L. juice phenolic compounds influence osteogenic differentiation in human osteosarcoma Saos-2 cells. *International Journal of Molecular Sciences*, **21**(14): 26 (2020b).

Zaklos-Szyda, M., Pawlik, N., Polka, D., Nowak, A., Koziolkiewicz, M., & Podsedek, A. Viburnum opulus fruit phenolic compounds as cytoprotective agents able to decrease free fatty acids and glucose uptake by caco-2 cells. *Antioxidants*, **8**(8): 25 (2019).

Zaklos-Szyda, M., Pietrzyk, N., Szustak, M., & Podsedek, A. Viburnum opulus L. juice phenolics inhibit mouse 3T3-L1 cells adipogenesis and pancreatic lipase activity. *Nutrients*, **12**(7): 29 (2020c).

Zarifikhosroshahi, M., Murathan, Z. T., Kafkas, E., & Okatan, V. Variation in volatile and fatty acid contents among Viburnum opulus L. Fruits growing different locations. *Scientia Horticulturae*, **264**: 6 (2020).

Zayachkivska, O. S., Gzhegotsky, M. R., Terletska, O. I., Lutsyk, D. A., Yaschenko, A. M., & Dzhura, O. R. Influence of Viburnum opulus proanthocyanidins on stress-induced gastrointestinal mucosal damage. *Journal of Physiology and Pharmacology*, **57**: 155–167 (2006).

29 *Viscum* sp.

Gökçe Şeker Karatoprak and Selen İlgün

INTRODUCTION

There are about 100 species of the genus *Viscum* L. (Santalaceae) in the world and distributed in Europe, Asia, Africa, and Australia. *Viscum* species is a dioecious, semi-parasitic, evergreen plant that grows on the branches of both deciduous and coniferous trees (Shahryar et al. 2012). *Viscum album* L. (mistletoe) and its three varieties are grown in Turkey. These subspecies are *Viscum album* subsp. *abietis* (Wiesb.) Abromerit, *Viscum album* subsp. *album*, and *Viscum album* subsp. *austriacum* (Wiesb.) Vollman (Güner and Aslan 2012). It has been determined that in Turkey these subspecies are generally growing on fruit trees such as apple, apricot, hawthorn, pear, and quince and on coniferous forest trees (Üstüner et al. 2015).

V. album L. has been an important part of human civilizations since ancient times. In general, mistletoe is used in different parts of the world to lower blood pressure, improve cardiac activity, relieve pain, treat constipation, and lower blood sugar. In Turkey, *V. album* is used in the eastern and southern regions of Anatolia, especially for the treatment of hypertension, and in Central Anatolia to lower blood sugar in diabetics (Karakaş et al. 2008).

V. album contains various biologically active compounds, and the main secondary metabolite groups are polypeptides, alkaloids, amines, flavonoids, acids, and terpenoids. Lectins and viscotoxins are known as the main groups of compounds that participate in cytotoxic biological activity (Ochocka and Piotrowski 2002). Several studies have shown that the phytochemical profile of mistletoe may vary according to the host plant (Sharma et al. 2018). The pharmacological activities of *V. album* extracts along with these isolated and identified compounds have been reported. These frequently studied activities consist of antioxidant, antimicrobial, anticancer, antidiabetic, immunomodulatory, blood pressure lowering, and hepatoprotective activities (Szurpnicka et al. 2020). Standardized extracts prepared from *V. album* such as Iscador® and Isorel® are used in cancer treatment, which emphasizes the importance of this plant.

Since only *V. album* and its subspecies are grown in Turkey, *V. album* is particularly emphasized in this section. The botanical properties, ethnobotanical use, phytochemicals, and biological activities of *V. album* are highlighted.

BOTANICAL FEATURES AND ETHNOBOTANICAL USAGE

Mistletoe (*Viscum*), a hemiparasitic genus, consists of approximately 100–150 species that survive parasitically on a wide variety of hosts. *Viscum* species are mainly distributed in Africa, Asia, and Europe (Petersen et al. 2015). It is a dioecious shrub that grows by wrapping the branches of deciduous trees as well as coniferous trees. These evergreen plants have leathery and opposite leaves. The inflorescences are axillary or terminal. Male flowers comprise four perigones. Anthers are implanted on perigones and dehisce by numerous pores. Fruit is white, ovoid, or elliptic, one-seeded berry, with bitter and unpleasant odor (Shahryar et al. 2012). The plant is generally accepted in the literature as belonging to the Loranthaceae and Viscaceae families; that is, a member of the Santalales order (Valle et al. 2021). The classification of *Viscum* species in plant systematics has caused numerous debates for many years due to the variability of the plant used as the host of the genus. Mainly, due to the differences in morphological characters such as the size and shape of the

DOI: 10.1201/9781003146971-29

leaves and the color and format of the fruits, the characteristics of the genus could not be determined (Ochocka and Piotrowski 2002). *Viscum* has long been considered to be related to the genus *Loranthus* Jacq and close allies, classified first in the Viscaceae and then in the Loranthaceae family (Nickrent et al. 2010). Later, the Santalalean group was studied in more detail, and it was noted that *Viscum*-related taxa were a monophyletic branch phylogenetically closer to Santalaceae, with molecular studies supported by morphological, cytological, embryological, and biogeographical data (Krasylenko et al. 2020).

V. album, which is widely growing in Europe, has different distributed varieties. In recent studies for this taxon, which is represented by a single species in Turkey, three varieties of the species have been reported. *Viscum album* subsp. *album* L. has been determined that it survives as a parasite in 452 species of deciduous dicotyledon trees in 96 genera and 44 families in Europe and up east to the Himalayas (Nepal) (Barney et al. 1998). *Viscum album* subsp. *abietis* (Wiesb.) Janch. Variety prefers *Abies* spp. as host. This taxon is generally distributed from the Pyrenees to Germany and through Eastern and Southern Europe to the Caucasus and Asia Minor (Anatolia) (Zuber 2004). In contrast, *Viscum album* subsp. *austriacum* (Wiesb.) Vollm. was observed in a few conifer taxa, mostly in *Pinus* and less frequently in *Larix* and *Picea*. (Dobbertin et al. 2005). This variety has survived in geography extending from Northwest Africa (Morocco) and the central Iberian Peninsula to the Caucasus and Asia Minor via Central and Southern Europe (Catalán and Aparicio 1997). Mainly, *V. album* often chooses apple tree (*Malus mali*), poplar (*Populus*), or oak (*Quercus*) species as its growing area, while it can colonize a large number of host trees such as elm (*Ulmus*), willow (*Salix alba*), pine (*Pinus*), fir (*Abies*), almond (*Prunus dulcis*), hawthorn (Crataegus), and ash (*Fraxinus*) (Valle et al. 2021).

Viscum album is generally referred to as "European mistletoe, Mistletoe, All-heal, Birdlime, Devil's fuge, Mystyldene". *Viscum album* L. is also known as evergreen and perennial plants. Plants obtain water and dissolved inorganic components straight from the xylem by living on woody trees. On the other hand, ripe pseudofruits' seeds are carried by birds and aid in pollination (Valle et al. 2021) Mistletoe is defined as shrub-like plants with linear and lanceolate leathery leaves that persist for several seasons. It has yellowish-green flowers and produces translucent and whitish fruits in autumn and winter. *V. album* has a vegetation period of 12 months. The plant, which completes its cycle without ever touching the soil, interestingly blooms in winter (Bussing 2000).

Viscum species (especially *V. album*) has been an important part of human culture since ancient times (Ramm 2015). Mistletoe has been used for centuries in traditional medicine to treat diseases, including cancer, inflammation, hypertension, arthritis, rheumatism, constipation, internal hemorrhages, stomach ulcers, and skin diseases. In ethnopharmacology, mistletoe was also used in the treatment of central nervous system ailments (Szurpnicka et al. 2019). Archaeological findings from the Neolithic period showed that this parasitic plant was used as fodder for animals, especially during the long winter periods. In addition, archaeological findings of the stems, leaves, and fruits of mistletoe have shown that mistletoe was used in rituals and for medicinal purposes. The first evidence that mistletoe was used for medicinal purposes began with the Greek physician Dioscorides (40–90 AD). However, Hippocrates used mistletoe to cure diseases of the spleen and menstruation-related diseases (Ramm 2015). In De Medicina, Celsus defined mistletoe as a component in various formulations; it is used in plasters or softeners by mixing with other substances. Mistletoe has also been used to alleviate pain or swelling, suppuration, or treating abscesses, carcinoids, or scrofula (Urech and Baumgartner 2015). Pliny stated that mistletoe is useful in the treatment of epilepsy, infertility, and ulcers (Ramm 2015). The use of mistletoe in the treatment of spleen and liver diseases has been recommended in ancient sources. The details about the use of mistletoe for treatment of illness, which especially grows as a host on oak trees, are explained in these books. In the 16th-century European herbal books, the warming, softening, and astringent properties of mistletoe were mentioned. It has also been noted that mistletoe and other herbs are used to treat bone fractures and labor pains (Bussing 2000). It is mentioned in the books that English chemist Sir Robert Doyle suggested taking mistletoe orally, dissolved in cherry juice for a few days during the full moon, as a cure for epilepsy. A German encyclopedia compiled in the middle of the 18th century wrote that oak mistletoe was used for leprosy, nosebleeds, roundworms, and other ailments. In addition, in 1720, the English physician Colbatch recommended mistletoe as a special therapy for epilepsy. In some literatures, it is stated that oak mistletoe is different from those grown on other trees, while some literatures state that there is no difference (Ramm 2015).

In Turkey, the fruits and leafy branches of the plant are used for constipation. It is also used as a diuretic, emetic, and blood pressure reducer. It is also used in infusion and decoction in cases such as stomach pain, hemorrhoid treatment, diarrhea, prostate, and urinary retention. Externally, it is used in the treatment of rheumatism pains and cuts (Demirezer 2011).

PHYTOCHEMICALS

The active phytochemical compounds of the *Viscum* genus are very diverse and therefore difficult to identify and classify. The complicated chemical profile of *Viscum* species displays a vast diversity of pharmacological activity, including the synergistic benefits. Mainly, bioactive phytochemicals of *Viscum* species contain nitrogen-containing compounds (amino acids, peptides, alkaloids, and nucleosides), carbohydrates, phenolic acids, flavonoids, terpenoids, fatty acids, lipids, viscotoxins, vesicles, and cyclitols (Song et al. 2021).

Viscotoxins are thionines and are classified as alpha and beta, and they are rich in cysteine and proteins. Viscotoxin variants A1, A2, A3, B, B2, B5, B6, B7, B8, C1, 1-PS, and U-PS were discovered in *Viscum album* L. (Song et al. 2021).

Dibutyl (2Z, 6Z)-octa-2, 6-dien-4-yne dioate and dibutyl (2E, 6E)-octa-2, 6-dien-4-yne dioate are conjugated acetylene compounds and were firstly identified by Duo *et al.* from *V. album* (Duo, Li-Qing et al. 2019).

Compounds reported as "alkaloid-like" include tyramine, phenylethylamine, choline, and acetylcholine. These components are not considered typical alkaloids and are present in the *V. album* extracts (Bussing 2000).

Three different types of mistletoe lectins (ML) were identified in *V. album*. These lectins are: MLI (115 kDa) – galactose, MLII (60 kDa) – galactose-and N-acetyl-D-galactosamine, and MLIII (60 kDa) – N-acetyl-D-galactosamine (Singh et al. 2016)

Carbohydrates such as homogalacturonan, pectin, arabinogalactan, and rhamnogalacturonan were detected in *V. album*. Polysaccharides detected in *V. album* are generally obtained from the fruit and green parts of the plant (leaf and stem). Highly esterified galacturonan, pectin, and arabinogalactan have been isolated especially from the berries and leaves. Rhamnogalacturonan, arabinogalactan, and small amounts of xyloglucan are compounds isolated from the fruits known to be particularly rich in polysaccharides (Jordan and Wagner 1986; Urech and Baumgartner 2015; Nazaruk and Orlikowski 2016).

5,7-dimethoxy-flavanone-4′-O-glucoside, 2′-hydroxy-4′,6′-dimethoxychalcone-4-O-glucoside, 5,7-dimethoxyflavanone-4′-O-[2″-O-(5″ ′-O-trans-cinnamoyl)-apiosyl]-glucoside), 2′-hydroxy-4′,6 ′-dimethoxychalcone-4-O-[2″-O-(5″ ′-O-trans-cinnamoyl)-apiosyl]-glucoside, 2′-hydroxy-3,4′,6′-trimethoxychalcone-4-O-glucoside, 2′-hydroxy-4′,6′-dimethoxychalcone-4-O-[apiosyl(1→2)]glucoside, 5,7-dimethoxyflavanone-4′-O-[apiosyl-(1→2)]-glucoside, and (2S)-3′,5,7-trimethoxyflavanone-4′-O-glucoside are the glycosidic form of flavonoids isolated from *V. album*, and they all have methoxyl groups in the chemical structures. Quercetin, kaempferol, and their mono-, di- and trimethyl ethers were also detected in the different subspecies of the European mistletoe (Haas et al. 2003). 4′-O-[β-apiosyl(1→2)]-β-glucosyl]-5-hydroxy-7-O-sinapylflavanone, 3-(4-acetoxy-3, 5-dimethoxy)-phenyl-2E-propenyl-β-glucoside, 3-(4-hydroxy-3,5-dimethoxy)-phenyl-2E-propenyl-β-glucoside, 5,7-dimethoxyflavanone-4′-O-β-glucoside, 4′,5-dimethoxy-7-hydroxy flavanone, and 5,7-dimethoxy-4′-hydroxy flavanone are other polyphenolic compounds that were isolated from

80% methanolic extract (Nazaruk and Orlikowski 2016). Viscumneoside XII, viscumneoside XIII, and viscumneoside XIV are new flavonoid glycosides that were isolated from the aerial part of *V. album* plant in 2019 (Jia-Kun et al. 2019).

Phenylpropanoids were isolated from *V. album* extract. The most popular and known components are coniferin, syringin, and syringenin (Panossian et al. 1998). Phytochemical studies on the subspecies of *V. album* showed that the phenylpropanoid content of the plant differs according to the host plant. It was stated that *V. album* spp. *album* had the highest syringenin and coniferin content. The presence of calopanaxin D (4-[2-0-(apiosyl)-β-d-glucosyloxy]-3-methoxycinnamyl alcohol) was detected in *V. album* spp. *album*. But calopanaxin D was not detected in *V. album* spp. *abietis*. Additionally, trace amounts of coniferin were detected in the extract of *V. album* spp. *Austriacum*, and syringin and coniferin were identified in *V. album* spp. *abietis* (Deliorman et al. 1999).

It has been determined that *V. album* is rich in triterpene content. Betulinic acid, ursanolinic, and lupeol acetate are the main compounds detected in the extract of this plant. Stigmasterol phytosteroid and beta-sitosterol compounds have also been identified in this plant. Lipophilic extracts of the plant contain saturated fats such as palmitic, arachidonic, lignoceric, and cerotic acids, and unsaturated fats such as linoleic and linolenic acids (Valle et al. 2021).

BIOLOGICAL ACTIVITIES

Anticancer and Immunomodulatory Activity

Lectin and visctoxins are the major compounds responsible for the anticancer activity of *Viscum* species. Although most of the anticancer activity studies focus on these compounds, subsequent studies have shown that the phenolic compounds, alkaloids, triterpene acids, and nonpolar compounds contained in *Viscum* species also have anticancer activity (Khwaja et al. 1986; Schaller et al. 1996; Thies et al. 2005, Eggenschwiler, von Balthazar et al. 2007; de Oliveira Melo et al. 2018). In a study conducted in 2019, it was stated that the *V. album* extract has stronger antitumor activity compared to the isolated compounds, and this effect may be due to the synergistic effect of the compounds in the extract (Twardziok et al. 2016; Kleinsimon et al. 2018; Felenda et al. 2019). More specifically, the activities of this species have been shown against osteosarcoma, leukemia, Ewing sarcoma, liver cancer, cervical cancer, breast cancer, and epidermal cancer (Bantel et al. 1999; Orhan et al. 2005; Sadeghi-Aliabadi et al. 2006; Vlad et al. 2016; Schötterl et al. 2017; Twardziok et al. 2017; de Oliveira Melo et al. 2018; Sharma et al. 2018; Yang et al. 2019; Kottireddy et al. 2020). When the studies on *Viscum* species are examined, it is found that *Viscum* extracts or their active compounds can arrest the cell cycle of cancer cells in certain phases, showing anticancer activity and preventing cell invasion. It has been shown in different cell lines that lectins can induce apoptosis in different ways. These mechanisms include the stimulation of caspases production or other proteins, down-regulation/up-regulation of certain genes involved in apoptotic suppression or induction (Yau et al. 2015). Molecular targets of the *Viscum* species are shown in Table 29.1.

The anticancer potential of *Viscum* species is associated with immunomodulatory activities, such as increased maturation and activation of dendritic cells, abolition of tumor-induced immunosuppression of dendritic cells, increased leukocytes, eosinophils, granulocytes, and lymphocytes, and the activity of natural killer cells. It also induces cytokines like IL-1, IL-2, IL-4, IL-5, IL-6, IL-8, IL10, IL-12, IFN-γ, and TNF-α (Thies et al. 2005; Szurpnicka et al. 2020).

Clinical studies were found on bladder, breast, colorectal, glioma, lung, and melanoma cancer patients. There are also studies reporting that *Viscum* has a synergistic effect with cancer treatments such as chemotherapy and radiotherapy (Szurpnicka et al. 2020). It was stated in the clinical study by Kim et al. (2012) that additional treatment with mistletoe was safe and increased the quality of life in gastric cancer patients. The use of Isorel® (mistletoe drug) on 70 patients with digestive tract

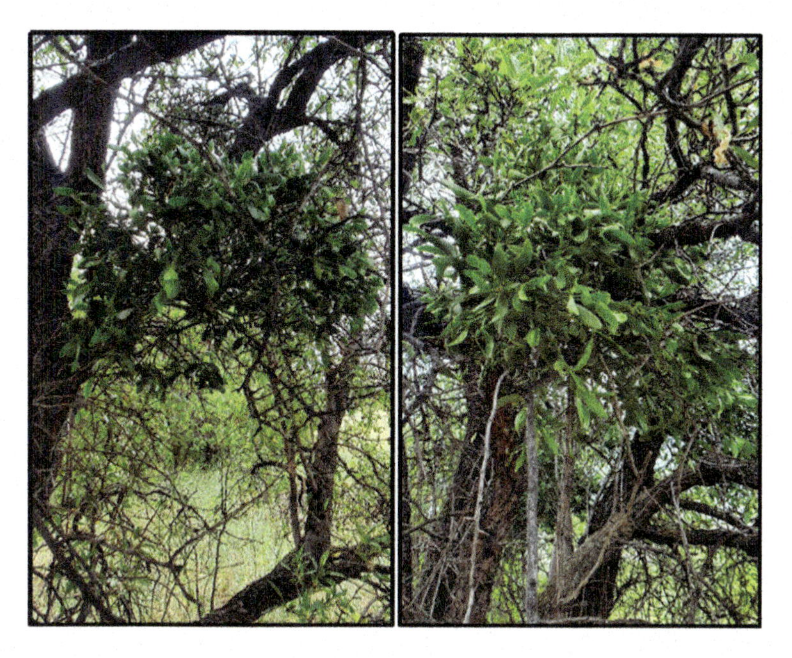

FIGURE 29.1 *Viscum album* colonized on *Prunus dulcis* (host tree).

TABLE 29.1
Molecular Targets of *Viscum* sp.

MAPK	Cytochrome c
PI3K/AKT	c-Myc
ERK,	MMP-2
p38	MMP-14
JNK	VEGF
Caspases	pAkt
Invasion and angiogenesis	β-Catenin
CCND1, CCNE, CCNA (Cyclins)	TGF-β
CDK4, CDK2 (cyclin-dependent protein kinases)	Bax
	Bcl-2

cancer has also demonstrated improved immune competence and improved overall health of cancer patients exposed to conventional anti-cancer treatment (Galun et al. 2015).

Activity on Cardiac System

The effects of *V. album* on the cardiac system have been reported, especially on hypertension in *in vitro* and *in vivo* studies, and its positive effects have been determined. Its effect on systolic and diastolic blood pressure reduction has been explained by a mechanism involving up-regulation of the nitric oxide (NO) pathway (Poruthukaren et al. 2014). The relationship between *V. album* and NO has also been proven in different clinical and *in vivo* studies (Mossalayi et al. 2006; Tenorio-Lopez et al. 2006). Since NO plays a significant role in heart failure pathophysiology, in a study conducted for this purpose, *V. album* exhibited a positive influence on left ventricular function in heart failure rats triggered by isoproterenol (Tang et al. 2014; Karagöz et al. 2016). The vasorelaxant activity of mistletoe in rabbit aortic rings has also been linked to the blockade of voltage-dependent Ca^{2+}

channel. It has been stated that this effect on Ca^{2+} channels is partially related to NO (Khan et al. 2016). Similar research is also executed in aortic rings of rats, and the extract (aqueous) of mistletoe has been reported to have a vasorelaxant effect that might be mediated by a non-specific and non-competitive inhibition of Ca^{2+} efflux (Mojiminiyi et al. 2008).

Antidiabetic Activity

In *in vivo* studies, *Viscum* species have been shown to reduce blood glucose levels and enhance insulin secretion, thus having anti-glycemic and insulinotropic effects. Gray and Flatt (1999) explained in their study that the aqueous extract of *V. album* increased the insulin release from clonal pancreatic β-cells of rats in a concentration-dependent manner; however, lectins are not responsible for the effect (Kim et al. 2014). It has also been reported that mistletoe increased insulin release from RINm5F cells (β cells of rat pancreas) without any toxicity. Furthermore, it has been shown in different *in vivo* studies that *V. album* reduces glucose levels in diabetic rats and has an antioxidant capacity to prevent the occurrence of diabetic complications (Orhan et al. 2005; Ahmed et al. 2019). Besides, the fact that *V. album* extracts have an α-amylase enzyme inhibitory effect has been supported by *in vitro* studies (Stefanucci et al. 2020; Orhan et al. 2014).

Hepatoprotective Activity

Although various *in vitro* and *in vivo* studies have tried to elucidate the mechanism of the hepatoprotective effect of *V. album*, the mechanism has not been fully elucidated. In an *in vivo* study, it was reported that the alkaloid fraction decreased the levels of TGF-1, TGF-β1 receptor, and phosphorylated Smad 2 protein and increased the level of Smad 7. In Jiang, Wang et al.'s (2014) *in vitro* study using rat hepatic stellate cell line (HSC-T6), it has been proven that the alkaloid fraction increased Smad7 expression and inhibited α-SMA, TGFβ1, and TGF-β1 receptors and Smad 2 and TIMP-1 expressions. *In vivo* studies performed with liver-damaged rats revealed that mistletoe extracts reduced the ALT, AST, and ALP levels (Ogbonnanya et al. 2010; Yusuf et al. 2013).

Effects on Central Nervous System (CNS)

Viscum species have different effects on the CNS. When *in vivo* studies are examined, it is observed that *Viscum* species have antiepileptic, sedative, analgesic, anti-anxiety, antidepressant, hypnotic, anti-stress, and antipsychotic activities (Gupta et al. 2012; Khatun et al. 2016; Kumar et al. 2016; Tsyvunin et al. 2016; Geetha et al. 2018). Effects on γ-aminobutyric acid receptors (GABA) and brain-derived neurotrophic factor (BDNF) levels were also reported for *Viscum* species (Ademola et al. 2016).

Other Activities

Numerous studies have shown that mistletoe extracts and lectin possess free radical scavenging activities and protective effects on oxidative stress caused by radicals (Sengul et al. 2009; Kim et al. 2010; Papuc et al. 2010; Kusi et al. 2015). In these studies, results stated that polar extracts are stronger antioxidants because they are rich in phenolic content, and the activity is also dependent on the host tree and the time of collection (ÖnayUçar et al. 2006; Vicaş et al. 2011; Pietrzak et al. 2017)

Few studies have also focused on wound healing activity. In one of these studies, lipophilic extract of *V. album* and triterpene–oleanolic acid content showed wound healing activity by promoting the fibroblast migration (Kuonen et al. 2013). Although there are many antibacterial activity studies, it can be concluded that their activity on Gram-positive bacteria is stronger according to the results, and antibiofilm activity is also one of the activities exhibited by mistletoe extracts (Ertürk et al. 2003; Muhammad et al. 2011; Kenar et al. 2016). Besides, few antiviral activity studies reported that mistletoe is active against HPIV-2 (human parainfluenza virus type 2), but weakly effective against the HIV-1 virus (Karagöz et al. 2003; Li et al. 2008).

CONCLUSION

Viscum album has been used by people for centuries for different purposes, and its three varieties were reported in literature as an important medicine in Turkey. The diversity of its bioactive compounds and different pharmacological activities make the plant important for pharmacognostic research. The popularity of the plant is increasing day by day and gains importance in *in vitro* and *in vivo* studies mainly due to its cytotoxic and immunomodulatory activities. It has been reported that it has dual activity when applied to cancer patients, as it stimulates the vital forces of the organism, improves immunity, and has a selective cytotoxic activity to cancer cells. At the same time, *V. album* is of great interest to the scientific community due to its satisfactory clinical research results.

This unique plant, which has been used as a remedy for many different diseases for several thousand years, has now proven its special potential as an element of integrative cancer treatment and supportive care. The results of more than 100 clinical studies have shown that mistletoe has many beneficial effects in the treatment of cancer. In complementary medicine in Europe, mistletoe extracts are among the most commonly used drugs for cancer treatment.

However, despite being used and known for a long time, scientific evidence to explain the molecular mechanisms of its effects is limited. Therefore, research on the isolation and identification of bioactive compounds, pharmacological activities, interactions and synergy among compounds, and possible mechanisms of action is a new field of study waiting to be explored by the scientific world. Considering all the clinical studies and the effects obtained, it can be concluded that the studies need to be further elaborated.

REFERENCES

Ademola, O., Edem, E., Olufunke, D., Oladunni, K. (2016). Cognitive-enhancing and neurotherapeutic prospects of *Viscum album* in experimental model of Alzheimer's disease. *African Journal of Cellular Pathology*, 7, 11–16.

Ahmed, A. K., Nihat, M., Handan, M. (2019). Investigation of the Antidiabetic effects of mistletoe (*Viscum album* L.) extract in experimental diabetes in rats. *Van Veterinary Journal*, 30(2), 121–125.

Bantel, H., Engels, I. H., Voelter, W., Schulze-Osthoff, K., Wesselborg, S. (1999). Mistletoe lectin activates caspase-8/FLICE independently of death receptor signaling and enhances anticancer drug-induced apoptosis. *Cancer Research*, 59(9), 2083–2090.

Barney, C. W., Hawksworth, F. G., Geils, B. W. (1998). Hosts of *Viscum album*. *European Journal of Forest Pathology*, 28(3), 187–208.

Bussing, A. (2000). *Mistletoe: The Genus Viscum*. CRC Press.

Catalán Rodríguez, P., Aparicio Martínez, A. (1997). *Viscum*. Flora Ibérica, nº 8.

de Oliveira Melo, M. N., Oliveira, A. P., Wiecikowski, A. F., Carvalho, R. S., de Lima Castro, J., de Oliveira, F. A. G., Holandino, C. (2018). Phenolic compounds from Viscum album tinctures enhanced antitumor activity in melanoma murine cancer cells. *Saudi Pharmaceutical Journal*, 26(3), 311–322.

Deliorman, D., Çaliş, I., Ergun, F., Tamer, U. (1999). The comparative studies on phenylpropanoid glycosides of Viscum album subspecies by high performance liquid chromatography. *Journal of Liquid Chromatography & Related Technologies*, 22(20), 3101–3114.

Demirezer, Ö. (2011). *FFD Monografları Tedavide Kullanılan Bitkiler.* MN Medikal and Nobel Tıp Kitapevi, Ankara.

Dobbertin, M., Hilker, N., Rebetez, M., Zimmermann, N. E., Wohlgemuth, T., Rigling, A. (2005). The upward shift in altitude of pine mistletoe (*Viscum album* ssp. austriacum) in Switzerland—The result of climate warming?. *International Journal of Biometeorology*, 50(1), 40–47.

Duo, C., Li-Qing, W., Xiao-Min, H., Hui-Rui, G., Meng, L., Ya-Hui, W., Jia-Kun, D. (2019). Two symmetrical unsaturated acids isolated from *Viscum album*. *Chinese Journal of Natural Medicines*, 17(2), 145–148.

Eggenschwiler, J., von Balthazar, L., Stritt, B., Pruntsch, D., Ramos, M., Urech, K., Viviani, A. (2007). Mistletoe lectin is not the only cytotoxic component in fermented preparations of Viscum album from white fir (Abies pectinata). *BMC Complementary and Alternative Medicine*, 7(1), 1–7.

Ertürk, Ö., Kati, H., Yayli, N., Demirbağ, Z. (2003). Antimicrobial activity of *Viscum album* L. subsp. abietis (Wiesb). *Turkish Journal of Biology*, 27(4), 255–258.

Felenda, J. E., Turek, C., Stintzing, F. C. (2019). Antiproliferative potential from aqueous *Viscum album* L. preparations and their main constituents in comparison with ricin and purothionin on human cancer cells. *Journal of Ethnopharmacology*, 236, 100–107.

Galun, D., Tröger, W., Milicevic, M. (2015). Cancer surgery and supportive mistletoe therapy: From scepticism to randomised clinical trials. In *Mistletoe: From Mythology to Evidence-Based Medicine* (Vol. 4, pp. 57–66). Karger Publishers.

Geetha, K. M., Murugan, V., Pavan Kumar, P., Wilson, B. (2018). Antiepileptic activity of Viscum articulatum Burm and its isolated bioactive compound in experimentally induced convulsions in rats and mice. *European Journal of Biomedical and Pharmaceutical Sciences*, 5, 311–318.

Gray, A. M., Flatt, P. R. (1999). Insulin-secreting activity of the traditional antidiabetic plant Viscum album (mistletoe). *Journal of Endocrinology*, 160(3), 409–414.

Gupta, G., Kazmi, I., Afzal, M., Rahman, M., Saleem, S., Ashraf, M. S., Anwar, F. (2012). Sedative, antiepileptic and antipsychotic effects of *Viscum album* L.(Loranthaceae) in mice and rats. *Journal of Ethnopharmacology*, 141(3), 810–816.

Güner, A., Aslan, S., Ekim, T., Vural, M., Babaç, M. T. (2012). *Türkiye bitkileri listesi.* Damarlı Bitkiler, Nezahat Gökyiğit Botanik Bahçesi ve Flora Araştırmaları Derneği Yayını, 262.

Haas, K., Bauer, M., Wollenweber, E. (2003). Cuticular waxes and flavonol aglycones of mistletoes. *Zeitschrift für Naturforschung C*, 58(7–8), 464–470.

Jia-Kun, D., Duo, C., Cui-Hua, L., Jing, G., Meng-Qing, L., Na, F., Meng-Yang, H. (2019). Three new bioactive flavonoid glycosides from *Viscum album. Chinese Journal of Natural Medicines*, 17(7), 545–550.

Jiang, Y., Wang, C., Li, Y. Y., Wang, X. C., An, J. D., Wang, Y. J., Wang, X. J. (2014). Mistletoe alkaloid fractions alleviates carbon tetrachloride-induced liver fibrosis through inhibition of hepatic stellate cell activation via TGF-β/Smad interference. *Journal of Ethnopharmacology*, 158, 230–238.

Jordan, E., Wagner, H. (1986). Structure and properties of polysaccharides from *Viscum album* (L.). *Oncology*, 43(Suppl. 1), 8–15.

Karagöz, A., Kesici, S., Vural, A., Usta, M., Tezcan, B., Semerci, T., Teker, E. (2016). Cardioprotective effects of *Viscum album* L. ssp. album (Loranthaceae) on isoproterenol-induced heart failure via regulation of the nitric oxide pathway in rats. *Anatolian Journal of Cardiology*, 16(12), 923.

Karagöz, A., Önay, E., Arda, N., Kuru, A. (2003). Antiviral potency of mistletoe (Viscum album ssp. album) extracts against human parainfluenza virus type 2 in Vero cells. *Phytotherapy Research*, 17(5), 560–562.

Karakaş, A., Serin, E., Guenduez, B., Tuerker, A. U. (2008). The effects of Mistletoe (*Viscum album* L. subsp. Album) extracts on isolated intestinal contractions. *Turkish Journal of Biology*, 32(4), 237–242.

Kenar, N., Erdonmez, D., Turkmen, K. E. (2016). Anti-quorum sensing and anti-biofilm activity of *Viscum album* L. on different pathogenic bacteria. *Journal of Biotechnology*, (231), S34.

Khan, T., Ali, S., Qayyum, R., Hussain, I., Wahid, F., Shah, A. J. (2016). Intestinal and vascular smooth muscle relaxant effect of *Viscum album* explains its medicinal use in hyperactive gut disorders and hypertension. *BMC Complementary and Alternative Medicine*, 16(1), 1–8.

Khatun, A., Rahman, M., Rahman, M. M., Hossain, H., Jahan, I. A., Nesa, M. L. (2016). Antioxidant, antinociceptive and CNS activities of *Viscum orientale* and high sensitive quantification of bioactive polyphenols by UPLC. *Frontiers in Pharmacology*, 7, 176.

Khwaja, T. A., Dias, C. B., Pentecost, S. (1986). Recent studies on the anticancer activities of mistletoe (Viscum album) and its alkaloids. *Oncology*, 43(Suppl. 1), 42–50.

Kim, B. K., Choi, M. J., Park, K. Y., Cho, E. J. (2010). Protective effects of Korean mistletoe lectin on radical-induced oxidative stress. *Biological and Pharmaceutical Bulletin*, 33(7), 1152–1158.

Kim, K. C., Yook, J. H., Eisenbraun, J., Kim, B. S., Huber, R. (2012). Quality of life, immunomodulation and safety of adjuvant mistletoe treatment in patients with gastric carcinoma–a randomized, controlled pilot study. *BMC Complementary and Alternative Medicine*, 12(1), 1–7.

Kim, K. W., Yang, S. H., Kim, J. B. (2014). Protein fractions from Korean mistletoe (Viscum album coloratum) extract induce insulin secretion from pancreatic beta cells. *Evidence-based Complementary and Alternative Medicine*, 1–8.

Kleinsimon, S., Longmuss, E., Rolff, J., Jäger, S., Eggert, A., Delebinski, C., Seifert, G. (2018). GADD45A and CDKN1A are involved in apoptosis and cell cycle modulatory effects of viscumTT with further inactivation of the STAT3 pathway. *Scientific Reports*, 8(1), 1–14.

Kottireddy, S., Koora, S., Selvaraj, J. (2020). Ameliorative effect of *Viscum album* on angiogenic markers in A431 human skin cancer cells in vitro. *Drug Invention Today*, 14(1).

Krasylenko, Y., Sosnovsky, Y., Atamas, N., Popov, G., Leonenko, V., Janošíková, K., Sytnyk, D. (2020). The European mistletoe (*Viscum album* L.): Distribution, host range, biotic interactions, and management worldwide with special emphasis on Ukraine. *Botany*, 98(9), 499–516.

Kumar, P., Kumar, D., Kaushik, D., Kumar, S. (2016). Screening of neuropharmacological activities of *Viscum album* and estimation of major flavonoid constituents using TLC densitometry. *International Journal of Toxicological and Pharmacological Research*, 8, 179–186.

Kuonen, R., Weissenstein, U., Urech, K., Kunz, M., Hostanska, K., Estko, M., Baumgartner, S. (2013). Effects of lipophilic extract of Viscum album L. and oleanolic acid on migratory activity of NIH/3T3 fibroblasts and on HaCat keratinocytes. *Evidence-Based Complementary and Alternative Medicine*, 1–7.

Kusi, M., Shrestha, K., Malla, R. (2015). Study on phytochemical, antibacterial, antioxidant and toxicity profile of *Viscum album* Linn associated with *Acacia catechu*. *Nepal Journal of Biotechnology*, 3(1), 60–65.

Li, Y., Zhao, Y. L., Huang, N., Zheng, Y. T., Yang, Y. P., Li, X. L. (2008). Two new phenolic glycosides from *Viscum articulatum*. *Molecules*, 13(10), 2500–2508.

Mojiminiyi, F. B., Owolabi, M. E., Igbokwe, U. V., Ajagbonna, O. P. (2008). The vasorelaxant effect of viscum album leaf extract is nediated by calcium-dependent mechanisms. *Nigerian Journal of Physiological Sciences*, 23(1–2).

Mossalayi, M. D., Alkharrat, A., Malvy, D. (2006). Nitric oxide involvement in the anti-tumor effect of mistletoe (*Viscum album* L.) extracts Iscador on human macrophages. *Arzneimittelforschung*, 56(6), 457–460.

Muhammad, A. H., Muhammad, Q. K., Nazar, H., Tariq, H. (2011). Antibacterial and antifungal potential of leaves and twigs of *Viscum album* L. *Journal of Medicinal Plants Research*, 5(23), 5545–5549.

Nazaruk, J., Orlikowski, P. (2016). Phytochemical profile and therapeutic potential of Viscum album L. *Natural Product Research*, 30(4), 373–385.

Nickrent, D. L., Malécot, V., Vidal-Russell, R., Der, J. P. (2010). A revised classification of Santalales. *Taxon*, 59(2), 538–558.

Ochocka, J. R., Piotrowski, A. (2002). Biologically active compounds from European mistletoe (*Viscum album* L.). *Canadian Journal of Plant Pathology*, 24(1), 21–28.

Ogbonnanya, A. E., Mounmbegna, E. P., Monago, C. C. (2010). Effect of ethanolic extract of mistletoe (*Viscum album* L.) leaves on paracetamol-induced hepatotoxicity in rats. *Journal of Pharmacy Research*, 3(8), 1888–1891.

ÖnayUçar, E., Karagöz, A., Arda, N. (2006). Antioxidant activity of *Viscum album* ssp. album. *Fitoterapia*, 77(7–8), 556–560.

Orhan, D. D., Aslan, M., Sendogdu, N., Ergun, F., Yesilada, E. (2005). Evaluation of the hypoglycemic effect and antioxidant activity of three *Viscum album* subspecies (European mistletoe) in streptozotocin-diabetic rats. *Journal of Ethnopharmacology*, 98(1–2), 95–102.

Orhan, N., Hoçbaç, S., Orhan, D., Asya, M., Ergun, F. (2014). Haziran Türkiye'nin bazı antidiyabetik bitkilerinin enzim inhibitör ve radikal süpürücü etkileri. *Journal of Basic Medical Sciences*, 17(6), 426–432.

Panossian, A., Kocharian, A., Matinian, K., Amroyan, E., Gabrielian, E., Mayr, C., Wagner, H. (1998). Pharmacological activity of phenylpropanoids of the mistletoe, *Viscum album* L., host: Pyrus caucasica Fed. *Phytomedicine*, 5(1), 11–17.

Papuc, C., Crivineanu, M., Goran, G., Nicorescu, V., Durdun, N. (2010). Free radicals scavenging and antioxidant activity of European mistletoe (*Viscum album*) and European birthwort (*Aristolochia clematitis*). *Revista de Chimie*, 61(7), 619–622.

Petersen, G., Cuenca, A., Møller, I. M., Seberg, O. (2015). Massive gene loss in mistletoe (Viscum, Viscaceae) mitochondria. *Scientific Reports*, 5(1), 1–7.

Pietrzak, W., Nowak, R., Gawlik-Dziki, U., Lemieszek, M. K., Rzeski, W. (2017). LC-ESI-MS/MS identification of biologically active phenolic compounds in mistletoe berry extracts from different host trees. *Molecules*, 22(4), 624.

Poruthukaren, K. J., Palatty, P. L., Baliga, M. S., & Suresh, S. (2014). Clinical evaluation of *Viscum album* mother tincture as an antihypertensive: A pilot study. *Journal of Evidence-based Complementary & Alternative Medicine*, 19(1), 31–35.

Ramm, H. (2015). Mistletoe through cultural and medical history: The all-healing plant proves to be a cancer-specific remedy. In *Mistletoe: From Mythology to Evidence-Based Medicine*. Karger Publishers. **4**: 1–10.

Sadeghi-Aliabadi, H., Ghasemi, N., & Fatahi, A. (2006). Cytotoxic effects of Iranian mistletoe extract on a panel of cancer cells. *Iranian Journal of Pharmaceutical Sciences*, 2(3), 157–162.

Schaller, G., Urech, K., & Giannattasio, M. (1996). Cytotoxicity of different viscotoxins and extracts from the European subspecies of Viscum album L. *Phytotherapy Research*, 10(6), 473–477.

Schötterl, S., Hübner, M., Armento, A., Veninga, V., Wirsik, N. M., Bernatz, S., ... Naumann, U. (2017). Viscumins functionally modulate cell motility-associated gene expression. *International Journal of Oncology*, 50(2), 684–696.

Sengul, M., Yildiz, H., Gungor, N., Cetin, B., Eser, Z., & Ercisli, S. (2009). Total phenolic content, antioxidant and antimicrobial activities of some medicinal plants. *Pakistan Journal of Pharmaceutical Sciences*, 22(1).

Shahryar, S. M., Robabeh, S. S., & Narges, G. (2012). Notes on the genus *Viscum* (Viscaceae) in Iran: A new combination based on morphological evidence. *African Journal of Agricultural Research*, 7(11), 1694–1702.

Sharma, S., Sharma, R. S., Sardesai, M. M., Mishra, V. (2018). Anticancer potential of leafless mistletoe (*Viscum angulatum*) from Western Ghats of India. *International Journal of Pharmaceutical Sciences and Research*, 9(5), 1902–1907.

Singh, B. N., Saha, C., Galun, D., Upreti, D. K., Bayry, J., & Kaveri, S. V. (2016). European Viscum album: A potent phytotherapeutic agent with multifarious phytochemicals, pharmacological properties and clinical evidence. *RSC Advances*, 6(28), 23837–23857.

Song, C., Wei, X. Y., Qiu, Z. D., Gong, L., Chen, Z. Y., Ma, Y., Yang, B. (2021). Exploring the resources of the genus Viscum for potential therapeutic applications. *Journal of Ethnopharmacology*, 277, 114233.

Stefanucci, A., Zengin, G., Llorent-Martinez, E. J., Dimmito, M. P., Della Valle, A., Pieretti, S., Mollica, A. (2020). *Viscum album* L. homogenizer-assisted and ultrasound-assisted extracts as potential sources of bioactive compounds. *Journal of Food Biochemistry*, 44(9), e13377.

Szurpnicka, A., Kowalczuk, A., Szterk, A. (2020). Biological activity of mistletoe: In vitro and in vivo studies and mechanisms of action. *Archives of Pharmacal Research*, 43(6), 593–629.

Szurpnicka, A., Zjawiony, J. K., Szterk, A. (2019). Therapeutic potential of mistletoe in CNS-related neurological disorders and the chemical composition of *Viscum* species. *Journal of Ethnopharmacology*, 231, 241–252.

Tang, L., Wang, H., & Ziolo, M. T. (2014). Targeting NOS as a therapeutic approach for heart failure. *Pharmacology & Therapeutics*, 142(3), 306–315.

Tenorio-Lopez, F. A., Valle Mondragon, L. D., Olvera, G. Z., Torres Narvaez, J. C., & Pastelin, G. (2006). Viscum album aqueous extract induces NOS-2 and NOS-3 overexpression in Guinea pig hearts. *Natural Product Research*, 20(13), 1176–1182.

Thies, A., Nugel, D., Pfüller, U., Moll, I., & Schumacher, U. (2005). Influence of mistletoe lectins and cytokines induced by them on cell proliferation of human melanoma cells in vitro. *Toxicology*, 207(1), 105–116.

Tsyvunin, V., Shtrygol, S., Prokopenko, Y., Georgiyants, V., Blyznyuk, N. (2016). Influence of dry herbal extracts on pentylenetetrazole-induced seizures in mice: Screening results and relationship chemical composition–pharmacological effect. *ScienceRise. Pharmaceutical Science*, 1(1), 18–28.

Twardziok, M., Kleinsimon, S., Rolff, J., Jäger, S., Eggert, A., Seifert, G., & Delebinski, C. I. (2016). Multiple active compounds from Viscum album L. synergistically converge to promote apoptosis in Ewing sarcoma. *PLoS One*, 11(9), e0159749.

Twardziok, M., Meierhofer, D., Börno, S., Timmermann, B., Jäger, S., Boral, S., Seifert, G. (2017). Transcriptomic and proteomic insight into the effects of a defined European mistletoe extract in Ewing sarcoma cells reveals cellular stress responses. *BMC Complementary and Alternative Medicine*, 17(1), 1–14.

Urech, K., Baumgartner, S. (2015). Chemical constituents of *Viscum album* L.: Implications for the pharmaceutical preparation of mistletoe. *Mistletoe: From Mythology to Evidence-Based Medicine. Journal of Biomedical and Translational Research*, 4, 11–23.

Üstüner, T., Düzenli, S., Kitiş, Y. E. (2015). Niğde Bölgesinde Ökse Otunun (*Viscum album*) konukçularında Oluşturduğu Enfeksiyon Şiddetinin Belirlenmesi. *Turkish Journal of Weed Science*, 18(1), 6–14.

Valle, A. C. V., de Carvalho, A. C., Andrade, R. V. (2021). *Viscum Album*-literature review. *International Journal of Science and Research*, 10, 63–71.

Vicaş, S. I., RuglnĂ, D., Leopold, L., Pintea, A., Socaciu, C. (2011). HPLC fingerprint of bioactive compounds and antioxidant activities of *Viscum album* from different host trees. *Notulae Botanicae Horti Agrobotanici Cluj-Napoca*, 39(1), 48.

Vlad, D. C., Popescu, R., Dumitrascu, V., Cimporescu, A., Vlad, C. S., Vágvölgyi, C., Horhat, F. G. (2016). Phytocomponents identification in mistletoe (*Viscum album*) young leaves and branches, by GC-MS and antiproliferative effect on HEPG2 and McF7 cell lines. *Farmacia Journal*, 64, 82–87.

Yang, P., Jiang, Y., Pan, Y., Ding, X., Rhea, P., Ding, J., Lee, R. T. (2019). Mistletoe extract Fraxini inhibits the proliferation of liver cancer by down-regulating c-Myc expression. *Scientific Reports*, 9(1), 1–12.

Yau, T., Dan, X., Ng, C. C. W., Ng, T. B. (2015). Lectins with potential for anti-cancer therapy. *Molecules*, 20(3), 3791–3810.

Yusuf, L., Oladunmoye, M. K., Ogundare, A. O. (2013). In-vivo antibacterial activities of mistletoe (*Viscum album*) leaves extract growing on cocoa tree in Akure North Nigeria. *European Journal of Biotechnology and Bioscience*, 1(1), 37–42.

Zuber, D. (2004). Biological flora of central Europe: *Viscum album* L. *Flora-Morphology, Distribution, Functional Ecology of Plants*, 199(3), 181–203.

Index